Ergebnisse der Mathematik und ihrer Grenzgebiete

3. Folge · Band 29

A Series of Modern Surveys in Mathematics

Anthony Joseph

Quantum Groups and Their Primitive Ideals

Springer-Verlag
Berlin Heidelberg New York
London Paris Tokyo
Hong Kong Barcelona
Budapest

Anthony Joseph
Department of Theoretical Mathematics
The Weizmann Institute of Science
76100 Rehovot
Israel
and
Laboratoire de Mathématiques Fondamentales
Université Pierre et Marie Curie
4 place Jussieu
F-75252 Paris Cédex 05
France

Mathematics Subject Classification (1991):
16W30, 17B35, 17B37, 81R50

With 2 Figures

ISBN-13:978-3-642-78402-6 e-ISBN-13:978-3-642-78400-2
DOI: 10.1007/978-3-642-78400-2

Library of Congress Cataloging-in-Publication Data.
Joseph, Anthony. Quantum groups and their primitive ideals/Anthony Joseph.
p. cm. – (Ergebnisse der Mathematik und ihrer Grenzgebiete;
3. Folge, Bd. 29)
Includes bibliographical references and index.
ISBN-13:978-3-642-78402-6

1. Quantum groups. 2. Mathematical physics. I. Title. II. Series.
QC20.7.G76J67 1995 512'.55–dc20 94-21758 CIP

Softcover reprint of the hardcover 1st edition 1995

Typesetting: Data conversion by Springer-Verlag
SPIN: 10087957 41/3140 - 5 4 3 2 1 0 – Printed on acid-free paper

To Denise

יעל, מיכאל, וענת

תפרחו כחבצלת

Contents

Introduction

A quantum group $\mathfrak{G}$ is at present a purely mythical being to which one nevertheless associates a Hopf algebra $R_q[G]$ resembling the algebra of regular functions $R[G]$ on a (genuine) affine algebraic group G. It is supposed that if $\mathfrak{G}$ were to exist it would have the same representations as those of G. This is expressed by requiring $R_q[G]$ and $R[G]$ to be isomorphic as coalgebras.

It does not seem that a quantum group should be viewed as possessing a Lie algebra. Yet it does admit a quantum analogue $U_q(\mathfrak{g})$ of the enveloping algebra $U(\mathfrak{g})$ of $\mathfrak{g} := \operatorname{Lie} G$. The main objective of this book is to give a detailed analysis of the structure of $U_q(\mathfrak{g})$ and of $R_q[G]$ for G semisimple and in particular to describe their primitive and prime spectra.

It is not our aim to attempt to define a quantum group nor even to give a systematic procedure for writing down $R_q[G]$. Yet it is appropriate to discuss some of the methods which might achieve this as have been proposed in the literature. Here we concentrate on the ideas rather than on a historical presentation.

First one may give an algebraic group G or more generally a Lie group an additional "Poisson" structure (A.4.2). Such objects are now rather well understood and lead to a Lie bialgebra structure on $\mathfrak{g}$. The latter is equivalent to extending the Lie algebra structure of $\mathfrak{g}$ to $\mathfrak{g}\oplus\mathfrak{g}^*$ so that the natural form is invariant (A.4.3). This defines a 1-cocycle which if represented as a coboundary gives a 2-cochain satisfying the classical Yang-Baxter equation (A.4.7) for $\mathfrak{g}$. Then $R_q[G]$ should be obtained by "quantizing the Poisson structure" through a procedure resembling the passage from classical to quantum mechanics. A basic tool for doing this is deformation theory; but this can be extremely cumbersome and it is not immediate that this should always be the correct recipe. In any case one expects the classical Yang-Baxter equation to be the limit of its quantum analogue (9.5.2) for $R_q[G]$. Moreover the orbit philosophy of Lie group representation theory should be realized. Concretely this means that the symplectic leaves of G should be in a natural bijection with the primitive ideals of $R_q[G]$. An achievement of the theory is that this holds (10.3 and A.4.5) for G semisimple.

A second rather attractive proposal (A.4.9) is to replace (ordinary) affine space by "quantum space" and then to view $\mathfrak{G}$ as a "group" of linear transformations of the latter. Concretely this means replacing the symmetric algebra by a more general quadratic algebra (possibly obtained by deformation) and then to derive $R_q[G]$ by requiring it to possess the latter as a comodule.

A third principle is to focus attention on the tensor structure of the category of $\mathfrak{G}$ modules. This means of course just defining an algebra structure on $R_q[G]$; but this is to be done in a very specific manner. Concretely the category is required to be braided and this forces (9.4.2) the existence of an "R-matrix" satisfying in particular the quantum Yang-Baxter equation and from which the algebra structure of $R_q[G]$ can be written down (9.4.5).

Finally there was a search for a perfectly self-dual model for $R_q[G]$ which would then be isomorphic to $U_q(\mathfrak{g})$. Apparently this failed; but V. G. Drinfeld found that it could be essentially made to work for the "Borel part" of $U_q(\mathfrak{g})$ denoted $U_q(\mathfrak{b})$ and further found a general construction (the Drinfeld double) mirroring a Lie bialgebra. This gives $U_q(\mathfrak{g})$ up to passage to a quotient.

One of the most remarkable aspects of the above superficially different approaches is their extraordinary intercoherence. In particular they essentially all lead for G semisimple to the same and hence "canonical", objects $R_q[G]$ and $U_q(\mathfrak{g})$, though this epithet may as yet be premature.

Returning to the main content of this book it should first be noted that the study of primitive ideals of enveloping algebras was a point of view emphasized during the seventies by the research groups of J. Dixmier and of P. Gabriel as an alternative to the seemingly hopeless task of classifying all the representations of a Lie group or its Lie algebra even those which are reasonable in some sense. A notable early success was the classification of primitive ideals in the solvable case mainly by W. Borho, P. Gabriel and R. Rentschler [BGR1] and here, just recently, O. Mathieu [M5] established the bicontinuity of the Dixmier map from coadjoint orbits (which one may now think of as symplectic leaves) to primitive ideals. The semisimple case proved to be particularly exciting and led to many unexpected relations with geometry especially that between Goldie rank polynomials [J17] and Springer's Weyl group representations [J13, Bo1]. This and related developments are particularly well-reviewed in the books of J. C. Jantzen [Ja2] and of W. Borho, J.-L. Brylinski and R. MacPherson [BBM1]. Finally the case of an arbitrary finite dimensional Lie algebra was settled by C. Moeglin and R. Rentschler [Mo1, MoR1,2].

The study of primitive ideals went into a regressive state during the early eighties partly as an aftermath of these achievements and partly because it had now become clearer which were the interesting representations to study. Nevertheless many key questions still remain open particularly concerning completely prime primitive ideals [Mo2], [McG1], Dixmier algebras [McG2], [V4], quantization of orbital varieties [J9] and multiplicities of the Jordan-Holder series of primitive quotients [J12]. Though these are not particularly the subject of the present text one can hope that in the longer term the theory of quantum groups will illucidate these questions.

The discovery (or invention) of quantum groups in the eighties gave a new impetus not only to the representation theory of Lie algebras; but also to primitive ideal theory. An underlying principle of this book is examine particularly those new features which quantum groups afford. "To those that have not read the story" it should be particularly exciting to see how much can be achieved in so apparently complex a situation. "Of such as have" it will be a refreshing experience to retread familiar paths in the quantum frame. Examples of these novelties which we call quantum phenomena are Drinfeld duality (concerning $U_q(\mathfrak{b})$ as noted above) and its consequences for example the Rosso form (3.3), the construction of centres (7.1.16–7.1.20) and the crystal limit (Chap. 5). Another important feature, though not one which will be considered here, is the possibility to take q a root of unity. This not only leads to an exciting representation theory mirroring in a remarkable fashion the positive characteristic case [L3, 5, 6, 12, KL2, AJS1, APK1,2]; but also has important applications to knot theory [Jo1, Re1, T1], invariants of 3-manifolds [RT1] and statistical physics [ShS1].

Concerning the central themes of this book the aim has been to achieve a self-contained presentation sparing neither detail nor precision. Thus Chapter 1 is a quite basic introduction to Hopf algebras though of course it mainly emphasizes the results needed for the subsequent analysis. Thus a significant amount of material concerns the nature of the adjoint action (1.3, 1.4.14) and the basic properties of the Hopf dual (1.4). Chapter 2 though tangential to the main argument is still important for understanding the origins (A.4) of quantum groups. Algebraic groups are treated from the point of view of cocommutative Hopf algebras. Though there is nothing new in this approach it is still quite difficult to find so compact an introduction to the subject. In particular the precise interrelation of the algebraic group to its Lie algebra is examined and this is done in a manner leaving no analytical aftertaste. Applications include Chevalley's classification (2.4.5) of algebraic Lie algebras and a new result (2.4.10) requiring, as Weyl might have said, the angel of topology to lie with the devil of abstract algebra.

The first part (3.1) of Chapter 3 constructs Drinfeld duality in the very general context of an arbitrary integer-valued matrix C. Perhaps a novel feature is the emphasis on skew derivations and the intrinsic definition of the Serre identities in these terms (3.1.5). This is more basic than the usual approach which starts from a skew-Hopf pairing. Section 3.2 describes the Drinfeld double and the resulting (3.2.9) Drinfeld-Jimbo quantized enveloping algebra $U_q(\mathfrak{g})$; the latter for C symmetric. Section 3.3 describes the Rosso form which is one of the new features of $U_q(\mathfrak{g})$ since the classical Killing form concerns $\mathfrak{g}$ rather than $U(\mathfrak{g})$. This leads to a Casimir-like semi-invariant (3.3.9) which lies only in a completion of $U_q(\mathfrak{g})$ but still in a smaller algebra than Drinfeld's invariant. In Section 3.4 a $q \longrightarrow 1$ limit is defined making Drinfeld duality become a non-degenerate pairing between $U(\mathfrak{n}^-)$ and $S(\mathfrak{n}^+)$. Assuming C symmetric this and the Casimir are used to show that the Shapovalev determinants can be written as a product of "linear" factors.

Chapter 4 describes the Jantzen filtration and sum formula (4.1). This is still an important technique of remarkable generality. Here it is used to establish the precise nature of the Shapovalev factors (4.1.16), a result of C. De Concini and V. Kac [DK1] for C of finite type. It is developed from [JL4] and is new in the present level of generality. Section 4.2 is a very rapid introduction to Kac-Moody Lie algebras and their highest weight modules particularly to their $\mathfrak{n}$-homology. This leads to the Weyl character formula (4.2.16) whose generalization by V. Kac to this situation was the first major success of the theory since it led to and generalized many marvelous formulae [Ka1, Chap. 12] of classical number theory. Section 4.3 reviews the corresponding theory for $U_q(\mathfrak{g})$ and is really a quite straightforward generalization. Demazure modules which have a significant geometric interpretation are described in Section 4.4. Uniqueness of Verma module embeddings is established in 4.4.15. This is more delicate in the Kac-Moody case and requires the Enright functors as interpreted by V. V. Deodhar [De1] through Ore localization. Finally the BGG theorem (4.4.14) concerning the possible simple subquotients of Verma modules is given a simple proof (due to Jantzen) through the Jantzen sum formula. Although this is now considered elementary it is still of some pedagological interest.

Chapters 5 and 6 develop certain results on bases needed for the subsequent structure theory of Chapters 7–10. These bases have several different sources. The first came from the following remarkable observation of C. M. Ringel [Ri1]. Consider the path algebra R_C over a field k defined by a Cartan matrix C of finite type. After P. Gabriel [Gab1] the isomorphism classes of finite dimensional R_C modules parametrize a PBW basis of $U(\mathfrak{n}^+)$ with the indecomposable modules corresponding to the root vectors. Now take k to be the field of q elements. Ringel introduced a Hall algebra for R_C whose structure constants $c^M_{N,N'}$ are the number of filtrations of M with quotient N and submodule N'. He observed that (up to some twist if C is not simply-laced) these give exactly the multiplication rules in $U_q(\mathfrak{n}^+)$. G. Lusztig [L7] reinterpreted Ringel's multiplication in geometric terms and applying the decomposition theory of perverse sheaves found a second basis, called Lusztig's canonical basis, for $U_q(\mathfrak{n}^+)$. This transformation was analogous to a situation familiar from Verma module theory. Moreover he found that this new basis was compatible with the integrable highest weight modules and indeed through a natural duality it gives an important "common basis theorem" (6.2.19). Moreover in terms of Lusztig's basis the structure constants are positive integers (in the simply-laced case) a fact which he has recently applied in [L15].

The second source was quantum statistical mechanics and in particular the zero temperature limit which in a suitable lattice model corresponds to the $q \longrightarrow 0$ limit. Indeed E. Date, M. Jimbo and T. Miwa observed (see [K2]) that the Gelfand-Tsetlin basis became monomial in the tensor algebra of the defining representation of $U_q(\mathfrak{sl}(n))$ at $q = 0$. Building on this elementary fact M. Kashiwara [K3] constructed his crystal basis for $U_q(\mathfrak{n}^+)$ which was similarly compatible with integrable highest weight modules. It also had an orthonormality property (6.1.2) which later allowed I. Grojnowski and

G. Lusztig [GrL1] to show that the globalized Kashiwara crystal basis coincided with Lusztig's canonical basis.

A third source was a conjecture independently formulated (see [M3]) by K. Baclawski and by I. M. Gelfand and A. V. Zelevinsky. The aim was to find for each integrable highest weight module a basis compatible with each subspace occurring in the tensor product decomposition formula of 6.3.20. This led O. Mathieu to construct a "good basis" [M3] with this property which incidentally also holds for the Kashiwara-Lusztig basis (6.5.5). There is here also an important connection [AP1], [M3, 5.5] with modules having a good filtration property [P1, M2, M4].

A fourth source was the Lakshmibai-Seshadri standard monomial theory [LS1] aimed at constructing bases compatible with the Demazure modules. Although their bases are a little different [La3], Kashiwara showed [K5] compatibility with the global crystal basis (6.3.10). Moreover their work motivated the Littelmann Path theory which ultimately (6.4.27) gives a quite explicit description of the crystal bases.

In the present work we have chosen to give a detailed description of the Kashiwara and of the Littelmann theory. This is because only elementary methods are needed and moreover it gives all the results we require for the subsequent structure theory thereby allowing a self-contained presentation.

Chapter 7 makes a detailed analysis of the structure of $U_q(\mathfrak{g}_C)$ mainly for C of finite type. Here the adjoint action is studied in some detail. Unlike $U(\mathfrak{g})$ only a subalgebra $F(U_q(\mathfrak{g}))$ is locally ad-finite. The latter admits a mock Peter-Weyl theorem (7.1.6) which must persist in $U(\mathfrak{g})$ but which had not been noted previously. This gives an elegant description of the centre (7.1.17) of $U_q(\mathfrak{g})$ and (7.1.20) of $U_q(\mathfrak{n}^+)$. Notably the proofs are easier than in the classical case. Using the Rosso form, the connection (due to P. Caldero [C1]) with the true Peter-Weyl theorem becomes apparent (7.1.23). Using the orthonormality of the crystal basis this is further deepened in section 7.2. Using the common basis theorem the quantum version of Kostant's separation of variables theorem is established (7.3). Finally (for C of finite type) the algebras under consideration (for example $F(U)$) are shown to be noetherian (7.4). This also uses the common basis theorem; but only with respect to fundamental modules.

Chapter 8 starts with the calculation of Poincaré series or generalized quantum exponents (8.1.9) which is a first step in computing Verma module annihilators (8.3.1). This surprisingly difficult result is obtained by comparing the factors of the quantum PRV determinants (8.2) with those of Shapovalev. Combined with the separation theorem this leads to an equivalence of the BGG and Harish-Chandra categories (8.4.11) analogous to the classical case and giving the quantum version of Duflo's theorem [Du1]. It may be noted that through the Drinfeld isomorphism (4.5.2) the left cell classification of primitive ideals of $U(\mathfrak{g})$ carries over to $U_q(\mathfrak{g})$. Finally a new result characterizes (8.2.2) the projective Verma modules even for C integrable.

Chapter 9 is devoted to the structure theory of $R_q[G]$. This is simplified by a judicious choice of Ore localization (9.1.10) and by $R_q[G]$ being the image (9.2.2) of an algebra built from subalgebras of invariants. A new result is that the latter is a free module over its centre (9.2.9). The coproduct and the Rosso form give an important embedding theorem (9.2.14) which can be used to show that all the primes of $R_q[G]$ are completely prime [J11]. The adjoint action is shown to have a surprisingly simple structure (9.3) even though the locally finite elements reduce to scalars. Finally the R-matrix is described (9.4) following a treatment of D. Gaitsgory [Ga1] and this is used to compute the second Hopf dual (9.4.9).

Chapter 10 starts with the theory of highest weight modules for $R_q[G]$. This leads to Soibelman's [S1] quantum Weyl group which also recovers the Lusztig automorphisms [L2, 6, 12]. In 10.2.8 their behaviour under the coproduct is directly related to the R-matrix. In 10.3 the prime and primitive spectra of $R_q[G]$ are classified. In 10.4 it is shown that Hopf algebra automorphisms come from only the torus action and Dynkin diagram automorphisms, a result found independently by A. Braverman [Br1] and by W. Chin and I. M. Musson [CM1]. Finally various twisted quantum Weyl algebras are briefly discussed. These are important for the quantum version of the Beilinson-Bernstein equivalence of categories [J9].

The first part (A.1) of the appendix gives some background material on Weyl groups. A more leisurely treatment of the first part may be found in [Ca1] and a more comprehensive one in [Bb1]. A rapid introduction to ring theory including the Jacobson and Artin-Wedderburn theorems, Ore localization, Goldie's theorem, the diamond lemma and Frobenius reciprocity for Hopf algebras is given in Section A.2. Miscellaneous results on combinatorial identities, filtrations, generic flatness, Gelfand-Kirillov dimension and the Hilbert-Samuel polynomial are reviewed in Section A.3. Finally some important topics which do not concern the main text are discussed in Section A.4. These include Poisson algebraic groups (A.4.2), the classical Yang-Baxter equation (A.4.6), Manin's construction of quantum groups (A.4.9) and an intrinsic derivation of the quantum determinant (A.4.10).

The bibliography for the main content of this book would come to about 40 key papers or books; but the peripheral material would involve several thousand. It would be a Herculean task to put all these many contributions into historical perspective and the rather small bibliography and comments at chapter endings are an unworthy contribution. One may recover some of the early history from Drinfeld's ICM address [Dr2]. The book of G. Lusztig [L12] gives an introduction to integrable modules, automorphisms and particularly to canonical bases giving also a historical perspective to this development. The book of S. Shnider and S. Sternberg [ShS1] gives a sweeping analysis of braided categories, deformation theory and applications to physics with more than 1250 references. Finally S. Montgomery's monograph [Mon2] describes recent developments in general Hopf algebra theory.

Acknowledgements. The author would like to thank the many people who have made this work possible. First, J. Dixmier whose book on enveloping algebras was a guideline of excellence. Second, those colleagues associated with the subject particularly J. Bernstein, W. Borho, M. Duflo, J. C. Jantzen, J. C. McConnell, R. Rentschler, M. Vergne and D. Vogan who have also been very good friends. I could not have learnt the subject nearly so efficiently without the enveloping algebra seminars of Paris 6 and the Weizmann Institute where we managed to persuade almost everybody in the field to speak, which in the last years also included many Russian mathematicians. Some of the results here represent a joint and very pleasant collaboration with G. Letzter. As the reader may discern much of this was inspired by work of B. Kostant whose ideas and those of A. W. Goldie have very much shaped my work. I would like to thank Miriam Abraham for painstakingly typing the manuscript and enduring the endless alterations I have inflicted upon her. Finally I would also like to thank Ruby Musrie for making the final corrections and ensuring the collation of the entire manuscript in good time. Several errors were corrected at the last minute by F. Millet who studied part of the manuscript.

Notes on Conventions. If unadorned, the symbols Hom and $\otimes$ refer to the base field. There is never more than one lemma, proposition, theorem or corollary in a given subsection and, apart from an occasional preamble, the proof begins directly after the assertion and terminates at the end of that subsection unless followed by a Remark. Since several different duals are involved, clarity and equity impose the use of asterisks of different pointage. The validity of certain symbols may be confined within designated chapters as defined in the Index of Notation. With a few exceptions having no particular significance, sources are accredited only at the end of each chapter. As the reader may independently realize, the spelling favours the Oxford tradition.

Anthony Joseph

Chapter I. Hopf Algebras

1.1 Axioms of a Hopf Algebra

1.1.1 Let k be a field. For any k-vector space V the symbol $1d_V$ (or simply, $1d$) denotes the identity map on V. An associative k-algebra (or simply, algebra) is a k-vector space A together with a linear map $\mu : A \otimes A \longrightarrow A$ such that the diagram

(i)

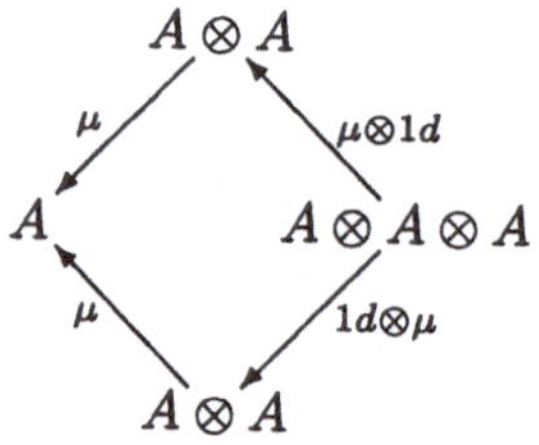

is commutative. Writing $\mu(a \otimes b) = ab$ this may be expressed as $(ab)c = a(bc)$ for all $a, b, c \in A$. An identity of an algebra A is a k-linear map $\eta : k \longrightarrow A$ such that the diagrams

$(\text{ii})_\ell$

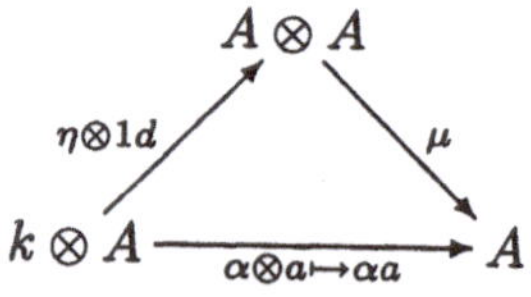

$(\text{ii})_\text{r}$

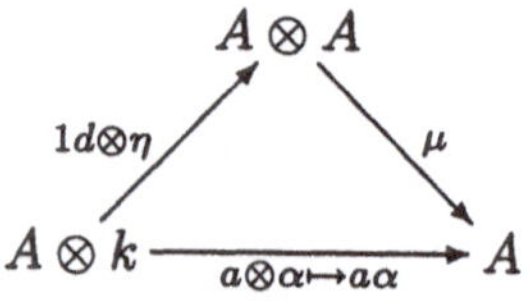

are commutative. By $(\text{ii})_\ell$ and $(\text{ii})_r$ it follows that $\eta(1) =: 1_A$ (or simply, 1) satisfies $1a = a1 = a$ for all $a \in A$. Moreover $\eta(\alpha)a = \alpha a$ and $\eta(\alpha\beta) = \alpha\eta(\beta) = \eta(\alpha)\eta(\beta)$ so η is an algebra homomorphism. It is often convenient to omit η. Unless otherwise specified the term algebra will mean associative algebra.

Let A be an algebra with identity. A (left) A module M is a vector space together with a linear map $\nu : A \otimes M \longrightarrow M$ such that the diagrams

(iii)$_\ell$

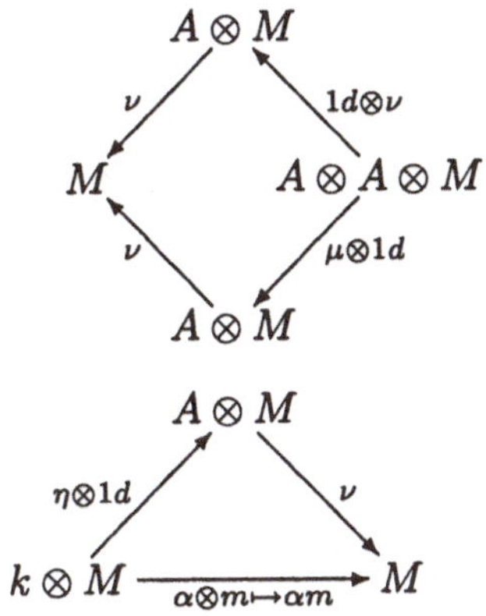

are commutative. Of course A is itself a (left) A module via μ and then a (left) A submodule of A is called a (left) ideal. Similarly right A modules may be defined. A linear subspace I of A which is both a left and a right ideal is simply called an ideal and in this case A/I inherits an algebra structure.

1.1.2 The corresponding notions of associative coalgebra and coidentity are immediately obtained by reversing arrows in the above. Thus a coproduct Δ on a vector space A is a linear map $\Delta : A \longrightarrow A \otimes A$ such that the diagram

(i)*

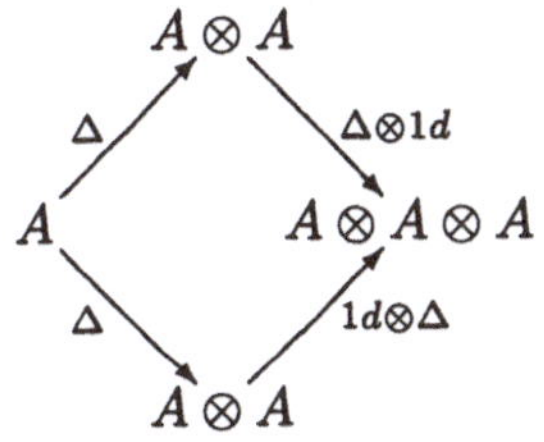

is commutative, making A a coalgebra. The common composed map $A \longrightarrow A \otimes A \otimes A$ is denoted Δ^2. A coidentity ε of A is a k-linear map such that the diagrams

(ii)$^*_\ell$

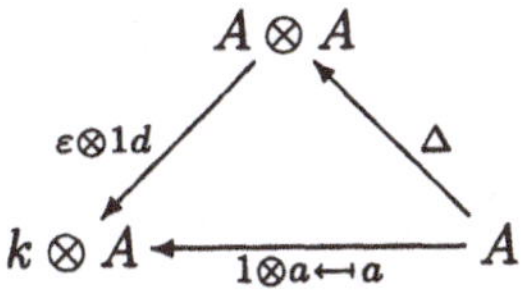

(ii)$^*_\mathrm{r}$

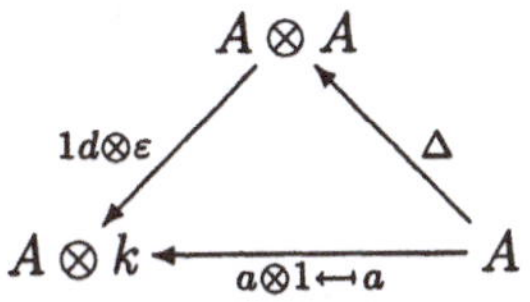

are commutative. If k itself is viewed as a coalgebra with coproduct defined by $\Delta(1) = 1 \otimes 1$, then ε is a coalgebra morphism, that is $(\varepsilon \otimes \varepsilon)\Delta = \Delta\varepsilon$.

Let A be a coalgebra with coidentity. Then a (left) A comodule M is a vector space together with a linear map $\lambda : M \longrightarrow A \otimes M$ such that the diagrams

(iii)$^*_\ell$

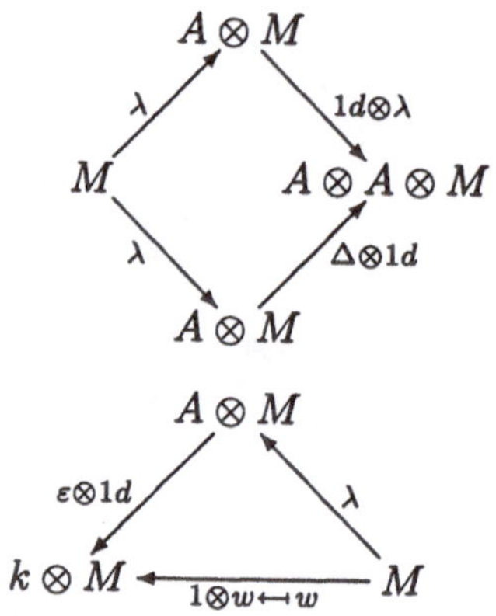

are commutative. A (left) subcomodule of A is called a (left) coideal. Notice however that it is appropriate to define a subspace I of A to be a coideal if just $\Delta(I) \subset I \otimes A + A \otimes I$ since this is what is required for A/I to inherit a coalgebra structure.

1.1.3 If A is both an algebra and a coalgebra, then it is called a bialgebra if the diagram (defining $\tau_{ij} \in \operatorname{End} A^{\otimes n}$ to interchange the i, j entries)

(iv)

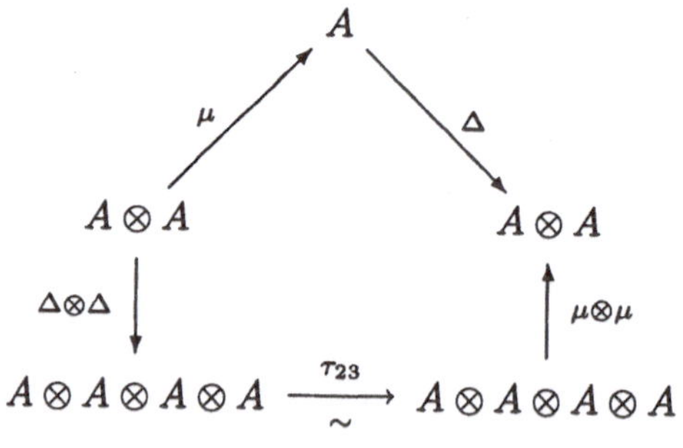

is commutative. Moreover this property is retained if we either replace μ by $\mu^{\text{opp}} := \mu\tau_{12}$ or Δ by $\Delta^{\text{opp}} := \tau_{12}\Delta$, (or both). Notice that (iv) is unchanged by reversal of arrows. It will generally be assumed that a bialgebra is equipped with an identity and coidentity. In this case one adds to (iv) the additional diagrams

(iv)′

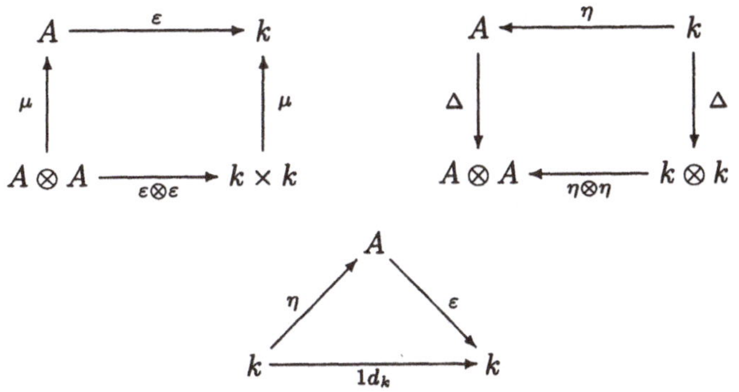

which are also assumed to be commutative. Notice that these are respectively equivalent to $\varepsilon(a)\varepsilon(b) = \varepsilon(ab)$, for all $a, b \in A$ and $\Delta(1) = 1 \otimes 1$, $\varepsilon(1) = 1$. One calls $\ker \varepsilon$ the augmentation ideal of A.

1.1.4 If A is an algebra, then the linear space $A \otimes A$ has an algebra structure through the product rule $(a \otimes b)(c \otimes d) = ac \otimes bd$, that is $A \otimes A$ is given the product $(\mu \otimes \mu)\tau_{23}$. Notice that for a bialgebra A the commutativity of (iv) (resp. and (iv)′) is equivalent to Δ (resp. and ε) being an algebra morphism, equivalently μ (resp. and η) being a coalgebra morphism. Now if M, N are (left) A modules the linear space $M \otimes N$ has an $A \otimes A$ module structure through the rule $(a \otimes b)(m \otimes n) = am \otimes bn$. Composing the resulting map $A \otimes A \longrightarrow \mathrm{End}(M \otimes N)$ with $\Delta : A \longrightarrow A \otimes A$ we conclude that $M \otimes N$ has an A module structure. This leads to the important conclusion that the category of (left) modules of a bialgebra A is closed under tensor product. Notice that $A \otimes A$ has $\tau_{23}(\Delta \otimes \Delta)$ as a coproduct.

1.1.5 For each vector space V, set $V^* = \mathrm{Hom}(V, k)$. Given vector spaces V, V' the map $\xi \otimes \xi' \mapsto (v \otimes v' \mapsto \xi(v)\xi'(v'))$ extends to a linear embedding of $V^* \otimes V'^*$ into $(V \otimes V')^*$, which is surjective if and only if either V or V' is finite dimensional. Let M be a left (resp. right) A module, then M^* is a right (resp. left) A module via $(\xi a)(m) = \xi(am)$ (resp. $(a\xi)(m) = \xi(ma)$).

1.1.6 Let A be a vector space equipped with a coproduct Δ. Then by transposition one has a linear map $\Delta^* : (A \otimes A)^* \longrightarrow A^*$ which by restriction (1.1.5) to $A^* \otimes A^*$ gives an algebra structure on A^*. Moreover if ε is a coidentity for A then ε^* is an identity for A^*. On the other hand, if A is an algebra with multiplication μ, then μ^* fails to give A^* a coalgebra structure unless A is finite dimensional. Nevertheless, A^* still inherits an A bimodule structure from A. Moreover if A is a bialgebra, giving $A^* \otimes A^*$ a left (resp. right) A module structure (1.1.4) via Δ makes $\Delta^* : A^* \otimes A^* \longrightarrow A^*$ a homomorphism of left (resp. right) A modules. More generally let B be an algebra which is a left (or right) A module. Then B is called an A-algebra if the multiplication $B \otimes B \longrightarrow B$ is a homomorphism of A modules and this rule is referred to as the (generalized) Leibnitz rule.

1.1.7 Let A be a bialgebra. The left action of A on itself, unlike that on A^*, does not satisfy the Leibnitz rule. To obtain an action which does have this property one needs an antipode σ on A. This is defined to be a linear map $\sigma : A \longrightarrow A$ such that the diagram

(v)

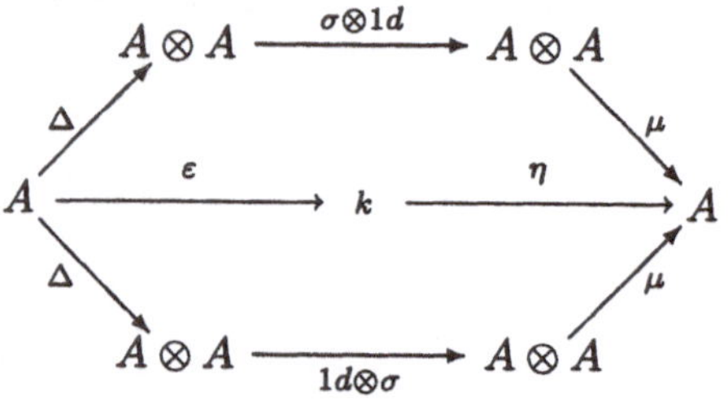

is commutative. A bialgebra A equipped with an antipode is called a Hopf algebra. An ideal of a Hopf algebra which is also a coideal, contained in the augmentation ideal and σ stable is called a Hopf ideal. The linear maps $\mu, \eta, \Delta, \varepsilon, \sigma$ in the definition of A as a Hopf algebra may be designated with an A subscript if more than one Hopf algebra is involved.

1.1.8 There comes a moment when the geometric elegance of diagrams must be forsaken for the banal realities of algebraic computation. Here coalgebras cause a particular difficulty. Thus for example $\Delta(a)$ is actually a sum and moreover the terms are only determined up to linear transformations. To Sweedler's well-known solution to this problem we have added the logical simplification of omitting both $\sum$ and the parentheses, writing this sum simply as $\Delta(a) = a_1 \otimes a_2$. The subscript is treated as a dummy index and for example $\Delta(a_i)$ is written as $a_i \otimes a_{i+1}$ where the dummy indices $j > i$ are increased by 1. Observe here that $\Delta^2(a) = a_1 \otimes a_2 \otimes a_3$ taking *either* branch of (i)*. Note that diagram (iv) gives $\Delta(a)\Delta(b) = \Delta(ab)$ which translates to $a_1b_1 \otimes a_2b_2 = (ab)_1 \otimes (ab)_2$. This can be interpreted as giving $a_1b_1 = (ab)_1$ and $a_2b_2 = (ab)_2$. However this is only a possible solution to the above equation. In these conventions (ii)* gives $a = \varepsilon(a_1)a_2 = a_1\varepsilon(a_2)$ and (v) that $\varepsilon(a) = \sigma(a_1)a_2 = a_1\sigma(a_2)$. Thus using ε or σ one may contract the *ordered* pair of dummy indices $i,\ i+1$. Difficulties which could in principle arise from different choices of representatives are avoided because of the linearity of all the transformations involved.

Let A be a bialgebra and B an A-algebra. Then one may form the smash product $B\#A$ which is $B \otimes A$ as a vector space and has multiplication given by

$$(b \otimes a)\ (b' \otimes a') = ba_1(b') \otimes a_2a' \quad \text{where} \quad \Delta(a) = a_1 \otimes a_2\ .$$

From the associativity of B and A, the coassociativity of A and the generalized Leibnitz rule, one checks that this is an associative multiplication.

Let A be a bialgebra and let M, N be A modules. A bilinear form $\psi : M \times N \longrightarrow k$ is said to be invariant if

$$\psi(a_1m, a_2n) = \varepsilon(a)\psi(m,n) \quad \textit{where} \quad \Delta(a) = a_1 \otimes a_2\ .$$

Lemma. *Suppose A is a Hopf algebra and let M, N be A modules. A bilinear form $\psi : M \times N \longrightarrow k$ is invariant if and only if $\psi(am, n) = \psi(m, \sigma(a)n)$, $\forall a \in A,\ m \in M,\ n \in N$.*

Suppose ψ is invariant. Then $\psi(m, \sigma(a)n) = \psi(m, \sigma(\varepsilon(a_1)a_2)n) = \varepsilon(a_1)$ $\psi(m, \sigma(a_2)n) = \psi(a_1 m, a_2\sigma(a_3)n) = \psi(a_1 m, \varepsilon(a_2)n) = \psi(am, n)$. For the converse observe that $\psi(a_1 m, a_2 n) = \psi(m, \sigma(a_1)a_2 n) = \varepsilon(a)\psi(m, n)$.

1.1.9 Let A be a coalgebra and B an algebra. Then the construction of 1.1.6 gives $\mathrm{Hom}(A,\ B)$ a multiplication called convolution, hence an algebra structure which in the notation of 1.1.8 can be expressed through the formula $(\xi_* \xi')(a) = \xi(a_1)\xi'(a_2)$. Now let A be a bialgebra. One checks that $\eta\varepsilon$ is the identity for $\mathrm{End}\, A$ (with respect to convolution). Then the commutative diagram (v) exactly states that $1d_A$ which is the identity for $\mathrm{End}\, A$ (with respect to composition) satisfies $\sigma_* 1d_A = 1d_{A*}\sigma = \eta\varepsilon$. We conclude that if A admits an antipode then it is unique. Then (1.1.6) $A \otimes A$ inherits a bialgebra structure and one checks that $\eta\varepsilon \otimes \eta\varepsilon$ is the identity for $\mathrm{Hom}(A \otimes A,\ A)$. Further assume that A has an antipode σ and define $\varphi, \psi \in \mathrm{Hom}(A \otimes A, A)$ through $\varphi(a \otimes b) = \sigma(ab)$, $\psi(a \otimes b) = \sigma(b)\sigma(a)$.

Lemma. *Let A be a Hopf algebra. Then in* $\mathrm{Hom}(A \otimes A, A)$ *one has*

(i) $\varphi_* \mu = 1$
(ii) $\mu_* \psi = 1$
(iii) *σ is an algebra antihomomorphism.*

Recalling the conventions and conclusions of 1.1.8 one obtains $(\varphi_* \mu)(a \otimes b) = \varphi(a_1 \otimes b_1)\mu(a_2 \otimes b_2) = \sigma(a_1 b_1)a_2 b_2 = \sigma((ab)_1)(ab)_2 = (\eta\varepsilon)(ab) = (\eta\varepsilon \otimes \eta\varepsilon)(a \otimes b)$. Hence (i). Again observing that $(\eta\varepsilon)(b)$ is a scalar gives $(\mu_* \psi)(a \otimes b) = \mu(a_1 \otimes b_1)\psi(a_2 \otimes b_2) = a_1 b_1 \sigma(b_2)\sigma(a_2) = a_1(\eta\varepsilon)(b)\sigma(a_2) = (\eta\varepsilon)(a)(\eta\varepsilon)(b) = (\eta\varepsilon \otimes \eta\varepsilon)(a \otimes b)$. Hence (ii). Combining (i) and (ii) gives (iii).

1.1.10 The result in 1.1.9 (iii) can be expressed by the commutativity of the diagram

(vi)

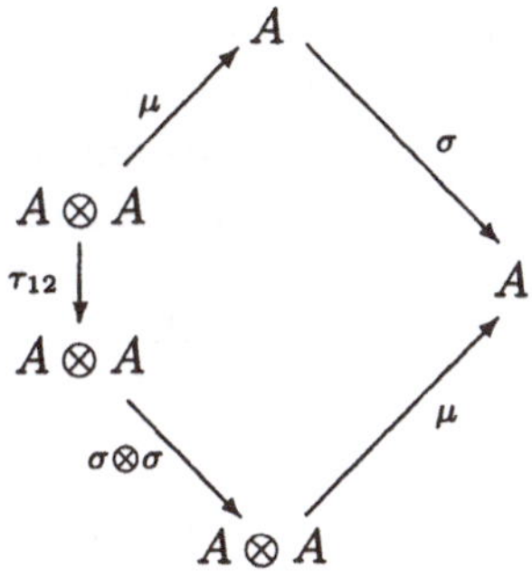

This immediately suggests the commutativity of the diagram

(vi)*

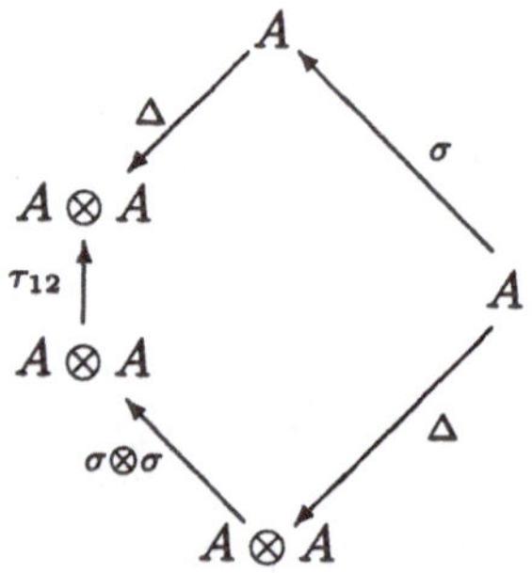

In the conventions (and within the limitations) of 1.1.8 this can be expressed by the following

Lemma. *Let A be a Hopf algebra. Writing $\Delta(a) = a_1 \otimes a_2$ one has $\sigma(a_1) = \sigma(a)_2$, $\sigma(a_2) = \sigma(a)_1$.*

One checks that $\Delta\eta\varepsilon$ is the identity of the algebra $\mathrm{Hom}(A, A\otimes A)$. Defining $\varphi = \Delta\sigma \in \mathrm{Hom}(A, A \otimes A)$ gives $\varphi_*\Delta = 1$. Defining $\psi = \tau_{12}(\sigma \otimes \sigma)\Delta \in \mathrm{Hom}(A, A \otimes A)$ gives $\Delta_*\psi = 1$. Hence $\varphi = \psi$ as required.

1.1.11 Let F be the free associative algebra with identity and generators $\{a^i\}_{i\in\mathcal{I}}$. Fix $\{r^j\}_{j\in\mathcal{J}} \subset F$ and let R be the ideal they generate in F. To define a coproduct Δ on $A := F/R$ it is enough to give the $\Delta(a^i) \in F \otimes F$ so that Δ extended to a homomorphism of F satisfies $\Delta(r^j) \in R \otimes F + F \otimes R$ and that $\Delta(1d \otimes \Delta)(a^i) = \Delta(\Delta \otimes 1d)(a^i)$ in $A \otimes A \otimes A$. To define a coidentity ε on A it is enough to give the $\varepsilon(a^i) \in k$ so that ε extended to a homomorphism of F satisfies $\varepsilon(r^j) \in R$ and $\varepsilon(a^i_1)a^i_2 = a^i_1\varepsilon(a^i_2) = a^i$ in A. To define an antipode σ on A it is enough to give the $\sigma(a^i) \in k$ so that σ extended to an antihomomorphism of F satisfies $\sigma(r^j) \in R$ and $\sigma(a^i_1)a^i_2 = a^i_1\sigma(a^i_2) = \varepsilon(a^i)$ in A. This Hopf algebra structure need not come from one on F. Often $\mathcal{I}$ and $\mathcal{J}$ are finite.

1.1.12 Let A be a Hopf algebra. By 1.1.3 one has $\Delta(1) = 1\otimes 1$ and $\varepsilon(1) = 1$. Then commutativity in (v) of 1.1.7 gives $\sigma(1) = 1$. On the other hand $a_1\sigma(a_2) = \varepsilon(a)$ where $\Delta(a) = a_1 \otimes a_2$. Thus if σ^{-1} exists it is an antiautomorphism and satisfies $a_2\sigma^{-1}(a_1) = \varepsilon(a)$ and similarly $\sigma^{-1}(a_2)a_1 = \varepsilon(a)$. One concludes that σ^{-1} is the antipode for A equipped with its opposed coproduct Δ^{opp} (1.1.3). Similarly σ^{-1} is the antipode for A equipped with the opposed product (and the same coproduct).

1.2 Group Algebras and Enveloping Algebras

In this section some basic examples of Hopf algebras are exhibited together with some considerations as to how they might be generalized.

1.2.1 Let G be a group. Then the group algebra kG is an associative k-algebra with identity. Since $(g \otimes g)(h \otimes h) = gh \otimes gh$, for all $g,\ h \in G$ it is immediate that the map $\Delta : g \mapsto g \otimes g$ of G into $kG \otimes kG$ extends linearly to a homomorphism of kG into $kG \otimes kG$. Moreover $\Delta^2(g) = g \otimes g \otimes g$ computed either way. Thus Δ defines a coproduct on G. One further checks that the map $\varepsilon(g) = 1$ (resp. $\sigma(g) = g^{-1}$) defined on G extends linearly to a coidentity (resp. antipode) on kG.

Let $\mathcal{F}(G)$ denote the set of functions on G with values in k. Then $\mathcal{F}(G)$ is a k-algebra with respect to scalar multiplication and pointwise operations. It is of course a fairly classical object which identifies with $(kG)^*$ given an algebra structure through Δ as in 1.1.6. Moreover $\mathcal{F}(G)$ is commutative and this results from $\Delta = \Delta^{\mathrm{opp}}$.

1.2.2 If G is a finite group, then $\mathcal{F}(G)$ is a Hopf algebra through the multiplication in G; but this is false in general. On the other hand suppose that G is also an affine algebraic variety and let $R[G]$ be the subalgebra of $\mathcal{F}(G)$ of regular functions on G. Then G is called an algebraic group if the transpose μ^* of multiplication μ in G restricted to $R[G]$ has its image in $R[G] \otimes R[G]$. Of course the latter is just $R[G \times G]$ and so this is what it means for $\mu : G \times G \longrightarrow G$ to be a morphism. The condition that G be affine ensures that $R[G]$ is not too small, or more precisely that it separates points of G. This means that the natural map $a \mapsto (f \mapsto f(a))$ of kG into $R[G]^*$ is an embedding and so we recover kG as a subalgebra of $R[G]^*$. One calls G irreducible if $R[G]$ is a domain. These concepts will be studied in more detail in sections 2.3, 2.4.

1.2.3 Suppose that k is algebraically closed in the example of 1.2.2. Then we may recover G from $R[G]$ as $\operatorname{Max} R[G]$. Indeed for each $g \in \operatorname{Max} R[G]$ we obtain an algebra homomorphism $e_g : R[G] \longrightarrow R[G]/g = k$ which in particular is an element of $R[G]^*$. Then the composed map $e_{g,h} : R[G] \xrightarrow{\mu^*} R[G] \otimes R[G] \xrightarrow{e_g \otimes e_h} R[G]/g \otimes R[G]/h \xrightarrow{\sim} k$ defines a multiplication $(g,h) \mapsto \ker e_{(g,h)}$ on $\operatorname{Max} R[G]$. This translates under the map $\operatorname{Max} R[G] \longrightarrow R[G]^*$ to the original multiplication on G viewed as a subgroup of $R[G]^*$ via the canonical map $g \mapsto (f \mapsto f(g))$. Moreover the identity of $\operatorname{Max} R[G]$ is just $\ker \eta^*$ and the inverse of g just $\ker(e_g \sigma^*)$.

1.2.4 It is useful (1.3.4) to observe that for any Hopf algebra A, replacing $\operatorname{Max} A$ by the subset $\operatorname{Max}^1 A$ of ideals of codimension 1, the above construction yields a subgroup $A^\wedge$ of A^*. Unlike the commutative case $A^\wedge$ will be too small to be separating.

1.2.5 Let V be a vector space. Then $\mathcal{T}(V) := k \oplus V \oplus V \oplus V \oplus \ldots$ has the structure of a graded associative algebra through the product $(u_1 \otimes u_2 \otimes \ldots \otimes u_m) \otimes (u_{m+1} \otimes \ldots \otimes u_{m+n}) \mapsto u_1 \otimes u_2 \otimes \ldots \otimes u_{m+n}$. It is called the tensor algebra over V. The linear map $v \mapsto v \otimes 1 + 1 \otimes v$ of V into $\mathcal{T}(V) \otimes \mathcal{T}(V)$ extends to an algebra homomorphism $\Delta : \mathcal{T}(V) \longrightarrow \mathcal{T}(V) \otimes \mathcal{T}(V)$ and is obviously a coproduct. Again set $\varepsilon(1) = 1$, $\varepsilon(v) = 0$ (resp. $\sigma(1) = 1$, $\sigma(v) = -v$) on V. Then ε (resp. σ) extends to an algebra homomorphism (resp. antihomomorphism) of $\mathcal{T}(V)$ into k (resp. $\mathcal{T}(V)$) and defines a coidentity (resp. antipode) on $\mathcal{T}(V)$. One may observe that taking a basis $\{v^i\}_{i\in\mathcal{I}}$ for V then $\mathcal{T}(V)$ is just the free associative algebra with generators $\{v^i\}_{i\in\mathcal{I}}$. The quotient of $\mathcal{T}(V)$ by the ideal generated by the commutators $v \otimes v' - v' \otimes v : v, v' \in V$ is called the symmetric algebra $S(V)$ over V.

1.2.6 A Lie algebra $\mathfrak{g}$ is a vector space given a (non-associative) product, called the Lie product which is a linear map $x \otimes y \mapsto [x, y]$ of $\mathfrak{g} \otimes \mathfrak{g}$ into $\mathfrak{g}$ satisfying

(i) $[x, y] + [y, x] = 0$ (antisymmetry)

(ii) $[x, [y, z]] + [y, [z, x]] + [z, [x, y]] = 0$ (Jacobi identity) .

Any associative algebra A defines a Lie algebra $\mathfrak{l}_A$ which is A as a vector space and whose Lie product is the commutator $xy - yx$. For example, let V be a vector space and set $A = \operatorname{End} V$. In this case $\mathfrak{l}_{\operatorname{End} V}$ is denoted $\mathfrak{gl}(V)$, or sometimes $\mathfrak{gl}(n)$ if $n = \dim V < \infty$. When $\operatorname{char} k = 0$ the set of trace zero matrices form a simple Lie subalgebra $\mathfrak{sl}(V)$ (or $\mathfrak{sl}(n)$ if $n = \dim V$) of $\mathfrak{gl}(V)$ having codimension one. A $\mathfrak{g}$ module V is a vector space with an action $\mathfrak{g} \times V \longrightarrow V$ such that the map $\rho : x \mapsto (v \mapsto xv)$ is a Lie algebra homomorphism of $\mathfrak{g}$ into $\mathfrak{gl}(V)$ and ρ is called a representation of $\mathfrak{g}$.

For each Lie algebra $\mathfrak{g}$ the set of pairs consisting of an associative algebra A and Lie algebra homomorphism $\mathfrak{g} \longrightarrow \mathfrak{l}_A$ admits a universal object denoted $(U(\mathfrak{g}), \mathfrak{i})$. It is immediate that $U(\mathfrak{g})$ is just the image of $\mathcal{T}(\mathfrak{g})$ by the ideal generated by the elements $x \otimes y - y \otimes x - [x, y] : x, y \in \mathfrak{g}$ with $\mathfrak{i}$ the composition of the canonical embedding $\mathfrak{g} \longrightarrow \mathcal{T}(\mathfrak{g})$ with the canonical projection $\mathcal{T}(\mathfrak{g}) \longrightarrow U(\mathfrak{g})$. It follows (A.2.14) from the Poincaré-Birkhoff-Witt theorem (PBW) that $\mathfrak{i}$ is an embedding (a fact which is otherwise not obvious). Moreover in $U(\mathfrak{g})$ the Lie product is exactly the commutator and so no loss of generality is obtained by viewing it in this fashion. Moreover it is immediate from universality that the category of $\mathfrak{g}$ modules is equivalent to the category of $U(\mathfrak{g})$ modules. Finally, observe that the gradation on $\mathcal{T}(V)$ induces a filtration $\mathcal{F}^n(U(\mathfrak{g})) := Im(k \otimes \mathfrak{g} \oplus \ldots \oplus \mathfrak{g}^{\otimes n}) : n \in \mathbb{N}$ on $U(\mathfrak{g})$. By the PBW theorem the associated graded algebra $gr_{\mathcal{F}} U(\mathfrak{g})$ is just the symmetric algebra $S(\mathfrak{g})$ over $\mathfrak{g}$. One calls $\mathcal{F}$ the canonical filtration of $U(\mathfrak{g})$.

An important and not quite trivial observation is that the Hopf structure on $\mathcal{T}(\mathfrak{g})$ given in 1.2.5 passes to $U(\mathfrak{g})$. Indeed in $\mathcal{T}(\mathfrak{g})$ one has

$$\Delta(xy) = (x \otimes 1 + 1 \otimes x)(y \otimes 1 + 1 \otimes y) = xy \otimes 1 + x \otimes y + y \otimes x + 1 \otimes xy$$

which implies that

$$\Delta(xy - yx - [x,y]) = (xy - yx - [x,y]) \otimes 1 + 1 \otimes (xy - yx - [x,y]) .$$

Consequently $\ker(\mathcal{T}(\mathfrak{g}) \longrightarrow U(\mathfrak{g}))$ is also a coideal. It is σ stable via the antisymmetry of the Lie product (and of the commutator).

1.2.7 It is worth noting that the examples of 1.2.1, 1.2.6 give not only Hopf algebras; but also Hopf algebra maps. Indeed a homomorphism $H \longrightarrow G$ (resp. $\mathfrak{h} \longrightarrow \mathfrak{g}$) of groups (resp. Lie algebras) yields a Hopf algebra map $kH \longrightarrow kG$ (resp. $U(\mathfrak{h}) \longrightarrow U(\mathfrak{g})$). In particular a representation $\mathfrak{g} \longrightarrow \mathfrak{gl}(V)$ gives a Hopf algebra map $U(\mathfrak{g}) \longrightarrow U(\mathfrak{gl}(V))$. This construction is less obvious for example 1.2.2 since one must ensure that both maps $H \longrightarrow G$ and $R[G] \longrightarrow R[H]$ are defined. The analogous requirement for quantized enveloping algebras (or for "quantum groups"), is even more difficult to satisfy (10.4.5).

1.2.8 Let G be an algebraic group. Then the tangent space T_e to the identity $e \in G$ has a Lie algebra structure and is denoted $\operatorname{Lie} G$. Set $\mathfrak{m} = \ker \eta^*$, where as in 1.2.2, 1.2.3, η^* is just the coidentity of $R[G]$. Then T_e identifies with $(\mathfrak{m}/\mathfrak{m}^2)^*$. More generally let A be any Hopf algebra and set $\mathfrak{m} = \ker \varepsilon$. Then $(\mathfrak{m}/\mathfrak{m}^2)^*$ has a Lie algebra structure. Moreover there is a canonical map $U((\mathfrak{m}/\mathfrak{m}^2)^*) \longrightarrow A^*$ which is injective if $\operatorname{char} k = 0$. If $A = R[G]$, G is irreducible and $\operatorname{char} k = 0$, then $U(\mathfrak{m}/\mathfrak{m}^2)^*)$ is separating; but in general this fails. This is studied in section 2.1.

1.2.9 Given a Lie algebra $\mathfrak{g}$ then one may construct from $U(\mathfrak{g})^*$ a Hopf algebra isomorphic to $R[G]$ for some irreducible algebraic group G. Ideally $\mathfrak{g}$ is recovered from $R[G]$ as in 1.2.8. Rather one obtains a finite dimensional image of $\mathfrak{g}$ occurring as an ideal of T_e. If $\mathfrak{g}$ is finite dimensional, then $R[G]$ can be chosen so that the map $\mathfrak{g} \longrightarrow T_e$ is injective and the latter is called the algebraic hull of $\mathfrak{g}$ relative to $R[G]$. This is studied in section 2.2.

1.2.10 Let A be an algebra (not necessarily associative). A derivation ∂ of A is a linear map satisfying the classical Leibnitz rule $\partial(ab) = (\partial a)b + a(\partial b)$. One easily checks that the set $\operatorname{Der} A$ of derivations of A is a Lie subalgebra of $\mathfrak{gl}(A)$. If the product rule $(x \otimes y) \mapsto [x,y]$ on A is antisymmetric then left multiplication $\ell_x : y \mapsto [x,y]$ is a derivation if and only if $[x,y]$ satisfies the Jacobi identity. In this case A is a Lie algebra and a further use of the Jacobi identity shows that the map $x \mapsto \ell_x$ is a Lie algebra homomorphism of A into $\operatorname{Der} A$ whose image is denoted $\operatorname{Int} A$, the set of all interior derivations of A. One may check that $\operatorname{Int} A$ is an ideal of $\operatorname{Der} A$.

It is clear that any derivation ∂ of A extends uniquely to a derivation, also denoted ∂, of $\mathcal{T}(A)$. For a ∂ stable ideal I of A, passage to A/I defines a derivation of A/I. If $\mathfrak{g}$ is a Lie algebra then this construction extends any $\partial \in Der(\mathfrak{g})$ to a derivation of $U(\mathfrak{g})$ leaving each $\mathcal{F}^n(U(\mathfrak{g}))$ stable.

A derivation ∂ of A is said to be locally finite (resp. nilpotent) if for all $a \in A$ one has $\partial^n a \in \sum_{m<n} k\partial^m a$ (resp. $\partial^n a = 0$) for n sufficiently large. For example if $\mathfrak{g}$ is a finite dimensional Lie algebra, then any $\partial \in \operatorname{Der} \mathfrak{g}$ extends to a locally finite derivation of $U(\mathfrak{g})$ since then a belongs to some $\mathcal{F}^n(U(\mathfrak{g}))$ which is finite dimensional and ∂ stable. When ∂ is locally nilpotent, $\exp \partial = 1+\partial+\partial^2/2!+\dots$, is defined as a linear map on A and a standard calculation shows that $\exp \partial$ is an automorphism of A. One may also note that if ∂ is locally finite and if k is a complete field, then $\exp \partial$ is defined on A and is an automorphism.

It is clear that a derivation ∂ of A is uniquely determined by its values $\partial a_i \in A$ on the generators $\{a_i\}_{i \in \mathcal{I}}$ of A. If these are free generators then the ∂a_i may be arbitrary. Otherwise they must be chosen so that $\partial r_j \in R := \langle r_j : j \in \mathcal{J} \rangle$ for each relation r_j. This holds trivially if the relations are all the commutators, that is if A is the polynomial algebra $k[a_i : i \in \mathcal{I}]$. We conclude that in this case $\operatorname{Der} A$ is the free A module on generators $\partial_i : i \in \mathcal{I}$ defined through $\partial_i a_j = \delta_{ij}$ (where δ is the Kronecker delta). As elements of $\operatorname{End} A$ the ∂_i, a_j satisfy exactly the relations

$$a_i a_j - a_j a_i = 0 \ , \quad \partial_i a_j - a_j \partial_i = \delta_{ij} \ , \quad \partial_i \partial_j - \partial_j \partial_i = 0 \ , \ \forall i, j \in \mathcal{I} \ .$$

When $|\mathcal{I}| = n < \infty$ this defines the nth Weyl algebra A_n which may also be viewed as the algebra of differential operators on affine n space. If G is an algebraic group, then the canonical map (1.2.8) gives an embedding $\operatorname{Lie} G \longrightarrow \operatorname{Der} R[G]$.

1.2.11 As the discussion in 1.2.10 indicates, algebraic groups are intimately related to Lie subalgebras of derivations of a commutative algebra. Our approach to quantum groups proceeds via a study of skew derivations of a not necessarily commutative algebra. In this let A be an algebra and $\theta \in Aut\ A$. A θ-leftskew (resp. θ-rightskew) derivation ∂ of A is a linear map satisfying the Leibnitz rule $\partial(ab) = (\partial a)b + \theta(a)\partial b$ (resp. $\partial(ab) = (\partial a)\theta(b) + a\partial b$). Either of these two properties is preserved under conjugation $\partial \mapsto \theta\partial\theta^{-1}$ and so it makes sense to further impose that $\theta\partial\theta^{-1} \in k^*\partial$. One may further verify that θ, ∂ generate a subalgebra B of $\operatorname{End} A$ which is equipped with a coproduct Δ_ℓ (resp. Δ_r) satisfying $\Delta_\ell(\theta) = \theta \otimes \theta$ and $\Delta_\ell(\partial) = \partial \otimes 1 + \theta \otimes \partial$ (resp. $\Delta_r(\partial) = \partial \otimes \theta + 1 \otimes \partial$) making A a B-algebra. Then $\Delta_\ell^{\mathrm{opp}} = \Delta_r \neq \Delta_\ell$. In particular B^* is not commutative and nor is B unless $\theta\partial = \partial\theta$. Thus we have a remarkably simple example of a bialgebra which is neither commutative nor cocommutative which combines a group-like element θ with an almost Lie algebra-like element ∂. As we shall see B is very nearly self-dual. It is the precursor of almost all quantized enveloping algebras (or their duals) that have been studied.

1.2.12 Let ∂ be a θ-leftskew derivation of an algebra A. One should expect a quantum group element to be some sort of exponential of ∂. The study of this has at least some interesting combinatorial aspects involving q-binomial coefficients. Indeed take A to be the polynomial algebra over $k(q)$, that is $A = k(q)[a]$, and define $\theta \in Aut\ A$ by $\theta(a) = q^2 a$. Define ∂ by setting $\partial a = 1$ and $\theta\partial\theta^{-1} = q^{-2}\partial$. An easy induction argument gives

$$\begin{aligned}\partial a^n &= (1 + q^2 + q^4 + \ldots + q^{2(n-1)})a^{n-1} \\ &= \left(\frac{1-q^{2n}}{1-q^2}\right) a^{n-1} = q^{n-1}\left(\frac{q^n - q^{-n}}{q - q^{-1}}\right) a^{n-1} .\end{aligned}$$

Set

$$[n]_q = \frac{q^n - q^{-n}}{q - q^{-1}} , \ [n!]_q = [n]_q[n-1]_q \ldots [1]_q , \ [0!]_q = 1$$

and

$$\begin{bmatrix} n \\ m \end{bmatrix}_q = \frac{[n!]_q}{[(n-m)!]_q [m!]_q} , \quad \text{for } 0 \leq m \leq n .$$

Then

$$\partial^m a^n = q^{mn - \frac{1}{2}m(m+1)} \frac{[n!]_q}{[(n-m)!]_q} a^{n-m} . \tag{1}$$

Thus ∂ is a natural q-analogue of differentiation. However when q is an nth root of unity, a new feature arises which parallels the case char $k > 0$, namely one has $\partial a^n = 0$. For this reason the representation theory of quantized enveloping algebras parallels the positive characteristic case. However this feature is not within the scope of our present text. Rather q is viewed as an indeterminate or as not being a root of unity other than 1.

1.2.13 The q-binomial coefficients satisfy the identity

$$q^m \begin{bmatrix} n \\ m \end{bmatrix}_q + q^{m-n-1} \begin{bmatrix} n \\ m-1 \end{bmatrix}_q = \begin{bmatrix} n+1 \\ m \end{bmatrix}_q , \quad \text{for all } n \geq m \geq 0 \tag{1}$$

which generalizes the Pascal triangle property for the ordinary binomial coefficients $\binom{n}{m}$. It leads naturally to two binomial theorems. The first concerns commuting indeterminates a, b and asserts that

$$\prod_{i=1}^{n} (a + q^{2(i-1)}b) = \sum_{m=0}^{n} \begin{bmatrix} n \\ m \end{bmatrix}_q a^{n-m}(q^{n-1}b)^m, \quad \text{for all } n \in \mathbb{N} . \tag{2}$$

The second concerns an algebra A containing elements x, y satisfying the relation $q^2 xy = yx$. In this case one has the identity

$$(x+y)^n = \sum_{m=0}^{n} \begin{bmatrix} n \\ m \end{bmatrix}_q q^{(n-m)m} x^{n-m} y^n . \tag{3}$$

1.2.14 There is more than one way to generalize the exponential of a derivation. However the most useful seems to be the following. Let A be an associative algebra over $k(q)$ and $\theta \in Aut\, A$. Let ∂ be a θ-leftskew derivation of A satisfying $\theta\partial\theta^{-1} = q^{-2}\partial$. Using 1.2.13 (1) an induction argument gives the identity

$$(1) \qquad \partial^n(fg) = \sum_{m=0}^{n} q^{-n(n-m)} \begin{bmatrix} n \\ m \end{bmatrix}_q \partial^m(\theta^{n-m} f)(\partial^{n-m} g)$$

for all $n \in \mathbb{N}$, $f, g \in A$. Now define the q-exponential $\exp_q \partial$ as the formal sum

$$\exp_q \partial = \sum_{n=0}^{\infty} \frac{q^{\frac{1}{2}n(n+1)}}{[n!]_q} \partial^n .$$

Under conditions similar to those discussed in 1.2.10 this defines a linear transformation of A. However it is not an automorphism. Rather, assuming that $f \in A$ satisfies $\theta(f) = q^s f$ for some $s \in \mathbb{Z}$, one obtains from (1) the identity

$$(2) \qquad (\exp_q \partial)(fg) = (\exp_q \partial) f (\exp_q q^s \partial) g , \quad \text{for all} \quad g \in A .$$

Thus the weight, namely s, of f modifies the anticipated automorphism property of $\exp_q \partial$.

A particularly interesting application of (2) occurs when A is the polynomial ring $k(q)[a]$, θ the automorphism defined by $\theta(a) = q^2 a$ and ∂ the θ-leftskew derivation defined by $\partial a = 1$, that is as in 1.2.12. Then the exponential of the ordinary derivation $\partial/\partial a$ is just translation and so sends a^n to $(a+1)^n$, whilst from (2) we obtain

$$(3) \qquad (\exp_q \partial) a^n = \prod_{i=1}^{n} (a + q^{2(i-1)}) .$$

This can be interpreted as translation with a spread and is therefore naturally a "quantum phenomenon". In any case whatever its philosophical meaning this identity has some useful computational value in deriving q-analogues of binomial identities. One example is given in A.3.1, where it recovers an identity needed in deriving the Lusztig-Soibelman automorphisms (10.2). Such identities are implicit in the representation theory of quantum groups. Ideally the machinery one develops should be sufficiently powerful so that such identities which are generally cumbersome, need never be explicitly used.

1.3 Adjoint Action

Throughout this section and the next it is assumed that A is a Hopf algebra over k with $(\mu, \eta, \Delta, \varepsilon, \sigma)$ the linear maps defined as in 1.1.

1.3.1 Given $a \in A$ define $\operatorname{ad} a \in \operatorname{End} A$ through $(\operatorname{ad} a)b = a_1 b \sigma(a_2)$ where $\Delta(a) = a_1 \otimes a_2$. By 1.1.4 and 1.1.9 the map $a \mapsto \operatorname{ad} a$ of A into $\operatorname{End} A$ is an algebra homomorphism. Thus *ad* defines an action of A on itself called the (left) adjoint action. It is sometimes useful to also consider the right adjoint action *Ad* defined by $(\operatorname{Ad} a)b = \sigma(a_1)ba_2$.

The adjoint action satisfies the Leibnitz rule, making A an A-algebra. That is

$$(1) \qquad (\operatorname{ad} a)bc = (\operatorname{ad} a_1)b(\operatorname{ad} a_2)c\ , \quad \forall\, a, b, c \in A\ .$$

Indeed $(\operatorname{ad} a)bc = a_1 bc\sigma(a_2) = a_1 \varepsilon(a_2) bc\sigma(a_3) = a_1 b \varepsilon(a_2) c\sigma(a_3)$ $= a_1 b\sigma(a_2)a_3 c\sigma(a_4) = (\operatorname{ad} a_1)b(\operatorname{ad} a_2)c$ illustrating nicely the computational rules indicated in 1.1.8. Now define $F(A)$ to be the set of elements of A on which the adjoint action is locally finite, that is

$$(2) \qquad F(A) := \{a \in A \mid \dim(\operatorname{ad} A)a < \infty\}\ .$$

It follows from (1) that $F(A)$ is a subalgebra of A.

1.3.2 We digress slightly to discuss the structure of End A as a *convolution algebra*. From the actions defined in 1.1.6 and 1.1.9 it follows that the natural embedding $A^* = \operatorname{Hom}(A, 1) \hookrightarrow \operatorname{Hom}(A, A)$ is an algebra homomorphism. Now for each $b \in A$ define $\varphi(b) \in \operatorname{End} A$ by $\varphi(b)(a) = (\operatorname{ad} a)b$. Then 1.3.1 (1) translates to give $\varphi(bc)(a) = (\varphi(b)_* \varphi(c))(a)$ and so $\varphi : b \mapsto \varphi(b)$ is an algebra homomorphism of A into $\operatorname{End} A$ which is injective since $\varphi(b)(1) = b$. Now let ψ be the linear map of $A^* \otimes A$ into $\operatorname{End} A$ defined by $\psi(\xi \otimes b) = \xi_* \varphi(b)$.

Lemma. *If either σ is injective or surjective, then ψ is injective.*

Suppose $\sum_i \xi_i \otimes b_i \in \ker \psi$. Take $a \in A$ and write $\Delta(a) = a_1 \otimes a_2$. Then substitution from 1.1.8 gives $\sum_i \xi_i(a_1)a_2 b_i = \sum_i \xi_i(a_1)a_2 b_i \varepsilon(a_3) =$ $\sum_i \xi_i(a_1)a_2 b_i \sigma(a_3)a_4 = \sum_i \xi_i(a_1)((\operatorname{ad} a_2)b_i)a_3 = (\sum_i (\xi_i * b_i)(a_1))a_2 = 0$. Applying σ and using 1.1.9 (iii) gives $\sum_i \xi_i(a_1)\sigma(b_i)\sigma(a_2) = 0$, which by a further reduction using 1.1.8 implies that $\sum_i \xi_i(a)\sigma(b_i) = 0$ for all $a \in A$. Injectivity of σ implies that $\sum_i \xi_i(a)b_i = 0$ for all a and hence that $\sum_i \xi_i \otimes b_i = 0$. Alternatively replacing a by $\sigma(a)$ in the first relation gives $\sum_i \xi_i(\sigma(a)_1)\sigma(a)_2 b_i = 0$. Then by 1.1.8 this can be expressed as $\sum_i \sigma(a_1)\xi_i(\sigma(a_2))b_i = 0$, which by 1.1.8 implies $\sum_i \xi_i(\sigma(a))b_i = 0$. Thus the injectivity of ψ also results from the surjectivity of σ.

Remark. In particular ψ is bijective if $A = kG$ with G a finite group. Moreover $\xi_* \varphi(b) = \varphi(b)_* \xi$ for all $\xi \in (kG)^* = \mathcal{F}(G)$, $b \in kG$. Thus the convolution algebra $\operatorname{End} kG$ is just the tensor product $\mathcal{F}(G) \otimes kG$. More generally if A is cocommutative, then $A^* \otimes A$ identifies with a subalgebra of the convolution algebra $\operatorname{End} A$.

1.3.3 Let $Z(A)$ denote the centre of A.

Lemma. $Z(A) = \{z \in A | (\operatorname{ad} a)z = \varepsilon(a)z, \ \forall a \in A\}$.

If $z \in Z(A)$ then $(\operatorname{ad} a)z = a_1 z\sigma(a_2) = a_1\sigma(a_2)z = \varepsilon(a)z$ by 1.1.8. Conversely $za = z\varepsilon(a_1)a_2 = \varepsilon(a_1)za_2 = ((\operatorname{ad} a_1)z)a_2 = a_1 z\sigma(a_2)a_3 = a_1 z\varepsilon(a_2) = a_1\varepsilon(a_2)z = az$ as required.

1.3.4 Let $A^\wedge$ denote the set of linear characters on A, equivalently the set of algebra homomorphisms of A into k satisfying $\chi(1) = 1$. For the convolution product on A^* it is a multiplicitively closed subset with identity ε. Moreover for each $\chi \in A^\wedge$ one has $\varepsilon(a) = \chi(\varepsilon(a)) = \chi(a_1)\chi(\sigma(a_2)) = (\chi_* \chi\sigma)(a)$ for all $a \in A$. This and a similar result show that χ has inverse $\chi\sigma$. In short convolution gives $A^\wedge$ a multiplicative group structure.

For each $\chi \in A^\wedge$ define $\theta_\chi \in \operatorname{End} A$ through $\theta_\chi(a) = \mu(1 \otimes \chi)\Delta(a) = a_1\chi(a_2)$. By 1.1.3, θ_χ is a homomorphism. Moreover given $\chi, \chi' \in A^\wedge$ then

$$\theta_{\chi'}(\theta_\chi(a)) = a_1\chi'(a_2)\chi(a_3) = a_1(\chi'\chi)(a_2) = \theta_{\chi'\chi}(a) \ .$$

Thus the map $\chi \mapsto \theta_\chi$ is a representation of $A^\wedge$. In particular θ_χ is an isomorphism with inverse $\theta_{\chi^{-1}}$. It is called a left winding automorphism of A. Consider moreover the canonical map $\varphi : A \longrightarrow (kA^\wedge)^*$ defined by $\varphi(a)(\chi) = \chi(a)$. Recall (1.1.6) the left $A^\wedge$ module structure on $(kA^\wedge)^*$ defined by $(\chi.\xi)(\chi') = \xi(\chi'\chi)$. Take $a \in A$ and $\xi = \varphi(a)$. Then $\varphi(a)(\chi'\chi) = (\chi'\chi)(a) = (\chi' \otimes \chi)\Delta(a) = \chi'(a_1)\chi(a_2)$ and so $\chi.\varphi(a) = \varphi(a_1\chi(a_2)) = \varphi(\theta_\chi(a))$, that is φ is a homomorphism of left $A^\wedge$ modules. Similarly φ is homomorphism of right $A^\wedge$ modules for the right winding automorphism $a \mapsto \mu(\chi \otimes 1)\Delta(a)$.

Suppose that $A = U(\mathfrak{g})$ for some Lie algebra $\mathfrak{g}$. Then $A^\wedge$ identifies with the orthogonal $[\mathfrak{g}, \ \mathfrak{g}]^\perp$ of $[\mathfrak{g}, \ \mathfrak{g}]$ in $\mathfrak{g}^*$. Moreover for each $\lambda \in [\mathfrak{g}, \ \mathfrak{g}]^\perp$ the winding automorphism θ_λ is defined on $\mathfrak{g}$ through $\theta_\lambda(x) = x + \lambda(x)$.

Each $\chi \in A^\wedge$ defines a one dimensional A module M_χ. It can happen that M_χ occurs as a submodule of A under the adjoint action, that is there exists $b \in A$ non-zero such that $(\operatorname{ad} a)b = \chi(a)b$ for all $a \in A$. If $M_{\chi'}$ occurs as the submodule kc of A, then by 1.3.1 (1) it follows that $(\operatorname{ad} a)bc = (\chi\chi')(a)bc$. Thus if A is an integral domain, the subset of $\chi \in A^\wedge$ for which M_χ occurs in A, is a semigroup. The action of the corresponding winding automorphisms on $\operatorname{Spec} A$ leads to the description of the latter as vertices of a directed graph. This situation has been particularly well-studied when $A = U(\mathfrak{g})$ with $\mathfrak{g}$ solvable and gives information on the localization theory at prime ideals.

1.3.5 In general $F(A)$ fails to be a Hopf (or even bi-) subalgebra of A. However it is often a left coideal. This will follow from the

Lemma. *Suppose $I \subset A$ is a left coideal, then so is* $(\operatorname{ad} A)I$.

Take $a \in A$, $b \in I$. It is convenient to make a slight modification in our summation convention writing $\Delta(b) = b' \otimes b''$. Then by 1.1.8 and 1.1.10

$$\begin{aligned}\Delta((\operatorname{ad} a)b) &= \Delta(a_1 b\sigma(a_2)) \\ &= (a_1 \otimes a_2)(b' \otimes b'')(\sigma(a_4) \otimes \sigma(a_3)) \\ &= a_1 b'\sigma(a_4) \otimes a_2 b''\sigma(a_3) \\ &= a_1 b'\sigma(a_3) \otimes (\operatorname{ad} a_2)b'' \ .\end{aligned}$$

Since $b'' \in I$ by hypothesis, the required result is obtained.

Remarks. If A is cocommutative, then one can interchange 2, 3 in the above to obtain $\Delta((\operatorname{ad} a)b) = (\operatorname{ad} a_1)b' \otimes (\operatorname{ad} a_2)b''$. Thus "left" can be replaced by "right" in the lemma. In many cases one can write $F(A)$ as $(\operatorname{ad} A)B$ where B is a sub-bialgebra of A.

1.3.6 Suppose $A = U(\mathfrak{g})$ with $\mathfrak{g}$ a finite dimensional Lie algebra. It is immediate consequence of the invariance of the canonical filtration (1.2.6) under adjoint action (1.2.10) that $F(A) = A$. This also holds (trivially) if $A = kG$ with G a finite group or if A is commutative. More interestingly for a quantized enveloping algebra, this equality fails but only slightly. However for the Hopf dual of such an algebra, $F(A)$ is very small. This difficulty can be overcome by twisting the adjoint action in the sense below.

Given $\tau \in Aut\ A$ define $ad_\tau a \in \operatorname{End} A$ through $(ad_\tau a)b = \tau(a_1)b\sigma(a_2)$, where $\Delta(a) = a_1 \otimes a_2$. As before $a \mapsto ad_\tau a$ is an algebra homomorphism of A into $\operatorname{End} A$. However the Leibnitz rule becomes rather unwieldly. Indeed given $\tau,\ \tau' \in Aut\ A$ one finds that

$$(ad_{\tau\tau'} a)(bc) = (\tau(ad_{\tau'} a_1)\tau^{-1} b)(ad_\tau a_2)c, \quad \forall b, c \in A \ .$$

However let us assume that $\tau,\ \tau'$ commute and that there exists an automorphism $\hat{\tau}$ of A such that

(2) $$\tau(1 \otimes \sigma)\Delta = (1 \otimes \sigma)\Delta\hat{\tau} \ .$$

Then

(3) $$\tau(ad_{\tau'} a)\tau^{-1} b = (ad_{\tau'} \hat{\tau}(a))b$$

and so (1) becomes

(4) $$(ad_{\tau\tau'} a)(bc) = (ad_{\tau'} \hat{\tau}(a_1)b)(ad_\tau a_2)c$$

which is better. Indeed define the linear subspace

$$F_\tau(A) = \{a \in A \mid \dim(ad_\tau A)a < \infty\} \ .$$

Then by (4) one has $F_\tau(A)F_{\tau'}(A) \subset F_{\tau\tau'}(A)$. Notice however one need not have $1 \in F_\tau(A)$. Indeed this would imply $F_{\tau'}(A) \subset F_{\tau\tau'}(A)$ and in particular if also $1 \in F_{\tau'}(A)$, then $1 \in F_{\tau\tau'}(A)$. Thus it is natural to consider a commutative semigroup $\mathcal{T}$ of automorphisms of A, satisfying (2) and $1 \in F_\tau(A)$ for all $\tau \in \mathcal{T}$, viewed as a directed set through $\tau, \tau' \leq \tau\tau'$ for all $\tau, \tau' \in \mathcal{T}$. Then $F_\tau(A) \subset F_{\tau'}(A)$ whenever $\tau \leq \tau'$ and one may form the direct limit

$$\mathcal{T}F(A) := \varinjlim \{F_\tau(A) \mid \tau \in \mathcal{T}\}$$

which is a subalgebra of A filtered by the $F_\tau(A)$. For A the algebra of functions on a quantum group, it will be seen (9.3.18) that $\mathcal{T}$ can be chosen so that $\mathcal{T}F(A)$ is a finite module over A and even a left coideal.

1.3.7 In the study of the prime spectrum (A.2.6) of a Hopf algebra A, one often first obtains information on the ad-invariant prime ideals of $F(A)$ (or of some $\mathcal{T}F(A)$). Transfering this information to $\operatorname{Spec} A$ requires the formalism, analogous to a Galois theory, described below.

1.3.8 Let A be graded by a finite additive group Γ whose identity is denoted by 0 and with A_γ the graded subspace corresponding to $\gamma \in \Gamma$. Assume

(i) $1 \in A_\gamma A_{-\gamma}\,, \quad \forall \gamma \in \Gamma$.

This implies $\varepsilon(A_\gamma) \neq 0$ and for all $\gamma, \gamma' \in \Gamma$ that $A_\gamma A_{\gamma'} = A_{\gamma+\gamma'}$. It follows that each graded left (or right) ideal is generated by its intersection with A_0. Assume that

(ii) A_γ *is* $\operatorname{ad} A$ *invariant,* $\forall \gamma \in \Gamma$

and that

(iii) A_γ *is a left coideal,* $\forall \gamma \in \Gamma$.

In particular A_0 is $\operatorname{ad} A$ invariant and so any two-sided ideal I of A intersecting A_0 is an $\operatorname{ad} A$ invariant two-sided ideal of A_0. Conversely, let J be an $\operatorname{ad} A$ invariant subspace of A_0. Then for all $a \in A$, $b \in J$ one has $ab = a_1\varepsilon(a_2)b = a_1 b\varepsilon(a_2) = a_1 b\sigma(a_2)a_3 = ((\operatorname{ad} a_1)b)a_2 \in JA$ and so JA is a two-sided ideal of A.

Identify the character group $\Gamma^\wedge$ of Γ with Γ. For each $\gamma \in \Gamma^\wedge$ define a linear map χ_γ on A by $\chi_\gamma(a) = \gamma(\gamma')\varepsilon(a)$, for all $a \in A_{\gamma'}$. It is immediate (1.3.4) that $\chi_\gamma \in A^\wedge$. Moreover for all $a \in A_{\gamma''}$ it follows from (iii) that $(\chi_\gamma * \chi_{\gamma'})(a) = \chi_\gamma(a_1)\chi_{\gamma'}(a_2) = \chi_\gamma(a_1)\gamma'(\gamma'')\varepsilon(a_2) = \chi_\gamma(a)\gamma'(\gamma'') = \gamma(\gamma'')\gamma'(\gamma'')\varepsilon(a) = \chi_{\gamma\gamma'}(a)$. Thus the map $\gamma \mapsto \chi_\gamma$ is a homomorphism of $\Gamma^\wedge$ into $A^\wedge$. It is injective by (i).

For each $\gamma \in \Gamma$, choose a linear combination e_γ of the $\gamma'' \in \Gamma^\wedge$ such that $e_\gamma(\gamma') = \delta_{\gamma,\gamma'}$. Let e_γ also denote the corresponding linear combination of the χ_γ. Then for all $a \in A_{\gamma'}$, we have $e_\gamma(a) = \delta_{\gamma,\gamma'}\varepsilon(a)$. In the notation of 1.3.4 let θ_{e_γ} denote the corresponding linear combination of θ_{χ_γ}. By (iii) it follows that θ_{e_γ} is just the projection of A onto A_γ defined by the gradation. Hence the

Lemma. *Under the above hypotheses*
(i) $A_0 = \{a \in A \mid \chi_\gamma(a) = \varepsilon(a), \quad \forall \gamma \in \Gamma^\wedge\}$.
(ii) *A subspace V of A is graded if and only if $\theta_{\chi_\gamma}(V) \subset V$ for all $\gamma \in \Gamma^\wedge$.*

1.3.9 Retain the hypotheses of 1.3.8 and further assume that A is left noetherian here and in the remainder of section 1.3. Since $AI \cap A_0 = I$ for any left ideal I of A_0 , this implies that A_0 is also left noetherian. View Γ as acting on ideals of A via the $\theta_{\chi_\gamma} : \gamma \in \Gamma$ and hence on Spec A.

Lemma

(i) *If P is a semiprime* ad A *invariant ideal of A_0 then PA is a semiprime ideal of A.*

(ii) *If $P \in$* Spec A_0 *and is* ad A *invariant, then the minimal primes Q_i over PA form a single Γ orbit and satisfy $P = A_0 \cap Q_i$, $\forall i$.*

(i) Recall that PA is a two-sided graded ideal of A and hence Γ invariant by 1.3.8 (ii). Its radical $\sqrt{PA}$ is hence also Γ invariant. By noetherianity there exists $n \in \mathbb{N}$ such that $(\sqrt{PA})^n \subset PA$ and so $(\sqrt{PA}\cap A_0)^n \subset (\sqrt{PA})^n \cap A_0 \subset PA \cap A_0 = P$. By the semiprimeness of P this implies that $\sqrt{PA} \cap A_0 = P$. Yet $\sqrt{PA}$ is graded by 1.3.8 (ii), hence generated by its intersection with A_0. Thus $\sqrt{PA} = (\sqrt{PA} \cap A_0)A = PA$, as required.

(ii) By (i) and noetherianity, PA is a finite intersection $\cap Q_i$ of the primes Q_i minimal over PA. Obviously $Q_i \cap A_0 \supset P$, for all i. Since $PA \cap A_0 = P$ we obtain $\cap(Q_i \cap A_0) = P$, which since P is prime implies that $Q_j \cap A_0 = P$ for some j. Let K be the intersection of the Γ translates of Q_j. Then K is Γ invariant and so $K = (K \cap A_0)A = (Q_j \cap A_0)A = PA$. Finally it is clear that $\theta_{\chi_\gamma}(Q_j)$ is minimal over PA and has intersection P with A_0. All this proves (ii).

1.3.10 Let I be a two-sided ideal of A_0 and set $\tau_\gamma(I) = A_\gamma I A_{-\gamma}$ which is a two-sided ideal of A_0 and equal to I if $\gamma = 0$. By 1.3.8 (i), one has $\tau_\gamma(\tau_{\gamma'}(I)) = \tau_{\gamma+\gamma'}(I)$. Again if I, J are two-sided ideals of A_0, then $\tau_\gamma(I)\tau_\gamma(J) = \tau_\gamma(IJ)$. It follows that τ_γ maps prime ideals to prime ideals. This defines an action of Γ on Spec A_0. Finally $\tau_\gamma(I) \subset I$ if and only if $A_\gamma I \subset IA_\gamma$.

Lemma. *Let Q be a semiprime ideal of A. Then $P_0 = Q \cap A_0$ is a semiprime ideal of A_0.*

It is clear that $\tau_\gamma(P_0) \subset P_0$ for all $\gamma \in \Gamma$ and even that equality holds. Then $(\tau_\gamma(\sqrt{P_0}))^n = \tau_\gamma((\sqrt{P_0})^n) \subset \tau_\gamma(P_0) \subset P_0$ for n sufficiently large and so $\tau_\gamma(\sqrt{P_0}) \subset \sqrt{P_0}$, for all $\gamma \in \Gamma$, and even equality holds. Thus $\sqrt{P_0}A = A\sqrt{P_0}$ which is hence a two-sided ideal of A. Moreover $(\sqrt{P_0}A)^n = (\sqrt{P_0})^n A \subset P_0 A$, for n sufficiently large. Thus $\sqrt{P_0}A \subset \sqrt{P_0 A} \subset \sqrt{Q} = Q$. In particular $\sqrt{P_0} \subset Q \cap A_0 = P_0$ as required.

1.3.11 Lemma. *Suppose Q is a prime ideal of A. Then $Q \cap A_0 = \bigcap_{\gamma\in\Gamma} \tau_\gamma(P)$ for some prime ideal P of A_0.*

By 1.3.10 we can write

$$Q \cap A_0 = \bigcap_{i=1}^{n} I_i \, , \quad \text{where } I_i = \bigcap_{\gamma \in \Gamma} \tau_\gamma(P_i) \text{ with } P_i \in \operatorname{Spec} A_0 \, .$$

Since $\tau_\gamma(I_i) \subset I_i$ for $\gamma \in \Gamma$ it follows that $I_i A = AI_i$. Then $\Pi_{i=1}^n (I_i A) = (\Pi_{i=1}^n I_i)A \subset (Q \cap A_0)A \subset Q$. Since Q is prime, this implies that $I_i A \subset Q$ for some i. Then $I_i \subset Q \cap A \subset I_i$, as required.

1.3.12 Lemma. *Take $P \in \operatorname{Spec} A_0$ and set $I = \cap_{\gamma \in \Gamma} \tau_\gamma(P)$. Then the primes Q_i minimal over IA form a single Γ orbit and satisfy $I = A_0 \cap Q_i$.*

By 1.3.9 (i) we have $\bigcap_i (Q_i \cap A_0) = (\bigcap_i Q_i) \cap A_0 = IA \cap A_0 = I$. Applying 1.3.11 to each Q_i implies that $Q_j \cap A_0 = I$ for some j. Conclude as in 1.3.9 (ii).

1.3.13 Combining 1.3.11, 1.3.12 gives a result which is slightly more complicated than that obtained when $A = K \otimes_k A_0$ for K a Galois extension of the base field, namely,

Theorem. *The map taking $\{\tau_\gamma(P)\}_{\gamma \in \Gamma}$ to the minimal primes over $(\bigcap_{\gamma \in \Gamma} \tau_\gamma(P))A$ is a bijection of the Γ orbits of* Spec A_0 *onto the Γ orbits of* Spec A.

1.3.14 Let I be an ad A invariant right ideal of A. Then (cf. 1.3.8) one has $A_\gamma I A_{-\gamma} \subset A_0 \cap IA = I$.

In many cases all two-sided ideals of A_0 are ad A invariant. Under this additional hypothesis Spec A_0 is in bijection with Γ orbits in Spec A via the map in theorem 1.3.13.

1.4 The Hopf Dual

Let A be a Hopf algebra. As pointed out in 1.1.5 the full algebraic dual A^* of A inherits an algebra structure; but, unless A is finite dimensional, it fails to inherit a coalgebra structure. Nevertheless, there is a canonical way to construct a subalgebra of A^* which does have this property and indeed is a Hopf algebra. It is called the Hopf dual of A and denoted $A^\star$. This is defined and studied below. Here the antipode σ is needed only to give $A^\star$ an antipode, namely σ^*.

1.4.1 Let $\mathcal{I}$ denote the set of all two-sided ideals of A of finite codimension. The most direct definition of the Hopf dual $A^\star$ is

$$A^\star := \{\xi \in A^* | \xi(I) = 0 \text{ for some } I \in \mathcal{I}\} \, .$$

Notice that if $\xi \in A^*$ vanishes on $I \in \mathcal{I}$, then with respect to the bimodule structure of A^*, it follows that $A.\xi.A$ vanishes on I and hence has dimension $\leq \operatorname{codim} I < \infty$. Conversely for any $\xi \in A^*$ the set $I = \{a \in A \mid (A.\xi.A)(a) = 0\}$ is a two-sided ideal of A and the map $(\eta,\ a) \mapsto \eta(a)$ of $A^* \times A \longrightarrow k$ factors to a non-degenerate pairing $A.\xi.A \times A/I \longrightarrow k$.

Consequently $\operatorname{codim} I = \dim A.\xi.A$ if the latter is finite. Again by the generalized Leibnitz rule (1.1.6) one has $A.\xi\eta.A \subset (A.\xi.A)(A.\eta.A)$ for all $\xi, \eta \in A^*$. All this proves the

Lemma

(i) $A^\star = \{\xi \in A^* \mid \dim A.\xi.A < \infty\}$.
(ii) $A^\star$ *is a subalgebra of* A^*.

1.4.2 The following result shows that $A^\star$ is a bialgebra and that it is the best possible choice.

Lemma. *Let* V *be a subspace of* A^*. *Consider*

(i) $\mu^*(V) \subset V \otimes V$.
(ii) $\dim A.\xi.A < \infty$ *for all* $\xi \in V$.

Then (i) $\Longrightarrow$ *(ii) and the converse holds if* V *is an* A *bisubmodule of* A^*.

Take $\xi \in A^*$, $a,\ b \in A$. Then $(\xi.a)(b) = \xi(a\,b) = \mu^*(\xi)(a \otimes b) = \xi_1(a)\xi_2(b)$ where $\mu^*(\xi) = \xi_1 \otimes \xi_2$ (cf. 1.1.8). It follows that $\xi.a = \xi_1(a)\xi_2$. Similarly $a.\xi = \xi_2(a)\xi_1$.

(i) $\Longrightarrow$ (ii). Take $\xi \in V$. Then $\mu^*(\xi) \in V \otimes V$ is a finite sum and so there exists a finite dimensional subspace U of V such that $\xi_1, \xi_2 \in U$. From the previous paragraph this gives $\xi.A \subset U$. Repeating this argument for a basis of $\xi.A$ shows that $\dim A.\xi.A < \infty$, as required.

Conversely take $\xi \in V$ and set $U = A.\xi.A$ which is finite dimensional. It is enough to show that $\mu^*(\xi) \in U \otimes U$. Let $\{u_i\}_{i=1}^n$ be a basis of U. Then $\xi.a = \sum v_i(a)u_i$ where the $v_i(a) \in k$ are uniquely determined and hence depend linearly on a, that is $v_i \in A^*$. Yet $\sum v_i(a)u_i(b) = (\xi.a)(b) = \xi(ab) = (b.\xi)(a)$ and so $b.\xi = \sum_i u_i(b)v_i \in U$, for all $b \in A$ by the definition of U. Since the $u_i \in A^*$ are linearly independent, there exists $b_j \in A$ such that $u_i(b_j) = \delta_{ij}$. Hence $v_i \in U$. Finally from the first paragraph above $\mu^*(\xi) = \sum_i v_i \otimes u_i \in U \otimes U$, as required.

1.4.3 Corollary. *The algebra* $A^\star$ is a Hopf algebra with respect to the linear maps $(\Delta^*, \varepsilon^*, \mu^*, \eta^*, \sigma^*)$.

It remains to consider σ^*. By 1.1.9 (iii) it follows that $a.\sigma^*(\xi).b = \sigma^*(\sigma(b).\xi.\sigma(a))$, $\forall\ a, b \in A$, $\xi \in A^*$. Thus σ^* leaves $A^\star$ invariant. It is an antipode for $A^\star$ as can be seen by reversal of arrows in 1.1.7 (v).

1.4.4 In this subsection A is just assumed to be an associative algebra with identity and V a left A module. For each $\xi \in V^*$, $v \in V$ define $c^V_{\xi,v} \in A^*$ (or simply, $c_{\xi,v}$) through $c^V_{\xi,v}(a) = \xi(av)$. Recalling 1.1.5, 1.1.6 one easily checks

(i) $a.\, c_{\xi,v} = c_{\xi,av}$, (ii) $c_{\xi,v}.b = c_{\xi b,v}$,

for all $a,\ b \in A$. Set $C^V = k\{c^V_{\xi,v} \mid \xi \in V^*,\ v \in V\}$.

Lemma. *Let V be a left A module.*

(i) *For each $\xi \in V^*$, the map $v \mapsto c_{\xi,v}$ of V into A^* is a homomorphism $\varphi(\xi)$ of left A modules. The map $\xi \mapsto \varphi(\xi)$ of V^* into* $\mathrm{Hom}_A(V, A^*)$ *is bijective.*

(ii) *For each $v \in V^*$, the map $\xi \mapsto c_{\xi,v}$ of V^* into A^* is a homomorphism $\psi(v)$ of right A modules. The map $v \mapsto \psi(v)$ of V into* $\mathrm{Hom}_A(V^*, A^*)$ *is injective. If* $\dim_k V < \infty$, *it is surjective.*

(i) The first part is (i) above. If $\xi \in \ker \varphi$, then $c_{\xi,v} = 0$, $\forall v \in V$. Then $0 = c_{\xi,v}(1) = \xi(v)$ and so $\xi = 0$. For surjectivity take $\theta \in \mathrm{Hom}_A(V, A^*)$ and let ξ_0 be the map $v \mapsto \theta(v)(1)$. Then $\xi_0 \in V^*$ and $c_{\xi_0,v}(a) = \xi_0(av) = \theta(av)(1) = (a.\theta(v))(1) = \theta(v)(a)$. Hence $c_{\xi_0,v} = \theta(v)$ and $\theta = \varphi(\xi_0)$.

(ii) The first part is (ii) above. If $v \in \ker \psi$, then $0 = c_{\xi,v}(1) = \xi(v)$, $\forall \xi \in V^*$ and so $v = 0$. For surjectivity, take $\theta \in \mathrm{Hom}_A(V^*, A^*)$ and let v_0 be the map $\xi \mapsto \theta(\xi)(1)$. Then $v_0 \in V^{**}$ which identifies with V if $\dim_k V < \infty$. Then $c_{\xi,v_0}(a) = \xi(av_0) = (\xi.a)(v_0) = \theta(\xi.a)(1) = (\theta(\xi).a)(1) = \theta(\xi)(a)$. Hence $c_{\xi,v_0} = \theta(\xi)$ and $\theta = \psi(v_0)$.

1.4.5 Let $\mathrm{Mod}_f\, A$ denote the category of all finite dimensional left A modules. Let F be the subspace of A^* spanned by the $C^V : V \in \mathrm{Mod}_f\, A$.

Corollary

(i) $F = \{\xi \in A^* \mid \dim A.\xi < \infty\}$.
(ii) $F = \{\xi \in A^* \mid \dim \xi.A < \infty\}$.
(iii) $F = A^*$.

(i) Take $\xi \in A^*$ such that $V := A.\xi$ if finite dimensional. Then by 1.4.4 (i) there exists $\eta \in V^*$ such that $A.\xi = \{c_{\eta,v} \mid v \in V\}$. Hence (i). Similarly (ii) follows from 1.4.4 (ii). Then (iii) follows from (i), (ii) and 1.4.1 (i).

1.4.6 The Hopf algebra structure of A^* can be very conveniently described in terms of the $C^V : V \in \mathrm{Mod}_f\, A$.

Lemma. *For all* $V,\ V' \in \mathrm{Mod}_f\, A$ *one has*

(i) $C^V + C^{V'} = C^{V \oplus V'}$.
(ii) $C^V\ C^{V'} = C^{V \otimes V'}$.

(i) By definitions it is immediate that $C^V + C^{V'}$ identifies with a subspace of $C^{V \otimes V'}$. Surjectivity follows from 1.4.4 (i) or (ii).

(ii) Given $a \in A$ write $\Delta(a) = a_1 \otimes a_2$. Then $c_{\xi\otimes\xi',v\otimes v'}(a) = (\xi \otimes \xi')(a(v \otimes v')) = (\xi \otimes \xi')(a_1 v \otimes a_2 v') = \xi(a_1 v)\xi'(a_2 v') = c_{\xi,v}(a_1)c_{\xi',v'}(a_2) = (c_{\xi,v}c_{\xi',v'})(a)$. Hence $c_{\xi\otimes\xi',v\otimes v'} = c_{\xi,v}c_{\xi',v'}$ which gives (ii).

Remarks. The sum $C^V + C^{V'}$ need not be direct. For example if $V = V'$, then $C^V = C^{V'}$. It is direct if V, V' are simple and non-isomorphic. The surjection $x \otimes y \mapsto xy$ of $C^V \otimes C^{V'}$ onto $C^V\, C^{V'}$ need not be injective. Indeed this is seldom the case.

1.4.7 For $V \in \mathrm{Mod}_f\, A$, let $\{v_i\}_{i=1}^n$ be a basis for V and $\{\xi_i\}_{i=1}^n$ the dual basis for V^*. Then for all $v \in V$, $\xi \in V^*$ one has

$$\mu^*(c_{\xi,v}) = \sum_{t=1}^{n} c_{\xi,v_t} \otimes c_{\xi_t,v} \,. \tag{1}$$

In particular $\mu^*(C^V) \subset C^V \otimes C^V$.

Lemma. *Let B be a bisubalgebra and left A submodule of A^*. If V is a finite dimensional submodule of B, then C^V occurs in B.*

By 1.4.4 (i), there exists $\xi_0 \in V^*$ such that $V = \{c_{\xi_0,v} \mid v \in V\}$. By (i) and the linear independence of the $c_{\xi_0,v} : v \in V$, it follows that the $c_{\xi_j,v}$ lie in B, as required.

1.4.8 To compute σ^*, give any left A module V a right A module structure V^σ through a linear isomorphism $\varphi : V \xrightarrow{\sim} V^\sigma$ satisfying $\varphi(\sigma(a)v) = \varphi(v)a$, $\forall a \in A$, $v \in V$. Let $\varphi^* : (V^\sigma)^* \xrightarrow{\sim} V^*$ be the transpose of φ. Then $(\varphi^*(\xi).\sigma(a))(v) = \varphi^*(\xi)(\sigma(a)v) = \xi(\varphi(\sigma(a)v)) = \xi(\varphi(v)a) = (a.\xi)(\varphi(v)) = (\varphi^*(a.\xi))(v)$, that is $\varphi^*(\xi).\sigma(a) = \varphi^*(a.\xi)$, $\forall a \in A, \xi \in (V^\sigma)^*$. Define $V^{*\sigma}$ to be V^* given a left A module structure by $a.\varphi^*(\xi) = \varphi^*(\xi).\sigma(a)$. This gives an isomorphism $(V^\sigma)^* \xrightarrow{\sim} (V^*)^\sigma$ of left A modules. The resulting module is called the contragredient module to V.

Lemma. *If $V \in \mathrm{Mod}_f\, A$ then $(V^\sigma)^* \in \mathrm{Mod}_f\, A$ and*

$$\sigma^*(c^V_{\varphi^*(\xi),v}) = c^{(V^\sigma)^*}_{\varphi(v),\xi}\,, \quad \forall v \in V, \quad \xi \in (V^\sigma)^* \,.$$

Indeed $\sigma^*(c_{\varphi^*(\xi),v})(a) = (c_{\varphi^*(\xi),v})(\sigma(a)) = \varphi^*(\xi)(\sigma(a)v) = \xi(\varphi(v)a) = c_{\varphi(v),\xi}(a)$.

1.4.9 Let $\mathrm{Mod}_f^0\, A$ be a subset $\mathrm{Mod}_f\, A$ and F_0 the subspace of A^* spanned by the $C^V : V \in \mathrm{Mod}_f^0\, A$. By 1.4.7, F_0 is a sub-coalgebra of A^*. If $\mathrm{Mod}_f^0\, A$ is closed under tensor products then by 1.4.6 it is also a subalgebra of A^*. If $\mathrm{Mod}_f^0\, A$ is further closed under $V \longrightarrow (V^\sigma)^*$ then F_0 is a Hopf subalgebra of A^*. It is often convenient to consider such Hopf subalgebras.

Let S be a subset of A^*. Then S is called separating if $\eta(a) = 0$, $\forall \eta \in S$ implies $a = 0$. Observe that $c^V_{\xi,v}(a) = 0$, $\forall \xi \in V^*$, $v \in V$ if and only if $a \in \operatorname{Ann} V$. Thus F_0 is separating if and only if

$$(*) \qquad \bigcap_{V \in \mathrm{Mod}^0_f A} \operatorname{Ann} V = 0 \,.$$

Moreover the left hand side is exactly the kernel of the canonical map $a \mapsto (\eta \mapsto \eta(a))$ of A into F_0^*. Notice that by 1.4.2 the image of A is contained in F_0^*. In particular if A^* is separating the canonical map gives an embedding $A \hookrightarrow A^{**}$.

When $A = U(\mathfrak{g})$ with $\mathfrak{g}$ finite dimensional. Then $(*)$ holds (for $\mathrm{Mod}_f\, U(\mathfrak{g})$) by Harish-Chandra's theorem (A.2.18).

1.4.10 Let $i : V \longrightarrow V'$ be an embedding of finite dimensional left A modules and let $C^{V,V'}$ denote the linear span of the $c^{V'}_{\xi,v} : \xi \in V'^*, v \in V$. It is clear that the map $c_{\xi,v} \mapsto c_{i^*(\xi),v}$ extends to an A bimodule isomorphism of $C^{V,V'}$ onto C^V. A similar result holds if i is a surjection. Taking account of 1.4.6 this gives the

Lemma. *Suppose $V \in \mathrm{Mod}_f A$ has the property that every indecomposable $V' \in \mathrm{Mod}_f A$ occurs either as a submodule or as a quotient of $T(V)$. Then C^V generates A^* as an algebra.*

1.4.11 Take $V \in \mathrm{Mod}_f A$. The following shows that V extends canonically to a left A^{**} module. For this adopt the notation of 1.4.7. Let $A \otimes V \otimes A^* \longrightarrow V$ (resp. $A^{**} \otimes V \otimes A^* \longrightarrow V$) be defined by the natural pairing.

Lemma. *The map $v \mapsto \sum_i v_i \otimes c_{\xi_i,v}$ defines the unique right A^* comodule structure on V which makes the diagram*

$$\begin{array}{ccc} A \otimes V & \longrightarrow & V \\ \big\downarrow & \nearrow & \\ A \otimes B \otimes A^* & & \end{array}$$

commutative. The map $x \otimes v \mapsto \sum x(c_{\xi_i,v})v_i$ defines the unique left A^{**} module structure on V which then makes the above diagram, with A replaced by A^{**}, commutative. Moreover if $f \in \mathrm{End}_A V$ then $f \in \mathrm{End}_{A^{**}} V$.

Writing the image of v as $\sum v_i \otimes c_i$ with $c_i \in A^*$, commutativity is equivalent to $c_i(a) = \xi_i(av) = c_{\xi_i,v}(a)$, for all $a \in A$, that is $c_i = c_{\xi_i,v}$. It is immediate from 1.4.7 (1) that this solution satisfies the comodule conditions $(\text{iii})^*_r$ of 1.1.2. This proves the first part. Commutativity in the second diagram determines the image of $x \otimes v$ as given. Moreover $x \otimes (y \otimes v) = xy \otimes v$

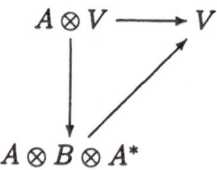

through the definition of the product in A^{**}. Finally if $f \in \mathrm{End}_A V$, then $\sum_i c_{\xi_i,fv}(a) \otimes v_i = 1 \otimes afv = 1 \otimes fav = \sum_i c_{\xi_i,v}(a) \otimes fv_i$ for all $a \in A$. Thus $xfv = \sum_i x(c_{\xi_i,fv})v_i = \sum_i x(c_{\xi_i,v})fv_i = fxv$ for all $x \in A^{**}$.

Remark. For some purposes one may need to replace A^* by a Hopf subalgebra A^+ and A^{**} by A^{+*}. If the canonical map $A \longrightarrow A^{+*}$ is injective, that is if A^+ is separating, then the above construction still extends the action of A on V to an action of A^{+*}.

1.4.12 It is false that the conclusion of 1.4.11 is the only way to extend the action of A. Let us digress slightly to examine this point.

Since (1.4.5) the left action of A on A^* is locally finite, one may take V to be A^* in 1.4.11. Now let $\psi : A^{**} \otimes A^* \longrightarrow A^*$ be an arbitrary left action of A^{**} on A^*. Then for each $y \in A^{**}$, the map $c \mapsto \psi(y \otimes c)(1)$ is an element φ_y of A^{**} and $\varphi : y \mapsto \varphi_y$ an element of $\mathrm{End}\, A^{**}$.

Suppose ψ to be defined by the conclusion of 1.14.11. Then

$$\varphi_x(c_{\xi,v}) = \psi(x \otimes c_{\xi,v})(1) = \sum_i x(c_{\xi,v_i})c_{\xi_i,v}(1)$$
$$= x(c_{\xi,\sum_i \xi_i(v)v_i}) = x(c_{\xi,v}) ,$$

that is $\varphi = Id_{A^{**}}$

Return to the general case.

Proposition

(i) *The map ψ extends the left action of A if and only if $\varphi_1(c) = c(1)$, $\varphi_{y.a} = \varphi_y.a$, $\varphi_{a.y}(c) = \psi(y \otimes c)(a)$ for all $a \in A$, $c \in A^*$, $y \in A^{**}$. Moreover in this case $\varphi|_A : A \longrightarrow A^{**}$ is the canonical map.*

(ii) *Suppose ψ extends the left action of A. Then ψ commutes with the right action of A if and only if $\varphi_{a.y} = a.\varphi_y$ for all $a \in A$, $y \in A^{**}$. Moreover in this case φ is an algebra homomorphism.*

(i) Suppose ψ extends the action of A. Then $\varphi_1(c) = c(1)$ and $\varphi_{y.a}(c) = \psi(y.a \otimes c)(1) = \psi(y \otimes \psi(a \otimes c))(1) = \psi(y \otimes a.c)(1) = \varphi_y(a.c) = (\varphi_y.a)(c)$, that is $\varphi_{y.a} = \varphi_y.a$. Finally $\varphi_{a.y}(c) = \psi(a.y \otimes c)(1) = \psi(a \otimes \psi(y \otimes c))(1) = (a.\psi(y \otimes c))(1) = \psi(y \otimes c)(a)$. Conversely the first two conditions imply $\varphi_a(c) = (\varphi_1.a)(c) = \varphi_1(a.c) = (a.c)(1) = c(a)$. Take $b \in A$. Then from the third condition $\psi(a \otimes c)(b) = \varphi_{ba}(c) = c(ba) = (a.c)(b)$, so ψ extends the action of A.

(ii) Suppose ψ commutes with the right action of A. Then $(a.\varphi_y)(c) = \varphi_y(c.a) = \psi(y \otimes c.a)(1) = (\psi(y \otimes c).a)(1) = \psi(y \otimes c)(a) = \varphi_{a.y}(c)$, that is $a.\varphi_y = \varphi_{a.y}$. The converse is similar.

For the last part of (ii), fix $c \in A^*$,. $y \in A^{**}$ and let C be a finite dimensional $A - A$ submodule of A^* containing c and $\psi(y \otimes c)$. Let $\{c_i\}$ be a basis for C. Since A separates elements of A^*, there exists $a_j \in A$ such that $c_i(a_j) = \delta_{ij}$. By 1.4.2, $\Delta(c) \in C \otimes C$ and one further checks that

$$\Delta(c) = \sum_j c_j \otimes c.a_j \ .$$

On the other hand $\psi(y \otimes c) = \sum_j c_j \psi(y \otimes c)(a_j) = \sum_j c_j \varphi_{a_j.y}(c) = \sum_j c_j(a_j.\varphi_y)(c) = \sum_j c_j \varphi_y(c.a_j)$. Thus for all $z \in A^{**}$ one obtains $\varphi_{zy}(c) = \psi(zy \otimes c)(1) = \psi(z \otimes \psi(y \otimes c))(1) = \varphi_z(\psi(y \otimes c)) = \sum_j \varphi_z(c_j)\varphi_y(c.a_j) = (\varphi_z \otimes \varphi_y)(\Delta(c)) = \varphi_z\varphi_y(c)$, as required.

Remark. Suppose $A = kG$ for some algebraic group G. Then $\operatorname{Lie} G$ is an $A - A$ submodule of A^{**} having zero as intersection with kG by say 2.1.8. Consequently ψ may satisfy both (i), (ii) above and yet φ may vanish on $\operatorname{Lie} G$.

1.4.13 Call A reductive if every $V \in \operatorname{Mod}_f A$ is semisimple. Let $S_f A$ denote the set of a simple finite dimensional A modules. If A is reductive, it is immediate from 1.4.6 (i) that

$$(*) \qquad A^* = \bigoplus_{V \in S_f} C^V \ .$$

Now C^V is isomorphic to $V \otimes V^*$ as an $A - A$ bimodule. If k is algebraically closed, $\operatorname{End}_A V = k$ by Schur's lemma. Take $c \in V \otimes V^*$ non-zero and write it in the form $c = \sum v_i \otimes \xi_i$ with the v_i linearly independent over k and $\xi_i \neq 0$, $\forall i$. By A.2.2 there exists $a \in A$ such that $av_1 = v_1$ and $av_i = 0$ for $i > 1$. Then $ac = v_1 \otimes \xi_1$ which by the simplicity of both V and V^*, is a cyclic vector for $V \otimes V^*$ as an $A - A$ bimodule. This proves that $V \otimes V^*$ is a simple $A - A$ bimodule. In particular C^V is a direct sum of $\dim V$ copies of V as a left A module. This provides an algebraic version of the Peter-Weyl theorem.

In the above situation it is usually easy to find $V \in \operatorname{Mod}_f A$ (and even in $S_f A$) such that every $V' \in S_f A$ occurs in $\mathcal{T}(V)$. Then by 1.4.10, A^* is generated by C^V.

Let G be a group. Then $V \in \operatorname{Mod}_f kG$ is faithful if and only if C^V separates points of G. Suppose V is faithful. If k has characteristic zero, then for any finite subset F of G, one may choose $c \in C^V$ to separate points of F. Then the powers of c separate points of kF. One concludes that the subalgebra $k[C^V]$ generated by C^V separates points of kG. If G is finite, this means that $k[C^V] = (kG)^* = (kG)^\star$, moreover by Maschke's theorem each $V \in \operatorname{Mod}_f kG$ is semisimple. If G is infinite, then $(kG)^\star$ is uncontrollably large. However take $k = \mathbb{C}$ and let G be a compact group with ν its Haar measure. Take $V \in \operatorname{Mod}_f \mathbb{C}G$ and equip $\operatorname{End}_{\mathbb{C}} V$ with the metric topology. Define $\operatorname{Mod}_{fc} \mathbb{C}G = \{V \in \operatorname{Mod}_f \mathbb{C}G \mid G \longrightarrow \operatorname{End}_{\mathbb{C}} V \text{ is continuous}\}$, and $S_{fc}\mathbb{C}G$ similarly. It is clear that

$$(\mathbb{C}G)^{\star}_{c} := \sum\{C^V \mid V \in \mathrm{Mod}_{fc}\mathbb{C}G\}$$

is a Hopf subalgebra of $(\mathbb{C}G)^{\star}$. Suppose $V \in \mathrm{Mod}_{fc}\mathbb{C}G$ is faithful. Since C^V separates points of G, it follows from the Stone-Weierstrass theorem that $\mathbb{C}[C^V]$ is dense in the space $L^2(G, \nu)$ of square integrable functions on G. Moreover every element of $\mathrm{Mod}_{fc}\mathbb{C}G$ is unitarizable, hence semisimple. By the classical Peter-Weyl theorem $L^2(G, \nu)$ is a Hilbert space direct sum of all $V \in S_{fc}\mathbb{C}G$ each occurring with multiplicity given by its dimension. Choosing an orthonormal basis in each isotypical component gives an orthonormal basis (in the Hilbert space sense) for $L^2(G, \nu)$. Orthogonality ensures that these basis vectors lie in the dense subspace $\mathbb{C}[C^V]$ which hence contains all their finite sums (with coefficients in $\mathbb{C}$). Moreover since $\mathbb{C}[C^V]$ is locally finite as a G module it coincides with this sum. This shows explicitly that C^V generates $(\mathbb{C}G)^{\star}_{c}$. By 1.4.6 it further implies that every object in $S_{fc}\mathbb{C}G$ occurs in $\mathcal{T}(V)$.

Now let $\mathfrak{g}$ be a complex semisimple Lie algebra. After Weyl, there exists a compact real group G such that every finite dimensional $\mathfrak{g}$ module and hence every $V' \in \mathrm{Mod}_f\, U(\mathfrak{g})$ admits the structure of a continuous G module via the exponential map. Thus if $V \in \mathrm{Mod}_f\, U(\mathfrak{g})$ is faithful as a G module (which is generally stronger than being faithful as a $\mathfrak{g}$ module) then C^V generates $U(\mathfrak{g})^*$ and this also coincides with $(\mathbb{C}G)^{\star}_{c}$. Though we need not make use of this argument to construct $U(\mathfrak{g})^*$, it serves as an important guide. Likewise it follows that $U(\mathfrak{g})$ is reductive and this also has a purely algebraic proof. Similar results will be obtained for $U_q(\mathfrak{g})$, and the analysis given also applies to $U(\mathfrak{g})$ over an arbitrary field of characteristic zero.

1.4.14 As a Hopf algebra $A^\star$ admits a (left) adjoint action which we recall is given by $(\mathrm{ad}\,\xi)\eta = \xi_1\eta\sigma^*(\xi_2)$ where $\mu^*(\xi) = \xi_1 \otimes \xi_2$. This is compatible with the right action of A as the following lemma shows.

Lemma. *For all $\xi, \eta \in A^\star$, $a, b \in A$ one has*

(i) $\mu^*(a.\xi.b) = \xi_1.b \otimes a.\xi_2$.
(ii) $\sigma^*(\sigma(a).\xi) = \sigma^*(\xi).a$.
(iii) $[(\mathrm{ad}\,\xi)\eta].a = (ad\sigma(a_3).\xi.a_1)\ (\eta.a_2)$ *where* $\Delta^2(a) = a_1 \otimes a_2 \otimes a_3$.

For all $c, d \in A$ one has $(a.\xi.b)(dc) = \xi(bdca) = \xi_1(bd)\xi_2(ca) = (\xi_1.b)(d)(a.\xi_2)(c)$, which gives (i).

For all $b \in A$ one has $\sigma^*(\sigma(a).\xi)(b) = (\sigma(a).\xi)(\sigma(b)) = \xi(\sigma(b)\sigma(a)) = \xi(\sigma(ab)) = \sigma^*(\xi)(ab) = (\sigma^*(\xi).a)(b)$, which gives (ii).

Finally

$$\begin{aligned}[(\mathrm{ad}\,\xi)\eta]\cdot a &= [\xi_1\eta\sigma^*(\xi_2)]\cdot a\ ,\\ &= (\xi_1.a_1)(\eta.a_2)(\sigma^*(\xi_2).a_3)\ ,\\ &= (\xi_1.a_1)(\eta.a_2)\sigma^*(\sigma(a_3).\xi_2)\ ,\quad \text{by (ii)}\ ,\\ &= (\mathrm{ad}(\sigma(a_3).\xi.a_1))(\eta.a_2)\ ,\quad \text{by (i)}\ .\end{aligned}$$

Remark. There is no corresponding compatibility for left action. However noting that $\sigma^*(\xi.\sigma(a)) = a.\sigma^*(\xi)$, $\forall a \in A$, $\xi \in A^*$ one may check that $a.(\mathrm{Ad}\,\xi)\eta = \mathrm{Ad}(a_3.\xi.\sigma(a_1))(a_2.\eta)$.

1.4.15 Let A be a Hopf algebra and B a Hopf subalgebra. Then A^* is a B bimodule. Define the set of left (resp. right) B invariants of A^* to be $(A^*)^B = \{\xi \in A^* \mid b.\xi = \varepsilon(b)\xi,\ \forall b \in B\}$ (resp. ${}^B(A^*) = \{\xi \in A^* \mid \xi.b = \varepsilon(b)\xi,\ \forall b \in B\}$) which is right (resp. left) A invariant. By the Leibnitz rule and noting that $\varepsilon(b_1)\varepsilon(b_2) = \varepsilon(b_1\varepsilon(b_2)) = \varepsilon(b)$, both are subalgebras of A^*.

Lemma

(i) $(A^*)^B$ *is a left coideal of* A^*,
(ii) ${}^B(A^*)$ *is a right coideal of* A^*,
(iii) $\sigma^*((A^*)^B) \subset {}^B(A^*)$,
(iv) $\sigma^*({}^B(A^*)) \subset (A^*)^B$.

For (i) take $\xi \in (A^*)^B$ and write $\mu^*(\xi) = \sum \xi_i \otimes \xi_i'$ with the $\xi_i \in A^*$ linearly independent. Since A separates A^* there exist $a_j \in A$ such that $\xi_i(a_j) = \delta_{ij}$. Then for all $a \in A$ one has $\varepsilon(a)\xi(a_j) = (a.\xi)(a_j) = \xi(a_j a) = \xi_j'(a)$. Then $(b.\xi_j')(a) = \xi_j'(ab) = \varepsilon(ab)\xi(a_j) = \varepsilon(b)\xi_j'(a)$, as required. The proof of (ii) is similar.

For (iii) take $\xi \in (A^*)^B$. Since $\varepsilon(\sigma(b)) = \varepsilon(b)$ by say 2.1.1 (iv), one has $(\sigma^*(\xi).b)(a) = \sigma^*(\xi)(ba) = \xi(\sigma(a)\sigma(b)) = (\sigma(b).\xi)\sigma(a) = \varepsilon(\sigma(b))\xi(\sigma(a)) = \varepsilon(b)(\sigma^*\xi)(a)$, as required. The proof of (iv) is similar.

1.4.16 The above result may be generalized to the algebras of semi-invariants. Namely for each $\chi \in B^\wedge$ one defines $A^*_\chi := \{\xi \in A^* \mid \xi.b = \varepsilon(\chi(b))\xi,\ \forall b \in B\}$. As in (i) above each A^*_χ is a left co-ideal of A^* and by the Leibnitz rule and the definition of the convolution product on B^* one obtains $A^*_\chi A^*_{\chi'} \subset A^*_{\chi * \chi'}$. The $A^*_\chi : \chi \in B^\wedge$ form a direct sum in A^* and the graded subalgebra of A^* so obtained is called the algebra of left B semi-invariants of A^*. One may remark that conditions (i), (iii) of 1.3.8 are satisfied by this graded subalgebra.

1.4.17 Let A and B be Hopf algebras. Then $A \otimes B$ is a Hopf algebra with multiplication $\mu_{A\otimes B} = (\mu_A \otimes \mu_B)\tau_{23}$ and comultiplication $\Delta_{A\otimes B} = \tau_{23}(\Delta_A \otimes \Delta_B)$. The canonical injection $A^* \otimes B^* \longrightarrow (A \otimes B)^*$ restricts to an isomorphism $A^\star \otimes B^\star \xrightarrow{\sim} (A \otimes B)^\star$. Indeed for the required surjectivity identify A (resp. B) with a subalgebra of $A \otimes B$ via the map $a \mapsto a \otimes 1$ (resp. $b \mapsto 1 \otimes b$). Then for any ideal I of $A \otimes B$ having finite codimension, $I \cap A$ (resp. $I \cap B$) is of finite codimension and $((A\otimes B)/I)^* \hookrightarrow (A \otimes B/A \cap I \otimes 1 + 1 \otimes B \cap I)^* \xrightarrow{\sim} (A/A \cap I \otimes B/B \cap I)^* \xrightarrow{\sim} (A/A \cap I)^* \otimes (B/B \cap I)^*$ which in view of 1.4.1 proves the required assertion.

1.4.18 Let $\mathcal{I}_f$ denote the set of (two-sided) ideals of A of finite codimension ordered by inclusion. It is immediate from 1.4.1 that

$$A^* = \varinjlim (A/I)^* , \quad A^{**} = \varprojlim A/I \quad (I \in \mathcal{I}_f) .$$

Again for all $M \in \mathrm{Mod}_f A$ there is a linear map $\varphi_M : M^* \otimes M \longrightarrow A^*$ defined by $\xi \otimes m \mapsto c^M_{\xi,m} : \xi \in M^*, \; m \in M$. This extends to a linear map φ of $\oplus\{M^* \otimes M \mid M \in \mathrm{Mod}_f A\}$ into A^* which by 1.4.5 (iii) is surjective.

Lemma. *The kernel of φ is the linear span K of the elements $g^*\xi \otimes m - \xi \otimes gm : M, N \in \mathrm{Mod}_f A, \; g \in \mathrm{Hom}_A(M,N), \; m \in M, \; \xi \in N^*$.*

It is immediate that $K \subset \ker \varphi$. If g is surjective, then C^N is generated by elements of the form $c^M_{g^*\xi,m} : m \in M, \; \xi \in N^*$. Thus it suffices to prove that K contains each $\ker \varphi_M$ for $M \in \mathrm{Mod}_f A$ which is a direct sum of cyclic modules. Using projectors one can assume M to be cyclic and even of the form A/I with $I \in \mathcal{I}_f$. Finally each $\xi \in (A/I)^*$ can be uniquely written as $c^{A/I}_{\xi,1}$, whilst for all $a \in A/I$ one has $c^{A/I}_{r_a^*\xi,1} = c^{A/I}_{\xi,a}$ where $r_a \in \mathrm{End}_A(A/I)$ denotes right multiplication by a.

Remark. This leads to the Grothendieck reconstruction lemma noted in the first part of 9.4.2.

1.5 Comments and Complements

1.5.1 The material in 1.1 is classical and can be found in a more fully developed form in the classical texts of M. E. Sweedler [Sw1] and E. Abe [Ab1]. Algebraic groups, briefly discussed in 1.2.1–1.2.3, are more fully developed in Chapter 2 following partly E. Abe [Ab1] and A. Borel [Bor1]. Further information on Lie algebras and their enveloping algebras (1.2.6–1.2.10) obtains from the classical texts of N. Jacobson [Jac1], J.-P. Serre [Se1] and J. Dixmier [Di2]. The subject matter of 1.2.11–1.2.13 is quite standard (see also [Ci1]), though 1.2.14 is more novel and developed from [J9]. For more details on differential operator rings one may consult [Bj1]. The material in 1.3.1–1.3.4 is fairly standard. Further discussions of winding automorphisms and their importance in the classification of prime ideals and localization theories for enveloping algebras can be found in [BGR1] and in [Bw1] respectively, see also [Ma1]. The material in 1.3.6 is taken from [J11, 5.4]. That of 1.3.7–1.3.14 is developed from [J11, 7.13–7.17] inspired in turn by Galois extension theory as described in [Di2, 3.4]. The material in 1.4.1–1.4.10 is standard and developed mainly from [Di2, 2.7]. I learnt 1.4.11 from R. Rentschler and developed 1.4.12 from these discussions. Concerning 1.4.13 a treatment of the representation theory of compact groups may be found in [Z1]. Finally 1.4.14 derives from [J11, 5.5].

Chapter 2.
Excerpts from the Classical Theory

2.1 Lie Algebras

Let A be a bialgebra. Here we examine part of the Hopf dual A^* of A and show that it admits a Lie subalgebra.

2.1.1 Recall 1.1.2 and set $\mathfrak{m} = \ker \varepsilon$. Let $\mathfrak{m}^0$ designate A. For each integer $j \geq 0$ let $1^{\otimes j}$ denote $1 \in k$ if $j = 0$ and $1 \otimes 1 \otimes \ldots \otimes 1$ (j times) if $j > 0$. Let S_r denote the group of permutations on r elements. Identify $\{\xi \in A^* \mid \xi(\mathfrak{m}^2) = 0\}$ with $(A/\mathfrak{m}^2)^*$.

Lemma. *For all integer $r > 0$, $a \in A$, $a_1, a_2, \ldots, a_r \in \mathfrak{m}$, $\xi_1, \xi_2, \ldots, \xi_r \in (A/\mathfrak{m}^2)^*$ one has*

(i) $\Delta(a) = -\varepsilon(a)(1 \otimes 1) + a \otimes 1 + 1 \otimes a \mod \mathfrak{m} \otimes \mathfrak{m}\,.$
(ii) $\Delta(\mathfrak{m}) \subset \mathfrak{m} \otimes A + A \otimes \mathfrak{m}\,.$
(iii) $\Delta(\mathfrak{m}^r) \subset \sum_{s=0}^{r} \mathfrak{m}^{r-s} \otimes \mathfrak{m}^s\,.$
(iv) *If σ is an antipode, then $\varepsilon\sigma(a) = \varepsilon(a)$. Moreover $\sigma(\mathfrak{m}) \subset \mathfrak{m}$ and $\sigma(a) + a \in \mathfrak{m}^2$ for all $a \in \mathfrak{m}$.*
(v) $\Delta^{r-1}(a_1 a_2 \ldots a_r) = \prod_{s=1}^{r} \sum_{t=1}^{r} (1^{\otimes t-1} \otimes a_s \otimes 1^{\otimes r-t}) \mod \sum_{t=1}^{r} A^{\otimes t-1} \otimes \mathfrak{m}^2 \otimes A^{\otimes r-t}\,.$
(vi) $\xi_1 \xi_2 \ldots \xi_r(a_1 a_2 \ldots a_r) = \sum_{\tau \in S_r} \prod_{t=1}^{r} \xi_t(a_{\tau(t)})\,.$

(i) Obviously $a - \varepsilon(a) \in \ker \varepsilon$ for all $a \in A$. Writing $\Delta(a) = a_1 \otimes a_2$ one has

$$\begin{aligned} a_1 - \varepsilon(a_1) \otimes a_2 - \varepsilon(a_2) &= a_1 \otimes a_2 - 1 \otimes \varepsilon(a_1)a_2 - a_1\varepsilon(a_2) \otimes 1 + \varepsilon(a_1)\varepsilon(a_2)1 \otimes 1 \\ &= \Delta(a) - 1 \otimes a - a \otimes 1 + \varepsilon(a)(1 \otimes 1) \end{aligned}$$

from which (i) results. (ii) follows from (i) and since Δ is a homomorphism (iii) follows from (ii). By (i) and 1.1.8

$$\varepsilon(a) = -\varepsilon(a) + \sigma(a) + a \qquad \mod \mathfrak{m}$$

and so $\varepsilon\sigma(a) = \varepsilon(2\varepsilon(a) - a) = \varepsilon(a)$, which is the first part of (iv). The last part of (iv) follows by the resubstitution in (i). Suppose $a \in \mathfrak{m}$, then repeated application of (i) gives

$$\Delta^{r-1}(a) = \sum_{t=1}^{r} 1^{\otimes t-1} \otimes a \otimes 1^{\otimes r-t} ,$$

modulo terms in which $\mathfrak{m}$ occurs in two factors. Since Δ^{r-1} is a homomorphism of algebras, $\Delta^{r-1}(a_1 a_2 \ldots a_r)$ equals the first term in the right hand side of (v) modulo $r-1+2=r+1$ occurrences of $\mathfrak{m}$ in r factors. Hence (v). (vi) follows from (v).

2.1.2 Let i denote the injection $\mathfrak{m} \hookrightarrow A$. Then i^* is a surjection of A^* onto $\mathfrak{m}^*$ whose kernel is $k\varepsilon = k1_{A^*}$. Given ξ, $\zeta \in \mathfrak{m}^*$ choose representatives $\xi' \in i^{*-1}(\xi)$, $\zeta' \in i^{*-1}(\zeta)$. Then $[\xi, \zeta] := i^*(\xi'\zeta' - \zeta'\xi')$ is independent of representatives and defines a Lie product on $\mathfrak{m}^*$.

Set $\mathfrak{g} = \{\xi \in \mathfrak{m}^* \mid \xi(\mathfrak{m}^2) = 0\}$ which may be identified with $(\mathfrak{m}/\mathfrak{m}^2)^*$. When A is the algebra of regular functions on an irreducible algebraic variety X defined over $\mathbb{C}$, the latter identifies with the tangent space $T_{\mathfrak{m},X}$ of X at $\mathfrak{m}$. It is convenient to identify $\mathfrak{g}$ with the subspace $(k1 + \mathfrak{m}^2)^\perp$ of A^*.

Lemma. *$\mathfrak{g}$ is a Lie subalgebra of $\mathfrak{m}^*$.*

One must show that $[\xi, \eta]\ (\mathfrak{m}^2) = 0$ for all $\xi, \eta \in \mathfrak{g}$. For this it suffices that $\xi\eta(ab)$ be symmetric in ξ, η for all $a,\ b \in \mathfrak{m}$. Yet by 2.1.1 (i) one has

$$\begin{aligned}\Delta(ab) = \Delta(a)\Delta(b) &= (a \otimes 1 + 1 \otimes a + \mathfrak{m} \otimes \mathfrak{m})\ (b \otimes 1 + 1 \otimes b + \mathfrak{m} \otimes \mathfrak{m}) \\ &= a \otimes b + b \otimes a + \mathfrak{m}^2 \otimes A + A \otimes \mathfrak{m}^2\end{aligned}$$

and so

$$\xi\eta(ab) = \xi(a)\eta(b) + \xi(b)\eta(a)$$

which is clearly symmetric.

2.1.3 For each $r \in \mathbb{N}$, let $\mathcal{F}^r_{\mathfrak{m}}(A)$ (or simply, $\mathcal{F}^r(A)$) be the subspace $\{\xi \in A^* \mid \xi(\mathfrak{m}^{r+1}) = 0\}$. Then $\mathcal{F}^0(A) = k1_{A^*}$ and it follows from 2.1.1 (iii) that $\mathcal{F}^r(A)\mathcal{F}^s(A) \subset \mathcal{F}^{r+s}(A)$ for all $r,\ s \in \mathbb{N}$.

Set

$$A^\times = \bigcup_{r \in \mathbb{N}} \mathcal{F}^r(A)$$

which is hence a subalgebra of A^* filtered by the $\mathcal{F}^r(A)$. Observe that $\mathcal{F}^1(A) = k1_{A^*} \oplus \mathfrak{g}$. With respect to the canonical filtration (1.2.6) of $U(\mathfrak{g})$ one obtains

Lemma. *For each $r \in \mathbb{N}$, the image of $\mathcal{F}^r(U(\mathfrak{g}))$ lies in $\mathcal{F}^r(A)$.*

2.1.4 For each $r \in \mathbb{N}$ the multiplication map $a \bmod \mathfrak{m}^r \otimes b \bmod \mathfrak{m}^2 \mapsto ab \bmod \mathfrak{m}^{r+1}$ defines a surjection of $\mathfrak{m}^{r-1}/\mathfrak{m}^r \otimes \mathfrak{m}/\mathfrak{m}^2$ onto $\mathfrak{m}^r/\mathfrak{m}^{r+1}$. Thus $\dim(\mathfrak{m}^r/\mathfrak{m}^{r+1}) \leq (\dim \mathfrak{g})^r$. In particular $\dim \mathcal{F}^r(A) < \infty$ for all $r \in \mathbb{N}$ if and only if $\dim \mathfrak{g} < \infty$. Now observe that $\ker \varepsilon \otimes \varepsilon = \mathfrak{m} \otimes A + A \otimes \mathfrak{m}$ and so $\mu(\ker \varepsilon \otimes \varepsilon)^{r+1} = \mathfrak{m}^{r+1}$. This gives

$$\mu^*(\mathcal{F}^r(A)) \subset \mathcal{F}^r(A \otimes A) := \{\xi \in (A \otimes A)^* \mid \xi(\ker \varepsilon \otimes \varepsilon)^{r+1} = 0\} .$$

On the other hand one has vector space isomorphisms

$$\begin{aligned}(A \otimes A)/(\ker \varepsilon \otimes \varepsilon)^{r+1} &= \bigoplus_{s=0}^{r} (\ker \varepsilon \otimes \varepsilon)^s/(\ker \varepsilon \otimes \varepsilon)^{s+1} \\ &= \bigoplus_{s=0}^{r} \bigoplus_{i=0}^{s} \mathfrak{m}^i \otimes \mathfrak{m}^{s-i}/\left(\mathfrak{m}^i \otimes \mathfrak{m}^{s-i+1} + \mathfrak{m}^{i+1} \otimes \mathfrak{m}^{s-i}\right) \\ &= \bigoplus_{s=0}^{r} \bigoplus_{i=0}^{s} (\mathfrak{m}^i/\mathfrak{m}^{i+1} \otimes \mathfrak{m}^{s-i}/\mathfrak{m}^{s-i+1}) \\ &= \sum_{s=0}^{r} (A/\mathfrak{m}^{r-s+1} \otimes A/\mathfrak{m}^{s+1}) .\end{aligned}$$

Consequently if $\mathcal{F}^r(A)$ is finite dimensional for all $r \in \mathbb{N}$, one may identify $\mathcal{F}^r(A \otimes A)$ with $\sum_{s=0}^{r} \mathcal{F}^{r-s}(A) \otimes \mathcal{F}^s(A)$. This proves the

Lemma. *Suppose* $\dim \mathfrak{g} < \infty$. *Then* $A^\times$ *is a bisubalgebra of* A^*. *If* A *has an antipode* $A^\times$ *is a Hopf subalgebra.*

2.1.5 Set $J = \cap_{r=0}^{\infty} \mathfrak{m}^r$. It is an ideal and a coideal of A. Moreover it is clear that J is just the kernel of the canonical map $A \longrightarrow A^{\times\times}$. Recall that by 2.1.3 there is a map $\theta : U(\mathfrak{g}) \longrightarrow A^\times$.

Proposition. ($\dim \mathfrak{g} < \infty$).

(i) θ *is homomorphism of bialgebras and of Hopf algebras if* A *has an antipode.*
(ii) *If* θ *is surjective, then* A/J *is commutative.*
(iii) *Suppose* $\operatorname{char} k = 0$. *Then* θ *is injective. If* A/J *is commutative, then* θ *is an isomorphism of* $U(\mathfrak{g})$ *onto* $A^\times$.

Obviously $(A/J)^\times \xrightarrow{\sim} A^\times$ and so replacing A by A/J one may assume $J = 0$. For the first part of (i) it suffices to show that $\mu^*(\xi) = \xi \otimes 1 + 1 \otimes \xi$, for all $\xi \in \mathfrak{g} = (\mathfrak{m}/\mathfrak{m}^2)^*$. Extend ξ to A^* by setting $\xi(1) = 0$. Then $\mu^*(\xi)(a \otimes b) = \xi((a-\varepsilon(a))(b-\varepsilon(b)) + \varepsilon(a)\xi(b) + \xi(a)\varepsilon(b)$. Recalling that $1_{A^*} = \varepsilon$, this gives the required assertion. The second part follows by the uniqueness of the antipode (1.1.9) or directly from 2.1.1 (iv).

If θ is surjective, then since $J = 0$ by assumption one has embeddings $A \hookrightarrow A^{\times\times} \hookrightarrow U(\mathfrak{g})^*$. Since $U(\mathfrak{g})^*$ is commutative, so is A.

Let $\{a_i\}_{i=1}^n$ be a basis for $\mathfrak{m}/\mathfrak{m}^2$ and $\{\xi_i\}_{i=1}^n$ the dual basis for $(\mathfrak{m}/\mathfrak{m}^2)^*$. For each $r \in \mathbb{N}^+$, one has $(\mathfrak{m}/\mathfrak{m}^2)^r = \mathfrak{m}^r/\mathfrak{m}^{r+1}$ and so the assumption that A/J is commutative implies that $\mathfrak{m}^r/\mathfrak{m}^{r+1}$ is spanned by the (ordered) monomials of degree r in the $a_i : i = 1, 2, \ldots, n$. Given $I = \{i_1, i_2, \ldots, i_n\} \in \mathbb{N}^n$, set $a^I = a_1^{i_1} \ldots a_n^{i_n}$ which has degree $|I| := \sum i_j$. Again since $\mathfrak{g}$ is a Lie algebra, the ordered monomials of degree r in ξ_i span $\mathfrak{g}^r$ and one sets $\xi^I = \xi_1^{i_1} \ldots \xi_n^{i_n}$. Clearly $\xi^I(a^J) = 0$ if $|I| < |J|$ whereas if $r = |I| = |J|$, then by 2.1.1 (vi)

$$\xi^I(a^J) = \sum_{\tau \in S_r} \prod_{s=1}^{r} \xi_{i_s}(a_{\tau(i_s)}) ,$$

$$= \begin{cases} |\operatorname{Stab}_{S_r} I| & : I = J , \\ 0 & : I \neq J . \end{cases}$$

By the assumption, $\operatorname{char} k = 0$, it follows that the matrix with entries $\xi^I(a^J)$ is upper triangular with non-zero scalars on the diagonal. This proves on the one hand the linear independence of ξ^I, hence the injectivity of θ and on the other hand if A/J is commutative that the $\xi^I : |I| \leq r$ *span* $\mathcal{F}^r(A)$, hence the surjectivity of θ.

Remark. Note that this also proves that the ξ^I are linearly independent. This proves the hard part of the PBW theorem for finite dimensional Lie algebras in characteristic zero of the form $(\mathfrak{m}/\mathfrak{m}^2)^*$. Historically Lie algebras were constructed as left invariant vector fields and hence essentially in this fashion. From this point of view the PBW theorem is rather easy and natural.

2.1.6 Suppose A is commutative and finitely generated (as an algebra) say by $a_1, a_2, \ldots, a_n$. Then $\mathfrak{m}/\mathfrak{m}^2$ is spanned by the $a_i - \varepsilon(a_i)$ and hence is finite dimensional. Let $A_\mathfrak{m}$ denote the localization of A at $\mathfrak{m}$ that is $A_\mathfrak{m} := \{s^{-1}a | s \in A \setminus \mathfrak{m},\ a \in A\}$ and $\varphi : A \longrightarrow A_\mathfrak{m}$ the canonical map. Then $A_\mathfrak{m}$ is a noetherian local ring with maximal ideal $\mathfrak{m}'$ generated by $\varphi(\mathfrak{m})$. Since $\mathfrak{m}J = J$ it follows that $\mathfrak{m}'J' = J'$ (where $J' = \varphi(J)A_\mathfrak{m})$) and hence by Nakayama's lemma (A.2.10) that $J' = 0$. Assuming that A is a domain, implies that φ is an embedding and so $J = 0$. Combined with 2.1.5 this proves the

Corollary. ($\operatorname{char} k = 0$). *Let A be a bialgebra which as an algebra is a finitely generated commutative domain. Then A is a subalgebra of $U(\mathfrak{g})^*$ where $\mathfrak{g} = (\mathfrak{m}/\mathfrak{m}^2)^*$ and $\mathfrak{m} = \ker \varepsilon$.*

2.1.7 Let A be a Hopf algebra. For all $g \in A^\wedge$, set $\mathfrak{m}_g = \ker g \in \operatorname{Max} A$ and recall that $\mathfrak{m} = \mathfrak{m}_e$ where e is the identity of $A^\wedge$.

Lemma. *For all $\xi \in A^*$, $r \in \mathbb{N}$, $g, g' \in A^\wedge$.*

(i) $\xi(\mathfrak{m}^r_{g'}) = 0$ *implies* $(g.\xi)(\mathfrak{m}^r_{gg'}) = 0$.

(ii) $\xi(\mathfrak{m}^r_g) = 0$ *implies* $(\xi.g')(\mathfrak{m}^r_{gg'}) = 0$.

(iii) $g.\mathfrak{g}.g^{-1} \subset \mathfrak{g}$.
(iv) $g^{-1}.(A/\mathfrak{m}_g^r)^* = (A/\mathfrak{m}_e^r)^* \subset \operatorname{Im}(U(\mathfrak{g}) \longrightarrow A^\times)$.

By definition $\mathfrak{m}_{gg'}$ is the kernel of the composed map $A \xrightarrow{\Delta} A \otimes A \longrightarrow A/\mathfrak{m}_g \otimes A/\mathfrak{m}_{g'}$. Consequently $\Delta(\mathfrak{m}_{gg'}) \subset \mathfrak{m}_g \otimes A + A \otimes \mathfrak{m}_{g'}$ and so for all $r \in \mathbb{N}$ one has

$$\Delta(\mathfrak{m}_{gg'}^r) \subset \sum_{s=0}^{r} \mathfrak{m}_g^{r-s} \otimes \mathfrak{m}_{g'}^s$$

from which (i), (ii) are immediate. Combined they give (iii) on taking $r = 2$. Finally (i) implies (iv).

2.1.8 Assume k to be algebraically closed and of characteristic zero. Assume A is a Hopf algebra which as a k-algebra is finitely generated and commutative. Under these hypotheses the Hopf dual $A^\star$ of A can be rather precisely described. First by 1.2.3 (or 1.3.4) the maximal spectrum Max A of A admits the structure of a multiplicative group which can also be viewed as the group of linear characters $A^\wedge \subset A^\star$ of A. Clearly $kA^\wedge \subset A^\star$. On the other hand it follows from 2.1.4 and 2.1.5 (iii) that $U(\mathfrak{g}) : \mathfrak{g} = (\mathfrak{m}/\mathfrak{m}^2)^*$ also identifies with a subalgebra of $A^\star$.

By 2.1.7 (iii) one has for each $g \in A^\wedge$ an endomorphism $a \mapsto g.a.g^{-1}$ of $U(\mathfrak{g})$ and these make $U(\mathfrak{g})$ a $kA^\wedge$ algebra. Recall (1.1.8) the definition of the smash product.

Theorem. *(k algebraically closed of characteristic zero). Let A be a Hopf algebra which as a k-algebra is finitely generated and commutative. Let $A^\wedge$ be the group of linear characters of A and $\mathfrak{g}$ the Lie algebra $(\mathfrak{m}/\mathfrak{m}^2)^*$ where $\mathfrak{m} = \ker \varepsilon$. Then $A^\star = U(\mathfrak{g})\#kA^\wedge$.*

Let $I \subset A$ be an ideal of finite codimension. Then by primary decomposition, there exists a finite subset $F \subset A^\wedge$ such that

$$I = \bigcap_{g \in F} P_g \quad \text{with} \quad \sqrt{P_g} = \mathfrak{m}_g \,.$$

By the Chinese remainder theorem

$$(A/I)^* = (\bigoplus_{g \in F} A/P_g)^* = \bigoplus_{g \in F} (A/P_g)^* \,.$$

Yet $(A/P_g)^* \subset g.U(\mathfrak{g})$ by 2.1.7 (iv). It is moreover clear that the $g.U(\mathfrak{g}) : g \in A^\wedge$ form a direct sum in $A^\star$ and by the above and the definition (1.4.1) of $A^\star$, this direct sum is $A^\star$.

2.1.9 The remainder of Section 2.1 is devoted to the study of $U(\mathfrak{g})^*$ with $\mathfrak{g}$ a finite dimensional Lie algebra.

Let $\{x_i\}_{i=1}^n$ be a basis for $\mathfrak{g}$. For each $I := \{i_1, i_2, \dots, i_n\} \in \mathbb{N}^n$ set $x^I = x_1^{i_1} x_2^{i_2} \dots x_n^{i_n} \in U(\mathfrak{g})$. By the PBW theorem the $x^I : I \in \mathbb{N}^n$ form a basis for $U(\mathfrak{g})$. Thus for each $I \in \mathbb{N}^n$, there exists $\xi_I \in U(\mathfrak{g})^*$ satisfying $\xi_I(x^J) = \delta_{I,J}$ (Kronecker delta). Since $\Delta(x^I) = \Pi \Delta(x_j^{i_j})$ and $\Delta(x^r) = \sum \binom{r}{s} (x^{r-s} \otimes x^s)$ it follows that

$$\xi_I \xi_J = \prod_{t=1}^{n} \binom{i_t + j_t}{i_t} \xi_{I+J} \, . \tag{1}$$

Finally each $\xi \in U(\mathfrak{g})^*$ can be expressed uniquely in the form

$$\xi = \sum_{I \in \mathbb{N}^n} \xi(x^I) \xi_I \, . \tag{2}$$

Thus $U(\mathfrak{g})^*$ is a commutative algebra and if $\operatorname{char} k > 0$ the ξ_I are nilpotent. Moreover $U(\mathfrak{g})^*$ cannot be finitely generated (unless $n = 0$).

Suppose here and in the remainder of Section 2.1 that $\operatorname{char} k = 0$. Set $\xi_{\{i\}} = \xi_i$. It is immediate from (1), (2) that $U(\mathfrak{g})^*$ is the algebra of formal power series $U(\mathfrak{g})^* = k[[\xi_1, \xi_2, \dots, \xi_n]]$. In particular $U(\mathfrak{g})^*$ is a domain. Its augmentation ideal $\mathfrak{m}$ is given by

$$\mathfrak{m} = \{\xi \in U(\mathfrak{g})^* \mid \xi(1) = 0\} \, .$$

Clearly $\mathfrak{m}^2$ vanishes on $\mathfrak{g}$ and $\mathfrak{m}/\mathfrak{m}^2$ is spanned by the images of the $\xi_i : i = 1, 2, \dots, n$. It follows that

$$\mathfrak{m}^2 = \{\xi \in \mathfrak{m} \mid \xi(\mathfrak{g}) = 0\} \, .$$

Thus $\mathfrak{g}^{**}$ and hence $\mathfrak{g}$ identifies with $(\mathfrak{m}/\mathfrak{m}^2)^*$.

Proposition. ($\operatorname{char} k = 0$). *Let A be a Hopf algebra which as a k-algebra is finitely generated and commutative. Then*

(i) *The canonical map $A \longrightarrow A^{**}$ is injective.*
(ii) *A has no nilpotent elements.*

(i) Take $0 \neq a \in A$ and set $\mathfrak{a} = \{b \in A \mid ba = 0\}$ which is a proper ideal of A. Let $\mathfrak{n}$ be a maximal ideal of A containing $\mathfrak{a}$. As in 2.1.6 it follows that $\cap(\mathfrak{n}A_\mathfrak{n})^n = \{0\}$. On the other hand the kernel of the canonical map $\varphi : A \longrightarrow A_\mathfrak{n}$ is given by $\ker \varphi = \{b \in A \mid sb = 0 \text{ for some } s \in A \setminus \mathfrak{n}\}$ and so does not contain a. Consequently there exists n such that $a \notin \mathfrak{n}^n$. By the nullstellensatz $\operatorname{codim} \mathfrak{n} < \infty$ and so $\operatorname{codim} \mathfrak{n}^n < \infty, \forall n \in \mathbb{N}$ (by the argument in 2.1.4). Thus a is not contained in the annihilator of the finite dimensional A module $A/\mathfrak{n}^n$ and hence not in $\ker(A \longrightarrow A^{**})$. Since a is arbitrary this proves (i).

For (ii) one may assume k algebraically closed. Then by 2.1.8 it follows that A^* is the direct sum of the $g.U(\mathfrak{g}) : g \in A^\wedge$. Consequently as a k-algebra A^{**} is a direct product of algebras isomorphic to $U(\mathfrak{g})^*$. Since $U(\mathfrak{g})^*$ is a domain, it follows that A^{**} has no nilpotent elements. Then (ii) follows from (i).

2.1.10 Set $\mathfrak{m}_0 = U(\mathfrak{g})^* \cap \mathfrak{m}$ which is the augmentation ideal of $U(\mathfrak{g})^*$. By Ado's theorem (A.2.16) there exists a finite dimensional faithful $\mathfrak{g}$ module V. Since $\{a \in U(\mathfrak{g}) \mid c(a) = 0, \forall c \in C^V\} = \operatorname{Ann} V$ it follows that $k + C^V + \mathfrak{m}^2 \supset \mathfrak{m}$. Consequently $\mathfrak{m}_0 + \mathfrak{m}^2 = \mathfrak{m}$ and so $\mathfrak{m}_0/\mathfrak{m}^2 \cap \mathfrak{m}_0 \xrightarrow{\sim} \mathfrak{m}/\mathfrak{m}^2$ which gives the embedding

$$\mathfrak{g} = (\mathfrak{m}/\mathfrak{m}^2)^* \hookrightarrow (\mathfrak{m}_0/\mathfrak{m}_0^2)^* =: \mathfrak{g}_0 \,.$$

Set $A = U(\mathfrak{g})^*$ which is a Hopf algebra with coproduct μ^* and antipode σ^*. Define the map $\varphi : A \longrightarrow A \otimes A$ by $\varphi(\xi) = \xi_1\sigma^*(\xi_3) \otimes \xi_2$ where $\mu^{*2}(\xi) = \xi_1 \otimes \xi_2 \otimes \xi_3$. For each $x \in A^*$ define the map $\varphi_x : A \longrightarrow A$ by composition of φ with the map $\xi \otimes \xi' \mapsto x(\xi)\xi'$ of $A \otimes A$ into A. Recall (2.1.2) that $\mathfrak{g}_0$ identifies with a Lie subalgebra of A^* by taking $y(1) = 0$ for all $y \in \mathfrak{g}_0$.

Proposition. (char $k = 0$).

(i) $x(\varphi_y(\xi)) = -[x, y](\xi)$ *for all* $x \in A^*$, $y \in \mathfrak{g}_0$, $\xi \in \mathfrak{m}_0^*$.

(ii) $\varphi(\mathfrak{m}_0) \subset A \otimes \mathfrak{m}_0$.

(iii) $\varphi(\mathfrak{m}^2 \cap \mathfrak{m}_0) \subset A \otimes (\mathfrak{m}^2 \cap \mathfrak{m}_0)$.

(iv) $\varphi_x(\mathfrak{m}^2 \cap \mathfrak{m}_0) \subset \mathfrak{m}^2 \cap \mathfrak{m}_0$, *for all* $x \in A^*$.

(v) $\mathfrak{g}$ *is an ideal of* $\mathfrak{g}_0$.

For (i) recall that η^* is the augmentation of A. Thus by 2.1.1 (iv) one has $\xi - \eta^*(\xi) \in \mathfrak{m}_0$, $\sigma^*(\xi) - \eta^*(\xi) \in \mathfrak{m}_0$. Since y vanishes on $\mathfrak{m}_0^2$ one obtains

$$\begin{aligned} x(\varphi_y(\xi)) &= y(\xi_1\sigma^*(\xi_3))x(\xi_2) \\ &= y(\sigma^*(\xi_3))\eta^*(\xi_1)x(\xi_2) + y(\xi_1)x(\xi_2)\eta^*(\xi_3) \\ &= y(\sigma^*(\xi_2))x(\xi_1) + y(\xi_1)x(\xi_2) \\ &= -[x, y](\xi) \,, \end{aligned}$$

since $y(\sigma^*(\xi_2)) = -y(\xi_2)$, by 2.1.1 (iv) again.

Take $\xi \in \mathfrak{m}_0$. Then

$$\varphi(\xi) = \xi_1\sigma^*(\xi_3) \otimes (\xi_2 - \eta^*(\xi_2)) + \xi_1\eta^*(\xi_2)\sigma^*(\xi_3) \otimes 1 \,.$$

The first term lies in $A \otimes \mathfrak{m}_0$, whereas the first factor in the second term equals $\xi_1\sigma^*(\xi_2) = \eta^*(\xi) = 0$. Hence (ii).

For all $a, b \in U(\mathfrak{g})$ one has

$$\varphi(\xi)(a \otimes b) = \xi_1\sigma^*(\xi_3)(a)\xi_2(b) = \xi(a_1 b \sigma(a_2)) = \xi((\operatorname{ad} a)b) \,.$$

Now take $b \in \mathfrak{g}$. By the remark in 1.3.5 one has

$$\begin{aligned}\Delta((\operatorname{ad} a)b) &= (\operatorname{ad} a_1)b \otimes (\operatorname{ad} a_2)1 + (\operatorname{ad} a_1)(1) \otimes (\operatorname{ad} a_2)b \\ &= (\operatorname{ad} a)b \otimes 1 + 1 \otimes (\operatorname{ad} a)b\ .\end{aligned}$$

It follows that for all $\xi, \eta \in \mathfrak{m}$, $a \in U(\mathfrak{g})$, $b \in \mathfrak{g}$ that

$$\begin{aligned}\varphi(\xi\eta)(a \otimes b) &= (\xi \otimes \eta)\Delta((\operatorname{ad} a)b) \\ &= \xi((\operatorname{ad} a)b)\eta(1) + \xi(1)\eta((\operatorname{ad} a)b) = 0\ .\end{aligned}$$

Combined with (ii), this proves (iii).
(iv) is immediate from (iii) and (v) follows from (i) and (iv).

2.1.11 Although $A := U(\mathfrak{g})^*$ is a commutative domain, it will not be finitely generated unless $\mathfrak{g}$ is semisimple. The difficulty already occurs when $\mathfrak{g}$ has dimension 1. Then $U(\mathfrak{g})^* = k[[\xi]]$ and the left (or right) action of $0 \neq x \in \mathfrak{g}$ is just differentiation. Then by 1.4.1 (i) any formal power series $f \in k[[\xi]]$ such that $\dim k[d/d\xi]f < \infty$ lies in A. In particular the $\exp n\xi : n \in \mathbb{Z}$ are algebraically independent and lie in A, which hence cannot be finitely generated.

2.1.12 Let $\mathfrak{g}$ be a semisimple Lie algebra. Then (c.f. 1.4.11 in the complex case) every finite dimensional $\mathfrak{g}$ module is semisimple. Again let V be the direct sum of the fundamental $\mathfrak{g}$ modules. Then every simple finite dimensional $\mathfrak{g}$ module occurs in $S(V)$. Finally

$$\bigcap_{V \in S_f U(\mathfrak{g})} \operatorname{Ann}_{U(\mathfrak{g})} V = 0\ .$$

By 1.4.10 it follows that $U(\mathfrak{g})^*$ is generated by C^V and hence is finitely generated. Moreover by (1), $U(\mathfrak{g})^*$ is separating. All these classical results will be established (4.3.10, 7.1.9) for the quantized enveloping algebra $U_q(\mathfrak{g})$ by proofs which also apply to $U(\mathfrak{g})$.

2.2 Algebraic Lie Algebras

Throughout this section it is assumed that k is algebraically closed of characteristic zero.

2.2.1 Let $\mathfrak{g}$ be a finite dimensional Lie algebra. By Ado's theorem (A.2.17) there exists a finite dimensional faithful $\mathfrak{g}$ module V. By definition this gives a Lie algebra embedding $\mathfrak{g} \hookrightarrow \mathfrak{gl}(V)$. Augmenting the dimension of V by 1 if necessary one may assume that this is an embedding into the simple Lie algebra $\mathfrak{sl}(V)$. Via the PBW theorem it is immediate that this extends to an embedding of $U(\mathfrak{g}) \hookrightarrow U(\mathfrak{sl}(V))$ which is moreover a Hopf algebra map. Duality gives a Hopf algebra map $U(\mathfrak{sl}(V))^* \longrightarrow U(\mathfrak{g})^*$. This is not necessarily surjective (because $W \in \text{Mod}_f\, U(\mathfrak{g})$ need not be a submodule of some finite dimensional $U(\mathfrak{sl}(V))$ module). Yet the image A of $U(\mathfrak{sl}(V))^*$ in $U(\mathfrak{g})^*$ is separating and moreover finitely generated. Since V is a faithful $SL(V)$ module it follows from 1.4.11 (for $k = \mathbb{C}$ and then in general through 4.3.6) that every simple finite dimensional $\mathfrak{sl}(V)$ module occurs in $\mathcal{T}(V)$. Since $U(\mathfrak{sl}(V))$ is reductive, 1.4.10 implies that $U(\mathfrak{sl}(V))^* = k[C^V] \twoheadrightarrow A$. In particular (the image of) C^V generates a Hopf algebra of $U(\mathfrak{g})^*$. If $n = \dim V$, the contragredient module to V identifies with the $(n-1)^{th}$ wedge product of V. More generally for an embedding $\mathfrak{g} \hookrightarrow \mathfrak{gl}(V)$ the nth wedge product of V is a non-trivial one dimensional $\mathfrak{g}$ module. Choosing a basis $\{v_i\}_{i=1}^n$ for V one has $x(v_1 \wedge \ldots \wedge v_n) = (det\; c_{\xi_j,v_i}(x))(v_1 \wedge \ldots \wedge v_n)$, for all $x \in U(\mathfrak{g})$. The element $x \mapsto det\; c_{\xi_j,x_i}(x)$ of $U(\mathfrak{g})^*$ is called the determinant polynomial and denoted by det (or simply, d). Clearly $d(1) = 1$, so $d \neq 0$. Whilst $k[C^V]$ is only a bialgebra; it may be made a Hopf algebra by inverting the determinant polynomial. This is because $\sigma^*(C^V) = d^{-1}C^{V'}$, where V' denotes the $(n-1)^{th}$ wedge product of V. The quantum analogue of this is A.4.10(∗).

2.2.2 Let $\mathfrak{g}$ be a finite dimensional Lie algebra. Although the assignment of $\mathfrak{g}_0$ defined in 2.1.10 to $\mathfrak{g}$ is canonical, this construction has the disadvantage that $\mathfrak{g}_0$ may be too big and indeed even infinite dimensional. Alternatively one may choose a faithful $\mathfrak{g}$ module V and replace $U(\mathfrak{g})^*$ by the Hopf subalgebra A_V generated by C^V (see 2.2.1). Setting $\mathfrak{m}_V = \mathfrak{m} \cap A_V$ and $\mathfrak{g}_V = (\mathfrak{m}_V/\mathfrak{m}_V^2)^*$, the argument in 2.1.10 shows that $\mathfrak{g}$ identifies with an ideal of $\mathfrak{g}_V$. Since A_V is finitely generated, $\mathfrak{g}_V$ is finite dimensional; yet it depends on V. To investigate this dependence one first recalls some elementary results concerning tangent spaces.

Let $\varphi : \mathcal{V} \longrightarrow \mathcal{W}$ be a morphism of irreducible affine algebraic varieties and assume $\mathcal{W} = \overline{\text{Im}\,\varphi}$. Then the comorphism $\varphi^* : R[\mathcal{W}] \longrightarrow R[\mathcal{V}]$ is an embedding of finitely generated integral domains and every such embedding so obtains.

Now suppose $R[\mathcal{V}]$ (resp. $R[\mathcal{W}]$) has generators $T_1, T_2, \ldots, T_m$ (resp. $S_1, S_2, \ldots, S_n$). These can be viewed as the co-ordinate functions on $\mathcal{V}$ (resp $\mathcal{W}$). For each $F \in k[T]$ define the differential d_vF at $v \in \mathcal{V}$ through

$$d_vF = \sum_{i=1}^{m} \frac{\partial F}{\partial T_i}(v)(T_i - v_i)\;,$$

where $v_i = T_i(v)$. The (classically defined) tangent space $T_{v,\mathcal{V}}$ of $\mathcal{V}$ at v is the linear space defined by the equations $d_vF_j = 0$ where the F_j runs through a set of generators of the ideal of definition $I(\mathcal{V})$ of $\mathcal{V}$. It follows that for each $f \in R[\mathcal{V}] \xleftarrow{\sim} k[T]/I(\mathcal{V})$, there is a uniquely determined linear form on $T_{v,\mathcal{V}}$ defined by

$$d_vf = d_vF|_{T_{v,\mathcal{V}}}$$

where $F \in k[T]$ has image f. One checks that $f \mapsto d_vf$ factors to a linear isomorphism of $\mathfrak{m}_v/\mathfrak{m}_v^2$ onto $T^*_{v,\mathcal{V}}$, where $\mathfrak{m}_v$ is the maximal ideal generated by the $T_i - v_i : i = 1, 2, \ldots, m$, that is corresponding to the point v. It is clear that $\dim T_{v,\mathcal{V}} = m - rk \frac{\partial F_j}{\partial T_i}(v)$, which hence takes its minimal value on a non-empty open subset $\mathcal{V}_0$ of $\mathcal{V}$. Since any such $\mathcal{V}$ is birationally isomorphic [Sh1, Thm. 6, Sect. 1.3] to a hypersurface in say $\mathbb{A}^m$, this minimal value equals $m - 1 = \dim \mathcal{V}$. One calls $\mathcal{V}_0$ the set of smooth points of $\mathcal{V}$.

Set $w = \varphi(v)$. The comorphism φ^* factors to a linear map $(d_v\varphi)^*$ of $\mathfrak{m}_w/\mathfrak{m}_w^2$ into $\mathfrak{m}_v/\mathfrak{m}_v^2$. By transposition this gives a linear map $d_v\varphi$ of $T_{v,\mathcal{V}}$ into $T_{\varphi(v),\mathcal{W}}$. One has $\xi((d_v\varphi)(x)) = x(\varphi^*\xi)$ for all $\xi \in \mathfrak{m}_w$, $x \in (\mathfrak{m}_v/\mathfrak{m}_v^2)^*$. Clearly $\varphi \mapsto d_v\varphi$ is a covariant functor and takes the identity morphism on $\mathcal{V}$ to the identity on $T_{v,\mathcal{V}}$. In particular if φ is isomorphism, then so is $d_v\varphi$. Since

$$(*) \qquad d_v(\varphi^*(S_j)) = \sum \frac{\partial \varphi^*(S_j)}{\partial T_i}(v)(T_i - v_i)\ , \quad d_w S_j = S_j - w_j$$

$(d_v\varphi)^*$ sends the linear function $S_j - w_j$ on $T_{w,\mathcal{W}}$ to the linear function $d_v(\varphi^*(S_j))$ on $T_{v,\mathcal{V}}$.

Some non-empty open subset $\mathcal{W}^0$ of $\mathcal{W}$ is isomorphic to a hypersurface in say $\mathbb{A}^n$ and restricted to this subset one may assume that the co-ordinate functions $S_1, S_2, \ldots, S_{n-1}$ are algebraically independent. Since φ^* is injective, the $\varphi^*(S_j) : j = 1, 2, \ldots, n-1$ are algebraically independent. Thus one may assume $T_j = \varphi^*(S_j) : j = 1, 2, \ldots, n-1$. Since some non-empty open set $\mathcal{V}^0$ of $\mathcal{V}$ may be assumed to be a hypersurface in say $\mathbb{A}^m$, then $\{T_j; j = 1, 2, \ldots, n-1\}$ may be completed to a set of co-ordinate functions on $\mathcal{V}^0$ satisfying the single relation $F(T_1, T_2, \ldots, T_m) = 0$. It follows that the differentials $d_v(\varphi^*(S_j))$ on $T_{v,\mathcal{V}}$ are linearly independent for any $v \in \mathcal{V}^0 \cap \varphi^{-1}(\mathcal{W}^0)$ satisfying $\partial F/\partial T_m(v) \neq 0$. Since $\varphi(\mathcal{V})$ contains a non-empty open subset of its closure $\mathcal{W}$, it follows that there exists a non-empty open subset $\mathcal{V}^{00}$ of $\mathcal{V}$ such that $\dim \operatorname{Im}(d_v\varphi)^* = n - 1 = \dim \mathcal{W}$ for all $v \in \mathcal{V}^{00}$. Yet $\dim(\mathfrak{m}_w/\mathfrak{m}_w^2) = \dim T_{w,\mathcal{W}} = \dim \mathcal{W}$ for all $w \in \mathcal{W}_0$ and so $(d_v\varphi)^*$ is injective on $\varphi^{-1}(\mathcal{W}_0) \cap \mathcal{V}^{00} =: \mathcal{U}$. This proves the

Lemma. *(k is algebraically closed of characteristic zero). Let $\varphi : \mathcal{V} \longrightarrow \mathcal{W}$ be a morphism of irreducible affine varieties and assume $\varphi(\mathcal{V}) = \mathcal{W}$. Then there exists a non-empty open subset $\mathcal{U}$ of $\mathcal{V}$ such that $d_v\varphi$ is a surjection of $T_{v,\mathcal{V}}$ onto $T_{\varphi(v),\mathcal{W}}$ for all $v \in \mathcal{U}$.*

2.2.3 Retain the notation and hypotheses of 2.2.2 and further assume that the comorphism $\varphi^* : R[\mathcal{W}] \longrightarrow R[\mathcal{V}]$ is a Hopf algebra map. Composing φ^* with a linear character $\chi : R[\mathcal{V}] \longrightarrow k$ gives a linear character $\chi\varphi^*$ on $R[\mathcal{W}]$. Then $\chi \mapsto \chi\varphi^*$ is a homomorphism $R[\mathcal{V}]^\wedge \longrightarrow R[\mathcal{W}]^\wedge$ of groups. Since $R[\mathcal{V}]^\wedge = \operatorname{Max} R[\mathcal{V}] = \mathcal{V}$, this map can be identified with φ. Let $\mathfrak{m}_g$ be a maximal ideal of $R[\mathcal{V}]$ such that $(d_g\varphi)$ is a surjection of $(\mathfrak{m}_g/\mathfrak{m}_g^2)^*$ onto $(\mathfrak{m}_{\varphi(g)}/\mathfrak{m}_{\varphi(g)}^2)^*$. Identifying g with an element of the group $R[\mathcal{V}]^\wedge = \mathcal{V}$ it follows from 2.1.7 (iv) that $d_g\varphi$ is surjective for all $g \in \mathcal{V}$ and in particular at the identity.

Now choose faithful $\mathfrak{g}$ modules V, V' and assume that V identifies with a submodule of V'. Then by 1.4.10, C^V identifies with a submodule of $C^{V'}$. This gives an embedding $\varphi^* : A_V \longrightarrow A_{V'}$ of finitely generated Hopf algebras which by 2.1.9 are furthermore integral domains.

Proposition. *Under the above hypotheses $d_e\varphi$ is a surjection of $\mathfrak{g}_{V'}$ onto $\mathfrak{g}_V$. In particular every $\mathfrak{g}_V$ is an image of $\mathfrak{g}_0$.*

The first part is already clear. Since $\mathfrak{m}_V = A_V \cap \mathfrak{m}_0$ there is an exact sequence

$$0 \longrightarrow (\mathfrak{m}_0^2 \cap \mathfrak{m}_V)/\mathfrak{m}_V^2 \longrightarrow \mathfrak{m}_V/\mathfrak{m}_V^2 \longrightarrow \mathfrak{m}_0/\mathfrak{m}_0^2 \longrightarrow \mathfrak{m}_0/\mathfrak{m}_V + \mathfrak{m}_0^2 \longrightarrow 0$$

which transposes to an exact sequence

$$0 \longleftarrow (\mathfrak{m}_0^2 \cap \mathfrak{m}_V/\mathfrak{m}_V^2)^* \longleftarrow \mathfrak{g}_V \longleftarrow \mathfrak{g}_0 \longrightarrow (\mathfrak{m}_0/\mathfrak{m}_V + \mathfrak{m}_0^2)^* \longleftarrow 0 .$$

Suppose $a \in \mathfrak{m}_0^2 \cap \mathfrak{m}_V$. Write $a = \sum a_i b_i : a_i, b_i \in \mathfrak{m}_0$. It is clear from 1.4.6(i) that the a_i, b_i can be assumed to belong to some $C^{V'}$ which may furthermore be assumed to contain C^V as a submodule. Then $a \in \mathfrak{m}_{V'}^2 \cap \mathfrak{m}_V$ and the surjectivity of $\mathfrak{g}_{V'} \longrightarrow \mathfrak{g}_V$ implies that a belongs to $\mathfrak{m}_V^2$, as required.

Remarks. Suppose all the indecomposable components of V' occur as submodules of $\mathcal{T}(V)$. Then by 1.4.10, φ^* is an isomorphism and hence so is $d_e\varphi$. From 2.1.10 it follows that $\mathfrak{g}$ is an ideal of $\mathfrak{g}_V$ which is called the algebraic hull of $\mathfrak{g}$ relative to the finite dimensional faithful $\mathfrak{g}$ module V. It is natural to call $\mathfrak{g}_0$ the universal hull of $\mathfrak{g}$.

2.2.4 It is already interesting to calculate $\mathfrak{g}_V$ when $\mathfrak{g}$ is one dimensional. Here recall (2.1.9) that $U(\mathfrak{g})^*$ is the algebra of formal power series in one variable ξ which can be chosen to satisfy $\xi(x^n) = \delta_{1,n}$ for some fixed nonzero $x \in \mathfrak{g}$. Then $\xi^n(x^m) = n!\delta_{n,m}$ and so left or right multiplication by x transposes to differentiation. Again multiplication in $U(\mathfrak{g})$ gives the coproduct in $k[[\xi]]$ defined by $\mu^*(\xi) = \xi \otimes 1 + 1 \otimes \xi$. Finally $\sigma^*(\xi) = -\xi$.

Now assume that V is a $\mathfrak{g}$ module in which x acts as a nilpotent linear transformation different from zero. Then x acts nilpotently on $k[C^V]$ and so the latter coincides with $k[\xi]$. Then $\mathfrak{m}_V = \langle \xi \rangle$ and so the embedding $\mathfrak{g} \hookrightarrow \mathfrak{g}_V$ is an isomorphism.

Now assume that x is a diagonal linear transformation on V, equivalently V is a direct sum of one dimensional subspaces V_i in which x has eigenvalue $\alpha_i \in k$. It is clear that C^{V_i} is spanned by the formal power series $\exp \alpha_i \xi$. Moreover $\sigma^*(\exp \alpha_i \xi) = \exp -\alpha_i \xi$. Enlarging V if necessary to V' it can be assumed that $\alpha_1, \alpha_2, \ldots, \alpha_n$ are linearly independent over $\mathbb{Q}$ and that the remaining $\alpha_j : j > n$ belong to the $\mathbb{Z}$ module generated by the $\alpha_i : i = 1, 2, \ldots, n$. Then the Hopf algebra A_V of $U(\mathfrak{g})^*$ generated by C^V is just the Laurent polynomial algebra on the (algebraically independent) generators

$\exp \alpha_i x : i = 1, 2, \ldots, n$. In this case $\mathfrak{m}_{V'} = \mathfrak{m} \cap A_{V'} = \langle 1 - \exp \alpha_i x : i = 1, 2, \ldots, n \rangle$ and so $\dim \mathfrak{g}_{V'} = n$. Since A_V is a finitely generated subalgebra of $A_{V'}$ admitting at least n algebraically independent generators it follows that $\dim \mathfrak{g}_V = \dim \mathfrak{g}_{V'}$ and so $\mathfrak{g}_{V'} \xrightarrow{\sim} \mathfrak{g}_V$.

Finally suppose that x is an arbitrary linear transformation on V. Let $x = x_n + x_s$ be the Jordan decomposition of V. Let V_n (resp. V_s) denote the $\mathfrak{g}$ module in which x acts by x_n (resp. x_s). Then x acts by $x_n + x_s$ on $V_n \otimes V_s$ and so V is a submodule of $V_n \otimes V_s$. Thus A_V is contained in the Hopf subalgebra generated by C^{V_n} and C^{V_s}.

Conversely let $V = \oplus V_i$ be a direct sum decomposition of V as a $\mathfrak{g}$ module corresponding to Jordan blocks. By 1.4.10, $k[C^V]$ is the subalgebra of $U(\mathfrak{g})^*$ generated by the $k[C^{V_i}]$. Applying the antipode it follows that A_V is generated by the A_{V_i}. For each i, there exists $\alpha_i \in k$ such that $x - \alpha_i 1d_V$ acts nilpotently. Then $d/d\xi - \alpha_i$ acts nilpotently on C^{V_i}. If this operator acts by zero, then C^{V_i} is spanned by $\exp \alpha_i \xi$. If not then C^{V_i} contains also $\xi \exp \alpha_i \xi$. Applying the antipode it follows that A_{V_i} is generated by ξ and $\exp \pm \alpha_i \xi$. Comparison with the analysis in the semisimple and nilpotent cases, it follows that A_V contains A_{V_n}, A_{V_s} and so is the algebra they generate. Consequently $\mathfrak{g}_V = \mathfrak{g}_{V_s} \times \mathfrak{g}_{V_n}$.

View the set $\text{Mod}_{ff}\, \mathfrak{g}$ of faithful finite dimensional $\mathfrak{g}$ modules as a directed set for inclusion. By 2.2.3 the map $V \mapsto \mathfrak{g}_V$ of $\text{Mod}_{ff}\, \mathfrak{g}$ into the set of finite dimensional Lie algebras is a contravariant functor. Further taking account of the last part of 2.2.3 one obtains

$$\mathfrak{g}_0 = \varprojlim \mathfrak{g}_V \ .$$

If $\dim \mathfrak{g} = 1$, it follows that $\mathfrak{g}_0$ is commutative and a direct product of a one dimensional Lie algebra (corresponding to the nilpotent part) and a possibly infinite dimensional Lie algebra (corresponding to the semisimple part) with basis indexed by a subset of a $\mathbb{Q}$ basis for k.

2.2.5 A Lie algebra of the form $\mathfrak{g}_V$ (notation 2.1.14) is called an algebraic Lie algebra. (Strictly speaking algebraic with respect to V). Let $\mathfrak{g}_V$ be an algebraic Lie algebra and $\mathfrak{h}$ a Lie subalgebra of $\mathfrak{g}$. Let $\mathfrak{h}_V$ denote the algebraic hull of $\mathfrak{h}$ relative to V. Note that each $x \in \mathfrak{g}_V$ may be considered as an element of $\mathfrak{gl}(V) = \text{End}\, V$ and hence has a Jordan decomposition.

Lemma. (*k algebraically closed of characteristic zero*).

(i) *$\mathfrak{h}_V$ identifies with a Lie subalgebra of $\mathfrak{g}_V$ containing $\mathfrak{h}$.*

(ii) *For all $x \in \mathfrak{g}$, the Jordan components x_n, x_s of x belong to $\mathfrak{g}_V$.*

(i) Let A_V denote the Hopf subalgebra of $U(\mathfrak{g}_V)^*$ generated by C^V. Then the Hopf subalgebra of $U(\mathfrak{h})^*$ used to construct $\mathfrak{h}_V$ is just the image of A_V under the restriction map $\xi \mapsto (a \mapsto \xi(a)) : a \in U(\mathfrak{h})$. This observation gives (i). Then (ii) follows from (i) and 2.2.4.

2.2.6 Let $\mathfrak{g}_V$ be an algebraic Lie algebra. By 2.2.4 and 2.2.5 (i), it follows that $\mathfrak{g}_V$ is spanned by the one dimensional algebraic Lie algebras it contains. The natural converse is less immediate; but will follow (2.4.4) from the subsequent section on Lie algebras of algebraic groups.

2.3 Algebraic Groups

We assume in this section that k is algebraically closed (but not for the moment necessarily of characteristic zero).

2.3.1 Let A be a Hopf algebra which as a k-algebra is commutative and finitely generated, with no nilpotent elements. (By 2.1.9 the last condition holds automatically if $\operatorname{char} k = 0$). Recall (1.2.3) that $G = \operatorname{Max} A$ has a group structure coming from the coproduct, coidentity and antipode on A. It is called an affine algebraic group, also linear algebraic group by some authors. By definition A separates points of G. Conversely G separates the elements of A since $\cap\{\mathfrak{m} \mid \mathfrak{m} \in \operatorname{Max} A\} = 0$, by the Hilbert nullstellensatz and the assumption that A has no nilpotent elements. Thus A embeds in $(kG)^\star$. However they need not be equal, indeed this would imply that every $V \in \operatorname{Mod}_f kG$ occurs in A viewed as a left G module (1.3.4). Consider for example the group algebra A of the integers $\mathbb{Z}$ with $k = \mathbb{C}$, that is $A = \mathbb{C}[a, a^{-1}]$. Then $A^\wedge = \operatorname{Max} A$ identifies with $\mathbb{C}$ as an additive group via the map $\alpha \mapsto \exp 2\pi i n\alpha =: \chi(a^n)$. The representations of $A^\wedge$ coming from A are just the integral characters $\alpha \mapsto \exp 2\pi i n\alpha : n \in \mathbb{Z}$. Moreover in this case $(kA^\wedge)^\star$ is not even finitely generated. On the other hand if G is compact and $k = \mathbb{C}$, then since A separates points of G, it follows from 1.4.11 that $A \xrightarrow{\sim} (kG)^\star_c$.

In the general situation described above, A is denoted $R[G]$ and is called the algebra of regular functions on G. However although A determines G, the group structure on G is by no means sufficient to determine $R[G]$. Consider for example $G = \mathbb{Z}$. For each complex number $\alpha \neq 0$, let c_α denote the matrix coefficient $c_\alpha(g^n) = \exp 2\pi i n\alpha : n \in \mathbb{Z}$ and let A_α denote the subalgebra of $(kG)^\star$ it generates. Then A_α is a Hopf algebra and isomorphic to $\mathbb{C}[a, a^{-1}]$ as an algebra. The A_α are generally distinct; but at least isomorphic. On the other hand $\mathbb{Z}$ also admits non-semisimple representations, in particular that given by $g^n \mapsto \begin{pmatrix} 1 & n \\ 0 & 1 \end{pmatrix} : n \in \mathbb{Z}$. In this the corresponding Hopf subalgebra of matrix coefficients is just $\mathbb{C}[a]$. The situation will become even worse for quantum groups which do not even exist. This "disfonctionnement" can be overcome by calling $R[G]$ an affine algebraic group and eventually its quantum analogue $R_q[G]$ – a quantum group.

2.3.2 Let G be an affine algebraic group. As a variety G is a finite union of its (irreducible) components which by definition are the closed irreducible subvarieties forming a finite non-redundant union of G. Let C be an irreducible component of G of maximal dimension and take $g \in G$. Since left multiplication in G is continuous (1.2.2) it follows that gC is again irreducible and that $\dim C \geq \dim gC \geq \dim g^{-1}(gC) = \dim C$. Thus the components of G have all the same dimension and are permuted by left multiplication, in particular they are pairwise disjoint. Set $G_0 = \text{Stab}_G\, C$.

Lemma. *G_0 is a normal subgroup of finite index of G. It coincides with the unique component C_0 of G containing the identity e of G.*

Finiteness is clear. Take $g \in C_0$. Then $g^{-1}C_0 \cap C_0 \supset \{e\} \neq \emptyset$ and so $g^{-1}C_0 = C_0$. Hence C_0 is a subgroup. As in the above C_0g is a component of G and so again $C_0g = C_0$. This proves that $C_0 \subset G_0 \subset \text{Stab}_G\, C_0 \subset C_0$, that is $C_0 = G_0$. Finally take $g \in G$. Then gC_0g^{-1} is a component of G containing e and so equal to C_0. Hence $G_0 = C_0$ is normal.

Remark. In particular G_0 is closed. It is called the identity component of G. The finite group G/G_0 is called the group of components of G. Note that an algebraic group is connected if and only if it is irreducible.

2.3.3 Let $R[G]$ be an affine algebraic group. By 2.1.2 the tangent space $T_{e,G} := (\mathfrak{m}/\mathfrak{m}^2)^*$ of G at the identity $e \in G$ is a Lie algebra and denoted $\text{Lie}\, G$. Taking $r = 1, 2$ in 2.1.7 (iv) gives $g.T_{e,G} = T_{g,G}$. In particular $\dim T_{g,G}$ is independent of $g \in G$, that is G is smooth and so $\dim G = \dim T_{g,G} = \dim \text{Lie}\, G$. Obviously $\text{Lie}\, G = \text{Lie}\, G_0$ in the notation of 2.3.2.

2.3.4 Let G be an affine algebraic group. A subgroup H of G which is also a closed subvariety is called an algebraic subgroup of G, for example G_0 of 2.3.2. This means that there exists a semiprime ideal I_H (or simply, I) of $R[G]$ such that $H = \{g \in G \mid f(g) = 0, \forall f \in I\}$. By the nullstellensatz $I = \{f \in R[G] \mid f(h) = 0, \forall h \in H\}$. Now for all $h, h' \in H$, $f \in I$ one has $\Delta(f)(h \otimes h') = f(hh') = 0$ and so $\Delta(I) \subset I \otimes R[G] + R[G] \otimes I$. Again from the definition of the antipode σ for $R[G]$ it follows that $\sigma(I) \subset I$, and $I \subset \mathfrak{m}$ since $e \in H$. Hence I is a Hopf ideal of $R[G]$.

Let H be an algebraic subgroup of G. For each $g, g' \in G$, $f \in R[G]$ one has $(g.f.g^{-1})(g') = f(g^{-1}g'g)$ and so f vanishes on H if and only if $g.f.g^{-1}$ vanishes on gHg^{-1}. Consequently gHg^{-1} is an algebraic subgroup of G and $I_{gHg^{-1}} = g.I_H.g^{-1}$. It is clear that $\text{Lie}\, H$ identifies with $(\mathfrak{m}/\mathfrak{m}^2 + I_H)^*$ and so $\text{Lie}\, gHg^{-1} = g.\text{Lie}\, H.g^{-1}$.

2.3.5 Let G be an affine algebraic group. An arbitrary subgroup H of G need not be closed; but its closure $\bar{H}$ is again a subgroup. Indeed since the multiplication map $\mu : G \times G \longrightarrow G$ is continuous, it follows that $\mu^{-1}(\bar{H}) \supset$

$\overline{\mu^{-1}(H)} \supset \overline{(H,H)} = (\bar{H},\bar{H})$ and so $\bar{H} \supset \underline{\mu(\bar{H},\bar{H})}$. Again the antipode σ : $g \mapsto g^{-1}$ is continuous and so $\sigma^{-1}(\bar{H}) \supset \overline{\sigma^{-1}(H)} = \bar{H}$, that is $\bar{H} \supset \bar{H}^{-1}$ which together with the first observation proves that $\bar{H}$ is a subgroup and hence an algebraic subgroup of G.

Recall that for an irreducible variety X an open subset U is both irreducible and dense.

Lemma. *Let G be an affine algebraic group. Let $H_1, H_2, \ldots, H_n$ be closed irreducible subvarieties of G. For each $i \in \{1,2,\ldots,n\}$ let U_i be non-empty open subsets of H_i and assume $e \in \bar{U}_i$ for all i. Then the subgroup H generated by the $U_i : i = 1,2,\ldots,n$ is a closed irreducible subvariety. Moreover for some $m \leq 2\dim G$ one has $H = V_1V_2\ldots V_m$ where the V_j are amongst the U_i or U_i^{-1}.*

One may assume $\dim U_i \geq 1$ for otherwise $\bar{U}_i = \{e\}$. Since the U_i are irreducible, the $\overline{V_1V_2\ldots V_m}$ are irreducible and since $e \in \bar{U}_i$ they form an ascending chain. Assume that this chain is strictly ascending. Then it must terminate at some maximal element $Z = \overline{V_1V_2\ldots V_m}$ with $m \leq \dim G$. Then $ZU_j \subset \overline{V_1V_2\ldots V_mU_j} = Z$ and similarly $ZU_j^{-1} \subset Z$. Hence $Z \supset H$. Finally $U = V_1V_2\ldots V_m$ contains an open subset of its closure Z and hence so does Uz for each $z \in Z$. Since Z is irreducible $U \cap Uz \neq \emptyset$ and so $z \in U^{-1}U$, as required.

2.3.6 Let G be an affine algebraic group and H, H' algebraic subgroups. Denote by $\langle H, H'\rangle$ the subgroup of G generated by H, H' and by $[H, H']$ the subgroup of G generated by all expressions of the form $hh'\,h^{-1}h'^{-1} : h \in H, h' \in H'$.

Lemma. *Assume H, H' are irreducible.*

(i) *$\langle H, H'\rangle$ is an irreducible algebraic subgroup of G.*
(ii) *$[H, H']$ is an irreducible algebraic subgroup of G.*
(iii) *$[H, H']$ is the smallest normal subgroup of $\langle H, H'\rangle$ for which H and H' commute in the quotient.*

(i) is immediate from 2.3.5. For (ii) consider the map $\psi : (h, h') \mapsto hh'h^{-1}h'^{-1}$ of $H \times H$ into $[H, H']$. Since ψ is a morphism and H, H' are irreducible, $\overline{\operatorname{Im}\psi}$ is irreducible and $\operatorname{Im}\psi$ contains a non-empty open subset U of $\overline{\operatorname{Im}\psi}$. Moreover $e \in \operatorname{Im}\psi \subset \bar{U}$. Applying 2.3.5 with $n = 1$, $H_1 = \overline{\operatorname{Im}\psi}$, $U_1 = U$, it follows that the subgroup generated by U is closed and irreducible. In particular it coincides with $[H, H']$ proving (ii). Normality in (iii) follows from the formula $h_1(hh'h^{-1}h'^{-1})h_1^{-1} = (h_1h)h'(h_1h)^{-1}h'^{-1}(h'h_1h'^{-1}h_1^{-1}) \in [H, H']$, where $h_1, h \in H$, $h' \in H'$, which shows that $[H, H']$ is stable under conjugation by H (and similarly by H'). In the quotient H and H' commute and $[H, H']$ must be contained in any normal subgroup with this property.

Remark. Suppose only H' is irreducible in (i) and let H_0 denote the component group of H. Then $\langle H_0, H'\rangle$ is an irreducible closed subgroup of $\langle H, H'\rangle$ and so are its (finitely many) conjugates under H which by 2.3.5 generate a normal subgroup of $\langle H, H'\rangle$ which is closed and irreducible. It follows that $\langle H, H'\rangle$ is the finite union $U\{h\langle H_0, H'\rangle h^{-1} \mid h \in H/H_0\}$ of closed sets and hence is also closed. On the other hand $\langle H, H'\rangle$ need not be closed if neither H nor H' are irreducible. Indeed it is enough that G contain two elements τ, τ' of finite order which together generate an infinite group. Then $H = \langle\tau\rangle$, $H' = \langle\tau'\rangle$ are closed; but $\langle H, H'\rangle$ is not.

2.3.7 Let V be an n dimensional vector space over k (which is assumed algebraically closed). Then the group $\mathrm{GL}(V)$ of linear bijections of V has a natural structure of a quasi-affine variety, namely as the open set of $\mathbb{A}^{n^2}$ of $n \times n$ matrices of non-vanishing determinant d. Then $R[\mathrm{GL}(V)]$ can be taken to be the algebra on n^2 indeterminates $c_{ij} : i, j = 1, 2, \dots, n$, localized at the determinant polynomial d, given a coproduct, coidentity and antipode through

$$\Delta(c_{ij}) = \sum_{t=1}^{n} c_{it} \otimes c_{tj} \ , \ \varepsilon(c_{ij}) = \delta_{ij} \ , \ \sigma(c_{ij}) = d^{-1} c^{ij} \ ,$$

where c_{ij} is the i, j cofactor of d.

Lemma. *Let $R[G]$ be an affine algebraic group. Then G admits a faithful finite dimensional module V such that there exists a Hopf algebra surjection $R[\mathrm{GL}(V)] \longrightarrow R[G]$, that is G is an algebraic subgroup of* $\mathrm{GL}(V)$.

Identify $R[G]$ with a finitely generated subalgebra and G submodule of $(kG)^*$. By 1.4.5 there exists a finite dimensional left G submodule V of $(kG)^*$ lying in and generating $R[G]$. By 1.4.7, one has $C^V \subset k[G]$ and a fortiori C^V generates $R[G]$, as required.

Remarks. All this may seem paradoxical; the point is that $c_{\xi,v} : \xi \in V^*$, $v \in V$ as functions on G will in general have many relations amongst themselves and may not even be linearly independent. Note also that $\mathrm{GL}(V)$ is itself affine viewed as the closed subvariety of A^{n^2+1} defined by $x_{n^2+1}d - 1 = 0$.

2.3.8 Let G be an affine algebraic group. An action of G on a finite dimensional vector space V may be viewed as a morphism $\psi : G \times V \longrightarrow V$ of varieties. One may then calculate $d_{e,v}\psi : T_{e,v,G\times V} \longrightarrow T_{v,V}$ at any point $v \in V$. Since V is already a vector space and ψ is linear in the second factor, the comorphism ψ^* restricts to the linear map $V^* \longrightarrow V^* \otimes R[G]$ given by $(\psi^*\xi)(g, v) = \xi(gv)$ which in turn determines ψ^*. Since $\xi(gv) = \sum \xi_i(v) c_{\xi,v_i}(g)$ one has $\psi^*\xi = \sum \xi_i \otimes c_{\xi,v_i}$.

Identify $T_{v,V}$ with V. Then $d_{e,v}\psi$ maps $T_{e,G} \times V = \mathrm{Lie}\, G \times V$ into V and from 2.2.2 one obtains

$$\xi((d_{e,v}\psi)(x,w)) = (w \otimes x)(\psi^*\xi) = \sum \xi_i(w)x(c_{\xi,v_i})$$

for all $x \in \operatorname{Lie} G$, $w \in V$. In particular, it is independent of $v \in V$. Omitting v in the subscript and rearranging gives

$$(d_e\psi)(x,w) = \sum x(c_{\xi_j,w})v_j \ .$$

Identifying $\operatorname{Lie} G$ with a subspace of $R[G]^*$ shows that this is exactly the canonical extension of actions defined in 1.4.11. In particular $d_e\psi$ is an action of $\operatorname{Lie} G$ on V.

2.3.9 Let G be an affine algebraic group. Then G acts on $\operatorname{Lie} G$ by conjugation (2.1.7 (iii)). According to 2.3.8, this defines an action of $\operatorname{Lie} G$ on itself. To compute this action write $\psi(g,x) = gxg^{-1}$, $\forall g \in G$, $x \in \mathfrak{g}$. Then $(gxg^{-1})(\xi) = g(\xi_1\sigma(\xi_3))x(\xi_2)$ for all $\xi \in \mathfrak{g}^*$ and so $\psi^*\xi = \xi_1\sigma(\xi_3) \otimes \xi_2$. Then as in 2.3.8

$$\begin{aligned}\xi((d_e\psi)(x,y)) &= x(\xi_1\sigma(\xi_3))y(\xi_2)\\ &= y\varphi_x(\xi) = [x,y](\xi)\ , \quad \text{by } \ 2.1.10\text{(i)}\ ,\end{aligned}$$

for all $x, y \in \mathfrak{g}$. Thus $(d_e\psi)(x,y) = [x,y]$, which is the Lie bracket.

Lemma. *Let G be an affine algebraic group, H an algebraic subgroup of G and $\mathfrak{s}$ a subspace of $\mathfrak{g} := \operatorname{Lie} G$. Then*

(i) *$\mathfrak{s}$ is H stable $\Longrightarrow$ $\mathfrak{s}$ is $\operatorname{Lie} H$ stable.*

(ii) *Suppose H is irreducible and $\operatorname{char} k = 0$. Then the converse of (i) holds.*

(i) is already clear. (ii) follows from the general considerations below.

Let A be a Hopf algebra and B a Hopf subalgebra of A^*. Set $I_B = \{\xi \in A \mid b(\xi) = 0,\ \forall b \in B\}$ which is a Hopf ideal of A. As in 2.1.10 define the map $\varphi : A \longrightarrow A \otimes A$ by $\varphi(\xi) = \xi_1\sigma(\xi_3) \otimes \xi_2$, and recall that $\varphi(\mathfrak{m}) \subset A \otimes \mathfrak{m}$. For each $b \in B$, set $\varphi_b(\xi) = b(\xi_1\sigma(\xi_3))\xi_2$. Let $\mathfrak{n}$ be a subspace of $\mathfrak{m}$. Then $\varphi_b(\mathfrak{n}) \subset \mathfrak{n}$, $\forall b \in B$ if and only if $\varphi(\mathfrak{n}) \subset A \otimes \mathfrak{n} + I_B \otimes \mathfrak{m}$. In particular if B, B' are Hopf subalgebras of A^* such that $I_B = I_{B'}$ then

$$(*) \qquad \varphi_b(\mathfrak{n}) \subset \mathfrak{n},\ \forall b \in B \Longleftrightarrow \varphi_{b'}(\mathfrak{n}) \subset (\mathfrak{n}), \quad \forall b' \in B' \ .$$

Now return to the hypotheses of (ii) and take $A = R[G]$ in the above. Since $R[H]$ is a domain and $\operatorname{char} k = 0$, it follows from 2.1.6 that $R[H]$ embeds in $U(\mathfrak{h})^*$ or equivalently that $I_{U(\mathfrak{h})} = I_H$. Thus $(*)$ applies with $\psi^* = \varphi$, $\mathfrak{n} := \{\xi \in \mathfrak{m} \mid \mathfrak{s}(\xi) = 0\} \supset \mathfrak{m}^2$. This gives (ii).

2.4 Lie Algebras of Algebraic Groups

2.4.1 In this section it is assumed that k is of characteristic zero and algebraically closed.

Let G be an affine algebraic group with Lie algebra $\mathfrak{g}$ and H, H' algebraic subgroups with Lie algebras $\mathfrak{h}$, $\mathfrak{h}'$. The subset $HH' = \{hh' \mid h \in H, h' \in H'\}$ is not a subgroup of G; but contains the identity e of G and is the $H \times H'$ orbit in G through e, so is an open subset of its closure. The ideal of definition of its closure is given by $I_{\overline{HH'}} = \{f \in R[G] \mid \mu^*(f) \in I_H \otimes R[G] + R[G] \otimes I_{H'}\}$. On the other hand $H \cap H'$ is an algebraic subgroup of G and $I_{H\cap H'} = \sqrt{I_H + I_{H'}}$ by the nullstellensatz.

Lemma

(i) $I_{H\cap H'} = I_H + I_{H'}$.

(ii) $\mathrm{Lie}(H \cap H') = \mathfrak{h} \cap \mathfrak{h}'$.

(iii) $T_{e,HH'} = \mathfrak{h} + \mathfrak{h}'$.

(i) Since $I_H, I_{H'}$ are both Hopf ideals so is $I_H + I_{H'}$. Thus $R[G]/I_H + I_{H'}$ is a Hopf algebra which is commutative and finitely generated as a k-algebra, hence has no nilpotent elements by 2.1.9 (ii). Consequently $I_H + I_{H'} = \sqrt{I_H + I_{H'}} = I_{H\cap H'}$, as required.

(ii) By definition $\mathrm{Lie}(H\cap H') = (\mathfrak{m}/\mathfrak{m}^2 + I_{H\cap H'})^* = (\mathfrak{m}/\mathfrak{m}^2 + I_H + I_{H'})^* = \{\xi \in \mathfrak{m}^* \mid \xi(\mathfrak{m}^2 + I_H + I_{H'}) = 0\} = \{\xi \in \mathfrak{m}^* \mid \xi(\mathfrak{m}^2 + I_H) = 0\} \cap \{\xi \in \mathfrak{m}^* \mid \xi(\mathfrak{m}^2 + I_{H'}) = 0\} = \mathrm{Lie}\, H \cap \mathrm{Lie}\, H' = \mathfrak{h} \cap \mathfrak{h}'$, as required.

(iii) Let η be the identity of kG, that is $\eta : k \longrightarrow kG$ with $\eta(1) = e$. For all $f \in R[G]$ one has $(\eta^* f)(1) = f(\eta(1)) = f(e)$. Then η^* identified with the element $f \mapsto f(e)$ of $R[G]^*$ is the identity of this algebra and moreover $\ker \eta^* = \mathfrak{m} \supset I_H$. Thus for all $\xi \in R[G]^*$, $f \in I_{\overline{HH'}}$ one has $\xi(f) = (\eta^* \xi)(f) = (\eta^* \otimes \xi)(\mu^*(f)) \in \eta^*(R[G]) \otimes \xi(I_{H'})$. Consequently $\xi(I_{H'}) = 0$ implies $\xi(I_{\overline{HH'}}) = 0$, that is $T_{e,HH'} \supset \mathrm{Lie}\, H' = \mathfrak{h}'$. Similarly from $\xi = \xi\eta^*$ one obtains $T_{e,HH'} \supset \mathfrak{h}$. Since $T_{e,HH'}$ is a linear space these combine to give $T_{e,HH'} \supset \mathfrak{h} + \mathfrak{h}'$, and for equality it suffices to establish equality of dimensions. By 2.3.2 one may assume H, H' irreducible.

Consider the morphism $(h, h') \mapsto hh'$ of $H \times H'$ into G. The fibre over hh' is just $(hg, g^{-1}h') : g \in H\cap H'$ and hence has dimension equal to $\dim(H \cap H')$. It follows that $\dim HH' = \dim(H \times H') - \dim H \cap H' = \dim H + \dim H' - \dim H \cap H' = \dim \mathfrak{h} + \dim \mathfrak{h}' - \dim(\mathfrak{h} \cap \mathfrak{h}') = \dim(\mathfrak{h} + \mathfrak{h}')$, by 2.3.4 and(ii). On the other hand HH' is smooth and so $\dim T_{e,HH'} = \dim HH' = dim(\mathfrak{h} + \mathfrak{h}')$, as required.

2.4.2 As noted in 2.3.1 an affine algebraic group is not determined by its group structure. However in characteristic zero it is determined by its Lie algebra in the sense described below. Here recall the notion of an algebraic Lie algebra $\mathfrak{g}_V$ defined in 2.2.5 and note (2.2.1) that V may be specified by viewing $\mathfrak{g}_V$ as a Lie subalgebra of $\mathfrak{gl}(V)$.

Proposition

(i) *To each algebraic Lie algebra* $\mathfrak{g}_V$ *there exists exactly one closed irreducible subgroup* G_V *of* $\mathrm{GL}(V)$ *such that* $\mathrm{Lie}\, G_V = \mathfrak{g}_V$.

(ii) $G_V \subset H_V \iff \mathrm{Lie}\, G_V \subset \mathrm{Lie}\, H_V$.

(iii) *Let* W *be a subspace of* V *stable by some Lie subalgebra* $\mathfrak{h}$ *of* $\mathfrak{gl}(V)$. *Then* W *is stable under* $\mathfrak{h}_V$ *and* H_V. *Moreover* W *is the trivial* $\mathfrak{h}$ *module if and only if it is the trivial* H_V *(or* $\mathfrak{h}_V$*) module.*

(i) Let A_V denote the Hopf algebra (2.2.2) used to define $\mathfrak{g}_V$. It is clear from 2.2.1 that $G_V := \mathrm{Max}\, A_V$ satisfies the conclusions of (i). Suppose H is an algebraic subgroup of $\mathrm{GL}(V)$ such that $\mathrm{Lie}\, H \supset \mathrm{Lie}\, G_V$. Then by 2.3.10 (ii), one has $\mathrm{Lie}(H \cap G_V) = \mathrm{Lie}\, H \cap \mathrm{Lie}\, G_V = \mathrm{Lie}\, G_V$. Now $H \cap G_V$ is a closed subgroup of the irreducible group G_V and so a strict inclusion $H \cap G_V \subsetneq G_V$ would imply $\dim \mathrm{Lie}(H \cap G_V) = \dim(H \cap G_V) < \dim G_V = \dim \mathrm{Lie}\, G_V$ which contradicts the previous equality. Hence $H \cap G_V = G_V$ and so $H \supset G_V$. In particular if H is irreducible, this forces $H = G_V$ and completes the proof of (i). For (ii) it remains to show that an inclusion $G \subset H$ of algebraic subgroups of $\mathrm{GL}(V)$ implies an inclusion $\mathrm{Lie}\, G \subset \mathrm{Lie}\, H$. This is immediate from 2.3.10 (ii). By 2.3.8, the action of $\mathfrak{gl}(V)$ on V is given by 1.4.11. Hence W is stable under $(\mathrm{Im}(A_V \longrightarrow U(\mathfrak{h})^*))^*$ which by 2.1.8 is $U(\mathfrak{h}_V)\# kH_V$. Hence (iii).

Remarks. For a given Lie subalgebra $\mathfrak{g}$ of $\mathfrak{gl}(V)$, it is immediate from (i) that $\mathfrak{g}_V$ is the unique smallest Lie subalgebra of $\mathfrak{gl}(V)$ containing $\mathfrak{g}$ which is the Lie algebra of a closed subgroup of $\mathrm{GL}(V)$. This is the classical definition of an algebraic Lie algebra. The proposition fails in positive characteristic.

2.4.3 Let G be an affine algebraic group, $H_1, H_2, \dots, H_n$ irreducible closed subgroups distinct from $\{e\}$. By 2.3.5 the group H generated by $H_1, H_2, \dots, H_n$ is an irreducible closed subgroup of G. Set $\mathfrak{h} = \mathrm{Lie}\, H$, $\mathfrak{h}_i = \mathrm{Lie}\, H_i$, : $i = 1, 2, \dots, n$. By 2.3.5 again there exists an integer $\dim H \geq m > 0$ such that $\overline{H_{i_1} H_{i_2} \dots H_{i_m}} = H$ for some $i_j \in \{1, 2, \dots, n\}$. Write $(H) = H_{i_1} \times H_{i_2} \times \dots \times H_{i_m}$. Given $h_i \in H : i = 1, 2, \dots, m$, write $(h) = (h_1, h_2, \dots, h_m)$ and $h = h_1 h_2 \dots h_m$. Let π_i denote the canonical projection $R[H] \longrightarrow R[H_i]$. Set $\pi = (\pi_{i_1}, \pi_{i_2}, \dots, \pi_{i_m})$, $\varphi = \mu^{m-1}$ (the $m-1$ fold) multiplication map of $H \times H \times \dots \times H \longrightarrow H$.

Proposition

(i) *The map* $\pi \mu^{*m-1} : R[H] \longrightarrow R[H_{i_1}] \otimes R[H_{i_2}] \otimes \dots \otimes R[H_{i_m}]$ *is injective.*

(ii) *There exist* $g_j \in H$ *such that* $\mathfrak{h} = \sum_{j=1}^{m} g_j \mathfrak{h}_{i_j} g_j^{-1}$.

(iii) $\mathfrak{h}$ *is the smallest* H *invariant subspace of* $\mathrm{Lie}\, G$ *containing the* $\mathfrak{h}_i : i = 1, 2, \dots, n$.

(iv) *As a Lie algebra* $\mathfrak{h}$ *is generated by the* $\mathfrak{h}_i : i = 1, 2, \dots, n$.

Set $\psi = \varphi \mid_{(H)}$. By the hypothesis $\overline{\operatorname{Im} \psi} = H$. Hence the comorphism
$\psi^* : R[H] \longrightarrow R[H_{i_1} \times H_{i_2} \times \ldots \times H_{i_m}] \xrightarrow{\sim} R[H_{i_1}] \otimes R[H_{i_2}] \otimes \ldots \otimes R[H_{i_m}]$
is injective. Yet $\psi^* = \pi \mu^{*m-1}$, hence (i).

Since $\overline{\operatorname{Im} \psi} = H$, it follows from 2.2.2 that there exist $h_{i_j} \in H_{i_j}$ such
that $d_{(h)}\psi : T_{(h),(H)} \longrightarrow T_{h,H}$ is surjective. By a slight abuse of notation,
H_{i_j} (resp. h_{i_j}) is simply written as H_j (resp. h_j) in what follows. Set $g_j = h_1 h_2 \ldots h_j \in H$, $H_j' = g_j H_j g_j^{-1}$, $(H') = H_1' \times H_2' \times \ldots \times H_m'$. Define $\lambda_{(h)} :$
$(H) \longrightarrow (H)$ by $\lambda_{(h)}(x_1, x_2, \ldots, x_m) = (h_1 x_1, h_2 x_2, \ldots, h_m x_m)$. Define $\gamma_{(g)} :$
$(H) \longrightarrow (H')$ by $\gamma_{(g)}(x_1, x_2, \ldots, x_m) = (g_1 x_1 g_1^{-1}, g_2 x_2 g_2^{-1}, \ldots, g_m x_m g_m^{-1})$.
Set $h = g_m$ and define $\rho_{h^{-1}} : H \longrightarrow H$ to be right multiplication by h^{-1}.
Finally set $\psi' = \varphi \mid_{(H')}$.

One checks that $(\rho_{h^{-1}} \varphi \lambda_{(h)})(x) = (\varphi \gamma_{(g)})(x)$ for all $(x) \in (H')$, that is
$\rho_{h^{-1}} \psi \lambda_{(h)} = \psi' \gamma_{(g)}$. Both map $(e) := (e, e, \ldots, e)$ to e. This gives a commutative diagram

$$
\begin{array}{ccc}
T_{(e),(H)} & \xrightarrow{\; d_{(e)}\gamma_{(g)} \;} & T_{(e),(H')} \\
\Big\downarrow {\scriptstyle d_{(e)}\lambda_{(h)}} & & \diagdown {\scriptstyle d_{(e)}\psi'} \\
& & T_{e,H} \\
& & \diagup {\scriptstyle d_h \rho_{h^{-1}}} \\
T_{(h),(H)} & \xrightarrow[\; d_{(h)}\psi \;]{} & T_{h,H}
\end{array}
$$

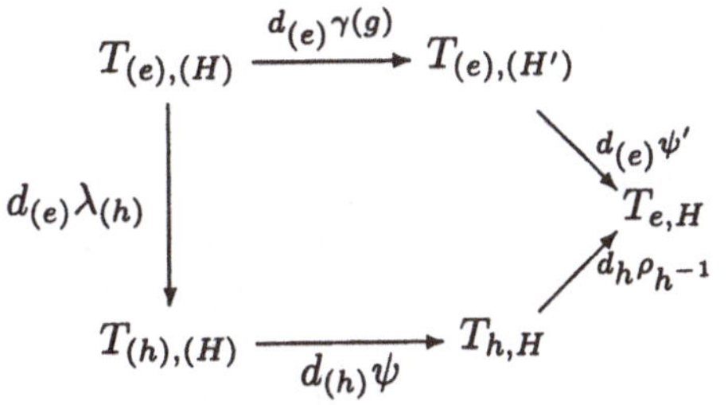

Since $\lambda_{(h)}$ and $\rho_{h^{-1}}$ are isomorphisms, so are $d_{(e)}\lambda_{(h)}$ and $d_h \rho_{h^{-1}}$. Then
since $d_{(h)}\psi$ is surjective, it follows that $d_{(e)}\psi'$ is surjective. Thus

$$\mathfrak{h} := T_{e,H} = (d_{(e)}\psi')(\bigoplus_{j=1}^{m} T_{e,H_j'})$$

$$= (d_{(e)}\psi')(\bigoplus_{j=1}^{m} g_j \mathfrak{h}_j g_j^{-1}) , \quad \text{by 2.3.4 ,}$$

$$= \sum_{j=1}^{m} (d_{(e)}\psi')(g_j \mathfrak{h}_j g_j^{-1}) , \quad \text{by linearity ,}$$

$$= \sum_{j=1}^{m} g_j \mathfrak{h}_j g_j^{-1} , \quad \text{viewing } g_j H_j g_j^{-1} \text{ as a subgroup of } H .$$

This is (ii). Assertion (iii) is immediate from (ii). Then (iv) follows from 2.3.9.

Remark. The map $d_{(e)}\psi$ is surjective if and only if the $\mathfrak{h}_i$ span $\mathfrak{h}$.

2.4.4 One now obtains the natural converse to 2.2.6.

Corollary. *Let $\mathfrak{g}$ be a Lie algebra of $\mathfrak{gl}(V)$. Then $\mathfrak{g}$ is algebraic if it is spanned (or generated) by algebraic subalgebras.*

2.4.5 From 2.2.4, 2.2.5, 2.4.4 it is quite easy to classify all algebraic subalgebras of $\mathfrak{gl}(V)$. Call a Lie subalgebra $\mathfrak{u}$ of $\mathfrak{gl}(V)$ *unipotent* if it is spanned by nilpotent elements (viewed as linear transformations of V). By 2.2.4 and 2.4.4 such a Lie subalgebra is algebraic. Any semisimple subalgebra $\mathfrak{s}$ of $\mathfrak{gl}(V)$ can be generated by nilpotent elements and hence is algebraic. Actually in this case one has by 2.1.12 the much better result that $\mathfrak{s}$ equals its universal hull $\mathfrak{s}_0$.

Now let $\mathfrak{g}$ be an arbitrary Lie subalgebra of $\mathfrak{gl}(V)$ and let $\mathfrak{r}$ denote its (solvable) radical. By Levi decomposition [Di2, 1.6.9] there exists a semisimple subalgebra (Levi factor) $\mathfrak{s}$ of $\mathfrak{g}$ such that $\mathfrak{g} = \mathfrak{s} \oplus \mathfrak{r}$. By 2.4.4 and the above $\mathfrak{g}$ is algebraic if and only if $\mathfrak{r}$ is algebraic. By Lie's theorem $\mathfrak{r}$ can be conjugated into the Lie subalgebra of $\mathfrak{gl}(V)$ of upper triangular matrices [Di2, 1.3.12]. Let $\mathfrak{u}$ denote the subalgebra of $\mathfrak{r}$ of those matrices which become strictly upper triangular. This is just the set of nilpotent elements of $\mathfrak{r}$ and is called the unipotent radical of $\mathfrak{g}$. Moreover $[\mathfrak{r}, \mathfrak{r}] \subset \mathfrak{u}$. Call a subalgebra $\mathfrak{k}$ of $\mathfrak{gl}(V)$ *reductive* if it is a product of a semisimple subalgebra and an abelian subalgebra $\mathfrak{a}$ of semisimple elements. Then V is a semisimple $\mathfrak{k}$ module and so is $\operatorname{End} V$ under adjoint action. Call $\mathfrak{k}$ *rationally reductive* if in addition $\mathfrak{a}$ admits a basis whose members have rational eigenvalues. Call $\mathfrak{g} \subset \mathfrak{gl}(V)$ *almost algebraic* if it contains the Jordan components of each of its elements.

Proposition. *A Lie subalgebra $\mathfrak{g}$ of $\mathfrak{gl}(V)$ is almost algebraic (resp. algebraic) if and only if its unipotent radical $\mathfrak{u}$ is complemented by a reductive (resp. rationally reductive) subalgebra $\mathfrak{k}$. Moreover $[\mathfrak{g}, \mathfrak{g}]$ is always algebraic.*

Let $\mathfrak{r}$ be the radical of $\mathfrak{g}$ and $\mathfrak{s}$ a Levi factor. Then by 2.4.4 and the above remarks $\mathfrak{g}_V = \mathfrak{s} \oplus \mathfrak{r}_V$ (notation 2.2.2). Let $G_V \subset \mathrm{GL}(V)$ be the unique irreducible algebraic group with Lie algebra $\mathfrak{g}_V$. Now $[\mathfrak{g}, \mathfrak{r}] \subset \mathfrak{r}$ and $[\mathfrak{g}_V, \mathfrak{g}] \subset \mathfrak{g}$ by 2.1.10 and 2.2.3, so $[\mathfrak{g}_V, \mathfrak{r}]$ is an ideal of $\mathfrak{g}$ contained in $\mathfrak{r}_V$ hence solvable. Consequently $[\mathfrak{g}_V, \mathfrak{r}] \subset \mathfrak{r}$. Then by 2.3.9 one has $g\mathfrak{r}g^{-1} \subset \mathfrak{r}$ for all $g \in G_V$. Since conjugation preserves nilpotence, this further implies that $g\mathfrak{u}g^{-1} \subset \mathfrak{u}$ for all $g \in G_V$. Hence G_V acts on $\mathfrak{r}/\mathfrak{u}$. Each $\bar{x} \in \mathfrak{r}/\mathfrak{u}$ can be viewed as a diagonal matrix and so the action of G_V can at most permute its eigenvalues. Thus the G_V orbit of x is finite, yet irreducible since G_V is and hence reduced to $\{\bar{x}\}$. Equivalently $x + \mathfrak{u}$ is G_V stable. By 2.3.9 again $kx + \mathfrak{u}$ is $\mathfrak{g}_V$ stable. In particular $k\bar{x}$ is a one dimensional $\mathfrak{g}$ module on which the radical $\mathfrak{r}$ of $\mathfrak{g}$ acts trivially. Since a semisimple Lie algebra has no non-trivial one dimensional modules, this proves that $\mathfrak{g}$ acts trivially on $k\bar{x}$. Consequently $[\mathfrak{g}, \mathfrak{r}] \subset \mathfrak{u}$. Moreover $[\mathfrak{g}, \mathfrak{g}] \subset \mathfrak{s} \oplus \mathfrak{u}$ and is hence algebraic.

Now assume that $\mathfrak{g}$ is almost algebraic. Let $\mathfrak{a}$ be an $\mathfrak{s}$ stable complement of $\mathfrak{u}$ in $\mathfrak{r}$. Then $[\mathfrak{s}, \mathfrak{a}] \subset \mathfrak{a} \cap [\mathfrak{g}, \mathfrak{u}] \subset \mathfrak{a} \cap \mathfrak{u} = 0$. Let $\mathfrak{a}_0$ be a maximal abelian subalgebra of $\mathfrak{a}$ of semisimple elements and set $\mathfrak{k}_0 = \mathfrak{s} \oplus \mathfrak{a}_0$. Let $\mathfrak{a}_1$ be a $\mathfrak{k}$ stable complement of $\mathfrak{a}_0 \oplus \mathfrak{u}$ in $\mathfrak{r}$. Then $[\mathfrak{k}, \mathfrak{a}_1] \subset \mathfrak{a}_1 \cap [\mathfrak{g}, \mathfrak{r}] = 0$. Given $x \in \mathfrak{a}_1$, let $x = x_s + x_n$ be its Jordan decomposition. Then $x_s \in \mathfrak{g}$ by hypotheses. Since x_s is polynomial in x one has $[\mathfrak{k}, x_s] = 0$, in particular $[\mathfrak{a}_0, x_s] = 0$ and so $x_s = 0$ by the maximality of $\mathfrak{a}_0$. Then $x = x_n \in \mathfrak{u}$ so $x = 0$, that is $\mathfrak{a}_0 = \mathfrak{a}$. Thus $\mathfrak{k} := \mathfrak{s} \oplus \mathfrak{a}$ is reductive and complements $\mathfrak{u}$. If $\mathfrak{g}$ is algebraic, $\mathfrak{k}$ is rationally reductive by 2.2.4. Conversely by 2.2.4, 2.4.4 and the above remarks any Lie algebra of this form is almost algebraic (resp. algebraic).

2.4.6 Let G be an affine algebraic group and H an algebraic subgroup. The notion of the quotient space G/H is not so easy as in the abstract case. For this one should consider the algebra of left kH semi-invariants of $R[G]$. That this algebra is sufficiently rich is provided by the following result. (Here k need not be assumed of characteristic zero).

Lemma. *There exists a finite dimensional left G submodule V of $R[G]$ and a line L of V such that $H = \{g \in G \mid gL = L\}$. If* $\operatorname{char} k = 0$, *then* $\operatorname{Lie} H = \{x \in \operatorname{Lie} G \mid xL \subset L\}$.

By noetherianity I_H admits a finite dimensional generating subspace W which can further be assumed to be a left H module. Let V_1 denote the left G module it generates in $R[G]$. This is again finite dimensional. Through the Leibnitz rule $gW = W$ for $g \in G$ implies $gI_H \subset I_H$. Evaluating at the identity implies $\xi(g) = 0$ for all $\xi \in I_H$ and so $g \in H$.

With $r = \dim W$, set $V = \wedge^r V_1$ which is a G module occurring in $R[G]$. It contains $L := \wedge^r W$ as a one dimensional H submodule. Clearly $gL = \wedge^r(gW) = L$ implies $gW = W$. The last part follows from the first and 2.4.2 (iii).

2.4.7 Define V as above. Then the $c^V_{\xi,\ell} : \xi \in V^*, \ell \in L$ are left H semi-invariants in $R[G]$. These separate the elements of G/H. Indeed suppose for some $\alpha \in k^*$ that $c^V_{\xi,\ell}(g) = \alpha c^V_{\xi,\ell}(g')$ for all $\xi \in V^*$. Then $g \in g'H$. However if $\alpha \neq 1$ these only define homogeneous functions of G/H, whilst the algebra of left H-invariants $R[G]^H$ may fail to be separating. This difficulty does not arise if H leaves the elements of L invariant. If V is the defining representation of G, that is G is viewed as a subgroup of $\mathrm{GL}(V)$, then this can be assured if

(*) *Every one dimensional H submodule of $T(V)$ is the trivial module.*

Indeed in this case every left H semi-invariant is invariant. For (*) to hold it is necessary and sufficient by 2.4.2 (iii) that the radical of $\operatorname{Lie} H$ be unipotent, for example if $\operatorname{Lie} H$ is a commutator subalgebra.

Assume (*) holds. A second difficulty, is that $R[G]^H$ need not be finitely generated. However this is less serious, as one may replace it by the subalgebra of $R[G]$ generated by the $c^V_{\xi,\ell} : \xi \in V^*$, which still separates points of G/H, is right G invariant and by 1.4.7 (1) is a left coideal. This latter algebra can be used to define the algebra $R[G/H]$ of regular functions on G/H and thereby defines the latter as an algebraic variety, namely $\operatorname{Spec} R[G/H]$. In particular the map $I \mapsto I \cap R[G/H]$ of $\operatorname{Spec} R[G]$ into $\operatorname{Spec} R[G/H]$ defines a morphism $\varphi : G \longrightarrow G/H$, whose comorphism φ^* is the original ring embedding $R[G/H] \hookrightarrow R[G]$. Moreover if $\xi \in R[G/H]$, then $\xi.g \in R[G/H]$ as noted above and $\varphi^*(\xi.g) = \varphi^*(\xi).g$, that is φ commutes with left multiplication. The separation property translates to give $\varphi^{-1}\varphi(g) = gH$ for all $g \in G$. From (A.3.10) one concludes that $\dim G/H = \dim G - \dim H$. Set $\mathfrak{g} = \operatorname{Lie} G$, $\mathfrak{h} = \operatorname{Lie} H$.

Lemma. *Under hypothesis (*) there is an exact sequence of vector spaces* $0 \longrightarrow \mathfrak{h} \longrightarrow \mathfrak{g} \longrightarrow T_{eH,G/H} \longrightarrow 0$.

Take $\xi \in R[G]^H \cap \mathfrak{m}$. Then $\xi(e) = 0$, since $\xi \in \mathfrak{m}$ and so $\xi(h) = \xi(eh) = \xi(e) = 0$, for all $h \in H$, since $\xi \in R[G]^H$. Consequently $\mathfrak{m}_{G/H} := \mathfrak{m} \cap R[G/H] \subset \mathfrak{m} \cap R[G]^H \subset I_H$.

By 2.2.2 there exists $g \in G$ such that $d_g\varphi : T_{g,G} \longrightarrow T_{gH,G/H}$ is surjective. Take $\xi \in R[G/H]$ and recall that $\xi.g^{-1} \in R[G/H]$. Given $x \in T_{g,H} = (\mathfrak{m}_g/\mathfrak{m}_g^2)^*$ one has $g^{-1}.x \in (\mathfrak{m}/\mathfrak{m}^2)^*$ by 2.1.7 (iv). Then by 2.2.2, $(\xi.g^{-1})((d_g\varphi)(x)) = x(\varphi^*(\xi.g^{-1})) = x(\varphi^*(\xi).g^{-1}) = (g^{-1}.x)(\varphi^*(\xi)) = \xi((d_e\varphi)(g^{-1}.x))$, that is $g^{-1}.(d_g\varphi) = (d_e\varphi).g^{-1}$. Consequently $d_e\varphi : T_{e,G} \longrightarrow T_{e,H,G/H}$ is also surjective.

Now $T_{eH,G/H} \subset \mathfrak{m}^*_{G/H}$, whilst $\mathfrak{h} = \{x \in \mathfrak{g} \mid x(I_H) = 0\}$ and so by the first part the image of the composed map $\mathfrak{h} \longrightarrow \mathfrak{g} \longrightarrow T_{eH,G/H}$ is zero. Then $\dim T_{eH,G/H} \geq \dim G/H = \dim G - \dim H = \dim \mathfrak{g} - \dim \mathfrak{h} = \dim \mathfrak{g}/\mathfrak{h}$ completes the proof.

2.4.8 Suppose in 2.4.7 that H is a normal subgroup of G. Then $R[G]^H = {}^H R[G]$. It then follows from 1.4.15 that $R[G]^H$ is a Hopf subalgebra of $R[G]$. If $R[G]^H$ is not finitely generated, one may replace it by a Hopf subalgebra which is finitely generated and still separates elements of G/H. This may be used to define G/H as an affine variety and hence as an algebraic group. From say 2.3.5, it follows that I_H is stable by conjugation G and hence so is $\mathfrak{h}$. Then by 2.3.9 (i), $\mathfrak{h}$ is an ideal of $\mathfrak{g}$. By 2.4.7, $\operatorname{Lie} G/H \xleftarrow{\sim} \mathfrak{g}/\mathfrak{h}$ and it is clear that this is an isomorphism of Lie algebras.

2.4.9 Let G be an affine algebraic group. Let H, H' be irreducible algebraic subgroups of G. By 2.4.3, $\operatorname{Lie}\langle H, H'\rangle$ is the Lie subalgebra $\langle \operatorname{Lie} H, \operatorname{Lie} H'\rangle$ of $\operatorname{Lie} G$ generated by $\operatorname{Lie} H$ and $\operatorname{Lie} H'$. By 2.3.6 (iii), $[H, H']$ is an irreducible algebraic group and so one may compute its Lie algebra to be the ideal of $\operatorname{Lie}\langle H, H'\rangle$ generated by $[\operatorname{Lie} H, \operatorname{Lie} H']$ as shown below in the case $H = H'$.

Lemma. *Let H be an irreducible algebraic subgroup of G. Then* $\operatorname{Lie}[H, H] = [\operatorname{Lie} H, \operatorname{Lie} H]$.

Set $\mathfrak{h} = \operatorname{Lie} H = (\mathfrak{m}/\mathfrak{m}^2)^*$. Let $\psi : H \times H \mapsto [H, H]$ be the morphism $\psi(h, h') = hh'h^{-1}h'^{-1}$. Then (see 2.3.9 (i)) for all $\xi \in R[[H, H]] = R[H]/I_{[H,H]}$ one has $\psi^*\xi = \xi_1\sigma(\xi_3) \otimes \xi_2\sigma(\xi_4)$. Now take $\xi \in \operatorname{Im} \mathfrak{m}$ and $x, y \in \mathfrak{h}$. A calculation similar to 2.1.10 (i) gives $(x \otimes y)(\psi^*\xi) = [x, y](\xi)$.

Recalling the definition (2.2.2) of $d_e\psi$ this gives $(d_e\psi)(x,y) = [x,y]$. Consequently $\mathrm{Lie}[H,H] \supset [\mathfrak{h},\mathfrak{h}]$.

Clearly $[\mathfrak{h},\mathfrak{h}]$ is an ideal of $\mathfrak{h}$. By 2.3.9 (ii) it is stable under conjugation by H. By 2.4.4 it is an algebraic Lie algebra and hence by 2.4.2 and the first part there exists a unique irreducible algebraic subgroup K of $\mathrm{Lie}[H,H]$ with $\mathrm{Lie}\, K = [\mathfrak{h},\mathfrak{h}]$. Since $\mathrm{Lie}\, h.K.h^{-1} = h.\,\mathrm{Lie}\, K.h^{-1} = h.[\mathfrak{h},\mathfrak{h}].h^{-1} = [\mathfrak{h},\mathfrak{h}]$ for all $h \in H$, it further follows from 2.4.2 that K is a normal subgroup of H. Since $\mathrm{Lie}\, K$ is a commutator subalgebra, 2.4.7, 2.4.8 apply and so H/K is defined as an irreducible algebraic group with Lie algebra $\mathfrak{h}/[\mathfrak{h},\mathfrak{h}]$. Since the latter is commutative, so is H/K. By 2.3.6 (iii) this implies $K = [H,H]$ as required.

2.4.10 Retain the notation of 2.4.3. One may ask if the conclusion of 2.4.3 (i) can be transposed to give that the multiplication map $U(\mathfrak{h}_{i_1}) \otimes U(\mathfrak{h}_{i_2}) \otimes \ldots \otimes U(\mathfrak{h}_{i_m}) \longrightarrow U(\mathfrak{h})$ is surjective. This is not immediate; but one may derive the even better result below.

Theorem. *Let $\mathfrak{g}$ be a Lie algebra of dimension $m < \infty$ and $x_1, x_2, \ldots, x_n$ a system of generators for $\mathfrak{g}$. Then there exist $i_1, i_2, \ldots, i_m \in \{1,2,\ldots,n\}$ such that the multiplication map $k[x_{i_1}] \otimes \ldots \otimes k[x_{i_m}] \longrightarrow U(\mathfrak{g})$ is surjective.*

Define a filtration $\mathcal{F}$ on $U(\mathfrak{g})$ by letting $\mathcal{F}^r(U(\mathfrak{g}))$ be the subspace of $U(\mathfrak{g})$ spanned by all monomials in the generators of length $\leq r$. Then $\mathcal{F}^r(\mathfrak{g}) := \mathfrak{g} \cap \mathcal{F}^r(U(\mathfrak{g}))$ defines a filtration of $\mathfrak{g}$ and it is immediate from this construction that $gr_{\mathcal{F}}(\mathfrak{g})$ generates $gr_{\mathcal{F}}(U(\mathfrak{g}))$ as a graded algebra. Universality then gives a surjection $U(gr_{\mathcal{F}}(\mathfrak{g})) \longrightarrow gr_{\mathcal{F}}(U(\mathfrak{g}))$ of graded algebras. (By the PBW theorem it is clear that this surjection is also bijective).

Consider the x_i as the (degree 1) generators of $gr_{\mathcal{F}}(\mathfrak{g})$. Suppose $i_1, i_2, \ldots, i_m \in \{1,2,\ldots,n\}$ are given such that the multiplication map $k[x_{i_1}] \otimes \ldots \otimes k[x_{i_m}] \longrightarrow U(gr_{\mathcal{F}}(\mathfrak{g}))$ is surjective. The composed map to $gr_{\mathcal{F}}(U(\mathfrak{g}))$ is also surjective and then by induction on filtration degree it follows that the x_i viewed as elements of $\mathfrak{g}$ satisfy (with the above choices of the i_j) the conclusion of the theorem. In other words the original problem has been reduced to the graded case.

Since $gr_{\mathcal{F}}(\mathfrak{g})$ is a graded Lie algebra with generators of degree 1, it is necessarily a nilpotent Lie algebra. Actually one may do better. Choose s sufficiently large so that $\mathfrak{g}$ is a subspace of $\mathcal{F}^s(U(\mathfrak{g}))$. Let I be the subspace of $U(gr_{\mathcal{F}}\mathfrak{g})$ spanned by elements of gradation degree $> s$. Obviously I is a two-sided ideal of finite codimension. Then $V := U(gr_{\mathcal{F}}\mathfrak{g}))/I$ is a finite dimensional graded $gr_{\mathcal{F}}\mathfrak{g}$ module which is faithful by choice of s. Then $\mathfrak{h} := gr_{\mathcal{F}}\mathfrak{g}$ can be viewed as a unipotent subalgebra of $\mathfrak{gl}(V)$. By 2.4.2, 2.4.4 there exist (unique) irreducible algebraic subgroups $H, H_1, H_2, \ldots, H_n$ of $\mathrm{GL}(V)$ such that $\mathfrak{h} = \mathrm{Lie}\, H$, $\mathfrak{h}_i := \mathrm{Lie}\, H_i$. Since $m = \dim \mathfrak{h} = \dim H$ one may choose $i_1, i_2, \ldots, i_m \in \{1,2,\ldots,n\}$ to satisfy the conclusion of 2.4.3 (i).

Set $\varphi = \pi\mu^{*m-1}$. Observe that $R[H]$ and the $R[H_i]$ inherit a gradation from $U(\mathfrak{h})$ and that φ may be viewed as the transpose of the multiplication map $U(\mathfrak{h}_{i_1}) \otimes \ldots \otimes U(\mathfrak{h}_{i_m}) \longrightarrow U(\mathfrak{h})$ and hence is a graded map of graded vector spaces. Generalizing the (Kunneth) formula in the first part of 2.1.4 to μ^{*m-1} gives for each integer $r \geq 1$, linear maps

$$R[H]/\mathfrak{n}_r \hookrightarrow \bigotimes_{i=1}^{m} R[H_i]/\varphi(\mathfrak{m}^r) \xrightarrow{\sim} \sum_{r_1+r_2+\cdot+r_m=r} \bigotimes_{i=1}^{m} R[H_i]/\pi_i(\mathfrak{m}^{r_i})$$

where $\mathfrak{n}_r := \varphi^{-1}\varphi(\mathfrak{m}^r) \supset \mathfrak{m}^r$. The $\mathfrak{n}_r$ are *graded* ideals satisfying $\cap \mathfrak{n}_r = 0$ by injectivity in 2.4.3 (i). Thus there exists $f : \mathbb{N} \longrightarrow \mathbb{N}$ satisfying $f(r) \longrightarrow \infty$ as $r \longrightarrow \infty$ (even $sf(r) \geq r$) such that $\mathfrak{n}_r \subset \mathfrak{m}^{f(r)}$ for all $r \in \mathbb{N}$. Since $(R[H_i]/\pi_i(\mathfrak{m}^{r_i}))^* = \mathcal{F}_c^{r_i}(U(\mathfrak{h}_i))$ (where $\mathcal{F}_c$ denotes the canonical filtration) and is finite dimensional, it follows that the multiplication map

$$\sum_{r_1+r_2+\ldots+r_m=r} \bigotimes_{i=1}^{m} \mathcal{F}_c^{r_i}(U(\mathfrak{h}_i)) \longrightarrow \mathcal{F}_c^{r}(U(\mathfrak{h}))$$

contains $\mathcal{F}_c^{f(r)}(U(\mathfrak{h}))$ in its image. This proves the theorem.

2.4.11 Though the conclusion of 2.4.10 does not quite give a basis of $U(\mathfrak{g})$ it can nevertheless be useful especially in the quantum situation (cf. 4.4.6(iii)) since $U_q(\mathfrak{g})$ does not possess a natural analogue of $\mathfrak{g}$. A typical application is the following.

Corollary. Let M be a $U(\mathfrak{g})$ module with cyclic vector m. *If* $\dim k[x_i]m < \infty$ *for a system of generators* $x_1, x_2, \ldots, x_n$ *of* $\mathfrak{g}$, *then* $\dim M < \infty$.

For all $a \in U(\mathfrak{g})$, one has $x_i am = ((\operatorname{ad} x_i)a)m + ax_i m$. Thus $k[x_i]am \subset (k[(\operatorname{ad} x_i)]a)m + ak[x_i]m$. The first term in the sum is finite dimensional by 1.2.10 and the second by hypothesis. It follows that $\dim k[x_i]m' < \infty$ for all $m' \in M$. Then the required conclusion is immediate from 2.4.10.

2.5 Comments and Complements

2.5.1 Excepting 2.4.10 the results of this chapter are well-known. Most have been reworked; but 2.1.9 and 2.3.5 have been taken from E. Abe [Ab1] and 2.4.3 and 2.4.6 from A. Borel [Bor1]. For more details on tangent spaces, see [Sh1, Chap. 2]. For a more detailed analysis of semisimple algebraic groups one may consult [Sp1]. For their representation theory see [Ja3]. Many authors have contributed to these remarkable results, however we specifically mention that 2.1.8 is due to B. Kostant and 2.4.5 to C. Chevalley.

2.5.2 The reduction to the nilpotent case in 2.4.10 is due to L. Makar-Limanov [M-L1] who then obtained a weaker result. Its main interest is pedagogical since one must use both the enveloping algebra and the algebraic group to obtain its conclusion. The conclusion of 2.4.11 is an unpublished result of B. Kostant. The idea of the present proof came from discussions with R. Rentschler. The latter is a corollary of a more general (and much more difficult) result of O. Gabber [G1] which asserts that the associated variety of any finitely generated $U(\mathfrak{g})$ module is involutive. Further discussion of this important topic can be found in [J9].

2.5.3 One cannot obtain 2.4.10 more directly (without first grading) in virtue of the following example pointed out to me by V. Berkovitch. Take $\varphi \in k[[x]]$ algebraically independent over x. Take $I = (x_1 - \varphi(x_2))k[[x_1, x_2]]$ and for each integer $\ell > 0$, set $\hat{\mathfrak{m}}_\ell = I + \mathfrak{m}^\ell$, where $\mathfrak{m}$ denotes the augmentation ideal $\langle x_1, x_2 \rangle$ of $k[[x_1, x_2]]$. Since $k[[x_1, x_2]]/I$ is a local noetherian ring whose maximal ideal is the image $\bar{\mathfrak{m}}$ of $\mathfrak{m}$, it follows as in 2.1.6 that $\cap \bar{\mathfrak{m}}^\ell = 0$ and hence that $\cap \hat{\mathfrak{m}}_\ell = I$. Set $\mathfrak{m}_\ell = \hat{\mathfrak{m}}_\ell \cap k[x_1, x_2]$. Then $\cap \mathfrak{m}_\ell = I \cap k[x_1, x_2] = 0$, by the choice of φ. Yet $\mathfrak{m}^2$ does not contain $\mathfrak{m}_\ell$ for any ℓ.

2.5.4 Let G be an affine algebraic group. How should one describe $R[G]^\times$ if $\operatorname{char} k > 0$? The calculations in 2.1.5 and 2.1.9 suggest introducing divided powers of the generators of $\mathfrak{g}$. However the problem is more subtle than describing $U(\mathfrak{g})^*$ since $U(\mathfrak{g})$ has a more complicated algebraic structure. For $\mathfrak{g}$ semisimple the problem was solved by B. Kostant [Ko3] using a Chevalley basis which in particular allows $\mathfrak{g}$ to be defined over $\mathbb{Z}$. Powers of root vectors are then replaced by divided powers and from the resulting $\mathbb{Z}$ form $U_{\mathbb{Z}}(\mathfrak{g})$ one obtains $R[G]^\times$ as $U_{\mathbb{Z}}(\mathfrak{g}) \otimes_{\mathbb{Z}} k$. The analogue of this construction for $U_q(\mathfrak{g})$ when $\operatorname{char} k = 0$, but q is a root of unity was given by G. Lusztig [L2]. Its study is however outside the scope of this present text. On the other hand, divided powers are used in globalizing the crystal basis (6.1.7).

2.5.5 T. Springer has informed me that the new edition of [Sp1] will treat algebraic groups over a field which is not necessarily algebraically closed.

Chapter 3. Encoding the Cartan Matrix

Unless otherwise indicated it is assumed in the remainder of this text that char $k = 0$ and that q is an indeterminate.

3.1 Quantum Weyl Algebras

Modifying the construction (1.2.10) of the Weyl algebras using skew derivations (1.2.11) not only generalizes these algebraic objects themselves, but also enables one to encode the information contained in the Cartan matrix which lies at the heart of the theory of semisimple Lie algebras. This is our present subject matter.

3.1.1 Let ℓ be an integer > 0 and $C = (c_{ij})_{i,j=1}^{\ell}$ a matrix of rank $r \leq \ell$ with integer entries. A realization of C is a $(2\ell - r)$ dimensional $\mathbb{Q}$ vector space $\mathfrak{h}_{\mathbb{Q}}$, a subset $\pi := \{\alpha_i\}_{i=1}^{\ell}$ of linearly independent elements of $\mathfrak{h}_{\mathbb{Q}}^*$ and a non-degenerate bilinear form $\{\alpha, \beta\} \mapsto (\alpha, \beta)$ on $\mathfrak{h}_{\mathbb{Q}}^*$ such that $c_{ij} = (\alpha_i, \alpha_j)$: $i, j = 1, 2, \ldots, \ell$.

Lemma. *Any matrix C of the above type admits a realization and moreover the form can be assumed to be symmetric if C is symmetric.*

Consider C as the matrix of the linear transformation of the n dimensional vector space $\mathbb{Q}^{\ell}$ with basis $\{\alpha_i\}_{i=1}^{\ell}$. By the diagonal action C induces a linear transformation $C^{(m)}$ on the mth wedge product $\wedge^m \mathbb{Q}^{\ell}$ of $\mathbb{Q}^{\ell}$ whose image is $\wedge^m (C\mathbb{Q}^{\ell})$. Consequently $C^{(r)}$ has rank 1.

Assume C is symmetric. Then $C^{(r)}$ is symmetric. It easily follows that the diagonal entries of $C^{(r)}$ cannot all be zero and so there is a relabelling of $\{\alpha_i\}_{i=1}^{\ell}$ such that $C^{(r)}$ does not vanish on $\alpha_1 \wedge \alpha_2 \wedge \ldots \wedge \alpha_r$. Then $(c_{ij})_{i,j=1}^{r}$ is non-degenerate. Augment the basis of $\mathbb{Q}^{\ell}$ to a basis $\{\alpha_i\}_{i=1}^{2\ell-r}$ of $\mathbb{Q}^{2\ell-r}$ and equip the latter with the symmetric bilinear form defined by

$$(*) \qquad (\alpha_i, \alpha_j) = \begin{cases} c_{ij} & : i, j \leq \ell \,, \\ 1 & : r < i \leq \ell, \ j = \ell - r + i \ \text{ or vise-versa} \,, \\ 0 & : \text{otherwise} \,, \end{cases}$$

which is easily checked to be non-degenerate. For the non-symmetric case permute both sets of indices for $i, j = 1, 2, \ldots, \ell$ so that $(c_{ij})_{i,j=1}^{r}$ is non-degenerate and extend C as in $(*)$. Finally apply the inverse permutations to recover C in the top left-hand corner.

Remark. The first part of the proof also applies if C is just quasi-symmetric (see below).

3.1.2 The rather clumsy notation of the previous paragraph arises from identifications which will eventually be made. In this further restrictions on C will be imposed. Call C quasi-symmetric if $c_{ij} = 0$ implies $c_{ji} = 0$ and specializable if $c_{ii} > 0$ and even for all i. Then one may write $\alpha_i^\vee = 2\alpha_i/(\alpha_i, \alpha_i)$ and the numbers $a_{ij} := 2c_{ij}/c_{ii} = (\alpha_i^\vee, \alpha_j)$ are defined and form the entries of the generalized Cartan matrix A in the sense of Kac-Moody. Call C integrable if $a_{ij} \in -\mathbb{N}$ for $i \neq j$. Such C give rise to the study of those (Kac-Moody) Lie algebras described in [Ka1]. Then $\mathfrak{h}_{\mathbb{Q}}$ is a Cartan subalgebra and π its set of simple roots. The form in the conclusion of 3.1.1 is called the Cartan form. The $\alpha_i^\vee : i = 1, 2, \ldots, \ell$ form the set $\pi^\vee$ of coroots. Since $(\alpha_i^\vee)^\vee = \alpha_i$ it follows that π is the set of coroots for the dual simple root system defined by $\pi^\vee$. Finally C is said to be of finite type if A is the Cartan matrix of a finite dimensional Kac-Moody Lie algebra (that is a semisimple Lie algebra). By [Ka1, Thm. 4.3], C integrable is of finite type if and only if it is positive definite. When C is integrable, π (or C) is said to be simply-laced if $a_{ij} < 1$ implies $a_{ij} = -1$.

The theory of semisimple and even of Kac-Moody Lie algebras is extremely rich. This is no less so for their quantum analogues. Much of this translates to remarkably deep but purely combinatorial questions which can be entirely described (but not at all solved!) in terms of C. This phenomenon is perhaps one of the most remarkable and inspiring aspects of the subject.

Slightly abusing standard terminology C itself will be called a Cartan matrix. Set $Q(\pi) = \mathbb{Z}\pi \subset \mathfrak{h}_{\mathbb{Q}}^*$ with respect to a realization of C. Set $\mathfrak{h}_{\mathbb{Z}}^* = \{\lambda \in \mathfrak{h}_{\mathbb{Q}}^* | (\alpha, \lambda) \in \mathbb{Z}, \forall\, \alpha \in \pi\}$. Set $Q^{\pm}(\pi) = \pm\mathbb{N}\pi$ and define an order relation on $Q(\pi)$ by $\mu \geq \nu$ if $\mu - \nu \in Q^+(\pi)$. Given $\nu \in Q^+(\pi)$, one may write

$$\nu = \sum_{\alpha \in \pi} k_\alpha \alpha : k_\alpha \in \mathbb{N}$$

and one defines $|\nu| = \sum k_\alpha$, called the order of ν.

3.1.3 Let C be a Cartan matrix. The quantum pre-Weyl algebra $\tilde{A}_C$ defined by C is the algebra over $k(q)$ with identity and generators $x_\alpha, y_{-\beta} : \alpha, \beta \in \pi$ satisfying the relations

$$(*) \qquad x_\alpha y_{-\beta} - q^{-(\alpha,\beta)} y_{-\beta} x_\alpha = \delta_{\alpha\beta} \,.$$

Let $\tilde{U}^+$ (resp. $\tilde{U}^-$) denote the subalgebra of $\tilde{A}_C$ generated by the x_α (resp. $y_{-\alpha}$) : $\alpha \in \pi$. Applying the diamond lemma (A.2.14), it follows that $\tilde{U}^+$ and

$\tilde{U}^-$ are freely generated and moreover the multiplication map $\tilde{U}^+ \otimes \tilde{U}^- \longrightarrow \tilde{A}_C$ is a linear bijection.

For each $\lambda \in \mathfrak{h}_{\mathbb{Z}}^*$, let τ_λ^- denote the automorphism of $\tilde{A}_C$ defined by $\tau_\lambda^-(x_\alpha) = q^{(\alpha,\lambda)}x_\alpha$, $\tau_\lambda^-(y_{-\beta}) = q^{-(\beta,\lambda)}y_{-\beta}, \forall\, \alpha, \beta \in \pi$. It is clear that for each monomial a in the generators there exists $\nu \in Q(\pi)$ such that $\tau_\lambda^-(a) = q^{(\nu,\lambda)}a$, $\forall\, \lambda \in \mathfrak{h}_{\mathbb{Z}}^*$. By the non-degeneracy of the Cartan form, ν is uniquely determined by this property. It is called the weight of a and one writes $a = a_\nu$. Moreover the set of elements of $\tilde{A}_C$ of a given weight ν form a subspace, called the ν-weight subspace and denoted $(\tilde{A}_C)_\nu$. Clearly

$$\tilde{A}_C = \bigoplus_{\nu \in Q(\pi)} (\tilde{A}_C)_\nu$$

defines a gradation of $\tilde{A}_C$. Similar considerations apply to $\tilde{U}^\pm$ which are graded by $Q^\pm(\pi)$.

For each $\beta \in \pi$ define $\delta'_{-\beta} \in \operatorname{End} \tilde{A}_C$ through

$$\delta'_{-\beta}(a_\nu) = a_\nu y_{-\beta} - q^{-(\nu,\beta)} y_{-\beta} a_\nu \; .$$

One checks easily that

$$\delta'_{-\beta}(a_\nu b_\mu) = (\delta'_{-\beta} a_\nu)\tau_{-\beta}^-(b_\mu) + a_\nu(\delta'_{-\beta} b_\mu)$$

and thus in the language of 1.2.11, $\delta'_{-\beta}$ is a $\tau_{-\beta}^-$-rightskew derivation of $\tilde{A}_C$. Moreover $\delta'_{-\beta}(x_\alpha) = \delta_{\alpha\beta}$ from the defining relation and so $\delta'_{-\beta}$ leaves $\tilde{U}^+$ stable.

Now consider the $\delta'_{-\beta} : \beta \in \pi$ just as endomorphisms of $\tilde{U}^+$. If V is a subspace of $\tilde{U}^+$ stable by these endomorphisms, then so are its graded components. It follows easily that there is a unique maximal proper ideal K_C^+, or simply K^+, of $\tilde{U}^+$ contained in the graded complement of $k(q)$ and stable by the $\delta'_{-\beta} : \beta \in \pi$. Moreover K^+ is graded. Set $U^+ = \tilde{U}^+/K^+$. On the other hand since $\tilde{U}^-$ is free the map sending 1 to $Id_{\tilde{U}^+}$ and $y_{-\beta}$ to $\delta'_{-\beta}$ extends to a homomorphism φ^- of $\tilde{U}^-$ into $\operatorname{End} \tilde{U}^+$. Let U^- denote its image. By construction the generators $\delta'_{-\beta}$ of U^- factor to endomorphisms of U^+. The quantum Weyl algebra A_C is defined to be the subalgebra of $\operatorname{End} U^+$ generated by the x_α and the $\delta'_{-\beta}$ acting on the *right*. When the entries of C are all zero or when $q = 1$, one recovers the Weyl algebra A_n described in 1.2.10. It is clear that the multiplication map $U^{+\,\mathrm{opp}} \otimes U^{-\,\mathrm{opp}} \longrightarrow A_C$ is a linear bijection.

Let θ be the antiautomorphism of the $\tilde{U}^\pm$ which is the identity on generators and set $\varphi_0^- = \varphi^-\theta$, $\tilde{U}_0^\pm := \tilde{U}^{\pm\,\mathrm{opp}}$. As endomorphisms of $\tilde{U}^+$ acting on the *right* one has $x_\alpha\delta'_{-\beta} - q^{-(\alpha,\beta)}\delta'_{-\beta}x_\alpha = \delta_{\alpha\beta}$. Thus $a \otimes b \mapsto (c \mapsto \varphi_0^-(b)(ca))$ defines an algebra homomorphism Φ^- of $\tilde{A}_C \cong \tilde{U}_0^+ \otimes \tilde{U}_0^-$ into $\operatorname{End} \tilde{U}^+$ whose image in $\operatorname{End} U^+$ coincides with A_C.

3.1.4 There is actually more structure in A_C than is immediately obvious. For each $\beta \in \pi$, define $\tau_\beta^+ \in Aut\, \tilde{A}_C$ by $\tau_\beta^+(b_\nu) = q^{(\beta,\nu)} b_\nu$ and $\delta''_{-\beta}$ to be the (unique) τ_β^+-rightskew derivation of $\tilde{U}^+$ satisfying $\delta''_{-\beta}(x_\alpha) = \delta_{\alpha\beta}$.

Consider the $\delta'_{-\beta}, \delta''_{-\beta} : \beta \in \pi$ as endomorphisms of $\tilde{U}^+$.

Lemma

(i) $\delta'_{-\beta}\delta''_{-\gamma} = q^{(\gamma,\beta)}\delta''_{-\gamma}\delta'_{-\beta}$, *for all* $\beta, \gamma \in \pi$.

(ii) K^+ *is* $\delta''_{-\gamma} : \gamma \in \pi$ *stable.*

(i) It is enough to show the relation holds on each weight vector a_ν. The proof is by induction on $|\nu|$. It is trivial for $\nu = 0$. Then it is enough to check that it holds on $a_\nu x_\alpha$ given that it holds on a_ν. One has

$$\begin{aligned}\delta'_{-\beta}\delta''_{-\gamma}(a_\nu x_\alpha) &= \delta'_{-\beta}(q^{(\gamma,\alpha)}(\delta''_{-\gamma}a_\nu)x_\alpha + a_\nu\delta_{\gamma\alpha}) \\ &= q^{(\gamma,\alpha)-(\alpha,\beta)}(\delta'_{-\beta}\delta''_{-\gamma}a_\nu)x_\alpha + q^{(\gamma,\alpha)}(\delta''_{-\gamma}a_\nu)\delta_{\beta\alpha} + (\delta'_{-\beta}a_\nu)\delta_{\gamma\alpha}\end{aligned}$$

whilst

$$\delta''_{-\gamma}\delta'_{-\beta}(a_\nu x_\alpha) = q^{(\gamma,\alpha)-(\alpha,\beta)}(\delta''_{-\gamma}\delta'_{-\beta}a_\nu)x_\alpha + q^{-(\alpha,\beta)}(\delta'_{-\beta}a_\nu)\delta_{\gamma\alpha} + (\delta''_{-\gamma}a_\nu)\delta_{\beta\alpha}\,,$$

hence the assertion.

(ii) Consider $K^+ + \delta''_{-\gamma}K^+ : \gamma \in \pi$. It is graded since K^+ is graded and so an ideal since $\delta''_{-\gamma}$ is a rightskew derivation. One has $1 \notin \delta''_{-\gamma}K^+$ for otherwise $x_\gamma \in K^+$ and then $K^+ \ni \delta'_{-\gamma}x_\gamma = 1$. On the other hand $\delta''_{-\gamma}K^+$ is $\delta'_{-\beta} : \beta \in \pi$ stable by (i). Hence $\delta''_{-\gamma}K^+ \subset K^+$ as required.

3.1.5 For each $\alpha \in \pi$ define $\delta'_\alpha \in \operatorname{End} \tilde{A}_C$, through

$$\delta'_\alpha(b_{-\nu}) = x_\alpha\, b_{-\nu} - q^{-(\alpha,\nu)} b_{-\nu}x_\alpha\ .$$

One easily checks that

$$\delta'_\alpha(a_{-\mu}b_{-\nu}) = \delta'_\alpha(a_{-\mu})b_{-\nu} + \tau_\alpha^+(a_{-\mu})\delta'_\alpha(b_{-\nu})$$

and so δ'_α is a τ_α^+-leftskew derivation of $\tilde{A}_C$. Since $\delta'_\alpha(y_{-\beta}) = \delta_{\alpha\beta}$ it leaves $\tilde{U}^-$ stable. As in 3.1.3 define K_C^-, or simply K^-, to be the unique maximal proper graded ideal of $\tilde{U}^-$ stable by the $\theta\delta'_\alpha\theta : \alpha \in \pi$.

Let δ''_α be the unique $\tau_{-\alpha}^-$-leftskew derivation of $\tilde{U}^-$ satisfying $\delta''_\alpha(y_{-\beta}) = \delta_{\alpha\beta}$. As in 3.1.4 one may check that K^- is $\theta\delta''_\alpha\theta : \alpha \in \pi$ stable.

Lemma. $K^- = \ker \varphi^-$.

Since Φ^- is an algebra homomorphism of $\tilde{U}_0^+ \otimes \tilde{U}_0^-$ into $\operatorname{End} \tilde{U}^+$ it follows that $\ker \Phi^- = \tilde{U}_0^+ \otimes \ker \varphi_0^-$ is a two-sided ideal of $\tilde{U}_0^+ \otimes \tilde{U}_0^-$. In particular $\ker \varphi_0^-$ is proper two-sided ideal of $\tilde{U}_0^-$ stable by the $\delta'_\alpha : \alpha \in \pi$. Hence $\ker \varphi^- \subset K^-$. Consider now the reverse inclusion.

Since K^- is graded it is enough to show that each weight vector $b_{-\nu} \in K^-$ lies in $\ker \varphi^-$. This is established by induction on $|\nu|$. For $\nu = 0$, it is trivial. Suppose $b_{-\nu} \in K^-$ with $|\nu| > 0$. Now $\varphi^-(b_{-\nu})a_\mu = 0$ for $\mu < \nu$ and in particular for $a_\mu = 1$. Through the action of $\delta'_\alpha(\theta(b_{-\nu}))$ on a_μ, one has

$$(*) \qquad \varphi^-(\theta\delta'_\alpha\theta(b_{-\nu}))a_\mu = \varphi^-(b_{-\nu})(a_\mu x_\alpha) - q^{-(\alpha,\nu)}\varphi^-(b_{-\nu})(a_\mu)x_\alpha \,.$$

Yet $\theta\delta'_\alpha\theta(b_{-\nu}) \in K^- \subset \ker \varphi^-$ by the induction hypothesis, so the left hand side above vanishes. Induction on $|\mu|$ then establishes that $\varphi^-(b_{-\nu})a_\mu = 0$ for all $\mu \in Q^+(\pi)$. Hence $\varphi^-(b_{-\nu}) = 0$, as required.

3.1.6 Consider the action of U^- on $\tilde{U}^+$ defined by the $\delta'_{-\beta} : \beta \in \pi$. Any invariant vector in $\tilde{U}^+$ is a sum of invariant weight vectors and clearly any non-scalar invariant weight vector lies in K^+. It is not obvious if K^+ is generated by its invariants though remarkably this can be established for C integrable (4.3.8). In any case U^+ can be viewed as being obtained by successively taking quotients by the ideal generated by non-scalar invariants. In particular

Lemma. *The set of U^- invariants of U^+ reduces to scalars.*

3.1.7 Let $\varepsilon^\pm$ denote the projection of $U^\pm$ onto $k(q)$ defined by the decomposition

$$U^\pm = \Big(\bigoplus_{\nu>0} U^\pm_\nu\Big) \oplus k(q)1 \,.$$

Obviously $\varepsilon^\pm$ is a homomorphism of algebras. Define a pairing $\varphi : U^- \times U^+ \longrightarrow k(q)$ by $\varphi(a,b) = \varepsilon^+(\varphi^-(a)b)$. It is clear that $\varphi(a_{-\nu}, b_\mu) = 0$ unless $\mu = \nu$.

Lemma. *The pairing φ is non-degenerate.*

Take $a_\mu \in U^+_\mu$ non-zero. Show that $\varphi(U^-_{-\mu}, a_\mu) \neq 0$ by induction on $|\mu|$. It is clear if $\mu = 0$. If $|\mu| > 0$, there exists $\alpha \in \pi$ such that $\varphi^-(y_{-\alpha})a_\mu \neq 0$ by 3.1.6. Then there exists $b_{-\mu+\alpha} \in U^-_{-\mu+\alpha}$ such that $\varphi(b_{-\mu+\alpha}, \varphi^-(y_{-\alpha})a_\mu) \neq 0$ by the induction hypothesis. Thus $\varphi(b_{-\mu+\alpha}y_{-\alpha}, a_\mu) \neq 0$ as required.

Take $b_{-\nu} \in U^-_{-\nu}$ non-zero. Show that $\varphi(b_{-\nu}, U^+_\nu) \neq 0$ by induction on $|\nu|$. It is clear if $\nu = 0$. If $|\nu| > 0$, there exists by the analogue of 3.1.6 some $\alpha \in \pi$ such that $\theta\delta'_\alpha\theta(b_{-\nu}) \neq 0$. Then by the induction hypothesis there exists $a_{\nu-\alpha} \in U^+_{\nu-\alpha}$ such that $\varphi^-(\theta\delta'_\alpha\theta(b_{-\nu}))a_{\nu-\alpha}$ is a non-zero scalar. Since $\varphi^-(b_{-\nu})a_{\nu-\alpha} = 0$, it follows from 3.1.5 (*) that $\varphi^-(b_{-\nu})(a_{\nu-\alpha}x_\alpha)$ is also a non-zero scalar. Thus $\varphi(b_{-\nu}, a_{\nu-\alpha}x_\alpha) \neq 0$, as required.

3.1.8 Let V be a graded vector space

$$V = \bigoplus_{\mu \in Q} V_\mu \, .$$

Then the graded dual of V is defined to be the graded vector space

$$V^{\succ} := \bigoplus_{\mu \in Q} V_\mu^*$$

where V_μ^* is identified with the subspace of V^* of linear functions vanishing on the $V_\nu : \nu \neq \mu$. It is useful to consider $V^{\succ}$ when the V_μ are all finite dimensional, since then $\dim V_\mu^* = \dim V_\mu$ and hence $V \xrightarrow{\sim} V^{\succ\succ}$.

3.1.9 Take $\mu \in Q^+(\pi)$. Then $\dim U_\mu^+$ and $\dim U_{-\mu}^-$ are finite since this already holds for the tildered quantities. By 3.1.7 they are equal, moreover U^- is just the graded dual of U^+. It follows that the product μ on U^+ defines a coproduct Δ on U^- by the rule

$$(*) \qquad \varphi(a, bc) = (\varphi \times \varphi)(\Delta(a), b \otimes c) \, .$$

Unfortunately this does not make U^- a bialgebra since it is not guaranteed (and is false) that $\Delta(ab) = \Delta(a)\Delta(b)$. On the other hand there is an action (given by φ^-) of U^- on U^+ which is moreover compatible with φ, that is

$$(**) \qquad \varphi(ab, c) = \varphi(a, \varphi^-(b)c) \, .$$

The interpretation of the action of the $\delta'_{-\beta} = \varphi^-(y_{-\beta})$ as a Leibnitz rule involves the additional quantities $\tau_{-\beta}^- : \beta \in \pi$ which generate an abelian subgroup of $\operatorname{End} U^+$. This suggests one should extend the above pairing to algebras augmented by this subgroup in order to recover the anticipated bialgebra structure. This is achieved below.

3.1.10 View $\mathfrak{h}_{\mathbb{Z}}^*$ as a multiplicative group by introducing the quantities $\tau^-(\lambda) : \lambda \in \mathfrak{h}_{\mathbb{Z}}^*$ satisfying $\tau^-(\lambda)\tau^-(\mu) = \tau^-(\lambda + \mu)$. The $\tau^-(\lambda)$ may be adjoined to any algebra with weight space decomposition having weights in $Q(\pi)$ by defining $\tau^-(\lambda)a_\nu = q^{(\nu,\lambda)}a_\nu\tau^-(\lambda)$. Similarly one may define $\tau^+(\lambda)$ requiring in this case that $\tau^+(\lambda)a_\nu = q^{(\lambda,\nu)}a_\nu\tau^+(\lambda)$. (Of course $\tau^-(\lambda) = \tau^+(\lambda)$ if the Cartan form is symmetric as will be eventually assumed).

Let $T^\pm$ denote the (free) abelian group on generators $\tau^\pm(\beta) : \beta \in \pi$. Then from the above construction the $\tilde{V}^\pm := T^\pm \tilde{U}^\pm$ acquire algebra structures. The action of $\tilde{U}^-$ on $\tilde{U}^+$ extends to an action of $\tilde{V}^-$ by setting $\varphi^-(\tau^-(\beta)) = \tau_\beta^-$. Then the coproduct Δ^- on $\tilde{V}^-$ compatible with $\delta'_{-\beta}$ being a $\tau_{-\beta}^-$-rightskew derivation and $\tau^-(\gamma)$ being an automorphism is given by

$$(-) \quad \Delta^-(y_{-\beta}) = y_{-\beta} \otimes \tau^-(-\beta) + 1 \otimes y_{-\beta} \, , \quad \Delta^-(\tau^-(\gamma)) = \tau^-(\gamma) \otimes \tau^-(\gamma)$$

for all $\beta \in \pi$, $\gamma \in Q(\pi)$. Of course using 1.1.11 one easily checks that this is indeed a coproduct and that combined with the coidentity ε^- given in 3.1.7 and extended by $\varepsilon^-(\tau(\gamma)) = 1$ this gives $\tilde{V}^-$ a bialgebra structure.

Similarly $\tilde{U}^+$ acts on $\tilde{V}^-$ through the map φ^+ defined by $\varphi^+(x_\alpha) = \delta'_\alpha$, $\varphi^+(\tau^+(\alpha)) = \tau^+_\alpha$ for all $\alpha \in \pi$. Then the coproduct Δ^+ on $\tilde{V}^+$ compatible with δ'_α being a τ^+_α-leftskew derivation and $\tau^+(\gamma)$ being an automorphism is given by

$$(+) \qquad \Delta^+(x_\alpha) = x_\alpha \otimes 1 + \tau^+(\alpha) \otimes x_\alpha \,, \quad \Delta^+(\tau^+(\gamma)) = \tau^+(\gamma) \otimes \tau^+(\gamma)$$

for all $\alpha \in \pi$, $\gamma \in Q(\pi)$. Combined with ε^+ extended by $\varepsilon^+(\tau^+(\gamma)) = 1$, this gives $\tilde{V}^+$ a bialgebra structure.

As a vector space $\tilde{V}^+$ (resp. $\tilde{V}^-$) is just a direct sum of copies $\tilde{U}^+\tau^+(\gamma)$ (resp. $\tilde{U}^-\tau^-(\gamma)$) $: \gamma \in Q(\pi)$ of $\tilde{U}^+$ (resp. $\tilde{U}^-$) and as an algebra is graded by them. It follows that the action of $\tilde{V}^-$ (resp. $\tilde{V}^+$) on $\tilde{U}^+$ (resp. $\tilde{U}^-$) can be extended to $\tilde{V}^+$ (resp. $\tilde{V}^-$) in a manner compatible with the Leibnitz rule by setting $\delta'_{-\beta}T^+ = 0$ (resp. $\delta'_\beta T^- = 0$) for all $\beta \in \pi$ and then (necessarily) setting $\tau^-_\alpha(\tau^+(\beta)) = q^{(\beta,\alpha)}$ (resp. $\tau^+_\alpha(\tau^-(\beta)) = q^{-(\alpha,\beta)}$) $: \alpha, \beta \in Q(\pi)$. By construction

$$(*) \qquad \varphi^-(y)(xx') = (\varphi^-(y_1)x)(\varphi^-(y_2)x') \;\text{ where }\; \Delta^-(y) = y_1 \otimes y_2 \,.$$

Extend the antiautomorphism θ of $\tilde{U}^-$ to $\tilde{V}^-$ by $\theta(\tau(\lambda)) = \tau(-\lambda)$ and set $\varphi_0^+(x)(y) = \theta(\varphi^+(x)(\theta(y)))$, for all $x \in \tilde{V}^+$, $y \in \tilde{V}^-$. It is immediate that

$$(**) \qquad \varphi_0^+(x)(yy') = (\varphi_0^+(x_2)y)(\varphi_0^+(x_1)y') \;\text{ where }\; \Delta^+(x) = x_1 \otimes x_2 \,.$$

Proposition. *There exists a unique bilinear form φ on $\tilde{V}^- \times \tilde{V}^+$ satisfying $\varphi(1,1) = 1$ and for all $y, y' \in \tilde{V}^-$ and all $x, x' \in \tilde{V}^+$,*

(i) $\varphi(yy', x) = \varphi(y, \varphi^-(y')x)$,
(ii) $\varphi(y, xx') = \varphi(\varphi_0^+(x')y, x)$,
Moreover for all $y \in \tilde{U}^-$, $x \in \tilde{U}^+$, $\gamma, \delta \in Q(\pi)$,
(iii) $\varphi(\tau^-(\gamma)y, \tau^+(\delta)x) = \varphi(y, x)q^{(\delta,\gamma)}$,
(iv) $\varphi(y, x) = \varepsilon^-(\varphi_0^+(x)y) = \varepsilon^+(\varphi^-(y)x)$,
In addition for all $y, y' \in \tilde{V}^-$ and all $x, x' \in \tilde{V}^+$,
(v) $\varphi(y, xx') = (\varphi \times \varphi)(\Delta^-(y), x \otimes x')$,
(vi) $\varphi(yy', x) = (\varphi \times \varphi)(y' \otimes y,\ \Delta^+(x))$,

Existence. Admitting the second equality in (iv), the first defines φ on $\tilde{U}^- \times \tilde{U}^+$ satisfying (i), (ii). To establish this second equality, it is enough to take $x \in \tilde{U}^+_\mu$, $y \in \tilde{U}^-_{-\mu}$ and to proceed by induction on $|\mu|$. It is trivial for $|\mu| = 0$. For $|\mu| > 0$ one may assume that $x = x_{\mu-\alpha}x_\alpha$, $y = y_{-(\mu-\beta)}y_{-\beta}$ for some $\alpha, \beta \in \pi$. Then

$$\begin{aligned}
\varepsilon^+(\varphi^-(y)x) &= \varepsilon^+(\varphi^-(y_{-(\mu-\beta)})(\delta'_{-\beta}(x_{\mu-\alpha})q^{-(\alpha,\beta)}x_\alpha + \delta_{\alpha\beta}x_{\mu-\alpha})) \\
&= q^{-(\alpha,\beta)}\varphi(y_{-(\mu-\beta)}, \delta'_{-\beta}(x_{\mu-\alpha})x_\alpha) + \delta_{\alpha\beta}\varphi(y_{-(\mu-\beta)}, x_{\mu-\alpha}) \\
&= q^{-(\alpha,\beta)}\varphi(\theta\delta'_\alpha\theta(y_{-(\mu-\beta)}), \delta'_{-\beta}(x_{\mu-\alpha})) + \delta_{\alpha\beta}\varphi(y_{-(\mu-\beta)}, x_{\mu-\alpha}) \,,
\end{aligned}$$

by the induction hypothesis. Similarly

$$\begin{aligned}\varepsilon^-(\varphi_0^+(x)y) &= \varepsilon^-(\varphi_0^+(x_{\mu-\alpha})(\delta_{\alpha\beta}y_{-(\mu-\beta)} + q^{-(\alpha,\beta)}\theta(y_{-\beta}\delta'_\alpha\theta(y_{-(\mu-\beta)}))) \\ &= \delta_{\alpha\beta}\varphi(y_{-(\mu-\beta)}, x_{\mu-\alpha}) + q^{-(\alpha,\beta)}\varphi(\theta\delta'_\alpha\theta(y_{-(\mu-\beta)}), \delta'_{-\beta}(x_{\mu-\alpha}))\ ,\end{aligned}$$

as required. It is clear that (iii) extends φ to $\tilde{V}^- \times \tilde{V}^+$. One may check that φ so defined satisfies (i), (ii) by reversing the uniqueness argument below.

Uniqueness. Assume that φ is a bilinear form on $\tilde{V}^- \times \tilde{V}^+$ satisfying (i). Then for all $y \in \tilde{U}^-_{-\nu}$, $x \in \tilde{U}^+$, $\delta, \gamma \in Q(\pi)$ it follows from $(-)$ that $\varphi(\tau^-(\gamma)y, \tau^+(\delta)x) = \varphi(\tau^-(\gamma), \varphi^-(y)\tau^+(\delta)x) = \varphi(\tau^-(\gamma), \tau^+(\delta)\varphi^-(y)x)$. Yet $\varphi^-(y)x = 0$ unless $\mu \geq \nu$. Further assume that (ii) holds. Then a similar reduction using $(+)$ shows that $\varphi(\tau^-(\gamma)y, \tau^+(\delta)x) = 0$ unless $\mu \leq \nu$. Finally if $\mu = \nu$, then $\varphi^-(y)x = \varepsilon^+(\varphi^-(y)x)$. This and $\varphi(1,1) = 1$ gives (iv) and that $\varphi(\tau^-(\gamma)y, \tau^+(\delta)x) = \varphi(\tau^-(\gamma), \tau^+(\delta))\varphi(y,x)$. Yet $\varphi(\tau^-(\gamma), \tau^+(\delta)) \overset{\text{(i)}}{=} \varphi(1, \tau_\gamma^-(\tau^+(\delta))) = q^{(\delta,\gamma)}\varphi(1, \tau^+(\delta)) \overset{\text{(ii)}}{=} q^{(\delta,\gamma)}$, hence (iii).

To show (v) take $y = 1$, $y' = y$ in (i) and replace x by xx'. By (iii), (iv) one has $\varphi(1, z) = \varepsilon^+(z)$ for all $z \in \tilde{V}^+$. Hence

$$\begin{aligned}\varphi(y, xx') &= \varphi(1, \varphi^-(y)xx') \\ &= \varepsilon^+(\varphi^-(y)xx') \\ &= \varepsilon^+(\varphi^-(y_1)x)\varepsilon^+(\varphi^-(y_2)x'), \quad \text{by } (*)\ , \\ &= \varphi(y_1, x)\varphi(y_2, x')\ , \\ &= (\varphi \times \varphi)(\Delta(y), x \otimes x'), \quad \text{as required}\ .\end{aligned}$$

Similarly (vi) follows from (ii) and $(**)$.

Remarks. Since $\Delta(1) = 1 \otimes 1$ only the normalization $\varphi(1,1) = 1$ is compatible with (v) (or with (vi)). To a large extent φ is determined by (i) alone. In particular one cannot have a non-zero form satisfying both (i) and $\varphi(y, xx') = \varphi(\varphi^+(x')y, x)$.

3.1.11 There is an antipode σ^- (resp. σ^+) on $\tilde{V}^-$ (resp. $\tilde{V}^+$) making it a Hopf algebra, namely $\sigma^-(\tau^-(\lambda)) = \tau^-(-\lambda)$, $\sigma^-(y_{-\alpha}) = -y_{-\alpha}\tau^-(\alpha)$; $\sigma^+(\tau^+(\lambda)) = \tau^+(-\lambda)$, $\sigma^+(x_\alpha) = -\tau^+(-\alpha)x_\alpha$ for all $\lambda \in Q(\pi)$, $\alpha \in \pi$. Assume for the moment that the Cartan form is symmetric. Then there is an isomorphism $\iota : \tilde{V}^- \xrightarrow{\sim} \tilde{V}^+$ of algebras defined by $\iota(y_{-\alpha}) = x_\alpha$, $\iota(\tau^-(\lambda)) = \tau^+(-\lambda)$ for all $\alpha \in \pi$, $\lambda \in Q(\pi)$. However ι is an antiautomorphism of Hopf algebras. Indeed let Δ_0^+ denote the opposed coproduct of $\tilde{V}^+$, that is $\Delta_0^+(x) = x_2 \otimes x_1$, where $\Delta^+(x) = x_1 \otimes x_2$. Then

$$\Delta_0^+\iota = (\iota \otimes \iota)\Delta^-\ , \quad \sigma_0^+\iota = \iota\sigma^-$$

where the second relation may viewed as defining σ_0^+. One has $\sigma_0^+(\tau^+(\lambda)) = \tau^+(-\lambda)$, $\sigma_0^+(x_\alpha) = -x_\alpha\tau^+(-\alpha)$ and from this or 1.1.12 it follows that σ_0^+

is the inverse of σ^+. Of course Δ_0^+, σ_0^+ can be defined without assuming the Cartan form to be symmetric. Then the result in 3.1.10 may be expressed through the

Corollary. *The bilinear form φ on $\tilde{V}^- \times \tilde{V}^+$ satisfies*

(i) $\varphi(1,x) = \varepsilon^+(x)$, $\varphi(y,1)) = \varepsilon^-(y)$
(ii) $\varphi(y,xx') = (\varphi \times \varphi)(\Delta^-(y), x \times x')$
(iii) $\varphi(yy',x) = (\varphi \times \varphi)(y \otimes y',\ \Delta_0^+(x))$
(iv) $\varphi(\sigma^-(y),x) = \varphi(y,\sigma_0^+(x))$
for all $y,y' \in \tilde{V}^-$, $x,x' \in \tilde{V}^+$.

3.1.12 Recall (3.1.3, 3.1.5) the definitions of $K^\pm$. It is clear that $L^\pm = K^\pm T^\pm$ is a two-sided ideal of $\tilde{V}^\pm$ and that the quotient identifies with $V^\pm := U^\pm T^\pm$. Moreover by 3.1.10 (iii), (iv) it follows easily that $L^- \times \tilde{V}^+ + \tilde{V}^- \times L^+$ is just $\ker \varphi$. This gives the

Corollary

(i) *L^- is a Hopf ideal of $\tilde{V}^-$.*
(ii) *L^+ is a Hopf ideal of $\tilde{V}^+$.*
(iii) *The induced pairing φ on $V^- \times V^+$ is non-degenerate.*

3.1.13 One may anticipate that A_C has a number of symmetries. Some of these derive from symmetries of $\tilde{A}_C$ and the canonical nature of the construction of A_C. Indeed let C' denote the transpose of C. Let ψ (resp. $\varkappa$) denote the isomorphism (resp. anti-isomorphism) of $\tilde{U}^-$ onto $\tilde{U}^+$ taking $y_{-\alpha}$ to x_α. Then θ restricted to $\tilde{U}^-$ identifies with $\psi^{-1}\varkappa = \varkappa^{-1}\psi$. Let ζ denote the automorphism $q \mapsto q^{-1}$ of $k(q)$ extended to the ζ–linear map on $\tilde{U}^- \otimes \tilde{U}^+$ fixing the basis given by monomials in $x_\alpha, y_{-\alpha} : \alpha \in \pi$. Examining the defining relation 3.1.3($*$), one obtains the

Lemma

(i) $\varkappa(\tilde{A}_C) \xrightarrow{\sim} \tilde{A}_{C'}$, *so* $\varkappa\theta(K_C^-) = K_{C'}^+$.
(ii) $\psi\zeta(\tilde{A}_C) \xrightarrow{\sim} \tilde{A}_{C'}$, *so* $\psi\zeta\theta(K_C^-) = K_{C'}^+$,
(iii) $\zeta\theta(K_C^-) = K_C^-$.

3.1.14 Further symmetries arise from the invariance properties of $K_C^\pm$.

Lemma

(i) $\zeta(K_C^\pm) = K_{C'}^\pm$.
(ii) $\psi\theta(K_C^-) = K_C^+$.
(iii) *If $C = C'$, then $K_C^\pm$ are invariant under ζ and θ.*

Observe that $\delta''_{-\alpha}$ obtains from $\delta'_{-\alpha}$ on replacing q by q^{-1} and C by C'. Thus $\zeta(K_C^+) = K_{C'}^+$. Hence (i). Combined with 3.1.13(ii) this gives (ii). Finally (iii) follows from (i), (ii) and 3.1.13(ii).

Remark. Notice one may also introduce, in a slightly more natural way, the quantum Weyl algebra A_C through the left action of $y_{-\beta}$ and δ'_α on $\tilde{U}^-$; though this would be slightly different if $C \neq C'$. This "duality" for $C = C'$ is of course present in A_n of 1.2.10 and there can be viewed as Fourier transform.

3.2 The Drinfeld Double

3.2.1 Let A, B be Hopf algebras. A Hopf pairing of A, B is a bilinear map $\varphi : A \times B$ satisfying

(i) $\varphi(1,b) = \varepsilon_B(b), \quad \varphi(a,1) = \varepsilon_A(a)$,
(ii) $\varphi(a,bb') = (\varphi \times \varphi)(\Delta_A(a),\ b \otimes b')$,
(iii) $\varphi(aa',b) = (\varphi \times \varphi)(a \otimes a',\ \Delta_B(b))$,
(iv) $\varphi(\sigma_A(a),b) = \varphi(a,\sigma_B(b))$,

for all $a \in A,\ b \in B$.

These axioms are not independent. Thus (ii)–(iv) imply that $\varepsilon(a)\varphi(1,b) = \varphi(a,1)\varepsilon(b)$ and hence (i) given $\varphi(1,1) = 1$. Conversely (i)–(iii) imply (iv). The two parts of (i) are equivalent under (ii), (iii) and the existence of an antipode.

Now assume that σ_B is invertible and recalling 1.1.12, let B^{opp}, (resp. B_{opp}) denote the Hopf algebra with the same algebra, (resp. coalgebra) structure as B but with product μ_B^{opp} (resp. coproduct Δ_B^{opp}) and antipode σ_B^{-1}. A skew-Hopf pairing φ of A, B is defined to be a Hopf pairing between A and B_{opp}. For example 3.1.11 shows that φ in its conclusion is a skew-Hopf pairing of $\tilde{V}^-$, $\tilde{V}^+$, whilst φ in the conclusion of 3.1.12 is a skew-Hopf pairing of V^-, V^+.

3.2.2 Let A, B be Hopf algebras admitting a skew-Hopf pairing φ.

Lemma. *There is an associative algebra structure on $A \otimes B$ such that for all $a \in A, b \in B$*

(i) $a \mapsto a \otimes 1$ *(resp. $b \mapsto 1 \otimes b$) is an algebra monomorphism of A (resp. B) into $A \otimes B$,*
(ii) $(a \otimes 1)(1 \otimes b) = a \otimes b$,
(iii) $(1 \otimes b)(a \otimes 1) = \varphi(a_1, \sigma_B(b_1))(a_2 \otimes b_2)\varphi(a_3, b_3)$,
where $\Delta_A^2(a) = a_1 \otimes a_2 \otimes a_3, \quad \Delta_B^2(b) = b_1 \otimes b_2 \otimes b_3$.

It is immediate that associativity is equivalent to the equality

$$[(1\otimes b)(a\otimes 1)(1\otimes b')](a'\otimes 1) = [(1\otimes b)(a\otimes 1)][(1\otimes b')(a'\otimes 1)] = (1\otimes b)[(a\otimes 1)(1\otimes b')(a'\otimes 1)] .$$

The central term equals

$$\begin{aligned}&\varphi(a_1,\sigma_B(b_1))\varphi(a_3,b_3)(a_2\otimes b_2)(a'_2\otimes b'_2)\varphi(a'_1,\sigma_B(b'_1))\varphi(a'_3,b'_3)\\ &\quad = \varphi(a_1,\sigma_B(b_1))\varphi(a_3,b_5)\varphi(a'_2,\sigma_B(b_2))(a_2a'_3\otimes b_3b'_2)\\ &\qquad \times \varphi(a'_4,b_4)\varphi(a'_1,\sigma_B(b'_1))\varphi(a'_5,b'_3) .\end{aligned}$$

The left hand side equals

$$\varphi(a_1,\sigma_B(b_1))(a_2\otimes b_2b')(a'\otimes 1)\varphi(a_3,b_3)$$
$$=\varphi(a_1,\sigma_B(b_1))\{\varphi(a'_1,\sigma_B(b_2b'_1))(a_2a'_2\otimes b_3b'_2)\varphi(a'_3,b_4b'_3)\}\varphi(a_3,b_5)\ .$$

Using 3.2.1 (ii) and 1.1.9 the curly bracketed term is replaced by

$$\varphi(a'_1,\sigma_B(b'_1))\varphi(a'_2,\sigma_B(b_2))(a_2a'_3\otimes b_3b'_2)\varphi(a'_4,b_4)\varphi(a'_5,b'_3)$$

and so the first equality is established.

The right hand side equals

$$\varphi(a'_1,\sigma_B(b'_1))(1\otimes b)(aa'_2\otimes b'_2)\varphi(a'_3,b'_3)$$
$$=\varphi(a'_1,\sigma_B(b'_1))\{\varphi(a_1a'_2,\sigma_B(b_1))(a_2a'_3\otimes b_2b'_2)\varphi(a_3a'_4,b_3)\}\varphi(a'_5,b'_3)\ .$$

Using 3.2.1 (iii) the curly bracketed term would *not* give the required result. However since φ is a skew-Hopf pairing this is replaced by $\varphi(a'a,b)=(\varphi\times\varphi)(a\otimes a',\Delta_B(b))$. Then, using 1.1.10, the curly bracketed term is replaced by

$$\varphi(a_1,\sigma_B(b_1))\varphi(a'_2,\sigma_B(b_2))(a_2a'_3\otimes b_3b'_2)\varphi(a_3,b_5)\varphi(a'_4,b_4)$$

and so the second equality is established.

Finally one notes from 3.2.1 (i) that (iii) is compatible with $1\otimes 1$ being the identity of $A\otimes B$.

Remark. For simplicity write $a\otimes b=ab$, $(1\otimes b)(a\otimes 1)=ba$. Set $\Delta_A(a)=a_1\otimes a_2$, $\Delta_B(b)=b_1\otimes b_2$. Then

$$\begin{aligned}b_2a_2\varphi(a_1,b_1)&=\varphi(a_1,b_1)\varphi(a_2,\sigma_B(b_2))a_3b_3\varphi(a_4,b_4)\\&=\varphi(a_1,b_1\sigma_B(b_2))a_2b_3\varphi(a_3,b_4),\quad\text{by 3.2.1 (iii)}\\&=\varphi(a_1,\varepsilon_B(b_1))a_2b_2\varphi(a_3b_3),\quad\text{by 1.1.8}\\&=a_1b_1\varphi(a_2,b_2),\quad\text{by 3.2.1 (i) and 1.1.8.}\end{aligned}$$

Conversely this identity implies 3.2.1 (iii).

3.2.3 Retain the notation and hypotheses of 3.2.2. It is immediate that $\Delta_{A\otimes B}:=\tau_{23}(\Delta_A\otimes\Delta_B)$ defines an associative coalgebra structure on $A\otimes B$ with coidentity $\varepsilon_A\otimes\varepsilon_B$. Set $\sigma_{A\otimes B}(a\otimes b)=(1\otimes\sigma_B(b))(\sigma_A(a)\otimes 1)$. Give $A\otimes B$ an algebra structure through 3.2.2.

Proposition. *$A\otimes B$ is a bialgebra with antipode $\sigma_{A\otimes B}$.*

To show that $\Delta:=\Delta_{A\otimes B}$ is an algebra homomorphism it suffices to show that $\Delta(a\otimes 1)(1\otimes b)=\Delta(a\otimes 1)\Delta(1\otimes b)$ and $\Delta(1\otimes b)(a\otimes 1)=\Delta(1\otimes b)\Delta(a\otimes 1)$ for all $a\in A$, $b\in B$. The first case results straightforwardly from 3.2.2 (ii). In the second the right hand side equals

$$(1\otimes b_1\otimes 1\otimes b_2)(a_1\otimes 1\otimes a_2\otimes 1)=(1\otimes b_1)(a_1\otimes 1)\otimes(1\otimes b_2)(a_2\otimes 1)$$
$$=\varphi(a_1,\sigma_B(b_1))(a_2\otimes b_2)\otimes(a_5\otimes b_5)\{\varphi(a_3,b_3)\varphi(a_4,\sigma_B(b_4))\}\varphi(a_6\otimes b_6)\ .$$

The term in curly brackets is just

$$\begin{aligned}&(\varphi\times\varphi)(a_3\otimes a_4, b_3\otimes\sigma(b_4))\\ &\quad=\varphi(a_3, b_3\sigma_B(b_4)),\ \text{ by } 3.2.1(\text{ii}),\\ &\quad=\varphi(a_3,\varepsilon_B(b_3)),\ \text{ by } 1.1.7(\text{v}),\\ &\quad=\varepsilon_A(a_3)\varepsilon_B(b_3),\ \text{ by } 3.2.1(\text{i}).\end{aligned}$$

Taking account of 1.1.2 (ii)$^*_\ell$ it follows that the previous expression simplifies to

$$\begin{aligned}&\varphi(a_1,\sigma_B(b_1))(a_2\otimes b_2)\otimes(a_3\otimes b_3)\varphi(a_4,b_4)\\ &\quad=\varphi(a_1,\sigma_B(b_1))\Delta(a_2\otimes b_2)\varphi(a_3,b_3)=\Delta((1\otimes b)(a\otimes 1)),\end{aligned}$$

as required. Finally that $\sigma_{A\otimes B}$ is the antipode is straightforward.

3.2.4 Retain the notation and hypotheses of 3.2.2. The Hopf algebra which results from the above conclusion is called the Drinfeld double $\mathcal{D}(A,B)$ of the pair A,B. Of course here A and B are related. Indeed the map $b\mapsto(a\mapsto\varphi(a,b))$ is Hopf algebra morphism of B_{opp} into the Hopf dual $A^\star$ of A. When this map is an isomorphism, B is determined and one simply writes $\mathcal{D}(A)$ for $\mathcal{D}(A,B)$.

3.2.5 One may describe $\mathcal{D}(\tilde{V}^-,\tilde{V}^+)$ (resp. $\mathcal{D}(V^-,V^+)$) as a Hopf algebra. The only new information is contained in its algebra structure which is described below.

Lemma. *As an algebra $\mathcal{D}(\tilde{V}^-,\tilde{V}^+)$ has generators $x_\alpha, y_{-\alpha}, \tau^\pm(\pm\alpha):\alpha\in\pi$ satisfying the relations*

(i) $\tau^\pm(\alpha)\tau^\pm(-\alpha)=\tau^\pm(-\alpha)\tau^\pm(\alpha)=1,\ \forall\,\alpha\in\pi$.
(ii) $\tau^+(\alpha),\tau^+(\beta),\ \tau^-(\alpha),\tau^-(\beta)$ *commute pairwise for all* $\alpha,\beta\in\pi$.
(iii) $\tau^+(\alpha)x_\beta\tau^+(-\alpha)=q^{(\alpha,\beta)}x_\beta,\ \tau^-(\alpha)y_{-\beta}\tau^-(-\alpha)=q^{-(\beta,\alpha)}y_{-\beta}$.
(iv) $\tau^-(\alpha)x_\beta\tau^-(-\alpha)=q^{(\beta,\alpha)}x_\beta,\ \tau^+(\alpha)y_{-\beta}\tau^+(-\alpha)=q^{-(\alpha,\beta)}y_{-\beta}$.
(v) $x_\alpha y_{-\beta}-y_{-\beta}x_\alpha=\delta_{\alpha\beta}(\tau^+(\alpha)-\tau^-(-\alpha))$.

The algebra $\mathcal{D}(V^-,V^+)$ is the quotient of $\mathcal{D}(\tilde{V}^-,\tilde{V}^-)$ defined by the additional relations

(iv) $K^-=0,\ K^+=0$.

The relations in (i), (iii) (resp. (vi)) are those from $\tilde{V}^-,\tilde{V}^+$ (resp. and V^-,V^+). Part of those in (ii) so arise and the remainder are clear. It remains to check (iv), (v). Consider (v). By ($\pm$) of 3.1.10 one has

$$(\Delta^+)^2(x_\alpha)=x_\alpha\otimes1\otimes1+\tau^+(\alpha)\otimes x_\alpha\otimes1+\tau^+(\alpha)\otimes\tau^+(\alpha)\otimes x_\alpha,$$

$$(\Delta^-)^2(y_{-\beta})=y_{-\beta}\otimes\tau^-(-\beta)\otimes\tau^-(-\beta)+1\otimes y_{-\beta}\otimes\tau^-(-\beta)+1\otimes1\otimes y_{-\beta}.$$

Thus by 3.2.2 (iii) and 3.1.10 (iii), (iv) one has

$$(1 \otimes x_\alpha)(y_{-\beta} \otimes 1) = \varphi(y_{-\beta}, -\tau^+(-\alpha)x_\alpha)(\tau^-(-\beta) \otimes 1)\varphi(\tau^-(-\beta), 1)$$
$$+ \varphi(1, \tau^+(-\alpha))(y_{-\beta} \otimes x_\alpha)\varphi(\tau^-(-\beta), 1)$$
$$+ \varphi(1, \tau^+(-\alpha))(1 \otimes \tau^+(\alpha))\varphi(y_{-\beta}, x_\alpha) .$$

So

$$x_\alpha y_{-\beta} = -\delta_{\alpha\beta}\tau^-(-\beta) + y_{-\beta}x_\alpha + \delta_{\alpha\beta}\tau^+(\alpha) ,$$

as required. Consider the first relation in (iv). One has

$$(1 \otimes x_\beta)(\tau^-(-\alpha) \otimes 1) = \varphi(\tau^-(-\alpha), \tau^+(-\beta)) \, (\tau^-(-\alpha) \otimes \tau_\beta)\varphi(\tau^-(-\alpha), 1) .$$

So

$$x_\beta \tau^-(-\alpha) = q^{-(\beta,\alpha)}\tau^-(-\alpha)x_\beta$$

as required. For the second relation one has

$$(1 \otimes \tau^+(\alpha))(y_{-\beta} \otimes 1) = \varphi(1, \tau^+(\alpha)) \, (y_{-\beta} \otimes \tau^+(\alpha))\varphi(\tau^-(-\beta), \tau^+(\alpha)) .$$

So

$$\tau^+(\alpha)y_{-\beta} = y_{-\beta}\tau^+(\alpha) \, q^{(\alpha,\beta)}$$

as required.

3.2.6 The algebras $\mathcal{D}(\tilde{V}^-, \tilde{V}^+)$ and $\mathcal{D}(V^-, V^+)$ have some symmetries. First the order 2 antiautomorphism $\varkappa$ sending $y_{-\alpha}$ to x_α which is the identity on T^+ and T^-. This interchanges the relations (iii) and (iv) and sends (i), (ii), (v) to themselves. If $C = C'$ then by 3.1.13(i) and 3.1.14(iii) it passes to an antiautomorphism of $\mathcal{D}(V^-, V^+)$. It is the quantum analogue of the Chevalley antiautomorphism, but requires modification (3.3.3) to also be an automorphism of coalgebras. Secondly the relations (i)–(v) are permuted amongst themselves under the order 2 antiautomorphism sending x_α to $-x_\alpha$, $y_{-\alpha}$ to $-y_{-\alpha}$, $\tau^+(\alpha)$ to $\tau^-(-\alpha)$ on replacing C by C'. By 3.1.13(iii) and 3.1.14(i) this also applies to (vi). Thus if $C = C'$ it defines an antiautomorphism of these algebras, and is the quantum analogue of the principal antiautomorphism. Composed with $\varkappa$ it defines ι which is an automorphism of algebras and an antiautomorphism of coalgebras and may be also viewed (8.3.6) as a product of the antipode and the modified Chevalley antiautomorphism.

3.2.7 It may seem strange that the relation $*$ of 3.1.3 on which the present construction was developed has now disappeared to be replaced by 3.2.5 (v). However all the skew derivations $\delta'_{\pm\alpha}, \delta''_{\pm\alpha} : \alpha \in \pi$ can be recovered from 3.2.5 (v). For the actual analysis of these algebras this is an important observation. Set $[a, b] = ab - ba$.

Lemma. *In the algebra $\mathcal{D}(\tilde{V}^-, \tilde{V}^+)$ one can write*

(i) $[x_\alpha, y] = (\theta\delta'_\alpha\theta)(y)\tau^+(\alpha) - (\theta\delta''_\alpha\theta)(y)\tau^-(-\alpha)$
(ii) $[x, y_{-\alpha}] = \delta''_{-\alpha}(x)\tau^+(\alpha) - \delta'_{-\alpha}(x)\tau^-(-\alpha)$

for all $x \in \tilde{U}^+$, $y \in \tilde{U}^-$. Moreover these formulae pass to $\mathcal{D}(V^-, V^+)$.

Consider for example (ii). Assume such a relation for some $\delta''_{-\alpha}, \delta'_{-\alpha} \in \operatorname{End}\tilde{U}^+$. Then 3.2.5 (v) gives $\delta'_{-\alpha}(x_\beta) = \delta''_{-\beta}(x_\beta) = \delta_{\alpha\beta}$. On the other hand expanding $[xx', y_{-\alpha}]$ and equating coefficients of $\tau^\pm(\pm\alpha)$ implies that $\delta'_{-\alpha}$ (resp. $\delta''_{-\alpha}$) is a $\tau^-_{-\alpha}$ (resp. τ^+_α) rightskew derivation of $\tilde{U}^+$. The proof of (i) is similar. The last part follows from the definition of $K^\pm$, 3.1.4 (ii) and 3.1.5.

3.2.8 There is a way to recover the defining relation of 3.1.3 from the relations in 3.2.5 which further shows that there are additional symmetries which the $K^\pm$ must satisfy. Set $g'_{-\beta} = -y_{-\beta}\tau^-(\beta)$. Then relation 3.2.5 (v) becomes

$$x_\alpha g'_{-\beta} - q^{-(\alpha,\beta)} g'_{-\beta} x_\alpha = \delta_{\alpha\beta}(1 - \tau^+(\alpha)\tau^-(\beta))$$

which is exactly the required relation if the second term on the right hand side is ignored. This may be achieved by introducing an appropriate filtration in $\mathcal{D}(\tilde{V}^-, \tilde{V}^+)$. First note that the latter is isomorphic to $\tilde{V}^- \otimes \tilde{V}^+$ as a vector space. Take the trivial gradation in $\tilde{V}^+$. Recall that $\tilde{V}^- = \tilde{U}^- \otimes k(q)T^-$, that $\tilde{U}^-$ is multigraded by $Q(\pi)$ and that $k(q)T^-$ is the Laurent polynomial ring on generators $\tau^-(\alpha) : \alpha \in \pi$. Thus the former can be graded by defining $y_{-\beta}$ to have degree one and the latter by defining $\tau^-(\alpha)$ to have degree minus one. The relations (i)–(iv) of 3.2.5 are homogeneous with respect to this gradation, whilst all the terms in (v) have degree one excepting $\tau^+(\alpha)$ which has degree zero. Consequently the above defines a filtration of $\mathcal{D}(\tilde{V}^-, \tilde{V}^+)$ such that in the associated graded algebra the relation

$$x_\alpha g'_{-\beta} - q^{-(\alpha,\beta)} g'_{-\beta} x_\alpha = \delta_{\alpha\beta}$$

holds. This construction and variants of it are important for the structure theory of quantized enveloping algebras (5.3, 7.1). For the moment recall that K^- is graded, multiply each vector $a_{-\nu}$ of weight $-\nu$ by $\tau^-(\nu)$, translate the resulting $\tau^-(\beta)$ factors to the left so that in each occurrence of $y_{-\beta}$ the latter may be replaced by $-y_{-\beta}\tau^-(\beta)$. Let $\hat{K}^-$ denote the ideal of $\tilde{U}^-$ in which each $g'_{-\beta}$ factor is further replaced by $y_{-\beta}$.

Lemma. $\hat{K}^- = K^-$.

Indeed it follows that $\hat{K}^-$ is $\theta\delta'_\alpha\theta : \alpha \in \pi$ stable and hence $\hat{K}^- \subset K^-$ by the above relation. Equality holds since both have the same weight space decomposition.

Remarks. Iterating this procedure the same applies to the variables $y_{-\beta}(\tau^-(\beta))^s : s \in \mathbb{Z}$. A similar argument applies to the $x_\alpha \tau^+(-\alpha)$.

3.2.9 Assume that C is symmetric. Then the $\tau^+(\alpha)\tau^-(-\alpha) : \alpha \in \pi$ are central in $\mathcal{D}(V^-, V^+)$. The quotient by the ideal generated by the $\tau^+(\alpha)\tau^-(-\alpha) - 1 : \alpha \in \pi$ is defined to be the (Drinfeld-Jimbo) quantized enveloping algebra $U_q(\mathfrak{g}_C)$, or simply, U. Here one writes $\tau^+(\gamma) = \tau^-(\gamma) = \tau(\gamma)$ for all $\gamma \in Q(\pi)$ and also $t_\alpha = \tau(\alpha) : \alpha \in \pi$. It is clear that $U_q(\mathfrak{g}_C)$ inherits a Hopf algebra structure.

Definition. The quantized enveloping algebra $U_q(\mathfrak{g}_C)$ associated to a symmetric Cartan matrix C is the Hopf algebra over $k(q)$ with identity 1 and generators $x_\alpha, y_{-\alpha}, t_\alpha, t_\alpha^{-1}$ satisfying the relations

(i) $t_\alpha t_\alpha^{-1} = t_\alpha^{-1} t_\alpha, \forall \alpha \in \pi$
(ii) $t_\alpha t_\beta = t_\beta t_\alpha, \forall \alpha, \beta \in \pi$
(iii) $t_\alpha x_\beta t_\alpha^{-1} = q^{(\alpha,\beta)} x_\beta$
(iv) $t_\alpha y_{-\beta} t_\alpha^{-1} = q^{-(\alpha,\beta)} y_{-\beta}$
(v) $x_\alpha y_{-\beta} - y_{-\beta} x_\alpha = \delta_{\alpha\beta}(t_\alpha - t_\alpha^{-1})$
(vi) $K^+ = K^- = 0$
(vii) $\Delta(x_\alpha) = x_\alpha \otimes 1 + t_\alpha \otimes x_\alpha$, $\Delta(y_{-\alpha}) = y_{-\alpha} \otimes t_\alpha^{-1} + 1 \otimes y_{-\alpha}$, $\Delta(t_\alpha^{\pm 1}) = t_\alpha^{\pm 1} \otimes t_\alpha^{\pm 1}$
(viii) $\varepsilon(x_\alpha) = \varepsilon(y_{-\alpha}) = 0$, $\varepsilon(t_\alpha^{\pm 1}) = 1$
(ix) $\sigma(x_\alpha) = -t_\alpha^{-1} x_\alpha$, $\sigma(y_{-\alpha}) = -y_{-\alpha} t_\alpha$, $\sigma(t_\alpha^{\pm 1}) = t_\alpha^{\mp 1}$.

Warning. When C is specializable, that is when $(\alpha,\alpha) > 0$ and is even for all $\alpha \in \pi$, it is convenient to set $q_\alpha = q^{\frac{(\alpha,\alpha)}{2}}$, to replace x_α by $x_\alpha / q_\alpha - q_\alpha^{-1}$, and α by $2\alpha/(\alpha,\alpha)$ in the definition of $\mathfrak{h}_{\mathbb{Z}}^*$. Then (v) becomes

$$x_\alpha y_{-\beta} - y_{-\beta} x_\alpha = \delta_{\alpha\beta} h_\alpha \quad \text{where} \quad h_\alpha = \frac{t_\alpha - t_\alpha^{-1}}{q_\alpha - q_\alpha^{-1}} .$$

The quantities h_α are natural analogues of the coroots $\alpha^\vee = 2\alpha/(\alpha,\alpha)$ lying in the Cartan subalgebra of the Kac-Moody Lie algebra $\mathfrak{g}$. It may be convenient to adjoin the $\tau(\gamma) : \gamma \in \mathfrak{h}_{\mathbb{Z}}^*$ to $U_q(\mathfrak{g}_C)$ in the obvious fashion, and to take $\pi = \{\alpha_i\}_{i=1}^{\ell}$ writing $x_{\alpha_i} = x_i$, $y_{-\alpha_i} = y_i$, $t_{\alpha_i} = t_i, h_{\alpha_i} = h_i, q_{\alpha_i} = q_i$. When C is integrable the generators of K^+, K^- can be computed explicitly (though they are practically never needed in an explicit form) and are known as the quantum Serre relations.

3.2.10 The Hopf subalgebra of $U_q(\mathfrak{g})$ generated by the x_α, t_α, t_α^{-1} (resp. $y_{-\alpha}$, t_α, t_α^{-1}) $: \alpha \in \pi$ is denoted by $U_q(\mathfrak{b}_C^+)$, or simply V^+, (resp. $U_q(\mathfrak{b}_C^-)$, or simply V^-). The subgroup generated by the $t_\alpha : \alpha \in \pi$ is called the (root) torus and denoted T and its group algebra $k(q)T$ by U^0. The (weight) torus $\check{T}$ is the multiplicative abelian group $\{\tau(\lambda) : \lambda \in \mathfrak{h}_{\mathbb{Z}}^*\}$ and one sets $\check{U}^0 := k(q)\check{T}$. The subalgebra generated by the x_α (resp. $y_{-\alpha}$) $: \alpha \in \pi$ is denoted $U_q(\mathfrak{n}_C^+)$, or simply U^+, (resp. $U_q(\mathfrak{n}_C^-)$, or simply U^-). By construction one has the vector space isomorphism

$$U^- \otimes U^0 \otimes U^+ \xrightarrow{\sim} U$$

defined by multiplication. It is called a triangular decomposition of U. One defines $\check{V}^{\pm} = U^{\pm} \otimes \check{U}^0$, $\check{U} = U^- \otimes \check{U}^0 \otimes U^+$ given Hopf algebra structures in the obvious way. One may remark that $U_q(\mathfrak{b}_C^+)$ is the quantum analogue of the enveloping algebra of a Borel subalgebra $\mathfrak{b}_C^+$ of $\mathfrak{g}_C$.

When C is of finite type $U_q(\mathfrak{g}_C)$ and $U(\mathfrak{g}_C)$ are very nearly isomorphic as algebras. Not surprisingly, therefore, their representation theory and primitive spectra very nearly coincide. Nevertheless one learns very new and interesting facts about $U_q(\mathfrak{g}_C)$ which in turn imply unsuspected properties of $U(\mathfrak{g}_C)$. Again the primitive spectra of $U_q(\mathfrak{b}_C^+)$ and $U(\mathfrak{b}_C^+)$ are very different and besides as shown by 3.1.12 the former is very nearly self-dual, whilst this is quite false of the latter. These effects are loosely referred to as "quantum phenomena" and form a main inspiration for this text.

3.2.11 Since $\tau_{-\gamma}^{+}\delta'_{-\beta} = q^{(\gamma,\beta)}\delta'_{-\beta}\tau_{-\gamma}^{+}$ for all $\beta, \gamma \in \pi$, it follows from 3.1.4 (i) that $\delta'_{-\beta}$ and $\tau_{-\gamma}^{+}\delta''_{-\gamma}$ commute. One should regard this as a consequence of commutation of left and right multiplication as indicated by the following

Lemma. *For all* $\alpha \in \pi$, $y \in \tilde{U}^-$, $x \in \tilde{U}^+$ *one has*

(i) $\varphi(yy_{-\alpha}, x) = \varphi(y, \delta'_{-\alpha}x)$
(ii) $\varphi(y_{-\alpha}y, x) = \varphi(y, \tau_{-\alpha}^{+}\delta''_{-\alpha}x)$.

(i) is just a special case of 3.1.10 (i). Noting that $\delta'_{-\alpha}x_\beta = \delta''_{-\alpha}x_\beta = \delta_{\alpha,\beta}$ which settles the case $y = 1$, (ii) obtains from (i) by induction on the order of the weight of y.

3.3 The Rosso Form and the Casimir Invariant

It will be assumed in this section that the Cartan matrix and form are symmetric.

3.3.1 Recall the notation and hypotheses of 3.2.4. It would be pleasant to interpret the Drinfeld double in the following way. First observe that $[(\operatorname{ad} b_1)a]b_2 = ba$. Thus the complicated product ba is determined by the easy product and the adjoint action of B on A. One might expect that the latter would be specified by requiring that φ be ad-invariant in the sense of 1.1.8. This doesn't quite work out – already the adjoint action of B in $\mathcal{D}(A,B)$ does not leave A invariant. Nevertheless one has the

Lemma. *The Rosso form* $R(a \otimes b, a' \otimes b') := \varphi(\sigma_A(a), b')\varphi(a', \sigma_B(b))$ *on the Drinfeld double* $\mathcal{D}(A,B)$ *is ad-invariant.*

By 3.2.2 (iii)

$$(\operatorname{ad} b'')(a \otimes b) = \varphi(a_1, \sigma_B(b_1''))\varphi(a_3, b_3'')(a_2 \otimes b_2'' b \sigma_B(b_4''))$$

and so by definition of R and φ this gives

$$R((\operatorname{ad} b'')(a \otimes b), a' \otimes b') = \varphi(a, \sigma_B(b_1'')\sigma_B^{-1}(b')b_3'')\varphi(a', \sigma_B(b_2'' b \sigma_B(b_4''))) .$$

On the other hand

$$\begin{aligned} &R(a \otimes b, \operatorname{ad} \sigma(b'')(a' \otimes b')) \\ &\quad = R(a \otimes b, a_2' \otimes \sigma_B(b_3'')b'\sigma_B^2(b_1''))\varphi(a_1', \sigma_B^2(b_4''))\varphi(a_3', \sigma_B(b_2'')) \\ &\quad = \varphi(a, \sigma_B(b_1'')\sigma_B^{-1}(b')b_3'')\varphi(a', \sigma_B(b_2'' b \sigma_B(b_4''))) , \end{aligned}$$

as required.

Similarly by 3.2.2 (iii),

$$(\operatorname{ad} a'')(a \otimes b) = \varphi(\sigma_A(a_2''), b_3)\varphi(a_4'', b_1)(a_1'' a \sigma_A(a_3'') \otimes b_2)$$

which by the definition of R and φ gives

$$R((\operatorname{ad} a'')(a \otimes b), a' \otimes b') = \varphi(\sigma_A(a_2'')\sigma_A^{-1}(a')a_4'', b)\varphi(\sigma_A(a_1'' a \sigma_A(a_3'')), b')$$

whilst

$$\begin{aligned} &R(a \otimes b, \operatorname{ad} \sigma(a'')(a' \otimes b')) \\ &\quad = R(a \otimes b, \sigma_A(a_4'')a'\sigma^2(a_2'') \otimes b_2')\varphi(\sigma_A(a_1''), b_1')\varphi(\sigma_A^2(a_3''), b_3') \\ &\quad = \varphi(\sigma_A(a_2'')\sigma_A^{-1}(a')a_4'', b)\varphi(\sigma_A(a_1'' a \sigma_A(a_3'')), b') , \end{aligned}$$

as required.

3.3.2 In the special case when $A = V^-$, $B = V^+$ the Rosso form on $\mathcal{D}(V^-, V^+)$ is non-degenerate and so cannot pass to a form on the quotient $U_q(\mathfrak{g})$. Yet one may modify the above formulae slightly to obtain an ad-invariant Rosso form R on $U_q(\mathfrak{g})$. This factorizes in a similar manner with respect to triangular decomposition (3.2.10), essentially coincides with φ on $U^- \otimes U^+$; but is modified on $T \times T$ by a factor $\frac{1}{2}$ which compensates for overcounting. The factor of $\frac{1}{2}$ is rather deep and runs through the whole theory of $U_q(\mathfrak{g})$ is an unavoidable fashion.

The considerations of 3.2.8 show that it may be appropriate to modify some variables replacing in particular $y_{-\beta}$ by $g_{-\beta} := \tau(\beta)y_{-\beta}$. Let G^- denote the subalgebra of $U_q(\mathfrak{g})$ generated by the $g_{-\beta} : \beta \in \pi$. By 3.2.8 it is isomorphic to U^-. Of course one still has triangular decompositions

$$\text{(i)} \quad G^- \otimes U^0 \otimes U^+ \xrightarrow{\sim} U , \qquad \text{(ii)} \quad U^+ \otimes U^0 \otimes G^- \xrightarrow{\sim} U .$$

The Rosso form is defined to respect one of these. From the formulae developed so far and for the Casimir invariant it is best to take (ii). However for discussing applications to highest weight modules and adjoint action (i) is better. The resulting forms are related by an automorphism.

For each $a \in U$, $\alpha \in \pi$ one has $(\operatorname{ad} y_{-\alpha})a = y_{-\alpha}at_\alpha - ay_{-\alpha}t_\alpha$. Thus

(1) $(\operatorname{ad} y_{-\alpha})g_{-\nu} = q^{(\alpha,\alpha)}(q^{(\alpha,\nu)}g_{-\alpha}g_{-\nu} - g_{-\nu}g_{-\alpha})$, $\forall\, g_{-\nu} \in G^-_{-\nu}$,

and by 3.2.7 (ii)

(2) $(\operatorname{ad} y_{-\alpha})a = \delta'_{-\alpha}(a) - \delta''_{-\alpha}(a)t^2_\alpha$, $\forall\, a \in U^+$.

3.3.3 Set $\varkappa'(x_\alpha) = \tau(\alpha)y_{-\alpha}$, $\varkappa'(y_{-\alpha}) = x_\alpha\tau(-\alpha)$ for all $\alpha \in \pi$. One checks that $\varkappa'$ extends to an involutory antiautomorphism of U which is the identity on T and an automorphism of coalgebras. Observe that $\varkappa'(U^+) = G^-$ and

(3) $(\operatorname{ad} y_{-\alpha})\varkappa'(a) = -q^{(\alpha,\alpha)}\varkappa'((\operatorname{ad} x_\alpha)a)$, $\forall \alpha \in \pi$, $a \in U$.

Fix $\rho \in \mathfrak{h}^*_{\mathbb{Z}}$ satisfying $2(\rho,\alpha) = (\alpha,\alpha)$ for all $\alpha \in \pi$. Extend $k(q)$ to include $q^{\frac{1}{2}}$.

Theorem. *The bilinear form R on $U_q(\mathfrak{g})$ defined by*

(i) $R(atg, a't'g') = R(a,g')R(t,t')R(g,a')$ *for all* $g, g' \in G^-$, $t, t' \in T$, $a, a' \in U^+$
(ii) $R(\tau(\lambda), \tau(\lambda')) = q^{\frac{1}{2}(\lambda,\lambda')}$, *for all* $\lambda, \lambda' \in Q(\pi)$
(iii) $R(g_{-\mu}, a_\mu) = \varphi(g_{-\mu}, a_\mu)$, *for all* $g_{-\mu} \in G^-_{-\mu}$, $a_\mu \in U^+_\mu$
(iv) $R(a_\mu, g_{-\mu}) = q^{-2(\rho,\mu)}R(g_{-\mu}, a_\mu)$ *for all* $g_{-\mu} \in G^-_{-\mu}$, $a_\mu \in U^+_\mu$

is ad $U_q(\mathfrak{g})$ *invariant.*

Furthermore R satisfies

(v) $R(a,b) = R(\varkappa'(b), \varkappa'(a))$ *for all* $a, b \in U_q(\mathfrak{g})$.
(vi) *R is non-degenerate on* $G^- \times U^+$.

Weight space considerations show that R is ad T invariant. Invariance under ad $y_{-\alpha}$ is equivalent by 3.1.10 (–) to

$$R((\operatorname{ad} y_{-\alpha})(atg), (\operatorname{ad} t_\alpha^{-1})(a't'g')) + R(atg, (\operatorname{ad} y_{-\alpha})a't'g')) = 0\ .$$

By 3.3.2 (1) and (2) one obtains

$$\begin{aligned}(\operatorname{ad} y_{-\alpha})(a_\nu\tau(\lambda)g_{-\mu}) &= q^{(\alpha,\mu)}(\delta'_{-\alpha}(a_\nu) - \delta''_{-\alpha}(a_\nu)t^2_\alpha)\tau(\lambda)g_{-\mu} \\ &\quad + q^{(\alpha,\alpha)}a_\nu\tau(\lambda)(q^{(\alpha,\mu+\lambda)}g_{-\alpha}g_{-\mu} - g_{-\alpha}g_{-\mu})\ .\end{aligned}$$

Substitution in the above gives eight terms which can be made to cancel pairwise if (ii) holds and if

$$(*)\qquad \begin{aligned} R(g_{-\alpha}g_{-\mu},\ a'_{\alpha+\mu}) &= R(g_{-\mu},\ \delta''_{-\alpha}(a'_{\alpha+\mu}))\ , \\ R(g_{-\mu}g_{-\alpha}, a'_{\alpha+\mu}) &= q^{(\alpha,\mu)}R(g_{-\mu}, \delta'_{-\alpha}(a'_{\alpha+\mu}))\ ,\end{aligned}$$

and two similar conditions hold equivalent to the first pair under (iv). Use of 3.1.10 (iii) and comparison with 3.2.11 gives (iii).

Under (i) and (iv) it suffices to prove (v) for $a \in G^-$, $b \in U^+$. Then by (iii), it suffices to prove that $\varphi(g,a) = \varphi(\varkappa'(a), \varkappa'(g))$ for all $g \in G^-$, $a \in U^+$. This is proved by induction on the order of weights. Since $\varkappa'(x_\beta) = g_{-\beta}$ and by (*) that $R(g_{-\alpha}, x_\beta) = \delta_{\alpha,\beta}$, it holds when the order is ≤ 1. Then

$$\begin{aligned}
\varphi(g_{-(\mu+\alpha)}, x_\mu x_\alpha) &= (\varphi \times \varphi)(\Delta g_{-(\mu+\alpha)}, x_\mu \otimes x_\alpha)\,, \quad \text{by 3.1.10 (v)}\,,\\
&= (\varphi \times \varphi)(\varkappa'(x_\mu) \otimes \varkappa'(x_\alpha),\ (\varkappa' \times \varkappa')\Delta(g_{-(\mu+\alpha)}))\,,\\
&\quad \text{by the induction hypothesis}\,,\\
&= (\varphi \times \varphi)(\varkappa'(x_\mu) \otimes \varkappa'(x_\alpha), \Delta\varkappa'(g_{-(\mu+\alpha)}))\,,\\
&\quad \text{since } \varkappa' \text{ is a coalgebra automorphism}\,,\\
&= \varphi(\varkappa'(x_\alpha)\varkappa'(x_\mu), \varkappa'(g_{-(\mu+\alpha)}))\,, \quad \text{by 3.1.10 (vi)}\,,\\
&= \varphi(\varkappa'(x_\mu x_\alpha), \varkappa'(g_{-(\mu+\alpha)}))\,, \quad \text{since } \varkappa' \text{ is an algebra antiautomorphism}\,,
\end{aligned}$$

which is the required result.

By (v) and (3) the $\operatorname{ad} y_{-\alpha}$ invariance of R implies its $\operatorname{ad} x_\alpha$ invariance. Finally (iv) is immediate from (iii) and 3.1.12 (iii).

3.3.4 A bilinear form satisfying 3.3.3 (i) is called triangular. By an appropriate extension of the base field $k(q)$ one can extend R to a triangular ad-invariant form on $\check{U}$ satisfying the appropriate analogue of 3.3.3 (ii). Moreover it is easy to verify that these conditions determine R uniquely. It is called the Rosso form on $U_q(\mathfrak{g})$ (resp. on $\check{U}_q(\mathfrak{g})$). By 3.3.3 (vi) and the non-degeneracy of the Cartan form, R is non-degenerate on $\check{U}_q(\mathfrak{g})$, though it may be degenerate on $U_q(\mathfrak{g})$.

One may ask if R is uniquely determined by just requiring it to be triangular and ad-invariant. This is not quite true. However under minor restrictions it is uniquely determined on a certain important subalgebra. Let T_2 denote the image under τ of $2Q(\pi)$.

Set $U_2^0 = k(q)T_2$ and $U_2 = G^- U_2^0 U^+$ which are subalgebras of U. One may check using (1)–(3) above that U_2 is ad-invariant. (Whilst U_2 is not a Hopf subalgebra, it is a left coideal of U. A deeper investigation of U_2 and certain important subalgebras of it, will be made in Chapter 7 for C integrable.) Taking representatives from T/T_2 one may write U as a direct sum of 2^ℓ free left U_2 modules each of which are $\operatorname{ad} U$ stable. Consequently there need be no relationship between the values of a triangular ad-invariant form on each pair of these submodules.

Now let R be a triangular ad-invariant form on U_2. Triangularity implies that $R(1,1) = R(1,1)^3$, that is $R(1,1) = \pm 1, 0$. Take $R(1,1) = 1$.

Proposition. *Assume C is specializable. Let R be a triangular ad-invariant form on U_2 satisfying $R(1,1) = 1$ and non-degenerate on $G^- \times U^+$. Then*

$$R(\tau(\lambda), \tau(\lambda')) = q^{\frac{1}{2}(\lambda,\lambda')}$$

for all $\lambda, \lambda' \in 2Q(\pi)$. In particular R is uniquely determined.

The ad x_α invariance of R gives in particular that $R(\operatorname{ad} x_\alpha(y_{-\alpha}\tau(\lambda)),1) = 0$ for all $\lambda \in 2Q(\pi)$. Expanding the first entry and using triangularity then gives $R((t_\alpha - t_\alpha^{-1})\tau(\lambda),1) = 0$. Since this holds for all $\alpha \in \pi$ induction then implies $R(\tau(\lambda),1) = R(1,1) = 1$ for all $\lambda \in 2Q(\pi)$.

The ad $y_{-\alpha}$ invariance of R gives in particular that

$$R((\operatorname{ad} y_{-\alpha})\tau(\lambda), (\operatorname{ad} t_\alpha^{-1})\tau(\lambda')x_\alpha) + R(\tau(\lambda), (\operatorname{ad} y_{-\alpha})\tau(\lambda')x_\alpha) = 0 .$$

The first term is just $q^{(\alpha,\alpha)}(1 - q^{-(\lambda,\alpha)})q^{-(\alpha,\alpha)}R(g_{-\alpha}\tau(\lambda),\tau(\lambda')x_\alpha) = q^{(\lambda+\lambda',\alpha)}(1-q^{-(\lambda,\alpha)})R(g_{-\alpha},x_\alpha)R(\tau(\lambda),\tau(\lambda'))$. The second term reduces by (2) and triangularity to give $q^{(\lambda',\alpha)}R(\tau(\lambda),\tau(\lambda')(1-t_\alpha^2))$. Thus one obtains

$$(*)\qquad \begin{aligned} &- q^{(\lambda,\alpha)}(1-q^{-(\lambda,\alpha)})R(\tau(\lambda),\tau(\lambda'))R(g_{-\alpha},x_\alpha) \\ &\quad = R(\tau(\lambda),\tau(\lambda')) - R(\tau(\lambda),t_\alpha^2\tau(\lambda')) , \end{aligned}$$

for all $\alpha \in \pi$; $\lambda, \lambda' \in 2Q(\pi)$.

Observe that $g_{-\alpha}^m\tau(-m\alpha) \in U_2$ for m even and that it is a multiple of $y_{-\alpha}^m$. Hence $(\operatorname{ad} y_\alpha)(g_{-\alpha}^m\tau(-m\alpha)) = 0$. Then ad $y_{-\alpha}$ invariance of R gives

$$\begin{aligned} 0 &= R(g_{-\alpha}^m\tau(-m\alpha), (\operatorname{ad} y_{-\alpha}(\tau(\lambda')x_\alpha^{m+1})) \\ &= R(g_{-\alpha}^m\tau(-m\alpha), \tau(\lambda')(\operatorname{ad} y_{-\alpha})x_\alpha^{m+1}) , \quad \text{using triangularity,} \\ &= R(g_{-\alpha}^m\tau(-m\alpha), \tau(\lambda')(\delta'_{-\alpha}(x_\alpha^{m+1}) - \delta''_{-\alpha}(x_\alpha^{m+1})t_\alpha^2)) , \quad \text{by (2)} . \end{aligned}$$

From their definition as skew derivations, it follows by induction that

$$\delta'_{-\alpha}(x_\alpha^{m+1}) = (1 + q^{-(\alpha,\alpha)} + \ldots + q^{-m(\alpha,\alpha)})x_\alpha^m = q^{-m(\alpha,\alpha)}\delta''_{-\alpha}(x_\alpha^{m+1}) .$$

Hence

$$R(g_{-\alpha}^m, x_\alpha^m)R(\tau(-m\alpha), \tau(\lambda') - q^{m(\alpha,\alpha)}t_\alpha^2\tau(\lambda')) = 0 ,$$

for all $m \in 2\mathbb{Z}$, $\lambda' \in 2Q(\pi)$. The non-degeneracy hypothesis implies that the second factor is zero. Taking $m = 2$, $\lambda' = 0$ gives $R(t_\alpha^{-2}, q^{2(\alpha,\alpha)}t_\alpha^2) = R(t_\alpha^{-2},1) = 1$, that is $R(t_\alpha^{-2},t_\alpha^2) = q^{-2(\alpha,\alpha)}$. Taking $\lambda = -2\alpha$, $\lambda' = 0$ in $(*)$ then gives $(1-q^{-2(\alpha,\alpha)})R(g_{-\alpha},x_\alpha) = 1 - q^{-2(\alpha,\alpha)}$. The hypothesis that C is specializable means that $(\alpha,\alpha) > 0$ and so one obtains $R(g_{-\alpha},x_\alpha) = 1$. Resubstitution in $(*)$ gives

$$R(\tau(\lambda), t_\alpha^2\tau(\lambda')) = q^{(\lambda,\alpha)}R(\tau(\lambda),\tau(\lambda'))$$

for all $\alpha \in \pi$, $\lambda, \lambda' \in 2Q(\pi)$. Induction and the first equality above gives the described result. For the last part recall the cancellation of the eight terms in the proof of theorem 3.3.3. By weight space considerations these must cancel in groups of four. Then using the above result and by varying λ (or λ') they must cancel in pairs. In particular $(*)$ of 3.3.3 must hold. However this was found to have a unique up to scalars solution given by 3.3.3 (iii). Via the assumed normalization it follows that R is uniquely determined on $G^- \times U^+$. Similarly from the remaining pair of cancellations one obtains uniqueness on $U^+ \times G^-$.

3.3.5 Let A be a Hopf algebra and M an A module admitting an invariant (1.1.8) non-degenerate bilinear form ψ. Recall (1.1.4) that $M \otimes M$ has an A module structure. If M is finite dimensional, ψ identifies M with M^* giving the latter the contragredient A module structure (1.4.3). From this one may construct an invariant element of $M \otimes M$. This is how the Killing form on a semisimple Lie algebra $\mathfrak{g}$ leads to an invariant element of $\mathfrak{g} \otimes \mathfrak{g}$ which can be interpreted as a quadratic central element of $U(\mathfrak{g})$ known as the Casimir invariant (or operator). This construction can be stretched to a Kac-Moody Lie algebra $\mathfrak{g}$ since the latter has a gradation by finite dimensional subspaces, though the resulting invariant does not lie in $U(\mathfrak{g})$ but only in a completion of it (defined by the gradation). It can almost be applied to $U_q(\mathfrak{g})$ using the Rosso form R. However here U^0 causes a difficulty because it is not a direct sum of finite dimensional subspaces pairwise orthogonal with respect to R. Some authors surmount this difficulty by augmenting U^0 to formal power series over $\mathfrak{h}$. However this may blur finer features of $U_q(\mathfrak{g})$. It also does not provide the factorization of the Shapovalev determinant given in 3.4.14.

The present construction of a "Casimir invariant" makes use of the restriction of R to $G^- \times U^+$. However this at once loses ad-invariance. Yet G^- is $\operatorname{ad} U^-$ invariant and U^+ is invariant under the skew-derivations $\delta'_{-\alpha}, \delta''_{-\alpha} : \alpha \in \pi$ which by 3.2.7(ii) reconstruct the $\operatorname{ad} y_{-\alpha}$ action. This leads to an invariant using the classical techniques mentioned above and made precise in the next section.

3.3.6 Let A be a Hopf algebra and M, N finite dimensional A-modules with $\psi : M \times N \longrightarrow k$ and $\psi' : N \times M \longrightarrow k$ non-degenerate invariant bilinear forms. Then $\psi \times \psi' : (m \otimes n,\ m' \otimes n') \mapsto \psi(m, n')\psi'(n, m')$ is a non-degenerate invariant bilinear form on $M \otimes N \times M \otimes N$. Let $\{m_j\}$ be a basis for M and $\{n_i\}$ a basis for N dual with respect to ψ' that is $\psi'(n_i, m_j) = \delta_{ij}$ and set $I = \sum(m_i \otimes n_i)$. Then $(\psi \times \psi')(I, m \otimes n) = \psi(m, n)$ and it follows easily that I is A invariant.

3.3.7 Let R denote the Rosso form restricted to $G^- \times U^+$. Instead of taking the Rosso form on $U^+ \times G^-$ it is slightly less confusing to take R' to be the form on $U^+ \times G^-$ defined by $R'(a, g) = R(g, a)$, $\forall g \in G^-$, $a \in U^+$. By 3.3.3 (iv) this makes no essential difference. For all $\gamma, \gamma' \in Q^+(\pi)$ one has $R(G^-_{-\gamma}, U^+_{\gamma'}) = 0$ unless $\gamma = \gamma'$. For each $\gamma \in Q^+(\pi)$, let R_γ (resp. R'_γ) denote the restriction of R (resp. R') to $G^-_{-\gamma} \times U^+_\gamma$ (resp. $U^+_\gamma \times G^-_{-\gamma}$) which is non-degenerate by the above and 3.3.3 (vi).

As in 3.3.6 there exists $I_\gamma \in G^-_{-\gamma} \otimes U^+_\gamma$ such that

$$(R_\gamma \times R'_\gamma)(I_\gamma, g \otimes a) = R_\gamma(g, a)$$

for all $g \in G^-_{-\gamma}$, $a \in U^+_\gamma$. Let I denote the sum of the $I_\gamma : \gamma \in Q^+(\pi)$, viewed as an element of $\prod_{\gamma \in Q^+(\pi)} G^-_{-\gamma} \otimes U^+_\gamma$. The anticipated invariance properties of I are expressed as follows.

Lemma. *For all $\alpha \in \pi$ one has*
(i) $(g_{-\alpha} \otimes 1)I = (1 \otimes \delta''_{-\alpha})I$,
(ii) $(1 \otimes \tau_{-\alpha})I(g_{-\alpha} \otimes 1) = (1 \otimes \delta'_{-\alpha})I$,

(i) Recalling $(*)$ of 3.3.3 one obtains

$$\begin{aligned}(R \times R')((g_{-\alpha} \otimes 1)I, g \otimes a) &= (R \times R')(I, g \otimes \delta''_{-\alpha}(a)) \\ &= R(g, \delta''_{-\alpha}(a)) = R(g_{-\alpha}g, a) = (R \times R')(I, g_{-\alpha}g \otimes a) \\ &= (R \times R')((1 \otimes \delta''_{-\alpha})I, g \otimes a) \ .\end{aligned}$$

Then (i) follows by the non-degeneracy of $R \times R'$. The proof of (ii) is similar.

3.3.8 Recall (3.1.2) the order relation of $Q^+(\pi)$. This gives it the structure of a directed set. For each $\gamma \in Q^+(\pi)$ set

$$U^\gamma = \bigoplus_{0 \leq \beta \leq \gamma} G^-_{-\beta} U^0 U^+_\beta \ .$$

From the relations in 3.2.9 one obtains $U^\beta U^\gamma \subset U^{\beta+\gamma}$ and so the $\{U^\gamma\}_{\gamma \in Q^+(\pi)}$ from an increasing exhaustive filtration of $U_q(\mathfrak{g})$. The direct limit

$$U_q^{\mathcal{O}}(\mathfrak{g}) := \varinjlim U^\gamma$$

is called the $\mathcal{O}$-completion of $U_q(\mathfrak{g})$. Recall $\mu : G^-_{-\beta} \otimes U^+_\beta \xrightarrow{\sim} G^-_{-\beta} U^+_\beta$. Then $\mu(I) \in U_q^{\mathcal{O}}(\mathfrak{g})$.

3.3.9 Set $J = \mu(I)$ which belongs to $U_q^{\mathcal{O}}(\mathfrak{g})$. To make the subsequent calculations easier to follow, it is convenient to write $I = a \otimes b$, though of course an infinite sum is involved. With this convention the relations of 3.3.7 become

$$(*) \qquad g_{-\alpha}a \otimes b = a \otimes \delta''_{-\alpha}(b), \ a \otimes \delta'_{-\alpha}(b) = a g_{-\alpha} \otimes \tau_{-\alpha}(b) \ .$$

Define ρ as in 3.3.3.

Proposition. *For all $\beta \in Q^+(\pi)$, $y_{-\beta} \in U^-_\beta$ one has*

$$J y_{-\beta} = q^{2(\rho,\beta)-(\beta,\beta)} y_{-\beta} J \tau(2\beta) \ .$$

The general case follows by induction on $|\beta|$ from the case $\beta \in \pi$. Take $\alpha \in \pi$. Relation 3.3.2 (2) can be expressed as

$$\delta''_{-\alpha}(b) = -(y_{-\alpha}b - by_{-\alpha})t^{-1}_{-\alpha} + \delta'_{-\alpha}(b)t^{-2}_\alpha$$

for all $b \in U^+$. Then from $(*)$ above

$$
\begin{aligned}
g_{-\alpha}J &= \mu(g_{-\alpha}a\otimes b) = \mu(a\otimes\delta''_{-\alpha}(b))\\
&= -a(y_{-\alpha}b - by_{-\alpha})t_\alpha^{-1} + \mu(a\otimes\delta'_{-\alpha}(b))t_\alpha^{-2}\\
&= -ay_{-\alpha}t_\alpha\tau_{-\alpha}(b)t_\alpha^{-2} + Jy_{-\alpha}t_\alpha^{-1} + \mu(ag_{-\alpha}\otimes\tau_{-\alpha}(b))t_\alpha^{-2}\\
&= Jy_{-\alpha}t_\alpha^{-1}\ .
\end{aligned}
$$

Since $g_{-\alpha}J = y_{-\alpha}t_\alpha J = y_{-\alpha}Jt_\alpha$, this gives the required result.

3.4 The Classical Limit and the Shapovalev Form

The aim of this section is to study the behaviour of $U^\pm$ in the limit $q \longrightarrow 1$.

3.4.1 Let A be an associative k-algebra and ∂ a derivation of A. The skew-polynomial extension $A\#_\partial k[x]$ (or simply, $A\#k[x]$) is defined to be the vector space $A\otimes k[x]$ with the multiplication $(a\otimes x)(b\otimes x^n) = a\partial(b)\otimes x^n + ab\otimes x^{n+1}$. This is an associative product and ∂ extends to the inner derivation $a\mapsto xa - ax$ of $A\#k[x]$.

Now assume that A admits an identity and that ∂ is a locally nilpotent derivation (1.2.10). Set $A^\partial = \{a\in A \mid \partial u = 0\}$. The following result is known as Taylor's lemma.

Lemma. *Assume that there exists $x\in A$ such that $\partial x = 1$. Then $b\mapsto xb - bx$ is a derivation ∂' of A^∂ and $A \xleftarrow{\sim} A^\partial\#_{\partial'}[x]$.*

If $b\in A^\partial$, then $\partial(xb - bx) = b - b = 0$, which gives the first assertion. Observe that $\partial^n x^n = n!$ for all $n\in\mathbb{N}^+$. It follows that the $x^n : n\in\mathbb{N}$ are linear independent over A^∂ and so the multiplication map $b\otimes x^n\mapsto bx^n$ of $A^\partial\#_{\partial'}k[x]\longrightarrow A$ is injective. For surjectivity take $a\in A$. By the hypothesis there exists $n\in\mathbb{N}$ such that $\partial^{n+1}a = 0$. Then $b := a - (\partial^n a)(x^n/n!)$ satisfies $\partial^n b = 0$. Since the second term on the right hand side belongs to $A^\partial\#k[x]$ an easy induction argument completes the proof.

3.4.2 It is useful to observe that 3.4.1 admits the following q-analogue. Let A be an associative $k(q)$-algebra, $\theta\in \mathit{Aut}\ A$ and ∂ a locally nilpotent θ-leftskew derivation of A satisfying $\theta\partial\theta^{-1} = q^{-2}\partial$. Assume there exists $x\in A$ such that $\theta(x) = q^2x$ and $\partial x = 1$. Then $\partial(xb - \theta^{-1}(b)x) = b + q^2x\partial b - q^{-2}(\partial b)x - b = q^2x\partial b - q^{-2}(\partial b)x = 0$ for all $b\in A^\partial$. Then $b\mapsto\partial' b := xb - \theta^{-1}(b)x$ is a θ^{-1}-leftskew derivation of A^∂. Then one may form the skew polynomial extension $A^\partial\#k[x]$ by the rule $xb - \theta^{-1}(b)x = \partial' b$. By 1.2.12 one has $\partial^n x^n = q^{\frac{1}{2}n(n-1)}[n!]_q$. Then as in 3.4.1 one obtains an isomorphism $A\xleftarrow{\sim} A^\partial\#k[x]$. Of course this fails if q is a root of unity, just as 3.4.1 would fail in positive characteristic.

Recall that a Lie algebra $\mathfrak{g}$ is nilpotent if its central descending series $\mathfrak{g}\supset C^1(\mathfrak{g}) := [\mathfrak{g},\mathfrak{g}]\supset\ldots\supset C^n(\mathfrak{g}) := [\mathfrak{g}, C^{n-1}(\mathfrak{g})]\supset\ldots$ terminates after finitely many steps at zero. By Engel's theorem [Di2, 1.3.15] this holds for any Lie algebra of nilpotent endomorphisms of a finite dimensional vector space.

Lemma. *Let $\mathfrak{g}$ be a finite dimensional Lie algebra of locally nilpotent endomorphisms of a vector space V. Then $\mathfrak{g}$ is nilpotent.*

If $\mathfrak{g}$ is not nilpotent, there exists $x \in \mathfrak{g}$ such that $\operatorname{ad}_{\mathfrak{g}} x$ is not a nilpotent derivation of $\mathfrak{g}$. Passing to the algebraic closure of k one may find $y \in \mathfrak{g}$ such that $(\operatorname{ad} x)y$ is a non-zero multiple of y. Let $\mathfrak{m}$ be the two-dimensional (solvable) Lie algebra spanned by x, y. By definition of $\mathfrak{g}$ there exists $v \in V$ such that $yv \neq 0$. Then the image of $\mathfrak{m}$ in $\operatorname{End} U(\mathfrak{m})v$ is $\mathfrak{m}$ and so is not nilpotent. Yet by the hypothesis and say 2.4.11 it follows that $U(\mathfrak{m})v$ is finite dimensional. This contradicts Engel's theorem.

3.4.3 Let A be an associative algebra with identity and $\mathfrak{g}$ a finite dimensional Lie algebra of locally nilpotent derivations of A. By 3.4.2, $\mathfrak{g}$ is nilpotent. Refine the central descending series of $\mathfrak{g}$ to a descending sequence $\mathfrak{g} = \mathfrak{g}_1 \supset \mathfrak{g}_2 \supset \ldots \supset \mathfrak{g}_{n+1} = 0$ of ideals with $\mathfrak{g}_{i+1}$ of codimension 1 in $\mathfrak{g}_i$. Set $A_i = A^{\mathfrak{g}_i}$. Obviously $A = A_{n+1}$, $A^{\mathfrak{g}} = A_1$ and $A_1 \subset A_2 \subset \ldots \subset A_{n+1}$. For each i choose $x_i \in \mathfrak{g}_i$, $x_i \notin \mathfrak{g}_{i+1}$. Given $A_j = A_{j+1}$ set $a_j = 0$.

Proposition. *Assume $A^{\mathfrak{g}}$ reduces to scalars. Then for all $i, j \in \{1, 2, \ldots, n\}$.*

(i) *Either $A_j = A_{j+1}$ or there exists $a_j \in A_{j+1}$ such that $x_j a_j = 1$. Moreover $A_{j+1} = A_j \# k[a_j]$ and a_j is uniquely determined up to an element of A_j.*
(ii) *The matrix with entries $x_i a_j$ is upper triangular.*
(iii) *If A is graded and $\mathfrak{g}$ consists of graded derivations then each a_j can be assumed homogeneous.*

(i) If $A_j \neq A_{j+1}$ there exists $a \in A_{j+1}$ such that $b := x_j a \neq 0$. For all $x \in \mathfrak{g}$ one has $[x, x_j] \in \mathfrak{g}_{j+1}$ and so $x_j(xa) = [x_j, x]a + x(x_j a) = xb$. Thus $x_j(U(\mathfrak{g})a) = U(\mathfrak{g})b$. Since $\mathfrak{g}$ acts by locally nilpotent endomorphisms $U(\mathfrak{g})b$ admits a non-zero element of $A^{\mathfrak{g}}$ and so by the hypothesis the first part of (i) results. Then the second part follows by 3.4.1.

(ii) follows from the construction in (i) and for (iii) it is enough to start with $a \in A_{j+1}$ homogeneous. Then applying to a an appropriate monomial in the $x_i : i = 1, 2, \ldots, n$ provides the required element.

3.4.4 Let Q^+ be a finitely generated additive semigroup with identity 0. Give Q^+ the order relation $\mu \geq \nu$ if there exists $\zeta \in Q^+$ such that $\mu = \nu + \zeta$. Assume that $\mu \leq 0$ implies $\mu = 0$. For example $Q^+(\pi)$ of 3.1.2 has these properties. Notice that Q^+ must be torsion-free and for $\alpha, \mu \in Q^+(\pi)$ the condition $\mu - m\alpha \in Q^+$, $\forall m \in \mathbb{N}$ implies $\alpha = 0$. Let A be an associative k-algebra with identity graded by Q^+ and assume that $\dim A_\mu < \infty$ for all

$\mu \in Q^+$ and that $A_0 = k1$. Let ε denote the projection of A onto A_0 defined by the gradation. Let $\mathfrak{g}$ be a possibly infinite dimensional Lie algebra of derivations of A graded by $-(Q^+ \setminus \{0\})$. Note that $\mathfrak{g}$ acts locally nilpotently on A. Give $U(\mathfrak{g})$ the induced gradation. Let φ be the bilinear form on $U(\mathfrak{g}) \times A$ defined by $\varphi(b, a) = \varepsilon(b(a))$. Let ψ denote the homomorphism of $U(\mathfrak{g})$ onto End A defined by the action. Set $U = \operatorname{Im} \psi$.

Theorem. *Assume that* $A^{\mathfrak{g}} = k1$. *Then*

(i) *A is a graded polynomial ring in possibly infinitely many variables.*
(ii) *Take $\nu \in Q^+$. Then φ restricts to a bilinear form φ_ν on $U_{-\nu} \times A_\nu$ which is non-degenerate on A_ν.*

Assume further that $\dim U_{-\nu} = \dim A_\nu$ *for all* $\nu \in Q^+$. *Then*

(iii) $\ker \psi = 0$.

(i) If A is not commutative, there exists $\mu, \nu \in Q^+$ such that $[a_\mu, b_\nu] \neq 0$. Assume that $\mu + \nu$ is minimal with this property. Take $x_{-\alpha} \in \mathfrak{g}_{-\alpha}$ non-zero. Since $x_{-\alpha}[a_\mu, b_\nu]$ is a sum of commutators lying in $A_{\mu+\nu-\alpha}$ minimality implies that $[a_\mu, b_\nu] \in A^{\mathfrak{g}} = k1$. Then $\mu + \nu = 0$ and so $\mu = \nu = 0$. Then $[a_\mu, b_\nu] = 0$ trivially. This contradiction shows that A is commutative.

For each $\mu \in Q^+$ set

$$A^\mu = \sum_{\nu \leq \mu} A_\nu \, .$$

Then A^μ is $\mathfrak{g}$ stable and finite dimensional. Let $\mathfrak{g}^\mu$ denote the image of $\mathfrak{g}$ in End A^μ. It is a finite dimensional Lie algebra of nilpotent derivations.

Since Q^+ is a finitely generated semigroup it follows from the definition of $\leq$ that one may choose a sequence $\mu_1 \leq \mu_2 \leq \ldots$ of elements of Q^+ such that for any $\nu \in Q^+$ one has $\nu \leq \mu_n$ for n sufficiently large. Then $\mathfrak{k}_i = \ker(\mathfrak{g} \longrightarrow \mathfrak{g}^{\mu_i})$ is a decreasing sequence of ideals of $\mathfrak{g}$ with each quotient $\mathfrak{g}/\mathfrak{k}_i$ nilpotent and finite dimensional. It follows as in 3.4.3 that this sequence may be refined to a possibly infinite decreasing sequence $\mathfrak{g} = \mathfrak{g}_1 \supseteq \mathfrak{g}_2 \supseteq \ldots$ of ideals of $\mathfrak{g}$ with $\dim \mathfrak{g}_i/\mathfrak{g}_{i+1} = 1$. Choose $x_i \in \mathfrak{g}_i$, $x_i \notin \mathfrak{g}_{i+1}$ homogeneous. Then the homogeneous elements $a_1, a_2, \ldots$ of A obtained from the construction of 3.4.3, commute because A is commutative, are algebraically independent by 3.4.3 (ii) and generate a subalgebra of A which for each $\nu \in Q^+$ contains A_ν and hence is A. This proves (i).

(ii) Suppose $0 \neq a \in A_\nu$ satisfies $\varphi(U_{-\nu}, a) = 0$. Then certainly $\nu > 0$ and by the hypothesis there exists $\mu \in Q^+$ and $b \in U_{-\mu}$ such that $b(a) \neq 0$. Then $\mu \leq \nu$. Assume μ maximal with the property that $b(a) \neq 0$. Then $(x_{-\alpha} b)(a) = 0$ for all $x_{-\alpha} \in \mathfrak{g}_{-\alpha}$ and all $\alpha > 0$, that is $b(a) \in A^{\mathfrak{g}} = k1$ and so $\varepsilon(b(a)) \neq 0$. This contradiction establishes (ii).

(iii) By (ii) one has $\dim U_{-\nu} \geq \dim A_{-\nu}$ and so equality means that φ_ν defines a non-degenerate pairing of $U_{-\nu}$ and $A_{-\nu}$. Hence for each ν and basis elements $x_j \in \mathfrak{g}_{-\nu}$ there exist $a'_j \in A_{-\nu}$ homogeneous satisfying $\varphi(x_i, a'_j) = \delta_{ij}$. Then $x_j a'_j = 1$ and it can be assumed that the matrix with entries $x_i a'_j$

is lower triangular. It is clear that the a'_j can be used in place of the a_j, as a system of homogeneous generators of A and that the derivations x_i of A can be expressed as differential operators of the form

$$x_i = \frac{\partial}{\partial a'_i} + \sum_{j<i} p_{ij}(a') \frac{\partial}{\partial a'_j}$$

for some polynomials p_{ij}. It follows that the ordered monomials in the x_i have linearly independent images in U. Hence (iii).

Remark. As in 2.1.5 the PBW theorem for $U(\mathfrak{g})$ becomes rather easy in the set-up of (iii) above.

3.4.5 Fix an arbitrary Cartan matrix C and recall the definitions of U^- and U^+ given in 3.1.3. To define the limit $q \longrightarrow 1$ certain choices must be made. First let A denote the localization $k[q]_{\langle q-1\rangle}$ of the polynomial ring $k[q]$ at the prime ideal $\langle q-1\rangle$. Let U_A^- denote the A-subring of U^- generated by the $y_{-\alpha} : \alpha \in \pi$. Note that with this definition U_A^- is automatically torsion-free as an A module and a direct sum of weight submodules $(U_A^-)_{-\mu} := \{a \in U_A^- \mid \tau^-(\lambda)a = q^{-(\mu,\lambda)}a\}$. Indeed $(U_A^-)_{-\mu} = U_A^- \cap U_{-\mu}^-$ and so

$$U_A^- = \bigoplus_{\mu \in Q^+(\pi)} (U_A^-)_{-\mu} ,$$

from the corresponding result for U^-. Again U_A^- can be viewed as a quotient of the free A-subring of $\tilde{U}^+$ generated by the $y_{-\alpha} : \alpha \in \pi$ and then $(U_A^+)_{-\mu}$ is the image of A submodule generated by the monomials in the $y_{-\alpha}$ where the exponents $k_\alpha \in \mathbb{N}$ satisfy

$$\sum_{\alpha \in \pi} k_\alpha \alpha = \mu .$$

Obviously only finitely many k_α satisfy this equation and so $(U_A^-)_\mu$ is a finitely generated A module. Since A is principal, it follows that each $(U_A^-)_\mu$ is a free A module and hence so is U_A^-.

Now set $W^- := U_A^- \otimes_A A/\langle q-1\rangle$ which may be viewed as a limit (specialization), of U^- as $q \longrightarrow 1$. Obviously W^- is a direct sum of its weight subspaces $W_{-\mu}^-$ and by the above

$$\dim_k W_{-\mu}^- = rk_A(U_A^-)_{-\mu} = \dim_{k(q)} U_{-\mu}^- .$$

3.4.6 It is clear that the analysis of 3.4.5 applies to U^+, but this is not particularly interesting. Indeed with this choice the non-degenerate pairing is not preserved (see 3.4.13 for an example). However one may choose a specialization of U^+ to force this property. Set $(S_A^+)_0 = A$ and then define by induction on $|\mu| : \mu \in Q^+$,

$$(S_A^+)_\mu = \{a \in U_\mu^+ \mid \delta'_{-\alpha} a \in (S_A^+)_{\mu-\alpha}, \ \forall \alpha \in \pi\}, \quad S_A^+ = \sum_{\mu \in Q^+(\pi)} (S_A^+)_\mu \ ,$$

where the latter is of course a direct sum. Then S_A^+ is an A–ring. Clearly each $(S_A^+)_\mu$ is torsion-free as an A module. Fix $|\mu| > 0$ and consider the map $a \mapsto \{\delta'_{-\alpha} a\}_{\alpha\in\pi}$ of $(S_A^+)_\mu$ into $\prod_{\alpha\in\pi}(S_A^+)_{\mu-\alpha}$. It is injective by 3.1.6. Since A is noetherian, it follows by induction on $|\mu|$ that $(S_A^+)_\mu$ is a finitely generated A module, hence free. Set

$$S^+ = S_A^+ \otimes_A A/\langle q-1\rangle$$

which also defines a limit of U^+ as $q \longrightarrow 1$. Obviously S^+ is a direct sum of its weight subspaces S_μ^+ and

$$\dim_k \ S_\mu^+ = rk_A (S_A^+)_\mu = \dim_{k(q)} \ U_\mu^+ \ .$$

Recalling that $\varphi^-(y_{-\alpha}) = \delta'_{-\alpha} : \alpha \in \pi$. it follows that the action of U^- on U^+ defined by φ^- passes to an action of W^- on S^+. Using $y_{-\alpha}$ to also denote the image of $y_{-\alpha}$ in W^-, the fact that $\delta'_{-\alpha}$ is a skew-derivation implies that $y_{-\alpha}$ acts as a derivation on S^+. Let $\mathfrak{n}_C^-$, or simply, $\mathfrak{n}^-$ denote the Lie algebra generated by these derivations.

Theorem

(i) $(S^+)^{\mathfrak{n}^-}$ *reduces to scalars.*
(ii) S^+ *is a graded polynomial ring with possibly infinitely many generators.*
(iii) $U(\mathfrak{n}^-) \xrightarrow{\sim} W^-$.

(i) It is clear that $(S^+)^{\mathfrak{n}^-} = \sum (S_\mu^+)^{\mathfrak{n}^-}$. Consider $|\mu| > 0$ and $\bar{a}_\mu \in (S_\mu^+)^{\mathfrak{n}^-}$. If $\bar{a}_\mu \neq 0$, there exists $a_\mu \in (S_A^+)_\mu$ whose image in S_μ^+ equals $\bar{a}_\mu$ and satisfies $\delta'_{-\alpha} a_\mu \in \langle q-1\rangle (S_A^+)_{\mu-\alpha}$ for all $\alpha \in \pi$. One may write $\delta'_{-\alpha} a_\mu = (q-1)^m b_{\mu-\alpha}$: $b_{\mu-\alpha} \in (S_A^+)_{\mu-\alpha}$ for some integer $m \geq 0$. By 3.1.6 the $b_{\mu-\alpha}$ cannot all vanish, so $m > 0$. Then $(q-1)^{-m} a_\mu \in (S_A^+)_\mu$ by definition of the latter. It follows that $\bar{a}_\mu = 0$. This contradiction proves (i).

(ii) follows from (i) and 3.4.4 (i). Since φ restricts to a non-degenerate pairing of $U_{-\mu}^- \times U_\mu^+$ one has $\dim_{k(q)} U_{-\mu}^- = \dim_{k(q)} U_\mu^+$ which by the previous dimension formulae implies the hypothesis of 3.4.4 (iii). From its conclusion and the fact that W^- is generated by the $y_{-\alpha} : \alpha \in \pi$, the assertion in (iii) follows.

3.4.7 For each $\mu \in \mathfrak{h}_\mathbb{Z}^*$, let e^μ denote the corresponding element in the associated multiplicative group so then $e^\mu e^\nu = e^{\mu+\nu}$ for all $\mu, \nu \in \mathfrak{h}_\mathbb{Z}^*$. Then the group ring $\mathbb{Z}\mathfrak{h}_\mathbb{Z}^*$ identifies with finite sums of the $e^\mu : \mu \in \mathfrak{h}_\mathbb{Z}^*$. For each T^- module M with weights in $\mathfrak{h}_\mathbb{Z}^*$ having finite dimensional weight spaces, define the formal character $\operatorname{ch} M$ to be the element of $\mathbb{Z}\mathfrak{h}_\mathbb{Z}^*$ given by $\operatorname{ch} M = \sum_{\mu\in\mathfrak{h}_\mathbb{Z}^*} (\dim M_\mu) e^\mu$. Suppose M, N are T^- modules whose weights lie in a finite number of sets of the form $\lambda + Q^+(\pi) : \lambda \in \mathfrak{h}_\mathbb{Z}^*$. Then $M \otimes N$ has this property and

$$(*) \qquad \operatorname{ch}(M \otimes N) = \operatorname{ch} M \operatorname{ch} N .$$

Let $\mathfrak{n}^+_C$, or simply, $\mathfrak{n}^+$ denote a graded subspace of S^+ paired to $\mathfrak{n}^-$ by the conclusion of 3.4.4 (ii). Call $\beta \in Q(\pi)$ a positive root if $\mathfrak{n}^+_\beta \neq 0$ and set $\operatorname{mult} \beta = \dim \mathfrak{n}^+_\beta$. Let $\Delta^+_{\mathrm{mult}}(C, q)$ (or simply, Δ^+_{mult}) denote the set of positive roots counted with their multiplicities. A similar definition applies to $\mathfrak{n}^-$ and it is clear that $\Delta^-_{\mathrm{mult}} := -\Delta^+_{\mathrm{mult}}$ defines the set of (negative) roots of $\mathfrak{n}^-$ counted with their multiplicities. The set of all non-zero roots is then defined to be $\Delta_{\mathrm{mult}} := \Delta^+_{\mathrm{mult}} \cup \Delta^-_{\mathrm{mult}}$. The subscript "mult" will be omitted if Δ_{mult} is just viewed as a set.

Since a polynomial ring is as a vector space the tensor product of polynomial rings in one variable it follows from $(*)$ and 3.4.6 that

Corollary

$$\operatorname{ch} U^\pm = \operatorname{ch} \; U(\mathfrak{n}^\pm) = \prod_{\beta \in \Delta^\pm_{\mathrm{mult}}} (1 - e^\beta)^{-1} =: D_\pm \; .$$

3.4.8 One may write

$$D_\pm = \sum_{\eta \in Q^+(\pi)} P(\eta) e^{\pm \eta}$$

for some $P(\eta) \in \mathbb{N}$. The function $\eta \mapsto P(\eta)$ is called the Kostant partition function. For each $\beta \in Q(\pi)$ set $P_\beta(\eta) := P(\eta - \beta)$, so then P_β is defined as a function on $Q(\pi)$.

Lemma

(i) *The* $P_\beta : \beta \in Q(\pi)$ *are linearly independent.*
(ii) *For all* $\eta \in Q(\pi)$ *one has*

$$P(\eta)\eta = \sum_{m=1}^{\infty} \sum_{\beta \in \Delta^+_{\mathrm{mult}}} P_{m\beta}(\eta) \beta \; .$$

(i) is immediate from the definition and the linear independence of the $e^\beta : \beta \in Q(\pi)$. For (ii) note that one may write

$$\sum_{\eta \in Q^+(\pi)} P(\eta) e^{t\eta} = \prod_{\beta \in \Delta^+_{\mathrm{mult}}} (1 - e^{t\beta})^{-1}$$

where t is considered as a free parameter. Differentiation with respect to t and setting $t = 1$ gives

$$\sum_{\eta \in Q^+(\pi)} \eta P(\eta) e^\eta = \sum_{\beta \in \Delta^+_{\mathrm{mult}}} \frac{\beta e^\beta}{(1 - e^\beta)} D_+ = \sum_{\eta \in Q^+(\pi)} \sum_{m=1}^{\infty} \sum_{\beta \in \Delta^+_{\mathrm{mult}}} \beta P(\eta) e^{\eta + m\beta} .$$

Then (ii) results on equating coefficients of e^η.

3.4.9 Here and in the remainder of section 3.4 it is assumed that C is symmetric. Then $U_q(\mathfrak{g}) : \mathfrak{g} = \mathfrak{g}_C$ is defined. The set $T^\wedge$ of linear characters on T identifies with $k(q)^{*\ell}$ through the map $\Lambda \mapsto \{\Lambda(t_i)\}_{i=1}^{\ell}$. One calls $\Lambda \in T^\wedge$ a (non-linear) weight. If Λ takes the form $\Lambda(t_i) = q^{(\alpha_i,\lambda)}$ for some $\lambda \in \mathfrak{h}^*_{\mathbb{Z}}$ one calls Λ a linear weight and one writes $\Lambda = q^\lambda$. Non-linear weights are important in studying the roots of the Shapovalev determinant defined below.

For each $\Lambda \in T^\wedge$ one may define a one-dimensional $U_q(\mathfrak{b}^+)$ module 1_Λ of weight Λ satisfying $x_\alpha 1_\Lambda = 0$, $\forall \alpha \in \pi$. Then the Verma module $M(\Lambda)$ with highest weight Λ is defined by

$$M(\Lambda) = U_q(\mathfrak{g}) \otimes_{U_q(\mathfrak{b}^+)} 1_\Lambda \ .$$

Observe that weights of $M(\Lambda)$ lie in the set $\Lambda q^{-\beta} : \beta \in Q^+(\pi)$. Thus Λ can be viewed as a highest weight of $M(\Lambda)$. The corresponding "highest weight vector" $1 \otimes 1_\Lambda$ is denoted by u_Λ and is called the canonical generator of $M(\Lambda)$. If Λ takes the form $\Lambda = q^\lambda$ one writes $M(\Lambda) = M(\lambda)$ and $u_\Lambda = u_\lambda$. The study of Verma modules gives deep information on the structure of $U_q(\mathfrak{g})$.

As a $U_q(\mathfrak{n}^-)$ module $M(\Lambda)$ is free of rank 1. A non-zero vector $u_{\Lambda'} \in M(\Lambda)$ of weight Λ' is called a primitive vector and Λ' a primitive weight if $x_\alpha u_{\Lambda'} = 0$ for all $\alpha \in \pi$. By universality of the tensor product construction and the above freeness it follows that $U_q(\mathfrak{g}) u_{\Lambda'}$ is isomorphic to $M(\Lambda')$. It is called a Verma submodule of $M(\Lambda)$.

Lemma. *If Λ' is a primitive weight of $M(\Lambda)$, then $\Lambda' = \Lambda q^{-\beta}$ for some $\beta \in Q^+(\pi)$ and*

$$q^{2(\rho,\beta)-(\beta,\beta)} \Lambda(\tau(2\beta)) = 1 \ .$$

For each $v \in M(\Lambda)$ one has $U^+_\mu v = 0$ for all μ sufficiently large. Thus the Casimir operator J given in 3.3.9 defines a linear endomorphism of $M(\Lambda)$. On a primitive vector $u_{\Lambda'}$ one has $J u_{\Lambda'} = u_{\Lambda'}$. On the other hand one must have $u_{\Lambda'} = y_{-\beta} u_\Lambda$ for some $\beta \in Q^+(\pi)$ and some $y_{-\beta} \in U^-_{-\beta}$. Then the asserted formula follows from 3.3.9.

3.4.10 Let U^-_+ (resp. U^+_+) denote the kernel of ε restricted to U^- (resp. U^+). The triangular decomposition (3.2.10) of $U_q(\mathfrak{g})$ gives the direct sum decomposition

$$U = U^0 \oplus (U^-_+ U + U U^+_+) \ .$$

Let $\mathcal{P}$ denote the projection of U onto U^0 defined by this decomposition. Restricted to the zero weight space of U it is an algebra homomorphism. Recall the antiautomorphism $\varkappa$ of U defined in 3.2.6. The map $(a,b) \mapsto \mathcal{S}(a,b) := \mathcal{P}(\varkappa(a)b)$ defines a bilinear form on $U_q(\mathfrak{g})$ called the Shapovalev form [Sha1]. It is symmetric since $\varkappa$ has order 2, is the identity on U^0 and commutes with $\mathcal{P}$. Its importance lies in the fact that it leads to a form on each Verma module. Indeed it is clear that $\mathcal{S}(a,b) = 0$ for all $b \in UU_+^+$. On the other hand for the canonical generator u_Λ of $M(\Lambda)$ one has $\operatorname{Ann}_U u_\Lambda \supset UU_+^+ + U \operatorname{Ann}_{U^0} u_\Lambda$. Moreover equality results since $U/UU_+^+ + U\operatorname{Ann}_{U^0} u_\Lambda \xrightarrow{\sim} U^-$ and $M(\Lambda)$ is a free U^- module. Now define the symmetric bilinear form $\mathcal{S}^\Lambda$ on $U_q(\mathfrak{g})$ by $(a,b) \mapsto \Lambda(\mathcal{P}(\varkappa(a)b))$, that is $\mathcal{S}^\Lambda = \Lambda\mathcal{S}$. It follows from the above remarks that $\mathcal{S}^\Lambda(a,b) = 0$ whenever a or b lie in $\operatorname{Ann}_U u_\Lambda$. Thus $(au_\Lambda, bu_\Lambda) \mapsto \mathcal{S}^\Lambda(a,b)$ defines a symmetric bilinear form on $M(\Lambda)$ which by a slight abuse of notation is also denoted by $\mathcal{S}^\Lambda$.

It is immediate from the definition of $\mathcal{S}$ that

$$\mathcal{S}(a, \varkappa(c)b) = \mathcal{S}(ca, b); \quad \mathcal{S}(a, cb) = \mathcal{S}(\varkappa(c)a, b)$$

where the second equality results from $\varkappa$ being of order 2. This property is known as contravariance. It obviously passes to $\mathcal{S}^\Lambda$. It implies the orthogonality of weight subspaces of $M(\Lambda)$ corresponding to different weights.

Lemma. $\operatorname{Ker}\mathcal{S}^\Lambda$ *is the unique maximal submodule of* $M(\Lambda)$. *The quotient* $V(\Lambda) := M(\Lambda)/\ker\mathcal{S}^\Lambda$ *is the unique simple* $U_q(\mathfrak{g})$ *module admitting a primitive vector of weight* Λ.

Contravariance implies that $\operatorname{Ker}\mathcal{S}^\Lambda$ is a submodule of $M(\Lambda)$. If the quotient is not simple it must itself admit a primitive vector $u_{\Lambda'}$ of weight $\Lambda' \in \Lambda - Q^+(\pi) \setminus \{\Lambda\}$. Yet for all $a \in U^-$ one has $\mathcal{S}^\Lambda(au_\Lambda, u_{\Lambda'}) = \mathcal{S}^\Lambda(u_\Lambda, \varkappa(a)u_{\Lambda'}) = \varepsilon(\varkappa(a))\mathcal{S}^\Lambda(u_\Lambda, u_{\Lambda'}) = 0$. Thus $u_{\Lambda'} \in \ker\mathcal{S}^\Lambda$ proving the first assertion. The second follows from the universality of $M(\Lambda)$.

3.4.11 The Shapovalev form can be easily computed from its restriction to U^-. Moreover $\mathcal{S}(U^-_{-\mu}, U^-_{-\nu}) = 0$ unless $\mu = \nu$ and one defines $\mathcal{S}_\nu$ to be the restriction of $\mathcal{S}$ to $U^-_{-\nu}$. Recall the notation of 3.4.5 and in particular the definition of A and U_A^-. For each $\nu \in Q^+(\pi)$ choose an A basis for the free finite rank A module $(U_A^-)_{-\nu}$. Obviously this serves as a $k(q)$ basis for $U^-_{-\nu}$ and then $\det\mathcal{S}_\nu$ is defined as the determinant of the matrix obtained by evaluating $\mathcal{S}_\nu$ on basis elements.

For each $\mu \in Q(\pi)$, $\alpha \in \pi$ write

$$h_{\alpha,\mu} = t_\alpha q^{-(\alpha,\mu)} - t_\alpha^{-1} q^{(\alpha,\mu)} .$$

Then $h_{\alpha,0} = h_\alpha := t_\alpha - t_\alpha^{-1}$. Set $h^+_{\alpha,\mu} = h_{\alpha,\mu}$ if $\mu \in Q^+(\pi)$ and $h^+_{\alpha,\mu} = 1$ otherwise. If C is specializable, it will always be assumed that relation (v) of 3.2.9 is modified by including $q_\alpha - q_\alpha^{-1}$ in the denominator of h_α which will also be assumed to be included in the denominators of the $h_{\alpha,\mu}, h^+_{\alpha,\mu}$. Let $\mathbb{Z}[\mathbf{h}^+]$ denote the $\mathbb{Z}$-ring generated by the $h^+_{\alpha,\mu}$. It inherits a filtration from the gradation on U^0 defined by taking each t_α of degree α.

Let U_A (resp. U_A^0) denote the AT subring of U (resp. U^0) generated by U_A^+ and U_A^- (resp. by the $h_{\alpha,\mu}$). It is immediate from the relations in 3.2.9, noting $a_\nu h_{\alpha,\mu} = h_{\alpha,\mu+\nu} a_\nu : a_\nu \in U_\nu$, that the multiplication map $U_A^- \otimes_A U_A^0 \otimes_A U_A^+ \longrightarrow U_A$ is surjective. It is injective since this is already true in U. If follows that the restriction of $\mathcal{P}$ to U_A has values in U_A^0. Since $\varkappa$ also restricts to an antiautomorphism of U_A, it follows that $\mathcal{S}$ has values in U_A^0.

Lemma. *Take $\nu \in Q^+(\pi)$. Then $c \det \mathcal{S}_\nu \in \mathbb{Z}[\mathbf{h}^+]$, for some non-zero $c \in A$, and has filtration degree $\leq \nu \dim U^-_{-\nu}$.*

Let $a_{-\mu}, b_{-\mu} \in U^-_{-\mu}$ be monomials in the $y_{-\alpha} : \alpha \in \pi$ and show that $\mathcal{S}(a_{-\mu}, b_{-\mu}) \in \mathbb{Z}[\mathbf{h}^+]$ is of filtration degree $\leq \mu$, by induction on $|\mu|$. It is obvious for $|\mu| = 0$. Take $|\mu| > 0$. Then there exists $\alpha \in \pi$ such that $\varkappa(a_{-\mu}) = \varkappa(a_{-(\mu-\alpha)})x_\alpha$ for some monomial $a_{-(\mu-\alpha)} \in U^-_{-(\mu-\alpha)}$ in the $y_{-\beta} : \beta \in \pi$. Write $b_{-\mu} = y_{-(\mu-\mu_s-\alpha)}y_{-\alpha}y_{-\mu_s}$ for each factor $y_{-\alpha}$ in $b_{-\mu}$ and $b^{(s)}_{-(\mu-\alpha)}$ for $b_{-\mu}$ with this factor eliminated. Then

$$\mathcal{S}(a_{-\mu}, b_{-\mu}) = \sum_s \mathcal{P}(\varkappa(a_{-(\mu-\alpha)})y_{-(\mu-\mu_s-\alpha)}[x_\alpha, y_{-\alpha}]y_{-\mu_s}) \, .$$

Each term in the above sum takes the form

$$\mathcal{P}(\varkappa(a_{-(\mu-\alpha)})y_{-(\mu-\mu_s-\alpha)}h_\alpha y_{-\mu_s}) = h^+_{\alpha,\mu_s}\varphi(a_{-(\mu-\alpha)}, b^{(s)}_{-(\mu-\alpha)})$$

and so the assertion obtains from the induction hypothesis.

Now choose monomials $a^{(i)}_{-\nu} \in U^-_{-\nu} : i = 1, 2, \ldots, \dim U^-_{-\nu}$ linearly independent over K. By definition of U^-_A these lie in $(U^-_A)_{-\nu}$. Thus $\det \mathcal{S}_\nu = c \det\{\mathcal{S}_\nu(a^{(i)}_{-\nu}, a^{(j)}_{-\nu})\}_{i,j}$ for some $c \in A$ non-zero (being the determinant of an appropriate transformation matrix). Yet by the above the determinant on the right hand side belongs to $\mathbb{Z}[\mathbf{h}^+]$, and has filtration degree $\leq \nu \dim U^-_{-\nu}$. This proves the lemma.

3.4.12 . By 3.4.7 and the definition of $P(\nu)$ one has $\dim U^-_{-\nu} = P(\nu)$. Then by 3.4.11 it follows that

$$\tau(\nu P(\nu)) \det \mathcal{S}_\nu \in k(q)[t_\alpha : \alpha \in \pi] \, .$$

For any $a \in k(q)[t_\alpha : \alpha \in \pi]$, let $a|_0$ denote the evaluation of a at $t_\alpha = 0 : \alpha \in \pi$, and set $det_0\ \mathcal{S}_\nu = \tau(\nu P(\nu)) \det \mathcal{S}_\nu \,|_0$.

Lemma. *For all $x_\nu \in U^+_\nu$, $y_{-\nu} \in U^-_{-\nu}$ one has*

$$\tau(\nu)\mathcal{S}_\nu(\varkappa(x_\nu), y_{-\nu}) \,|_0 = (-1)^{|\nu|} q^{\frac{1}{2}(\nu,\nu)-(\rho,\nu)} \varphi(y_{-\nu}, x_\nu) \, .$$

In particular $det_0\ \mathcal{S}_\nu \neq 0$.

Observe first that $\tau(\nu)\mathcal{S}_\nu(\varkappa(x_\nu), y_{-\nu}) = \mathcal{P}(x_\nu y_{-\nu}\tau(\nu))$. Then the assertion is proved by induction on $|\nu|$. It is clear for $|\nu| = 0$. If $|\nu| > 0$ one may write $y_{-\nu} = y_{-\alpha}y_{-(\nu-\alpha)}$ for some $\alpha \in \pi$. Then

$$\mathcal{P}(x_\nu y_{-\nu}\tau(\nu))\mid_0 = \mathcal{P}([x_\nu, y_{-\alpha}]y_{-(\nu-\alpha)}\tau(\nu))\mid_0$$
$$= -q^{(\alpha,\nu-\alpha)}\mathcal{P}(\delta'_{-\alpha}(x_\nu)y_{-(\nu-\alpha)}\tau(\nu-\alpha))\mid_0 , \quad \text{by 3.2.7 (ii).}$$

Yet by 3.1.10 (i) the right hand side satisfies a similar recurrence relation. Hence the assertion. The last part follows from 3.1.7.

3.4.13 It is instructive to compute $\det \mathcal{S}_{\alpha+\beta} : \alpha, \beta \in \pi$. If $(\alpha,\beta) \neq 0$, then $y_{-\alpha}y_{-\beta}, y_{-\beta}y_{-\alpha}$ form an A basis for $(U_A^-)_{-(\alpha+\beta)}$. Following say the calculation in 3.4.11 one obtains

$$\det \mathcal{S}_{\alpha+\beta} = h_\alpha h_\beta (h_{\alpha,\beta}h_{\beta,\alpha} - h_\alpha h_\beta) .$$

The term in brackets reduces to

$$(q^{-2(\alpha,\beta)} - 1)(t_\alpha t_\beta - q^{2(\alpha,\beta)}t_\alpha^{-1}t_\beta^{-1}) .$$

In particular $det_0\ \mathcal{S}_{\alpha+\beta} = 1 - q^{2(\alpha,\beta)}$ which vanishes at $q = 1$. By 3.4.12 this corresponds to the pairing φ degenerating at $q = 1$ as mentioned in 3.4.6. However if C is specializable, then introducing the denominators $q_\alpha - q_\alpha^{-1}$ means that $\det \mathcal{S}_{\alpha+\beta}$ takes the form

$$\det \mathcal{S}_{\alpha+\beta} = \frac{t_\alpha - t_\alpha^{-1}}{q_\alpha - q_\alpha^{-1}}\,\frac{t_\beta - t_\beta^{-1}}{q_\beta - q_\beta^{-1}}\,\frac{t_\alpha t_\beta - q^{2(\alpha,\beta)}t_\alpha^{-1}t_\beta^{-1}}{q_\alpha - q_\alpha^{-1}}\,\frac{q^{-2(\alpha,\beta)} - 1}{q_\beta - q_\beta^{-1}} .$$

Evaluating t_α, t_β on a linear weight, say $q^\lambda : \lambda \in \mathfrak{h}_{\mathbb{Z}}^*$, gives an expression lying in A which is generally non-zero. Thus $\det \mathcal{S}_{\alpha+\beta}$ has a good specialization at $q = 1$. This does not rescue the pairing φ which in the present conventions acquires poles at $q = 1$.

3.4.14 The factorization of $\det \mathcal{S}_\nu$ described in 3.4.13 (*) for $\nu = \alpha + \beta$ holds quite remarkably in complete generality. For this set

$$Q_{irr}^+(\pi) = \{\gamma \in Q^+(\pi) \mid \gamma = r\beta : \beta \in Q^+(\pi),\ r \in \mathbb{N}^+ \text{ implies } r = 1\} .$$

For each $s \in \mathbb{N}^+$, let Γ_s denote the set of primitive s roots of unity. Set $\Gamma^m = \bigcup_{s|m} \Gamma_s$, $\Gamma = \bigcup \Gamma_s$.

Lemma. *Take $\nu \in Q^+(\pi)$. Then $\det \mathcal{S}_\nu$ is a product of factors of the form*

$$\prod_{\zeta \in \Gamma_s} (\tau(\beta) - \zeta q^{-2(\rho,\beta)+m(\beta,\beta)}\tau(-\beta))$$

where $\beta \in Q_{irr}^+(\pi)$, $m \in \mathbb{N}^+$ satisfy $m\beta \leq \nu$ and s is a divisor of m.

Replace the base field $k(q)$ by its algebraic closure $\overline{k(q)}$. By 3.4.11, 3.4.12, $\det \mathcal{S}_\nu$ is a non-zero element of the Laurent polynomial ring $\overline{k(q)}[t_\alpha, t_\alpha^{-1} :$

$\alpha \in \pi]$. Let $\Lambda \in \overline{k(q)}^{*\ell}$ be a zero of $\det S_\nu$. Then by 3.4.10 the corresponding Verma module $M(\Lambda)$ is not simple. More precisely its unique maximal submodule $\overline{M(\Lambda)} := \ker(M(\Lambda) \longrightarrow V(\Lambda))$ must satisfy $\overline{M(\Lambda)}_{\Lambda q^{-\nu}} \neq 0$. Thus there exists $\gamma \in Q^+(\pi) \setminus \{0\}$ with $\gamma \leq \nu$ and minimal with the property that $\overline{M(\Lambda)}_{\Lambda q^{-\gamma}} \neq 0$. Then $u \in \overline{M(\Lambda)}_{\Lambda q^{-\gamma}}$ is a primitive vector and so by 3.4.9, Λ is a root of the polynomial

$$(*) \qquad \tau(2\gamma) - q^{-2(\rho,\gamma)+(\gamma,\gamma)} .$$

One may write $\gamma = m\beta$ for some $\beta \in Q^+_{irr}(\pi)$, $m \in \mathbb{N}^+$. Then the irreducible factors of $(*)$ take the form $\tau(\beta) - \zeta q^{-(\rho,\beta)+\frac{m}{2}(\beta,\beta)}$ where ζ is a $2m^{th}$ root of unity. Let $\mathcal{H}_{\beta,m,\zeta}$ be the hypersurface of zeros such a factor defines.

Let $\mathcal{H}$ be an irreducible component of the zero variety of $\det S_\nu$. By the above $\mathcal{H}$ lies in a union of the $\mathcal{H}_{\beta,m,\zeta}$ with $m\beta \leq \nu$ and $\zeta^{2m} = 1$. However the set of triples β, m, ζ satisfying the above conditions is finite and so $\mathcal{H}$ must coincide with one of the $\mathcal{H}_{\beta,m,\zeta}$. Combined with 3.4.11 this gives the conclusion of the lemma.

3.4.15 Not all the factors described in the conclusion of 3.4.14 occur. Their description, which can be almost completely given, needs the more advanced techniques of the next chapter.

3.5 Comments and Complements

3.5.1 The Drinfeld double construction is of course due to V. G. Drinfeld [Dr2, Sect.13] who also noted the very important duality between $U(\mathfrak{b}_C^-)$ and $U(\mathfrak{b}_C^+)$ at least for C of finite type. The very general formulation (for any C) given here is taken partly from unpublished lecture notes of M. Rosso and [R3]. However Rosso needed to make some choices in particular for the coproduct. The present development motivated by a question of I. M. Gelfand [Dr5, Qu.8.3] and some computations of M. Kashiwara [K3, Sect.3] is extracted from an unpublished note of the author and makes the quantum Weyl algebra as its starting point. Hopefully the reader is convinced that the whole theory of quantized Kac-Moody enveloping algebras derives from 3.1.3 $(*)$ in a natural manner free of further choices. The existence of a skew-Hopf pairing between V^- and V^+ was also observed by T. Tanisaki [Ta1].

3.5.2 The Rosso form (3.3.1) on the Drinfeld double was pointed out to me by M. Rosso. The Rosso form on $U_q(\mathfrak{g})$ is older and was first given by Rosso in his thesis and then in [R2]. Only a weaker uniqueness result was obtained by Rosso and indeed 3.3.4 is taken from [JL4]. The construction of the Casimir invariant and the factorization of the Shapovalev determinant is also taken from the same paper, the latter inspired by work of V. G. Kac and D. A. Kazhdan [KK1]. A different Casimir invariant defined for a slightly larger algebra than $U_q(\mathfrak{g})$ was constructed by V. G. Drinfeld [Dr3, Sect.5].

3.5.3 Proposition 3.4.3 is taken from [J1] where it is moreover shown that the hypotheses even imply that A is a Weyl algebra over a polynomial ring [J1, Sect. 3]! The specialization of U^- described in 3.4.5 first occurred in the work of G. Lusztig [L2]. The specialization of U^+ described in 3.4.6 is derived from [JL2].

Chapter 4. Highest Weight Modules

The base field k is assumed of characteristic zero, with $\overline{k(q)}$ the algebraic closure of $k(q)$. The Cartan matrix is assumed symmetric, so then $U_q(\mathfrak{g})$ is defined.

4.1 The Jantzen Filtration and Sum Formula

This is an advanced technique which allows one to almost completely determine the factors of the Shapovalev determinants. Though this is not the easiest way to obtain the desired result for C of finite type, its beauty lies in its startling generality – no further restrictions on C being necessary.

4.1.1 Set $B = \overline{k(q)}[S]_{\langle S-1 \rangle}$ and $Q = \operatorname{Fract} B$. One views S as being q^s for some indeterminate s. In particular for each $\lambda \in \mathfrak{h}^*_{\mathbb{Z}}$ let $q^{s\lambda}$ denote the linear character on T with values in B sending $\tau(\alpha)$ to $q^{s(\alpha,\lambda)} := S^{(\alpha,\lambda)}$. Let $\rho^\vee \in \mathfrak{h}^*_{\mathbb{Z}}$ have the property that $(\alpha, \rho^\vee) = 1$ for all $\alpha \in \pi$.

Replacing k by $\overline{k(q)}$ and A by B in 3.4.5 defines the B rings $U_B^\pm$ which are free as B modules. Set $U_B^0 = BT$. Similarly replacing $\overline{k(q)}$ by Q the Q-algebras $U_Q^\pm$, U_Q^0, U_Q are defined. Let U_B denote the image of $U_B^- \otimes U_B^0 \otimes U_B^+ \longrightarrow U_Q$ under the multiplication map (which is injective). It is immediate from 3.2.9 that U_B is a B-subring of U_Q.

For each linear character on T with values in B, the Verma module $M_B(\Lambda)$ with canonical generator u_Λ is defined exactly as in 3.4.9, namely $M_B(\Lambda) = U_B \otimes_{U_B^+ T} 1_\Lambda$. By the above triangular decomposition, it is a free U_B^- module of rank 1. Replacing B by Q defines $M_Q(\Lambda)$. Since $M_B(\Lambda)$ is a free B module $M_B(\Lambda) \otimes_B Q \xrightarrow{\sim} M_Q(\Lambda)$. Triangular decomposition allows one to construct a Shapovalev form S on U_B with values in U_B as in 3.4.10. Evaluation at Λ defines S^Λ as a bilinear contravariant form on $M_B(\Lambda)$ which extends by scalars to $M_Q(\Lambda)$. Let $T^\wedge$ denote the set of linear characters on T with values in $\overline{k(q)}$.

Lemma. *Take $\Lambda_0 \in T^\wedge$ and set $\Lambda = \Lambda_0 q^{s\rho^\vee}$. Then $M_Q(\Lambda)$ is simple. In particular S^Λ is non-degenerate on $M_B(\Lambda)$.*

Since $(\beta, \rho^\vee) \in \mathbb{N}^+$ for all $\beta \in Q^+(\pi) \setminus \{0\}$, it follows from 3.4.9 that u_Λ is the unique up to scalars primitive vector of $M_Q(\Lambda)$. The second part follows from 3.4.10 (or 3.4.14).

4.1.2 Take Λ_0, Λ as in 4.1.1. For each $i \in \mathbb{N}$ define

$$M_B^i(\Lambda) = \{m \in M_B(\Lambda) \mid \mathcal{S}(M_B(\Lambda), m) \in (S-1)^i B\} .$$

By contravariance, the $M_B^i(\Lambda)$ form a decreasing sequence of U_B submodules of $M_B(\Lambda)$. By 4.1.1 one has

$$\bigcap_{i \in \mathbb{N}} M_B^i(\Lambda) = 0 .$$

Set $M^i(\Lambda_0) = M_B^i(\Lambda) \otimes_B B/\langle S-1 \rangle$. Then the $M^i(\Lambda_0)$ form a decreasing sequence of U submodules of $M(\Lambda_0)$ with zero intersection. It is called the Jantzen filtration of $M(\Lambda_0)$. Since B is principal and each weight submodule $M_B(\Lambda)$ is a free finite rank B module, the same holds for each $M_B^i(\Lambda)$. In particular

$$(*) \qquad rk_B\, M_B^i(\Lambda)_{\Lambda q^{-\mu}} = \dim_{\overline{k(q)}} M^i(\Lambda_0)_{\Lambda_0 q^{-\mu}}$$

for all $\mu \in Q^+(\pi)$. For each $\mu \in Q^+(\pi)$ it is immediate that $M^i(\Lambda_0)_{\Lambda_0 q^{-\mu}} = 0$ for i sufficiently large.

4.1.3 Recall 3.4.14 and let $d_{\beta,\zeta,m}(\nu) : \nu \in Q^+(\pi)$, $\beta \in Q^+_{\text{irr}}(\pi)$, $m \in \mathbb{N}^+$, $\zeta \in \Gamma^m$ denote the multiplicity of the factor $\tau(\beta) - \zeta q^{-2(\rho,\beta)+(\beta,\beta)}\tau(-\beta)$ occurring in $\det \mathcal{S}_\nu$. Given $\Lambda_0 \in T^\wedge$, set $\mathcal{T}(\Lambda_0) = \{(\beta, m, \zeta) \mid \Lambda_0(\tau(2\beta)) = \zeta q^{-2(\rho,\beta)+m(\beta,\beta)}\}$.

Proposition. *Take $\Lambda_0 \in T^\wedge$. Then*

$$\sum_{i=1}^{\infty} \dim_{\overline{k(q)}} M^i(\Lambda_0)_{\Lambda_0 q^{-\nu}} = \sum_{(\beta,m,\zeta) \in \mathcal{T}(\Lambda_0)} d_{\beta,m,\zeta}(\nu) .$$

The assertion results by calculating the largest power r of $S-1$ dividing $\det \mathcal{S}_\nu(\Lambda)$ in two different ways. First $\Lambda(\tau(2\beta)) = S^{2(\beta,\rho^\vee)}\Lambda_0(\tau(2\beta))$ and so exactly one such power occurs in $\det \mathcal{S}_\nu(\Lambda)$ whenever $(\beta, m, \zeta) \in \mathcal{T}(\Lambda_0)$. Thus the right hand side above equals r.

With respect to a basis $\{a_i\}_{i=1}^n$ for the free B module $M_B(\Lambda)_{\Lambda q^{-\nu}}$ consider $\mathcal{S}_\nu$ as the element of End B^n with matrix coefficients $\mathcal{S}_\nu(a_i, a_j)$. Since B is a noetherian principal ideal domain there exists bases $\{b_i\}$, $\{c_i\}$ of B^n and $s_i \in B$ such that $\mathcal{S}_\nu(b_i, c_j) = s_j \delta_{ij}$. Since B is moreover a local ring with maximal ideal $\langle S-1 \rangle$, there exists integers $r_j \geq 0$ such that $s_j = (S-1)^{r_j}$ up to units. Then by definition of $M_B^i(\Lambda)$ one has

$$rk_B M^i(L)_{\Lambda q^{-\nu}} = |\{j \mid r_j \geq i\}| ,$$

where $|\cdot|$ denotes cardinality.

Summing over i and using 4.1.2 (*) gives

$$\sum_{i=1}^{\infty} \dim_{\overline{k(q)}} M^i(\Lambda_0)_{\Lambda_0 q^{-\nu}} = \sum_{i=1}^{\infty} |\{j \mid r_j \geq i\}| = \sum_{j=1}^{n} r_j = r ,$$

as required.

4.1.4 A $U_q(\mathfrak{g})$ module M is said to belong to the $\mathcal{O}$ category if

(i) *M is a direct sum of its T weight subspaces $M_\Lambda : \Lambda \in T^\wedge$.*
(ii) $\dim M_\Lambda < \infty$ *for all $\Lambda \in T^\wedge$.*
(iii) *The set $\Omega(M) := \{\Lambda \in T^\wedge \mid M_\Lambda \neq 0\}$ of weights of M lies in a set of the form $\{\Lambda q^{-\mu} : \mu \in Q^+(\pi)\}$.*

Clearly $\mathcal{O}$ is an abelian category. Again each $M \in Ob\mathcal{O}$ admits a formal character defined by

$$\operatorname{ch} M = \sum (\dim_K M_\Lambda)\Lambda .$$

This is additive on exact sequences. Moreover given $M, N \in Ob\mathcal{O}$, then $M \otimes N \in Ob\mathcal{O}$ and

$$\operatorname{ch}(M \otimes N) = \operatorname{ch} M \operatorname{ch} N .$$

Given $M \in Ob\mathcal{O}$, it follows from (i) and (iii) that M is generated as a U^- module by a T stable complement to $U^-_+ M$.

Finally for each $\Lambda \in \Omega(M)$ it follows from (iii) that $U^+_\beta M_\Lambda = 0$ for β sufficiently large. Thus the $\mathcal{O}$-completion (3.3.8) of $U_q(\mathfrak{g})$ acts on objects in $\mathcal{O}$.

The category $\mathcal{O}$ admits an exact contravariant functor δ (the $\mathcal{O}$ dual) taking simples to simples defined as follows. Given $M \in Ob\mathcal{O}$, consider M^* as a left $U_q(\mathfrak{g}_C)$ module via the antiautomorphism $\varkappa$. Let δM be the largest submodule of M^* on which the $t_\alpha : \alpha \in \pi$ act locally finitely. Viewing M as a vector space graded by weight space decomposition, it is immediate that δM is just the graded dual (3.1.8) of M and in particular finite dimensionality of weight spaces gives an isomorphism $M \xrightarrow{\sim} \delta(\delta M)$. Finally the contravariant form (3.4.10) on $M(\Lambda)$ gives a $U_q(\mathfrak{g}_C)$ homomorphism $M(\Lambda) \longrightarrow \delta M(\Lambda)$ and hence an isomorphism $V(\Lambda) \xrightarrow{\sim} \delta V(\Lambda)$.

For each $\Lambda \in T^\wedge$ define $\mathcal{O}^\Lambda$ to be the full subcategory of $\mathcal{O}$ whose objects M satisfy $\Omega(M) \subset \{\Lambda q^{-\mu} : \mu \in Q(\pi)\}$. Given $M \in Ob\mathcal{O}$ and $\Lambda \in T^\wedge$ set

$$M^\Lambda = \bigoplus_{\mu \in Q(\pi)} M_{\Lambda q^{-\mu}} .$$

It is immediate that M^Λ is a $U_q(\mathfrak{g})$ direct summand of M and belongs to $\mathcal{O}^\Lambda$. Of course $\mathcal{O}^\Lambda$ is an abelian category; but it is not closed under tensor products, except in the case $\Lambda = q^\mu$ for some $\mu \in Q(\pi)$. Clearly $M(\Lambda) \in Ob\mathcal{O}^\Lambda$ and so is its simple quotient $V(\Lambda)$.

Lemma. *Suppose $M \in Ob\mathcal{O}^\Lambda$. Then for each $\lambda \in Q(\pi)$ there exists a chain $M = M_1 \supsetneq M_2 \supsetneq \cdots \supsetneq M_{n+1} = 0$ of U submodules of M such that either $M_i/M_{i+1} \cong V(\Lambda q^{\lambda_i})$ for some $\lambda_i \geq \lambda$ or $\Omega(M_i/M_{i+1}) \cap \Lambda q^{\lambda+Q^+(\pi)} = \emptyset$.*

The proof is by induction on

$$r(M) := \sum_{\mu \geq \lambda} \dim M_{\Lambda q^\mu}$$

which by hypotheses (ii) and (iii) of 4.1.4 is finite. If $r(M) = 0$ the chain $M \supset 0$ suffices. Otherwise choose $\Lambda' = \Lambda q^\mu \in \Omega(M)$ with $\mu \in \lambda + Q^+(\pi)$ maximal. Then any non-zero $u \in M_{\Lambda'}$ is a primitive vector and by universality of $M(\Lambda')$ one obtains a surjection $M(\Lambda') \longrightarrow Uu =: M'$. Recalling 3.4.10 let M'' be the image of the unique maximal submodule of $M(\Lambda')$. Then $M'/M'' \cong V(\Lambda')$. Yet $r(M) = r(M/M') + r(M'/M'') + r(M'')$ and $r(M'/M'') \geq 1$, so applying the induction hypothesis to M/M' and M'' completes the proof.

4.1.5 Given $M \in Ob\mathcal{O}$ one may define the multiplicity $[M : V(\Lambda)]$ of $V(\Lambda)$ in M as follows. First one may assume $M \in Ob\mathcal{O}^\Lambda$ by direct sum decomposition. Applying 4.1.4 with $\lambda \leq 0$, let $[M : V(\Lambda)]$ denote the number of times M_i/M_{i+1} is isomorphic to $V(\Lambda)$. Using ch it is immediate that the above expression is independent of λ and the chain. Moreover one has

$$\operatorname{ch} M = \sum_{\Lambda \in \Omega(M)} [M : V(\Lambda)] \operatorname{ch} V(\Lambda)$$

in which one may also note that the $\operatorname{ch} V(\Lambda)$ are linearly independent.

A similar formula results if $[M]$ denotes the image of M in the resulting Grothendieck group $\mathcal{G}$ of $\mathcal{O}$. Applied to $M(\Lambda)$ one obtains

$$[M(\Lambda)] = [V(\Lambda)] + \sum_{\mu < 0} [M(\Lambda) : V(\Lambda q^{-\mu})] \; [V(\Lambda q^{-\mu})] \, .$$

It follows that the $[M(\Lambda)]$ also form a basis for $\mathcal{G}$. Applying the Casimir invariant one obtains as in 3.4.9.

Lemma. *Let M be a submodule of $M(\Lambda)$. There exist $b_\gamma \in \mathbb{N}$, $c_\gamma \in \mathbb{Z}$ such that*

$$[M] = \sum b_\gamma [V(\Lambda q^{-\gamma}] = \sum c_\gamma [M(\Lambda q^{-\gamma})]$$

where the sums extend over all $\gamma \in Q^+(\pi)$ satisfying $q^{2(\rho,\gamma)-(\gamma,\gamma)} \Lambda(\tau(2\gamma)) = 1$.

4.1.6 Call $\beta \in Q^+(\pi)$ isotropic if $(\beta, \beta) = 0$. Extending k one may assume it to contain the complex field $\mathbb{C}$.

Lemma. *Fix $m \in \mathbb{N}^+$, $\zeta \in \Gamma$. Then for each $\alpha \in Q^+_{\text{irr}}(\pi)$ isotropic (resp. non-isotropic) there exists $\Lambda_0 \in T^\wedge$ such that $\Lambda_0(\tau(2\beta)) = \zeta q^{-2(\rho,\beta)+m(\beta,\beta)}$ implies $\beta \in \mathbb{N}^+\alpha$ (resp. $\beta = \alpha$).*

Assume first $\zeta = 1$. Up to a numbering of π one may write

$$\alpha = \sum_{i=1}^{n} r_i \alpha_i \quad : \quad r_i \in \mathbb{N}^+, \quad n \leq \ell \, .$$

Choose $\lambda \in \mathfrak{h}_{\mathbb{Z}}^*$ such that

$$\frac{1}{m}(\lambda + \rho, \alpha_i) = \frac{1}{2}(\alpha_i, \alpha_i) r_i + \sum_{j=i+1}^{n} (\alpha_i, \alpha_j) r_j \in \mathbb{Z} \, .$$

Then $2(\lambda + \rho, \alpha) = m(\alpha, \alpha)$. Take $u_i \in \mathbb{R} : i = 1, 2, \ldots, \ell$ linearly independent over $\mathbb{Q}$ and set $u_0 = u_n$. Defining $\Lambda_0 \in T^\wedge$ through

$$\Lambda_0(t_i) = \begin{cases} q^{(\lambda, \alpha_i)} exp \, \frac{1}{r_i} \, (u_i - u_{i-1}) & : 1 \leq i \leq n \, , \\ q^{(\lambda, \alpha_i)} exp \; u_i & : i > n \, . \end{cases}$$

gives the required result. Inserting into $\Lambda_0(t_i)$ the factor ζ^{s_i} where $\sum r_i s_i = 1$ settles the general case.

4.1.7 Recall the notation of 3.4.8 and 4.1.3.

Lemma. *For all $\beta \in Q^+_{\text{irr}}(\pi)$, $m \in \mathbb{N}^+$, $\zeta \in \Gamma$, there exists $c_{\beta,m,\zeta} \in \mathbb{Z}$ such that*

(i) $\sum_i \operatorname{ch} M^i(\Lambda_0) = \sum_{(\beta,m,\zeta) \in \mathcal{T}(\Lambda_0)} c_{\beta,m,\zeta} \operatorname{ch} M(\Lambda_0 q^{-m\beta})$ *for all* $\Lambda_0 \in T^\wedge$.
(ii) $\sum_{m=1}^\infty d_{\beta,m,\zeta} = \sum_{m=1}^\infty c_{\beta,m,\zeta} P_{m\beta}$ *if β is isotropic.*
(iii) $d_{\beta,m,\zeta} = c_{\beta,m,\zeta}$ *if β is non-isotropic.*

Take $\Lambda_0 \in T^\wedge$ and $M = M^i(\Lambda_0)$ in 4.1.5. Let $c^{\Lambda_0,i}_{\beta,m,\zeta}$ denote the coefficient of $[M(\Lambda_0 q^{-m\beta})]$ in $[M^i(\Lambda_0)]$ and set $c^{\Lambda_0}_{\beta,m,\zeta} = \sum_i c^{\Lambda_0,i}_{\beta,m,\zeta}$.

By 4.1.3 and 4.1.5 one has

$$\begin{aligned} \sum d_{\beta,m,\zeta}(\nu) &= \sum c^{\Lambda_0}_{\beta,m,\zeta} \; \dim M(\Lambda_0 q^{-m\beta})_{\Lambda_0 q^{-\nu}} \\ &= \sum c^{\Lambda_0}_{\beta,m,\zeta} \; P_{m\beta}(\nu) \, , \end{aligned}$$

where the sums are over $\mathcal{T}(\Lambda_0)$. Taking Λ_0 as in 4.1.6 successively for each $\beta \in Q^+_{\text{irr}}(\pi)$, noting that the $P_{m\beta}$ are linearly independent (3.4.8 (i)) and that $d_{\beta,m,\zeta}(\nu)$ are independent of Λ_0 gives the required assertions.

4.1.8 Retain the notation of 4.1.7 and let $\operatorname{mult} \alpha$ denote the number of times $\alpha \in Q^+(\pi)$ occurs as a weight of $\mathfrak{n}^+$.

Lemma. *For all $\beta \in Q^+_{\text{irr}}(\pi)$, $m \in \mathbb{N}^+$ one has*

$$\sum_{\zeta \in \Gamma^m} c_{\beta,m,\zeta} = \sum_{s|m} s(\operatorname{mult} s\beta) \, .$$

By 3.4.12 the leading coefficient of $\det S_\nu$ equals $\tau(\nu P(\nu))$. From the definition of $d_{\beta,m,\zeta}$ and 3.4.8 (ii) this gives

$$\sum_{m=1}^{\infty} \sum_{\zeta \in \Gamma^m} \sum_{\beta \in Q^+_{\text{irr}}(\pi)} \beta d_{\beta,m,\zeta}(\nu) = \sum_{n=1}^{\infty} \sum_{\alpha \in \Delta^+_{\text{mult}}} \alpha P_{n\alpha}(\nu) \,.$$

Substitution from 4.1.7 (ii), (iii), writing $\alpha = s\beta : \beta \in Q^+_{\text{irr}}(\pi)$ and using the linear independence of $P_{m\beta}$ gives the required assertion.

4.1.9 The above formulae do not quite determine the irreducible factors in the Shapovalev determinants. To resolve these ambiguities the leading term in their $(q-1)$ expansion is computed in the following sections.

Recall (3.4.6) the Lie algebra $\mathfrak{n}^-$ generated by the images of the $y_{-\alpha}$: $\alpha \in \pi$ in $U_A^- \otimes_A A/\langle q-1\rangle$. Obviously $\mathfrak{n}^-$ admits a basis formed from their commutators. Let the $y_{-\nu} : \nu \in \Delta^+_{\text{mult}}$ denote such a set of commutators but with the $y_{-\alpha} : \alpha \in \pi$ viewed as elements of U_A^-. Lift the ordering on Δ^+_{mult} induced by $Q^+(\pi)$ to a total ordering. Then the images of the standard monomials $Y_{-\nu} = y_{-\nu_1} y_{-\nu_2} \cdots y_{-\nu_r} : r \in \mathbb{N},\ \nu_1 \leq \nu_2 \leq \ldots \leq \nu_r,\ \sum \nu_i = \nu$ form a basis for $U(\mathfrak{n}^-)$, so by 3.4.6 (iii) and the dimensionality formulae in 3.4.5 they form a K basis for U^-.

Now let $y_{-\nu} : \nu \in Q^+(\pi)$ denote any A linear combination of commutators in the $y_{-\alpha} : \alpha \in \pi$ of weight $-\nu$. For each $Y_{-\nu} \in U_A^-$ of weight $-\nu$ let $\deg Y_{-\nu}$ be the smallest degree of $Y_{-\nu}$ as a polynomial (not necessarily ordered) in the $y_{-\nu}$. Obviously $\deg Y_{-\nu} \leq |\nu|$. Note that for the $Y_{-\nu}$ of the previous paragraph one has $\deg Y_{-\nu} \leq r$ and the inequality might be strict. For each $i \in \mathbb{N}$ let $\mathcal{D}^i(U_A^+)$ denote the A linear span of the $(q-1)^{i-(|\nu|-\deg Y_{-\nu})} Y_{-\nu}$, so in particular $Y_{-\nu} \in \mathcal{D}^{|\nu|-\deg Y_{-\nu}}(U_A^-)$ and $y_{-\alpha} \in \mathcal{D}^0(U_A^-)$. It follows that $\{\mathcal{D}^i(U_A^-)\}_{i\in\mathbb{N}}$ is a decreasing filtration of U_A^- with $\mathcal{D}^0(U_A^-) = U_A^-$. Extend it to a filtration of $U_A^- T$ by setting $\mathcal{D}^i(U_A^- T) = \mathcal{D}^i(U_A^-)T$. Applying the antiautomorphism $\varkappa$ gives similar filtrations for U_A^+ and $U_A^+ T$.

For each $\mu \in Q^+(\pi)$, set $h_\mu^+ = \tau(\mu) - \tau(\mu)^{-1}$, $h_\mu^- = \tau(\mu) + \tau(\mu)^{-1}$. Viewing $h_\alpha^\pm$ to be of weight α let ℓ_μ denote any sum of products of the $h_\alpha^\pm$ of weight μ. By 3.2.9(v) one has $[x_\alpha, y_{-\alpha}] = h_\alpha^+$ for all $\alpha \in \pi$. Define $(\pm)^r$ to be $+$ if r is even and $-$ if r odd. One has

$$\begin{aligned} [h_\mu^+, Y_{-\nu}] &= Y_{-\nu}[(q^{-(\mu,\nu)} - 1)\tau(\mu) - (q^{(\mu,\nu)} - 1)\tau(\mu)^{-1}] \\ &= -(\mu,\nu)(q-1) Y_{-\nu} h_\mu^- \bmod \mathcal{D}^{|\nu|-\deg Y_{-\nu}+2}(U_A^- T) \,. \end{aligned} \tag{$*$}$$

Similarly

$$[h_\mu^-, Y_{-\nu}] = -(\mu,\nu)(q-1) Y_{-\nu} h_\mu^+ \bmod \mathcal{D}^{|\nu|-\deg Y_{-\nu}+2}(U_A^-) \,.$$

Lemma. *For each $\alpha \in \pi$, $\nu \in Q^+(\pi)$ and each commutator $y_{-\nu}$ (in the $y_{-\beta} : \beta \in \pi$) one has*

$$[x_\alpha, y_{-\nu}] = \begin{cases} 0 & : \text{if } \nu \not\geq \alpha \\ h_\alpha^+ & : \text{if } \nu = \alpha \\ \sum_{r=1}^{\infty} h_\alpha^{(\pm)^r} (q-1)^r y_{-\nu_1} y_{-\nu_2} \cdots y_{-\nu_r} \bmod \mathcal{D}^{|\nu|}(U_A^- T) & : \text{if } \nu > \alpha \end{cases}$$

where in each summand the $y_{-\nu_i}$ are $\mathbb{Z}$ linear combinations of commutators and $\sum \nu_i = \nu - \alpha$.

The proof is an easy induction on commutator length. Write $y_{-\nu} = [y_{-\beta}, y_{-\mu}] : \beta \in \pi,\ \mu = \nu - \beta$. Then

$$[x_\alpha, y_{-\nu}] = \delta_{\alpha\beta}[h_\alpha^+, y_{-\mu}] + [y_{-\beta}, [x_\alpha, y_{-\mu}]] \ .$$

The first term reduces to the asserted form by $(*)$ above. For the second term the induction hypothesis is applied and then it suffices to observe that we can take

$$[y_{-\beta}, h_\alpha^{(\pm)^r} Y_{-(\mu-\alpha)}] = h_\alpha^{(\pm)^r}[y_{-\beta}, Y_{-(\mu-\alpha)}] + (q-1)(\alpha,\beta)h_\alpha^{(\pm)^{r+1}} y_{-\beta}Y_{-(\mu-\alpha)} \ .$$

4.1.10 Retain the above conventions and consider $x_\mu Y_{-\nu} \bmod UU_+^+$ viewed as an element of $U_A^- T$. By 4.1.9 it lies in $\mathcal{D}^{|\nu|-\deg Y_{-\nu}}(U_A^- T)$. Then one may calculate

$$(*) \qquad (x_\mu Y_{-\nu} \bmod UU_+^+) \bmod \mathcal{D}^{|\nu|-\deg Y_{-\nu}+1}(U_A^- T) \ .$$

In this one may remark that if $y_{-\nu_1} y_{-\nu_2} \cdots y_{-\nu_r}$ had in fact degree $< r$, then it would lie in a strictly higher filtration subspace than expected and so can be ignored. Now let us write $Y_{-\nu} = y_{-\nu_1} y_{-\nu_2} \cdots y_{-\nu_n} : n = \deg Y_{-\nu}$ and assume that $\mu \not> \nu_i$ for all i. If $\mu \neq \nu_i$ for all i, every term in $(*)$ takes the form

$$(**) \qquad (q-1)^{|\mu| + \deg Y_{-(\nu-\mu)} - \deg Y_{-\nu}} \ell_\mu \, Y_{-(\nu-\mu)}$$

with $\deg Y_{-(\nu-\mu)} \geq \deg Y_{-\nu}$, whilst if $\mu = \nu_i$ for some i there may also be a term of degree given by $\deg Y_{-\nu} - 1$.

Now consider $\mathcal{P}(X_\nu\ Y_{-\nu})$ with $Y_{-\nu}$ as before and $X_\nu = x_{\mu_1} x_{\mu_2} \ldots x_{\mu_m}$: $m = \deg X_\nu$ which lies in AT. Write $X_\nu \sim Y_{-\nu}$ if $m = n$ and if up to permutation $\mu_i = \nu_i$ for all i. Write $X_\nu \not\sim Y_{-\nu}$ otherwise.

Lemma

$$\mathcal{P}(X_\nu Y_{-\nu}) \in \begin{cases} \mathcal{D}^{|\nu|-\deg X_\nu}(AT) & : \textit{if } X_\nu \sim Y_{-\nu} \\ \mathcal{D}^{|\nu|-\deg X_\nu + 1}(AT) & : \textit{if } \deg X_\nu = \deg Y_{-\nu} \textit{ but } X_\nu \not\sim Y_{-\nu} \\ \mathcal{D}^{|\nu|-\min(\deg X_\nu, \deg Y_{-\nu})}(AT) & : \textit{in general}\ . \end{cases}$$

The proof is by induction on $|\nu|$ and within a given $|\nu|$ value by induction on $\deg X_\nu + \deg Y_{-\nu}$. One may assume one of the μ_i or one of ν_i are minimal and both cases are equivalent. Suppose μ_i is minimal. One may push x_{μ_i} to be a right hand factor of X_ν introducing commutators which strictly decrease $\deg X_\nu$. If $\deg X_\nu \leq \deg Y_{-\nu}$, then by the induction hypothesis these terms gain a $(q-1)$ factor and so satisfy the conclusion. If $\deg X_\nu > \deg Y_{-\nu}$, then one is only requiring the last conclusion to hold. This is again true by the induction hypothesis since it is enough to consider $\deg Y_{-\nu}$. Consequently one may assume $\mu := \mu_n$ minimal. Write $X_{\nu-\mu} = x_{\mu_1}x_{\mu_2}\dots x_{\mu_{n-1}}$. Consider (∗) above.

Suppose $\deg X_\nu \leq \deg Y_{-\nu}$. If $\mu \neq \nu_i$ for all i, then by (∗∗), $\deg X_{\nu-\mu} < \deg Y_{-(\nu-\mu)}$ for each term in the resulting sum and so by the induction hypothesis the power of $(q-1)$ dividing $\mathcal{P}(X_\nu Y_{-\nu})$ is at least $|\mu_n| + |\nu - \mu_n| - (\deg X_\nu - 1) = |\nu| - \deg X_{-\nu} + 1$ as required. If $\mu = \nu_i$ for some i, there can be a term $Y_{-(\nu-\mu)}$ with $\deg Y_{-(\nu-\mu)} = \deg Y_{-\nu} - 1 \geq \deg X_{\nu-\mu}$ multiplied by a factor of $(q-1)^{|\mu|-1}$. Again the result follows by the induction hypothesis.

Suppose $\deg X_\nu > \deg Y_{-\nu}$. If $\mu = \nu_i$ for some i then one may have a term $Y_{-(\nu-\mu)}$ with $\deg Y_{-(\nu-\mu)} = \deg Y_{-\nu} - 1 < \deg X_{-(\nu-\mu)}$ multipled by a factor of $(q-1)^{|\mu|-1}$. Then the result obtains by the induction hypothesis. Finally one may have a term $Y_{-(\nu-\mu)}$ with $\deg Y_{-(\nu-\mu)} \geq \deg Y_{-\nu}$. Then by the induction hypothesis the $(q-1)$ power dividing $\mathcal{P}(X_\nu Y_{-\nu})$ is at least

$$\begin{aligned}|\mu| + \deg Y_{-(\nu-\mu)} - \deg Y_{-\nu} + |\nu - \mu| - \min(\deg X_{\nu-\mu}, \deg Y_{-(\nu-\mu)}) \\ \geq |\nu| - \deg Y_{-\nu} \ ,\end{aligned}$$

as required.

4.1.11 Let $Y^i_{-\nu} : i = 1, 2, \dots, \dim U^-_{-\nu}$ denote the basis of $U^-_{-\nu}$ described in 4.1.9. Set $\det \mathcal{S}'_\nu = \det\{\mathcal{S}_\nu(Y^i_{-\nu}, Y^j_{-\nu})\}$. Write $Y^i_{-\nu} \sim Y^j_{-\nu}$ if $\varkappa(Y^i_{-\nu}) \sim Y^j_{-\nu}$. This is obviously an equivalence relation on the above set and so the elements may be relabelled $Y^{i,j}_{-\nu}$ with $Y^{i,j}_{-\nu} \sim Y^{i',j'}_{-\nu}$ if and only if $j = j'$. For each j let $\det \mathcal{S}^j_\nu$ denote the determinant of the $n_j \times n_j$ matrix $\mathcal{S}_\nu(Y^{i,j}_{-\nu}, Y^{i',j}_{-\nu})$ and let m_j denote the common degree of the $Y^{i,j}_{-\nu}$ viewed as a monomial in the commutators (specializing to a basis for $\mathfrak{n}^-$). By the first conclusion in 4.1.10 it follows that $(q-1)^{(|\nu|-m_j)}$ divides $\mathcal{S}_\nu(Y^{i,j}_{-\nu}, Y^{i',j}_{-\nu})$. (As before it does not matter if $\deg Y^{i,j}_{-\nu} < m_j$. This would only result in a possibly higher power of $(q-1)$ factoring out). Thus $(q-1)^{n_j(|\nu|-m_j)}$ divides $\det \mathcal{S}^j_\nu$. Set $d_\nu = \sum_j n_j(|\nu| - m_j)$.

Corollary

(i) $\prod_j \det \mathcal{S}^j_\nu \in \mathcal{D}^{d_\nu}(AT)$.

(ii) $\det \mathcal{S}'_\nu = \prod_j \det \mathcal{S}^j_\nu \mod \mathcal{D}^{d_\nu+1}(AT)$.

4.1.12 It remains to calculate $\mathcal{P}(X_\nu Y_{-\nu})$ when $X_\nu \sim Y_{-\nu}$. For this the Drinfeld double formula turns out to be a powerful computational technique. In the following it is convenient to write $X_{\nu-\mu} \otimes X_\mu$ for a sum of tensor products of such weight vectors.

Lemma. *Let $x_\nu, y_{-\nu} : \nu \in Q^+(\pi)$ be commutators. Then one may write*

(i) $\Delta(x_\nu) = x_\nu \otimes 1 + \sum_{0<\mu<\nu} (q-1)^{\deg X_{\nu-\mu} + \deg X_\mu - 1} X_{\nu-\mu} \otimes X_\mu + \tau(\nu) \otimes x_\nu$

(ii) $\Delta(y_{-\nu}) = y_{-\nu} \otimes \tau(-\nu) +$
$+ \sum_{0<\mu<\nu} (q-1)^{\deg Y_{-(\nu-\mu)} + \deg Y_{-\mu} - 1} Y_{-(\nu-\mu)} \tau(-(\nu-\mu)) \otimes Y_{-\mu} + 1 \otimes y_{-\nu}$

(iii) $x_\nu y_{-\nu} - y_{-\nu}x_\nu - (\tau(\nu) - \tau(\nu)^{-1})\varphi(y_{-\nu}, x_\nu) = \sum_{0<\mu<\nu}(q-1)^{d_{\mu,\nu}}$ $\{Y_{-(\nu-\mu)}\tau(-(\nu-\mu))X_{\nu-\mu}\varphi(Y_{-\mu}, X_\mu) - X_\mu Y_{-\mu}\varphi(Y_{-(\nu-\mu)}\tau(-(\nu-\mu)), X_{\nu-\mu})\}$, *where* $d_{\mu,\nu} = \deg X_{\nu-\mu} + \deg Y_{-(\nu-\mu)} + \deg X_\mu + \deg Y_{-\mu} - 2$.

Since Δ is an algebra homomorphism $\Delta[x_\alpha, x_\nu] = [\Delta(x_\alpha), \Delta(x_\nu)]$. Then (i) follows from 3.1.10 (+) by induction on $|\nu|$. Similarly (ii) follows from 3.1.10 (−). Then (iii) follows from (i), (ii) and remark 3.2.2 noting that the pairing φ respects weight space decomposition.

4.1.13 Consider 4.1.12 (iii). One would like to show that the terms on the right hand side can be ignored in calculating $\mathcal{P}(X_\nu Y_{-\nu})$ to its lowest $(q-1)$ power. By 3.1.3 and 3.4.12, the power of $q-1$ dividing $\varphi(Y_{-\nu-\mu}\tau(-(\nu-\mu)), X_{\nu-\mu})$ is at least that dividing $\mathcal{S}_\nu(\varkappa(X_{-\nu-\mu}), Y_{-(\nu-\mu)})$. By 4.1.10 this provides a $(q-1)$ exponent $\geq |\nu-\mu| - \min(\deg X_{\nu-\mu}, \deg Y_{-(\nu-\mu)})$. Again by 4.1.10 reordering the contribution of the term $X_\mu\, Y_{-\mu}$ gives a further $(q-1)$ exponent $\geq |\mu| - \min(\deg X_\mu, \deg Y_{-\mu})$. Since all terms have degree ≥ 1, the overall $(q-1)$ exponent coming from the second term in the summation is $\geq |\nu|$. Again by 4.1.10 a similar conclusion holds for the first term. On the other hand the term on the left hand side of 4.1.12 (iii) coming from $\varphi(y_{-\nu}, x_\nu)$ will in general only have an exponent in $(q-1)$ of $|\nu| - 1$. This gives the

Proposition. *For all* $\nu \in Q^+(\pi)$ *the coefficient of* $(q-1)^{d_\nu}$ *in* $\det \mathcal{S}'_\nu$ *equals*

$$a_\nu \prod_{m=1}^{\infty} \prod_{\beta\in\Delta^+_{\text{mult}}} (\tau(\beta) - \tau(\beta)^{-1})^{P(\nu-m\beta)}$$

for some $a_\nu \in A$.

4.1.14 Of course the scalar a_ν occurring in the conclusion of 4.1.13 could be divisible by $q-1$. This is shown not to be so for C integrable. Assume first that C is specializable and divide each $x_\alpha : \alpha \in \pi$ by $q_\alpha - q_\alpha^{-1}$ (which equals $q-1$ times a unit in A). This brings a denominator of $(q-1)^{|\nu|P(\nu)}$ into $\det \mathcal{S}'_\nu$. By 3.4.8 (ii)

$$|\nu|P(\nu) = \sum_{m=1}^{\infty} \sum_{\beta\in\Delta^+_{\text{mult}}} |\beta|P(\nu - m\beta) .$$

On the other hand d_ν equals

$$\sum_{m=1}^{\infty} \sum_{\beta\in\Delta^+_{\text{mult}}} (|\beta| - 1)P(\nu - m\beta) .$$

One concludes that the leading term of $\det \mathcal{S}'_\nu$ becomes up to a unit in A,

$$a_\nu \prod_{m=1}^{\infty} \prod_{\beta\in\Delta^+_{\text{mult}}} \left(\frac{\tau(\beta) - \tau(\beta)^{-1}}{(q-1)}\right)^{P(\nu-m\beta)} .$$

4.1.15 Recall the notation of 3.4.11 and set

$$U_1 = U_A \otimes_A A/\langle q-1\rangle, \quad U_1^0 = U_A^0 \otimes_A A/\langle q-1\rangle .$$

Then by 3.4.6 (iii) and the triangular decomposition in 3.4.11 one obtains the triangular decomposition

$$U(\mathfrak{n}^-) \otimes U_1^0 \otimes U(\mathfrak{n}^+) \xrightarrow{\sim} U_1 .$$

Moreover U_1^0 contains the images of the $(t_\alpha - t_\alpha^{-1})/(q_\alpha - q_\alpha^{-1})$ which will also be denoted by h_α. Thus $t_\alpha^2 = 1$ in U_1 and t_α is easily seen to be central in U_1. Set $\bar{U}_1 = U_1/\langle t_\alpha - 1 : \alpha \in \pi\rangle$, $\bar{U}_1^0 = U_1^0/\langle t_\alpha - 1 : \alpha \in \pi\rangle$. One easily checks that $\bar{U}_1^0$ is generated by the $h_\alpha : \alpha \in \pi$ which are moreover algebraically independent and commute pairwise. Let $\mathfrak{h}^\vee$ denote the commutative Lie algebra they generate. One has $x_\alpha y_{-\beta} - y_{-\beta} x_\alpha = \delta_{\alpha,\beta} h_\alpha$, since this already holds before specialization. Writing $\alpha^\vee = 2\alpha/(\alpha,\alpha)$ for all $\alpha \in \pi$ one further checks that

$$[h_\alpha, x_\beta] = (\alpha^\vee, \beta) x_\beta \ , \ [h_\alpha, y_{-\beta}] = -(\alpha^\vee, \beta) y_{-\beta}$$

for all $\alpha, \beta \in \pi$. One concludes that $\mathfrak{g} := \mathfrak{n}^- \oplus \mathfrak{h}^\vee \oplus \mathfrak{n}^+$ admits a Lie algebra structure and $\bar{U}_1 \xrightarrow{\sim} U(\mathfrak{g})$. Again the factors $(\tau(\beta) - \tau(\beta)^{-1})/(q-1) : \beta \in \Delta^+_{\text{mult}}$ lie in $\mathfrak{h}^\vee$.

4.1.16 Just as for $U_q(\mathfrak{g}_C)$ the enveloping algebra $U(\mathfrak{g})$ admits the calculation of Shapovalev determinants. One says that $U_q(\mathfrak{g}_C)$ admits good specialization if all these are non-vanishing. (This condition means that $\mathfrak{g}$ is essentially the Kac-Moody Lie algebra $\mathfrak{g}_C$ constructed from C). Since the $Y^i_{-\nu}$ used in the definition of $\det S'_\nu$ specialize to a basis of $U(\mathfrak{n}^-)$, the above condition also means that the a_ν in the conclusion of 4.1.13 are units in A and then $\det S'_\nu = \det S_\nu$ up to a unit in A. Combined with 3.4.14, 4.1.7 and 4.1.8, the ambiguity in the $c_{\beta,m,\zeta}$ is resolved and one obtains

Theorem. *Assume that C is specializable and that $U_q(\mathfrak{g}_c)$ admits good specialization. Then up to a unit in A*

(i) $\det S_\nu = \prod_{m=1}^{\infty} \prod_{\beta \in \Delta^+_{\text{mult}}} (\tau(\beta) - q^{-2(\rho,\beta)+m(\beta,\beta)}\tau(-\beta))^{P(\nu - m\beta)}$
or equivalently
(ii) $c_{\beta,m,\zeta} = \sum_{r|m, \zeta^r=1} \operatorname{mult} r\beta$, *for all $\beta \in Q^+_{\text{irr}}(\pi)$, $m \in \mathbb{N}^+$, $\zeta \in \Gamma^m$.*

Remark. It is clear that the conclusion $\det S_\nu = \det S'_\nu$ up to a unit in A means that the $Y_{-\nu} : \nu \in \Delta^+_{\text{mult}}$ constructed in 4.1.9 actually form a free A basis for $(U_A^-)_{-\nu}$. This also holds independently by Nakayama's lemma.

4.1.17 That the hypothesis holds is indicated by the following general result which also has an analogue for $U(\mathfrak{g}_C)$.

Proposition. *The ideal K_C^- (resp. K_C^+) is generated by weight vectors $y_{-\mu}$ (resp. x_μ) satisfying $2(\rho,\mu) = (\rho,\rho)$ and $\mu \notin \pi$.*

Using $\varkappa$ it suffices to prove the assertion for K_C^-. Let $\tilde{U}_q(\mathfrak{g}_C)$ (or simply, $\tilde{U}$) denote the quotient of the Drinfeld double $\mathcal{D}(\tilde{V}^-, \tilde{V}^+)$ by the ideal generated by the $\tau^+(\alpha)\tau^-(-\alpha) - 1 : \alpha \in \pi$. It is the Hopf algebra given by the relations of 3.2.9 with (vi) omitted. Let $p : \tilde{U} \longrightarrow U$ denote the canonical projection.

Identify $\tilde{U}^-$ with the Verma module $\tilde{M}(0)$ for $\tilde{U}_q(\mathfrak{g})$ with highest weight 0. Observe that $y_{-\alpha}\tilde{M}(0) : \alpha \in \pi$ identifies with the Verma submodule $\tilde{M}(-\alpha)$ of $\tilde{M}(0)$ with highest weight $-\alpha$ and that the sum $\tilde{S} = \sum_\alpha \tilde{M}(-\alpha)$ in $\tilde{M}(0)$ is direct since $\tilde{U}$ is freely generated by the $y_{-\alpha} : \alpha \in \pi$. Moreover K^- lies in $\tilde{S}$. Set $S = U \otimes_{\tilde{U}} \tilde{S}$ which on the one hand is isomorphic to $\tilde{S}/(\ker p)\tilde{S}$ and on the other hand to the direct sum of the Verma modules $M(-\alpha) : \alpha \in \pi$ for U of highest weight $-\alpha$. Let $\lambda : K^- \longrightarrow S$ be the composed map. Since $M(-\alpha)$ is a free U^- module it easily follows that $\ker \lambda = \sum_\alpha K^- y_{-\alpha}$. Thus $M := K^-/\sum_\alpha K^- y_{-\alpha}$ is a U submodule of S. Use of the Casimir operator and 4.1.4 shows that the simple subquotients occurring in M take the form $V(-\mu)$ with $2(\rho, \mu) = (\mu, \mu)$ and then that every weight $-\mu$ of the "homology space" $M/U_+^- M$ also satisfies this relation. Yet $M \in Ob\mathcal{O}$ so by 4.1.4 it is generated over U^- by a T stable complement N to $U_+^- M$ which has necessarily the same set of weights. Finally it is clear that the preimage of $U_+^- M$ is K^- is just $\sum_\alpha (y_{-\alpha} K^- + K^- y_{-\alpha})$, so in fact N forms a generating subspace for K^- as an ideal of $\tilde{U}^-$. Trivially $y_{-\alpha} \notin K^-$, so one cannot have $\mu \in \pi$.

4.2 Kac-Moody Lie Algebras

To each Cartan matrix C one may associate a Kac-Moody Lie algebra $\mathfrak{g}_C$. Naturally the structure and representation theory of its enveloping algebra $U(\mathfrak{g}_C)$ parallels that of $U_q(\mathfrak{g}_C)$. Whereas the aim of the present text is to develop the theory of $U_q(\mathfrak{g}_C)$ in an independent fashion and to concentrate on properties not apparent in $U(\mathfrak{g}_C)$, this is not always practical nor possible. Thus the present section briefly develops those aspects of $U(\mathfrak{g}_C)$ required for the study $U_q(\mathfrak{g}_C)$, particularly to show it has good specialization for C integrable.

4.2.1 Let C be a specializable Cartan matrix and set $\alpha^\vee = 2\alpha/(\alpha, \alpha)$, $\forall \alpha \in \pi$. (The denominators in h_α of 3.2.9 could be omitted in the general case; but this leads to a degenerate situation of no particular interest here). By virtue of the Cartan form, these may be viewed as elements of $\mathfrak{h}_\mathbb{Q}$. Set $\mathfrak{h}_\mathbb{Z} = \{h \in \mathfrak{h}_\mathbb{Q} \mid (h, \beta) \in \mathbb{Z}, \forall \beta \in \pi\}$ and view $\mathfrak{h} := \mathfrak{h}_\mathbb{Z} \otimes_\mathbb{Z} k$ as a commutative Lie algebra. (It will shortly be assumed that $(\alpha^\vee, \beta) \in \mathbb{Z}$, $\forall \alpha, \beta \in \pi$ and so $\alpha^\vee \in \mathfrak{h}_\mathbb{Z}$.)

The pre Kac-Moody Lie algebra $\tilde{\mathfrak{g}}_C$ defined by C is the k-Lie algebra with generators $x_\alpha, y_{-\beta} : \alpha, \beta \in \pi$; $h \in \mathfrak{h}$ satisfying the relations

(i) $[h, h'] = 0$,

(ii) $[h, x_\alpha] = (h, \alpha)x_\alpha$,

(iii) $[h, y_{-\beta}] = -(h, \beta) y_{-\beta}$,
(iv) $[x_\alpha, y_{-\beta}] = \delta_{\alpha,\beta} \alpha^\vee$,

for all $\alpha, \beta \in \pi$, $h, h' \in \mathfrak{h}$. Using the diamond lemma (A.2.13) one may show that the Lie algebra $\tilde{\mathfrak{n}}^+$ (resp. $\tilde{\mathfrak{n}}^-$) generated by the x_α (resp. $y_{-\alpha}$) : $\alpha \in \pi$ is a free Lie algebra graded by $Q^+(\pi)$ (resp. $Q^-(\pi)$). As in the quantum case (3.1.3) there exists a unique maximal proper ideal $\mathfrak{k}_C^+$ (resp. $\mathfrak{k}_C^-$) of $\tilde{\mathfrak{n}}^+$ (resp. $\tilde{\mathfrak{n}}^-$) stable by the $\operatorname{ad} y_{-\alpha}$ (resp. $\operatorname{ad} x_\alpha$) : $\alpha \in \pi$ and moreover $\mathfrak{k}^+$ (resp. $\mathfrak{k}^-$) is graded. Set $\mathfrak{n}_C^\pm = \tilde{\mathfrak{n}}^\pm / \mathfrak{k}_C^\pm$. As in 3.4.7 the set of weights of $\mathfrak{n}_C^+$ counted with their multiplicities is denoted $\Delta^+_{\text{mult}}(C)$ with $\Delta_{\text{mult}}(C)$ and $\Delta(C)$ similarly defined. The corresponding dual root system obtained from $\pi^\vee$ is denoted $\Delta^\vee(C)$. It is clear that $\mathfrak{g}_C := \mathfrak{n}_C^- \oplus \mathfrak{h} \oplus \mathfrak{n}_C^+$ acquires a Lie algebra structure. It is called the Kac-Moody Lie algebra defined by C. Assume C symmetric and let $\mathfrak{g}$ denote the Lie algebra obtained in 4.1.15 by specializing $U(\mathfrak{g}_C)$. It is immediate that the identity on the $x_\alpha, y_{-\beta} : \alpha, \beta \in \pi$ extends to a Lie algebra homomorphism $\mathfrak{g} \longrightarrow \mathfrak{g}_C$. It will be shown that this map is injective (equivalently that $U_q(\mathfrak{g}_C)$ admits good specialization) when C is integrable. Observe $\varkappa(x_\alpha) = y_{-\alpha}$, $\varkappa(y_{-\alpha}) = x_\alpha$ extends to an antiautomorphism of $\tilde{\mathfrak{g}}_C$ (and hence of $\mathfrak{g}_C$) which is the identity on $\mathfrak{h}$. It is called the Chevalley antiautomorphism.

4.2.2 Assume from now on that C is symmetric and that $\operatorname{char} k = 0$. Then C defines a non-degenerate symmetric bilinear form $(\ ,\ )$ on $\mathfrak{h}$ called its Cartan form (3.1.1, 3.1.2).

Proposition. *The Cartan form extends uniquely to an ad-invariant symmetric bilinear form $K(\ ,\ )$ on $\tilde{\mathfrak{g}}_C$. Moreover $K(\ ,\ )$ factors to a non-degenerate form on $\mathfrak{g}_C$.*

It is clear that any such form must respect weight space decomposition (that is $K(z_\mu, z_\nu) = 0$ unless $\mu + \nu = 0$). Moreover for all $h \in \mathfrak{h}$

$$K(x_\alpha, [y_{-\beta}, h]) = -K([y_{-\beta}, x_\alpha], h) = \delta_{\alpha,\beta} K(\alpha^\vee, h)$$

so then

$$(*) \qquad K(x_\alpha, y_{-\beta}) = 2\delta_{\alpha,\beta}/(\alpha, \alpha)\ .$$

Since the $\operatorname{ad} x_\alpha : \alpha \in \pi$ are derivations and $\tilde{\mathfrak{n}}^+$ is a free graded Lie algebra, it follows that $(*)$ extends to a unique bilinear form on $\tilde{\mathfrak{n}}^+ \times \tilde{\mathfrak{n}}^-$ respecting weight space decomposition and satisfying

$$K([x_\alpha, x_\nu], y_{-\mu}) = -K(x_\nu, [x_\alpha, y_{-\mu}])$$

for all $\mu, \nu \in Q^+(\pi)$, $\nu \neq 0$. An easy induction argument on $|\nu|$ using the relations (ii)–(iv) of 4.2.1 then shows that

$$K(x_{\nu+\alpha}, [y_{-\alpha}, y_{-\mu}]) = -K([y_{-\alpha}, x_{\nu+\alpha}], y_{-\mu})$$

for all $\mu, \nu \in Q^+(\pi)$, $\nu \neq 0$. Finally by the symmetry of C one has

$$K([x_\alpha, h], y_{-\beta}) = -(h, \alpha)K(x_\alpha, y_{-\beta}) = -\delta_{\alpha,\beta}K(h, \alpha^\vee) = -K(h, [x_\alpha, y_{-\beta}]) ,$$

which completes the verifications needed to establish the first part. It is clear from the definition of $\mathfrak{k}^\pm$ that $K(,)$ factors to $\mathfrak{g}_C$ and an easy induction argument on $|\nu|$ establishes the non-degeneracy.

4.2.3 Suppose $\dim \mathfrak{g}_C < \infty$. Then the construction of 3.3.6 applied to 4.2.3 gives a (quadratic) element $J \in U(\mathfrak{g}_C)$ which is ad-invariant hence central. In particular when $|\pi| = 1$ and $(\alpha, \alpha) = 2$, this recovers the classical Casimir invariant. Moreover generally as noted in 3.3.7 one obtains an element

$$I \in \prod_{\gamma \in Q^+(\pi)} \mathfrak{n}^-_{-\gamma} \otimes \mathfrak{n}^+_\gamma$$

satisfying

$$(*) \qquad (K \times K)(I, (x \otimes y)) = K(x, y)$$

for all $x \in \mathfrak{n}^+$, $y \in \mathfrak{n}^-$. Set $\mathfrak{m}^\pm = \sum_{|\gamma|>1} \mathfrak{n}^\pm_{\pm\gamma}$.

Lemma. *For all* $\alpha \in \pi$

(i) $(\operatorname{ad} x_\alpha)I = \frac{(\alpha,\alpha)}{2} (\alpha^\vee \otimes x_\alpha)$

(ii) $(\operatorname{ad} y_{-\alpha})I = \frac{-(\alpha,\alpha)}{2} (y_{-\alpha} \otimes \alpha^\vee)$.

From the construction of I and 4.2.2 $(*)$ one has

$$I = \sum_{\beta \in \pi} \frac{(\beta, \beta)}{2} (y_{-\beta} \otimes x_\beta) \operatorname{mod}(\mathfrak{m}^- \otimes \mathfrak{m}^+) .$$

On the other hand from $(*)$ and the ad-invariance of K it follows that $(K \times K)((\operatorname{ad} x_\alpha)I, \mathfrak{n}^+ \otimes \mathfrak{m}^-) = 0$. Hence

$$(\operatorname{ad} x_\alpha)I = \sum_{\beta \in \pi} \frac{(\beta, \beta)}{2} [x_\alpha, y_{-\beta}] \otimes x_\beta = \frac{(\alpha, \alpha)}{2} (\alpha^\vee \otimes x_\alpha)$$

which is (i). The proof of (ii) is similar.

4.2.4 Recall 3.3.8, 4.1.4. Define the $\mathcal{O}$ category and the $\mathcal{O}$-completion $U^{\mathcal{O}}(\mathfrak{g})$ of $U(\mathfrak{g})$ similarly. As before $\mu(I) \in U^{\mathcal{O}}(\mathfrak{g})$. Let $\{h_i\}$ be a basis for $\mathfrak{h}$ orthonormal with respect to K. Choose $h_\rho \in \mathfrak{h}$ such that $(h_\rho, \alpha) = (\rho, \alpha) = \frac{1}{2} (\alpha, \alpha)$ for all $\alpha \in \pi$. Then

$$(\operatorname{ad} x_\alpha)(\frac{1}{2} \sum h_i^2 + h_\rho) = -\frac{1}{2} \sum (h_i, \alpha)(x_\alpha h_i + h_i x_\alpha) - \frac{(\alpha, \alpha)}{2} x_\alpha$$
$$= - \frac{(\alpha, \alpha)}{2} \alpha^\vee x_\alpha .$$

Set $J = \mu(I) + \frac{1}{2} \sum h_i^2 + h_\rho$. From the above, a similar expression for $\operatorname{ad} y_{-\alpha}$ and 4.2.3, it follows that J is an ad-invariant element of $U^{\mathcal{O}}(\mathfrak{g})$. It is called the Casimir invariant. By 1.3.3, the above result may be expressed by the

Proposition. *The element $J \in U^{\mathcal{O}}(\mathfrak{g})$ is central and acts on modules in the $\mathcal{O}$ category.*

4.2.5 The simplest example of a Kac-Moody Lie algebra is obtained when $|\pi| = 1$ and $C = (2)$. It admits a faithful simple module if dimension 2 which expressed in matrix notation is given by

$$x_\alpha = \begin{pmatrix} 0 & 1 \\ 0 & 0 \end{pmatrix}, \qquad \alpha^\vee = \begin{pmatrix} 1 & 0 \\ 0 & -1 \end{pmatrix}, \qquad y_{-\alpha} = \begin{pmatrix} 0 & 0 \\ 1 & 0 \end{pmatrix}.$$

Thus in this case $\mathfrak{g}_C \cong \mathfrak{sl}(V)$, when $\dim V = 2$. It is immediate that the nth symmetric power $V^{\circledS n}$ is a simple $\mathfrak{sl}(V)$ module (this holds irrespective of $\dim V$) and has dimension $n+1$. Verma module theory (detailed for the quantum analogue in 4.3.1) shows that this construction exhausts all the simple finite dimensional $\mathfrak{sl}(2)$ modules and furthermore using in particular J above that every finite dimensional $\mathfrak{sl}(2)$ module is semisimple. (This also results from the considerations in 1.4.13).

For any Kac-Moody Lie algebra and for any $\alpha \in \pi$, the subspace $\mathfrak{sl}_\alpha$ spanned by $x_\alpha, y_{-\alpha}, \alpha^\vee$ is isomorphic to $\mathfrak{sl}(2)$. Let M be a $U(\mathfrak{g}_C)$ module admitting a formal character M (defined as in 3.4.7). Let $s_\alpha \in Aut\ \mathfrak{h}^*$ be the reflection defined by α, that is $s_\alpha\lambda = \lambda - (\alpha^\vee, \lambda)\alpha$. Then $s_\alpha^2 = 1d$ and $(s_\alpha\lambda, s_\alpha\lambda) = (\lambda, \lambda)$. The Weyl group $W(C)$ (or simply, W) is defined to be the subgroup of $Aut\ \mathfrak{h}^*$ generated by the $s_\alpha : \alpha \in \pi$. Observe that one may write $s_\alpha\lambda = \lambda - (\alpha, \lambda)\alpha^\vee$, so s_α may also be viewed as the reflection defined by the coroot $\alpha^\vee$.

Lemma. *Assume that $U(\mathfrak{sl}_\alpha)$ acts locally finitely on M. Then $s_\alpha \operatorname{ch} M = \operatorname{ch} M$.*

By assumption M is a sum of its finite dimensional $\mathfrak{sl}_\alpha$ modules and hence a direct sum of its simple finite dimensional $\mathfrak{sl}_\alpha$ modules. It remains to observe that the above property holds for $V^{\circledS n}$, since it holds for V.

4.2.6 One may anticipate that the easiest $U(\mathfrak{g}_C)$ modules to describe are those which admit a formal character and are locally finite for each $\mathfrak{sl}_\alpha$. They are called integrable modules. Nevertheless unless $\dim \mathfrak{g}_C < \infty$, these have not as yet been classified. In any case one doesn't get too far without some further assumptions on C. Some results do hold in general however as noted below. Set $\mathfrak{b}^+ = \mathfrak{h} + \mathfrak{n}^+$. Then $\lambda \in \mathfrak{h}^*$ defines a one dimensional $\mathfrak{h}$ module 1_λ of weight λ which extends to a $U(\mathfrak{b}^+)$ module by setting $x_\alpha 1_\lambda = 0$ for all $\alpha \in \pi$. As in 3.4.9 for $U_q(\mathfrak{g})$ the Verma module $\mathcal{M}(\lambda)$ with highest weight λ is defined by

$$\mathcal{M}(\lambda) = U(\mathfrak{g}) \otimes_{U(\mathfrak{b}^+)} 1_\lambda\ .$$

As before the highest weight vector $1 \otimes 1_\lambda$ of $\mathcal{M}(\lambda)$ is denoted by u_λ. For each $\alpha \in \pi$, $\mu \in \mathfrak{h}^*$ set $s_\alpha.\mu := s_\alpha(\mu + \rho) - \rho$. This defines the translated action of W on $\mathfrak{h}^*$.

Lemma. *Consider $\alpha \in \pi$. Suppose $(\alpha^\vee, \lambda) = r \in \mathbb{N}$. Then $y_{-\alpha}^{r+1}u_\lambda$ is a primitive vector of $\mathcal{M}(\lambda)$ and generates a submodule of $\mathcal{M}(\lambda)$ isomorphic to $\mathcal{M}(s_\alpha.\lambda)$.*

Suppose $\beta \in \pi$ and $\beta \neq \alpha$. It is immediate that $x_\beta y_{-\alpha}^{r+1}u_\lambda = y_{-\alpha}^{r+1}x_\beta u_\lambda = 0$. It remains to show that $x_\alpha y_{-\alpha}^{r+1}u_\lambda = 0$ which can be verified by direct computation. Alternatively consider $k[y_{-\alpha}]u_\lambda$. It is a Verma module for $\mathfrak{sl}_\alpha$ with highest weight λ. By 4.2.5 there exists a simple dimensional $\mathfrak{sl}_2$ module whose highest weight coincides with $\lambda|_{\alpha^\vee}$. By universality of Verma modules the latter is a quotient of $k[y_{-\alpha}]u_\lambda$. This forces $k[y_{-\alpha}]y_{-\alpha}^{r+1}u_\lambda$ to be a submodule and hence $y_{-\alpha}^{r+1}u_\lambda$ to be a primitive vector relative to $\mathfrak{sl}_\alpha$, as required.

4.2.7 Assume $(\alpha^\vee, \beta) \in -\mathbb{N}$ for some $\alpha, \beta \in \pi$ distinct. The significance of this condition derives from the previous lemma. Indeed consider $U(\mathfrak{g})$ as a module for the adjoint action of $\mathfrak{sl}_\alpha$. Then $y_{-\beta}$ has weight $-\beta$. Thus it is a primitive vector and hence so is $(\operatorname{ad} y_{-\alpha})^{-(\alpha^\vee,\beta)+1}y_{-\beta}$. One checks easily that this element commutes with x_β and hence with all the $x_\gamma : \gamma \in \pi$. It follows in particular that

$$(*) \quad (\operatorname{ad} y_{-\alpha})^{-(\alpha^\vee,\beta)+1}y_{-\beta} \in \mathfrak{k}^- \quad \text{and similarly} \quad (\operatorname{ad} x_\alpha)^{-(\alpha^\vee,\beta)+1}x_\beta \in \mathfrak{k}^+ .$$

The following is immediate from $(*)$ and the PBW theorem applied to $U(\mathfrak{sl}_\alpha)$.

Lemma. *Fix $\alpha \in \pi$ and assume that $(\alpha^\vee, \beta) \in -\mathbb{N}$ for all $\beta \in \pi \setminus \{\alpha\}$. Then*

(i) *$U(\mathfrak{g})$ is locally finite under the adjoint action of $\mathfrak{sl}_\alpha$.*
(ii) *Suppose $\lambda \in \mathfrak{h}^*$ satisfies $(\alpha^\vee, \lambda) \in \mathbb{N}$. Then $\mathcal{M}(\lambda)/\mathcal{M}(s_\alpha.\lambda)$ is locally finite under the action of $\mathfrak{sl}_\alpha$. In particular any subquotient M of $\mathcal{M}(\lambda)/\mathcal{M}(s_\alpha.\lambda)$ satisfies*

$$s_\alpha \operatorname{ch} M = \operatorname{ch} M .$$

4.2.8 The significance of C being integrable can be readily appreciated from 4.2.7. Thus this condition provides for each $\alpha, \beta \in \pi$ distinct, a generator $y^{\alpha,\beta}$ (resp. $x^{\alpha,\beta}$) of $\mathfrak{k}^-$ (resp. $\mathfrak{k}^+$) and these will be shown to provide a complete set. Again set

$$P(\pi) = \{\lambda \in \mathfrak{h}^* \mid (\alpha^\vee, \lambda) \in \mathbb{Z}, \forall \alpha \in \pi\}$$

and

$$P^+(\pi) = \{\lambda \in \mathfrak{h}^* \mid (\alpha^\vee, \lambda) \in \mathbb{N}, \forall \alpha \in \pi\} .$$

Given $\lambda \in P^+(\pi)$ it follows from 4.2.6 that $\sum_{\alpha\in\pi} \mathcal{M}(s_\alpha \cdot \lambda)$ is a submodule of $\mathcal{M}(\lambda)$. Let $\mathcal{V}'(\lambda)$ denote the resulting quotient. In general it is not known if $\mathcal{V}'(\lambda)$ equals its simple quotient $\mathcal{V}(\lambda)$. However assume C integrable. Then $\mathcal{V}'(\lambda)$ is integrable. Moreover

Proposition. *Take $\lambda \in P^+(\pi)$.*

(i) *Suppose $\mu \in \lambda - \mathbb{N}\pi$ satisfies $(\mu+\rho, \mu+\rho) = (\lambda+\rho, \lambda+\rho)$. If $\mu+\rho \in P^+(\pi)$ then $\mu = \lambda$. If $w.\mu \leq \lambda$ for all $w \in W$, then equality holds for some $w \in W$.*

Assume C integrable.

(ii) *Take $\alpha \in \pi$. Then $\Delta^+_{\mathrm{mult}}(C) \setminus \{\alpha\}$ is s_α stable. In particular $\Delta_{\mathrm{mult}}(C)$ is $W(C)$ stable. Similarly $\Delta^\vee_{\mathrm{mult}}(C)$ is $W(C)$ stable.*

(iii) *$\mathcal{V}'(\lambda)$ is simple.*

(iv) $\operatorname{ch}\mathcal{V}(\lambda) = \sum_{w\in W}(-1)^{\ell(w)}\operatorname{ch}\mathcal{M}(w.\lambda)$, *where $\ell(w)$ denotes the reduced length of $w \in W$.*

The relation in (i) can be rewritten as $(\lambda-\mu, \mu+\rho)+(\lambda+\rho, \lambda-\mu) = 0$. The hypotheses $\lambda, \mu+\rho \in P^+(\pi)$, $\lambda-\mu \in \mathbb{N}\pi$ imply that both terms are positive and so must separately vanish. Yet $(\lambda+\rho, \lambda-\mu) = 0$ implies $\lambda-\mu = 0$ as required. Finally choose $\mu_0 \in W(\mu+\rho)-\rho = W.\mu$ so that $|\lambda-\mu_0|$ is minimal. It is immediate $\mu_0+\rho \in P^+(\pi)$, so from the first part $\mu_0 = \lambda$ as required.

Fix $\alpha \in \pi$ and consider the Lie subalgebra of $\mathfrak{n}^-$ of codimension 1 given by

$$\mathfrak{m}^-(\alpha) := \bigoplus_{\beta\in\Delta^+\setminus\{\alpha\}} \mathfrak{n}^-_{-\beta} .$$

It is immediate that $\mathfrak{m}^-(\alpha)$ is an $\mathfrak{sl}_\alpha$ module which moreover is integrable by the hypothesis on C. This implies that $\Delta^+_{\mathrm{mult}} \setminus \{\alpha\}$ is s_α stable.

Not (iii) would imply that $\mathcal{V}'(\lambda)$ admits a primitive vector u_μ with $\mu < \lambda$. Take $\alpha \in \pi$. By 4.2.7 (ii), $\mathcal{V}'(\lambda)$ is $\mathfrak{sl}_\alpha$ locally finite and so 4.2.5 implies that $(\alpha^\vee, \mu) \in \mathbb{N}$. Consequently $\mu \in P^+(\pi)$. On the other hand as in the quantum case, the Casimir invariant can be applied to $\mathcal{M}(\lambda)$ and to any of its subquotients. On any primitive vector of weight ν it acts by the scalar $\frac{1}{2}\sum_i(h_i,\nu)^2+(\rho,\nu) = \frac{1}{2}(\nu+2\rho,\nu)$. Taking $\nu = \mu, \lambda$ gives the identity in the hypothesis of (i) and so $\mu = \lambda$ which is a contradiction.

By 4.2.7 (ii) it is immediate that $\operatorname{ch}\mathcal{V}(\lambda)$ is W stable. The enveloping algebra version of 4.1.5 gives $\operatorname{ch}\mathcal{V}(\lambda) = \sum a_\mu \operatorname{ch}\mathcal{M}(\mu)$ where $a_\lambda = 1, a_\mu \in \mathbb{Z}\setminus\{0\}, \mu \leq \lambda$, $(\lambda+2\rho,\lambda) = (\mu+2\rho,\mu)$ in the sum. Applied to the formula for $\operatorname{ch} U(\mathfrak{n}^-_C)$ which results from the enveloping algebra analogue of 3.4.7 it follows that $s_\alpha \operatorname{ch}\mathcal{M}(w.\lambda) = -\operatorname{ch}\mathcal{M}(s_\alpha w.\lambda)$ and so $w\operatorname{ch}\mathcal{M}(\lambda) = (-1)^{\ell(w)}\operatorname{ch}\mathcal{M}(w.\lambda)$. Then the linear independence of the $\operatorname{ch}\mathcal{M}(\mu)$ implies that $|a_{w.\mu}|$ is independent of $w \in W$ in the sum. In particular $w.\mu \leq \lambda$ and so (iv) obtains from (i).

4.2.9 In the integrable case one may compute the homology groups $H_*(\mathfrak{n}^-, \mathcal{V}(\lambda)) : \lambda \in P^+(\pi)$. This makes the result in 4.2.8 (iv) more precise and natural. Taking $\mathcal{V}(\lambda)$ to be the trivial module it also determines $\mathfrak{k}^-$. Applying $\varkappa$ this determines $\mathfrak{k}^+$.

Let $\mathfrak{n}$ be a Lie algebra and M an $\mathfrak{n}$ module. The rth homology space $H_r(\mathfrak{n}, M)$ with coefficients in M is defined as follows. Set $d_0 m = 0$ for all $m \in M$ and for $r > 0$ define a linear map $d_r : \wedge^r\mathfrak{n} \otimes M \longrightarrow \wedge^{r-1}\mathfrak{n} \otimes M$ by

$$d_r(x_1 \wedge x_2 \dots x_r \otimes m) = \sum_{i=1}^{r} (-1)^{r-i}(x_1 \dots \wedge \hat{x}_i \dots \wedge x_r) \otimes x_i m$$

$$+ \sum_{i<j=1}^{r} (-1)^{i+j}(x_1 \dots \wedge \hat{x}_i \dots \wedge \hat{x}_j \dots x_r \wedge [x_i, x_j]) \otimes m ,$$

where $\hat{x}$ means omitting that argument. One checks that $d_r d_{r+1} = 0$ and sets $H_r(\mathfrak{n}, M) := \ker d_r / \operatorname{Im} d_{r+1}$. For example $H_0(\mathfrak{n}, M) \xrightarrow{\sim} M/\mathfrak{n}M$. If $\mathfrak{n}$ is an ideal of $\mathfrak{b}$ and M is a $\mathfrak{b}$ module, taking the adjoint action of $\mathfrak{b}$ on $\mathfrak{n}$ combined with the coproduct gives each $\wedge^r \mathfrak{n} \otimes M$ a $\mathfrak{b}$ module structure. One checks that this action of $\mathfrak{n}$ commutes with d_r. Since $\operatorname{Im} d_r \supset \mathfrak{n}(\wedge^{r-1}\mathfrak{n} \otimes M)$ this gives $H_r(\mathfrak{n}, M)$ a $\mathfrak{b}/\mathfrak{n}$ module structure. If M is free, in particular for $M = U(\mathfrak{n})$, then $H_i(\mathfrak{n}, M) = 0$ for $i > 0$. Thus setting $Y^r := U(\mathfrak{n}) \otimes \wedge^r \mathfrak{n}$ makes (Y^*, ε) a free (hence projective) resolution of the trivial $U(\mathfrak{n})$ module 1 and one defines the cohomology spaces

$$H^r(\mathfrak{n}, M) := H^*(\operatorname{Hom}_{U(\mathfrak{n})}(Y^r, M)) = H^*(\operatorname{Hom}(\wedge^r \mathfrak{n}, M)) .$$

By Frobenius reciprocity $(\wedge^r \mathfrak{n} \otimes M)^* \xrightarrow{\sim} \operatorname{Hom}(\wedge^r \mathfrak{n}, M^*)$. Consequently

(1) $\quad H_r(\mathfrak{n}, M)^* = (\ker d_r / \operatorname{Im} d_{r+1})^* = \ker d_{r+1}^* / \operatorname{Im} d_r^* = H^r(\mathfrak{n}, M^*) .$

Now take $\mathfrak{n} = \mathfrak{n}_C^+$, $\mathfrak{b} = \mathfrak{b}_C^+ := \mathfrak{h} \oplus \mathfrak{n}_C^+$. Take $\lambda \in \mathfrak{h}^*$ and defining 1_λ as in 4.2.6, identify Y^r with $U(\mathfrak{b}_C^+) \otimes_{U(\mathfrak{h})} (\wedge^r \mathfrak{n}^+ \otimes 1_\lambda)$. Then (Y^*, ε) is a projective resolution of the $U(\mathfrak{b}_C^+)$ module 1_λ and

$$H^r(\mathfrak{n}_C^+, M)_\lambda = H^*(\operatorname{Hom}_{U(\mathfrak{b}_C^+)}(Y^r, M)) .$$

By PBW the functor $U(\mathfrak{g}_C) \otimes_{U(\mathfrak{b}_C^+)} -$ is exact and so taking $Z^r := U(\mathfrak{g}_C) \otimes_{U(\mathfrak{b}_C^+)} Y^r$ makes $(Z^*, 1 \otimes \varepsilon)$ a projective resolution of $\mathcal{M}(\lambda)$. Now assume that M is a $U(\mathfrak{g}_C)$ module. Then $\operatorname{Hom}_{U(\mathfrak{g}_C)}(Z^r, M) = \operatorname{Hom}(\Lambda^r \mathfrak{n}_C^+, M)$ and consequently

(2) $$\operatorname{Ext}^r(\mathcal{M}(\lambda), M) = H^r(\mathfrak{n}_C^+, M)_\lambda .$$

Observe that since each Z^r is a sum of its $\mathfrak{h}$ weight spaces one may replace M in (2) by its locally $\mathfrak{h}$ finite submodule. Now assume that $M \in Ob\mathcal{O}$. Recalling the definition of the $\mathcal{O}$ dual (4.1.4) it follows that one may replace M^* by δM in (1). However in doing this the latter is given a left module structure via $\varkappa$ which interchanges $\mathfrak{n}_C^-$ and $\mathfrak{n}_C^+$ and is the identity on $\mathfrak{h}$. Finally $\delta^2 = Id$ on $\mathcal{O}$. Thus (1), (2) translate to give

Lemma. *Suppose $M \in Ob\mathcal{O}$. Then for all $\lambda, \mu \in \mathfrak{h}^*$*

(i) $\operatorname{Ext}^r(\mathcal{M}(\lambda), M) = H^r(\mathfrak{n}_C^+, M)_\lambda = H_r(\mathfrak{n}_C^-, \delta M)_\lambda$
(ii) $H_r(\mathfrak{n}_C^-, \mathcal{V}(\mu))_\lambda = 0$ *unless* $(\lambda + \rho, \lambda + \rho) = (\mu + \rho, \mu + \rho)$.

Clearly (ii) obtains from (i) using the Casimir invariant.

4.2.10 Take $\mathcal{V}(\lambda)$ to be the trivial $U(\mathfrak{g}_C)$ module 1 in 4.2.9 (ii), that is take $\lambda = 0$. Since $H_2(\mathfrak{n}_C^-, 1)$ determines a set of generators for the relation ideal $\mathfrak{k}_C^-$, it follows that $\mathfrak{k}_C^-$ is generated as a Lie algebra by weight vectors $-\mu$ satisfying $2(\mu, \rho) = (\rho, \rho)$. In particular $\ker((U(\tilde{\mathfrak{n}}^-) \longrightarrow U(\tilde{\mathfrak{n}}_C^-))$ is generated as a two-sided ideal by weight vectors having this property. Thus the former result is a (slightly finer) version of 4.1.17 obtained for the quantum case.

4.2.11 When C is integrable, the conclusion of 4.2.9 (ii) can be markedly improved. This obtains from the following general result. Let $\mathfrak{p}$ be a Lie algebra and $\mathfrak{n} \supset \mathfrak{m}$ ideals in $\mathfrak{p}$. Let M be a $\mathfrak{p}$ module. Then $H_r(\mathfrak{n}/\mathfrak{m}, H_s(\mathfrak{m}, M))$ is defined and admits a $\mathfrak{p}/\mathfrak{n}$ module structure. The embedding $\mathfrak{m} \hookrightarrow \mathfrak{n}$ gives a $\mathfrak{p}$ module map $\varphi_r : \wedge^r\mathfrak{m} \otimes M \longrightarrow \wedge^r\mathfrak{n} \otimes M$ satisfying $\varphi_{r-1}d_r = d_r\varphi_r$. Thus φ_r factors to a $\mathfrak{p}$ module map $H_r(\mathfrak{m}, M) \longrightarrow H_r(\mathfrak{n}, M)$. One checks that the latter further factors to a $\mathfrak{p}$ module map $\bar{\varphi}_r : H_0(\mathfrak{n}/\mathfrak{m}, H_r(\mathfrak{m}, M)) \longrightarrow H_r(\mathfrak{n}, M)$ which is an embedding if $\dim \mathfrak{n}/\mathfrak{m} = 1$. Moreover in this case one may write $\mathfrak{n} = kx \oplus \mathfrak{m}$ and then each $a \in \wedge^r\mathfrak{n} \otimes M$ may be uniquely expressed in the form $a = x \otimes b + b'$ with $b \in \wedge^{j-1}\mathfrak{m} \otimes M$, $b' \in \wedge^j\mathfrak{m} \otimes M$. Define $\psi_r : \wedge^r\mathfrak{n} \otimes M \longrightarrow \mathfrak{n}/\mathfrak{m} \otimes \wedge^{r-1}\mathfrak{m} \otimes M$ by $\psi_r(a) = x \otimes b$. Obviously ψ_r is surjective, satisfies $\ker \psi_r = \operatorname{Im} \varphi_r$ and is a map of $\mathfrak{p}$ modules. Now $d_r a = xb - x \otimes d_{r-1}b + d_r b'$, so $d_r a = 0$ if and only if $d_{r-1}b = 0$ and $xb + d_r b' = 0$, whilst $\psi_{r-1}(d_r a) = -x \otimes d_{r-1}b$. Thus ψ_r factors to a $\mathfrak{p}$ module map $\bar{\psi}_r$ of $H_r(\mathfrak{n}, M)$ into $\mathfrak{n}/\mathfrak{m} \otimes H_{r-1}(\mathfrak{m}, M)$. Moreover $x \otimes b \in \operatorname{Im} \bar{\psi}_r$ satisfies $xb = 0$, that is $\operatorname{Im} \bar{\psi}_r \subset H_1(\mathfrak{n}/\mathfrak{m}, H_{r-1}(\mathfrak{m}, M))$. One checks that equality holds and $\ker \bar{\psi}_r = \operatorname{Im} \bar{\varphi}_r$. This proves the

Lemma. *Let* $\mathfrak{m} \subset \mathfrak{n}$ *be ideals of a Lie algebra* $\mathfrak{p}$ *and suppose* $\dim \mathfrak{n}/\mathfrak{m} = 1$. *Let* M *be a* $\mathfrak{p}$ *module. Then for each* $r \in \mathbb{N}$ *there is an exact sequence*

$$0 \longrightarrow H_0(\mathfrak{n}/\mathfrak{m}, H_r(\mathfrak{m}, M)) \longrightarrow H_r(\mathfrak{n}, M) \longrightarrow H_1(\mathfrak{n}/\mathfrak{m}, H_{r-1}(\mathfrak{m}, M)) \longrightarrow 0$$

of $\mathfrak{p}/\mathfrak{n}$ *modules.*

4.2.12 Fix $\alpha \in \pi$. Define $\mathfrak{m}^-(\alpha)$ as in 4.2.8 and set $\mathfrak{p}^-(\alpha) = \mathfrak{n}^- \oplus \mathfrak{h} \oplus kx_\alpha$. Then the conclusion of 4.2.11 applies to any $U(\mathfrak{g}_C)$ module M taking $\mathfrak{p} = \mathfrak{p}^-(\alpha)$, $\mathfrak{n} = \mathfrak{n}^-$, $\mathfrak{m} = \mathfrak{m}^-(\alpha)$.

Assume C integrable and take $M = \mathcal{V}(\lambda) : \lambda \in P^+(\pi)$. Then since $\mathfrak{m}^-(\alpha)$ and $\mathcal{V}(\lambda)$ are integrable $\mathfrak{sl}_\alpha + \mathfrak{h} \cong \mathfrak{p}(\alpha)^-/\mathfrak{m}(\alpha)^-$ modules so is $H_r(\mathfrak{m}_\alpha^-, \mathcal{V}(\lambda))$ which is hence (4.2.5) semisimple. Let N be a simple submodule with highest weight μ. Then 4.2.5 gives $(\alpha^\vee, \mu) \in \mathbb{N}$ and that N has lowest weight $s_\alpha\mu$. Let $u_\mu, u_{s_\alpha\mu}$ denote the corresponding weight vectors. Then $H_0(\mathfrak{n}^-/\mathfrak{m}^-(\alpha), N) = N/y_{-\alpha}N = ku_\mu$ whilst $H_1(\mathfrak{n}^-/\mathfrak{m}^-(\alpha), N) = k(y_{-\alpha} \otimes u_{s_\alpha\mu})$.

Now suppose $H_r(\mathfrak{n}^-, \mathcal{V}(\lambda))_\nu \neq 0$ for some $\nu \in \mathfrak{h}^*$. Trivially $\nu \in \lambda - \mathbb{N}\pi \subset P(\pi)$. Moreover by 4.2.11 either $(\alpha^\vee, \nu) \in \mathbb{N}$ and ν is a highest weight of $H_r(\mathfrak{m}^-(\alpha), \mathcal{V}(\lambda))$ or $(\alpha^\vee, \nu) \in -\mathbb{N}^+$ and $\nu + \alpha$ is a lowest weight of $H_{r-1}(\mathfrak{m}^-(\alpha), \mathcal{V}(\lambda))$. This gives the following key

Proposition. *Assume C integrable and take $\lambda \in P^+(\pi)$. Suppose $H_r(\mathfrak{n}^-, \mathcal{V}(\lambda))_\nu \neq 0$. Then $\nu \leq \lambda$. Moreover for each $\alpha \in \pi$ either $(\alpha^\vee, \nu) \in \mathbb{N}$ or $(\alpha^\vee, \nu) \in -\mathbb{N}^+$. In the latter case $r \geq 1$ and there is an isomorphism $H_r(\mathfrak{n}^-, \mathcal{V}(\lambda))_\nu \xrightarrow{\sim} H_{r-1}(\mathfrak{n}^-, \mathcal{V}(\lambda))_{s_\alpha.\nu}$ of vector spaces.*

4.2.13 Assume C integrable in the remainder of this section and take $\lambda \in P^+(\pi)$. It follows from 4.2.12 using induction on $|\lambda-\mu|$ that $H_*(\mathfrak{n}^-, \mathcal{V}(\lambda))$ is completely determined by $H_*(\mathfrak{n}^-, \mathcal{V}(\lambda))_\mu$ for $\mu \in P^+(\pi) \cap \lambda - \mathbb{N}\pi$. Yet by 4.2.10 (ii) it follows from 4.2.8 (i) that $\mu = \lambda$ in this case. Now $H_r(\mathfrak{n}^-, \mathcal{V}(\lambda))_\lambda \neq 0$ forces $r = 0$ whilst $H_0(\mathfrak{n}^-, \mathcal{V}(\lambda)) = \mathcal{V}(\lambda)/\mathfrak{n}^-\mathcal{V}(\lambda) = ku_\lambda$. This proves the

Theorem. *Take $\lambda \in P^+(\pi)$. For all $\nu \in \mathfrak{h}^*, r \in \mathbb{N}$ one has*

$$\dim H_r(\mathfrak{n}_C^-, \mathcal{V}(\lambda))_\nu = \begin{cases} 1 & : \nu = w.\lambda : w \in W \quad \textit{with} \quad r = \ell(w) \\ 0 & : \textit{otherwise} . \end{cases}$$

4.2.14 Take $\lambda = 0$ in 4.2.13. Then $H_2(\mathfrak{n}_C^-, 1)_{s_\alpha s_\beta .0}$ is one dimensional for each pair $\alpha, \beta \in \pi$ distinct. Yet $-s_\alpha s_\beta.0 = \beta - (\alpha^\vee, \beta)\alpha$ so this module determines a generator of $\mathfrak{k}_C^-$ of weight $-(\beta - (\alpha^\vee, \beta)\alpha)$ which has already been found to be $y^{\alpha,\beta}$. Hence the

Corollary. *The $y^{\alpha,\beta}$ (resp. $x^{\alpha,\beta}$) : $\alpha, \beta \in \pi$, $\alpha \neq \beta$, generate $\mathfrak{k}_C^-$ (resp. $\mathfrak{k}_C^+$).*

4.2.15 The formal character of $\mathcal{V}(\lambda)$ may also be computed from 4.2.13. Indeed for each $r \in \mathbb{N}$ one has

$$\begin{aligned} \operatorname{ch}(\wedge^r \mathfrak{n}^-) \operatorname{ch} \mathcal{V}(\lambda) &= \operatorname{ch}(\wedge^r \mathfrak{n}^- \otimes \mathcal{V}(\lambda)) \\ &= \operatorname{ch} \operatorname{Im} d_r + \operatorname{ch} \ker d_r \\ &= \operatorname{ch} \operatorname{Im} d_r + \operatorname{ch} \operatorname{Im} d_{r+1} + \operatorname{ch} H_r(\mathfrak{n}^-, \mathcal{V}(\lambda)) . \end{aligned}$$

Taking the alternating sum (which converges in the Krull topology) gives

$$\operatorname{ch} \mathcal{V}(\lambda) = \frac{\sum_{r=0}^{\infty}(-1)^r ch\ H_r(\mathfrak{n}^-, \mathcal{V}(\lambda))}{\sum_{r=0}^{\infty}(-1)^r \operatorname{ch} \wedge^r \mathfrak{n}^-} .$$

The denominator may be evaluated by noting that $\mathcal{V}(0)$ is the trivial module and so $\operatorname{ch} \mathcal{V}(0) = 1$. Combined with 4.2.13 this gives the

Corollary. *For all $\lambda \in P^+(\pi)$,*

$$\operatorname{ch} \mathcal{V}(\lambda) = \frac{\sum_{w \in W}(-1)^{\ell(w)} e^{w(\lambda+\rho)-\rho)}}{\sum_{w \in W}(-1)^{\ell(w)} e^{w\rho-\rho}} .$$

4.2.16 It is immediate that

$$\left(\sum_{r=0}^{\infty}(-1)^r \operatorname{ch} \wedge^r \mathfrak{n}^-\right)^{-1} = \prod_{\alpha\in\Delta^+_{\text{mult}}} (1-e^{-\alpha})^{-1} ,$$

which may also recall is the formal character of $U(\mathfrak{n}^-)$ or of $\mathcal{M}(0)$. This recovers 4.2.8 (iii). Either from 4.2.8 (iii) or the above the following identity is obtained.

Corollary

$$\sum_{w\in W} (-1)^{\ell(w)} e^{w\rho-\rho} = \prod_{\alpha\in\Delta^+_{\text{mult}}} (1-e^{-\alpha}) .$$

4.3 Integrable Modules for $U_q(\mathfrak{g}_C)$

In this section and in the remainder of the text it will be assumed that C is integrable. It will be seen that the theory of highest weight modules for $U_q(\mathfrak{g}_C)$ closely parallels that for $U(\mathfrak{g}_C)$.

4.3.1 The simplest example of a quantized enveloping algebra of the form $U_q(\mathfrak{g}_C)$ is obtained when $|\pi| = 1$ and $C = (2)$. By analogy with 4.2.5 it is denoted as $U_q(\mathfrak{sl}(2))$. As a $k(q)$-algebra it is defined by generators $x_\alpha, y_{-\alpha}, t_\alpha, t_\alpha^{-1}$ and relations

$$t_\alpha t_\alpha^{-1} = t_\alpha^{-1} t_\alpha = 1 , \quad t_\alpha x_\alpha t_\alpha^{-1} = q^2 x_\alpha , \quad t_\alpha y_{-\alpha} t_\alpha^{-1} = q^{-2} y_{-\alpha} ,$$

$$x_\alpha y_{-\alpha} - y_{-\alpha} x_\alpha = \frac{t_\alpha - t_\alpha^{-1}}{q-q^{-1}} .$$

In what follows it will be convenient to omit the α-subscript. As for $\mathfrak{sl}(2)$ it is easy to write down a two-dimensional simple $U_q(\mathfrak{sl}(2))$ module V (with highest weight $\alpha/2$) and then using the Hopf algebra structure to show that $V^{\otimes n}$ is a simple $U_q(\mathfrak{sl}(2))$ module of dimension $n+1$ and highest weight $n\alpha/2$. Alternatively consider the identity

$$
\begin{aligned}
[x, y^r] &= \sum_{s=0}^{r-1} y^s [x, y] y^{r-s-1} \\
(*) \qquad &= \frac{1}{(q-q^{-1})} \sum_{s=0}^{r-1} y^s (t-t^{-1}) y^{r-s-1} \\
&= \frac{y^{r-1}}{(q-q^{-1})} [r]_q \{q^{-(r-1)}t - q^{(r-1)}t^{-1}\}
\end{aligned}
$$

(notation 1.2.12). This implies that $y^r u_\Lambda$ is a primitive vector of the Verma module $M(\Lambda)$ (notation 3.4.9) if and only if $\Lambda(t)^2 = q^{2(r-1)}$. This has two solutions. Either $\Lambda(t) = q^{r-1}$ and in this case Λ is a linear weight (3.4.9) of the form q^λ with $(\alpha^\vee, \lambda) = r-1$, or Λ is the non-linear weight $-q^\lambda$. By universality of Verma modules, this proves the

Lemma. *For each integer $r > 0$ there exist up to isomorphism exactly two simple r dimensional simple $U_q(\mathfrak{sl}(2))$ modules. These have respectively highest weights $\pm q^{\frac{1}{2}(r-1)\alpha}$.*

Remarks. The conclusion changes dramatically for q a root of unity. However this situation is beyond the scope of the present text. Note that if $(\alpha^\vee, \lambda) \in \mathbb{N}$, then $M(q^{s_\alpha \cdot \lambda})$ is a submodule of $M(q^\lambda)$. Following 3.4.9 one simply writes $M(\lambda)$ for $M(q^\lambda)$.

4.3.2 It follows in particular from 4.3.1 that $U_q(\mathfrak{sl}(2))$ admits two-isomorphic simple one dimensional modules. Tensoring by the non-trivial module interchanges modules with highest weight $\pm q^\lambda$. For this reason the family in 4.3.1 with non-linear highest weight can be ignored. However this doubling is an essential feature of $U_q(\mathfrak{g}_C)$ and persists in many related forms.

4.3.3 A new feature of $U_q(\mathfrak{g}_C)$ is that the adjoint action is no longer locally finite (c.f. 1.2.10). This will be studied in detail in 7.1. It leads to some remarkable simplifications which aid the understanding of even $U(\mathfrak{g}_C)$. For the moment consider just $\mathfrak{sl}(2)$ retaining the notation of 4.3.2. Write $U = U_q(\mathfrak{sl}(2))$.

Lemma. *The action of* $\operatorname{ad} x$ *(resp.* $\operatorname{ad} y$*) on* $t^r : r \in \mathbb{Z}$ *is locally finite if and only if* $r \leq 0$. *In this case* $(\operatorname{ad} U)t^r$ *is finite dimensional.*

From 3.2.9 one has $(\operatorname{ad} x)a = xa - tat^{-1}x$. Thus $(\operatorname{ad} x)t^r x^s = xt^r x^s - q^{2s}t^r x^{s+1} = (q^{-2r} - q^{2s})t^r x^{s+1}$. Again $(\operatorname{ad} y)a = yat - ayt$ and so $(\operatorname{ad} y)t^r(yt)^s = (q^{2(r+s)} - 1)t^r(yt)^{s+1}$. In both cases the coefficient exactly vanishes when $r + s = 0$. This proves the first part. For the second part it is enough by 1.3.1 to take $r = -1$. Then by triangular decomposition one is reduced to checking that $(\operatorname{ad} y)^3(\operatorname{ad} x)t^{-1} = 0$ which is straightforward.

4.3.4 Take $r \in \mathbb{N}$. It will be shown as part of a general result for $U_q(\mathfrak{g}_C)$ that $(\operatorname{ad} U)t^{-r}$ is isomorphic to the endomorphism algebra of the unique simple $(r+1)$ dimensional $U_q(\mathfrak{sl}(2))$ module with highest weight. This gives a powerful method for constructing centres, not available for enveloping algebras! Indeed the identity of the endomorphism ring transforms like the trivial $U_q(\mathfrak{g}_C)$ module. The corresponding element in $U_q(\mathfrak{g}_C)$ must also have this property and is hence (1.3.3) central. Applied here this means that $(\operatorname{ad} U)t^{-1}$ is four dimensional and admits the trivial one dimensional module as a submodule. This can of course be checked directly and the resulting central element z found to be

$$z = yx + \frac{qt + q^{-1}t^{-1}}{(q - q^{-1})^2} .$$

It is immediate that z acts on the Verma module $M(\Lambda)$ by the scalar $(q - q^{-1})^{-2}(q\Lambda(t) + q^{-1}\Lambda(t)^{-1}) = \pm(q - q^{-1})^{-2}(q^r + q^{-r})$ when $\Lambda =$

$\pm q^{\frac{1}{2}(r-1)\alpha}$. In particular z separates the simple finite dimensional modules. This will give the

Lemma. *Every finite dimensional $U_q(\mathfrak{sl}(2))$ module is semisimple.*

It remains to exclude self-extensions. Let V be an extension of m isomorphic simple modules of dimension r. Then V splits if and only if its generalized highest weight subspace

$$V' = \{v \in V \mid (t^2 - q^{-(r-1)})^s v = 0 \text{ for } s \text{ sufficiently large}\}$$

splits. Now $xV' = 0$ and $y^r V' = 0$, so 4.3.1 (*) gives $y^{r-1}(q^{-(r-1)}t - q^{(r-1)}t^{-1})V' = 0$. Yet $\dim y^{r-1}V' = \dim V'$ since this holds for each simple factor. Hence $V' \subset \ker(t^2 - q^{-(r-1)})$ as required.

4.3.5 Take $\alpha \in \pi$. It is clear from 3.2.9 that the subalgebra $U_q(\mathfrak{sl}_\alpha)$ (or simply, U_α) of $U_q(\mathfrak{g}_C)$ generated by $x_\alpha, y_{-\alpha}t_\alpha, t_\alpha^{-1}$ is isomorphic to $U_q(\mathfrak{sl}(2))$ as a Hopf algebra. Call a $U_q(\mathfrak{g}_C)$ module admitting a formal character integrable if it is locally finite for each $U_\alpha : \alpha \in \pi$. By triangular decomposition of U_α any such module is a sum of its finite dimensional U_α submodules and hence by 4.3.4 a direct sum of simple finite dimensional U_α modules.

The condition that C be integrable has a similar consequence as it does in the enveloping algebra case (4.2.7). Indeed suppose $\alpha, \beta \in \pi$ are distinct and satisfy $(\alpha^\vee, \beta) \in -\mathbb{N}$. Then $y_{-\beta}t_\beta$ has weight $-\beta$. One checks that it is a primitive vector for the adjoint action of U_α on U. Hence by 4.3.1 it follows that $(\operatorname{ad} y_{-\alpha})^{-(\alpha^\vee,\beta)+1}(y_{-\beta}t_\beta)$ is a primitive vector. By say the calculation in 4.3.3 it can be expressed in the form $y_q^{\alpha,\beta}t_\alpha^{-(\alpha^\vee,\beta)+1}t_\beta$ for some $y_q^{\alpha,\beta} \in U^-$ of weight $-(\beta + (1-(\alpha^\vee,\beta))\alpha)$. Then the property that this vector be primitive translates to $[x_\alpha, y_q^{\alpha,\beta}] = 0$. Again $(\operatorname{ad} x_\beta)y_{-\beta}t_\beta = h_\beta t_\beta = \frac{1}{(q_\beta - q_\beta^{-1})}(t_\beta^2 - 1)$, whilst with respect to $\operatorname{ad} y_{-\alpha}$ one may observe that t_β^2 behaves like $t_\alpha^{2(\alpha,\beta)/(\alpha,\alpha)} = t_\alpha^{-(-(\alpha^\vee,\beta))}$ and so by the calculation in 4.3.3 one has $((\operatorname{ad} y_{-\alpha})^{-(\alpha^\vee,\beta)+1})t_\beta^2 = 0$. Thus

$$(\operatorname{ad} x_\beta)(y_q^{\alpha,\beta}t_\alpha^{-(\alpha^\vee,\beta)+1}) = (\operatorname{ad} x_\beta)((\operatorname{ad} y_{-\alpha})^{-(\alpha^\vee,\beta)+1}y_{-\beta}t_\beta) = 0 ,$$

which translates to give $[x_\beta, y_q^{\alpha,\beta}] = 0$. Trivially $[x_\gamma, y_q^{\alpha,\beta}] = 0$ for $\gamma \in \pi \setminus \{\alpha,\beta\}$. From all this one concludes that $y_q^{\alpha,\beta} \in K^-$.

Now take $\lambda \in P(\pi)$ and identify q^λ with the (linear) weight $t_\alpha \mapsto q^{(\alpha^\vee,\lambda)}$. Fix $\alpha \in \pi$ and assume $(\alpha^\vee, \lambda) \in \mathbb{N}$. It follows by 4.3.1 (*) that $y_{-\alpha}^{(\alpha^\vee,\lambda)+1}u_\lambda$ is a primitive vector of $M(\lambda)$ and generates a submodule isomorphic to $M(s_\alpha.\lambda)$.

Lemma

(i) *$M(\lambda)/M(s_\alpha.\lambda)$ is U_α locally finite and is the largest such quotient.*

(ii) *Any subquotient M of $M(\lambda)/M(s_\alpha.\lambda)$ satisfies $s_\alpha \operatorname{ch} M = \operatorname{ch} M$.*

Note that $\operatorname{ad} y_{-\alpha}$ is locally finite on $y_{-\alpha}$ and on the $y_{-\beta}t_\beta : \beta \in \pi \setminus \{\alpha\}$. Moreover one may view these elements as generating $M(\lambda)$ over u_λ. Then (i) follows from repeated use of the identity $y_{-\alpha}at_\alpha\bar{u}_\lambda = (\operatorname{ad} y_{-\alpha})a\bar{u}_\lambda + ay_{-\alpha}t_\alpha\bar{u}_\lambda$ and the fact that the image $\bar{u}_\lambda$ of u_λ in $M(\lambda)/M(s_\alpha.\lambda)$ is annihilated by $y_{-\alpha}^{(\alpha^\vee,\lambda)+1}$ whilst no power of $y_{-\alpha}$ annihilates the image of $u_{s_\alpha.\lambda}$ in the unique simple quotient $V(s_\alpha.\lambda)$ of $M(s_\alpha.\lambda)$. Then (ii) follows from (i) and 4.3.1.

4.3.6 Take $\lambda \in P^+(\pi)$. Then $\sum_{\alpha\in\pi} M(s_\alpha.\lambda)$ is a submodule of $M(\lambda)$. Let $V'(\lambda)$ denote the resulting quotient which is integrable by 4.3.5 (i) and is the largest such quotient.

Theorem. *Take* $\lambda \in P^+(\pi)$.

(i) $V'(\lambda)$ *is a simple* $U_q(\mathfrak{g}_C)$ *module.*
(ii) *Take* $\mu = -\sum k_\alpha\alpha : k_\alpha \in \mathbb{N}$. *If* $(\alpha^\vee,\lambda) \geq k_\alpha$ *for all* $\alpha \in \pi$, *then* $M(\lambda)_{\lambda-\mu} \xrightarrow{\sim} V(\lambda)_{\lambda-\mu}$.
(iii) $s_\alpha \operatorname{ch} V(\lambda) = \operatorname{ch} V(\lambda)$ *and* $s_\alpha \operatorname{ch} M(\mu) = -\operatorname{ch} M(s_\alpha.\mu)$ *for all* $\alpha \in \pi$, $\mu \in P(\pi)$.
(iv) $\operatorname{ch} V(\lambda) = \sum_{w\in W}(-1)^{\ell(w)} \operatorname{ch} M(w.\lambda)$.

(i) follows from 3.4.9, 4.3.5 (i) and 4.2.8 (i) exactly as in 4.2.8 (iii). For (ii) note that if $\lambda-\mu$ is a weight of $\sum_{\alpha\in\pi} M(s_\alpha.\lambda)$ then $\mu \in \bigcup_{\alpha\in\pi}\{((\alpha^\vee,\lambda)+1)\alpha + \mathbb{N}\pi\}$. The first part of (iii) follows from 4.3.5 (ii). From the locally nilpotent skew derivation δ'_α of U^- which satisfies $\delta'_\alpha y_{-\alpha} = 1$, it follows from 3.4.2 that $y_{-\alpha}$ is a non-zero divisor in U^-. Then if N^- is a T stable complement to $U^-y_{-\alpha}$ in U^- the sum $\sum_\alpha N^-y_{-\alpha}^r$ is direct and equals U^-. Consequently $\operatorname{ch} U^- = \operatorname{ch} N^-(1-e^{-\alpha})^{-1}$. Yet $N^- \cong U^-/U^-y_{-\alpha} \cong M(0)/M(s_\alpha.0)$ so $s_\alpha \operatorname{ch} N^- = \operatorname{ch} N^-$ by 4.3.5 (ii). Then $s_\alpha \operatorname{ch} M(\mu) = e^{s_\alpha\mu} \operatorname{ch} N^-(1-e^\alpha)^{-1} = -\operatorname{ch} M(s_\alpha.\mu)$ as required. Then (iv) follows from 4.1.5 and 4.2.8 (i) exactly as in 4.2.8 (iv).

4.3.7 The conclusion of 4.3.6 (iv) may also be obtained from 4.2.8 (iv) by specialization in the sense of 4.1.15. Set $V_A(\lambda) = U_A^- u_\lambda = U_A^-/U_A^- \cap \operatorname{Ann}_{U^-} u_\lambda$ which identifies with an A submodule of $V(\lambda)$ and so is torsion free. It follows as in 3.4.5 that each weight A submodule $V_A(\lambda)_{\lambda-\mu}$ is free and then that $\dim_{k(q)} V(\lambda)_{\lambda-\mu} = rk_A V_A(\lambda)_{\lambda-\mu} = \dim_k \mathcal{V}(\lambda)_{\lambda-\mu}$, where $\mathcal{V}(\lambda) := V_A(\lambda) \otimes_A A/\langle q-1\rangle$ is just the corresponding quotient of the Verma module $\mathcal{M}(\lambda)$ for $U(\mathfrak{g}_C)$.

4.3.8 It is immediate from the definition of $y_q^{\alpha,\beta}$ that it specializes (in the sense of 4.1.15) to $y^{\alpha,\beta}$. Thus U^- specializes to $U(\mathfrak{n}_C^-)$ and in particular the $y_q^{\alpha,\beta} : \alpha, \beta \in \pi$ distinct, form a set of generators for U^-. Applying $\varkappa$ the analogous result follows for U^+. Consequently (if C is integrable) $U_q(\mathfrak{g}_C)$ admits good specialization. By 4.1.16 this completely determines the factors of the Shapovalev determinants in the integrable case.

4.3.9 It follows from the defining relations (3.4.9) that the group Γ of one dimensional $U_q(\mathfrak{g}_C)$ modules is generated by the characters $\chi_\alpha : \alpha \in \pi$, when $\chi_\alpha(t_\beta) = 1,\ \alpha \neq \beta,\ \chi_\alpha(t_\alpha) = -1$ and is hence isomorphic to $\mathbb{Z}_2^{|\pi|}$. Via 1.3.4 it acts on the subcategory of integrable (resp. integrable and highest weight) modules taking simples to simples. From 4.3.1 it follows that

Lemma. *Let V be a simple integrable highest weight module. Then there exists $\gamma \in \Gamma$, $\lambda \in P^+(\pi)$ such that $V = \gamma \otimes V(\lambda)$, up to isomorphism.*

4.3.10 Let V, V' be $U = U_q(\mathfrak{g}_C)$ modules admitting a weight space decomposition. If $\operatorname{Ext}^1(V, V') \neq 0$ then there must be a weight vector of U taking some weight subspace of V to a weight subspace of V'. In particular $\operatorname{Ext}^1(\gamma \otimes V(\lambda), \gamma' \otimes V(\mu)) \neq 0$ implies $\gamma = \gamma'$ and then that $\operatorname{Ext}^1(V(\lambda), V(\mu)) \neq 0$.

Theorem. *For all $\lambda, \mu \in P^+(\pi)$ one has $\operatorname{Ext}^1(V(\lambda), V(\mu)) = 0$. In particular every integrable highest weight module is simple and every integrable module in $\mathcal{O}$ is semisimple.*

Let M be an extension of $V(\mu)$ by $V(\lambda)$. Take $\alpha \in \pi$. Then M is by definition a sum of finite dimensional U_α modules and so by 4.3.1 and 4.3.2, t_α acts locally semisimply on M. Thus M is a weight module so in particular belongs to $\mathcal{O}$ and is integrable. Thus if $\lambda = \mu$ it is immediate that M splits. If $\lambda \neq \mu$ one may assume applying $\mathcal{O}$ duality (4.1.17) that $\lambda \not> \mu$, that is λ is not a weight of $V(\mu)$. Let u_λ be a non-zero vector in M_λ. Since M_λ is one dimensional, u_λ maps to the canonical generator of $V(\lambda)$. Thus either Uu_λ is isomorphic to $V(\lambda)$ and M splits, or $M = Uu_\lambda$ and so is a quotient of $M(\lambda)$. Yet by 4.3.6 (i), $V(\lambda)$ is the largest integrable quotient of $M(\lambda)$ and so only the first possibility applies proving the theorem.

Remark. Of course a similar result applies to $U(\mathfrak{g}_C)$.

4.3.11 An important property of the $x_q^{\alpha,\beta}$ and the $y_q^{\alpha,\beta}$ is that they behave well under the coproduct Δ for $\tilde{U}_q(\mathfrak{g}_C)$. This is expressed as follows

Lemma. *For all $\alpha, \beta \in \pi$ distinct one has*

(i) $\Delta x_q^{\alpha,\beta} = x_q^{\alpha,\beta} \otimes 1 + \tau(s_\alpha.\beta) \otimes x_q^{\alpha,\beta}$,
(ii) $\Delta y_q^{\alpha,\beta} = y_q^{\alpha,\beta} \otimes \tau(-s_\alpha.\beta) + 1 \otimes y_q^{\alpha,\beta}$.

(i) follows from 3.1.10 (+) assuming that the intermediate terms vanish. These take the form $X_{\nu-\mu} \otimes X_\mu$ where $X_{\nu-\mu} \in (\tilde{U}^+T)_{\nu-\mu}$, $X_\mu \in \tilde{U}_\mu^+$ in which $\nu = s_\alpha.\beta$ and $0 < \mu < \nu$. Yet K^+T is a coideal (3.1.12 (ii)). Thus it is enough to show that $K_\mu^+ = 0$ for μ in the above range, that is $\mu = \alpha$ or $\mu = \beta + r\alpha$ with $0 \leq r < -(\alpha^\vee, \beta)$. By 4.1.17 it can be assumed that $\mu \notin \pi$

and $(\mu,\mu) = 2(\rho,\mu)$. This latter has only the solutions $r = 0$ or $r = -(\alpha^\vee,\beta)$ which are excluded. Similarly (ii) follows from 3.1.10 (−), 3.1.12 (i) and 4.1.17.

4.3.12 Let A be a $U_q(\mathfrak{g})$ algebra and S a T stable left Ore subset of A. Assume either that S consists of regular elements of A or that A satisfies *acc* on left annihilators. In either case $S^{-1}A$ is defined and can be viewed as the algebra with identity 1 and generators $\{a\}_{a\in A}$, $\{s^{-1}\}_{s\in S}$ satisfying those relations coming from A together with $ss^{-1} = s^{-1}s = 1 : s \in S$. For each $s \in S$ the coproduct and coidentity on $U_q(\mathfrak{g})$ uniquely extend the action of $\tau(\mu), x_\alpha, y_{-\alpha} : \mu \in \mathbb{Z}\pi, \alpha \in \pi$ to s^{-1} via the relation $s^{-1}s = 1$. The resulting formulae are

$$(*)\qquad \begin{aligned} \tau(\mu)(s^{-1}) = (\tau(\mu)s)^{-1},\ x_\alpha(s^{-1}) &= -(\tau(\alpha)s)^{-1}(x_\alpha s)s^{-1}\ , \\ y_{-\alpha}(s^{-1}) &= -s^{-1}(y_{-\alpha}s)(\tau(-\alpha)s)^{-1}\ . \end{aligned}$$

One easily checks that these are compatible with the additional relation $ss^{-1} = 1$.

Lemma. *The relations (∗) extend the $U_q(\mathfrak{g})$ subalgebra structure of A to $S^{-1}A$.*

It remains to show that relations (i)–(vi) of 3.2.9 hold with respect to the images of the $\tau(\mu)$, $x_\alpha, y_{-\alpha}$ in End $S^{-1}A$ defined by (∗). Since $U_q(\mathfrak{g})$ is a bi-algebra, it is enough to show each relation in $U_q(\mathfrak{g})$ annihilates each s^{-1}. This is quite easy for relations (i)–(iv). For (v) set $r^{\alpha,\beta} = x_\alpha y_{-\beta} - y_{-\beta}x_\alpha - \delta_{\alpha,\beta}h_\alpha$. One checks that $\Delta(r^{\alpha,\beta}) = r^{\alpha,\beta} \otimes \tau(-\beta) + \tau(\alpha) \otimes r^{\alpha,\beta}$ up to the relations (i)–(vi), and $\varepsilon(r^{\alpha,\beta}) = 0$. Since Δ, ε are algebra homomorphisms, it follows from (∗) that

$$r^{\alpha,\beta}(s^{-1})\tau(-\beta)s + (\tau(\alpha)s)^{-1}r^{\alpha,\beta}(s) = \varepsilon(r^{\alpha,\beta}) = 0\ .$$

Yet $r^{\alpha,\beta}(s) = 0$ because A is an $U_q(\mathfrak{g})$ module. Multiplying on the right by $(\tau(-\beta)s)^{-1}$ then gives $r^{\alpha,\beta}(s^{-1}) = 0$ as required. Using 4.3.11 which by definition holds up to (i)–(v), a similar argument applies to (vi). This proves the lemma.

Remark. Let H be a Hopf algebra. Call $g \in H$ group-like if $\Delta(g) = g \otimes g$. Necessarily $\varepsilon(g) = 1$ in this case. Call $x \in H$ primitive if $\Delta(x) = x \otimes 1 + 1 \otimes x$ and quasi-primitive if $\Delta(x) = x \otimes 1 + g \otimes x$ for some group-like element g. Necessarily $\varepsilon(x) = 0$ in this case. It is easy to generalize the above result to the case when all the generators and the relations of H (as an algebra) are group-like or quasi-primitive, modulo generators and relations already disposed of. For example this holds for $U_q(\mathfrak{g}_C)$ assuming C just symmetric via the first part of 3.1.6 and the argument in 4.3.11.

4.3.13 Explicit formulae for the $x_q^{\alpha,\beta}, y_q^{\alpha,\beta}$ ought never to be needed. Nevertheless they can be obtained as follows.

Lemma. *Fix* $\alpha, \beta \in \pi$ *satisfying* $m = m(\alpha,\beta) := 1 - (\alpha^\vee, \beta) \in \mathbb{N}^+$. *Then*

(i) $x_q^{\alpha,\beta} = \sum_{n=0}^m (-1)^n \begin{bmatrix} m \\ n \end{bmatrix}_{q_\alpha} x_\alpha^{m-n} x_\beta x_\alpha^n$

(ii) $y_q^{\alpha,\beta} = \sum_{n=0}^m (-1)^n \begin{bmatrix} m \\ n \end{bmatrix}_{q_\alpha} y_{-\alpha}^{m-n} y_{-\beta} y_{-\alpha}^n$.

Recall that $(\operatorname{ad} y_{-\alpha})g = (y_{-\alpha}g - gy_{-\alpha})\tau(\alpha) = g'_{-\alpha}(\tau_\alpha^{-1}g) - gg'_{-\alpha}$, where $g'_{-\alpha} = y_{-\alpha}t_\alpha$. Define $\ell_{-\alpha}, r_{-\alpha} \in \operatorname{End} U$ by $\ell_{-\alpha}(g) = g'_{-\alpha}\tau_\alpha^{-1}(g)$, $r_{-\alpha}(g) = gg'_{-\alpha}$. Then $\operatorname{ad} y_{-\alpha} = \ell_{-\alpha} - r_{-\alpha}$. Moreover $\ell_{-\alpha}r_{-\alpha} = q_\alpha^2 r_{-\alpha}\ell_{-\alpha}$. It follows from 1.2.13 (3) and the invariance of the q-binomial coefficients under $q \longrightarrow q^{-1}$ that

$$
\begin{aligned}
(\operatorname{ad} y_{-\alpha})^m g &= (\ell_{-\alpha} - r_{-\alpha})^m g \\
&= \sum_{n=0}^m (-1)^n \begin{bmatrix} m \\ n \end{bmatrix}_{q_\alpha} q_\alpha^{-(m-n)n} \ell_{-\alpha}^{m-n} r_{-\alpha}^n (g) \\
&= \left(\sum_{n=0}^m (-1)^n \begin{bmatrix} m \\ n \end{bmatrix}_{q_\alpha} q_\alpha^{-(m-1)n} y_{-\alpha}^{m-n} g y_{-\alpha}^n \right) t_\alpha^m .
\end{aligned}
$$

Replacing g by $y_{-\beta}t_\beta$, recalling (4.3.5) that $(\operatorname{ad} y_{-\alpha})^m y_{-\beta}t_\beta = y_q^{\alpha,\beta} t_\alpha^m t_\beta$ gives (ii). Then (i) obtains from (ii) applying $\varkappa$.

4.3.14 One may remark that $x_q^{\alpha,\beta}$, $y_q^{\alpha,\beta}$ are invariant under $q \to q^{-1}$ as would be expected by 3.1.14 (i). One may also explicitly check that they are invariant under the transformation described in 3.2.8.

The relations $x_q^{\alpha,\beta} = 0$, $y_q^{\alpha,\beta} = 0$ are known as the quantum Serre relations. Introduce the divided powers $y_{-\alpha}^{(m)} := y_{-\alpha}^m/[m!]_{q_\alpha}$, $x_\alpha^{(m)} := x_\alpha^m/[m!]_{q_\alpha}$. Then it is useful to observe that the second relation can be written in the form

$$\sum_{n=0}^m (-1)^n y_{-\alpha}^{(m-n)} y_{-\beta} y_{-\alpha}^{(n)} = 0$$

with a similar expression for the first.

4.4 Demazure Modules and Product Formulae

Throughout this section it is assumed that C is integrable.

4.4.1 Take $\lambda \in P^+(\pi)$. For each $w \in W$ one has $\dim V(\lambda)_{w\lambda} = 1$ by 4.3.6 (iv). Choose $u_{w\lambda} \in V(\lambda)_{w\lambda}$ non-zero and set $V_w(\lambda) = U^+ u_{w\lambda}$. It is called the Demazure module for the pair λ, w. Clearly $V_w(\lambda)$ is a TU^+ module with finite dimensional weight spaces. It therefore makes sense to define $\operatorname{ch} V_w(\lambda)$.

The above construction makes sense and indeed was first studied for $U(\mathfrak{g}_C)$. Identify $\mathcal{V}(\lambda)$ with $V_A(\lambda) \otimes_A A/\langle q-1\rangle$ as in 4.3.7, choose $u_{w\lambda} \in V_A(\lambda)_{w\lambda}$ and let $\bar{u}_{w\lambda}$ denote its image in $\mathcal{V}(\lambda)$. Then the corresponding module is $\mathcal{V}_w(\lambda) := U(\mathfrak{n}^+)\bar{u}_{w\lambda}$. Clearly the image of $\operatorname{Ann}_{U_A^+} u_{w\lambda}$ in $U(\mathfrak{n}^+)$ is contained in $\operatorname{Ann}_{U(\mathfrak{n}^+)} \bar{u}_{w\lambda}$. Moreover equality holds if and only if $\operatorname{ch} V_w(\lambda) = \operatorname{ch} \mathcal{V}_w(\lambda)$ since the argument of 4.3.7 shows $U_A^+ u_{w\lambda}$ to be a free A module. It is known that $\operatorname{ch} \mathcal{V}_w(\lambda)$ satisfies the Demazure character formula and this will also be shown for $V_w(\lambda)$ in Chapter 6.

4.4.2 The module $\mathcal{V}_w(\lambda)$ has an important geometric meaning, namely it is the dual to the space of global sections of a certain sheaf $\mathcal{L}_\lambda$ on the Schubert variety defined by w. In particular the inclusion relations for these varieties is given [BGG1] by the inclusion relations for the $\mathcal{V}_w(\lambda)$ whenever λ is regular, that is when $\lambda \in P^+(\pi)+\rho$. Sweeping geometry aside one may simply analyze the inclusion relations for the $V_w(\lambda)$. This is done below.

4.4.3 Retain the notation of 4.3.5, 4.4.1 and A.1.1. Set $U_\alpha^- = k[y_{-\alpha}]$.

Lemma. *Take $\lambda \in P^+(\pi)$, $w \in W$, $\alpha \in \pi$ such that $\ell(s_\alpha w) < \ell(w)$. Set $r = -(\alpha^\vee, w\lambda)$.*

(i) *If $V(\lambda)_\mu \neq 0$ then $(\mu,\mu) \leq (\lambda,\lambda)$.*
(ii) *$y_{-\alpha}u_{w\lambda} = 0$, $x_\alpha u_{s_\alpha w\lambda} = 0$.*
(iii) *One has $r \in \mathbb{N}$ and up to non-zero scalars $x_\alpha^r u_{w\lambda} = u_{s_\alpha w\lambda}$, $y_{-\alpha}^r u_{s_\alpha w\lambda} = u_{w\lambda}$.*
(iv) *$U^+U_\alpha = U_\alpha U^+$.*
(v) *$V_w(\lambda) = U_\alpha V_{s_\alpha w}(\lambda) = U_\alpha^- V_{s_\alpha w}(\lambda)$.*

Suppose $(\mu,\mu) > (\lambda,\lambda)$ and $V(\lambda)_\mu \neq 0$. By 4.3.6 (iv) both still hold for any $\nu \in W\mu$. Assume $|\lambda-\nu|$ is minimal for these properties. Then $\nu \in P^+(\pi)$ and so $0 > (\lambda,\lambda)-(\nu,\nu) = (\lambda-\nu,\lambda)+(\nu,\lambda-\nu) \geq 0$. This contradiction proves (i).

One has $r = -(\alpha^\vee, w\lambda) = -((w^{-1}\alpha)^\vee, \lambda) \in \mathbb{N}$ by A.1.1 (iv). Hence $(w\lambda-\alpha, w\lambda-\alpha) = (\lambda,\lambda)+(\alpha,\alpha)-2(\alpha,w\lambda) > (\lambda,\lambda)$ which by (i) gives (ii). Then (iii) follows from 4.3.1 applied to the U_α module generated by the highest weight vector $u_{s_\alpha w\lambda}$. Assertion (iv) is an easy verification. Then $V_w(\lambda) = U^+U_\alpha u_{s_\alpha w\lambda} = U_\alpha U^+ u_{s_\alpha w\lambda} = U_\alpha V_{s_\alpha w}(\lambda) = U_\alpha^- V_{s_\alpha w}(\lambda)$ which is (v).

4.4.4 Lemma. *Take $\alpha \in \pi$ and suppose $w, w' \in W$ satisfy $\ell(s_\alpha w) < \ell(w)$ and $\ell(s_\alpha w') < \ell(w')$. Then $V_w(\lambda) \supset V_{w'}(\lambda)$ implies $V_{s_\alpha w}(\lambda) \supset V_{s_\alpha w'}(\lambda)$.*

The hypothesis implies that $u_{w'\lambda} \in V_w(\lambda)$ and this by 4.4.3 (ii), (v) is a lowest weight vector of a simple U_α submodule M of $V_w(\lambda)$. By 4.3.4 there exists a TU_α stable complement of M in $V_w(\lambda)$. Let p denote the projection of $V_w(\lambda)$ onto M defined by this complement. It follows from 4.4.3 (v) that $p(V_{s_\alpha w}(\lambda)) \neq 0$. Since $V_{s_\alpha w}(\lambda)$ is x_α stable the latter contains the highest weight vector of $p(M)$, namely $p(u_{s_\alpha w'\lambda})$. Yet $\dim V_w(\lambda)_{s_\alpha w'\lambda} = 1$ and so in fact $u_{s_\alpha w'\lambda} \in V_{s_\alpha w}(\lambda)$, which gives the required assertion.

Remark. A common error is to assume that $V_{s_\alpha w}(\lambda)$ contains the highest weight vectors of every simple U_α submodule of $V_w(\lambda)$. Although this is false, it does hold in the $q \longrightarrow 0$ limit by 6.3.12.

4.4.5 Let $\leq$ denote the Bruhat order (A.1.7) on W.

Proposition. *Take $\lambda \in P^+(\pi) + \rho$. Then for all $w, w' \in W$ one has $w' \leq w \iff V_{w'}(\lambda) \subset V_w(\lambda)$.*

The assertion $\Longrightarrow$ is an immediate consequence of 4.4.3 (v) and A.1.8. For $\Longleftarrow$, view the inclusion $V_w(\lambda) \subset V_{w'}(\lambda)$ as an order relation $\preceq$ on W. Then it is enough to show that $\preceq$ satisfies (i)–(iv) of A.1.7. In this (ii) is immediate from 4.4.3 (v). Again by 4.4.3 (v) one has $V_{s_\alpha w'}(\lambda) \subset U_\alpha V_{w'}(\lambda) \subset U_\alpha V_w(\lambda)$ which equals either $V_w(\lambda)$ or $V_{s_\alpha w}(\lambda)$, hence (iii). Similarly (iv) follows from 4.4.4.

Suppose $V_e(\lambda) \supset V_w(\lambda)$. If $w \neq e$ there exists $\alpha \in \pi$ such that $\ell(ws_\alpha) < \ell(w)$. Then $V_{s_\alpha}(\lambda) \subset V_w(\lambda)$ by 4.4.3 (v) which by the hypothesis gives $u_{s_\alpha\lambda} = u_\lambda$. Hence $(\alpha^\vee, \lambda) = 0$ in contradiction with the supposed regularity of λ.

Remark. Suppose $|W| < \infty$. Then one may check that $w \preceq w_0 \Longrightarrow w = w_0$ and thereby avoid the use of 4.4.4.

4.4.6 One may deduce a number of combinatorial consequences from the above. For example, take $w \in W$, let $w = s_{\alpha_1} s_{\alpha_2} \dots s_{\alpha_r}$ be a reduced decomposition and set $U_w^- = U_{\alpha_1}^- U_{\alpha_2}^- \dots U_{\alpha_r}^-$.

Corollary

(i) $U_w^- u_\lambda = V_w(\lambda)$, $\forall \lambda \in P^+(\pi)$.

(ii) *U_w^- is independent of the reduced decomposition of w.*

(iii) *Suppose $|W| < \infty$. Then $U^- = U_{w_0}^-$.*

(i) is immediate from 4.4.3 (v). Then (ii) and (iii) follow from 4.3.6 (ii).

Remarks. (iii) is a direct generalization of the corresponding result for $U(\mathfrak{n}^-)$. Since $\ell(w_0) = \dim \mathfrak{n}^-$ it exemplifies 2.4.10. Actually in this case the product of the corresponding one parameter groups coincide with the algebraic group corresponding to $\mathfrak{n}^-$, by say Bruhat decomposition [Ca1].

4.4.7 Further combinatorial relations may be obtained from Verma module theory some of which is developed below. Recall the Bruhat order $\leq$.

Lemma. *Take $\mu \in P^+(\pi)$ and $w, y \in W$ with $w \geq y$. Then $M(w.\mu)$ is a submodule of $M(y.\mu)$.*

The proof is by induction on $\ell(w)$. If $w = e$ the assertion is obvious. Otherwise choose $\alpha \in \pi$ such that $s_\alpha w < w$. Then $M(w.\lambda)$ is a submodule of $M(s_\alpha w.\lambda)$ by 4.3.1 $(*)$. By A.1.7 (iii) either $s_\alpha w \geq y$ or $s_\alpha w \geq s_\alpha y$ and one can assume $y \geq s_\alpha y$. In the first case the assertion obtains by the induction hypothesis. In the second case consider the image $\bar{u}_{w.\lambda}$ of $u_{w.\lambda}$ in $M(s_\alpha y.\lambda)/M(y.\lambda)$ under the composed map $M(w.\lambda) \hookrightarrow M(s_\alpha w.\lambda) \hookrightarrow M(s_\alpha y.\lambda) \longrightarrow M(s_\alpha y.\lambda)/M(y.\lambda)$ which is defined by the induction hypothesis and 4.3.1 $(*)$. By 4.3.5 (ii) one has $y_{-\alpha}^r \bar{u}_{w.\lambda} = 0$ for r sufficiently large. Yet from 4.3.1 $(*)$ since $(\alpha^\vee, w.\lambda) < 0$ it follows that $x_\alpha^r y_{-\alpha}^r \bar{u}_{w.\lambda}$ is a non-zero multiple of $\bar{u}_{w.\lambda}$ which must hence be zero. Consequently $M(w.\lambda)$ is a submodule of $M(y.\lambda)$ as required.

4.4.8 Lemma. *Take $\beta \in Q^+(\pi)$, $r \in \mathbb{N}$. Then*

$$\Omega_{\beta,r} := \{\Lambda \in T^\wedge \mid \dim_U \operatorname{Hom}(M(\Lambda q^{-\beta}), M(\Lambda)) = r\}$$

is an algebraic subset of $T^\wedge$.

For each $\gamma \in Q^+(\pi)$ fix a basis $\{y_{-\gamma}^{(i)}\}$ for $U_{-\gamma}^-$. Take $\alpha \in \pi$. From 3.2.9 it is immediate that there exist Laurent polynomials $p_{i,j}^\alpha \in U^0$ such that $[x_\alpha, y_{-\beta}^{(i)}] = \sum y_{-\beta+\alpha}^{(j)} p_{i,j}^\alpha$. Write $y_{-\beta} = \sum c_i y_{-\beta}^{(i)} : c_i \in k(q)$. Then

$$x_\alpha y_{-\beta} u_\Lambda = [x_\alpha, y_{-\beta}] u_\Lambda = \sum c_i y_{-\beta+\alpha}^{(j)} p_{i,j}^\alpha(\Lambda) \, .$$

Since $M(\Lambda)$ is a free U^- module, it follows that $y_{-\beta} u_\Lambda$ is a generator of a Verma submodule $M(\Lambda q^{-\beta})$ if and only if $\sum c_i p_{i,j}^\alpha(\Lambda) = 0$ for all α. Hence $\Omega_{\beta,r}$ can be expressed as the common zeros of certain minors of the matrices $\{p_{i,j}^\alpha\}$.

4.4.9 The above result combined with 4.2.6 may be used to simplify the proof of 4.1.16 in the case when C is of finite type.

Lemma. *Suppose $\gamma \in \Delta_{re}^+$, $m \in \mathbb{N}$ and $q^{2(\rho,\gamma)-m(\gamma,\gamma)}\Lambda(\tau(2\gamma)) = 1$. Then $M(\Lambda q^{-m\gamma})$ is a submodule of $M(\Lambda)$.*

The set $H_\gamma := \{\lambda \in \mathfrak{h}_{\mathbb{Q}}^* \mid (\lambda + \rho, \gamma^\vee) = m\}$ is a hyperplane in $\mathfrak{h}_{\mathbb{Q}}^*$ which by say the calculation in 4.1.6 has non-null and hence Zariski dense intersection with $P(\pi)$. Hence by 4.4.8 it suffices to show that $M(s_\gamma.\lambda)$ is a submodule of $M(\lambda)$ for all $\lambda \in H_\gamma \cap P(\pi)$. Choose $w \in W$ such that $\mu + \rho := w^{-1} s_\gamma(\lambda + \rho) \in$

$P^+(\pi)+\rho$. Then $(w^{-1}\gamma^\vee, \mu+\rho) = -(\gamma^\vee, \lambda+\rho) = -m$, so $w^{-1}\gamma \in \Delta^-$. Then by A.1.5 and A.1.8 it follows that $s_\gamma w < w$ for the Bruhat order, so the assertion follows from 4.4.7.

4.4.10 By 4.1.6, 4.4.8 and 4.4.9 it follows that $c_{\beta,m,1} = 1$ for all $\beta \in \Delta^+_{re}$, $m \in \mathbb{N}^+$. Now for $\beta \in \Delta^+_{re}$, the sum in the right hand side 4.1.8 equals 1. In view of 4.1.7 (iii) this gives $c_{\beta,m,\zeta} = 0$ for $\zeta \neq 1$. Thus the factorization of $\mathcal{S}_\nu$ is completely determined when $\Delta^+_{re} = \Delta^+$, that is when C is of finite type. Conversely from 4.1.6, 4.1.16 and 4.4.8 one obtains the following general result. Here it is helpful to note by [Ka1, Prop. 5.1 b) and Prop. 5.5] that Δ^+ is generated by its intersection Δ^+_{irr} with $Q^+_{\text{irr}}(\pi)$, that is $\Delta^+_{\text{irr}} = \Delta^+ \cap \{m\gamma \mid m \in \mathbb{N}, \gamma \in \Delta^+ \cap Q^+_{\text{irr}}(\pi)\}$.

Proposition. *Take $\Lambda \in T^\vee, \beta \in Q^+(\pi)$. Consider*

(i) *$M(\Lambda q^{-\beta})$ is a submodule of $M(\Lambda)$.*
(ii) $q^{2(\rho,\beta)-(\beta,\beta)}\Lambda(\tau(2\beta)) = 1$, $\beta = m\gamma$, $m \in \mathbb{N}$, $\gamma \in \Delta^+_{\text{irr}}$.

Then (ii) $\Longrightarrow$ (i). *Conversely* (i) *implies* (ii) *or that* (i) *holds for some $\beta' \in Q^+(\pi) : 0 < \beta' < \beta$.*

4.4.11 Take Δ, γ as in 4.4.10 (i) and suppose $(\gamma,\gamma) > 0$. The $c_{\gamma,m,1} = 1$ for all $m \in \mathbb{N}^+$ and so one should expect that $\dim \operatorname{Hom}_U(M(\Lambda q^{-m\gamma}), M(\Lambda)) = 1$. This can be resolved for $\Lambda = q^\lambda : \lambda \in P(\pi)$ through the Enright functors $C_\alpha : \alpha \in \pi$ described below.

4.4.12 Fix $\alpha \in \pi$. It has already been noted in 4.3.6 (iii) that $y_{-\alpha}$ is a non-zero divisor in U^- and hence by triangular decomposition it is a non-zero divisor in U. Set $Y_{-\alpha} = \{y^r_{-\alpha}\}_{r\in\mathbb{N}}$. From the relations in 3.2.9 using 4.3.13 (ii) or the more intrinsic analysis of 4.3.5, it is easily checked that $Y_{-\alpha}$ is an Ore subset of U. Similarly $X_\alpha := \{x^r_\alpha\}_{r\in\mathbb{N}}$ is an Ore subset of U.

Set $D_\alpha M = Y^{-1}_{-\alpha}U \otimes_U M$. Then $M \longrightarrow D_\alpha M$ is an exact functor on the category of left U modules. Let $C_\alpha M$ denote the torsion submodule of $D_\alpha M$ with respect to the Ore subset X_α. Let C_α denote the covariant functor $M \longrightarrow C_\alpha M$ on the category of left U modules which are X_α torsion.

Some elementary properties of C_α are listed below.

Lemma

(i) *C_α is left exact and commutes with direct sums.*
(ii) *$C_\alpha M = 0$ if M is $Y_{-\alpha}$ torsion.*
(iii) *The map $m \mapsto 1 \otimes m$ defines a homomorphism $M \longrightarrow C_\alpha M$ whose kernel is the $Y_{-\alpha}$ torsion submodule of M.*

(iv) *Given a homomorphism* $N \longrightarrow M$ *of* U *modules, functoriality and* (iii) *give a commutative diagram*

$$\begin{array}{ccc} N & \longrightarrow & M \\ \downarrow & & \downarrow \\ C_\alpha N & \longrightarrow & C_\alpha M \end{array}$$

(v) $C_\alpha M \xrightarrow{\sim} C_\alpha(\mathrm{Im}(M \longrightarrow C_\alpha M))$.
(vi) $C_\alpha M \xrightarrow{\sim} C_\alpha^2 M$.
(vii) *As a functor on* U_α *modules,* C_α *commutes with tensoring by integrable* U_α *modules.*

(i)–(iv) are clear. Set $N = \mathrm{Im}(M \longrightarrow C_\alpha M)$, $K = \mathrm{Ker}(M \longrightarrow C_\alpha M)$. Then $D_\alpha K = 0$, so $D_\alpha M \xrightarrow{\sim} D_\alpha N$ by the exactness of D. Hence $C_\alpha M \xrightarrow{\sim} C_\alpha N$ by the definition of C_α which is (v). Since $C_\alpha M/N$ is $Y_{-\alpha}$ torsion, (vi) follows from (i) and (v).

Let V be an integrable U_α module. By the first paragraph of 4.3.5 and (i) one may assume V finite dimensional in proving (vi). Let M be a U_α module which is X_α torsion and let N be its $Y_{-\alpha}$ torsion submodule. From 3.2.9 (vii) it follows that $V \otimes M$ (resp. $V \otimes N$) is X_α (resp. and $Y_{-\alpha}$) torsion. Let $\varphi : V \otimes M \longrightarrow V \otimes D_\alpha M$ be the canonical map $v \otimes m \mapsto v \otimes (1 \otimes m)$. It passes to an injection $D_\alpha(V \otimes M) \overset{\varphi'}{\hookrightarrow} D_\alpha(V \otimes D_\alpha M)$ of $Y_\alpha^{-1}U$ modules. From 3.2.9 (vii) and the $y_{-\alpha}$ local finiteness of V one checks that the canonical map $V \otimes D_\alpha M \longrightarrow D_\alpha(V \otimes D_\alpha M)$ is bijective and moreover that this module coincides with $Y_\alpha^{-1} \mathrm{Im}\, \varphi = \mathrm{Im}\, \varphi'$. Finally from 3.2.9 (vii) and the x_α local finiteness of V one deduces that φ' restricts to an isomorphism of $V \otimes C_\alpha M$ onto $C_\alpha(V \otimes M)$. Hence (vii).

4.4.13 Fix $\alpha \in \pi$ and take $\beta \in \pi \setminus \{\alpha\}$. In 4.3.5 it was noted that $\mathrm{ad}\, U_\alpha(y_{-\beta} t_\beta)$ is a simple $1 - (\alpha^\vee, \beta)$ dimensional U_α submodule of G^-. Let G_α^- denote the subalgebra of G^- generated by these submodules. It is $\mathrm{ad}\, U_\alpha$ locally finite. Recall that $g_{-\alpha} = t_\alpha y_{-\alpha}$.

Lemma. *The multiplication map* $G_\alpha^- \otimes k(q)[g_{-\alpha}] \longrightarrow G^-$ *is an isomorphism of vector spaces.*

Let $\bar{u}_0$ denote the image of the canonical generator u_0 of $M(0)$ in $M(0)/M(s_\alpha.0)$. Then $y_{-\alpha}\bar{u}_0 = 0$. It follows easily from the definition of G_α^- and the formula (3.3.2 (1)) for $\mathrm{ad}\, y_{-\alpha}$ that $G_\alpha^- \bar{u}_0 \supset U^- \bar{u}_0 = M(0)/M(s_\alpha.0)$. Since $\mathrm{Ann}_{U^-} \bar{u}_0 = U^- y_{-\alpha}$, surjectivity follows easily. For injectivity apply $\mathrm{ad}\, x_\alpha$ to an element in the kernel of the map. Since $(\mathrm{ad}\, x_\alpha) g_{-\alpha}$ is proportional to $t_\alpha^2 - 1$ it follows that the resulting expression has a coefficient of t_α^2 which is also in the kernel. Repeated application shows that the coefficient of the highest power of $g_{-\alpha}$ must vanish and so this gives injectivity.

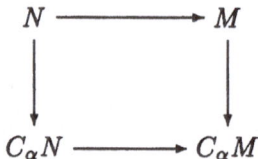

Remark. It is clear that G_α^- specializes in the sense of 4.1.15 to $U(\mathfrak{m}^-(\alpha))$.

4.4.14 Fix $\alpha \in \pi$. Take $\Lambda \in T^\wedge$ such that $\Lambda(\tau(2\alpha)) = q^{m(\alpha,\alpha)-(\rho,\alpha)}$ for some $m \in \mathbb{N}$. Then from 4.4.9 it follows that $M(\Lambda q^{-m\alpha})$ is a submodule of $M(\Lambda)$.

Lemma

(i) $C_\alpha M(\Lambda) \xrightarrow{\sim} M(\Lambda)$

(ii) $C_\alpha M(\Lambda q^{-m\alpha}) \xrightarrow{\sim} M(\Lambda)$.

Let $M_\alpha(\Lambda)$ denote the Verma module for U_α defined by Λ. Then $M_\alpha(\Lambda)$ has basis $y_{-\alpha}^j u_\Lambda : j \in \mathbb{N}$ and so $D_\alpha M_\alpha(\Lambda)$ has basis $y_{-\alpha}^j u_\Lambda : j \in \mathbb{Z}$. Then $C_\alpha M_\alpha(\Lambda)/M_\alpha(\Lambda)$ which is X_α torsion, hence a finite dimensional U_α module, has weights in the set $\Lambda + j\alpha : j \in \mathbb{N}^+$. By the assumption on Λ and 4.3.1 it follows that $M_\alpha(\Lambda) \xrightarrow{\sim} C_\alpha M_\alpha(\Lambda)$. Finally by 4.4.13 the multiplication map $G_\alpha^- \otimes M_\alpha(\Lambda) \longrightarrow M(\Lambda)$ is an isomorphism of U_α modules and so (i) results from 4.4.12 (vii).

As in 4.3.5 (i) it follows that $M(\Lambda)/M(\Lambda q^{-m\alpha})$ is $Y_{-\alpha}$ torsion. Then $D_\alpha M(\Lambda q^{-m\alpha}) \xrightarrow{\sim} D_\alpha M(\Lambda)$ and so (ii) results from (i).

4.4.15 The following complements 4.4.7.

Lemma. *Take* $\mu \in P^+(\pi)$ *and* $w, y \in W$. *Then* $\dim \operatorname{Hom}_U(M(w.\mu), M(y.\mu)) \leq 1$.

The proof is by induction on $\ell(w)$. If $\ell(w) = 0$ the assertion is obvious. Otherwise choose $\alpha \in \pi$ such that $s_\alpha w < w$. A homomorphism $\varphi : M(w.\mu) \longrightarrow M(y.\mu)$ gives by functoriality a homomorphism $C_\alpha\varphi : C_\alpha M(w.\mu) \longrightarrow C_\alpha M(y.\mu)$. By 4.4.12 (iv), $C_\alpha\varphi = 0$ implies that $Im\varphi \subset \ker(M(y.\mu) \longrightarrow C_\alpha M(y.\mu)) = 0$ using 4.4.12 (iii). Hence $\varphi = 0$. Consequently $\dim \operatorname{Hom}_U(M(w.\mu), M(y.\mu)) \leq \dim \operatorname{Hom}_U(C_\alpha M(w.\mu), C_\alpha M(y.\mu))$. Finally $C_\alpha M(w.\mu) = M(s_\alpha w.\mu)$ and $C_\alpha M(y.\mu) = M(y'.\mu)$ for some $y' \in W$ by 4.4.14, so the assertion follows by the induction hypothesis.

4.4.16 Consider the case when $\mathfrak{g} = \mathfrak{sl}(3)$, that is $\pi = \{\alpha, \beta\}$ and $(\alpha,\alpha) = (\beta,\beta) = 2$, $(\alpha,\beta) = -1$. Then $w_0 = s_\alpha s_\beta s_\alpha = s_\beta s_\alpha s_\beta$. Writing the generators of the submodule $M(w_0.0)$ of $M(0)$ in the two resulting ways gives $y_{-\alpha} y_{-\beta}^2 y_{-\alpha} = y_{-\beta} y_{-\alpha}^2 y_{-\beta}$ up to scalar which can easily seen to be 1 using the Shapovalev form. This can also be checked from the quantum Serre relations (4.3.13). Similar relations hold in other rank 2 cases of finite type.

4.4.17 The Jantzen sum formula simplifies considerably for weights of the form $y.\mu : y \in W$, $\mu \in P^+(\pi)$. Indeed

Proposition. *For all $\mu \in P^+(\pi)$, $y \in W$ one has*

$$\operatorname{ch} M^i(y.\mu) = \sum_{\gamma \in \Delta^+_{re} \setminus S(y^{-1})} \operatorname{ch} M(s_\gamma y.\mu) \ .$$

Suppose $\gamma \in \Delta^+$, $m \in \mathbb{N}^+$ satisfies $\Lambda(\tau(2\gamma)) = q^{-2(\rho,\gamma)+m(\gamma,\gamma)}$ where $\Lambda = q^{y.\mu}$. Then $2(y(\mu+\rho),\gamma) = m(\gamma,\gamma)$. If $\gamma \in \Delta^+_{im}$, then $y^{-1}\gamma \in \Delta^+_{im}$ by A.1.3 (i). Then $2(\mu+\rho, y^{-1}\gamma) > 0$ which contradicts $(\gamma,\gamma) \leq 0$. Thus $\gamma \in \Delta^+_{re}$ and $y^{-1}\gamma \in \Delta^+$, that is $\gamma \in \Delta^+ \setminus S(y^{-1})$. Finally observe that $y.\mu - m\gamma = s_\gamma y.\mu$. Then the assertion follows from 4.1.7 and 4.1.16.

4.4.18 Corollary. *Take $\mu \in P^+(\pi)$ and $y \in W$. Then V is a simple subquotient of $M(y.\mu)$ if and only if $V \cong V(w.\mu)$ with $w \geq y$.*

Sufficiency follows from 4.4.7. Conversely suppose that V is a simple subquotient of $M(y.\mu)$. Then either $V \cong V(y.\mu)$ or V is a simple subquotient of $M^1(y.\mu)$. Moreover $V \cong V(\nu)$ for some $\nu \leq y.\mu$. Then by 4.4.17 it follows that V is a subquotient of $M(s_\gamma y.\mu)$ for some $\gamma \in \Delta^+_{re} \setminus S(y^{-1})$ so then $s_\gamma y > y$ by A.1.8. Moreover $\nu \leq s_\gamma y.\mu < y.\mu$ and in particular $|s_\gamma y.\mu - \nu| < |y.\mu - \nu|$. Thus this process must eventually terminate giving $\nu = w.\mu$ with $w \geq y$.

4.5 Comments and Complements

4.5.1 The Jantzen filtration and sum formula were first described by J.-C. Jantzen [Ja1] in the case of the enveloping algebra of a semisimple Lie algebra. Jantzen suggested that it should behave well under Verma module embeddings. In [GJ2] it was shown that this would imply the Kazhdan-Lusztig conjectures and that the quotients would be semisimple. Conversely R.S. Irving [I1] showed that truth of the Kazhdan-Lusztig conjecture determines the semisimple factors in the socle filtration of a Verma module. Finally the Jantzen filtration has been stated to be Gabber's weight filtration by J. Bernstein and A.A. Beilinson and has been further studied particularly in [BGS1]. The Jantzen sum formula was also studied for Weyl modules of Chevalley groups by Jantzen himself and by H.H. Andersen [A2].

4.5.2 Sections 4.1.1–4.1.8 follow [JL4], with 4.1.3 taken from [Ja1] and 4.1.4 from [BGG3] whilst 4.1.9–4.1.16 answers (for C integrable, 4.3.8) a conjecture of [JL4]. The intended application was to show that $V(\lambda)$ for $\lambda \in \mathfrak{h}^*_{\mathbb{Z}}$ remain simple on specialization to answer a question of V.G. Drinfeld [Dr5]. The idea was that the Jantzen filtration for $M(\lambda)$ should be coarser than for $\mathcal{M}(\lambda)$ so coincidence of the character formulae for the sum forces equality and in particular that $\operatorname{ch} V(\lambda) = \operatorname{ch} \mathcal{V}(\lambda)$. However difficulties occur as two indeterminates are involved. The factorization of the Shapovalev form for

$U_q(\mathfrak{g}_C)$ for C of finite type was first given by C. De Concini and V.G. Kac [DK1]. They also determined the scalar (a_ν of 4.1.13). This gives a finer result discussed briefly in Chapter 10.5.2.

4.5.3 The result in 4.1.17 was related to me by M. Rosso. A proof has been reconstructed following the enveloping case described in [Ka1, 9.11]. (Incidentally [Ka1, Remark 9.3] used in [Ka1, 9.11] is incorrect. One should replace primitive vectors by representatives of $\mathfrak{n}^-$ homology). The result in 4.2.11 can be derived from the Hochshild-Serre spectral sequence [HSe1] known already in 1954. However its first use in the present (Lie algebra) context seems to be in [J6]. The result in 4.2.13 for C is finite type is due to B. Kostant [Ko2] being itself developed from the well-known Bott formula. Further applications of 4.2.11 are described in [J9]. One may avoid use of the Casimir invariant if C is of finite type because of vanishing of homology in dimension greater than $\dim \mathfrak{n}^-$.

4.5.4 The Casimir invariant for Kac-Moody Lie algebra is due to V.G. Kac as is 4.2.8 and 4.2.16 which generalize the character formula and denominator formula of H. Weyl. Formula 4.2.16 was the first major success of Kac-Moody Lie algebras as it proved and generalized formulae of I.G. Macdonald which itself generalized formulae of Euler, Gauss and Jacobi. This theory is described in [Ka1].

4.5.5 The result in 4.3.6 (iv) was first obtained by G. Lusztig [L2] using specialization (as discussed in 4.3.7). Assertion 4.3.10 is taken from [JL1, 5.12] extended a result of M. Rosso [R1, R2] who considered just the case C of finite type. The aim following a suggestion of A. Borel was to eliminate completely the use of the centre. This can be achieved for C of finite type using homological methods replacing 4.2.10–4.2.13 and recalling the last remark in 4.5.3. Again the quantum Kac-Moody version of the BGG resolution can be easily developed by calculating $\mathrm{Ext}^i(M(w.\mu), V(\mu)) : \mu \in P^+(\pi), w \in W$ or by specialization. The analysis mirrors the enveloping algebra case (see [GJ1, Ku3] for example) so has been omitted.

4.5.6 The problem discussed in 4.3.12 for an arbitrary Hopf algebra H acting on a semi-prime left Goldie (A.2.7) ring A has been studied by M. Cohen [Co1] and S. Montgomery [Mon1, 2, 3]. Let $\mathcal{I}$ be the set of essential (A.2.7) left ideals of A. Assume that the action of H on A is $\mathcal{I}$ continuous, that is for each $I \in \mathcal{I}$ and $h \in H$ there exists $J \in \mathcal{I}$ such that $hJ \subset I$. Assume further that the antipode is bijective. Then the action of H on A extends uniquely to its ring of fractions $S^{-1}A$ as follows. Given $x \in S^{-1}A$ choose $I \in \mathcal{I}$ such that $Ix \subset A$. Then given $h \in H$ choose $J \in \mathcal{I}$ such that $\sigma^{-1}(h_1)J \subset I$ for all h_1 occurring in the sum $\Delta(h) = h_1 \otimes h_2$. Given $s \in J \cap S$, one defines $h \cdot x \in S^{-1}A$ by

$$h \cdot x = s^{-1}(h_2(\sigma^{-1}(h_1)s)x) .$$

For example let $G(H)$ denote the set of group-like elements of H. Then $\sigma(g)g = g\sigma(g)$ for all $g \in G(H)$. Assume that H is pointed, that is its simple subcoalgebras are one dimensional, hence of the form $kg : g \in G(H)$. This holds in particular for $H = U_q(\mathfrak{g}_C)$ with C symmetric. One has the

Lemma. *If H is pointed, then the action of H on A is $\mathcal{I}$ continuous.*

By the Taft-Wilson theorem [TW1] each $h \in H \setminus G(H)$ is quasi-primitive up to a multiple of $g \in G(H)$ and modulo the coradical filtration. Since $\mathcal{I}$ is closed under intersection and products, one is reduced to establishing continuity for quasi-primitive elements. Assume x quasi-primitive and take $I \in \mathcal{I}$. It is enough to show that $J := \{a \in I \mid x(a) \in I\} \in \mathcal{I}$. Let L be any non-zero left ideal of A. Take $s \in I \cap S$. Then $(L \cap I)s^{-1} \cap A$ is a non-zero left ideal L' of A. Writing $\Delta(x) = x \otimes 1 + g \otimes x$ one has $x(L's) = x(L')s + g(L')x(s) = g(L')x(s) \bmod I$. Since $g(L')x(s)$ is a left ideal of A, there exists $\ell \in L's$ non-zero such $x(\ell) \in I$. Then $\ell \in L \cap J$ as required.

4.5.7 M. Demazure [D1] first gave the character formula for the modules $\mathcal{V}_w(\lambda)$ described in 4.4.1. An error in the proof was noted by V.G. Kac and this generated a spate of correct proofs [A1, Ku1, M1] and further works [Ku3, M2, M4, P1]. The proof of 4.4.5 is taken from J. Bernstein, I. Gelfand and S. Gelfand [BGG1] who established (correctly!) a slightly weaker version of Demazure's formula – at the time thought to be equivalent. The result in 4.4.6 is taken from [JL1] which also establishes the corresponding result for q a root of unity using divided powers.

When C is of finite type the result in 4.4.15 obtains from easy standard arguments using the fact that $U_q(\mathfrak{g}_C)$ has no zero divisors and is noetherian (see A.3.7 (iv) and [Di2, 7.6]). The Enright functor is of course due to T.J. Enright. The development here extends work of V. Deodhar [De1] for $U(\mathfrak{g}) : \mathfrak{g}$ semisimple and W. Neidhardt [N1]. The identities given as an exercise in 4.4.16 are due to D.N. Verma in the enveloping case. The method used here is that noted by J. Dixmier [Di2, 7.8.8] which generalizes without change to the quantum case. Notice that the noetherianity of $U_q(\mathfrak{g}_C)$ which holds ([DK1] and 7.5.7) for C of finite type and whose proof is more delicate than for $U(\mathfrak{g}_C)$, is not needed here.

4.5.8 The result in 4.4.18 for $U(\mathfrak{g}) : \mathfrak{g}$ semisimple is due to J. Bernstein, I. Gelfand and S. Gelfand [BGG2]. The present proof for this case is due to J.-C. Jantzen and is *much* shorter. It is remarkable that it extends to the present case. Of course one now has much information on the multiplicities themselves [KT1].

Chapter 5. The Crystal Basis

Here and in chapter 6, C is assumed integrable and one writes $\mathfrak{g}_C$ simply as $\mathfrak{g}$.

The advantage of $U_q(\mathfrak{g})$ over $U(\mathfrak{g})$ is that the former contains a free parameter which in particular may be given specific values. This leads to information on $U_q(\mathfrak{g})$ which can even yield new information on $U(\mathfrak{g})$ itself. In this chapter the $q \longrightarrow 0$ limit is studied. This process is called crystalization by virtue of the interpretation of q as the absolute temperature.

A Note on Conventions. The original Drinfeld-Jimbo quantization of $U(\mathfrak{g})$ had a little extra symmetry. However it soon became replaced by the present version which is a Hopf subalgebra of index 2. The trouble is that the latter has two natural forms for its generators and relations. No doubt in a really well-planned text only one form would appear throughout. This will not be the case here. Recall (3.1.13) the algebra automorphism $\zeta \varkappa$ of $U_q(\mathfrak{g})$ interchanging x_α, $y_{-\alpha}$, and sending $q \longrightarrow q^{-1}$. It is convenient to let e_α (resp. $f_{-\alpha}$) denote the image of x_α (resp. $y_{-\alpha}$). With respect to 3.2.9 only Δ and σ are altered. However the relations will be given again below. This new notation will also help to recall the denominator in 3.2.9 (v) which has further replaced 3.1.3 (*). Moreover in this and the next chapter the base field is taken to be the rational field $\mathbb{Q}$ and k is used to designate an arbitrary integer.

5.1 Operators in the Crystal Limit

5.1.1 Set $\ell = |\pi|$ and $q_i = q^{(\alpha_i,\alpha_i)/2} : i = 1, 2, \ldots, \ell$. Then $U_q(\mathfrak{g})$ is the Hopf algebra over $\mathbb{Q}(q)$ with identity and generators $e_i, f_i, t_i, t_i^{-1} : i = 1, 2, \ldots, \ell$ satisfying the relations

(i) $t_i t_i^{-1} = t_i^{-1} t_i = 1$
(ii) $t_i t_j = t_j t_i$
(iii) $t_i e_j t_i^{-1} = q_i^{(\alpha_i^\vee,\alpha_j)} e_j$
(iv) $t_i f_j t_i^{-1} = q_i^{-(\alpha_i^\vee,\alpha_j)} f_j$
(v) $e_i f_j - f_j e_i = \delta_{ij} \left(\frac{t_i - t_i^{-1}}{q_i - q_i^{-1}}\right)$

(vi) $\sum_{n=0}^{m}(-1)^n e_i^{(n)} e_j e_i^{(m-n)} = 0$, $\sum_{n=0}^{m}(-1)^n f_i^{(n)} f_j f_i^{(m-n)} = 0$ *where* $m = 1 - (\alpha_i^\vee, \alpha_j)$ *and* $e_i^{(n)} = e_i^n/[n!]_{q_i}$, $f_i^{(n)} = f_i^n/[n!]_{q_i}$

(vii) $\Delta(e_i) = e_i \otimes t_i^{-1} + 1 \otimes e_i$, $\Delta(f_i) = f_i \otimes 1 + t_i \otimes f_i$, $\Delta(t_i^{\pm 1}) = t_i^{\pm 1} \otimes t_i^{\pm 1}$

(viii) $\varepsilon(e_i) = \varepsilon(f_i) = 0$, $\varepsilon(t_i^{\pm 1}) = 1$

(ix) $\sigma(e_i) = -e_i t_i$, $\sigma(f_i) = -t_i^{-1} f_i$, $\sigma(t_i^{\mp 1}) = t_i^{\pm 1}$.

Let $U_q(\mathfrak{s}_i)$ be the Hopf subalgebra generated by the e_i, f_i, t_i, t_i^{-1}.

5.1.2 Let M be an integrable $U_q(\mathfrak{g})$ module. One defines for each i, operators $\tilde{e}_i$, $\tilde{f}_i$ leaving each $U_q(\mathfrak{s}_i)$ submodule invariant. For this one may assume that M is a simple $U_q(\mathfrak{s}_i)$ module and drop the i subscript.

Recall (4.3.1) that if M is simple, then $\ker e$ is one dimensional and $M = \oplus\{f^{(n)}(\ker e) | n = 0, 1, 2, \ldots, \dim M - 1\}$. Then define for all $u \in \ker e$, $n = 0, 1, 2, \ldots, \dim M - 1$, taking $f^{(-1)} = 0$,

$$\tilde{f}(f^{(n)}u) := f^{(n+1)}u$$
$$\tilde{e}(f^{(n)}u) := f^{(n-1)}u\,.$$

More generally these operators can be defined for any module M satisfying for each i that $M = \oplus(f_i^{(n)} \ker e_i)$. However if f_i is not injective then the definition of $\tilde{e}_i$ can depend on how $\ker e_i$ is expressed as a sum of one dimensional subspaces. For an integrable module first take a decomposition of $\ker e_i$ into weight subspaces $(\ker e_i)_\mu$. Since

$$\dim f_i^{(n)}(\ker e_i)_\mu = \begin{cases} \dim(\ker e_i)_\mu & : n \le (\alpha_i^\vee, \mu) \\ 0 & : \text{otherwise} \end{cases}$$

no further ambiguity arises.

5.1.3 Let A denote the localization of $\mathbb{Q}[q]$ at $q = 0$. Let M be a module of the type described in 5.1.2. A *crystal basis* for M is a pair (L, B) satisfying

(A1) *L is a free A module such that* $M \cong \mathbb{Q}(q) \otimes_A L$.

(A2) *B is a basis for* L/qL.

(A3) *L is stable by the* $\tilde{e}_i$, $\tilde{f}_i$, $\forall i$.

(A4) $\tilde{e}_i B$, $\tilde{f}_i B \subset B \cup \{0\}$, $\forall i$.

(A5) $L = \bigoplus L_\lambda$, $B = \amalg B_\lambda$ *where* $L_\lambda = L \cap M_\lambda$, $B_\lambda = B \cap (L_\lambda/qL_\lambda)$.

(A6) *For* $b, b' \in B$ *one has* $b' = \tilde{f}_i b \iff \tilde{e}_i b' = b$.

The main task of this chapter is to prove the existence of crystal bases for certain modules of the above type (in particular all integrable ones). Their properties are studied in the succeeding chapter.

5.1.4 Recall (4.3.9) that any integrable module M is a direct sum of simple highest weight modules and that every simple highest weight module $V(\lambda)$ is determined by its highest weight $\lambda \in P^+(\pi)$. Fix a highest weight vector

u_λ of $V(\lambda)$ and define a lattice $L(\lambda)$ in $V(\lambda)$ by taking all A linear combinations of vectors of the form $\tilde{f}_{i_1}\tilde{f}_{i_2}\cdots\tilde{f}_{i_n}u_\lambda : i_j \in \{1,2,\ldots,\ell\}$. Define $B(\lambda) \subset L(\lambda)/qL(\lambda)$ to be the images of all such monomials. A basic result is that $(L(\lambda), B(\lambda))$ is a crystal basis for $V(\lambda)$. Admit this result and consider uniqueness of a crystal basis for M integrable.

5.1.5 Lemma. *Take $\lambda \in P^+(\pi)$. Then*

(i) $\{u \in L(\lambda)/qL(\lambda) | \tilde{e}_i u = 0, \ \forall i\} = \mathbb{Q}u_\lambda$
(ii) $\{u \in V(\lambda) | \tilde{e}_i u \in L(\lambda), \ \forall i\} = L(\lambda) + V(\lambda)_\lambda$.

(i) It is enough to show that if $u \in (L(\lambda)/qL(\lambda))_\mu : \mu \neq \lambda$ satisfies $\tilde{e}_i u = 0, \forall i$ then $u = 0$. By (A4), (A5) one can assume $u \in B_\mu$ without loss of generality. Then the assertion follows from (A6) and the *definition* of $B(\lambda)$.

(ii) One can assume $u \in L(\lambda)_\mu : \mu \neq \lambda$. Let n be the smallest integer such that $u' := q^n u \in L(\lambda)$. If $n > 0$ then $\tilde{e}_i u' = 0 \bmod qL(\lambda)$, $\forall i$ and so $u' = 0 \bmod qL(\lambda)$ by (i), which is a contradiction.

5.1.6 A lattice L in $V(\lambda)$ is determined by its highest weight vector u_λ and (A3). This is shown by the

Lemma. *Assume L is an A submodule of $V(\lambda)$ satisfying $L = \bigoplus_{\mu \in P(\pi)} L_\mu$ and $L_\lambda = Au_\lambda$. Then*

(i) $\tilde{f}_i L \subset L, \ \forall i \Longrightarrow L(\lambda) \subset L$
(ii) $\tilde{e}_i L \subset L, \ \forall i \Longrightarrow L \subset L(\lambda)$.

(i) is clear. For (ii) it is enough to show that $L_\mu \subset L(\lambda)_\mu$ given $L_{\mu+\alpha_i} \subset L(\lambda)_{\mu+\alpha_i}$, $\forall i$. The latter implies $\tilde{e}_i L_\mu \subset L(\lambda)$ and so the assertion follows from 5.1.5(ii).

5.1.7 Proposition. *Let M be integrable with $\lambda \in P^+(\pi)$ its highest weight. Let (L, B) be a crystal basis for M and set $M = N_1 \oplus N_2 : N_1 = U_q(\mathfrak{g})M_\lambda$; $L_i = N_i \cap L$, $B_i = B \cap L_i/qL_i : i = 1, 2$. Then*

(i) $L = L_1 \oplus L_2$, $B = B_1 \amalg B_2$
(ii) $(L_1, B_1) \cong (L(\lambda), B(\lambda))^{\dim M_\lambda}$.

It is clear that N_1 admits a crystal basis $(\tilde{L}, \tilde{B}) \cong (L(\lambda), B(\lambda))^{\dim M_\lambda}$ such that $\tilde{L}_\lambda = (L_1)_\lambda$. Then 5.1.6 gives $L_1 = \tilde{L}$. Again let $p : M \longrightarrow N_1$ be the projection defined by the decomposition $M = N_1 \oplus N_2$. One checks that $L' := p(L)$ satisfies the hypothesis of 5.1.6 and so again $L' = \tilde{L} = L_1$. Consequently $L = L_1 \oplus L_2$. Now show that $B_\mu = \tilde{B}_\mu \amalg (B_2)_\mu$, $\forall \mu \in P(\pi)$. It is trivial for $\mu = \lambda$ so one can assume $\mu \neq \lambda$ and $B_{\mu+\alpha_i} \subset \tilde{B} \amalg B_2$, $\forall i$. For $b \in B_\mu$ the above decomposition of L gives $b = u_1 + u_2 : u_j \in L_j/qL_j$. If $u_1 = 0$ then $b = u_2 \in B_2$. By 5.1.5(i) if $u_1 \neq 0$, then $\tilde{e}_i u_1 \neq 0$ for some i. Then

$\tilde{e}_i b = \tilde{e}_i u_1 + \tilde{e}_i u_2 \in B_{\mu+\alpha_i} \subset \tilde{B}_{\mu+\alpha_i} \cup (B_2)_{\mu+\alpha_i}$. This forces $\tilde{e}_i u_1 \in \tilde{B}_{\mu+\alpha_i}$ and $\tilde{e}_i u_2 = 0 \pmod{qL_2}$. Then $\tilde{e}_i b \in \tilde{B}_{\mu+\alpha_i}$ and so by (A6) one has $b = \tilde{f}_i \tilde{e}_i b \in \tilde{B}$. Thus $B \subset \tilde{B} \amalg B_2$. From the *definition* of $\tilde{B}$ it is clear that $\tilde{B} \subset B_1$, whilst $B_1 \subset B$ trivially. Hence $B = B_1 \amalg B_2$ as required.

5.1.8 It is clear from 5.1.7 that

Theorem. *Let M be integrable and (L, B) a crystal basis for M. Then there is an isomorphism $\varphi : M \xrightarrow{\sim} \oplus V(\lambda_j) : \lambda_j \in P^+(\pi)$ taking (L, B) onto $\oplus(L(\lambda_j), B(\lambda_j))$.*

Of course all this assumes that $(L(\lambda), B(\lambda))$ is a crystal basis of $V(\lambda)$. Otherwise it is not obvious that a crystal basis of $V(\lambda)$ has to satisfy the conclusion of 5.1.5(i).

5.1.9 Already the theory of crystal bases has some interesting properties at the level of $U_q(\mathfrak{sl}(2))$. Here the existence of a crystal basis is quite obvious. Simply fix a $\mathbb{Q}(q)$ basis $\{u_i\}$ for $\ker e$ formed from weight vectors and set

$$L = \sum_{i,n \in \mathbb{N}} A f^{(n)} u_i \, , \quad B = \{f^{(n)} u_i \bmod qL\} \, .$$

In particular, the conclusion of 5.1.8 applies.

Suppose M is a simple $U_q(\mathfrak{sl}(2))$ module of dimension $\ell + 1$. Choose $u \in \ker e$ non-zero and write $u = u_\ell$. Then set $u_{\ell-2m} = f^{(m)} u_\ell$ for $m = 0, 1, 2, \ldots, \ell$. From 4.3.1 $(*)$ one has

$$[e, f^{(m)}] = f^{(m-1)} (q^{-(m-1)} t - q^{(m-1)} t^{-1})/(q - q^{-1}) \, .$$

Hence

$$e u_{\ell-2m} = [\ell - (m-1)] u_{\ell-2(m-1)} \, , \quad t u_{\ell-2m} = q^{\ell-2m} u_{\ell-2m} \, ,$$
$$f u_{\ell-2m} = [m+1] u_{\ell-2(m+1)} \, .$$

The unique (up to isomorphism) crystal basis for M is given by $L = \bigoplus_{j=0} A u_{\ell-2j}$ with B being the set of images of the $\{u_{\ell-2j}\}_{j=0}^{\ell}$ in L/qL.

5.1.10 Let M be an integrable module. Given $u \in M_\mu$ one can write $u = \sum_{n=0}^m f^{(n)} u_n$ with $u_n \in \ker e$. Moreover one can suppose $u_n \in M_{\mu - n\alpha}$ and $0 \leq n \leq (\mu, \alpha)$ for otherwise $f^{(n)} u_n = 0$. *In what follows this will always be assumed.*

Lemma. *Assume (L, B) is a crystal basis for M and $u \in L$ a weight vector.*

(i) *Suppose $u = \sum_{n=0}^m f^{(n)} u_n \in L$ with $u_n \in \ker e$. Then $u_n \in L$, $\forall n$.*
(ii) *If moreover $u \in B \bmod qL$, then $u = f^{(n_0)} u_{n_0} \in B \bmod qL$ where $n_0 = \max\{n | \tilde{e}^n u \notin qL\}$.*

(i) is proved by induction on m. It is trivial for $m = 0$. By (A3), the definition of $\tilde{e}$ and the assumption on the sum one has $L \ni \tilde{e}u = \sum_{n=1}^{m} f^{(n-1)}u_n$. Thus $u_n \in L : n \geq 1$ by the induction hypothesis. For $n \geq 1$ this gives $f^{(n)}u_n = \tilde{f}^n u_n \in L$ and so $u_0 \in L$ also.

(ii) By (A4) one has $\tilde{e}^{n_0}u \in B \bmod qL$. Then by (A6) one has $u = \tilde{f}^{n_0}\tilde{e}^{n_0}u = \tilde{f}^{n_0}u_{n_0} \bmod qL$ by definition of n_0.

5.1.11 Let (L, B) be a crystal basis for an integrable $U_q(\mathfrak{g})$ module M. Define for all $b \in B$

$$(*) \qquad \varepsilon_i(b) = \max\{n | \tilde{e}_i^n b \neq 0\}$$

$$(**) \qquad \varphi_i(b) = \max\{n | \tilde{f}_i^n b \neq 0\} \ .$$

Given $b \in B_\lambda$, set $wt(b) = \lambda$ and $wt_i(b) = (\alpha_i^\vee, wt(b))$.

Consider M as an $U_q(\mathfrak{sl}(2))$ module by restriction. Applying 5.1.8 it is immediate that

Lemma. $wt_i(b) = \varphi_i(b) - \varepsilon_i(b)$.

5.1.12 Undoubtedly, the crucial property of a crystal basis is its good behaviour under tensor product. This is expressed as follows.

Theorem. *Let $M_j : j = 1, 2$ be integrable $U_q(\mathfrak{g})$ modules with crystal bases (L_j, B_j). Set $L = L_1 \otimes_A L_2$, $B = \{b_1 \otimes b_2 : b_j \in B_j\}$. Then*

(i) *(L, B) is a crystal basis for $M_1 \otimes M_2$.*

(ii) $$\tilde{f}_i(b_1 \otimes b_2) = \begin{cases} \tilde{f}_i b_1 \otimes b_2 & \text{if } \varphi_i(b_1) > \varepsilon_i(b_2) \\ b_1 \otimes \tilde{f}_i b_2 & \text{otherwise.} \end{cases}$$

$$\tilde{e}_i(b_1 \otimes b_2) = \begin{cases} \tilde{e}_i b_1 \otimes b_2 & \text{if } \varphi_i(b_1) \geq \varepsilon_i(b_2) \\ b_1 \otimes \tilde{e}_i b_2 & \text{otherwise.} \end{cases}$$

Since it is enough to check the statements for each i this reduces to a purely $\mathfrak{sl}(2)$ calculation and the i subscript is dropped. Then by 5.1.8 it can be assumed that M_1, M_2 are simple $U_q(\mathfrak{sl}(2))$ modules given their unique crystal bases.

Assume first that M_1 is two-dimensional. Write $B_1 = \{\bar{u}_1, \bar{u}_{-1}\}$, $B_2 = \{\bar{v}_\ell, \bar{v}_{\ell-2,\ldots}, \bar{v}_{-\ell}\}$. One easily checks that $\ker e$ on $M := M_1 \otimes M_2$ is spanned by $u_1 \otimes v_\ell$ and $u_1 \otimes v_{\ell-2} - q^\ell[\ell]u_{-1} \otimes v_\ell$. Now setting $U = U_q(\mathfrak{sl}(2))$

$$\begin{aligned} \Delta(f^m) &= (f \otimes 1 + t \otimes f)^m \\ &= t^m \otimes f^m + \sum_{j=0}^{m-1} t^{m-j} f t^{j-1} \otimes f^{m-1} \bmod U f^2 \otimes U \ , \\ &= t^m \otimes f^m + q^{-(m-1)}[m] f t^{m-1} \otimes f^{m-1} \bmod U f^2 \otimes U \ , \end{aligned}$$

which gives

$$f^{(m)}(u_1 \otimes v_\ell) = q^m u_1 \otimes v_{\ell-2m} + u_{-1} \otimes v_{\ell-2m+2} : m = 0, 1, \dots, \ell+1$$

and

$$\begin{aligned} f^{(m)}&(u_1 \otimes v_{\ell-2} - q^\ell[\ell]u_{-1} \otimes v_\ell) \\ &= q^m[m+1]u_1 \otimes v_{\ell-2-2m} + [m]u_1 \otimes v_{\ell-2m} - q^{\ell-m}[\ell]u_1 \otimes v_{\ell-2m} \\ &= \frac{(1-q^{2(m+1)})}{(1-q^2)} u_1 \otimes v_{\ell-2(m+1)} - q^\ell[\ell-m]u_{-1} \otimes v_{\ell-2m} \,, \\ &\text{for } m = 0, 1, \dots, \ell-1 \,. \end{aligned}$$

Noting that $q^\ell[\ell-m] = q^{m+1}(1-q^{2(\ell-m)})/(1-q^2)$, these formulae imply that $u_1 \otimes v_{\ell-2m} : m = 0, 1, 2, \dots, \ell$ form an A basis for $L_1 \otimes L_2$. Then (i) follows from the definition (1.1.2) of $\tilde{e}$, $\tilde{f}$. Moreover, $\bmod qL_1 \otimes L_2$ the formulae of (ii) hold. Notice also that the crystal basis is compatible with the direct sum decomposition of $M_1 \otimes M_2$ into its two components.

Return to the general case which will be established by induction on $m = \dim M_1 - 1$. Write V_m for the unique simple module of dimension $m+1$. By the induction hypothesis $(L_{m-1} \otimes L_\ell,\ B_{m-1} \otimes B_\ell)$ is a crystal basis for $V_{m-1} \otimes V_\ell$. Then by the above $(L_1 \otimes (L_{m-1} \otimes L_\ell),\ B_1 \otimes (B_{m-1} \otimes B_\ell))$ is a crystal basis for $V_1 \otimes (V_{m-1} \otimes V_\ell) = (V_1 \otimes V_{m-1}) \otimes V_\ell$, using the associativity of Δ. Moreover $V_1 \otimes V_{m-1} \cong V_m \oplus V_{m-2}$ and $L_1 \otimes L_{m-1} = L_m \oplus L_{m-2}$, $B_1 \otimes B_{m-1} = B_m \amalg B_{m-2}$ taking $\ell = m-1$ in the above. Hence the direct summand $(L_m \otimes L_\ell,\ B_m \otimes B_\ell)$ is a crystal basis for $V_m \otimes V_\ell$. This proves (i). Using the induction hypothesis and the result for $m = 1$, verification of (ii) results from (ii) being an associative rule (as shown in 5.2.5).

5.2 Crystals

One may abstract the notion of a crystal basis and obtain some interesting generalizations corresponding to limit cases when dimension goes to infinity. It leads to a purely set theoretic or combinatorial problem with the Cartan matrix playing a fundamental role. This is studied below.

5.2.1 Give $\mathbb{Z} \cup -\infty$ its linear order with $-\infty$ the smallest element and an additive structure via $-\infty + n = -\infty$. Here the set I indexes the simple roots of $\mathfrak{g}$ with P its lattice of weights. Recall that $(\alpha_i^\vee, \omega) \in \mathbb{Z}$ for all $\omega \in P$.

Definition. A crystal B is a set
with maps

(1) $\quad wt : B \longrightarrow P \,, \quad \varepsilon_i : B \longrightarrow \mathbb{Z} \cup \{-\infty\} \,, \quad \varphi_i : B \longrightarrow \mathbb{Z} \cup \{-\infty\} \,,$

and operations,

(ii) $\quad \tilde{e}_i : B \longrightarrow B \cup \{0\} \,, \quad \tilde{f}_i : B \longrightarrow B \cup \{0\} \,, \quad \tilde{e}_i 0 = \tilde{f}_i 0 = 0 \,,$

satisfying

(C1) *$\varphi_i(b) = \varepsilon_i(b) + (\alpha_i^\vee, wtb)$ (which defines φ_i).*
(C2) *If $b \in B$, $\tilde{e}_i b \in B$ (resp. $\tilde{f}_i b \in B$) then $wt(\tilde{e}_i b) = wtb + \alpha_i$, $\varepsilon_i(\tilde{e}_i b) = \varepsilon_i(b) - 1$, $\varphi_i(\tilde{e}_i b) = \varphi_i(b) + 1$ (resp. $wt(\tilde{f}_i b) = wtb - \alpha_i$, $\varepsilon_i(\tilde{f}_i b) = \varepsilon_i(b) + 1$, $\varphi_i(\tilde{f}_i b) = \varphi_i(b) - 1$).*
(C3) *For $b, b' \in B$ one has $b' = \tilde{e}_i b \Longleftrightarrow b = \tilde{f}_i b'$.*
(C4) *For $b \in B$, if $\varphi_i(b) = -\infty$ then $\tilde{e}_i b = \tilde{f}_i b = 0$.*

A crystal is called upper (resp. lower) normal if 5.1.11(*) (resp. (**)) is satisfied. If both hold, a crystal is called normal. Set $wt_i(b) = (wtb, \alpha_i^\vee)$.

A morphism ψ of crystals is a map $B_1 \cup \{0\} \longrightarrow B_2 \cup \{0\}$ with $\psi(0) = 0$ satisfying

(i) *ψ commutes with wt, ε_i, φ_i, $\forall i \in I$*
(ii) *$\psi(\tilde{e}_i b) = \tilde{e}_i \psi(b)$ if $\tilde{e}_i b \neq 0$, $\forall i \in I$*
(iii) *$\psi(\tilde{e}_i b) = \tilde{f}_i \psi(b)$ if $\tilde{f}_i b \neq 0$, $\forall i \in I$.*

A morphism is called strict if ψ commutes with $\tilde{e}_i$, $\tilde{f}_i$, $\forall i \in I$.

Clearly a morphism is strict if it is an isomorphism or if both B_1, B_2 are normal.

5.2.2 Given crystals B_1, B_2 one may form their direct sum $B_1 \oplus B_2$ whose underlying set is $B_1 \amalg B_2$ with the obvious actions. A morphism ψ is called an embedding if it is injective and its image is called a subcrystal. Any subset B' of a crystal B has the structure of a subcrystal. An embedding is said to be full if $\tilde{e}_i \psi(b) \in B_2 \Longrightarrow \tilde{e}_i b \in B_1$. By (C3) a strict embedding is full and $B_2 = B_1 \oplus (B_2 \backslash \psi(B_1))$ in this case, that is $\psi(B_1)$ is a component of B_2 viewed as a graph. For any morphism $\psi : B_1 \longrightarrow B_2$ one has $B_1 = \psi^{-1}(B_2) \oplus \psi^{-1}(0)$.

5.2.3 Examples

(i) Let (L, B) be a crystal basis of an integrable $U_q(\mathfrak{g})$ module (if it exists!). Recalling (A1)–(A6) and 5.1.11, it is clear that B is a normal crystal.

(ii) For each $\lambda \in P$ define T_λ to be the crystal consisting of a single element t_λ satisfying $wt(t_\lambda) = \lambda$, $\varepsilon_i(t_\lambda) = \varphi_i(t_\lambda) = -\infty$, $\tilde{e}_i t_\lambda = \tilde{f}_i t_\lambda = 0$.

(iii) For $\forall i \in I$ the crystal C_i defined as follows

$$C_i = \{c_i(n) | n \in \mathbb{Z}\}, \ wt\ c_i(n) = n\alpha_i, \ \varphi_i(c_i(n)) = n, \ \varphi_j(c_i(n)) = -\infty \ j \neq i$$

with the action

$$\tilde{e}_i(c_i(n)) = c_i(n+1)\ , \quad \tilde{f}_i(c_i(n)) = c_i(n-1) \quad \tilde{e}_j c_i(n) = \tilde{f}_j c_i(n) = 0 : j \neq i\ .$$

5.2.4 Given crystals B_1, B_2 define their tensor product $B_1 \otimes B_2$ to be $B_1 \times B_2$ (writing (b_1, b_2) as $b_1 \otimes b_2$ and identifying $(0, b)$, $(b, 0)$ with 0) by the following rules.

(1) *The formulae* 5.1.12(ii) *hold.*
(2) $wt(b_1 \otimes b_2) = wt\ b_1 + wt\ b_2$, $\varepsilon_i(b_1 \otimes b_2) = \max\{\varepsilon_i(b_1), (\varepsilon_i(b_2) - \varphi_i(b_1)) + \varepsilon_i(b_1)\}$, $\varphi_i(b_1 \otimes b_2) = \max\{\varphi_i(b_2), (\varphi_i(b_1) - \varepsilon_i(b_2)) + \varphi_i(b_2)\}$.

Lemma. *$B_1 \otimes B_2$ is a crystal.*

(C1) is easily checked. The assertion for wt in (C2) is obvious. For the assertion in (C2) concerning ε_i observe that

$$\varepsilon_i(b_1 \otimes b_2) = \begin{cases} \varepsilon_i(b_1) : \varphi_i(b_1) \geq \varepsilon_i(b_2) \\ \varepsilon_i(b_2) - wt_i(b_1) : \varphi_i(b_1) < \varepsilon_i(b_2) \end{cases}$$

as the assertion is immediate if $\varphi_i(b_1) \neq \varepsilon_i(b_2)$ and one again checks it when equality holds noting that in this case

$$\begin{aligned} \varepsilon_i(\tilde{f}_i b_1 \otimes b_2) &= \varepsilon_i(b_1 \otimes \tilde{f}_i b_2) \\ &= \varepsilon_i(\tilde{f}_i b_2) - \varphi_i(b_1) + \varepsilon_i(b) \\ &= 1 + \varepsilon_i(b_2) - \varphi_i(b_1) + \varepsilon_i(b_1) = 1 + \varepsilon_i(b_1) \\ &= 1 + \varepsilon_i(b_1 \otimes b_2)\ , \quad \text{as required}\ . \end{aligned}$$

For (C3) one observes that

$$\tilde{e}_i \tilde{f}_i(b_1 \otimes b_2) = \begin{cases} \tilde{e}_i \tilde{f}_i b_1 \otimes b_2 & \text{if } \varphi_i(b_1) > \varepsilon_i(b_2) \\ b_1 \otimes \tilde{e}_i \tilde{f}_i b_2 & \text{if } \varphi_i(b_1) \leq \varepsilon_i(b_2) \end{cases}$$

and

$$\tilde{f}_i \tilde{e}_i(b_1 \otimes b_2) = \begin{cases} \tilde{f}_i \tilde{e}_i b_1 \otimes b_2 & \text{if } \varphi_i(b_1) \geq \varepsilon_i(b_2) \\ b_1 \otimes \tilde{f}_i \tilde{e}_i b_2 & \text{if } \varphi_i(b_1) < \varepsilon_i(b_2)\ . \end{cases}$$

Thus if $b \in B$, then $\tilde{e}_i b \neq 0$ implies $\tilde{f}_i \tilde{e}_i b = b$ and $\tilde{f}_i b \neq 0$ implies $\tilde{e}_i \tilde{f}_i b = b$ at least when either $\varphi_i(b_1)$ or $\varepsilon_i(b_2)$ are finite. If $\varphi_i(b_1) = \varepsilon_2(b_2) = -\infty$ then (C4) for B_1, B_2 may be invoked to establish (C3) for $B_1 \otimes B_2$.

Finally $\varphi_i(b_1 \otimes b_2) = -\infty$ implies $\varphi_i(b_1) = \varphi_i(b_2) = -\infty$ and so $\tilde{e}_i$ and $\tilde{f}_i$ are zero on $b_1 \otimes b_2$. Hence (C4) holds for $B_1 \otimes B_2$.

5.2.5 Associativity of the tensor product obtains from the following formulae for $\tilde{e}_i(b_1 \otimes \ldots \otimes b_n)$ and $\tilde{f}_i(b_1 \otimes \ldots \otimes b_n)$ which are found to be independent of partitioning. Here it is convenient to assume that $-\infty$ is replaced by a negative number strictly smaller than all the finite r_j given below.

Let $B_k : k = 1, 2, \ldots, n$ be crystals and $b_k \in B_k$. Define r_k^i, or simply r_k, by

$$r_k = \varepsilon_i(b_k) - \sum_{1 \leq j < k} wt_i(b_j)\ .$$

One has

$$\begin{aligned} r_k - r_{k+1} &= \varepsilon_i(b_k) - \varepsilon_i(b_{k+1}) + wt_i(b_k) \\ &= \varphi_i(b_k) - \varepsilon_i(b_{k+1})\ . \end{aligned}$$

Now set $r_k^- = \max\{r_j | j \le k\}$, $r_k^+ = \max\{r_j | j \ge k\}$.

Notice that $\varepsilon_i((b_1 \otimes b_2) \otimes b_3) = \varepsilon_i(b_1 \otimes (b_2 \otimes b_3)) = \max\{r_j\}$. Since wt_i is trivially associative, so is φ_i.

Lemma

(i) $\varphi_i(b_1 \otimes b_2 \otimes \ldots \otimes b_j) - \varphi_i(b_j) = r_j^- - r_j$.

(ii) $\varepsilon_i(b_{j+1} \otimes b_{j+2} \otimes \ldots \otimes b_n) - \varepsilon_i(b_{j+1}) = r_{j+1}^+ - r_{j+1}$.

(iii) $\varphi_i(b_1 \otimes \ldots \otimes b_j) - \varepsilon_i(b_{j+1} \otimes \ldots \otimes b_n) = r_j^- - r_{j+1}^+$.

(iv) $\tilde{e}_i(b_1 \otimes \ldots \otimes b_n) = b_1 \otimes \ldots \otimes \tilde{e}_i b_k \otimes \ldots \otimes b_n$ *where* $k = \min\{j | r_j^- \ge r_{j+1}^+\} = \min\{j | r_\nu < r_j,\ r_{\nu'} \le r_j \textit{ for } \nu < j \le \nu'\}$.

(v) $\tilde{f}_i(b_1 \otimes \ldots \otimes b_n) = b_1 \otimes \ldots \otimes \tilde{f}_i b_k \otimes \ldots b_n$ *where* $k = \min\{j | r_j^- > r_{j+1}^+\} = \min\{j | r_\nu \le r_j,\ r_{\nu'} < r_j \textit{ for } \nu \le j < \nu'\}$.

(i) is clearly true for $j = 1$. It is established in the general case by induction on j. From 5.2.4(2)

$$\begin{aligned}\varphi_i(b_1 \otimes b_2 \otimes \ldots \otimes b_j) - \varphi_i(b_j) &= \max\{0, \varphi_i(b_1 \otimes b_2 \otimes \ldots \otimes b_{j-1}) - \varepsilon_i(b_j)\} \\ &= \max\{0, \varphi_i(b_1 \otimes b_2 \otimes \ldots \otimes b_{j-1}) - \varphi_i(b_{j-1}) + r_{j-1} - r_j\} \\ &= \max\{0, r_{j-1}^- - r_j\} = r_j^- - r_j \ .\end{aligned}$$

(ii) is clearly true for $j + 1 = n$. It is established in the general case by decreasing induction on j. From 5.2.4(2)

$$\begin{aligned}&\varepsilon_i(b_{j+1} \otimes \ldots \otimes b_n) - \varepsilon_i(b_{j+1}) \\ &\quad = \max\{0, \varepsilon_i(b_{j+2} \otimes \ldots \otimes b_n) - \varepsilon_i(b_{j+2}) + r_{j+2} - r_{j+1}\} \\ &\quad = \max\{0, r_{j+2}^+ - r_{j+1}\} = r_{j+1}^+ - r_{j+1} \ .\end{aligned}$$

(iii) follows trivially from (i), (ii). By 5.2.4(i) and (iii)

$$\begin{aligned}&\tilde{e}_i((b_1 \otimes \ldots \otimes b_j) \otimes (b_{j+1} \otimes \ldots \otimes b_n)) \\ &\quad = \begin{cases} \tilde{e}_i(b_1 \otimes \ldots \otimes b_j) \otimes (b_{j+1} \otimes \ldots \otimes b_n) & \text{if } r_j^- \ge r_{j+1}^+ \\ (b_1 \otimes \ldots \otimes b_j) \otimes \tilde{e}_i(b_{j+1} \otimes \ldots \otimes b_n) & \text{if } r_j^- < r_{j+1}^+ \end{cases} \\ &\quad = b_1 \otimes \ldots \otimes \tilde{e}_i b_k \otimes \ldots b_n\end{aligned}$$

where $k = \min\{j | r_j^- \ge r_{j+1}^+\}$. Moreover this result is independent of where one places the parentheses in the tensor product. Similar reasoning gives (v).

5.2.6 If $\varepsilon_i(b_1)$, $\varepsilon_i(b_2) \in \mathbb{Z}$, then

$$\tilde{e}_i^a(b_1 \otimes b_2) = \tilde{e}_i^x b_1 \otimes \tilde{e}_i^y b_2$$

where $x + y = a$ and $y = \max\{0, \varepsilon_i(b_2) - \varphi_i(b_1)\} = \varepsilon_i(b_1 \otimes b_2) - \varepsilon_i(b_1)$. Now if B_1 is lower normal, then $\varphi_i(b_1) \ge 0$ and so if B_2 is upper normal, one has

$\tilde{e}_i^y b_2 \neq 0$. Then if B_1 is upper normal one has $\tilde{e}_i^a(b_1 \otimes b_2) = 0 \iff \tilde{e}_i^x b_1 = 0$. Since $x - \varepsilon_i(b_1) = a - \varepsilon_i(b_1 \otimes b_2)$, one concludes that

(*) *B_1 normal and B_2 upper normal $\implies B_1 \otimes B_2$ is upper normal .*

Again if $\varphi_i(b_1),\ \varphi_i(b_2) \in \mathbb{Z}$, then

$$\tilde{f}_i^a(b_1 \otimes b_2) = \tilde{f}_i^x b_1 \otimes \tilde{f}_i^y b_2$$

where $x+y=a$ and $x = \max\{0, \varphi_i(b_1) - \varepsilon_i(b_2)\} = \varphi_i(b_1 \otimes b_2) - \varphi_i(b_2)$. Now if B_2 is upper normal, then $\varepsilon_i(B_2) \geq 0$ and so if B_1 is lower normal, one has $\tilde{f}_i^x b_1 \neq 0$. Then if B_2 is lower normal one has $\tilde{f}_i^a(b_1 \otimes b_2) = 0 \iff \tilde{f}_i^y b_2 = 0$. Since $y - \varphi_i(b_2) = a - \varphi_i(b_1 \otimes b_2)$, one concludes that

(**) *B_1 lower normal and B_2 normal $\implies B_1 \otimes B_2$ is lower normal .*

Consequently

Lemma. *If $B_1,\ B_2$ are normal, then so is $B_1 \otimes B_2$.*

5.2.7 Define $s_i \in Aut\ \mathfrak{h}^*$ by $s_i\lambda = \lambda - (\alpha_i^\vee, \lambda)\alpha_i$.

Lemma. *The map $\psi : c_i(n) \otimes t_\lambda \mapsto t_{s_i\lambda} \otimes c_i(n + (\alpha_i^\vee, \lambda))$ is an isomorphism of $C_i \otimes T_\lambda$ onto $T_{s_i\lambda} \otimes C_i$.*

First observe that $wt(c_i(n) \otimes t_\lambda) = \lambda + n\alpha_i = s_i\lambda + (n + (\alpha_i^\vee, \lambda))\alpha_i = wt(t_{s_i\lambda} + c_i(n + (\alpha_i^\vee, \lambda)))$. Again $\varepsilon_i(b \otimes t_\lambda) = \varepsilon_i(b)$ whilst $\varepsilon_i(t_{s_i\lambda} \otimes b) = \varepsilon_i(b) - wt_i(t_{s_i\lambda}) = \varepsilon_i(b) + (\alpha_i^\vee, \lambda)$. Consequently $\varepsilon_i(c_i(n) \otimes t_\lambda) = -n = -(n + (\alpha_i^\vee, \lambda)) + (\alpha_i^\vee, \lambda) = \varepsilon_i(t_{s_i\lambda} \otimes c_i(n + (\alpha_i^\vee, \lambda))$ and moreover $\varphi_i(c_i(n) \otimes t_\lambda) = n + (\alpha_i^\vee, \lambda) = \varphi_i(t_{s_i\lambda} \otimes c_i(n + (\alpha_i^\vee, \lambda)))$. This proves that ψ satisfies (i) of 5.2.1. Since say $\tilde{e}_i(b \otimes t_\lambda) = \tilde{e}_i b \otimes t_\lambda$ and $\tilde{e}_i(t_\mu \otimes b) = t_\mu \otimes \tilde{e}_i b$ one concludes that ψ commutes with $\tilde{e}_i$ and similarly with $\tilde{f}_i$. This proves the lemma.

5.2.8 Take $r, s \in \mathbb{Z}$ and set $m(r,s) = \max\{r, s\}$.

Lemma. *The map $\psi : c_i(-r) \otimes c_i(s) \mapsto c_i(s - r - m(r,s)) \otimes t_{m(r,s)\alpha_i}$ induces an isomorphism of $C_i \otimes C_i$ onto $\oplus C_i \otimes T_{k\alpha_i}$.*

Obviously ψ commutes with wt. Now $\varepsilon_i(c_i(-r) \otimes c_i(s)) = \max\{r, -s+2r\}$, whilst $\varepsilon_i(c_i(x) \otimes t_{k\alpha_i}) = -x$. Hence taking $x = -\max\{r, -s+2r\} = s - r - m(r,s)$ and $k = s - r - x = m(r,s)$ shows that ψ commutes with ε_i and hence with φ_i. Notice also that $\varphi_i(c_i(-r)) - \varepsilon_i(c_i(s)) = s - r$. Hence

$$\tilde{e}_i(c_i(-r) \otimes c_i(s)) = \begin{cases} c_i(-r+1) \otimes c_i(s) : s \geq r \\ c_i(-r) \otimes c_i(s+1) : s < r\ . \end{cases}$$

Thus if $s \geq r$ one has $\psi(\tilde{e}_i(c_i(-r) \otimes c_i(s)) = c_i(-r+1) \otimes t_{s\alpha_i}$ whilst $\tilde{e}_i\psi(c_i(-r) \otimes c_i(s)) = \tilde{e}_i(c_i(-r) \otimes t_{s\alpha_i}) = c_i(-r+1) \otimes t_{s\alpha_i}$ and if $s < r$

one has $\psi(\tilde{e}_i(c_i(-r)\otimes c_i(s))=\psi(c_i(-r)\otimes c_i(s+1))=c_i(s-2r+1)\otimes t_{r\alpha_i}$ whilst $\tilde{e}_i\psi(c_i(-r)\otimes c_i(s))=\tilde{e}_i(c_i(s-2r)\otimes t_{r\alpha_i}=c_i(s-2r+1)\otimes t_{r\alpha_i}$. A similar argument shows that $\tilde{f}_i$ commutes with ψ. Finally if $\psi(c_i(-r)\otimes c_i(s))=c_i(x)\otimes t_{k\alpha_i}$, then

$$k=\max\{r,s\}\ ,\qquad x=\min\{-r,\ -2r+s\}$$

and

$$r=-\max\{-k,\ x\}\ ,\qquad s=\min\{k,2x+k\}$$

which shows that ψ is bijective.

5.3 Ad-invariant Filtrations, Twisted Actions and the Crystal Basis for $U_q(\mathfrak{n}^-)$

5.3.1 The relations in 5.1.1 give

$$(\operatorname{ad} e_i)a=e_ia\sigma(t_i^{-1})+a\sigma(e_i)=e_iat_i-ae_it_i=[e_i,a]t_i$$

and similarly

$$(\operatorname{ad} f_i)a=f_ia-t_iat_i^{-1}f_i\ ,\quad (\operatorname{ad} t_i)a=t_iat_i^{-1}\ .$$

Taking account of the triangular decomposition (3.2.10) of $U_q(\mathfrak{g})$ into graded subalgebras one may define a filtration $\mathcal{F}$ of $U_q(\mathfrak{g})$ in which e_i has degree 1, f_i has degree 0 and t_i has degree -1. By the formulae above this filtration is ad-invariant and consequently there is an induced action of $U_q(\mathfrak{g})$ on $gr_{\mathcal{F}}U_q(\mathfrak{g})$. The relation

$$(\operatorname{ad} e_i)f_j=[e_i,f_j]t_i=\delta_{ij}\left(\frac{t_i-t_i^{-1}}{q_i-q_i^{-1}}\right)t_i=\frac{-\delta_{ij}}{(q_i-q_i^{-1})}$$

in $gr_{\mathcal{F}}\ U_q(\mathfrak{g})$, implies that $gr_{\mathcal{F}}U_q(\mathfrak{n}^-)$ is ad-invariant. Since $gr_{\mathcal{F}}U_q(\mathfrak{n}^-)$ identifies with $U_q(\mathfrak{n}^-)$ as an algebra this gives an action of $U_q(\mathfrak{g})$ on $U_q(\mathfrak{n}^-)$ compatible with multiplication in the latter. All this has some important consequences. First define $\tau_i\in Aut\ U_q(\mathfrak{g})$ by $\tau_i(a)=t_iat_i^{-1}$. Then from 3.2.7 or direct computation one checks that there exists a τ_i-leftskew (resp. τ_i^{-1}-leftskew) derivation e_i' (resp. e_i'') of $U_q(\mathfrak{n}^-)$ such that

$$[e_i,\ b]=\frac{t_ie_i''(b)-t_i^{-1}e_i'(b)}{q_i-q_i^{-1}}\ ,\qquad \forall b\in U_q(\mathfrak{n}^-)\ .$$

Viewing f_j as the endomorphism of $U_q(\mathfrak{n}^-)$ given by left multiplication, one has the commutation rules

$$(*)\qquad e_i''f_j=q^{(\alpha_i,\alpha_j)}f_je_i''+\delta_{ij}\ ,\quad e_i'f_j=q^{-(\alpha_i,\alpha_j)}f_je_i'+\delta_{ij}\ .$$

On the other hand identifying $U_q(\mathfrak{n}^-)$ with its image in $gr_{\mathcal{F}}U_q(\mathfrak{g})$ gives $e_i'=-(q_i-q_i^{-1})\operatorname{ad} t_ie_i$. Since (3.2.8) the $-(q_i-q_i^{-1})t_ie_i$ satisfy the same quantum

Serre relations as the e_i and ad is a homomorphism it follows that the e_i' also satisfy the quantum Serre relations. This is of course also true of the f_j. Denote by $A_q(\mathfrak{g})$ the algebra generated by the e_i', f_j subject to the above three sets of relations. This is just the quantum Weyl algebra A_C of 3.1.3 for C is integrable and $\mathfrak{g} = \mathfrak{g}_C$. The above analysis shows that $U_q(\mathfrak{n}^-)$ is also an $A_q(\mathfrak{g})$ module. Yet it is also a $U_q(\mathfrak{g})$ module and inspection shows that the action of the $U_q(\mathfrak{n}^+)$ subalgebra is essentially the same as under the $A_q(\mathfrak{g})$ action. In particular

Lemma. *View $U_q(\mathfrak{n}^-)$ as an $A_q(\mathfrak{g})$ or an* $\operatorname{ad} U_q(\mathfrak{g})$ *module. Then for all $i \in \{1, 2, \ldots, \ell\}$ one has*

$$\ker(\operatorname{ad} e_i) = \ker e_i' \ .$$

5.3.2 A similar analysis may be used to discuss e_i''. Though this is not so necessary it leads to a better understanding of the structure of the objects defined. First take the new coproduct Δ_+ and antipode σ_+ on $U_q(\mathfrak{g})$ to be defined by 3.2.9 (vii) - (ix). Then the right adjoint action (1.3.1) satisfies

$$(\operatorname{Ad} e_i)a = -t_i^{-1} e_i a + t_i^{-1} a e_i = -t_i^{-1}[e_i, a] \ ,$$

$$(\operatorname{Ad} f_i)a = -f_i t_i a t_i^{-1} + a f_i \ , \quad (\operatorname{Ad} t_i)a = t_i^{-1} a t_i \ .$$

One obtains an Ad invariant filtration $\mathcal{F}'$ of $U_q(\mathfrak{g})$ by now taking t_i to have degree 1. Then in $gr_{\mathcal{F}'} U_q(\mathfrak{g})$

$$(\operatorname{Ad} e_i) f_j = -\delta_{ij} t_i^{-1} \frac{(t_i - t_i^{-1})}{(q_i - q_i^{-1})} = \frac{-\delta_{ij}}{(q_i - q_i^{-1})}$$

which gives $U_q(\mathfrak{n}^-)$ the structure of *right* $U_q(\mathfrak{g})$ module. Observe that

$$e_i'' = -(q_i - q_i^{-1}) \operatorname{Ad} e_i \ .$$

Since Ad is an antihomomorphism it follows that the e_i'' satisfy the quantum Serre relations. Of course this result and that in 5.3.1 is implicit in 3.2.7.

5.3.3 The above considerations do not have an immediate analogue for $U(\mathfrak{g})$. However they can be understood in the following way. Consider the Verma module $\mathcal{M}(\lambda)$ of highest weight λ. One has

$$\mathcal{M}(\lambda) := U(\mathfrak{g}) \otimes_{U(\mathfrak{b})} \mathbb{Q}_\lambda \cong U(\mathfrak{n}^-) \cong S(\mathfrak{n}^-)$$

where the last isomorphism obtains from the symmetrization map. The left action of $U(\mathfrak{g})$ on $\mathcal{M}(\lambda)$ translates to an action of $U(\mathfrak{g})$ on $S(\mathfrak{n}^-)$ by differential operators. On the other hand there are two further actions on $S(\mathfrak{n}^-)$, namely multiplication by functions and the action by differential operators induced by the right action of $U(\mathfrak{n}^-)$ on itself. Moreover these actions under Fourier transform and the Chevalley antiautomorphism give an action of

$U(\mathfrak{g})$ on $\delta\mathcal{M}(\lambda)$ in which (for $\lambda = 0$) the elements of $\mathfrak{g}$ become vector fields, and a further action of $U(\mathfrak{n}^+)$ commuting with the first. Notice also that $\delta\mathcal{M}(\lambda)$ may be identified with the sections of the standard invertible sheaf $\mathcal{L}_\lambda$ on the flag manifold restricted to the open Bruhat cell. This observation suggests the following should hold.

1) That e_i', e_j'' commute possibly up to factors in q.
2) That $U_q(\mathfrak{n}^-)$ is isomorphic to $\delta M(0)$ viewed as an $\operatorname{ad} U_q(\mathfrak{g})$ module.
3) That one may twist the action of $U_q(\mathfrak{g})$ on $U_q(\mathfrak{n}^-)$, *without altering the* $U_q(\mathfrak{n}^+)$ *action*, so that it becomes isomorphic to $\delta M(\lambda)$ for any chosen λ. Then the unique simple quotient $V(\lambda)$ of $M(\lambda)$ can be viewed as a submodule of $U_q(\mathfrak{n}^-)$. Notice here that the conclusion of 5.3.1 will remain valid.

These three assertions are established below. Their relationship to the enveloping algebra theory obtains from the specialization of 3.4.6.

5.3.4 By 3.1.4 or by direct computation one obtains the

Lemma. *As endomorphisms of* $U_q(\mathfrak{n}^-)$ *one has*

$$e_i'e_j'' = q^{(\alpha_i,\alpha_j)}e_j''e_i' \ .$$

Remark. We may define e_i''' through $[e_i, b] = \frac{t_i c - d' t_i^{-1}}{q_i - q_i^{-1}}$ and $e_i'''(b) = d' =: t_i^{-1} d t_i$. Then $e_i''' = -(q_i - q_i^{-1}) \operatorname{ad} e_i = e_i' \operatorname{ad} t_i^{-1}$ and so $e_i'''e_j'' = e_i'(\operatorname{ad} t_i^{-1})e_j'' = q^{-(\alpha_i,\alpha_j)}e_i'e_j''(\operatorname{ad} t_i^{-1}) = e_j''e_i'''$. Thus the above result expresses the fact that $\operatorname{ad} e_i$ and $\operatorname{Ad} e_j$ commute on $U_q(\mathfrak{n}^-)$.

5.3.5 Translated to the present situation, 3.1.5 gives the

Propositión. *View* $U_q(\mathfrak{n}^-)$ *as an* $A_q(\mathfrak{g})$ *module. Then*

$$\bigcap_{i=1}^{\ell} \ker e_i' = \mathbb{Q}(q)1 \ .$$

5.3.6 Recall the action of $U_q(\mathfrak{g})$ on $U_q(\mathfrak{n}^-)$ defined in 5.3.1 and the $\mathcal{O}$ dual (4.1.4).

Corollary. *As a* $U_q(\mathfrak{g})$ *module* $U_q(\mathfrak{n}^-)$ *is isomorphic to* $\delta M(0)$.

By 5.3.5 and 5.3.4 it follows that $\delta U_q(\mathfrak{n}^-)$ is generated over $U_q(\mathfrak{n}^-)$ by a weight vector of weight 0, hence it is an image of $M(0)$. Yet $\operatorname{ch} \delta U_q(\mathfrak{n}^-) = \operatorname{ch} U_q(\mathfrak{n}^-) = \operatorname{ch} M(0)$ and so $\delta U_q(\mathfrak{n}^-)$ is isomorphic to $M(0)$. Hence the assertion.

5.3.7 Corollary. *As an* $A_q(\mathfrak{g})$ *module* $U_q(\mathfrak{n}^-)$ *is simple.*

Indeed any non-zero $A_q(\mathfrak{g})$ submodule must contain a non-zero multiple of 1 by 5.3.5. Yet $U_q(\mathfrak{n}^-)$ is generated by 1 over the $U_q(\mathfrak{n}^-)$ subalgebra of $A_q(\mathfrak{g})$.

5.3.8 Recall 3.1.3 that the multiplication map $U_q(\mathfrak{n}^-) \otimes U_q(\mathfrak{n}^+) \longrightarrow A_q(\mathfrak{g})$ is bijective. Let $\mathbb{Q}(q)$ denote the trivial $U_q(\mathfrak{n}^+)$ module (i.e., in which e_i' acts by zero). Then $U_q(\mathfrak{n}^-)$ identifies as an $A_q(\mathfrak{g})$ module with $A_q(\mathfrak{g}) \otimes_{U_q(\mathfrak{n}^+)} \mathbb{Q}(q)$.

5.3.9 Observe that the symmetry of (α_i, α_j) implies that $\varkappa(e_i') = f_i$, $\varkappa(f_i) = e_i'$ extends to an antiautomorphism of $A_q(\mathfrak{g})$. View $M := \operatorname{Hom}(U_q(\mathfrak{n}^-), \mathbb{Q}(q))$ as a left $U_q(\mathfrak{g})$ module via $\varkappa$. Define $\varphi_0 \in M$ by $\varphi_0(1) = 1$ and $\varphi_0(\sum_i f_i U_q(\mathfrak{n}^-)) = 0$. Then $e_i'\varphi_0 = 0$, $\forall i$ and so there is a homomorphism

$$\psi : U_q(\mathfrak{n}^-) \cong A_q(\mathfrak{g}) \otimes_{U_q(\mathfrak{n}^+)} \mathbb{Q}(q) \text{ to } M$$

sending 1 to φ_0. Consequently one may define a bilinear form $(\,,)$ on $U_q(\mathfrak{n}^-)$ by

$$(u, v) = \psi(u)(v) .$$

It is immediate that $(1,1) = \psi(1)(1) = \varphi_0(1) = 1$ and that $(au, v) = \psi(au)(v) = \psi(u)(\varkappa(a)v) = (u, \varkappa(a)v)$ for all $a \in A_q(\mathfrak{g})$. Such a form is unique since for any $f, g \in U_q(\mathfrak{n}^-)$ one has $(f,g) = (1, \varkappa(f)g) = (\varkappa^{-1}(\varkappa(f)g)1, 1)$ and $\varkappa^{-1}(\varkappa(f)g)1$ is a multiple of 1 whose value determines (f,g). Since $\varkappa^2 = 1d$ it follows that the bilinear form $(u,v)' := \psi(v)u$ has the same properties as (u,v) and hence coincides with (u,v). Thus $(\,,)$ is symmetric. Finally the kernel of $(\,,)$ is obviously a proper submodule of $U_q(\mathfrak{n}^-)$, so by 5.3.7 the form is non-degenerate.

5.3.10 One may define an action of $U_q(\mathfrak{g})$ on $U_q(\mathfrak{n}^-)\tau(\lambda)$ by the Leibnitz rule and the given adjoint action of $U_q(\mathfrak{g})$ on $U_q(\mathfrak{n}^-)$ if it can be specified how generators should act on $\tau(\lambda)$. For this set

$$(\operatorname{ad}_\lambda e_i)\tau(\lambda) = 0 , \quad (\operatorname{ad}_\lambda t_i)\tau(\lambda) = q^{-1/2(\lambda,\alpha_i)}\tau(\lambda)$$

and

$$(\operatorname{ad}_\lambda f_i)\tau(\lambda) = (1 - q^{-(\lambda,\alpha_i)}) f_i \tau(\lambda) .$$

It remains to verify that $\operatorname{ad}_\lambda$ defines a homomorphism of $U_q(\mathfrak{g})$. For this it is enough to verify that the relations on generators expressed in terms of $\operatorname{ad}_\lambda$ vanish on $\tau(\lambda)$. This is quite trivial except for the relation $e_i f_i - f_j e_i = \delta_{ij}(t_i - t_1^{-1})/(q_i - q_i^{-1})$. In fact

$$\begin{aligned}
((\operatorname{ad}_\lambda e_i)(\operatorname{ad}_\lambda f_j) - (\operatorname{ad}_\lambda f_j)(\operatorname{ad}_\lambda e_i))\tau(\lambda) &= (\operatorname{ad}_\lambda e_i)(\operatorname{ad}_\lambda f_j)\tau(\lambda) \\
&= (1 - q^{-(\lambda,\alpha_i)}) \operatorname{ad}_\lambda e_i (f_j \tau(\lambda)) \\
&= (1 - q^{-(\lambda,\alpha_i)})(\operatorname{ad}_\lambda e_i) f_j (\operatorname{ad}_\lambda t_i^{-1})\tau(\lambda) , \quad \text{by the Leibnitz rule} \\
&= (q^{\frac{1}{2}(\lambda,\alpha_i)} - q^{-\frac{1}{2}(\lambda,\alpha_i)}) \frac{(-\delta_{ij})\tau(\lambda)}{(q_i - q_i^{-1})} = \delta_{ij} \left(\frac{\operatorname{ad}_\lambda t_i - \operatorname{ad}_\lambda t_i^{-1}}{q_i - q_i^{-1}} \right) \tau(\lambda)
\end{aligned}$$

as required.

Proposition. *With the above action, $U_q(\mathfrak{n}^-)\tau(\lambda)$ is isomorphic to $\delta M(-\frac{1}{2}\lambda)$.*

Since the e_i action is unchanged it follows by 5.3.6 that $\delta(U_q(\mathfrak{n}^-)\tau(\lambda))$ is generated over $U_q(\mathfrak{n}^-)$ by a weight vector of weight $-\frac{1}{2}\lambda$. As before this gives an isomorphism $\delta(U_q(\mathfrak{n}^-)\tau(\lambda)) \xrightarrow{\sim} M(-\lambda/2)$. Hence the assertion of the proposition.

Remark. Of course one may identify $U_q(\mathfrak{n}^-)\tau(\lambda)$ with $U_q(\mathfrak{n}^-)$ and this gives the required twisted action on the latter.

5.3.11 Consider $U_q(\mathfrak{n}^-)$ as an $A_q(\mathfrak{g})$ module and fix $i \in \{1, 2, \ldots, n\}$. It is clear that e_i' is a nilpotent endomorphism of $U_q(\mathfrak{n}^-)$. From $e_i'(f_i) = 1$ and the Leibnitz rule one obtains $e_i'(f_i^n) = q_i^{-(n-1)}[n]f_i^{n-1}$ and so

$$e_i'^{(n)}(f_i^{(n)}) = q_i^{-\frac{1}{2}n(n-1)}$$

It then results from the Leibnitz rule and induction on nilpotence degree that

Lemma

$$U_q(\mathfrak{n}^-) = \oplus_{n\geq 0} f_i^{(n)} \ker e_i' \ .$$

5.3.12 By 5.3.11 one may define endomorphisms $\tilde{e}_i, \tilde{f}_i$ of $U_q(\mathfrak{n}^-)$ by

$$\tilde{e}_i(f_i^{(n)}u) = f_i^{(n-1)}u \ , \qquad \tilde{f}_i(f_i^{(n)}u) = f_i^{(n+1)}u$$

for all $u \in \ker e_i'$. Obviously $\tilde{e}_i\tilde{f}_i = 1$.

5.3.13 A crystal basis for $U_q(\mathfrak{n}^-)$ is by definition a pair (L, B) satisfying conditions (A1)-(A6) of 5.1.3. However one should note that these axioms simplify in the present situation. Thus the relation $\tilde{e}_i\tilde{f}_i = 1$ and (4) forces $\tilde{f}_iB \subset B$. Again one does not need weight space decomposition in developing the analogue of 5.1.8 for $U_q(\mathfrak{n}^-)$. Then the crystal basis for $U_q(\mathfrak{n}^-)$ determined in the next section shows that axiom (A5) may be omitted. Again the proviso in the analogue of 5.1.10 for $U_q(\mathfrak{n}^-)$ is unnecessary.

Take A as in 5.1.3. Define a lattice $L(\infty)$ in $U_q(\mathfrak{n}^-)$ to be all A linear combinations of the $\tilde{f}_{i_1}\tilde{f}_{i_2}\cdots\tilde{f}_{i_n}.1 : n \in \mathbb{N}, i_1, i_2, \ldots, i_n \in \{1, 2, \ldots, \ell\}$, and let $B(\infty)$ denote the set of their images in $L(\infty)/qL(\infty)$. The main result of the next section is that $(L(\infty), B(\infty))$ is a crystal basis for $U_q(\mathfrak{n}^-)$. At the same time it is shown that $V(\lambda) : \lambda \in P^+(\pi)$ admits $(L(\lambda), B(\lambda))$ as a crystal basis and that these crystal bases are compatible with the canonical surjection $U_q(\mathfrak{n}^-) \cong U(\lambda) \xrightarrow{\pi_\lambda} V(\lambda)$. More precisely one has

(i) $\pi_\lambda(L(\infty)) = L(\lambda)$ *which hence factors to a map* $\bar{\pi}_\lambda : L(\infty)/qL(\infty) \twoheadrightarrow L(\lambda)/qL(\lambda)$

(ii) $B(\lambda) \xleftarrow{\sim} \{b \in B(\infty) \mid \bar{\pi}_\lambda(b) \neq 0\}$

(iii) $\tilde{f}_i$ *commutes with* $\bar{\pi}_\lambda$.
(iv) *For all* $b \in B(\infty)$ *satisfying* $\bar{\pi}_\lambda(b) \neq 0$ *one has* $\tilde{e}_i\bar{\pi}_\lambda(b) = \bar{\pi}_\lambda(\tilde{e}_i b)$.

Let u_∞ denote the image of 1 in $B(\infty)$ and u_λ a generator of $L(\lambda)_\lambda$ (as an A module) which is also identified as the unique element of $B(\lambda)$ of weight λ. Observe that $B(\infty)$ becomes an upper normal crystal by taking u_∞ to have weight 0. In the language of crystals (ii)-(iv) express the fact that there is an embedding $\psi : B(\lambda) \longrightarrow B(\infty) \otimes T_\lambda$ defined by $\psi(u_\lambda) = u_\infty \otimes t_\lambda$ commuting with the $\tilde{e}_i$. Indeed by (ii) given $b' \in B(\lambda)$ one can write $b' = \bar{\pi}_\lambda(b)$ for a unique $b \in B(\infty)$. Set $\psi(b') = b \otimes t_\lambda$. Then ψ is obviously injective and (iv) expresses the fact that ψ commutes with $\tilde{e}_i$. Again if $0 \neq \tilde{f}_i b' = \tilde{f}_i \bar{\pi}_\lambda(b) = \bar{\pi}_\lambda(\tilde{f}_i b)$ by (iii) one has $\psi(\tilde{f}_i b') = \tilde{f}_i b \otimes t_\lambda = \tilde{f}_i(b \otimes t_\lambda) = \tilde{f}_i \psi(b')$ as required. Actually one may do better. Let S_λ denote the crystal consisting of a single element s_λ of weight λ satisfying $\varepsilon_i(s_\lambda) = -(\alpha_i^\vee, \lambda)$, $\tilde{e}_i s_\lambda = \tilde{f}_i s_\lambda = 0$, for all i.

Corollary. *Take* $\lambda \in P^+(\pi)$. *Then the map* $\psi : B(\lambda) \longrightarrow B(\infty) \otimes S_\lambda$ *defined by* $\psi(\bar{\pi}_\lambda(b)) = b \otimes s_\lambda$ *when* $\bar{\pi}_\lambda(b) \neq 0$, *is a strict embedding.*

Suppose $\bar{\pi}_\lambda(b) \neq 0$. Then $\varphi_i(b) := \varepsilon_i(b) + wt_i(b) = \varepsilon_i(\bar{\pi}_\lambda(b)) + wt_i(\bar{\pi}_\lambda(b)) - (\alpha_i^\vee, \lambda) = \varphi_i(\bar{\pi}_\lambda(b)) - (\alpha_i^\vee, \lambda)$, so $\varphi_i(b) - \varepsilon_i(s_\lambda) = \varphi_i(\bar{\pi}_\lambda(b)) \geq 0$. Hence $\varepsilon_i(b \otimes s_\lambda) = \max\{\varepsilon_i(b), \varepsilon_i(s_\lambda) - \varphi_i(b) + \varepsilon_i(b)\} = \varepsilon_i(b)$. Moreover $\tilde{e}_i(b \otimes s_\lambda) = \tilde{e}_i b \otimes s_\lambda$ whilst $\tilde{f}_i(b \otimes s_\lambda) = \tilde{f}_i b \otimes s_\lambda$ unless $\varphi_i(\bar{\pi}_\lambda(b)) = 0$ in which case it equals $b \otimes \tilde{f}_i s_\lambda = 0$. Yet $\varphi_i(\bar{\pi}_\lambda(b)) = 0 \iff \tilde{f}_i(\bar{\pi}_\lambda(b)) = 0$ and so $\tilde{e}_i$, $\tilde{f}_i$ commute with ψ.

5.3.14 Take $\lambda \in P^+(\pi)$. Then for the twisted action $\operatorname{ad}_{-2\lambda}$ on $U_q(\mathfrak{n}^-)$ it follows from 5.3.10 that $V(\lambda)$ identifies with $(\operatorname{ad}_{-2\lambda} U_q(\mathfrak{g}))1$. Asymptotically the action of $\operatorname{ad}_{-2\lambda} f_i$ becomes multiplication by f_i. This may be expressed more precisely as follows. Let $L'(\infty)$ denote the A linear span of monomials in the $f_j : j = 1, 2, \ldots, \ell$.

Lemma. *Fix* $i \in \{1, 2, \ldots, \ell\}$ *and* $a_\xi \in L'(\infty)$ *of weight* ξ. *Fix* $k \in \mathbb{N}$ *and assume that* $r := (\alpha_i^\vee, \xi + \lambda - k\alpha_i) \geq 0$. *Then*

$$(\operatorname{ad}_{-2\lambda} f_i)^{n+1} a_\xi = f_i^{n+1} a_\xi \bmod q_i^r L'(\infty)$$

for all $n \in \{0, 1, 2, \ldots, k\}$.

By definition of the twisted action,

$$\begin{aligned}(\operatorname{ad}_{-2\lambda} f_i) a_\xi &= (\operatorname{ad} f_i) a_\xi + (t_i a_\xi t_i^{-1})(\operatorname{ad}_{-2\lambda} f_i)1 \\ &= (\operatorname{ad} f_i) a_\xi + q_i^{(\alpha_i^\vee, \xi)}(1 - q_i^{(\alpha_i^\vee, \lambda)}) a_\xi f_i \,.\end{aligned}$$

Whilst

$$f_i a_\xi = (\operatorname{ad} f_i) a_\xi + q_i^{(\alpha_i^\vee, \xi)} a_\xi f_i \,.$$

Hence

$$(\mathrm{ad}_{-2\lambda} f_i)a_\xi = f_i a_\xi - q_i^{(\alpha_i^\vee,\xi+\lambda)} a_\xi f_i = f_i a_\xi \bmod q_i^r L'(\infty) ,$$

from which the assertion results by an easy induction argument.

5.3.15 From 5.3.1 and 5.3.14, the compatibility of crystal bases discussed in 5.2.13 can be expected to hold in some asymptotic sense. To make this precise first write $V(\infty) = U_q(\mathfrak{n}^-)$ and $Q^-(\pi) = -\mathbb{N}\pi$. Given $\xi = \sum k_i\alpha_i \in Q^-(\pi)$ set $|\xi| = -\sum k_i$. Then for each $k \in \mathbb{N}$, set $Q_k^- = \{\xi \in Q^-(\pi) \,\big|\, |\xi| \le k\}$ and define

$$L_k'(\infty) = \bigoplus_{\xi\in Q_k^-} L'(\infty) , \quad V_k(\infty) = \bigoplus_{\xi\in Q_k^-} U_q(\mathfrak{n}^-)_\xi .$$

Similarly define $L_k(\infty), V_k(\lambda), L_k(\lambda), B_k(\infty), B_k(\lambda)$.

Fix $k \in \mathbb{N}$. It is clear that $e_iV_k(\infty) \subset V_k(\infty)$ for all i. Now $V_k(\infty)$ is finite dimensional and generated over $\mathbb{Q}(q)$ by either $L_{k+1}'(\infty)$ or $L_{k+1}(\infty)$. Hence there is a positive integer s such that $q^s L_{k+1}'(\infty) \subset L_{k+1}(\infty) \subset q^{-s}L_{k+1}'(\infty)$ and such that in the decomposition 5.3.11 of $a \in L_{k+1}'(\infty)$, namely

$$(*) \qquad a = \sum f_i^{(n)} a_n : a_n \in \ker e_i'$$

one has $a_n \in q^{-s}L_{k+1}'(\infty)$. Note also that $\mathrm{ad}_{-2\lambda} e_i$ is independent of λ, whilst by 5.3.14 for λ satisfying $(\alpha_i^\vee, \xi + \lambda - k\alpha_i) \ge 0$, $\forall \xi \in Q_k^-$ one has $(\mathrm{ad}_{-2\lambda} f_i)L_k'(\infty) \subset L_{k+1}'(\infty)$. Thus one may also choose s so that $(\mathrm{ad}_{-2\lambda} x)L_k(\infty) \subset q^{-s+1}L_{2k}(\infty)$ for all $\lambda \in P^+(\pi)$ and all x monomial of degree $\le k$ in the e_i, f_i.

Now let $\tilde{e}_{i,\lambda}$, $\tilde{f}_{i,\lambda}$ denote the linear maps of $V(\lambda) \hookrightarrow U_q(\mathfrak{n}^-)$ defined as in 5.3.12 but with respect to the twisted adjoint action $\mathrm{ad}_{-2\lambda}$ on $U_q(\mathfrak{n}^-)$. Since $\mathrm{ad}_{-2\lambda} e_i$ is independent of λ it follows from 5.3.1 that these operators differ from $\tilde{e}_i$, $\tilde{f}_i$ only through the dependence of $\mathrm{ad}_{-2\lambda} f_i$ on λ.

Lemma. *For each $a \in L_k(\infty)$ and each $i \in \{1, 2, \dots, \ell\}$ one has*

(i) $\tilde{f}_{i,\lambda} a = \tilde{f}_i a \bmod qL_{k+1}(\infty)$,
(ii) $\tilde{e}_{i,\lambda} a = \tilde{e}_i a \bmod qL_{k+1}(\infty)$,

given that $(\alpha_i^\vee, \lambda) > 3s + 4k$.

Since $(\alpha_i^\vee, \alpha_j) \le 0$ if $j \ne i$ and equals 2 otherwise, it follows that $r := (\alpha_i^\vee, \lambda+\xi-k\alpha_i) \ge (\alpha_i^\vee, \lambda) - 4k > 3s$. Consider (i) and take $a \in L_k(\infty)$. Then $a \in q^{-s}L_k'(\infty)$ and so in the decomposition of $(*)$ one has $a_n \in q^{-2s}L_k'(\infty)$. Then by 5.3.14

$$\begin{aligned}\tilde{f}_{i,\lambda}(f_i^{(n)}a_n) &= \tilde{f}_{i,\lambda}(\mathrm{ad}_{-2\lambda} f_i)^{(n)} a_n \bmod q^{r-2s} L_{k+1}'(\infty) ,\\ &= (\mathrm{ad}_{-2\lambda} f_i)^{(n+1)} a_n \bmod q^{r-2s} L_{k+1}'(\infty) ,\\ &= f_i^{(n+1)} a_n \bmod q^{r-2s} L_{k+1}'(\infty) ,\\ &= \tilde{f}_i(f_i^{(n)}a_n) \bmod qL_{k+1}(\infty) ,\end{aligned}$$

since $r - 3s > 0$. Summing over n gives the required assertion. The proof of (ii) is the same.

5.3.16 View $V(\lambda)$ as a quotient of $U_q(\mathfrak{n}^-)$ via the map $\pi_\lambda : a \mapsto au_\lambda$. Write $\lambda \gg 0$ if $(\alpha_i^\vee, \lambda) \gg 0, \forall i$.

Corollary. *Fix $k \in \mathbb{N}$. If $\lambda \gg 0$, then for all $a \in L_k(\infty)$,*

(i) $\tilde{f}_i(au_\lambda) = (\tilde{f}_i a)u_\lambda \mod q\pi_\lambda(L_{k+1}(\infty))$,
(ii) $\tilde{e}_i(au_\lambda) = (\tilde{e}_i a)u_\lambda \mod q\pi_\lambda(L_k(\infty))$,
(iii) $\pi_\lambda(L_{k+1}(\infty)) = L_{k+1}(\lambda)$.

(i) results from 5.3.15 given that $\tilde{f}_i(au_\lambda) = (\tilde{f}_{i,\lambda}a)u_\lambda \mod q\pi_\lambda(L_{k+1}(\infty))$. For this it suffices to show that $x(au_\lambda) = ((\mathrm{ad}_{-2\lambda}\, x)a)u_\lambda \mod q(L_{2k}(\infty))u_\lambda$ for any monomial x in the e_i, f_j of degree $\leq k$. Let $\theta_\lambda : V(\lambda) \longrightarrow U_q(\mathfrak{n}^-)$ be the embedding defined by 5.3.10. Writing $(\mathrm{ad}_{-2\lambda}\, a)1 = \psi_\lambda(a)$ one has $\theta_\lambda(xau_\lambda) = (\mathrm{ad}_{-2\lambda}\, x)\psi_\lambda(a) = \theta_\lambda((\psi_\lambda^{-1}((\mathrm{ad}_{-2\lambda}\, x)\psi_\lambda(a)))u_\lambda)$, assuming ψ_λ invertible on $L_{2k}(\infty)$. As in 5.3.15 there exists $s \in \mathbb{N}$ satisfying $(\mathrm{ad}_{-2\lambda}\, x)L_k(\infty) \subset q^{-s+1}L_{2k}(\infty)$, for all $\lambda \in P^+(\pi)$. Again choosing $\lambda \gg 0$ one can ensure for all $b \in L_{2k}(\infty)$ that $\psi_\lambda(b) = b \mod q^m L_{2k}(\infty)$. Then

$$\psi_\lambda^{-1}((\mathrm{ad}_{-2\lambda}\, x)\psi_\lambda(a)) = (\mathrm{ad}_{-2\lambda}\, x)a \mod qL_{2k}(\infty)$$

giving the required assertion. One also obtains $\tilde{e}_i(au_\lambda) = (\tilde{e}_{i,\lambda}a)u_\lambda \mod q\pi_\lambda(L_k(\infty))$ and hence (ii).

Finally set $M = \pi_\lambda(L_{k+1}(\infty))$, $N = L_{k+1}(\lambda)$. Then (i) gives $M = N \mod qM$, that is $M \subset N + qM$, $N \subset M$. Since M is finitely generated over the local ring A, Nakayama's lemma gives $M = N$. Hence (iii).

5.3.17 Since π_λ becomes injective on $L_k(\infty)$ for $\lambda \gg 0$ it is clear that 5.3.16 gives

Corollary. *Fix $k \in \mathbb{N}$, then for $\lambda \gg 0$ one has*

(i) $L_k(\infty) \xrightarrow{\sim} L_k(\lambda)$,
(ii) $B_k(\infty) \xrightarrow{\sim} B_k(\lambda)$.

5.3.18 Observe that $\tilde{\varkappa}(e_i) = q_i^{-1}t_i^{-1}f_i$, $\tilde{\varkappa}(f_i) = q_i^{-1}t_ie_i$, $\tilde{\varkappa}(t_i) = t_i$ extends to an involutory antiautomorphism of $U_q(\mathfrak{g})$. Fixing $u_\lambda \in V(\lambda)_\lambda$ there is a unique symmetric form satisfying $(u_\lambda, u_\lambda) = 1$ and $(au_\lambda, bu_\lambda) = (u_\lambda, \tilde{\varkappa}(a)bu_\lambda)$ for all $a, b \in U_q(\mathfrak{g})$ defined as in 3.4.10. Its kernel being a submodule of $V(\lambda)$ is zero.

Lemma. *For all $\xi = -\sum n_i\alpha_i \in Q^-(\pi)$ and $a, b \in U_q(\mathfrak{n}^-)$ there exists a polynomial f in the $q_i^{(\alpha_i^\vee,\lambda)}$ with coefficients in $\mathbb{Q}(q)$ such that*

(i) $(au_\lambda, bu_\lambda) = \frac{f}{\Pi_i(1-q_i^2)^{n_i}}$
(ii) *For $\lambda \gg 0$, $f = (a, b)$ up to a unit in A equal to 1 at $q = 0$.*

Both are obvious for $|\xi| = 0$. When $|\xi| = k > 0$ one can write $b = f_i c$ with $c \in V_{k-1}(\lambda)$ and then

$$\begin{aligned}(au_\lambda, bu_\lambda) &= q_i^{-1}(t_i e_i a u_\lambda, cu_\lambda)\\ &= q_i^{-1}\left(\frac{t_i^2 e_i''(a) - e_i'(a)}{(q_i - q_i^{-1})} u_\lambda, cu_\lambda\right)\\ &= (1-q_i^2)^{-1}\left[(e_i'(a)u_\lambda, cu_\lambda) - q_i^{2(\alpha_i^\vee, \lambda+\xi+\alpha_i)}(e_i''(a)u_\lambda, cu_\lambda)\right]\end{aligned}$$

from which the first part follows by induction. Again if $\lambda \gg 0$, then

$$f|_{q=0} = (e_i'(a)u_\lambda, cu_\lambda)|_{q=0} = (e_i'(a), c) = (a, f_i c) = (a, b)$$

which establishes the second part.

5.3.19 Set

$$\begin{aligned}L(\infty)^* &= \{a \in U_q(\mathfrak{n}^-) | (a, L(\infty)) \subset A\}\\ L(\lambda)^* &= \{u \in V(\lambda) | (u, L(\lambda)) \subset A\}\ .\end{aligned}$$

Since $L(\lambda)$ is a free, finite rank A module it follows that $L(\lambda) = \{u \in V(\lambda) | (u, L(\lambda)^*) \subset A\}$. By weight space decomposition a similar result holds for $L(\infty)$.

Lemma. *Take $k \in \mathbb{N}$. Then for $\lambda \in P^+(\pi)$ sufficiently dominant $\pi_\lambda(L(\infty)^*_\xi) = L(\lambda)^*_{\lambda+\xi}$, for all $\xi \in \mathbb{Q}_k^-$.*

This is immediate from 5.3.17 and 5.3.18.

5.3.20 Recall that $V(\lambda+\mu)$ is a direct summand of $V(\lambda) \otimes V(\mu)$ and let $\Phi(\lambda, \mu) : V(\lambda+\mu) \longrightarrow V(\lambda) \otimes V(\mu)$, $\Psi(\lambda, \mu) : V(\lambda) \otimes V(\mu) \longrightarrow V(\lambda+\mu)$ be the $U_q(\mathfrak{g})$ homomorphisms defined by $\Phi(\lambda, \mu)(u_{\lambda+\mu}) = u_\lambda \otimes u_\mu$ and $\Psi(\lambda, \mu)(u_\lambda \otimes u_\mu) = u_{\lambda+\mu}$. Because they are homomorphisms they commute with $\tilde{e}_i, \tilde{f}_i$. Define $S(\lambda, \mu)$: $V(\lambda) \otimes V(\mu) \longrightarrow V(\lambda)$ to be the linear map satisfying

$$S(\lambda, \mu)(u \otimes v) = \begin{cases} u & \text{if } v = v_\mu \\ 0 & \text{if } v \in \sum f_i V(\mu)\ . \end{cases}$$

Since $f_i(u \otimes v) = f_i u \otimes v + t_i u \otimes f_i v$ it follows that $S(\lambda, \mu)$ commutes with f_i. However it may fail to commute with $\tilde{f}_i$. (See however 5.4.7).

Take the product contravariant form on $V(\lambda) \otimes V(\mu)$. Then

$$(*) \qquad (\Psi(\lambda, \mu)(w), u) = (w, \Phi(\lambda, \mu)u)$$

for all $w \in V(\lambda) \otimes V(\mu)$, $u \in V(\lambda+\mu)$ since this holds for $w = u_\lambda \otimes u_\mu$, $u = u_{\lambda+\mu}$ and by contravariance in general.

5.4 The Grand Loop

5.4.1 The existence of crystal bases for $V(\lambda)$; $\lambda \in P^+(\pi)$ and $U_q(\mathfrak{n}^-)$ together with their compatibility is proved by induction on $|\xi|$ as ξ runs over $Q^-(\pi)$. This involves a number of steps described below. Each of these are assumed to apply to all $i = 1, 2, \dots, \ell$ and all $\lambda, \mu \in P^+(\pi)$ where appropriate. Moreover λ^ε designates ∞ if $\varepsilon = \infty$ and $\lambda \in P^+(\pi)$ if $\varepsilon = 0$.

Let $(L(\lambda) \otimes L(\mu))_k$ (resp. $(B(\lambda) \otimes B(\mu))_k$) denote

$$\sum_{j=0}^{k} L_{k-j}(\lambda) \otimes L_j(\mu) \quad (\text{resp. } \bigcup_{j=0}^{k} B_{k-j}(\lambda) \otimes B_j(\mu)) .$$

Define the following induction steps.

$S_1^\varepsilon(k)$ $\tilde{e}_i(L_k(\lambda^\varepsilon)) \subset L(\lambda^\varepsilon)$.

$S_2(k)$ $\pi_\lambda(L_k(\infty)) = L_k(\lambda)$.

$S_3^\varepsilon(k)$ $B_k(\lambda^\varepsilon)$ *is a basis for* $L_k(\lambda)^\varepsilon / qL_k(\lambda^\varepsilon)$.

$S_4(k)$ $\tilde{f}_i(au_\lambda) = (\tilde{f}_i a)u_\lambda \bmod qL_k(\lambda)$ *for all* $a \in L_{k-1}(\infty)$.

Note that $S_4(k)$ immediately gives $\pi_\lambda(L_k(\infty)) \subset L_k(\lambda)$.

$S_5^\varepsilon(k)$ $\tilde{e}_i B_k(\lambda^\varepsilon) \subset B_{k-1}(\lambda^\varepsilon) \cup \{0\}$.

$S_6^1(k)$ $\Phi(\lambda,\mu)L_k(\lambda+\mu) \subset (L(\lambda) \otimes L(\mu))_k$.

$S_6^2(k)$ $\Phi(\lambda,\mu)B_k(\lambda+\mu) \subset (B(\lambda) \otimes B(\mu))_k$.

$S_7^1(k)$ $\Psi(\lambda,\mu)(L(\lambda) \otimes L(\mu))_k \subset L_k(\lambda+\mu)$.

$S_7^2(k)$ $\Psi(\lambda,\mu)(B(\lambda) \otimes B(\mu))_k \subset B_k(\lambda+\mu) \cup \{0\}$.

$S_8(k)$ *Let* $\bar{\pi}_\lambda$ *be the surjection* $L_k(\infty)/qL_k(\infty) \twoheadrightarrow L_k(\lambda)/qL_k(\lambda)$ *defined by* $S_4(k)$. *Then* $\bar{\pi}_\lambda$ *induces a bijection of* $B_k(\infty) \setminus \bar{\pi}_\lambda^{-1}(0)$ *onto* $B_k(\lambda)$.

$S_9^+(k)$ $\tilde{e}_i \bar{\pi}_\lambda(b) = \bar{\pi}_\lambda(\tilde{e}_i b)$ *for all* $b \in B_k(\infty) \setminus \bar{\pi}_\lambda^{-1}(0)$.

$S_9^-(k)$ $\tilde{f}_i \bar{\pi}_\lambda(b) = \bar{\pi}_\lambda(\tilde{f}_i b)$ *for all* $b \in B_k(\infty)$.

$S_{10}^\varepsilon(k)$ *For all* $b \in B_k(\lambda^\varepsilon)$, $b' \in B_{k-1}(\lambda^\varepsilon)$ *one has* $b = \tilde{f}_i b' \iff b' = \tilde{e}_i b$.

It is perhaps worth recalling that by definition of $B(\lambda^\varepsilon)$ one has $\tilde{f}_i B(\lambda^\varepsilon) \subset B(\lambda^\varepsilon)$. As noted in 5.3.13 the relation $\tilde{e}_i \tilde{f}_i = 1$ being valid in $L(\infty)$ forces $\tilde{f}_i$ to be injective on $B(\infty)$. Note that $S_2(k)$, $S_4(k)$ imply $S_9^-(k)$. Again $S_1^\infty(k)$ results from $S_1^0(k)$ and $S_2(k)$ by 5.3.16 and 5.3.17 (see 5.4.10).

In the proof one may assume that the fundamental weights ϖ_j, defined by $(\alpha_i^\vee, \varpi_j) = \delta_{ij}$, all belong to $P^+(\pi)$. Given $u \in L_k(\lambda^\varepsilon)$ let $\bar{u}$ denote its image in $L_k(\lambda^\varepsilon)/qL_k(\lambda^\varepsilon)$. If $u \in L_k(\lambda^\varepsilon)$ then $\bar{u} \in B_k(\lambda^\varepsilon)$ is sometimes written without the bar. For the induction to proceed one needs $2k \geq k+1$. Yet $S_1 - S_{10}$ are easily checked for $k \leq 1$.

5.4.2 It is convenient to supplement $S_1 - S_{10}$ by further steps of the induction loop. These summarize to some extent what has already been observed. Then the proof becomes almost a matter of constructing the appropriate flow

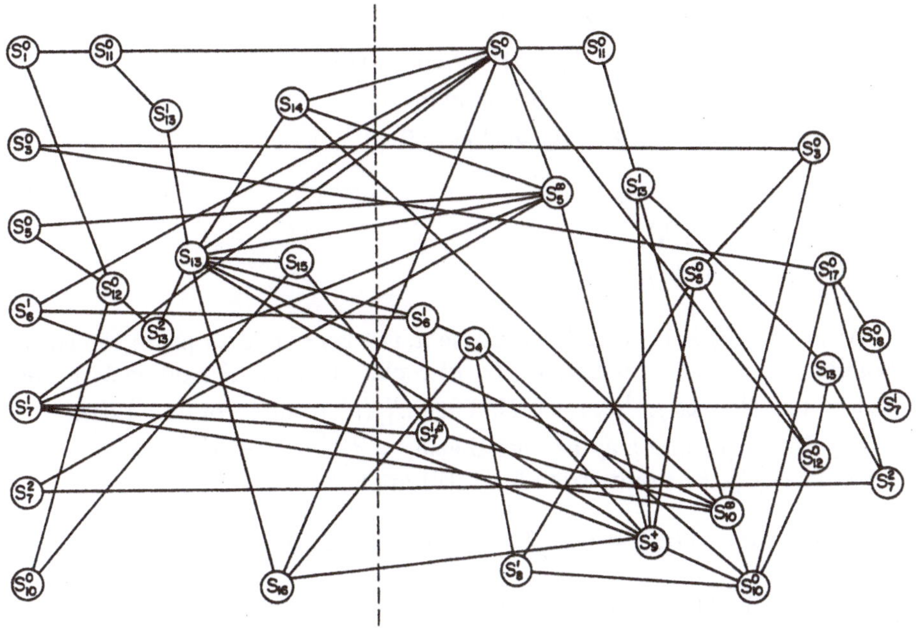

Fig. 1. Flow Chart for the Grand Loop. The induction proceeds from left to right with k increasing by 1 as the vertical dashed line is crossed

chart (namely, Fig. 1 above). Here some steps not needed in the induction process have been omitted, for example S_6^2.

Recall 5.3.11 (resp. 5.1.2) and write $u \in L_k(\lambda^\varepsilon)$ as $u = \sum f_i^{(n)} u_n : u_n \in \ker e_i' \cap U_q(\mathfrak{n}^-)$ (resp. $u_n \in \ker e_i \cap V(\lambda)$). Then

$S_{11}^\varepsilon(k)$ *$u_n \in L(\lambda^\varepsilon)$ for all n.*

$S_{12}^\varepsilon(k)$ *If $\bar{u} \in B_k(\lambda^\varepsilon)$ then $u = f_i^{(n_0)} u_{n_0} \bmod qL(\lambda^\varepsilon)$ where $n_0 = \max\{n \in \mathbb{N} | \tilde{e}_i^n u \notin qL(\lambda^\varepsilon)\}$.*

Note that in S_{11}^0, S_{12}^0 one must take care to choose the u_n as prescribed in 5.1.10. Observe that the argument of 5.1.10 gives the

Lemma

(i) $S_1^\varepsilon(k) \Longrightarrow S_{11}^\varepsilon(k)$.

(ii) $S_j^\varepsilon(k) : j = 1, 5, 10 \Longrightarrow S_{12}^\varepsilon(k)$.

5.4.3 Recall (5.1.11) the definitions of $\varphi_i(b)$, $\varepsilon_i(b)$ for $b \in B(\lambda)$. Let $S_{13}(k)$ denote the following collection of steps.

$S_{13}^1(k)$ *$\tilde{e}_i, \tilde{f}_i$ map $L_k(\lambda) \otimes L_k(\mu)$ to $L(\lambda) \otimes L(\mu)$.*

$S_{13}^2(k)$ *The relations of* 5.1.12 (ii) *hold for $b_1 \in B_k(\lambda)$, $b_2 \in B_k(\mu)$.*

Take $b_1 \in B_k(\lambda)$, $b_2 \in B_k(\mu)$. Note that $S^2_{13}(k)$ implies (cf. 5.2.4) that

$S^3_{13}(k)$ $\tilde{e}_i(b_1 \otimes b_2) \neq 0 \Longrightarrow b_1 \otimes b_2 = \tilde{f}_i \tilde{e}_i(b_1 \otimes b_2)$
$S^4_{13}(k)$ *If* $\tilde{e}_i(b_1 \otimes b_2) = 0$ *for all* i, *then* $b_1 = u_\lambda$.
$S^5_{13}(k)$ $\tilde{f}_i(b_1 \otimes u_\mu) = \tilde{f}_i b_1 \otimes u_\mu$ *or* $\tilde{f}_i b_1 = 0$.

Lemma
(i) $S^0_{11}(k) \Longrightarrow S^1_{13}(k)$,
(ii) $S^0_{12}(k) \Longrightarrow S^2_{13}(k)$.

(i) By $S^0_{11}(k)$ it is enough to choose $u, v \in \ker e_i$ and to prove the corresponding assertions with $L_k(\lambda) \otimes L_k(\mu)$ replaced by the A module M generated by the $f_i^{(r)} u \otimes f_i^{(s)} v : r, s \in \mathbb{N}$. Then $S^1_{13}(k)$ results from 5.1.12(i).

(ii) By $S^0_{12}(k)$ one may assume that $b_1 = f_i^{(r)} u$, $b_2 = f_i^{(s)} v$ with u, v as in (i). Then $S^2_{13}(k)$ results from 5.1.12(ii) applied to M.

5.4.4 Lemma. *Take* $j = 1, 2$. *Then* $S^j_6(k)$, $S^j_{13}(k)$ *imply* $S^j_6(k+1)$.

Take $j = 1$. Then by definition

$$L_{k+1}(\lambda + \mu) = \sum_{i=1}^{\ell} \tilde{f}_i L_k(\lambda + \mu) \, .$$

Since $\Phi(\lambda, \mu)$ commutes with the $\tilde{f}_i$ by 5.3.20 the result follows. The case $j = 2$ is the same.

5.4.5 Take $i_1, i_2, \ldots, i_{k+1} \in \{1, 2, \ldots, \ell\}$.

$S_{14}(k)$ $\tilde{f}_{i_1} \tilde{f}_{i_2} \ldots \tilde{f}_{i_{k+1}} (u_{\omega_{i_k}} \otimes u_\mu) \in B_k(\omega_{i_k}) \otimes B_k(\mu) \cup \{0\}$.

Lemma. $S^2_{13}(k) \Longrightarrow S_{14}(k)$.

Set $\omega_{i_k} = \omega$. If $i_{k+1} \neq i_k$, then $\tilde{f}_{i_{k+1}} u_\omega = 0$ so

$$\tilde{f}_{i_{k+1}}(u_\omega \otimes u_\mu) = f_{i_{k+1}}(u_\omega \otimes u_\mu) = t_{i_{k+1}} u_\omega \otimes f_{i_{k+1}} u_\mu = u_\omega \otimes \tilde{f}_{i_{k+1}} u_\mu \, .$$

Since $\varphi_{i_k}(u_\omega) = 1$, whilst $\varepsilon_{i_k}(\tilde{f}_{i_{k+1}} u_\mu) = 0$ it follows from $S^2_{13}(1)$ that

$$(*) \qquad \tilde{f}_{i_k} \tilde{f}_{i_{k+1}} (u_\omega \otimes u_\mu) = \tilde{f}_{i_k} u_\omega \otimes \tilde{f}_{i_{k+1}} u_\mu \, .$$

Similar reasoning shows that $(*)$ also holds when $i_{k+1} = i_k$. Applying $S^2_{13}(k)$ to $(*)$ gives the required assertion.

5.4.6 Now consider

$S_{15}(k)$ $(L(\lambda) \otimes L(\mu))_{k+1} = \sum_{i=1}^{\ell} \tilde{f}_i (L(\lambda) \otimes L(\mu))_k + u_\lambda \otimes L_{k+1}(\mu)$.

Lemma. $S^0_{10}(k), S_{13}(k) \Longrightarrow S_{15}(k)$.

Let L (resp. L') denote the left (resp. right) hand side in $S_{15}(k)$. By $S^1_{13}(k)$ one has $L' \subset L$. Take $b_1 \in B_k(\lambda) \setminus \{u_\lambda\}$. Then by definition of $B(\lambda)$ and $S^0_{10}(k)$ there exists $i \in \{1, 2, \dots, \ell\}$ such that $\tilde{e}_i b_1 \neq 0$. Take $b_2 \in B_k(\mu)$ and set $b = b_1 \otimes b_2$. By $S^4_{13}(k)$ one has $\tilde{e}_i b = \tilde{e}_i(b_1 \otimes b_2) \neq 0$. Then by $S^3_{13}(k)$ one has $b = \tilde{f}_i \tilde{e}_i b$. By choice of b_1, this gives

$$L \subset L' + L_{k+1}(\mu) \otimes u_\mu + qL \; .$$

Yet by $S^5_{13}(k)$ one also has $L_{k+1}(\mu) + u_\mu \subset L'$ and so $L \subset L' + qL$. Then Nakayama's lemma implies $L = L'$ as required.

5.4.7 It is obvious that $S(\lambda, \mu)(L(\lambda) \otimes L(\mu)) = L(\lambda)$. Assume $S^1_{13}(k)$ holds.

$S_{16}(k)$ *The diagram*

$$\begin{array}{ccc} (L(\lambda) \otimes L(\mu))_k / q(L(\lambda) \otimes L(\mu))_k & \overset{S(\lambda,\mu)}{\longrightarrow} & L_k(\lambda)/qL_k(\lambda) \\ \downarrow \tilde{f}_i & & \downarrow \tilde{f}_i \\ (L(\lambda) \otimes L(\mu))_{k+1} / q(L(\lambda) \otimes L(\mu))_{k+1} & \overset{S(\lambda,\mu)}{\longrightarrow} & L_{k+1}(\lambda)/qL_{k+1}(\lambda) \end{array}$$

commutes.

Lemma. $S_{13}(k) \Longrightarrow S_{16}(k)$.

Take $w \in (L(\lambda) \otimes L(\mu))_k$. It is enough to show that $\tilde{f}_i S(\lambda, \mu) w = S(\lambda, \mu) \tilde{f}_i w \bmod qL(\lambda)$. Define M as in the proof of 5.4.3(i). One may assume $w = f_i^{(n)} u \otimes f_i^{(m)} v \in M$. Then by $S^2_{13}(k)$ one has $\tilde{f}_i w = f_i^{(n+1)} u \otimes f_i^{(m)} v$ or $f_i^{(n)} u \otimes f_i^{(m+1)} v \bmod qM$. Thus both sides above are zero unless $m = 0$ and $v = u_\mu$. In this last case the assertion follows $S^5_{13}(k)$.

5.4.8 Proposition. $S^1_6(k+1),\ S_{16}(k) \Longrightarrow S_4(k+1)$.

Take $a \in L_k(\infty)$. By 5.3.16 taking $\mu \gg 0$ one has

$$(\tilde{f}_i a) u_{\lambda+\mu} = \tilde{f}_i (a u_{\lambda+\mu}) \bmod qL(\lambda + \mu) \; .$$

Apply $\Phi(\lambda+\mu)$ to both sides (which belong to $L_{k+1}(\lambda+\mu)$). Then $S^1_6(k+1)$ combined with $S_{16}(k)$ gives $S_4(k+1)$.

5.4.9 Consider the following weakened version of $S_8(k)$.

$S'_8(k)$ $\bar{\pi}_\lambda(B_k(\infty)) \setminus \{0\} = B_k(\lambda)$.

Lemma
(i) $S_4(k+1) \Longrightarrow S_2(k+1)$.
(ii) $S_4(k+1) \Longrightarrow S'_8(k+1)$.

(i) follows using Nakayama's lemma as in 5.3.16 and then (ii) is immediate.

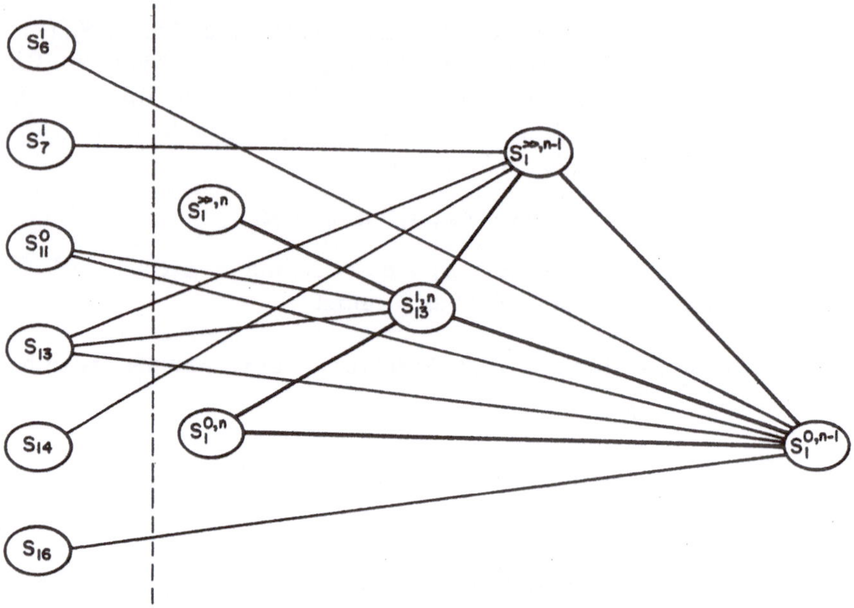

Fig. 2. Flow Chart for the Small Loop. The conventions of Fig. 1 are retained. The input on the extreme left comes from the Grand Loop. The Small Loop indicated by the boldfaced lines gives the conclusion of 5.4.14

5.4.10 To obtain $S_1^0(k+1)$ a "small" induction loop is performed whose flow chart is given in Fig. 2 above.

Let n be an integer ≥ 0. Consider the following refinements of $S_1^\varepsilon(k)$.

$S_1^{\infty,n}(k)$ $\tilde{e}_i L_k(\infty) \subset q^{-n}L(\infty)$.
$S_1^{0,n}(k)$ $\tilde{e}_i L_k(\lambda) \subset q^{-n}L(\lambda)$.
$S_1^{\gg,n}(k)$ $\tilde{e}_i L_k(\mu) \subset q^{-n}L(\mu)$ *for* $\mu \gg 0$.

Lemma. *Assume* $S_2(k+1)$. *Then*

$$S_1^{\infty,n}(k+1) \Longleftrightarrow S_1^{\gg,n}(k+1)\ .$$

Consider $\Longrightarrow$. One has

$$\begin{aligned}\tilde{e}_i L_{k+1}(\mu) &\overset{S_2(k+1)}{=} \tilde{e}_i \pi_\mu(L_{k+1}(\infty)) \overset{5.3.16}{=} \pi_\mu(\tilde{e}_i L_{k+1}(\infty)) \bmod qL(\mu)\\ &\subset q^{-n}\pi_\mu(L_k(\infty)) \bmod qL(\mu),\ \textit{by}\ S_1^{\infty,n}(k+1)\\ &= q^{-n}L_k(\mu),\ \textit{by}\ S_2(k)\ \text{again}\ .\end{aligned}$$

Conversely

$$\tilde{e}_i \pi_\mu(L_{k+1}(\infty)) \overset{S_2(k+1)}{=} \tilde{e}_i L_{k+1}(\mu) \subset q^{-n}L_k(\mu) \overset{S_2(k)}{=} q^{-n}\pi_\mu(L_k(\infty))$$

from which $S_1^{\infty,n}(k+1)$ follows by applying 5.3.17.

5.4.11 One may similarly refine part of $S^1_{13}(k)$ by

$S^{1,n}_{13}(k)$ $\tilde{e}_i(L(\lambda)\otimes L(\mu))_k \subset q^{-n}L(\lambda)\otimes L(\mu)$ *for* $\mu \gg 0$.

Lemma. $S^{\gg,n}_1(k+1), S^{0,n}_1(k+1),\ S^0_{11}(k), S^1_{13}(k) \Longrightarrow S^{1,n}_{13}(k+1)$.

Take $u \in L_r(\lambda)$, $v \in L_s(\mu)$ with $r+s \leq k+1$. It must be shown that $\tilde{e}_i(u\otimes v) \in q^{-n}L(\lambda)\otimes L(\mu)$. When $r,s \leq k$ this is $S^1_{13}(k)$. When $s = k+1$ one may assume that $u = u_\lambda$. Then $\tilde{e}_i v \in q^{-n}L(\mu)$ by $S^{\gg,n}_1(k+1)$ and 5.4.10. Write $v = \sum f_i^{(m)} v_m$ with $v_m \in \ker e_i$. Then $\tilde{e}_i v = \sum f_i^{(m-1)} v_m$, and so by $S^0_{11}(k)$ one obtains $v_m \in q^{-n}L(\mu)$ for $m \geq 1$. Yet

$$\tilde{e}_i(u\otimes v) = \sum_{m\geq 1} \tilde{e}_i(u_\lambda \otimes f_i^{(m)} v_m)$$

and so now the proof may proceed as in 5.4.3(i). The case $r = k+1$ is similar using $S^{0,n}_1(k+1)$ instead of $S^{\gg,n}_1(k+1)$.

5.4.12 Lemma. $S^1_7(k),\ S^1_{13}(k),\ S_{14}(k),\ S^{1,n}_{13}(k+1) \Longrightarrow S^{\gg,n-1}_1(k+1)$.

It must be shown that $\tilde{e}_i\tilde{f}_{i_1}\tilde{f}_{i_2}\dots\tilde{f}_{i_{k+1}}u_\mu \in q^{1-n}L(\mu)$ for $\mu \gg 0$. Set $\omega = \omega_{i_k}$. Then by $S_{14}(k)$ one has

$$w := \tilde{f}_{i_1}\dots\tilde{f}_{i_{k+1}}(u_\omega \otimes u_{\mu-\omega}) = v\otimes v' \bmod q(L(\omega)\otimes L(\mu-\omega))_{k+1}$$

for some $v \in L_k(\omega)$, $v' \in L_k(\mu-\omega)$. Then by $S^1_{13}(k)$ and $S^{1,n}_{13}(k+1)$ one obtains

$$\tilde{e}_i w \in q^{1-n}(L(\omega)\otimes L(\mu-\omega))_k\ .$$

Applying $\Psi(\omega,\mu-\omega)$ the required assertion follows from $S^1_7(k)$.

5.4.13 Lemma. $S^1_6(k),\ S^0_{11}(k),\ S^5_{13}(k),\ S_{16}(k),\ S^{\gg,n-1}_1(k+1),\ S^{0,n}_1(k+1),\ S^{1,n}_{13}(k+1) \Longrightarrow S^{0,n-1}_1(k+1)$.

Take $w = \tilde{f}_{i_1}\tilde{f}_{i_2}\dots\tilde{f}_{i_{k+1}}u_\lambda : \lambda \in P^+(\pi)$ and show that $\tilde{e}_i w \in q^{1-n}L(\lambda)$. By $S^{0,n}_1(k+1)$ one can assume $w \notin qL(\lambda)$. Then by $S^5_{13}(k)$ one has

$$\tilde{f}_{i_1}\tilde{f}_{i_2}\dots\tilde{f}_{i_{k+1}}(u_\lambda\otimes u_\mu) = w\otimes u_\mu \bmod q(L(\lambda)\otimes L(\mu))_{k+1}\ .$$

Now assume $\mu \gg 0$. Then from $S^{\gg,n-1}_1(k+1)$

$$\begin{aligned}\tilde{e}_i\tilde{f}_{i_1}\dots\tilde{f}_{i_{k+1}}(u_\lambda\otimes u_\mu) &= \Phi(\lambda,\mu)(\tilde{e}_i\tilde{f}_{i_1}\dots\tilde{f}_{i_{k+1}}u_{\lambda+\mu})\\ &\in q^{1-n}\Phi(\lambda,\mu)L_k(\lambda+\mu)\\ &\subset q^{1-n}L(\lambda)\otimes L(\mu)\ ,\quad \text{by}\quad S^1_6(k)\ .\end{aligned}$$

Combined with the previous relation and $S^{1,n}_{13}(k+1)$, one deduces that

$$(*)\qquad \tilde{e}_i(w\otimes u_\mu) \in q^{1-n}(L(\lambda)\otimes L(\mu))\ .$$

Write $w = \sum f_i^{(m)} w_m$ with $w_m \in \ker e_i$. Then $\sum f_i^{(m-1)} w_m = \tilde{e}_i w \in q^{-n}L_k(\lambda)$ by $S^{0,n}_1(k+1)$. Hence by $S^0_{11}(k)$ one obtains $w_m \in q^{-n}L(\lambda)$ for

$m > 0$. Let M be the A module generated by the $f_i^{(r)} w_m \otimes f_i^{(s)} u_\mu$ for $m > 0$, $r, s \in \mathbb{N}$. Then by 5.1.12

$$\begin{aligned}\tilde{e}_i(w \otimes u_\mu) &= \sum_{m>0} \tilde{e}_i(f_i^{(m)} w_m \otimes u_\mu) = \sum_{m>0} f_i^{(m-1)} w_m \otimes u_\mu \bmod qM \\ &= \tilde{e}_i w \otimes u_\mu \bmod q^{1-n}(L(\lambda) \otimes L(\mu)) .\end{aligned}$$

Applying $S(\lambda, \mu)$ using $S_{16}(k)$ one concludes from the above and $(*)$ that $\tilde{e}_i w \in q^{1-n} L(\lambda)$ as required.

5.4.14 Take $k \in \mathbb{N}$. Then $S_1^{\infty,n}(k)$, $S_1^{0,n}(k)$ hold for $n \gg 0$, whilst $S_1^{0,0}(k)$ is just $S_1^0(k)$. Hence the

Proposition. $S_6^1(k)$, $S_7^1(k)$, $S_{11}^0(k)$, $S_{13}(k)$, $S_{14}(k)$, $S_{16}(k) \implies S_1^0(k+1)$.

5.4.15 The small loop having been completed, it remains to terminate the grand loop.

Lemma. $S_1^0(k+1)$, $S_5^0(k)$, $S_7^1(k)$, $S_7^2(k)$, $S_{13}^2(k)$, $S_{14}(k) \implies S_5^\infty(k+1)$.

Take $w = \tilde{f}_{i_1} \dots \tilde{f}_{i_{k+1}}.1$ and set $\omega = \omega_{i_k}$. By $S_{14}(k)$ one can write

$$\tilde{f}_{i_1} \dots \tilde{f}_{i_{k+1}}(u_\omega \otimes u_\mu) = b_1 \otimes b_2 \bmod q(L(\omega) \otimes L(\mu))_{k+1}$$

where $b_1 \in B_k(\omega)$, $b_2 \in B_k(\mu) \cup \{0\}$. Then by $S_{13}^2(k)$ and $S_1^0(k+1)$

$$\tilde{e}_i \tilde{f}_{i_1} \dots \tilde{f}_{i_{k+1}}(u_\omega \otimes u_\mu) = \tilde{e}_i b_1 \otimes b_2 \text{ or } b_1 \otimes \tilde{e}_i b_2 \text{ or } 0 \bmod q(L(\omega) \otimes L(\mu))_k .$$

Thus applying $\Psi(\omega, \mu)$ one obtains from $S_5^0(k)$, $S_7^1(k)$, $S_7^2(k)$ that

$$\tilde{e}_i \tilde{f}_{i_1} \dots \tilde{f}_{i_{k+1}} u_{\omega+\mu} \in B_k(\omega + \mu) \cup \{0\} .$$

Now take $\mu \gg 0$ and apply 5.3.16. Then $\pi_{\omega+\mu}(\tilde{e}_i w) = (\tilde{e}_i w) u_{\omega+\mu} = \tilde{e}_i(w u_{\omega+\mu}) \in B_k(\omega + \mu) \cup \{0\}$. Applying 5.3.17 gives the required conclusion.

5.4.16 Lemma. $S_4(k+1)$, $S_5^\infty(k+1)$, $S_6^1(k)$, $S_{13}^5(k)$, $S_{13}^1(k+1)$, $S_{16}(k) \implies S_9^+(k+1)$.

By $S_4(k+1)$ the induced map $\bar{\pi}$ is defined on $B_{k+1}(\infty)$. Take $a = \tilde{f}_{i_1} \tilde{f}_{i_2} \cdots \tilde{f}_{i_{k+1}}.1$ which one may view as representing an arbitrary element of $b \in B_{k+1}(\infty)$. Assume that $\bar{\pi}_\lambda(b) \neq 0$, equivalently that $w := a u_\lambda \neq 0$. Then by $S_{13}^5(k)$ one has

$$(*) \qquad a(u_\lambda \otimes u_\mu) = w \otimes u_\mu \bmod q(L(\lambda) \otimes L(\mu))_{k+1} .$$

Take $\mu \gg 0$. Then by 5.3.16

$$(\tilde{e}_i a) u_{\lambda+\mu} = \tilde{e}_i(a u_{\lambda+\mu}) \bmod q L_k(\lambda + \mu)$$

so applying $\Phi(\lambda, \mu)$ one obtains from $S_6^1(k)$ that

$$(**) \qquad (\tilde{e}_i a)(u_\lambda \otimes u_\mu) = \tilde{e}_i(a(u_\lambda \otimes u_\mu)) \bmod q L_k(\lambda + \mu) .$$

Then

$$\begin{aligned}\tilde{e}_i w \otimes u_\mu &= \tilde{e}_i(w \otimes u_\mu) \operatorname{mod} q(L(\lambda) \otimes L(\mu))_k \quad \text{by} \quad S^1_{13}(k+1) \ , \\ &= (\tilde{e}_i a)(u_\lambda \otimes u_\mu) \operatorname{mod} q(L(\lambda) \otimes L(\mu))_k \ ,\end{aligned}$$

by $S^1_{13}(k+1)$ and $(*)$, $(**)$ above. Yet $\tilde{e}_i a \in B_k(\infty)$ by $S^\infty_5(k+1)$ so applying $S^5_{13}(k)$, the right hand side above becomes $\tilde{e}_i a u_\lambda \otimes u_\mu \operatorname{mod} qL(\lambda) \otimes L(\mu)$ and so by $S_{16}(k)$ one obtains

$$\tilde{e}_i w = (\tilde{e}_i a) u_\lambda \operatorname{mod} qL(\lambda) \ .$$

Equivalently $\tilde{e}_i \bar{\pi}_\lambda(b) = \bar{\pi}_\lambda(\tilde{e}_i b)$ as required.

5.4.17 Lemma. $S^\infty_5(k+1)$, $S'_8(k+1)$, $S^+_9(k+1) \implies S^0_5(k+1)$.

Indeed $\tilde{e}_i B_{k+1}(\lambda) = \tilde{e}_i \bar{\pi}_\lambda(B_{k+1}(\infty)) = \pi_\lambda(\tilde{e}_i B_{k+1}(\infty)) \subset \pi_\lambda(B_{k+1}(\infty)) = B_{k+1}(\lambda) \cup \{0\}$.

5.4.18 Consider the following asymptotic version of $S^1_7(k)$, namely

$S^{1,a}_7(k)$ $\Psi(\lambda,\mu)(L(\lambda) \otimes L(\mu))_k \subset L_k(\lambda+\mu)$ *for* $\mu \gg 0$.

Lemma. $S^1_6(k+1)$, $S^1_7(k)$, $S_{15}(k) \Longrightarrow S^{1,a}_7(k+1)$.

By $S_{15}(k)$ one has

$$(*) \qquad (L(\lambda) \otimes L(\mu))_{k+1} = \sum \tilde{f}_i (L(\lambda) \otimes L(\mu))_k + u_\lambda \otimes L_{k+1}(\mu) \ .$$

Yet for all $u \in L(\lambda+\mu)^*_{\lambda+\mu+\xi} : \xi \in Q^-_{k+1}$ it follows from 5.3.20$(*)$ and $S^1_7(k)$ that

$$(**) \qquad \begin{aligned}(\Phi(\lambda,\mu)(u), \tilde{f}_i(L(\lambda) \otimes L(\mu))_k) &= (u, \tilde{f}_i \Psi(\lambda,\mu)(L(\lambda) \otimes L(\mu))_k) \\ &\subset (u, \tilde{f}_i L_k(\lambda+\mu)) \subset A \ .\end{aligned}$$

Now assume $\mu \gg 0$. Then by 5.3.19 there exists $a \in L(\infty)^*_\xi$ such that $u = a u_{\lambda+\mu}$. Writing $\xi = -\sum n_i \alpha_i$ one has

$$\Delta(a) = \Pi t_i^{n_i} \otimes a \operatorname{mod} \sum f_i U_q(\mathfrak{g}) \otimes U_q(\mathfrak{n}^-)$$

and so

$$\Phi(\lambda,\mu) a u_{\lambda+\mu} = (\Pi q_i^{n_i(\alpha_i^\vee,\lambda)}) u_\lambda \otimes a u_\mu \operatorname{mod}(\sum f_i V(\lambda)) \otimes V(\mu) \ .$$

Therefore recalling 5.3.18 and that $\pi_\mu(L(\infty)^*_\xi) = L(\mu)^*_{\mu+\xi}$ by 5.3.19 one obtains

$$(***) \qquad ((\Phi(\lambda,\mu)(u)), u_\lambda \otimes L_{k+1}(\mu)) \subset (\Pi q_i^{n_i(\alpha_i^\vee,\lambda)})(a u_\mu, L_{k+1}(\mu)) \subset A \ .$$

Consequently

$$\begin{aligned}&(L(\lambda+\mu)^*_{\lambda+\mu+\xi}, \Psi(\lambda,\mu)(L(\lambda)\otimes L(\mu))_{k+1})\\&\quad=(\Phi(\lambda,\mu)L(\lambda+\mu)^*_{\lambda+\mu+\xi},(L(\lambda)\otimes L(\mu))_{k+1})\subset A\end{aligned}$$

by (*), (**) and (***) above. By the characterization of $L(\lambda)$ in 5.3.19 one concludes that

$$\Psi(\lambda,\mu)(L(\lambda)\otimes L(\mu))_{k+1}\subset L_{k+1}(\lambda+\mu)\ .$$

Since $\Psi(\lambda,\mu)\Phi(\lambda,\mu)=1d_{L(\lambda+\mu)}$ the opposite inclusion follows from $S_6^1(k+1)$.

5.4.19 Lemma. $S_7^1(k)$, $S_7^{1,a}(k+1)$, $S_{13}^1(k+1)$, $S_{13}^3(k)$, $S_{14}(k)\Longrightarrow S_{10}^{\infty}(k+1)$.

Take $b,b'\in B_{k+1}(\infty)$ with $\tilde{e}_ib=b'$. One can write $b=a_{k+1}.1 \bmod qL_{k+1}(\infty)$ where $a_{k+1}=\tilde{f}_{i_1}\tilde{f}_{i_2}\dots\tilde{f}_{i_{k+1}}$. Set $\omega=\omega_{i_k}$. Then by $S_{14}(k)$ one has

$$a_{k+1}(u_\omega\otimes u_\mu)=b_1\otimes b_2 \bmod q(L(\omega)\otimes L(\mu))_{k+1}$$

for some $b_1\in B_k(\omega)$, $b_2\in B_k(\mu)\cup\{0\}$. Now by $S_{13}^1(k+1)$ and $S_7^1(k)$ one has

$$\Psi(\omega,\mu)\tilde{e}_i(L(\omega)\otimes L(\mu))_{k+1}\subset L_k(\omega+\mu)$$

and so

$$\tilde{e}_ia_{k+1}u_{\omega+\mu}=\tilde{e}_i\Psi(\omega,\mu)(b_1\otimes b_2)\bmod qL_k(\omega+\mu)\ .$$

Now take $\mu\gg 0$. Then by 5.3.17 and 5.3.16 one has $\bmod qL(\omega+\mu)$ that

$$0\neq\pi_{\omega+\mu}(b')=\tilde{e}_ibu_{\omega+\mu}=\tilde{e}_ia_{k+1}u_{\omega+\mu}$$

and consequently $\tilde{e}_i(b_1\otimes b_2)\neq 0$. By $S_{13}^3(k)$ one concludes that

$$a_{k+1}(u_\omega\otimes u_\mu)=\tilde{f}_i\tilde{e}_ia_{k+1}(u_\omega\otimes u_\mu)\bmod q(L(\omega)\otimes L(\mu))_{k+1}\ .$$

Applying $S_7^{1,a}(k+1)$ one can replace $u_\omega\otimes u_\mu$ by $u_{\omega+\mu}$ and then 5.3.16 gives

$$\bar{\pi}_{\omega+\mu}(b)=\bar{\pi}_{\omega+\mu}(\tilde{f}_i\tilde{e}_ib)\ .$$

From this 5.3.17 implies that $b=\tilde{f}_i\tilde{e}_ib$. Since already $\tilde{e}_i\tilde{f}_i=1$ on $B(\infty)$, this completes the proof of 5.4.19.

5.4.20 Lemma. $S_4(k+1)$, $S_8'(k+1)$, $S_9^+(k+1)$, $S_{10}^{\infty}(k+1)\Longrightarrow S_{10}^0(k+1)$.

Take $b\in B_{k+1}(\lambda)$ such that $\tilde{e}_ib\neq 0$. By $S_8'(k+1)$ there exists $b'\in B_{k+1}(\infty)$ such that $b=\pi_\lambda(b')$. Then $S_9^+(k+1)$ implies $\bar{\pi}_\lambda(\tilde{e}_ib')=\tilde{e}_ib\neq 0$. Hence $\tilde{e}_ib'\neq 0$. Then by $S_{10}^{\infty}(k+1)$ one obtains $b'=\tilde{f}_i\tilde{e}_ib'$ which gives

$$\tilde{f}_i\tilde{e}_ib=\tilde{f}_i\bar{\pi}_\lambda(\tilde{e}_ib')\overset{S_4(k+1)}{=}\bar{\pi}_\lambda(\tilde{f}_i\tilde{e}_ib')=\bar{\pi}_\lambda(b')=b\ .$$

Now take $b\in B_{k+1}(\lambda)$ such that $\tilde{f}_ib\neq 0$. By $S_8'(k+1)$ there exists $b'\in B_{k+1}(\infty)$ such that $b=\pi_\lambda(b')$. Then $S_4(k+1)$ implies $\bar{\pi}_\lambda(\tilde{f}_ib')=\tilde{f}_ib\neq 0$ and hence by $S_9^+(k+1)$ one obtains

$$\tilde{e}_i\tilde{f}_ib = \tilde{e}_i\bar{\pi}_\lambda(\tilde{f}_ib') = \bar{\pi}_\lambda(\tilde{e}_i\tilde{f}_ib') = \bar{\pi}_\lambda(b') = b$$

as required.

5.4.21 Lemma. $S_3^\varepsilon(k)$, $S_5^\varepsilon(k+1)$, $S_{10}^\varepsilon(k+1) \Longrightarrow S_3^\varepsilon(k+1)$.

Take $\varepsilon = \infty$. Assume $\sum_{b\in B_{k+1}(\infty)} a_b b = 0$ and show $a_b = 0$. For any i the $\tilde{e}_ib$ which are non-zero lie in $B_k(\infty)$ by $S_5^\infty(k+1)$ and are distinct by $S_{10}^\infty(k+1)$, hence are linearly independent by $S_3^\infty(k)$. One concludes that $a_b = 0$ if there exists i such that $\tilde{e}_ib \neq 0$. By $S_{10}^\infty(k)$ again this only leaves a_{u_λ} which must therefore also vanish. The case $\varepsilon = 0$ is the same.

5.4.22 Recall 5.3.15. Set $V_k(\lambda^\varepsilon) = V_k(\infty)$ if $\varepsilon = \infty$ and $V_k(\lambda)$ if $\varepsilon = 0$. Define

$S_{17}^\varepsilon(k)$ $\{u \in L_k(\lambda^\varepsilon)/qL_k(\lambda^\varepsilon) | \tilde{e}_iu = 0, \forall i\} = 0$.
$S_{18}^\varepsilon(k)$ $\{u \in V_k(\lambda^\varepsilon) | \tilde{e}_iu \in L(\lambda^\varepsilon), \forall i\} = L_k(\lambda^\varepsilon)$.

As in 5.1.5 one obtains

Lemma
(i) $S_3^\varepsilon(k)$, $S_{10}^\varepsilon(k+1) \Longrightarrow S_{17}^\varepsilon(k+1)$,
(ii) $S_{17}^\varepsilon(k+1) \Longrightarrow S_{18}^\varepsilon(k+1)$.

5.4.23 Lemma. $S_7^1(k)$, $S_{18}^0(k+1) \Longrightarrow S_7^1(k+1)$.

Indeed

$$\begin{aligned}\tilde{e}_i\Psi(\lambda,\mu)(L(\lambda)\otimes L(\mu))_{k+1} &\subset \Psi(\lambda,\mu)(L(\lambda)\otimes L(\mu))_k \\ &\subset L_k(\lambda+\mu)\ , \quad \text{by} \quad S_7^1(k)\ ,\end{aligned}$$

so the assertion follows from $S_{18}^0(k+1)$.

5.4.24 Lemma. $S_5^\infty(k+1)$, $S_8(k)$, $S_8'(k+1)$, $S_9^+(k+1)$, $S_{10}^\varepsilon(k+1) \Longrightarrow S_8(k+1)$.

By $S_8'(k+1)$ it suffices to show that for $b, b' \in B_{k+1}(\infty)$ the relation $\bar{\pi}_\lambda(b) = \bar{\pi}_\lambda(b') \neq 0$ implies $b = b'$. By $S_{10}^0(k+1)$ one can find i such that $\tilde{e}_i\bar{\pi}_\lambda(b) \neq 0$. By $S_9^+(k+1)$ one obtains $\bar{\pi}_\lambda(\tilde{e}_ib) = \bar{\pi}_\lambda(\tilde{e}_ib') \neq 0$ and so $\tilde{e}_ib = \tilde{e}_ib'$ by $S_5^\infty(k+1)$ and $S_8(k)$. Then $S_{10}^\infty(k+1)$ gives $b = b'$.

5.4.25 Lemma. $S_7^2(k)$, $S_{13}(k+1)$, $S_{17}^0(k+1) \Longrightarrow S_7^2(k+1)$.

Take $b \in (B(\lambda)\otimes B(\mu))_{k+1}$. Suppose there exists i such that $\tilde{e}_ib \neq 0$. Then by $S_{13}(k+1)$ and $S_7^2(k)$ one has $\Psi(\lambda,\mu)b = \Psi(\lambda,\mu)(\tilde{f}_i\tilde{e}_ib) = \tilde{f}_i\Psi(\lambda,\mu)\tilde{e}_ib \subset \tilde{f}_i(B_k(\lambda+\mu)\cup\{0\}) \subset B_{k+1}(\lambda+\mu)\cup\{0\}$. Otherwise $\tilde{e}_i\Psi(\lambda,\mu)b = 0$ for all i and then $\Psi(\lambda,\mu)b = 0$ by $S_{17}^0(k+1)$.

5.4.26 The main theorems proved in this chapter are summarized below. Recall 5.3.13 the definitions of $L(\infty)$, $B(\infty)$.

Theorem. $(L(\infty), B(\infty))$ *is a crystal basis for* $U_q(\mathfrak{n}^-)$. *Moreover up to isomorphism any crystal basis has this form.*

The first part is just the content of S_1^∞, S_3^∞, S_5^∞, S_{10}^∞. Notice also that (A5) holds. The second part follows from S_{17}^∞, S_{18}^∞ as in the proof of 5.1.8.

Remark. Any $A_q(\mathfrak{g})$ module M satisfying $\dim U_q(\mathfrak{n}^+)m < \infty, \forall m \in M$ is a direct sum of the $U_q(\mathfrak{n}^-)$ and moreover the analogue of 5.1.8 extends to give a uniqueness theorem for a crystal basis of M. However this does not seem to have any particular importance.

5.4.27 Recall (5.1.4) the definitions of $L(\lambda), B(\lambda)$.

Theorem. $(L(\lambda), B(\lambda))$ *is a crystal basis for* $V(\lambda)$. *(It is unique by 5.1.8).*

This is just S_1^0, S_3^0, S_5^0, S_{10}^0 .

5.4.28 The compatibility of these crystal bases was discussed in 5.3.13. Here one notes that S_2, S_8, S_9^-, S_9^+ give 5.3.13 (i)–(iv) respectively.

5.5 Comments and Complements

5.5.1 The main results of this chapter are due to M. Kashiwara [K2–K5]. Although the induction steps of the grand loop have been modified, the only significant change has been to remove the asymptotic results 5.3.16, 5.3.17 from the induction loop. Actually inspection of the flow chart for the grand loop show that the only induction steps concerning $U_q(\mathfrak{n}^-)$ appealed to are S_5^∞ and S_{10}^∞. As in 5.4.10 these are equivalent to an asymptotic version for λ very dominant. In this fashion one may first deduce theorem 5.4.27 and obtain theorem 5.4.26 as a corollary.

5.5.2 The twisted action described in 5.2.9 is taken from [JL2] as is also the construction of an ad-invariant filtration in 5.3.1. This leads to a slight simplification of Kashiwara's analysis. However its main importance is realized only in subsequent chapters discussing the structure of $U_q(\mathfrak{g})$. The crystal S_λ together with its use in 5.3.13, 6.1.13 and 6.4.21 are new.

5.5.3 The appearance of the quantum Weyl algebra in 5.3 motivated partly the development in section 3.1. This is why some of Kashiwara's results seem obvious. In any case these results and the twisted adjoint action can be regarded as being really a consequence of Drinfeld duality (3.1.10).

Chapter 6. The Global Bases

The compatibility of $B(\lambda)$ with the tensor product (5.2.4) already has some combinatorial consequences. A crucial additional fact is that the *antiautomorphism* $\star$ of $U_q(\mathfrak{n}^-)$ defined by $f_i^\star = f_i$ induces an involution of $B(\infty)$. Again the crystal bases $B(\lambda)$, $B(\infty)$ can be lifted to global bases of $V(\lambda)$ and $U_q(\mathfrak{n}^-)$ respectively. This leads to a common basis theorem (6.2.19) which is central to the structure theory of $U_q(\mathfrak{g})$. It also gives the Demazure character formula (6.3.15). In the last section Littelmann's theory is described and compared to Kashiwara's. The global bases, obtained here are now known to coincide with Lusztig's canonical bases (6.5.1) and there are ideas here common to those of Lusztig particularly the use of $\star$, see [L7].

6.1 The $\star$ Operation and the Embedding Theorem

6.1.1 Let $(\ ,\ )$ denote the form on $V(\lambda)$ defined in 5.3.18.

Lemma. *Fix $i \in \{1, 2, \ldots, \ell\}$. Let $u_\mu, v_\nu \in \ker e_i \cap L(\lambda)$ be weight vectors and set $r_i = (\alpha_i^\vee, \nu)$. Then for all $m, n \in \mathbb{N}$ one has*

(i) $(f_i^{(m)} u_\mu, f_i^{(n)} v_\nu) = 0$ *unless* $m = n \leq r_i$ *and* $\mu = \nu$.
(ii) $(f_i^{(m)} u_\mu, f_i^{(m)} v_\nu) = (u_\mu, v_\nu) \bmod qA$, *for all* $m \leq r_i$.
(iii) $(L(\lambda), L(\lambda)) \subset A$.

Drop the subscript i. Recalling that $\tilde{\varkappa}(f) = q^{-1} te$ one has

$$(f^{(m)} u_\mu, f^{(n)} v_\nu) = \frac{1}{[m]_q} (f^{(m-1)} u_\mu, q^{-1} te f^{(n)} v_\nu)$$

which equals zero if $n = 0$. If $n > 0$ the formula in 5.1.9 gives

$$\begin{aligned} e f^{(n)} v_\nu &= [f^{(n-1)}(q^{-(n-1)} t - q^{n-1} t^{-1})/(q - q^{-1})] v_\nu \\ &= [r - (n-1)]_q f^{(n-1)} v_\nu \end{aligned}$$

and then by resubstitution

$$(*) \quad (f^{(m)} u_\mu, f^{(n)} v_\nu) = q^{r-2n+1} \frac{[r - (n-1)]_q}{[m]_q} (f^{(m-1)} u_\mu, f^{(n-1)} v_\nu) .$$

It is immediate from the above that the left hand side vanishes unless $n \geq m$ so by symmetry one must have $n = m$. Since q is an indeterminate (or just not a root of unity) the form is orthogonal on distinct weight spaces. All this gives (i). Finally (ii) results from $(*)$ because $q^m[m+1]_q$ is a unit in A whose value at $q = 0$ equals 1. (iii) follows from (ii).

6.1.2 Let $(\,,\,)_0$ denote the $\mathbb{Q}$-valued inner product on $L(\lambda)/qL(\lambda)$ induced by $(\,,\,)|_{q=0}$ on $L(\lambda)$.

Proposition

(i) $(\tilde{e}_i u, v)_0 = (u, \tilde{f}_i v)_0$, $\forall\, u, v \in L(\lambda)/qL(\lambda)$, $\forall\, i$.

(ii) *$B(\lambda)$ is an orthonormal base with respect to $(\,,\,)_0$. In particular $(\,,\,)_0$ is positive definite.*

(iii) $L(\lambda)^* = L(\lambda)$.

(iv) $L(\lambda) = \{u \in V(\lambda) | (u, u) \in A\}$.

(i) is immediate from 6.1.1 (ii). Combined with S^0_{10} it gives (ii). By 6.1.1(i) one has $L(\lambda) \subset L(\lambda)^*$ and by (ii) that $L(\lambda)^* \subset L(\lambda) + qL(\lambda)^*$, so (iii) follows by Nakayama's lemma. For (iv) take $u \in U(\lambda)$ satisfying $(u, u) \in A$. If $u \neq 0$ there exists an integer $n \geq 0$ such that $v := q^n u \in L(\lambda)$ and $\bar{v} \neq 0$ in $L(\lambda)/qL(\lambda)$. Yet if $n > 0$ one has $(\bar{v}, \bar{v})_0 = q^{2n}(u, u)|_{q=0} = 0$ contradicting (ii).

6.1.3 Let $(\,,\,)$ denote the form on $U_q(\mathfrak{n}^-)$ defined in 5.3.9. Let a, b be monomials in the $\tilde{f}_i$. By 5.3.18 taking $\lambda \gg 0$ one has $(a, b) = (au_\lambda, bu_\lambda)$ up to a unit in A taking the value 1 at $q = 0$. It is then immediate that the conclusions of 6.1.2 apply to $(L(\infty), B(\infty))$. These are designated 6.1.3 (i)–(iv) in what follows.

6.1.4 Define $e_i^\backprime$ by $[e_i, b] = (q_i - q_i^{-1})^{-1}(e_i^\backprime(b)t_i - t_i^{-1}e_i'(b))$, for all $b \in U_q(\mathfrak{n}^-)$. Then $e_i^\backprime = (\operatorname{ad} t_i)e_i''$. The following may be compared with 3.2.11.

Lemma. *For all $i, j \in \{1, 2, \dots, \ell\}$; $a, b \in U_q(\mathfrak{n}^-)$ one has*

(i) $e_i^\backprime e_j' = e_j' e_i^\backprime$.

(ii) $(af_i, b) = (a, e_i^\backprime b)$.

(i) By 5.3.4, one has $(\operatorname{ad} t_i)e_i'' e_j' = q^{-(\alpha_i, \alpha_j)}(\operatorname{ad} t_i)(e_j' e_i'') = e_j'(\operatorname{ad} t_i)e_i''$.

(ii) If $a = 1$, it is enough to check this assertion with $b = f_i$ and this is quite easy. It is then enough to show that (ii) implies that this assertion holds with a replaced by $f_j a$. Yet

$$\begin{aligned}
(f_j a f_i, b) &= (af_i, e_j'(b)) \,, && \text{by definition of the form ,}\\
&= (a, e_i^\backprime e_j'(b)) \,, && \text{by the induction hypothesis ,}\\
&= (a, e_j' e_i^\backprime(b)) \,, && \text{by (i) ,}\\
&= (f_j a, e_i^\backprime(b)) \,, && \text{by definition of the form .}
\end{aligned}$$

6.1.5 Let $\star$ denote the antiautomorphism of $U_q(\mathfrak{g})$ defined by $e_i^\star = e_i$, $f_i^\star = f_i$, $t_i^\star = t_i^{-1}$. It is immediate from the definition of $e_i^\backprime$ and e_i' that

Lemma. $e_i'(a^\star) = (e_i^\backprime(a))^\star$.

6.1.6 Proposition

(i) $(a^\star, b^\star) = (a, b)$, $\forall\, a, b \in U_q(\mathfrak{n}^-)$
(ii) $L(\infty)^\star = L(\infty)$.

(i) is clear for $a = 1$. Thus it is enough to show that (i) implies this assertion with a replaced by af_i. Yet

$$\begin{aligned}((af_i)^\star, b^\star) &= (f_i a^\star, b^\star)\,, \quad \text{by definition of } \star\,,\\ &= (a^\star, e_i'(b^\star))\,, \quad \text{by definition of the form}\,,\\ &= (a, (e_i'(b^\star))^\star)\,, \quad \text{by the induction hypothesis}\,,\\ &= (a, e_i^\backprime(b))\,, \quad \text{by 6.1.5}\,,\\ &= (af_i, b)\,, \quad \text{by 6.1.4(ii)}\,.\end{aligned}$$

(ii) follows from (i) and 6.1.3(iv).

6.1.7 Let $U_q^{\mathbb{Z}}(\mathfrak{n}^-)$ denote the $\mathbb{Z}[q, q^{-1}]$ subalgebra of $U_q(\mathfrak{n}^-)$ generated by the $f_i^{(m)} : i = 1, 2, \ldots, \ell;\ m \in \mathbb{N}$. Note that the Serre relations for the f_i are defined in $U_q^{\mathbb{Z}}(\mathfrak{n}^-)$.

Lemma. *For all $i, j \in \{1, 2, \ldots, m\}$ one has*

(i) $e_i'(f_j^{(m)}) = \delta_{ij} q_j^{-(m-1)} f_j^{(m-1)}$.
(ii) $U_q^{\mathbb{Z}}(\mathfrak{n}^-)$ *is stable by* e_i'.
(iii) $U_q^{\mathbb{Z}}(\mathfrak{n}^-)$ *is stable by* $\tilde{e}_i$, $\tilde{f}_i$.
(iv) $U_q^{\mathbb{Z}}(\mathfrak{n}^-)$ *is the free* $\mathbb{Z}[q, q^{-1}]$ *module generated by all the distinct* $\tilde{f}_{i_1} \cdots \tilde{f}_{i_r} u_\infty$.

(i) is proved by induction on m recalling that e_i' is a skew derivation. Indeed one has

$$e_i'(f_j^{(m)}) = \frac{1}{[m]_{q_i}}\, e_i'(f_j f_j^{(m-1)}) = \frac{1}{[m]_{q_i}}\, (\delta_{ij} f_j^{(m-1)} + q_i^{-(\alpha_i^\vee, \alpha_j)} f_j e_i'(f_j^{(m-1)}))\,.$$

This gives (i) for $i \neq j$. For $i = j$ drop the i subscript. Then

$$\begin{aligned}e'(f^{(m)}) &= \frac{1}{[m]_q}\, (f^{(m-1)} + q^{-2} f e'(f^{(m-1)}))\,,\\ &= \frac{1}{[m]_q}\, (1 + q^{-m}[m-1]_q) f^{(m-1)}\,, \quad \text{by the induction hypothesis}\\ &= q^{-(m-1)} f^{(m-1)}\,, \quad \text{as required}\,.\end{aligned}$$

(ii) follows from (i) by e_i' being a skew derivation. Then the conclusion of 5.3.11 applies to $U_q^{\mathbb{Z}}(\mathfrak{n}^-)$ and this gives (iii). Finally (ii), (iii) imply (iv).

6.1.8 Set $L^{\mathbb{Z}}(\infty) = L(\infty) \cap U_q^{\mathbb{Z}}(\mathfrak{n}^-)$ which by 6.1.7 is stable by $\tilde{e}_i$, $\tilde{f}_i$. Consequently

$$B(\infty) \subset L^{\mathbb{Z}}(\infty)/qL^{\mathbb{Z}}(\infty) \subset L(\infty)/qL(\infty)$$

where the last inclusion follows from $qL(\infty) \cap L^{\mathbb{Z}}(\infty) = qL(\infty) \cap U_q^{\mathbb{Z}}(\mathfrak{n}^-) = qL^{\mathbb{Z}}(\infty)$.

Let $A^{\mathbb{Z}}$ denote the $\mathbb{Z}$ subalgebra of $\mathbb{Q}(q)$ generated by q and the $(1 - q^{2n})^{-1} : n \geq 1$. Set $K^{\mathbb{Z}} = A^{\mathbb{Z}}[q^{-1}]$. Then $A^{\mathbb{Z}} = A \cap K^{\mathbb{Z}}$ (because q is prime to the denominators in A whilst $\mathbb{Z}[q]$ is a unique factorization domain). Now for all $a, b \in U^{\mathbb{Z}}(\mathfrak{n}^-)$, $m \in \mathbb{N}$, $i \in \{1, 2, \dots, \ell\}$ one has

$$(f_i^{(m)}a, b) = \frac{1}{[m]_{q_i}} \, (f_i^{(m-1)}a, e_i'(b)) \, .$$

Since the coefficient in the right hand side belongs to $K^{\mathbb{Z}}$ it follows from 6.1.7(ii) and the obvious induction on weights that

$$(U^{\mathbb{Z}}(\mathfrak{n}^-),\ U^{\mathbb{Z}}(\mathfrak{n}^-)) \subset K^{\mathbb{Z}} \, .$$

Consequently 6.1.3(iii) gives

$$(L^{\mathbb{Z}}(\infty), L^{\mathbb{Z}}(\infty)) \subset A^{\mathbb{Z}} \, .$$

Since $f(0) \in \mathbb{Z}$ for any $f \in A^{\mathbb{Z}}$ it follows that the form $(\,,\,)_0$ is $\mathbb{Z}$-valued on $L^{\mathbb{Z}}(\infty)/qL^{\mathbb{Z}}(\infty)$.

Lemma

(i) $L^{\mathbb{Z}}(\infty)/qL^{\mathbb{Z}}(\infty)$ *is a free* $\mathbb{Z}$ *module with base* $B(\infty)$.
(ii) $B(\infty) \cup -B(\infty) = \{u \in L^{\mathbb{Z}}(\infty)/qL^{\mathbb{Z}}(\infty) \mid (u, u)_0 = 1\}$.

(i) is immediate from 6.1.3(ii). Again if $u = \sum a_b b \in L^{\mathbb{Z}}(\infty)/qL^{\mathbb{Z}}(\infty)$ satisfies $(u, u)_0 = 1$ then $\sum a_b^2 = 1$. Since $a_b \in \mathbb{Z}$ by (i), this gives (ii).

6.1.9 Corollary

(i) $L^{\mathbb{Z}}(\infty)^\star = L^{\mathbb{Z}}(\infty)$.
(ii) $B(\infty)^\star \subset B(\infty) \cup -B(\infty)$.

Since $U_q^{\mathbb{Z}}(\mathfrak{n}^-)$ is trivially $\star$ stable, (i) follows from 6.1.6(ii). Then (ii) follows from 6.1.8(ii) and 6.1.6(i).

6.1.10 The aim of this and the next two sections is to show that $B(\infty)^\star = B(\infty)$. This will lead to some useful additional operators on $B(\infty)$.

Lemma. *Take* $a \in U_q(\mathfrak{n}^-)$ *non-zero. Fix* i *and assume* $e_i'(a) = 0$. *Consider* $\lambda \in P^+(\pi)$ *such that* $(\alpha_i^\vee, \lambda) = 0$ *and assume* $(\alpha_j^\vee, \lambda) \gg 0$ *for* $j \neq i$. *Then*

(i) $au_\lambda \neq 0$.
(ii) *If* $a \in L(\infty)$ *and* $a \bmod qL(\infty) \in B(\infty)$, *then* $au_\lambda \in B(\lambda)$.

(i) One can assume that a is a weight vector of weight $\xi \in Q^-(\pi)$. By 4.3.6 (i)

$$\mathrm{Ann}_{U_q(\mathfrak{n}^-)}\, u_\lambda = \sum_{i=1}^{\ell} U_q(\mathfrak{n}^-) f_i^{(\alpha_i^\vee,\lambda)+1} \,.$$

Thus for $(\alpha_j^\vee, \lambda) \geq -(\alpha_j^\vee, \xi) : j \neq i$ one has

$$(\mathrm{Ann}_{U_q(\mathfrak{n}^-)}\, u_\lambda)_\xi = (U_q(\mathfrak{n}^-) f_i)_\xi \,.$$

Consequently if $au_\lambda = 0$, one can write $a = bf_i^m$ for some $m > 0$, $b \in U_q(\mathfrak{n}^-)$. Now by 6.1.5 the hypothesis on a implies that $e_i'(a^\star) = 0$, whilst $b^\star$ can be written in the form $\sum f_i^n b_n : b_n \in \ker e_i'$. Then

$$\sum f_i^{m+n} b_n = a^\star \in \ker e_i' \,.$$

This contradicts the direct sum decomposition in 5.3.11. Hence (i). Then (ii) follows from (i) and S_8.

6.1.11 Take a, λ as in 6.1.10(ii).

Proposition. *Take* $m \in \mathbb{N}$ *and* $\mu \in P^+(\pi)$ *with* $(\alpha_i^\vee, \mu) \gg 0$. *Then*

(i) $af_i^{(m)}(u_\lambda \otimes u_\mu) = au_\lambda \otimes f_i^{(m)} u_\mu \bmod qL(\lambda) \otimes L(\mu)$,
(ii) $af_i^{(m)} \bmod qL(\infty) \in B(\infty)$.

(i) By choice of λ, μ one has

$$f_i^{(m)}(u_\lambda \otimes u_\mu) = u_\lambda \otimes f_i^{(m)} u_\mu \neq 0 \,.$$

Now $\Delta(a) = a \otimes 1 \bmod U_q(\mathfrak{g}) \otimes U_q(\mathfrak{n}^-)_+$, where $+$ denotes the augmentation. Hence by the preceeding equality

$$(*) \qquad af_i^{(m)}(u_\lambda \otimes u_\mu) = au_\lambda \otimes f_i^{(m)} u_\mu \bmod \sum_{\xi \neq \mu - m\alpha_i} V(\lambda) \otimes V(\mu)_\xi \,.$$

Since $a \bmod qL(\infty) \in B(\infty)$, one obtains $a^\star \bmod qL(\infty) \in \pm B(\infty)$ by 6.1.9. Yet $e_i'(a^\star) = 0$ by the hypothesis on a, so $f_i^{(m)} a^\star \bmod qL(\infty) \in \pm B(\infty)$ by the definition of $B(\infty)$. Hence $af_i^{(m)} \bmod qL(\infty) \in \pm B(\infty)$ by 6.1.9 again. Then since $\lambda + \mu \gg 0$ one has $af_i^{(m)}(u_\lambda \otimes u_\mu) \bmod q(L(\lambda) \otimes L(\mu)) \in \pm B(\lambda) \otimes B(\mu)$ by S_6, S_8. On the other hand by 6.1.10(ii) the first term in the right hand side of $(*)$ lies $\bmod\, q(L(\lambda) \otimes L(\mu))$ in $B(\lambda) \otimes B(\mu)$. Comparison shows that the remaining terms must lie in $q(L(\lambda) \otimes L(\mu))$. Hence (i). It further forces the $+$ sign to hold above which hence gives (ii).

6.1.12 Take a as in 6.1.11 and let b be the corresponding element in $B(\infty)$. It was shown that $b^\star \in B(\infty)$, $\tilde{e}_i b^\star = 0$, $af_i^{(m)} \bmod qL(\infty) \in B(\infty)$ for all $m \in \mathbb{N}$. However although $e_i^\prime(a) = 0$ it is in general *false* that $af_i^{(m)} \bmod qL(\infty) = b\tilde{f}_i^m$. On the other hand by 6.1.9(i) one may define the operators $\tilde{e}_i^\star = \star\tilde{e}_i\star$, $\tilde{f}_i^\star := \star\tilde{f}_i\star$ on $L(\infty)$ and then the above result translates as follows. Take $b \in B(\infty)$ such that $b^\star \in B(\infty)$ and $\tilde{e}_i b^\star = 0$. Then $b^\star$ admits a representative $a^\star$ in $L(\infty)$ satisfying $e_i^\prime a^\star = 0$, hence $e_i^\prime a = 0$. By 6.1.9.(i), one has $a \bmod qL(\infty) = b$ and so $\tilde{f}_i^{\star m} a \bmod qL(\infty) = \tilde{f}_i^{\star m} b$ for all $m \in \mathbb{N}$. Yet $\tilde{f}_i^{\star m} a = (\tilde{f}_i^m a^\star)^\star = (f_i^{(m)} a^\star)^\star = af_i^{(m)}$ and so $\tilde{f}_i^{\star m} b \in B(\infty)$ by the above. It is clear that this proves the

Corollary. $B(\infty)^\star = B(\infty)$. *Moreover each $b \in B(\infty)$ may be uniquely written in the form $b = \tilde{f}_i^{\star m} b_0$ with $\tilde{e}_i^\star b_0 = 0$, $m \in \mathbb{N}$.*

6.1.13 Fix $i \in \{1, 2, \ldots, m\}$. Recall the crystal C_i defined in 5.2.3(ii). Represent $b \in B(\infty)$ in the form given by 6.1.12. Then one has the following important embedding theorem.

Theorem. *The map $\Psi_i : b \mapsto b_0 \otimes c_i(-m)$ is a strict embedding of $B(\infty)$ into $B(\infty) \otimes C_i$.*

Take $\lambda, \mu \in P^+(\pi)$ with $(\alpha_i^\vee, \lambda) = 0$, $(\alpha_i^\vee, \mu) \gg 0$ and $(\alpha_j, \lambda - \mu) \gg 0$ for $j \neq i$ (relative to the given b). Then $\mu + \lambda \gg 0$ and so $\bar{\pi}_{\lambda+\mu}(b) \neq 0$. By 6.1.11(i) and 6.1.12 one has $\Phi_{\lambda,\mu}(\bar{\pi}_{\lambda+\mu}(b)) = b_0 u_\lambda \otimes \tilde{f}_i^m u_\mu$ which under the strict embedding ψ of 5.3.13 is further sent by $\psi \otimes 1d$ to $b_0 \otimes s_\lambda \otimes \tilde{f}_i^m u_\mu$. Since $\tilde{f}_i^m(s_\lambda \otimes u_\mu) = s_\lambda \otimes \tilde{f}_i^m u_\mu$ and

$$\varepsilon_j(s_\lambda \otimes \tilde{f}_i^m u_\lambda) = \max\{-(\alpha_j^\vee, \lambda), \varepsilon_j(\tilde{f}_i^m u_\mu) - (\alpha_j^\vee, \lambda)\} = \begin{cases} m & : j = i \\ << 0 & : j \neq i \end{cases}$$

it follows from the tensor product formula (5.2.4) that one may replace $s_\lambda \otimes \tilde{f}_i^m u_\mu$ by $c_i(-m)$ without losing the property that $(\psi \otimes 1d)\Phi_{\lambda,\mu}\bar{\pi}_{\lambda+\mu}$ commutes with the $\tilde{e}_i$, $\tilde{f}_j$ (for appropriately large λ, μ). Hence the assertion of the theorem.

6.1.14 Corollary *Take $i \neq j$. Each of the pairs $\tilde{e}_i, \tilde{f}_j^\star$; $\tilde{f}_i, \tilde{f}_j^\star$; $\tilde{e}_i^\star, \tilde{f}_j$; $\tilde{e}_i, \tilde{e}_j^\star$ commute.*

Consider $\tilde{e}_i$, $\tilde{f}_j^\star$. Write $b \in B(\infty)$ in the form $\tilde{f}_j^{\star m} b_0$ with $\tilde{e}_j^\star b_0 = 0$, $m \in \mathbb{N}$. Then by 6.1.13

$$\Psi_j(\tilde{e}_i \tilde{f}_j^\star b) = \tilde{e}_i \Psi_j(\tilde{f}_j^\star b) = \tilde{e}_i(b_0 \otimes c_j(-(m+1)) = \tilde{e}_i b_0 \otimes c_j(-(m+1)) ,$$

since $i \neq j$. Yet $\tilde{e}_i \tilde{f}_j^\star b \in B(\infty)$ so 6.1.13 forces one to conclude that $\tilde{e}_j^\star(\tilde{e}_i b_0) = 0$ and that the last expression above equals $\Psi_j(\tilde{f}_j^{\star m+1} \tilde{e}_i b_0)$. Again from 6.1.13 and the first conclusion $\Psi_j(\tilde{f}_j^{\star m}(\tilde{e}_i b_0)) = \tilde{e}_i b_0 \otimes c_j(-m) = \tilde{e}_i(b_0 \otimes c_j(-m)) = \tilde{e}_i \Psi_j(b) = \Psi_j(\tilde{e}_i b)$. Since Ψ_j is an embedding this gives the two equalities

$$\tilde{e}_i \tilde{f}_j^\star b = \tilde{f}_j^{\star m+1} \tilde{e}_i b_0 \ , \quad \tilde{e}_i b = \tilde{f}_j^{\star m} \tilde{e}_i b_0$$

from which the first of the commutation relations follows. The second is exactly the same. The third obtains by multiplying the first by $\star$ on both sides. For the fourth observe for $m > 0$ that $\tilde{e}_j^\star \tilde{e}_i b = \tilde{e}_j^\star \tilde{e}_i \tilde{f}_j^{\star m} b_0 = \tilde{e}_j^\star \tilde{f}_j^{\star m} \tilde{e}_i b_0 = \tilde{f}_j^{\star m-1} \tilde{e}_i b_0 = \tilde{e}_i \tilde{f}_j^{\star m-1} b_0 = \tilde{e}_i \tilde{e}_j^\star b$. On the other hand one already has $\tilde{e}_j^\star \tilde{e}_i b_0 = 0 = \tilde{e}_i \tilde{e}_j^\star b_0$ and so the last commutation relation is obtained .

6.1.15 The embedding theorem leads to a purely combinatorial description of $B(\infty)$. Let $I = \{i_n, i_{n-1}, \dots, i_1\}$ be an ordered subset of $\{1, 2, \dots, \ell\}^n$ and $\varphi : I \longrightarrow \mathbb{N}^n$. Set

$$\tilde{f}^\varphi = \tilde{f}_{i_n}^{k_n} \tilde{f}_{i_{n-1}}^{k_{n-1}} \dots \tilde{f}_{i_1}^{k_1} : k_j = \varphi(i_j) \ .$$

Define $\Psi_I : B(\infty) \longrightarrow B(\infty) \otimes C_{i_n} \otimes C_{i_{n-1}} \otimes \dots \otimes C_{i_1}$ inductively through $\Psi_I = (\Psi_{I'} \otimes 1d)\Psi_{i_1}$, where the prime denotes that i_1 has been omitted from I. By 6.1.14, Ψ_I is a strict embedding. Moreover it is easy to check that

$$\Psi_I(\tilde{f}^\varphi u_\infty) = u_\infty \otimes c_{i_n}(-k'_n) \otimes \dots \otimes c_{i_1}(-k'_1) \ ,$$

for some integers $k'_1, k'_2, \dots, k'_n$ (in general different to the $k_1, k_2, \dots, k_n$). Thus $B(\infty)$ may be viewed as subcrystal of the limit $u_\infty \otimes \dots \otimes C_{i_2} \otimes C_{i_1}$, in which each index appears infinitely many times. In particular, using the tensor product theorem one can determine $\varepsilon_i(\tilde{f}^\varphi)$ for each choice of φ and each i. By S_{10}^∞ this determines when two expressions $\tilde{f}^{\varphi_1}, \tilde{f}^{\varphi_2}$ represent the same element. (Here the sets I_1, I_2 do not need to coincide). Notice that the only data needed to obtain $B(\infty)$ is the Cartan matrix $(\alpha_i, \alpha_j) : i, j = \{1, 2, \dots, \ell\}$. One must take heed that the map omitting u_∞ is a morphism; but not strict. This difficulty may be avoided by taking $\tilde{e}_i c_i(0) = 0$.

Example. Take $\mathfrak{g}$ of type A_2 and $I = \{1, 2, 1\}$. One checks that

$$\Psi_I(\tilde{f}_1^r \tilde{f}_2^s \tilde{f}_1^t u_\infty) = \begin{cases} u_\infty \otimes c_1(0) \otimes c_2(-s) \otimes c_1(-r-t) & : t > s \\ u_\infty \otimes c_1(-r) \otimes c_2(-s) \otimes c_1(-t) & : r \leq s-t \\ u_\infty \otimes c_1(-(s-t)) \otimes c_2(-s) \otimes c_1(-t-(r-(s-t))) & : r \geq s-t \geq 0 . \end{cases}$$

In particular $\operatorname{Im} \Psi_I = u_\infty \otimes c_1(-u) \otimes c_1(-v) \otimes c_1(-w)$ with $0 \leq u \leq v$, $0 \leq v, w$. Moreover all these elements are distinct. One may go on to check that this defines a subcrystal of $C_1 \otimes C_2 \otimes C_1$ (i.e., is stable under $\tilde{f}_2$ also) and hence coincides with $B(\infty)$. Write the above element simply as (u, v, w). By 5.2.5(ii)

$$\varepsilon_1(u, v, w) = \max\{u, w - v + 2u\} \ , \quad \varepsilon_2(u, v, w) = v - u$$

$$wt(u, v, w) = -(u + w)\alpha_1 - v\alpha_2 \ .$$

In general for $\mathfrak{g}$ finite dimensional one may embed $B(\infty)$ in a fixed $C_{i_n} \otimes C_{i_{n-1}} \otimes \ldots \otimes C_{i_1}$ where $s_{i_n} s_{i_{n-1}} \cdots s_{i_1}$ is a reduced decomposition of the longest element w_0 of W. (This embedding is not strict).

6.1.16 Retain the above notation. Define φ to be normal if

$$(*) \quad \tilde{e}_{i_m} \tilde{f}_{i_{m+1}}^{k_{m+1}} \cdots \tilde{f}_{i_n}^{k_n} u_\infty = 0 \quad \text{for all} \quad m = 1, 2, \ldots, n \quad \text{with} \quad k_j = \varphi(i_j) .$$

and set

$$b_\varphi = \tilde{f}_{i_1}^{k_1} \cdots \tilde{f}_{i_n}^{k_n} u_\infty .$$

One easily checks from the embedding theorem that

$$\Psi_I(b_\varphi^\star) = u_\infty \otimes c_{i_n}(-k_n) \otimes \ldots \otimes c_{i_1}(-k_1) .$$

According to 6.1.12 there exists (I_1, φ_1) such that $\Psi_{I_1}(f^{\varphi_1} u_\infty) = \Psi_I(b_\varphi^\star)$, and then $b_\varphi^\star = f^{\varphi_1} u_\infty$. However the calculation of (I_1, φ_1), is a very difficult combinatorial problem. It appears that even to prove its existence would be very difficult by just combinatorial considerations.

Example. Take $\mathfrak{g}$ of type A_2 and I as in 6.1.15. Then $\varphi(1,2,1) = \{u, v, w\}$ is normal exactly when $u \leq v$. In this case one concludes that

$$(\tilde{f}_1^w \tilde{f}_2^v \tilde{f}_1^u u_\infty)^\star = \begin{cases} \tilde{f}_1^u \tilde{f}_2^v \tilde{f}_1^w u_\infty & : v \geq u + w , \\ \tilde{f}_1^{u+a} \tilde{f}_2^v \tilde{f}_1^{w-a} u_\infty & : a = u + w - v \geq 0 . \end{cases}$$

Notice that $(w - a, v, u + a) = (v - u, v, 2u + w - v)$ is also normal. Thus $\star$ does not take a particularly simple form even in type A_2.

6.1.17 Retain the notation of 6.1.16. A natural question that arises is whether one can choose $I_1 = \{i_1, i_2, \ldots, i_n\}$. This is obtained from sections 6.2, 6.3.

6.1.18 It is interesting to note the following consequence of 6.1.15 which follows easily from 5.2.5(ii).

Lemma. *Take $i, j \in \{1, 2, \ldots, \ell\}$ distinct. Then $\varepsilon_i(b) \leq \varepsilon_i(\tilde{f}_j b) \leq \varepsilon_i(b) + (\alpha_i^\vee, \alpha_j)$.*

6.2 Globalization

6.2.1 Recall the definition (6.1.17) of $U_q^{\mathbb{Z}}(\mathfrak{n}^-)$. For each $i \in \{1, 2, \ldots, \ell\}$ and $n \in \mathbb{N}$ set

$$(f_i^n U_q(\mathfrak{n}^-))^{\mathbb{Z}} := f_i^n U_q(\mathfrak{n}^-) \cap U_q^{\mathbb{Z}}(\mathfrak{n}^-) .$$

Lemma. $(f_i^n U_q(\mathfrak{n}^-))^{\mathbb{Z}} = \sum_{k \geq n} f_i^{(k)} U_q^{\mathbb{Z}}(\mathfrak{n}^-)$.

By 5.3.11 applied to 6.1.7(ii) every element of $U_q^{\mathbb{Z}}(\mathfrak{n}^-)$ can be expressed uniquely as

$$(*) \qquad \sum f_i^{(k)} u_k \text{ with } u_k \in \ker e_i' \cap U_q^{\mathbf{Z}}(\mathfrak{n}^-)$$

and so clearly this belongs to $f_i^n U_q(\mathfrak{n}^-)$ if and only if $u_k = 0$ for $k < n$. Hence the assertion.

6.2.2 Define $U_q^{\mathbf{Z}}(\mathfrak{n}^+)$ as in 6.1.7 interchanging e_i, f_i. Recall the definition (3.2.9) of τ and let $U_q^{\mathbf{Z}}(\mathfrak{g})$ denote the subring of $U_q(\mathfrak{g})$ generated by $U_q^{\mathbf{Z}}(\mathfrak{n}^-)$, $U_q^{\mathbf{Z}}(\mathfrak{n}^+)$ and the $\tau(\lambda) : \lambda \in P(\pi)$. For $\lambda \in P^+(\pi)$ let u_λ be a generator of weight λ of $V(\lambda)$ and set $V^{\mathbf{Z}}(\lambda) := U_q^{\mathbf{Z}}(\mathfrak{n}^-)u_\lambda$ which is clearly a $U_q^{\mathbf{Z}}(\mathfrak{g})$ module. Set $L^{\mathbf{Z}}(\lambda) = V^{\mathbf{Z}}(\lambda) \cap L(\lambda)$ which by S_2 equals $\pi_\lambda(U_q^{\mathbf{Z}}(\mathfrak{n}^-)) \cap \pi_\lambda(L(\infty)) \supset \pi_\lambda(L^{\mathbf{Z}}(\infty))$. Consequently $B(\lambda) = \pi_\lambda(B(\infty)) \subset L^{\mathbf{Z}}(\lambda)/qL^{\mathbf{Z}}(\lambda) \subset L(\lambda)/qL(\lambda)$, by S_8 and 6.1.8. It is immediate from 6.1.2(ii) that $L^{\mathbf{Z}}(\lambda)/qL^{\mathbf{Z}}(\lambda)$ is a free $\mathbb{Z}$ module with base $B(\lambda)$. For each $i \in \{1, 2, \ldots, \ell\}$, $n \in \mathbb{N}$ set

$$(*) \qquad (f_i^n V(\lambda))^{\mathbf{Z}} := (f_i^n U_q(\mathfrak{n}^-))^{\mathbf{Z}} u_\lambda = \sum_{k \geq n} f_i^{(k)} V^{\mathbf{Z}}(\lambda) ,$$

where the last equality obtains from 6.2.1.

6.2.3 Restricting ζ of 3.1.13 to $U_q(\mathfrak{n}^-)$ gives a ring homomorphism $a \mapsto \bar{a}$ of $U_q(\mathfrak{n}^-)$ defined by $\bar{q} = q^{-1}$, $\bar{f_i} = f_i$. Since by 4.3.6 (i)

$$\mathrm{Ann}_{U_q(\mathfrak{n}^-)} u_\lambda = \sum_{i=1}^{\ell} U_q(\mathfrak{n}^-) f_i^{\langle \alpha_i^\vee, \lambda \rangle + 1} ,$$

one may define an isomorphism $v \mapsto v^-$ of $V(\lambda)$ by $(au_\lambda)^- = \bar{a}u_\lambda$. Since $\overline{f_i^{(n)}} = f_i^{(n)}$ it follows that $U_q^{\mathbf{Z}}(\mathfrak{n}^-)$, hence $V^{\mathbf{Z}}(\lambda)$ and $(f_i^n V(\lambda))^{\mathbf{Z}} : n \in \mathbb{N}$ are stable by $v \mapsto v^-$.

In what follows $L^{\mathbf{Z}}(\lambda) \cap L^{\mathbf{Z}}(\lambda)^-$ is shown to map isomorphically to $L^{\mathbf{Z}}(\lambda)/qL^{\mathbf{Z}}(\lambda)$. Denoting the inverse isomorphism by G_λ, this provides a global basis, namely $G_\lambda(b) : b \in B(\lambda)$ for $V^{\mathbf{Z}}(\lambda)$. A particularly important corollary is the compatibility of these bases for different λ.

6.2.4 Consider the following general problem. Let V be a vector space over $\mathbb{Q}(q)$, M a $\mathbb{Z}[q, q^{-1}]$ submodule, L_+ (resp. L_-) a free A (resp. $\bar{A} := \zeta(A)$) submodule of V satisfying $V \xleftarrow{\sim} \mathbb{Q}(q) \otimes_A L_\pm$. Set $L = L_+$. Given a basis B for L/qL actually lying in $M \cap L/M \cap qL$ can one lift it to a basis for V lying in M? This is answered by studying when $M \cap L \cap L_- \longrightarrow M \cap L/M \cap qL$ is an isomorphism.

Lemma. *Let E be a $\mathbb{Z}$ submodule of V such that $E \longrightarrow L/qL$ is injective. Then*

(i) *E has no $\mathbb{Z}$ torsion.*
(ii) *$A \otimes_{\mathbb{Z}} E \longrightarrow L$, $\mathbb{Q}(q) \otimes_{\mathbb{Z}} E \longrightarrow V$ are injective.*
(iii) *$(\mathbb{Q}(q) \otimes_{\mathbb{Z}} E) \cap L = A \otimes_{\mathbb{Z}} E$.*

(i) is immediate. For the first part of (ii), suppose $\sum f_i(q) \otimes e_i$ belongs to the kernel of the map. By (i) one can assume the e_i to be linearly independent over $\mathbb{Q}$. Then $\sum f_i(0)e_i \in E \cap qL = 0$ and this implies the $f_i(q)$ to be all divisible by q. Dividing by q eventually gives a contradiction. The second part of (ii) and (iii) follow by eliminating powers of q in denominators.

6.2.5 Lemma. *Assume that $M \cap L \cap L_- \longrightarrow M \cap L / M \cap qL$ is an isomorphism. Then*

(i) $M \cap L \overset{\sim}{\longleftarrow} \mathbb{Z}[q] \otimes_{\mathbb{Z}} (M \cap L \cap L_-)$.
(ii) $M \cap L_- \overset{\sim}{\longleftarrow} \mathbb{Z}[q^{-1}] \otimes_{\mathbb{Z}} (M \cap L \cap L_-)$.
(iii) $M \overset{\sim}{\longleftarrow} \mathbb{Z}[q, q^{-1}] \otimes_{\mathbb{Z}} (M \cap L \cap L_-)$.
(iv) $M \cap L \cap L_- \overset{\sim}{\longrightarrow} M \cap L_- / M \cap qL_-$.
(v) $A \otimes_{\mathbb{Z}} (M \cap L \cap L_-) \overset{\sim}{\longrightarrow} M' \cap L$ *where* $M' = \mathbb{Q}(q) \otimes_{\mathbb{Z}[q,q^{-1}]} M$.
(vi) $(\mathbb{Q} \otimes_{\mathbb{Z}} M) \cap L \cap L_- \overset{\sim}{\longrightarrow} M' \cap L / M' \cap qL$, *with* M' *as in* (v).

Set $E = M \cap L \cap L_-$. Note that $M \cap q^k L = q^k(M \cap L), \ \ \forall\, k \in \mathbb{N}$. By the surjectivity hypothesis $M \cap L \subset E + q(M \cap L)$ and then by induction on n one has

$$M \cap L \subset \sum_{k=0}^{n} q^k E + q^{k+1}(M \cap L)\ .$$

Again $\sum_{k=0}^{n} q^k E \subset \sum_{k=0}^{n} q^k L_- = q^n L_-$. Thus

$$\sum_{k=0}^{n} q^k E \subset M \cap L \cap q^n L_- \subset \left(\sum_{k=0}^{\infty} q^k E + q^{n+1}(M \cap L) \right) \cap q^n L_-$$
$$= \sum_{k=0}^{\infty} q^k E + q^n (M \cap qL \cap L_-) = \sum_{k=0}^{\infty} q^k E$$

by the injectivity hypothesis. Then 6.2.4(ii) gives (i) and (iii) follows from (i). Again

$$M \cap L_- = \left(\bigcup_{n \in \mathbb{N}} q^{-n}(M \cap L) \right) \cap L_-$$
$$= \bigcup_{n \in \mathbb{N}} q^{-n}(M \cap L \cap q^n L_-)$$
$$= \mathbb{Z}[q^{-1}]E\ , \quad \text{by the above.}$$

Then (ii) follows from 6.2.4(ii) replacing q by q^{-1}. Now (iii) implies that $M \cap L_- \subset E + q^{-1}(M \cap L_-)$, whilst one already has $M \cap L \cap q^{-1} L_- = q^{-1}(M \cap qL \cap L_-) = 0$. Hence (iv).

Set $S = \{f(q) \in \mathbb{Z}[q] | f(0) \neq 0\}$, so $A = S^{-1}\mathbb{Z}[q]$ by definition. Then $M' \cap L = S^{-1} M \cap L = S^{-1}(M \cap L) \overset{\sim}{\longleftarrow} A \otimes_{\mathbb{Z}} E$, by (i). Hence (v). Finally by (v) one has $M' \cap L / q(M' \cap L) \overset{\sim}{\longrightarrow} \mathbb{Q} \otimes_{\mathbb{Z}} E$, hence (vi).

6.2.6 Lemma. *Let E be a $\mathbb{Z}$ module and $\varphi : E \longrightarrow M \cap L \cap L_-$ a homomorphism. Suppose that*

(a) $E \longrightarrow L/qL$, $E \longrightarrow L_-/qL_-$ *are injective.*
(b) $M = \mathbb{Z}[q, q^{-1}]\varphi(E)$.

Then $E \xrightarrow{\sim} M \cap L \cap L_- \xrightarrow{\sim} M \cap L/M \cap qL$.

By (a), φ is injective and so one may view E as a submodule of $M \cap L \cap L_-$. By (b) there is a surjective map $\mathbb{Z}[q, q^{-1}] \otimes_{\mathbb{Z}} E \longrightarrow M$ which is injective by (a) and 6.2.4(ii). Then 6.2.4(iii) gives $\mathbb{Z}[q] \otimes_{\mathbb{Z}} E = (\mathbb{Z}[q, q^{-1}] \cap A) \otimes_{\mathbb{Z}} E = \mathbb{Z}[q, q^{-1}] \otimes_E A \cap A \otimes_{\mathbb{Z}} E \xrightarrow{\sim} M \cap L$. Similarly $\mathbb{Z}[q^{-1}] \otimes_{\mathbb{Z}} E \xrightarrow{\sim} M \cap L_-$. These give the asserted conclusions.

6.2.7 Lemma. *Let* N *be a* $\mathbb{Z}[q, q^{-1}]$ *submodule of* M *such that*

(a) $N \cap L \cap L_- \xrightarrow{\sim} N \cap L/N \cap qL$

Let F *be a* $\mathbb{Z}$ *module and* $\varphi : F \longrightarrow M \cap (L+N) \cap (L_- + N)$ *a homomorphism such that*

(b) $M = \mathbb{Z}[q, q^{-1}]\varphi(F) + N$
(c) *The homomorphisms* $\psi_\pm : F \longrightarrow (L_\pm + \mathbb{Q} \otimes_{\mathbb{Z}} N)/(q^{\pm 1} L_\pm + \mathbb{Q} \otimes_{\mathbb{Z}} N)$ *induced by* φ *are injective.*

Then

(i) $M \cap L \cap L_- \xrightarrow{\sim} M \cap L/M \cap qL$.
(ii) $0 \longrightarrow N \cap L/N \cap qL \longrightarrow M \cap L/M \cap qL \xrightarrow{g} (L + \mathbb{Q} \otimes_{\mathbb{Z}} N)/(qL + \mathbb{Q} \otimes_{\mathbb{Z}} N)$ *is exact.*
(iii) $\operatorname{Im} g = \operatorname{Im} \psi_+$.

By (c), φ is injective so one may view F as a submodule of $M \cap (L + N) \cap (L_- + N)$. Applying 4.2.2(i)–(iii) to (a) gives $N = N \cap L + N \cap L_-$, so $L_- + N = L_- + N \cap L$ and then

$$\begin{aligned} M \cap (L+N) \cap (L_- + N) &= (N + M \cap L) \cap (L_- + N) \quad \text{since } N \subset M \\ &= N + M \cap L \cap (L_- + N)\ , \\ &= N + M \cap L \cap (L_- + N \cap L), \quad \text{by the above} \\ &= N + M \cap L \cap L_-\ . \end{aligned}$$

Thus one may just take F to be a submodule of $M \cap L \cap L_-$. By (c), F has no $\mathbb{Z}$ torsion and $F \cap \mathbb{Q} \otimes_{\mathbb{Z}} N = 0$. Hence $F \cap N = 0$ and so one may view $F \oplus (N \cap L \cap L_-)$ as a submodule of $M \cap L \cap L_-$ which in turn maps to L/qL. Again $\ker(L \longrightarrow (L + \mathbb{Q} \otimes_{\mathbb{Z}} N)/(qL + \mathbb{Q} \otimes_{\mathbb{Z}} N)) = qL + L \cap (\mathbb{Q} \otimes_{\mathbb{Z}} N)$. This gives a commuting diagram

$$\begin{array}{ccccccccc} 0 & \longrightarrow & N \cap L \cap L_- & \longrightarrow & F \oplus (N \cap L \cap L_-) & \xrightarrow{m \mapsto (m+N)/N} & F & \longrightarrow & 0 \\ & & \downarrow \beta & & \downarrow \alpha & & \downarrow \psi_+ & & \\ 0 & \longrightarrow & \mathbb{Q} \otimes_{\mathbb{Z}} (N \cap L \cap L_-) & \longrightarrow & L/qL & \xrightarrow{\gamma} & (L + \mathbb{Q} \otimes_{\mathbb{Z}} N)/(qL + \mathbb{Q} \otimes_{\mathbb{Z}} N) & & \end{array}$$

Moreover $\ker\gamma = (qL + L\cap \mathbb{Q}\otimes_{\mathbb{Z}} N)/qL = L\cap(\mathbb{Q}\otimes_{\mathbb{Z}} N)/qL\cap\mathbb{Q}\otimes_{\mathbb{Z}} N = \mathbb{Q}\otimes_{\mathbb{Z}}(N\cap L/N\cap qL)$ so the last row is exact by (a). Now ψ_+ is injective by (c) and β is injective since L is defined over $\mathbb{Q}$. Consequently α is injective. Similarly $F\oplus(N\cap L\cap L_-)\longrightarrow L_-/q^{-1}L_-$ is injective. Applying 6.2.6 with $E = F\oplus(N\cap L\cap L_-)$ one concludes that $F\oplus(N\cap L\cap L_-)\xrightarrow{\sim} M\cap L\cap L_-\xrightarrow{\sim} M\cap L/M\cap qL$. Hence (i). Finally observe that $g=\gamma\alpha$ so $\ker g = N\cap L\cap L_- = N\cap L/N\cap qL$ and $\operatorname{Im} g = \operatorname{Im}\psi_+$.

6.2.8 Recall the conventions of 5.3.15 and 5.4.1. Set $V^{\mathbb{Z}}(\infty) = U_q^{\mathbb{Z}}(\mathfrak{n}^-)$. Observe that $V_k^{\mathbb{Z}}(\lambda^\varepsilon)\cap L(\lambda^\varepsilon) = L_k^{\mathbb{Z}}(\lambda^\varepsilon)$. Consider the conditions

$G_1^\varepsilon(k)$ *The map* $V_k^{\mathbb{Z}}(\lambda^\varepsilon)\cap L(\lambda^\varepsilon)\cap L(\lambda^\varepsilon)^-\longrightarrow L_k^{\mathbb{Z}}(\lambda^\varepsilon)/qL_k^{\mathbb{Z}}(\lambda^\varepsilon)$ *is an isomorphism.*

Given $G_1^\varepsilon(k)$, let G_{λ^ε} denote the inverse isomorphism. Set $G_{\lambda^\infty} = G$. Observe from 6.1.8 and 6.2.2 that

$$(\varepsilon)\qquad b = G_{\lambda^\varepsilon}(b) \bmod qL_k^{\mathbb{Z}}(\lambda^\varepsilon), \quad \textit{for all} \quad b\in B_k(\lambda^\varepsilon)\ .$$

$G_2(k)$ *For all* $\xi\in Q_k^-$, $n\in\mathbb{N}$, $b\in\tilde f_i^n B(\infty)_{\xi+n\alpha_i}$ *one has* $G(b)\in f_i^n V(\infty)$.

$$G_3^\varepsilon(k)\qquad V_k^{\mathbb{Z}}(\lambda^\varepsilon) = \bigoplus_{b\in B_k(\lambda^\varepsilon)}\mathbb{Z}[q,q^{-1}]G_{\lambda^\varepsilon}(b)\ .$$

It follows from 6.2.5 (iii), 6.1.8 (i) and 6.2.2 that

Lemma. $G_1^\varepsilon(k)\Longrightarrow G_3^\varepsilon(k)$.

6.2.9 Assume that $G_1^\varepsilon(k)$ hold. Then S_4 gives a commutative diagram

$$\begin{array}{ccc} V_k^{\mathbb{Z}}(\infty)\cap L(\infty)\cap L(\infty)^- & \xleftarrow{G} & L_k^{\mathbb{Z}}(\infty)/qL_k^{\mathbb{Z}}(\infty)\\ \downarrow \pi_\lambda & & \downarrow \bar\pi_\lambda\\ V_k^{\mathbb{Z}}(\lambda)\cap L(\lambda)\cap L(\lambda)^- & \xleftarrow{G_\lambda} & L_k^{\mathbb{Z}}(\lambda)/qL_k^{\mathbb{Z}}(\lambda)\ . \end{array}$$

In particular, $G_1^\varepsilon(k)$ imply

$$G_4(k)\qquad G(b)u_\lambda = \pi_\lambda G(b) = G_\lambda(\bar\pi_\lambda b)\ \text{ for all }\ b\in B_k(\infty)\ .$$

6.2.10 Observe that injectivity in $G_1^\infty(k)$ means that $V_k^{\mathbb{Z}}(\infty)\cap qL(\infty)\cap L(\infty)^- = 0$. Thus given $G_1^\infty(k)$ one has for all $b\in L_k^{\mathbb{Z}}(\infty)/qL_k^{\mathbb{Z}}(\infty)$ that

$$\begin{aligned}(q-q^{-1})^{-1}(G(b)-\overline{G(b)}) &= -q(1-q^2)^{-1}(G(b)-\overline{G(b)})\\ &= q^{-1}(1-q^{-2})^{-1}(G(b)-\overline{G(b)})\in qL(\infty)\cap q^{-1}L(\infty)\ .\end{aligned}$$

On the other hand for any $a\in\mathbb{Z}[q,q^{-1}]$ one has $(q-q^{-1})^{-1}(a-\bar a)\in\mathbb{Z}[q,q^{-1}]$ and so $(q-q^{-1})^{-1}(G(b)-\overline{G(b)})$ also lies in $V_k^{\mathbb{Z}}(\infty)$ and hence vanishes. That is $G(b)=\overline{G(b)}$.

6.2.11 Fix $k \in \mathbb{N}$, $n \in \mathbb{N}^+$. Consider

$G_5^{0,n}(k)$ *For all* $\xi \in Q_k^-$, *one has isomorphisms*

$$(f_i^n V(\lambda))_{\lambda+\xi}^{\mathbb{Z}} \cap L(\lambda) \cap L(\lambda)^- \xrightarrow{\sim} (f_i^n V(\lambda))_{\lambda+\xi}^{\mathbb{Z}} \cap L(\lambda)/(f_i^n V(\lambda))_{\lambda+\xi}^{\mathbb{Z}} \cap qL(\lambda)$$
$$\xrightarrow{\sim} \bigoplus_{b \in B(\lambda)_{\lambda+\xi} \cap \tilde{f}_i^n B(\lambda)} \mathbb{Z}b \,.$$

Proposition. $G_1^\infty(k)$, $G_2(k)$, $G_3^0(k)$, $G_4(k)$, $G_k^{0,n+1}(k+1) \Longrightarrow G_5^{0,n}(k+1)$.

By 6.2.2(∗)

$$(*) \qquad M^n := (f_i^n V(\lambda))_{\lambda+\xi}^{\mathbb{Z}} = f_i^{(n)} V^{\mathbb{Z}}(\lambda)_{\lambda+\xi+n\alpha_i} + (f_i^{n+1} V(\lambda))_{\lambda+\xi}^{\mathbb{Z}} \,.$$

Since $n \geq 1$ by hypothesis, $G_3^0(k)$, $G_4(k)$ give

$$V^{\mathbb{Z}}(\lambda)_{\lambda+\xi+n\alpha_i} = \bigoplus_{b \in S} \mathbb{Z}[q, q^{-1}] G(b) u_\lambda \,,$$

where $S = \{b \in B(\infty)_{\xi+n\alpha_i} | \bar{\pi}_\lambda(b) \neq 0\}$.

Take $b \in S$. If $\tilde{e}_i b \neq 0$, then S_{10}^∞ gives $b = \tilde{f}_i B(\infty)$ and so $G(b) \in (f_i V(\infty))^{\mathbb{Z}}$ by $G_1^\infty(k)$ and $G_2(k)$. Yet

$$(**) \qquad \begin{aligned} f_i^{(n)}(f_i V(\infty))^{\mathbb{Z}} &= f_i^{(n)}(f_i V(\infty) \cap V^{\mathbb{Z}}(\infty)) \\ &= f_i^{n+1} V(\infty) \cap V^{\mathbb{Z}}(\infty) = (f_i^{n+1} V(\infty))^{\mathbb{Z}} \,. \end{aligned}$$

Consequently by S_9^+ one can just take

$$S = \bar{\pi}_\lambda^{-1}\{b \in B(\lambda)_{\lambda+\xi+n\alpha_i} | \tilde{e}_i b = 0\}$$

in the above. Now take $V = V(\lambda)_{\lambda+\xi}$, $M = M^n$, $N = M^{n+1}$, $L = L(\lambda)_{\lambda+\xi}$, $L_- = L(\lambda)_{\lambda+\xi}^-$ and

$$F = \bigoplus_{b \in S} \mathbb{Z} f_i^{(n)} G(b) u_\lambda$$

in 6.2.7. Hypothesis (a) of 6.2.7 is just $G_5^{0,n+1}(k+1)$. Let P_i be the projector onto $\ker e_i'$ defined by the decomposition $V(\infty) = \ker e_i' \oplus f_i V(\infty)$. Then 6.1.7 gives $V^{\mathbb{Z}}(\infty) = \ker e_i' \oplus (f_i V(\infty))^{\mathbb{Z}}$ and so by (∗∗) that

$$f_i^{(n)} G(b) = f_i^{(n)} P_i G(b) \bmod (f_i^{n+1} V(\infty))^{\mathbb{Z}} \,.$$

Again writing

$$G(b) = P_i G(b) + \sum_{s>0} f_i^{(s)} u_s : u_s \in V^{\mathbb{Z}}(\infty) \cap \ker e_i'$$

gives

$$\tilde{f}_i^n G(b) = f_i^{(n)} P_i G(b) + \sum_{s>0} f_i^{(n+s)} u_s = f_i^{(n)} P_i G(b) \bmod (f_i^{n+1} V(\infty))^{\mathbb{Z}}$$

by 6.2.1. Thus by $G_1^\infty(k)$ and 6.2.8 (∞) one obtains

$$(***) \qquad \tilde{f}_i^n(b) = f_i^{(n)} G(b) \bmod qL^{\mathbb{Z}}(\infty) + (f_i^{n+1} V(\infty))^{\mathbb{Z}}$$

for all $s \in S$. From S_2 one concludes that $F \subset (L+N) \cap (L_- + N)$. Moreover the definition of F gives $F \subset M$ and from $(*)$ that hypothesis (b) of 6.2.7 is satisfied. To check hypothesis (c) set

$$\begin{aligned} H &= (L + \mathbb{Q} \otimes_{\mathbb{Z}} N)/(qL + \mathbb{Q} \otimes_{\mathbb{Z}} N) \\ &= (L/qL)/((\mathbb{Q} \otimes_{\mathbb{Z}} N) \cap L)/(\mathbb{Q} \otimes_{\mathbb{Z}} N) \cap qL) \\ &= \bigoplus_{b \in S'} \mathbb{Q}b \,, \quad \text{by} \quad G_5^{0,n+1}(k+1) \,, \end{aligned}$$

where $S' = B(\lambda)_{\lambda+\xi} \setminus (B(\lambda)_{\lambda+\xi} \cap \tilde{f}_i^{n+1} B(\lambda))$. Now for $b \in S$ one has by $(***)$ that the image of $f_i^{(n)} G(b) u_\lambda$ in H is just $\bar{\pi}_\lambda(\tilde{f}_i^n b)$. Yet $\bar{\pi}_\lambda(S) = B(\lambda)_{\lambda+\xi+n\alpha_i} \setminus \tilde{f}_i B_{\lambda+\xi+(n+1)\alpha_i}$ by definition and S_{10}^0. Thus $\bar{\pi}_\lambda(\tilde{f}_i^n S) = \tilde{f}_i^n B(\lambda)_{\lambda+\xi+n\alpha_i} \setminus \tilde{f}_i^{n+1} B_{\lambda+\xi+(n+1)\alpha_i} \subset B(\lambda)_{\lambda+\xi} \setminus (B(\lambda)_{\lambda+\xi} \cap \tilde{f}_i^{n+1} B(\lambda)) = S'$. This proves the injectivity of ψ_+ and that of ψ_- is obtained by applying ζ. Then 6.2.7(i) gives the first part of $G_5^{0,n}(k+1)$. The second part follows from the second part of $G_5^{0,n+1}(k+1)$, the above formula for $\bar{\pi}_\lambda(\tilde{f}_i^n S)$ and 6.2.7 (iii).

6.2.12 Since $G_5^{0,n}(k+1)$ holds trivially for $n \gg 0$ one obtains

Corollary. $G_1^\infty(k)$, $G_2(k)$, $G_3^0(k)$, $G_4(k) \Longrightarrow G_5^{0,n}(k+1)$ *for all* $n \geq 1$.

6.2.13 Replacing λ by ∞ in $G_5^{0,n}(k)$ gives an assertion which is called $G_5^{\infty,n}(k)$. By 5.3.17

Lemma. $G_5^{0,n}(k+1) \Longrightarrow G_5^{\infty,n}(k+1)$.

6.2.14 Use of $G_5^{\epsilon,1}(k+1)$ gives a commutative diagram

$$\begin{array}{ccc} f_i V_{k+1}^{\mathbb{Z}}(\infty) \cap L(\infty) \cap L(\infty)^- & \xrightarrow{\sim} & \bigoplus_{\tilde{f}_i B(\infty) \cap B_{k+1}(\infty)} \mathbb{Z}b \\ \downarrow \pi_\lambda & & \downarrow \bar{\pi}_\lambda \\ f_i V_{k+1}^{\mathbb{Z}}(\lambda) \cap L(\lambda) \cap L(\lambda)^- & \xrightarrow{\sim} & \bigoplus_{\tilde{f}_i B(\lambda) \cap B_{k+1}(\lambda)} \mathbb{Z}b \,. \end{array}$$

Let G_i denote the inverse of the map in the top row.

Lemma. *Given* $b \in \tilde{f}_i B(\infty) \cap \tilde{f}_j B(\infty) \cap B_{k+1}(\infty)$ *one has* $G_i(b) = G_j(b)$.

One may write $b = \tilde{f}_{i_1} \ldots \tilde{f}_{i_r}.1$. Choose $\lambda \in P^+(\pi)$ such that $(\alpha_{i_r}^\vee, \lambda) = 0$ and $(\alpha_s^\vee, \lambda) \gg 0$ for $s \neq i_r$. Then

$$V_{k+1}(\lambda) \cong V_{k+1}(\infty)/V_k(\infty) f_{i_r} \,.$$

Now $(\alpha_{i_r}^\vee, \lambda) = 0$ implies $\tilde{f}_{i_r} u_\lambda = 0$ and hence $\bar{\pi}_\lambda(b) = 0$. Through the above commutative diagram one obtains $G_i(b)u_\lambda = 0$ which by $(*)$ gives $G_i(b) \in V_k(\infty) f_{i_r}$. Combined with a similar assertion for j and 6.2.8 (∞) one obtains

$$Q := G_i(b) - G_j(b) = G_i(b) - b - (G_j(b) - b) \in V_{k+1}(\infty) f_{i_r} \cap qL(\infty) \cap L(\infty)^- .$$

Then by 6.1.3 (iii) and injectivity in $G_5^{\infty,1}(k+1)$

$$Q^* \in f_{i_r} V_{k+1} \cap qL(\infty) \cap L(\infty)^- = 0$$

and so $Q = Q^{**} = 0$.

6.2.15 Corollary

(i) $G_5^{\infty,1}(k+1)$, $G_5^{0,1}(k+1) \Longrightarrow G_1^\infty(k+1)$, $G_1^0(k+1)$.

(ii) $G_5^{\infty,n}(k+1) : \forall n \in \mathbb{N}^+$, $G_5^{0,1}(k+1) \Longrightarrow G_2(k+1)$.

(i) Obviously, $B_{k+1}(\infty) = u_\infty \cup \bigcup_{i=1}^{\ell} \tilde{f}_i B_k(\infty)$. By 6.2.1

$$M := V_{k+1}^{\mathbb{Z}}(\infty) = Z[q, q^{-1}]u_\infty + \sum_{i=1}^{\ell} f_i V_k^{\mathbb{Z}}(\infty) .$$

Then $G_5^{\varepsilon,1}(k+1)$ and 6.2.14 give a homomorphism G of $\mathbb{Z}$ modules

$$E := \bigoplus_{b \in B_{k+1}(\infty)} \mathbb{Z}b \longrightarrow M \cap L(\infty) \cap L(\infty)^-$$

defined by setting $Gu_\infty = u_\infty$ and $G(b) = G_i(b)$ when $b \in \tilde{f}_i B_k(\infty)$. By 6.2.5 (iii) applied to $G_5^{\infty,1}(k+1)$ one has

$$M = \mathbb{Z}[q, q^{-1}]u_\infty + \sum_{i=1}^{\ell} \sum_{b \in \tilde{f}_i B_k(\infty)} Z[q, q^{-1}]G_i(b) = \sum_{b \in B_{k+1}(\infty)} Z[q, q^{-1}]G(b) .$$

Taking $\varphi = G$ in 6.2.6 this establishes hypothesis (b) of 6.2.6. Hypothesis (a) obtains from S_3^∞. The conclusion of 6.2.6 then gives $G_1^\infty(k+1)$.

Take $b \in B_{k+1}(\infty)$ and suppose $\bar{\pi}_\lambda(b) = 0$. Then $b \in \tilde{f}_i B_k(\infty)$ for some i, and $G(b)u_\lambda = G_i(b)u_\lambda = 0$ as already seen in 6.2.14. Recalling S_8 one may define a $\mathbb{Z}$ module homomorphism G_λ from $E_\lambda := \oplus_{b \in B_{k+1}(\lambda)} \mathbb{Z}b \longrightarrow Mu_\lambda \cap L(\lambda) \cap L(\lambda)^-$ by $G_\lambda(\bar{\pi}_\lambda(b)) = G(b)u_\lambda$. Since $V_{k+1}^{\mathbb{Z}}(\lambda) = V_{k+1}^{\mathbb{Z}}(\infty)u_\lambda = Mu_\lambda$ by the definition of $V^{\mathbb{Z}}(\lambda)$, one concludes recalling S_3^0 that the hypotheses of 6.2.6 are satisfied. Then $G_1^0(k+1)$ obtains from its conclusion.

(ii) Define G as above. Applying 6.2.5 (iii) to $G_5^{\infty,n}(k+1)$ gives

$$(f_i^n V(\infty))_\xi^{\mathbb{Z}} = \bigoplus_{b \in \tilde{f}_i^n B(\infty) \cap B(\infty)_\xi} \mathbb{Z}[q, q^{-1}]G(b) : n \geq 1 ,$$

from which $G_2(k+1)$ is immediate.

6.2.16 Set $U_q^{\mathbb{Q}}(\mathfrak{n}^-) = \mathbb{Q} \otimes_{\mathbb{Z}} U_q^{\mathbb{Z}}(\mathfrak{n}^-)$ which is just the $\mathbb{Q}[q, q^{-1}]$ subring of $U_q(\mathfrak{n}^-)$ generated by the $f_i^{(n)}$: $n \in \mathbb{Z}$, $i \in \{1, 2, \ldots, \ell\}$. Then $V^{\mathbb{Q}}(\lambda) := \mathbb{Q} \otimes_{\mathbb{Z}} V^{\mathbb{Z}}(\lambda) = U_q^{\mathbb{Q}}(\mathfrak{n}^-)u_\lambda$. Since the $G_1^{\varepsilon}(k)$ are obvious for $k = 0$ it follows from 6.2.8–6.2.15 that they hold for all k. Through 6.2.5 (v) their conclusion gives a commutative diagram

$$\begin{array}{ccc} U_q^{\mathbb{Q}}(\mathfrak{n}^-) \cap L(\infty) \cap L(\infty)^- & \overset{G}{\underset{\sim}{\longleftarrow}} & L(\infty)/qL(\infty) \\ \downarrow \pi_\lambda & & \downarrow \bar{\pi}_\lambda \\ V^{\mathbb{Q}}(\lambda) \cap L(\lambda) \cap L(\lambda)^- & \overset{G_\lambda}{\underset{\sim}{\longleftarrow}} & L(\lambda)/qL(\lambda) \, . \end{array}$$

Moreover 6.2.5 (iii) gives

$$U_q^{\mathbb{Q}}(\mathfrak{n}^-) = \bigoplus_{b \in B(\infty)} \mathbb{Q}[q, q^{-1}]G(b) \, , \qquad V^{\mathbb{Q}}(\lambda) = \bigoplus_{b \in B(\lambda)} \mathbb{Q}[q, q^{-1}]G_\lambda(b) \, .$$

Further tensoring over $\mathbb{Q}(q)$ shows that $\{G(b)\}_{b \in B(\infty)}$ is a basis for $U_q(\mathfrak{n}^-)$ with the property that for each $\lambda \in P^+(\pi)$, the subset $\{G(b)u_\lambda | G(b)u_\lambda \neq 0\}$ is a basis for $V(\lambda)$. All this may be expressed in a dual version as follows.

6.2.17 Recall (3.2.6) the antiautomorphism $\varkappa$ of $U_q(\mathfrak{g})$ defined by $\varkappa(e_i) = f_i$, $\varkappa(f_i) = e_i$, $\varkappa(t_i) = t_i$. Let $\varepsilon : U_q(\mathfrak{g}) \longrightarrow \mathbb{Q}(q)$ be the augmentation and recall the adjoint action of $U_q(\mathfrak{n}^+)$ on $U_q(\mathfrak{n}^-)$ defined in 5.3.1. Recall (5.3.9) the form $(\,,)$ defined on $U_q(\mathfrak{n}^-)$.

Lemma. *For all $\mu, \nu \in \mathbb{N}\pi$ one has*

$$\varepsilon(\operatorname{ad} \varkappa(a_{-\mu})b_{-\nu}) = \frac{q^{(\mu,\mu)/2}}{\Pi_i(1 - q_i^2)^{k_i}} (a_{-\mu}, b_{-\nu}), \quad \textit{where} \quad \mu = \sum k_i \alpha_i \, .$$

The proof is by induction on $|\mu|$. It is clear for $|\mu| = 0$. Note that one may assume $\mu = \nu$ without loss of generality. Denote the scale factor in the right hand side above by $q(\mu)$ and take $a_{-\mu} = f_i a_{-\mu+\alpha_i}$. Then by the induction hypothesis

$$\begin{aligned} \varepsilon(\operatorname{ad} \varkappa(a_{-\mu})b_{-\mu}) &= \varepsilon(\operatorname{ad} \varkappa(a_{-\mu+\alpha_i})(\operatorname{ad} e_i)b_{-\mu}) \\ &= q(\mu - \alpha_i)(a_{-\mu+\alpha_i}, (\operatorname{ad} e_i)b_{-\mu}) \, . \end{aligned}$$

Yet 5.3.1 gives $(\operatorname{ad} e_i)a_{-\mu} = -(q_i - q_i^{-1})^{-1}(\operatorname{ad} t_i^{-1})e_i' b_{-\mu} = -q_i^{-2+(\alpha_i^\vee,\mu)}(q_i - q_i^{-1})^{-1}(e_i' b_{-\mu})$. Recalling 5.3.9 it remains to observe that

$$-q(\mu - \alpha_i)q_i^{-2+(\alpha_i^\vee,\mu)}(q_i - q_i^{-1})^{-1} = q(\mu) \, .$$

6.2.18 For each subspace $S \subset U_q(\mathfrak{n}^-)$ let $S^\perp$ denote its orthogonal with respect to the form $(\,,)$ defined in 5.3.9. Identifying $U_q(\mathfrak{n}^-)$ with $U_q(\mathfrak{n}^-)\tau(-2\lambda)$

gives via 5.3.10 an embedding $V(\lambda) \hookrightarrow U_q(\mathfrak{n}^-)$ of $U_q(\mathfrak{g})$ modules for the twisted action of $U_q(\mathfrak{g})$. Since $\operatorname{ad}_{-2\lambda} x = \operatorname{ad} x$ for all $x \in U_q(\mathfrak{n}^+)$, this is a $U_q(\mathfrak{n}^+)$ homomorphism for the adjoint action.

Lemma. *For each $\lambda \in P^+(\pi)$ one has*

$$V(\lambda) = (\operatorname{Ann}_{U_q(\mathfrak{n}^-)} u_\lambda)^\perp .$$

Let $F = \mathcal{S}^\lambda$ denote the contravariant form (3.4.10) on $V(\lambda)$ defined by $F(u_\lambda, u_\lambda) = 1$ and $\varkappa$. Recall that in the above embedding u_λ identifies with 1. Then for all $b \in V(\lambda)$, $a \in U_q(\mathfrak{n}^-)$ one has $\varepsilon(\operatorname{ad} \varkappa(a)b) = F(\operatorname{ad} \varkappa(a)b, u_\lambda) = F(b, au_\lambda)$. Then by 6.2.17 and the non-degeneracy of F one obtains $V(\lambda)^\perp = \operatorname{Ann}_{U_q(\mathfrak{n}^-)} u_\lambda$. Since $(\ ,\ )$ restricts to a non-degenerate form on each weight space and since weight spaces are finite dimensional, this gives the lemma.

6.2.19 Let $H(b) : b \in B(\infty)$ be a basis of $U_q(\mathfrak{n}^-)$ satisfying $(H(b), G(b')) = \delta_{bb'}$, that is to say a dual basis to $G(b) : b \in B(\infty)$ relative to the form $(\,,)$.

Theorem. *For each $\lambda \in P^+(\pi)$ the subset $\{H(b) : b \in B(\lambda)\}$ is a basis for the image (6.2.18) of $V(\lambda)$ in $U_q(\mathfrak{n}^-)$.*

6.2.20 The following immediate consequence of 6.2.19 has important implications for the structure theory of $U_q(\mathfrak{g})$.

Corollary. *The $V(\lambda) : \lambda \in P^+(\pi)$ generate a distributive lattice of subspaces of $U_q(\mathfrak{n}^-)$.*

6.3 The Demazure Property

6.3.1 The following reformulates slightly 6.2.4–6.2.6.

Let V be a $\mathbb{Q}(q)$ vector space. Let $V^\mathbb{Q}$ (resp. L, L_-) be a $\mathbb{Q}[q, q^{-1}]$ (resp. A, $\bar{A}$) submodule of V such that

$$V \xleftarrow{\sim} \mathbb{Q}(q) \otimes_{\mathbb{Q}[q,q^{-1}]} V^\mathbb{Q} , \quad V \xleftarrow{\sim} \mathbb{Q}(q) \otimes_A L , \quad V \xleftarrow{\sim} \mathbb{Q}(q) \otimes_{\bar{A}} L_- .$$

Lemma

(i) $L \xleftarrow{\sim} A \otimes_{\mathbb{Q}[q]} (V^\mathbb{Q} \cap L)$

(ii) $L_- \xleftarrow{\sim} A \otimes_{\mathbb{Q}[q^{-1}]} (V^\mathbb{Q} \cap L_-)$.

Both parts are similar, so just consider (i). Injectivity is immediate from the hypothesis. For surjectivity, take $u \in L \subset V$. Then from the hypothesis there exists $0 \neq f(q) \in \mathbb{Q}[q]$ such that $f(q)u \in V^\mathbb{Q}$. One can suppose $f(0) \neq 0$. Then $f(q)^{-1} \in A$ and so $u \in A(V^\mathbb{Q} \cap L)$.

6.3.2 Retain the above hypotheses and notation. Set $E = V^\mathbb{Q} \cap L \cap L_-$.

Lemma. *The following three conditions are equivalent*

(i) $E \xrightarrow{\sim} L/qL$.
(ii) $E \xrightarrow{\sim} L_-/q^{-1}L_-$.
(iii) $A \otimes_{\mathbb{Q}} E \xrightarrow{\sim} L$, $\bar{A} \otimes_{\mathbb{Q}} E \xrightarrow{\sim} L_-$.

Obviously (iii) $\Longrightarrow$ (i), (ii). Consider (i) $\Longrightarrow$ (iii). The map $L \cap V^{\mathbb{Q}} \longrightarrow L/qL$ factors to an injection $L \cap V^{\mathbb{Q}}/qL \cap V^{\mathbb{Q}} \hookrightarrow L/qL$ which by 6.3.1 (i) is surjective. Taking $M = V^{\mathbb{Q}}$ in 6.2.5 one obtains the first (resp. second) part of (iii) from 6.2.5 (i) and 6.3.1 (i) (resp. 6.2.5 (ii) and 6.3.1 (ii)). Similarly (ii) $\Longrightarrow$ (iii).

6.3.3 One calls a triple $\{V^{\mathbb{Q}}, L, L_-\}$ satisfying one the equivalent conditions of 6.3.2 *balanced*. Notice that for a balanced triple

(i) $\mathbb{Q}[q] \otimes_{\mathbb{Q}} E \xrightarrow{\sim} V^{\mathbb{Q}} \cap L$ so $A \otimes_{\mathbb{Q}} E \xrightarrow{\sim} L$,
(ii) $\mathbb{Q}[q^{-1}] \otimes_{\mathbb{Q}} E \xrightarrow{\sim} V^{\mathbb{Q}} \cap L_-$ so $\bar{A} \otimes_{\mathbb{Q}} E \xrightarrow{\sim} L^-$,

and by 6.2.5 (iii)

(iii) $\mathbb{Q}[q, q^{-1}] \otimes_{\mathbb{Q}} E \xrightarrow{\sim} V^{\mathbb{Q}}$.
In particular, $V^{\mathbb{Q}}$, L, L_- are free over the appropriate rings.

6.3.4 Here and in 6.3.5–6.3.8, it is assumed that $\ell = 1$ and the i subscript is omitted. Let V be an integrable $U_q(\mathfrak{g})$ module and $V^{\mathbb{Q}}$ a $U_q^{\mathbb{Q}}(\mathfrak{g}) := \mathbb{Q} \otimes_{\mathbb{Z}} U_q^{\mathbb{Z}}(\mathfrak{g})$ submodule of V. Let (L, B) be a crystal basis for V and assume that $(V^{\mathbb{Q}}, L, L^-)$ form a balanced triple. Let G denote the inverse of the isomorphism $E := V^{\mathbb{Q}} \cap L \cap L^- \xrightarrow{\sim} L/qL$. Obviously G respects weight space decomposition. In particular for each $\xi \in Q(\pi)$, $b \in B_\xi$ one has $b - G(b) \in qL_\xi$. Now choose ξ such that $L_{\xi+\alpha} = 0$. In particular $e_i L_\xi = e_i L_\xi^- = 0$, so $\tilde{f}^n L_\xi = f^{(n)} L_\xi$ and $\tilde{f}^n L_\xi^- = f^{(n)} L_\xi^-$. Consequently $f^{(n)} E_\xi \subset E_{\xi - n\alpha}$ for all $n \in \mathbb{N}$. For a similar reason $\tilde{f}^n b - f^{(n)} G(b) \in qL_{\xi-n\alpha}$, $\tilde{f}^n b - G(\tilde{f}^n b) \in qL_{\xi-n\alpha}$. Hence $f^{(n)} G(b) - G(\tilde{f}^n b) \in qL_{\xi-n\alpha} \cap E_{\xi-n\alpha} = 0$. Consequently

$$(*) \qquad f^{(r)} G(\tilde{f}^s b) = \begin{bmatrix} r+s \\ r \end{bmatrix} G(\tilde{f}^r(\tilde{f}^s b)) .$$

Now set

$$B_{\max} = \{\tilde{f}^n b : b \in B_\xi,\ n \in \mathbb{N}\} , \quad B' = B \setminus B_{\max} ,$$

$$L' = \sum_{b \in B'} AG(b) , \quad V'^{\mathbb{Q}} = \sum_{b \in B'} \mathbb{Q}[q, q^{-1}] G(b) , \quad W = \sum_{b \in B_{\max}} \mathbb{Q}(q) G(b) .$$

It is clear that W is a $U_q(\mathfrak{g})$ submodule of V and that (L', B') is a crystal base for V/W with $(V'^{\mathbb{Q}}, L', L'^-)$ a balanced triple. Thus the above process may be repeated and gives the lemma below.

For each $r \in \mathbb{N}$, set $I^r(B) = \{b \in B \mid \varepsilon(b) + \varphi(b) = r\}$ and $W^r(B) = \bigcup_{s \geq r} I^s(B)$. Let $I^r(V)$ denote the sum of all $r+1$ dimensional submodules of V and $W^r(V) = \bigoplus_{s \geq r} I^s(V)$. Then

Lemma. *For each* $r \in \mathbb{N}$

(i) $W^r(V) = \bigoplus_{b \in W^r(B)} \mathbb{Q}(q)G(b)$

(ii) $f^{(k)}G(b) = \begin{bmatrix} \varepsilon(b)+k \\ k \end{bmatrix} G(\tilde{f}^k b) \bmod W^{r+1}(V)$

(iii) $e^{(k)}G(b) = \begin{bmatrix} \varphi(b)+k \\ k \end{bmatrix} G(\tilde{e}^k b) \bmod W^{r+1}(V)$.

Indeed (i) is immediate. In (ii), (iii) one can assume by (i) that $W^s(V) = 0$ for $s > r$ without loss of generality. Then (ii) follows from $(*)$ above. Finally by 5.1.9 one has $e^{(t)} f^{(s)} u_r = \begin{bmatrix} r-s+t \\ t \end{bmatrix} u_{r-2s+2t}$, $0 \leq t \leq s \leq r$. Combined with $(*)$ this gives (iii).

6.3.5 Here and in 6.3.6–6.3.8 let N be a $U_q(\mathfrak{n}^+)$ submodule of V having the form

$$N = \bigoplus_{b \in B_N} \mathbb{Q}(q)G(b)$$

for some subset $B_N \subset B$.

Lemma. $\tilde{e}B_N \subset B_N \cup \{0\}$.

If $\tilde{e}b : b \in B_N$ is non-zero, then by 6.3.4 (iii) the expansion of $eG(b)$ has $G(\tilde{e}b)$ as a non-zero coefficient and so the assertion follows from 6.3.4 $(*)$.

6.3.6 Set $M := U_q(\mathfrak{g})N = \sum_{k \geq 0} f^k N$ and $B_M = \bigcup_{k \geq 0} (\tilde{f}^k B_N \setminus \{0\})$. Then $W^r(M)$, $W^r(B_M)$ are defined. Set $W^r(N) = N \cap W^r(M)$, $W^r(B_N) = B_N \cap W^r(B_M)$. It is immediate from 6.3.4 (i) that

Lemma. $W^r(N) = \bigoplus_{b \in W^r(B_N)} \mathbb{Q}(q)G(b)$.

6.3.7 Lemma. $W^r(M) = U_q(\mathfrak{n}^-)W^r(N)$.

It is clear that $W^r(M) \supset U_q(\mathfrak{n}^-)W^r(N)$. For the converse it is enough to show that if u_ξ has weight ξ and $u_\xi \in W^r(M) \cap \ker e$ then $u_\xi \in U_q(\mathfrak{n}^-)W^r(N)$. One can write $u_\xi = \sum_k f^k v_{\xi+k\alpha} : v_{\xi+k\alpha} \in N$. Since $r \leq (\alpha^\vee, \xi) \leq (\alpha^\vee, \xi + k\alpha)$, it follows that $v_{\xi+k\alpha} \in W^r(M)$. Hence $v_{\xi+k\alpha} \in W^r(N)$ as required.

Remark. Notice one does not need $v_{\xi+k\alpha} \in \ker e$. Indeed it is generally false that $N \supset M \cap \ker e$. On the other hand by 6.3.5 and the definition B_M one has $B_N \supset \{b \in B_M \mid \tilde{e}b = 0\}$.

6.3.8 Proposition. $W^\ell(M) = \bigoplus_{b \in W^\ell(B_M)} \mathbb{Q}(q)G(b)$.

This is proved by descending induction on ℓ. It is obvious for $\ell \gg 0$ since both sides are zero. Assume the assertion for $\ell+1$. Then both sides contain $W^{\ell+1}(M)$. Taking account of 6.3.4(i) one may assume that $W^{\ell+1}(M)=0$ by passing to $M/W^{\ell+1}(M)$. One may then replace M by $W^{\ell}(M)/W^{\ell+1}(M)$. Since $\{b \in B_M \mid \tilde{e}b = 0\} \subset B_N$ the assertion results from 6.3.4 (ii) in this case.

6.3.9 For each $i \in \{1,2,\dots,\ell\}$ let $U_q(\mathfrak{g}_i)$ denote the Hopf subalgebra of $U_q(\mathfrak{g})$ generated by e_i, f_i, t_i, t_i^{-1} over $\mathbb{Q}(q)$. These are all isomorphic to $U_q(\mathfrak{sl}(2))$. Fix $\lambda \in P^+(\pi)$. By 6.2.16 one has

$$\cdot\, V(\lambda) = \bigoplus_{b\in B(\lambda)} \mathbb{Q}(q)G(b) \,.$$

Define $u_{w\lambda} : w \in W$ inductively on length as follows. Set $u_{w\lambda} = u_\lambda$ if $w=1$. Otherwise there exists $i \in \{1,2,\dots,\ell\}$ such that $s_iw < w$. Then $u_{s_iw\lambda}$ is defined and one sets $u_{w\lambda} = f_i^{(m)}u_{s_iw\lambda}$ where $m = (\alpha_i^\vee, s_iw\lambda)$. Note that $e_iu_{s_iw\lambda} = 0$ under the hypothesis, so $m \geq 0$ and furthermore $u_{w\lambda} = \tilde{f}_i^m u_{s_iw\lambda}$. In particular $u_{w\lambda} \in L(\lambda)$ and $u_{w\lambda} \bmod qL(\lambda) \in B(\lambda)$. Set

$$V_w(\lambda) := U_q(\mathfrak{n}^+)u_{w\lambda} \,.$$

Since $\dim V(\lambda)_{w\lambda} = 1$ it follows that $V_w(\lambda)$ coincides with the module defined in 4.4.1 and in particular is independent of the reduced decomposition of w.

6.3.10 Let $w = s_{i_n}s_{i_{n-1}}\dots s_{i_1}$ be a reduced expression for w. Set $w' = s_{i_{n-1}}\dots s_{i_1}$ and put $i = i_n$. Define for $\varepsilon = 0,\ \infty$, the set $B_w(\lambda^\varepsilon)$ through

$$B_w(\lambda^\varepsilon) = \bigcup_{k\geq 0} \tilde{f}_i^k B_{w'}(\lambda^\varepsilon) \,.$$

By 4.4.3 (v)

$$V_w(\lambda^\varepsilon) = \sum_{k_i\in\mathbb{N}} f_{i_n}^{k_n} f_{i_{n-1}}^{k_{n-1}} \dots f_{i_1}^{k_1} u_{\lambda^\varepsilon}$$

if $\varepsilon = 0$. It defines $V_w(\infty)$ when $\varepsilon = \infty$.

Proposition

(i) $V_w(\lambda^\varepsilon) = \bigoplus_{b\in B_w(\lambda^\varepsilon)} \mathbb{Q}(q)G_{\lambda^\varepsilon}(b)$.

(ii) $\tilde{e}_iB_w(\lambda^\varepsilon) \subset B_w(\lambda^\varepsilon)\bigcup\{0\}$.

(iii) *$B_w(\lambda^\varepsilon)$ is independent of the choice of reduced decomposition. In particular $\tilde{f}_jB_w(\lambda^\varepsilon) \subset B_w(\lambda^\varepsilon)$ if $s_jw < w$.*

(iv) $\bar{\pi}_\lambda(B_w(\infty)) = B_w(\lambda)$.

Take $\varepsilon = 0$. Then (i) results from 6.3.8 and (ii) from 6.3.5 by induction on k. Then (iii) results from (i) and the independence of $V_w(\lambda)$ on reduced decomposition. Finally (iv) follows from 5.3.17 and 6.2.16.

Remarks. Via 6.1.15 assertions (ii) and (iii) for $\varepsilon = \infty$ can be expressed purely combinatorially. However it seems to be extremely difficult to establish either in this fashion. Notice that $\mathfrak{g}$ is finite dimensional if and only if $|W| < \infty$ and so admits a unique longest element w_0. Then by 6.1.15, $B(\infty)$ embeds in $C_{i_n} \otimes C_{i_{n-1}} \cdots \otimes C_{i_1}$ where $w_0 = s_{i_n} s_{i_{n-1}} \cdots s_{i_1}$ is a reduced decomposition. Note that this is not a strict embedding because u_∞ was omitted.

6.3.11 The following answers the question raised in 6.1.17.

Proposition. $B_w(\infty)^\star = B_{w^{-1}}(\infty)$.

This results from 6.1.12, 6.3.10 (i) and the fact that $V_w(\infty)^\star = V_{w^{-1}}(\infty)$, the latter being immediate from definitions.

6.3.12 First note the following consequence of 6.3.10 (ii), (iv) and the definition of $B_w(\infty)$.

Lemma. *Suppose $s_i w < w$. If $b \in B(\infty)$ satisfies $\tilde{e}_i b = 0$ and $\tilde{f}_i^k b \in B_w(\infty)$ for some $k \in \mathbb{N}$ one has $b \in B_{s_i w}(\infty)$.*

6.3.13 Theorem. *Fix $w \in W$. If $b \in B(\infty)$ satisfies $\tilde{f}_i b \in B_w(\infty)$, then $\tilde{f}_i^k b \in B_w$ for all $k \in \mathbb{N}$.*

Replacing b by $\tilde{e}_i b$ if the latter is non-zero one may assume that $\tilde{e}_i b = 0$. Now $(\tilde{f}_i^k b)^\star = \tilde{f}_i^{\star k} b^\star$, so replacing b by $b^\star$ and w by w^{-1} it is enough to show that

$$(*) \quad \text{If } \ \tilde{e}_i b^\star = 0 \ \text{ and } \ \tilde{f}_i^\star b \in B_w(\infty), \ \text{ then } \ \tilde{f}_i^{\star k} b \in B_w(\infty), \ \ \forall k \geq 0 .$$

This is established by induction on the length $\ell(w)$ of w. It is obvious if $\ell(w) = 0$. If $\ell(w) > 0$ one can find $j \in \{1, 2, \dots, \ell\}$ such that $s_j w < w$. Recall (6.1.13) the definition of Ψ_i and set $c_i = c_i(0)$. The condition $\tilde{e}_i(b^\star) = 0$ is equivalent to $\tilde{e}_i^\star(b) = 0$ and so $\Psi_i(b) = b \otimes c_i$.

If $j \neq i$ write $b = \tilde{f}_j^t b'$ with $\tilde{e}_j b' = 0$. Then by 6.1.14 one has $\tilde{e}_j(\tilde{f}_i^\star b') = 0$, whilst $\tilde{f}_j^t \tilde{f}_i^\star b' = \tilde{f}_i^\star b \in B_w(\infty)$. Thus $\tilde{f}_i^\star b' \in B_{s_j w}(\infty)$ by 6.3.12. Again by 6.1.14 one has $0 = \tilde{e}_i^\star(b) = \tilde{f}_j^t(\tilde{e}_i^\star b')$, so $\tilde{e}_i^\star b' = 0$ by the injectivity of $\tilde{f}_j$. Then $\tilde{f}_i^{\star k} b' \in B_{s_j w}(\infty)$ for all $k \geq 0$ by the induction hypothesis. Then by 6.3.10 (iii) and 6.1.14 one has $\tilde{f}_i^{\star k} b = \tilde{f}_j^t \tilde{f}_i^{\star k} b' \in B_w(\infty)$ for all $k \geq 0$, as required.

Finally suppose $j = i$. If $\varphi_i(b) \leq \varepsilon_i(\tilde{f}_i c_i) = 1$, then 6.1.13 gives $\Psi_i(\tilde{f}_i^{k-1} \tilde{f}_i^\star b) = \tilde{f}_i^{k-1}(b \otimes \tilde{f}_i c_i) = b \otimes \tilde{f}_i^k c_i = \Psi_i(\tilde{f}_i^{\star k} b)$. Then the injectivity of Ψ_i gives $\tilde{f}_i^{\star k} b = \tilde{f}_i^{k-1} \tilde{f}_i^\star b$ which is contained in $B_w(\infty)$ by the hypothesis and 6.3.10 (iii). Hence one may assume $\varphi_i(b) > 1$. Write $b = \tilde{f}_i^t b'$ with $\varepsilon_i(b') = 0$. Then $\varphi_i(b') = \varphi_i(b) + t > 1$ so $\varepsilon_i(b' \otimes \tilde{f}_i c_i) = 0$ and moreover $\tilde{f}_i^t(b' \otimes \tilde{f}_i c_i) = \tilde{f}_i^t b' \otimes \tilde{f}_i c_i = b \otimes \tilde{f}_i c_i = \Psi_i(\tilde{f}_i^\star b) \in \Psi_i(B_w(\infty))$ by the hypothesis. One concludes that $b' \otimes \tilde{f}_i c_i \in \Psi_i(B_{s_i w}(\infty))$ using 6.3.12. Yet 6.1.13

gives $\tilde{e}_i^\star(b')=0$ and $b'\otimes\tilde{f}_i c_i=\Psi_i(\tilde{f}_i^\star b')$. Then by the induction hypothesis $b'\otimes\tilde{f}_i^k c_i=\Psi_i(\tilde{f}_i^{\star k}b')\in\Psi_i(B_{s_iw}(\infty))$, for all $k\geq 0$. Yet $\Psi_i(\tilde{f}_i^{\star k}b)=b\otimes\tilde{f}_i^k c_i=\tilde{f}_i^t b'\otimes\tilde{f}_i^k c_i=\tilde{f}_i^p(b'\otimes\tilde{f}_i^s b_i)\in\tilde{f}_i^p\Psi_i(B_{s_iw}(\infty))\subset\Psi_i(B_w(\infty))$, by 6.1.13 and 6.3.10 (iii). Hence $\tilde{f}_i^{\star k}b\in B_w(\infty)$ for all $k\geq 0$ as required.

6.3.14 Fix $i\in\{1,2,\dots,\ell\}$. Call an i-string of a crystal B any subset of the form $\{\tilde{e}_i^k b,\tilde{f}_i^k b|k\in\mathbb{N}\}$ for some $b\in B$. By (C3) any crystal B is a disjoint union of its i-strings. Then 6.3.13 can be reformulated as saying that for any i-string S of $B(\infty)$ one has either $S\cap B_w(\infty)=\emptyset$, $S\subset B_w(\infty)$ or $S\cap B_w(\infty)=S\cap\ker\tilde{e}_i$, which of course is reduced to a single element denoted s. By 6.3.10 (iv) and S_8 the corresponding result holds for $B_w(\lambda)$. That is one has the

Corollary. *Take $\lambda\in P^+(\pi)$, $w\in W$, $i\in\{1,2,\dots,\ell\}$. Then for any i-string S of $B(\lambda)$ one of the following three conclusions hold*

(i) $S\cap B_w(\lambda)=\emptyset$,
(ii) $S\subset B_w(\lambda)$,
(iii) $S\cap B_w(\lambda)=\{s\}$.

Remark. If $s_iw<w$, then by 6.3.9 and 6.3.10 (i) either (i) or (ii) must hold.

6.3.15 In view of 6.3.10 one has

$$\operatorname{ch}V_w(\lambda):=\sum_{\xi\in P(\pi)}(\dim V_w(\lambda)_\xi)e^\xi=\sum_{\xi\in P(\pi)}|B_w(\lambda)_\xi|e^\xi=:\operatorname{ch}B_w(\lambda)\,.$$

Observe that $\operatorname{ch}V_w(\lambda)$ can then be computed inductively through 6.3.14. Define the ith Demazure operator $\Delta_i\in\operatorname{End}_{\mathbb{Z}}\mathbb{Z}[P(\pi)]$ through

$$\Delta_i e^\xi=\frac{e^\xi-e^{s_i\xi-\alpha_i}}{1-e^{-\alpha_i}}\,.$$

One notes in particular that if $(\alpha_i^\vee,\xi)>0$, then

$$\Delta_i e^\xi=e^\xi+e^{\xi-\alpha_i}+\dots+e^{-\xi}\quad\text{and}\quad\Delta_i(e^\xi+e^{s_i\xi-\alpha_i})=0\,.$$

Consequently Δ_i acts like the identity on the character of any i string S of $B(\lambda)$, whilst $\operatorname{ch}S=\Delta_i\operatorname{ch}s$ with $s\in S$ the unique element of $S\cap\ker\tilde{e}_i$. Thus $\Delta_i\operatorname{ch}V_{s_iw}(\lambda)=\operatorname{ch}V_w(\lambda)$ if $s_iw<w$. Now fix a reduced decomposition $w=s_{i_n}s_{i_{n-1}}\cdots s_{i_1}$ and set

$$\Delta_w:=\Delta_{i_n}\Delta_{i_{n-1}}\cdots\Delta_{i_1}\,.$$

Theorem

(i) $\operatorname{ch}V_w(\lambda)=\Delta_w e^\lambda$.
(ii) *Δ_w is independent of the choice of reduced decomposition.*

6.3.16 Corollary. *For each $\mu \in \lambda - \mathbb{N}\pi$ there exists $w \in W$ such that $\dim V(\lambda)_\mu$ equals the coefficient of e^μ in $\Delta_w(e^\lambda)$. In particular if $|W| < \infty$ then $\operatorname{ch} V(\lambda) = \Delta_{w_0}(e^\lambda)$.*

By 4.4.3 (v) the subspace $\sum_{y \in W} V_y(\lambda)$ of $V(\lambda)$ is U_α stable for each $\alpha \in \pi$ and hence coincides with $V(\lambda)$. Since any weight subspace $V(\lambda)_\mu$ is finite dimensional one has $V(\lambda)_\mu \subset \sum_{y \in F} V_y(\lambda)_\mu$ for some finite subset F of W. By A.1.10 there exists $w \in W$ such that $w \geq y$ for all $y \in F$. The required assertion then follows from 4.4.5 and 6.3.15.

The last assertion has also a more elementary proof avoiding the use of 6.3.15 (i). Set $\Delta_\alpha = \Delta_i$ if $\alpha = \alpha_i$. One checks that $\Delta_\alpha^2 = \Delta_\alpha$ and using A.1.9 (i) the purely combinatorial result 6.3.15 (ii). This gives in particular that $\Delta_\alpha \Delta_{w_0} = \Delta_{w_0}$ which translates to give $s_\alpha \Delta_{w_0}(e^\lambda) = \Delta_{w_0}(e^\lambda)$, that is $\Delta_{w_0}(e^\lambda)$ is W invariant. Now set $\Delta_\alpha^0 = e^\rho \Delta_\alpha e^{-\rho}$, $\Delta_w^0 = e^\rho \Delta_w e^{-\rho}$. Then $\Delta_\alpha^0 = (1 - e^{-\alpha})^{-1}(1 - s_\alpha)$ and so by A.1.1 (iv) and A.1.8

$$(*) \qquad \Delta_w^0 = (-1)^{\ell(w)} \Bigg(\prod_{\beta \in S(w^{-1})} (1 - e^{-\beta}) \Bigg) w + \sum_{w' < w} a(w') w'$$

for some $a(w') \in \operatorname{Fract} \mathbb{Z}[P(\pi)]$. Now set $J = \sum_{w \in W} (-1)^{\ell(w)} w$ which is proportional to the projector onto the sign representation of w. By $(*)$ and 4.2.16 one obtains for all $\lambda \in P(\pi)$ that

$$J(e^\rho) \Delta_{w_0}(e^\lambda) = \sum_{w \in W} b(w) w(e^{\lambda + \rho})$$

for some $b(w) \in \operatorname{Fract} \mathbb{Z}[P(\pi)]$ with $b(w_0) = (-1)^{\ell(w_0)} w_0$. Since the left hand side transforms like the sign representation, it follows that the right hand side equals $J(e^{\lambda+\rho})$, as required.

6.3.17 Let V be an integrable module with a crystal basis (L, B). Fix $i \in \{1, 2, \ldots, \ell\}$, $\mu \in P(\pi)$. It is clear that $\ker e_i \cap L_\mu = \ker \tilde{e}_i \cap L_\mu$. Again by applying 5.1.8 to V and with respect to the $U_q(\mathfrak{sl}(2))$ subalgebra generated by e_i, f_i, t_i, t_i^{-1} it follows that specialization maps $\ker \tilde{e} \cap L_\mu$ surjectivity to $\ker \tilde{e}_i \cap (L/qL)_\mu$. Since $\tilde{e}_i$ preserves $B \cup \{0\}$ it follows that the latter admits $\{b \in B_\mu \mid \tilde{e}_i b = 0\}$ as a basis. Since A is a principal ideal domain and L_μ a free finite rank A module, it follows that $\ker \tilde{e}_i \cap L_\mu$ is free and then

$$\begin{aligned} \dim_{\mathbb{Q}(q)}(\ker e_i \cap V_\mu) &= rk_A(\ker \tilde{e}_i \cap L_\mu) \\ &= \dim_{\mathbb{Q}}(\overline{\ker \tilde{e}_i \cap L_\mu}) = \#\{b \in B_\mu \mid \tilde{e}_i b = 0\} \,. \end{aligned}$$

Since specialization commutes with intersection this gives

Lemma. *For any subset $I \subset \{1, 2, \ldots, \ell\}$ and any $\mu \in P(\pi)$ one has*

$$\dim_{\mathbb{Q}(q)} \Bigg(\bigcap_{i \in I} \ker e_i \Bigg) \cap V_\mu = \#\{b \in B_\mu \mid \tilde{e}_i b = 0, \ \forall i \in I\} \,.$$

6.3.18 Fix $\lambda, \mu \in P^+(\pi)$ and recall that $V(\lambda) \otimes V(\mu)$ is a direct sum of simple integrable modules. Let $N^\xi_{\lambda,\mu}$ denote the multiplicity of $V(\xi)$ in $V(\lambda)\otimes V(\mu)$.

Proposition

$$N^\xi_{\lambda,\mu} = \#\{b \in B(\mu)_{\xi-\lambda} \,\big|\, \tilde{e}_i^{(\lambda,\alpha_i^\vee)+1} b = 0, \ \forall i \in \{1,2,\ldots,\ell\}\} .$$

Clearly $N^\xi_{\lambda,\mu} = \dim_{\mathbb{Q}(q)}((V(\lambda) \otimes V(\mu))_\xi \cap \bigcap_{i=1}^\ell \ker e_i)$, so 6.3.17 gives $N^\xi_{\lambda,\mu} = \#\{b \in (B(\lambda) \otimes B(\mu))_\xi \mid \tilde{e}_i b = 0, \ \forall i\}$. Write $b = b_1 \otimes b_2 : b_1 \in B(\lambda)$, $b_2 \in B(\mu)$. If $\tilde{e}_i b_2 = 0$ then $\varepsilon_i(b_2) = 0 \le \varphi_i(b_1)$. Thus by 5.1.12 (ii) one has $\tilde{e}_i b = 0 \iff \tilde{e}_i b_1 = 0$ and $\varphi_i(b_1) \ge \varepsilon_i(b_2)$. Now the condition $\tilde{e}_i b_1 = 0$, $\forall i$ implies $b_1 = u_\lambda$. Since $\varphi_i(u_\lambda) = (\lambda, \alpha_i^\vee)$ the conclusion is obtained.

6.3.19 Fix $\lambda, \mu \in P^+(\pi)$ and set $B^\lambda(\mu) = \{b \in b(\mu) \mid \varepsilon_i(b) \le (\alpha_i^\vee, \lambda)\}$. As shown in 6.3.18 the specialization at $q = 0$ of the A submodule $I_\xi^{\lambda,\mu}$ of highest weight vectors of $L(\lambda) \otimes L(\mu)$ of weight ξ has basis $u_\lambda \otimes b : b \in B^\lambda(\mu)_{\xi-\lambda}$. Now $U_q(\mathfrak{n}^-)I_\xi^{\lambda,\mu}$ is just the isotypical component $V_\xi^{\lambda,\mu}$ of $V(\lambda) \otimes V(\mu)$ corresponding to highest weight ξ. Let $L_\xi^{\lambda,\mu}$ (resp. $B_\xi^{\lambda,\mu}$) denote the A linear span of the subset of $L(\lambda) \otimes L(\mu)$ (resp. the subset of $B(\lambda) \otimes B(\mu)$) generated by the action of the $\tilde{f}_i : i \in \{1,2,\ldots,\ell\}$ on $I_\xi^{\lambda,\mu}$ (resp. $u_\lambda \otimes B^\lambda(\mu)_{\xi-\lambda}$). Then by 5.4.27 it follows that $(L_\xi^{\lambda,\mu}, B_\xi^{\lambda,\mu})$ is a crystal basis for $V_\xi^{\lambda,\mu}$.

6.3.20 Following the reasoning of 6.3.17 one may deduce from 6.3.18 that

$$N^\xi_{\lambda,\mu} = \dim\{a \in V(\mu)_{\xi-\lambda} \,\big|\, e_i^{(\lambda,\alpha_i^\vee)+1} a = 0, \ \forall i\} .$$

However this result is not particularly deep and may be obtained as follows. Fix $\lambda, \mu \in P^+(\pi)$ and observe that $V(\mu)^*$ viewed as a left $U_q(\mathfrak{g})$ module via the antipode has lowest weight $-\mu$.

Lemma. *The map* $\varphi : u_\lambda \otimes u_{-\mu} \longrightarrow \varphi(u_\lambda \otimes u_{-\mu})$ *of* $\mathrm{Hom}_{U_q(\mathfrak{g})}(V(\lambda) \otimes V(\mu)^*, V(\xi))$ *into* $V(\xi)$ *is injective and has image* $V^\mu(\xi)_{\lambda-\mu} := \{a \in V(\xi)_{\lambda-\mu} \,\big|\, e_i^{(\mu,\alpha_i^\vee)+1} a = 0, \ \forall i\}$.

Since $u_\lambda \otimes u_{-\mu}$ is a generator of the $U_q(\mathfrak{g})$ module $V(\lambda) \otimes V(\mu)^*$ it is immediate that φ is injective. From the form of the coproduct one easily deduces that $\mathrm{Im}\,\varphi \subset V^\mu(\xi)_{\lambda-\mu}$. For injectivity recall as above that $U_q(\mathfrak{n}^+)u_{-\mu} = V(\mu)^*$ and (4.3.6 (i)) that

$$\mathrm{Ann}_{U_q(\mathfrak{n}^+)}\, u_{-\mu} = \sum_{i=1}^{\ell} U_q(\mathfrak{n}^+) e_i^{(\mu,\alpha_i^\vee)+1} \subset \mathrm{Ann}_{U_q(\mathfrak{n}^+)}\, a \ ,$$

for all $a \in V^\mu(\xi)_{\lambda-\mu}$. Thus for each $a \in V^\mu(\xi)_{\lambda-\mu}$ one may define a map $\psi_a \in \mathrm{Hom}_{\mathbf{Q}(q)}(V(\mu)^*, V(\xi))$ by $\psi_a(xu_{-\mu}) = xa$ for all $x \in U_q(\mathfrak{n}^+)$. Now for any $\psi \in \mathrm{Hom}_{\mathbf{Q}(q)}(V(\mu)^*, V(\xi))$ one has $(e_i.\psi)(u) = e_i\psi(\sigma(t_i^{-1})u) + \psi(\sigma(e_i)u) = e_i\psi(t_iu) - \psi(e_it_iu)$ and $(t_i.\psi)(u) = t_i\psi(t_i^{-1}u)$. Hence ψ_a is a highest weight vector of weight λ in the integrable module $\mathrm{Hom}_{\mathbf{Q}(q)}(V(\mu)^*, V(\xi))$. This gives $\theta \in \mathrm{Hom}_{U_q(\mathfrak{g})}(V(\lambda), \mathrm{Hom}_{\mathbf{Q}(q)}(V(\mu)^*, V(\xi)))$ defined by $\theta(u_\lambda) = \psi_a$. In particular $\theta(u_\lambda)(u_{-\mu}) = a$. Then the required map φ is the image of θ under the Frobenius reciprocity

$$\mathrm{Hom}_{U_q(\mathfrak{g})}(V(\lambda), \mathrm{Hom}_{\mathbf{Q}(q)}(V(\mu)^*, V(\xi))) \xrightarrow{\sim} \mathrm{Hom}_{U_q(\mathfrak{g})}(V(\lambda) \otimes V(\mu)^*, V(\xi)) .$$

This proves surjectivity.

Remark. Set $\mu_0 = -w_0\mu$. The lemma shows that $N^\xi_{\lambda,\mu_0} = \dim V^\mu(\xi)_{\lambda-\mu}$, which had previously been shown to be equal to $N^\lambda_{\mu,\xi}$. Yet the former is just the multiplicity of $V(0)$ in $V(\lambda) \otimes V(\mu)^* \otimes V(\xi)^*$ and the latter the multiplicity of $V(0)$ in $V(\mu) \otimes V(\xi) \otimes V(\lambda)^*$. Since $V(0)$ is self dual and multiplicity is independent of the order in the tensor product one does in fact have $N^\xi_{\lambda,\mu_0} = N^\lambda_{\mu,\xi}$.

6.4 Littelmann's Path Crystals

6.4.1 Recall A.1.6. Call $\{w_1, w_2, \ldots;\ \beta_1, \beta_2, \ldots : w_j \in W,\ \beta_j \in \Delta^+_{re}\}$ a Bruhat sequence if $w_1 = e$ and $w_j \xleftarrow{\beta_j} w_{j+1}$ for all $j = 1, 2, \ldots$. If $|W| < \infty$ every Bruhat sequence is finite and can be taken to terminate in the unique longest element w_0 of W. In general it is convenient to consider only finite sequences terminating in some $w_{r+1} \in W$, $r \in \mathbb{N}$ though of course neither r nor w_{r+1} will be considered fixed. If $w_{r+1} = w \in W$ one says that the Bruhat sequence terminates in w. Given $\alpha_i \in \pi$ one writes $s_i = s_{\alpha_i}$. Note that in defining a Bruhat sequence it suffices to give the w_j.

6.4.2 Fix $\lambda \in P^+(\pi)$. To each partition $\sum_{j=1}^{r+1} b_j = 1$ of 1 by positive rational numbers b_j and to each Bruhat sequence $\{w_1, w_2, \ldots, w_{r+1} : w_j \in W\}$ associate the sum $\sum_{j=1}^{r+1} b_j w_j\lambda$. This may be considered as an ordered partition of the element it defines.

Set $b_j = a_{j-1} - a_j$ with $a_0 = 1,\ a_{r+1} = 0$ in the above. Then

$$\lambda - \sum_{j=1}^{r+1} b_j w_j \lambda = \sum_{j=2}^{r+1} a_{j-1}(w_{j-1}\lambda - w_j\lambda) = \sum_{j=1}^{r} a_j(\beta_j^\vee, w_j\lambda)\beta_j .$$

A partition of μ is said to be a Lakshmibai-Seshadri, or simply, an LS partition, for λ if the positive rational numbers $n_j := a_j(\beta_j^\vee, w_j\lambda)$ are integers. The set of LS partitions for λ is denoted by $C(\lambda)$. In enumerating $C(\lambda)$ it should be noticed that different sequences can give the same partition if λ is not regular or if the a_j are not strictly decreasing.

6.4.3 To each partition for λ one may associate a continuous piecewise linear path $b : [0,1] \mapsto \mathbb{R}P(\pi)$ as follows. For each $s \in \{1, 2, \ldots, r+1\}$ let $s_+ \geq s$ be maximal such that $a_{s_+} = a_s$. Then

$$b(t) = \sum_{j=s+1}^{r+1} (a_{j-1} - a_j) w_j \lambda + (t - a_s) w_{s_+} \lambda \quad \text{for} \quad t \in [a_{s-1}, a_s] \; .$$

One calls b an LS path if the corresponding partition is LS. Observe that $b(0) = 0$ and the original partition is recovered from the linear components of b. Thus these notions are equivalent. In either case, one may omit j in the sum for which $a_{j-1} - a_j = 0$. Reindexing, one can assume $s = s_+$. Again if λ is not regular one may omit some w_j; however in both cases the property that $\ell(w_{j+1}) = \ell(w_j) + 1 = j$ is lost.

6.4.4. More generally let $\mathbb{P}$ denote the set of all continuous paths $b : [0,1] \longrightarrow \mathbb{R}P(\pi)$ satisfying $b(0) = 0$, $b(1) \in P(\pi)$. For each $b \in \mathbb{P}$, set $wt(b) = b(1)$. Fix $i \in \{1, 2, \ldots, \ell\}$ and set $h_b^i(t) = -(\alpha_i^\vee, b(t))$ and $m_b^i = \max\{h_b^i(t) \cap \mathbb{Z} \mid t \in [0,1]\}$ which is a positive integer. In what follows the quantities $e_\pm^i(b), f_\pm^i(b)$ are defined. The i superscript and b will often be omitted.

Let e_+ (resp. f_+) be the minimal (resp. maximal) element in $[0,1]$ such that $h(e_+) = m$ (resp. $h(f_+) = m$). If $h(e_+) \geq 1$ (resp. $h(1) - h(f_+) \leq -1$) let $e_- < e_+$ (resp. $f_- > f_+$) be maximal (resp. minimal) such that $h(e_-) = h(e_+) - 1$ (resp. $h(f_-) = h(f_+) - 1$) and set

$$(\tilde{e}_i b)(t) = \begin{cases} b(t) & : t \in [0, e_-] \\ b(e_-) + s_i(b(t) - b(e_-)) & : t \in [e_-, e_+] \\ b(t) + \alpha_i & : t \in [e_+, 1] \end{cases}$$

respectively

$$(\tilde{f}_i b)(t) = \begin{cases} b(t) & : t \in [0, f_+] \\ b(f_+) + s_i(b(t) - b(f_+)) & : t \in [f_+, f_-] \\ b(t) - \alpha_i & : t \in [f_-, 1] \end{cases}$$

Otherwise set $\tilde{e}_i b = 0$ (resp. $\tilde{f}_i b = 0$). In this construction one may note that $(\alpha_i^\vee, b(e_+) - b(e_-)) = (\alpha_i^\vee, b(f_+) - b(f_-)) = -1$. Consequently $\tilde{e}_i b$, $\tilde{f}_i b$ are continuous. Finally they satisfy the required boundary conditions and hence belong to $\mathbb{P}$. The following is straightforward, though rather tiresome.

Lemma. *Take* $b \in \mathbb{P}$.

(i) *Suppose* $\tilde{e}_i b \neq 0$, *then*

a) $m_{\tilde{e}_i b} = m_b - 1$ *and* $h_{\tilde{e}_i b}(1) = h_b(1) - 2$,
b) $f_+(\tilde{e}_i b) = e_-(b)$ *and* $f_-(\tilde{e}_i b) = e_+(b)$.

In particular $\tilde{f}_i\tilde{e}_ib = b$.

c) $wt\tilde{e}_ib = b + \alpha_i$

(ii) *Suppose* $\tilde{f}_ib \neq 0$, *then*

a) $m_{\tilde{f}_ib} = m_b + 1$ *and* $h_{\tilde{f}_ib}(1) = h_b(1) + 2$,

b) $e_+(\tilde{f}_ib) = f_-(b)$ *and* $e_-(\tilde{f}_ib) = f_+(b)$.

In particular $\tilde{e}_i\tilde{f}_ib = b$.

c) $wt\tilde{f}_ib = b - \alpha_i$.

6.4.5 For each $b \in \mathbb{P}$, $i \in \{1, 2, \ldots, \ell\}$ let $\varepsilon_i(b)$ (resp. $\varphi_i(b)$) be the largest integer $k \geq 0$ such that $\tilde{e}_i^k b \neq 0$ (resp. $\tilde{f}_i^k b \neq 0$).

Lemma

(i) $\varepsilon_i(b) = m_b^i$

(ii) $\varphi_i(b) = \varepsilon_i(b) + (\alpha_i^\vee, wtb)$.

By 6.4.4, one may assume $\tilde{e}_ib = 0$. Then by definition $m_b^i = 0$, hence (i). Again writing $b' = \tilde{f}_i^k b$, the condition $\tilde{f}_i(b') = 0$ gives $m_{b'}^i = r_{b'}^i(1) = -(\alpha_i^\vee, wtb')$. By 6.4.4 (ii(a)) the latter translates to $m_b^i + k = -(\alpha_i^\vee, wtb) + 2k$, that is $(\alpha_i^\vee, wtb) = k = \varphi_i(b) - \varepsilon_i(b)$, hence (ii).

6.4.6 Fix $\lambda \in P^+(\pi), b \in C(\lambda)$, $i \in \{1, 2, \ldots, \ell\}$. Let $[t'_-, t'_+]$ (resp. $[t''_-, t''_+]$) be the leftmost (resp. rightmost) maximal connected interval in which $h(t)$ takes its maximal value. Primes are omitted when assertions are common to both. Clearly $t_- = a_s$, $t_+ = a_u$ for some $s \geq u$ and one may assume s maximal and u minimal with this property. Then either $s = r + 1$ or $(\alpha_i^\vee, w_{s+1}\lambda) < 0$ and in this case $s_iw_{s+1} < w_{s+1}$. Similarly either $u = 1$ or $(\alpha_i^\vee, w_u\lambda) > 0$ and in both cases $s_iw_u > w_u$.

Suppose $s < r + 1$. Then there exists k' maximal (resp. k'' minimal) with $s \geq k \geq u$ such that $s_iw_{k+1} < w_{k+1}$, $s_iw_k > w_k$. As the sequence is Bruhat this forces $\beta_k = \alpha_i$ and then

$$b(a_k) = \sum_{j=k+1}^{r+1} (a_{j-1} - a_j)w_j\lambda = \sum_{j=k+1}^{r+1} a_{j-1}(w_j\lambda - w_{j-1}\lambda) + a_kw_k\lambda$$

and so

$$(*) \qquad b(a_k) = -\sum_{j=k}^{r} n_j\beta_j + a_kw_k\lambda \,,$$

which gives

$$(**) \qquad h(a_k) = \sum_{j=k}^{r} n_j(\alpha_i^\vee, \beta_j) - n_k \,.$$

Lemma. *Suppose $b \in C(\lambda)$. Then $m_b^i = \max\{h_b^i(t) \mid t \in [0,1]\}$.*

If $s = r+1$ then $\max\{h_b^i(t) \mid t \in [0,1]\} = h(a_{r+1}) = 0 = m$. If $s < r+1$, then $(*)$ above holds and so $h(a_k)$ is an integer. Together these give the assertion.

Remarks. The corresponding assertion for $\min\{h_b^i(t)\}$ is false. One may note that for k chosen as above $h^i(a_k)$ and r_k^i of 5.2.5 coincide.

6.4.7 Retain the notation of 6.4.6.

Lemma. *Assume $b \in C(\lambda)$ and $\tilde{e}_i b \neq 0$. Then $\tilde{e}_i b \in C(\lambda)$. Moreover there exist integers $r+1 \geq v > k \geq 1$ such that $e_+ = a_k$,*

$$(w_1, \ldots, w_k = s_i w_{k+1}, \ldots, s_i w_v,\ w_v, \ldots;$$
$$\beta_1, \ldots, \beta_{k-1}, s_i\beta_{k+1}, \ldots, s_i\beta_{v-1}, \alpha_i, \beta_v, \ldots)$$

is a Bruhat sequence and $\tilde{e}_i b$ is the partition

$$(*)\quad \begin{aligned} \tilde{e}_i b = &\sum_{j=1}^{k-1} (a_{j-1} - a_j) w_j \lambda + (a_{k-1} - a_{k+1}) w_k \lambda + \sum_{j=k+2}^{v-1} (a_{j-1} - a_j) s_i w_j \lambda \\ &+ (a_{v-1} - e_-) s_i w_v \lambda + (e_- - a_v) w_v \lambda + \sum_{j=v+1}^{r+1} (a_{j-1} - a_j) w_j \lambda\ . \end{aligned}$$

Take $k = k'$ in 6.4.6. Then $e_+ = a_k$ by 6.4.6 and moreover $s < r+1$. Take $v \geq s+1$ such that $a_{v-1} > e_- \geq a_v$. Then $-(a_{v-1} - a_v)(\alpha_i^\vee, w_r\lambda) > 0$ by definition of e_-. In particular $s_i w_v < w_v$. Recall also that $-(a_s - a_{s+1})(\alpha_i^\vee, w_{s+1}\lambda) > 0$ by choice of s.

Suppose $s_i w_m > w_m$ for some $m \in \{v, v-1, \ldots, k+1\}$ and let m be maximal with this property. Then $v > m > s+1$ by definition of v, k, s. Consequently $s_i w_{m+1} < w_{m+1}$ and this forces $\alpha_i = \beta_m$. Then 6.4.6 $(**)$ holds with $k = m$ and implies that $h(a_m) \in \mathbb{Z}$. On the other hand the above inequalities imply that $m_b = h(e_+) > h(a_m) > h(e_-) = m_b - 1$, which is absurd.

Recall that $w_k = s_i w_{k+1}$ and $s_i w_v \overset{\alpha_i}{\longleftarrow} w_v$. By the previous paragraph and A.1.6 one has $s_i w_m \overset{s_i\beta_m}{\longleftarrow} s_i w_{m+1}$ for $m = \{k+1, k+2, \ldots, v-1\}$. This establishes the claim on the Bruhat sequence. Noting that

$$b(e_+) - b(e_-) = \sum_{m=k+1}^{v-1} (a_{m-1} - a_m) w_m \lambda + (a_{v-1} - e_-) w_v \lambda$$

$(*)$ of the lemma results. Finally to establish the integrality condition it is enough to observe that $((s_i\beta_m)^\vee, s_i w_m \lambda) a_m = (\beta_m^\vee, w_m\lambda) a_m \in \mathbb{Z}$ for $m \in \{k+1, k+2, \ldots, v-1\}$, whilst $e_-(\alpha_i^\vee, s_i w_v \lambda) = h(e_-) - \sum_{j=v}^{r} n_j(\alpha_i^\vee, \beta_j)$ is again an integer.

6.4.8 Retain the notation of 6.4.6.

Lemma. *Assume $b \in C(\lambda)$ and $\tilde{f}_i b \neq 0$. Then $\tilde{f}_i b \in C(\lambda)$. Moreover there exist integers $1 \leq v < k \leq r+1$ such that $f_+ = a_k$,*

$$(w_1, \dots, w_v, s_i w_v, \dots, s_i w_k = w_{k+1}, w_{k+2}, \dots; \\ \beta_1, \dots, \beta_{v-1}, \alpha_i, s_i\beta_v, \dots, s_i\beta_{k-1}, \beta_{k+1}, \dots)$$

is a Bruhat sequence and $\tilde{f}_i b$ is the partition

$$(*) \quad \begin{aligned} \tilde{f}_i b &= \sum_{j=1}^{v-1} (a_{j-1} - a_j) w_j \lambda + (a_{v-1} - f_-) w_v \lambda + (f_- - a_v) s_i w_v \lambda \\ &+ \sum_{j=v+1}^{k-1} (a_{j-1} - a_j) s_i w_j \lambda + (a_{k-1} - a_{k+1}) w_{k+1} \lambda + \sum_{j=k+2}^{r+1} (a_{j-1} - a_j) w_j \lambda \,, \end{aligned}$$

where $w_{k+1} := s_i w_k$ when $k = r+1$.

Take $k = k''$ in 6.4.6. Then $f_+ = a_k$ by 6.4.6 and moreover $u > 1$. Take $v \leq u$ such that $a_{v-1} \geq f_- > a_v$. Then $-(a_{v-1} - a_v)(\alpha_i^\vee, w_v\lambda) < 0$ by definition of f_-. In particular $s_i w_v > w_v$. Recall also that $-(a_{u-1} - a_u)(\alpha_i^\vee, w_u\lambda) < 0$ by choice of u. As in 6.4.7 it follows that $s_i w_m > w_m$ for all $m \in \{v, v+1, \dots, k\}$. By A.1.6 this establishes the claim on the Bruhat sequence and then $(*)$ above follows. For the integrality condition observe that $((s_i\beta_m)^\vee, s_i w_m \lambda) a_m = (\beta_m^\vee, w_m\lambda) a_m \in \mathbb{Z}$ for all $m \in \{v, v+1, \dots, k-1\}$ whilst $f_-(\alpha_i^\vee, w_v\lambda) = \sum_{j=v}^{r} n_j(\alpha_i^\vee, \beta_j) - h(f_-)$ is again an integer.

6.4.9 Recall 5.2.1. The rules in 6.4.4 and 6.4.5 endow $\mathbb{P}$ with the structure of a normal crystal. View $C(\lambda) : \lambda \in P^+(\pi)$ as a subcrystal of $\mathbb{P}$. By 6.4.7 and 6.4.8

Theorem. *For each $\lambda \in P^+(\pi)$ the embedding $C(\lambda) \hookrightarrow \mathbb{P}$ is strict.*

6.4.10 For $w \in W$ let $C_w(\lambda)$ denote the subset of $C(\lambda)$ obtained by imposing that every Bruhat sequence terminates in w.

Lemma

(i) $\tilde{e}_i C_w(\lambda) \subset C_w(\lambda) \cup \{0\}$ *for all $i \in \{1, 2, \dots, \ell\}$.*
(ii) $\tilde{f}_i C_w(\lambda) \subset C_w(\lambda) \cup \{0\}$ *for all $i \in \{1, 2, \dots, \ell\}$ satisfying $s_i w < w$.*

These assertions follow immediately from 6.4.7 and 6.4.8 where for (ii) one notes that $k = r+1$ implies $s_i w_{r+1} > w_{r+1}$.

6.4.11 Let b_λ denote the path $b(t) = \lambda t$. Fix $w \in W$ and let $w = s_{i_1} s_{i_2} \dots s_{i_r}$ be a reduced decomposition.

Lemma

(i) *Assume* $s_i w < w$. *Then* $\{b \in C_w(\lambda) \mid \tilde{e}_i b = 0\} \subset C_{s_i w}(\lambda)$.
(ii) b_λ *is the unique element of* $C(\lambda)$ *for which* $\tilde{e}_i b_\lambda = 0$ *for all* i.
(iii) $C_w(\lambda) = \{\tilde{f}_{i_1}^{k_1} \tilde{f}_{i_2}^{k_2} \dots \tilde{f}_{i_r}^{k_r} b_\lambda : k_i \in \mathbb{N}\}$.

Let $\{w_1, w_2, \dots, w_{r+1}\}$ be a Bruhat sequence terminating in w. Assume $s_i w < w$ and let k be maximal such that $s_i w_k > w_k$. This forces $\beta_k = \alpha_i$ and $s_i w \geq w_k > w_{k-1} > \dots$ by A.1.7(iv). On the other hand $s_i w_j < w_j$ for all $j \geq k = 1$ and so

$$\sum_{j=k+1}^{r+1} (a_{j-1} - a_j)(\alpha_i^\vee, w_j \lambda) \geq 0$$

if and only if each term in the sum vanishes. Now a strict inequality implies by 6.4.6 that $\tilde{e}_i b \neq 0$. Thus $\tilde{e}_i b = 0$ implies $a_{j-1} = a_j$ or $s_i w_j \lambda = w_j \lambda$ for each $j \in \{k+1, \dots, r+1\}$. In the first case $w_j \lambda$ can be omitted from the partition sum, whilst in the second case it can be replaced by $s_i w_j \lambda$. One concludes that the given Bruhat sequence can be replaced by $\{w_1, w_2, \dots, w_k, s_i w_{k+2}, \dots, s_i w_{r+1}\}$ which terminates in $s_i w$. This proves (i).

(ii) is immediate from (i) and the fact that $s_i w > w$, $\forall\, i$ implies $w = e$. Finally by 6.4.4 (i(b)) one has setting $i = i_1$ that

$$C_w(\lambda) = \bigcup_{k \in \mathbb{N}} \tilde{f}_i^k \{b \in C_w(\lambda) \,\big|\, \tilde{e}_i b = 0\} = \bigcup_{k \in \mathbb{N}} \tilde{f}_i^k C_{s_i w}(\lambda) ,$$

by (i). Hence (iii) follows by induction on $\ell(w)$.

6.4.12 Recall the notion (6.3.14) of an i-string S_i of a crystal B.

Lemma. *The conclusion of* 6.3.14 *holds for any* i*-string of* $C(\lambda)$.

If $s_i w < w$, then by 6.4.10 only (i) or (ii) of 6.3.14 can hold. Suppose $s_i w > w$. Then it is enough to show that if $b \in C_w(\lambda)$ satisfies $\tilde{e}_i b = 0$ and $\tilde{f}_i b \in C_w(\lambda)$, then $\tilde{f}_i^j b \in C_w(\lambda) \cup \{0\}$ for all $j \in \mathbb{N}$. Set $b' = \tilde{f}_i^j b : j \geq 1$ which can be assumed non-zero. Then $m_{b'}^i = m_b^i + j \geq 1$ by 6.4.4 (ii(a)). Consequently in the notation of 6.4.6 the integer k defined for b' must satisfy $k < r+1$. Then by 6.4.8 the Bruhat sequence defined for $\tilde{f}_i b'$ terminates in w, as required.

6.4.13 Define

$$\operatorname{ch} C_w(\lambda) := \sum_{\mu \in P(\pi)} \#\{b \in C_w(\lambda) \,\big|\, wt b = \mu\} e^\mu .$$

Recall (6.3.16) the Demazure operator Δ_w. Just as in 6.3.15 one obtains from 6.4.12 the

Corollary. $\operatorname{ch} C_w(\lambda) = \Delta_w(e^\lambda)$.

6.4.14 Define

$$\operatorname{ch} C(\lambda) = \sum_{\mu \in P(\pi)} \#\{b \in C(\lambda) \,\big|\, wtb = \mu\} e^\mu \, .$$

Theorem. *For all $\lambda \in P(\pi)^+$ one has*

$$\operatorname{ch} V(\lambda) = \operatorname{ch} C(\lambda) \, .$$

Take $\mu \in \lambda - \mathbb{N}\pi$. It is enough to show that

$$\#\{b \in C(\lambda) \,\big|\, wtb = \mu\} = \dim V(\lambda)_\mu \, .$$

For each $y \in W$ set $C_y(\lambda)_\mu = \{b \in C_y(\lambda) \,\big|\, wtb = \mu\}$. Let w be as in the conclusion of 6.3.16. It is immediate that $C_y(\lambda) \supset C_w(\lambda)$ if $y \geq w$ and so from 6.3.16 one concludes that $C_y(\lambda)_\mu = C_w(\lambda)_\mu$ for all $y \geq w$ and that this set has cardinality $\dim V(\lambda)_\mu$. Noting that each $b \in C(\lambda)$ of weight μ lies in some $C_y(\lambda)_\mu$ completes the proof.

6.4.15 Take $\lambda, \mu \in P^+(\pi)$. Then $b \in C(\mu)$ is said to be λ dominant if $(\alpha_i^\vee, \lambda + b(t)) \geq 0$ for all $t \in [0,1]$ and all $i \in \{1, 2, \ldots, \ell\}$. Set $C(\mu)^\lambda = \{b \in C(\mu) \mid b \text{ is } \lambda \text{ dominant}\}$.

Lemma. *Take $b \in C(\lambda)$. The following are equivalent:*

(i) $b \in C(\mu)^\lambda$.
(ii) $\tilde{e}_i^{(\alpha_i^\vee, \lambda)+1} b = 0$ *for all* i.

Indeed $(\alpha_i^\vee, \lambda + b(t)) = (\alpha_i^\vee, \lambda) - r_i^b(t) \geq 0 \iff m \leq (\alpha_i^\vee, \lambda) \iff \tilde{e}_i^{m+1} b = 0$ by 6.4.4 (i(a)) and definitions. Hence the assertion of the lemma.

6.4.16 A significant achievement of Littelmann's theory is the following tensor product decomposition theorem. Moreover at least for $|W| < \infty$, the proof is, after admitting the Weyl character formula, purely combinatorial.

Theorem. *For all $\lambda, \mu \in P(\pi)^+$ one has*

$$V(\lambda) \otimes V(\mu) = \sum_{b \in C(\mu)^\lambda} V(\lambda + wtb) \, .$$

Let $\Omega(V(\mu))$ denote the set of weights of $V(\mu)$ counted with their multiplicities. Recall that by 4.3.10 one has (notation 6.3.18) that

$$V(\lambda) \otimes V(\mu) = \bigoplus_{\xi \in P^+(\pi)} N_{\lambda,\mu}^\xi V_\xi$$

and equate the characters of both sides using 4.2.15 (cf. 4.3.7). This gives

$$\sum_{\xi\in P^+(\pi)} N^\xi_{\lambda,\mu} \sum_{w\in W}(-1)^{\ell(w)}e^{w(\xi+\rho)} = \sum_{w\in W}\sum_{\nu\in\Omega(V(\mu))}(-1)^{\ell(w)}e^{w(\lambda+\rho)}e^\nu$$

$$= \sum_{w\in W}\sum_{\nu\in\Omega(V(\mu))}(-1)^{\ell(w)}e^{w(\lambda+\nu+\rho)}$$

by the W invariance (4.3.6 (iii)) of $\operatorname{ch} V(\mu)$. Then by A.1.11 and 6.4.14 one concludes that

$$(*) \qquad N^\xi_{\lambda,\mu} = \sum_{b\in C(\mu)|\lambda+wtb\in W.\xi} sg(\lambda+wtb+\rho)\ .$$

Define the following equivalence relation on $C(\mu)\backslash C(\mu)^\lambda$. Given $b\in C(\mu)\backslash C(\mu)^\lambda$ there exist $i\in\{1,2,\ldots,\ell\}$; $s\in[0,1]$ such that $(\alpha_i^\vee,\lambda)-h_b^i(s)<0$ and then by 6.4.6 that $(\alpha_i^\vee,\lambda)-h_b^i(s)\le -1$. Set $s_b=\min\{s\in[0,1]\mid (\alpha_i^\vee,\lambda+\rho)\le h_b^i(s)$, for some $i\}$. Let I_b be the subset of all $i\in\{1,2,\ldots,\ell\}$ such that $(\alpha_i^\vee,\lambda+\rho)\le h_b^i(s_b)$. Obviously $(\alpha_i^\vee,\lambda+\rho)\ge h_b^i(t)$ for $t\in[0,s_b]$ with equality if and only if $t=s_b$ and $i\in I_b$. Define $b,b'\in C(\mu)\setminus C(\mu)^\lambda$ to be equivalent if $s_b=s_{b'}$, $I_b=I_{b'}$ and b, b' coincide on $[0,s_b]$. Let $C_b(\mu)$ denote the class containing b. Now write simply $I=I_b$, $s=s_b$.

For each $b'\in C_b(\mu)$, $i\in I$ one has $f_+^i(b)\ge s$. Consequently $\tilde f_ib'$, b' coincide on $[0,s]$ and so $\tilde f_ib'\in C_b(\mu)$. In particular $C_b(\mu)$ is $\tilde f_i$ stable for all $i\in I$. Fix $i\in I$. Since $\tilde f_ib'=\tilde f_ib''\ne 0$ implies $b'=b''$ by 6.4.4 (ii) one may write $C_b(\mu)$ as a disjoint union of $\tilde f_i$ stable subsets. Any such subset takes the form $\{b',\tilde f_ib',\ldots,\tilde f_i^rb'\}$ where $n=h_{b'}(f_+(b'))-h_{b'}(1)\ge h_{b'}(s)-h_{b'}(1)=(\alpha_i^\vee,\lambda+\nu+\rho)$ and $\nu=wt(b')$. Moreover one has an inequality only if $h_{b'}(f_+(b'))\ge h_{b'}(s)+1$. Then $e_-(b')\ge s$ and so $\tilde e_ib'\in C_b(\mu)$ contradicting the choice of b'. Thus equality holds and then

$$(**) \qquad s_{\alpha_i}(\lambda+wt(\tilde f_i^jb')+\rho)=\lambda+wt(\tilde f_i^{n-j}b')+\rho : j=1,2,\ldots,n\ .$$

From $(**)$ the contributions in $(*)$ from $\tilde f_i^jb'$ and $\tilde f_i^{n-j}b'$ cancel. Consequently the contributions in $(*)$ from $C(\mu)\backslash C(\mu)^\lambda$ cancel. Finally if $b\in C(\mu)^\lambda$ then $\lambda+wtb\in P^+(\pi)$ and so

$$N^\xi_{\lambda,\mu}=\#\{b\in C(\mu)^\lambda \mid \lambda+wtb=\xi\}$$

as required.

Remark. By 6.4.27 and 6.4.15 the above result is equivalent to 6.3.18.

6.4.17 Call $b \in C(\mu)$ almost λ dominant if $\lambda + b(a_j) \in P^+(\pi)$ for all $a_i \neq 1$. In this case $b \in C(\mu)^\lambda$ if and only if $\lambda + wtb \in P^+(\pi)$. For example every path of the form $b(t) = tw\mu$ is almost λ dominant and it is λ dominant if $w\mu = \mu$.

Lemma. *Suppose $b \in C(\mu)$ is almost λ dominant. Then there exists $b' \in C(\mu)^\lambda$ such that $\lambda + wtb \in W(\lambda + wtb')$.*

The proof is by induction on $|\mu - wtb|$ (notation 5.3.15). Write b as in 6.4.2 and choose j_0 minimal such that $a_{j_0} < 1$. Choose $s \in [0,1]$ maximal such that $\lambda + b(t)$ is dominant for all $t \in [0,s]$. Then $s \geq a_{j_0}$ by the hypothesis. If b is not λ dominant, $s < 1$ and there exists $i \in \{1,2,\ldots,\ell\}$ such that $n(t) := -(\alpha_i^\vee, \lambda + b(t))$ satisfies $n(s) = 0$ and $(dn(t)/dt)_{t=s} > 0$. By the assumption of b being almost λ dominant this forces $(dn(t)/dt) > 0$ for all $t \in [s,1]$ and in particular that $n := n(1)$ be a strictly positive integer.

Consider $\tilde{e}_i^n b$ which is non-zero since $m_b^i \geq (\alpha_i^\vee, \lambda) + n \geq n$. Since $n(t)$ is increasing in $[s,1]$ it follows that $e_-^i(\tilde{e}_i^{n-1} b) = s$ and then from 6.4.7 that viewed as a partition

$$\tilde{e}_i^n b = (1-s)s_i w_{j_0}\lambda + (s - a_{j_0})w_{j_0}\lambda + \sum_{j=j_0+1}^{r+1} (a_{j-1} - a_j)w_j\lambda$$

which by definition of s is again almost λ dominant. (Note this may fail for the $\tilde{e}_i^m b$ if $1 \leq m < n$). Moreover $\lambda + wt\tilde{e}_i^n b = \lambda + wtb + n\alpha_i = s_{\alpha_i}(\lambda + wtb)$ by definition of n. Finally it is clear that $|\mu - wt\tilde{e}_i^n b| = |\mu - wtb| - n < |\mu - wtb|$ and so the assertion obtains from the induction hypothesis.

6.4.18 Recall A.1.11.

Corollary. *Take $\lambda, \mu \in P^+(\pi)$, $w \in W$. Choose $y \in W$ such that $\xi := y(\lambda + w\mu) \in P^+(\pi)$. Then $V(\xi)$ is a component of $V(\lambda) \otimes V(\mu)$.*

This follows from 6.4.16, 6.4.17 and the remark preceeding the latter.

6.4.19 One of the remarkable aspects of Littelmann's paths crystals is that the tensor product can be defined by concatenation of paths. Indeed given $b_1, b_2 \in \mathbb{P}$ choose $s : 0 < s < 1$ and define $b_1 \otimes b_2 \in \mathbb{P}$ through

$$(b_1 \otimes b_2)(t) = \begin{cases} b_1(t/s) & : 0 \leq t \leq s\,, \\ b_1(1) + b_2\left(\frac{t-s}{1-s}\right) & : s \leq t \leq 1\,. \end{cases}$$

Then

$$m_{b_1 \otimes b_2}^i = \max\{m_{b_1}^i, m_{b_2}^i - (\alpha_i^\vee, wtb_1)\}\,.$$

From this and 6.4.4 it is immediate that the tensor product rules of 5.2.4 hold. In particular as a crystal $b_1 \otimes b_2$ is independent of the choice of s.

6.4.20 Take $\lambda, \mu \in P^+(\pi)$, $b \in C(\mu)^\lambda$ and set $\nu = b_\lambda(1)+b(1) = wt(b_\lambda \otimes b) \in P^+(\pi)$. It is clear that $\tilde{e}_i(b_\lambda \otimes b) = 0$, for all i, from the definition of $C(\mu)^\lambda$. Let $\tilde{C}(\nu)$ denote the subcrystal of $C(\lambda) \otimes C(\mu)$ generated by $b_\lambda \otimes b$. One would like to show that $\tilde{C}(\nu)$ is isomorphic to $C(\nu)$. Actually it is enough by 6.4.21 to show that this holds when $b = b_\mu$, and the latter results from 6.4.26. For the moment note that by 6.4.15 and the argument in 6.3.18 one obtains the

Lemma. *For all $\lambda, \mu \in P^+(\pi)$ one has*

$$C(\lambda) \otimes C(\mu) = \bigoplus_{b \in C(\mu)^\lambda} \tilde{C}(\lambda + wt\ b) .$$

6.4.21 A crystal D is said to be of highest weight λ if

(i) *There exists $d \in D$ of weight λ such that $\tilde{e}_i d = 0$ for all i.*
(ii) *D is generated by the $\tilde{f}_i : i = 1, 2, \ldots, \ell$ acting on d.*

It is clear that d is the unique element of D of weight λ and one writes $D = D(\lambda)$ and $d = d_\lambda$. Again there can be only finitely many elements of $D(\lambda)$ of a given weight and the set of weights of $D(\lambda)$ must be contained in $\lambda - \mathbb{N}\pi$.

Let $\mathcal{D} = \{D(\lambda) : \lambda \in P^+(\pi)\}$ be a family of highest weight crystals with $D(\lambda)$ of highest weight λ. Suppose that $d_\lambda \otimes d_\mu : \lambda, \mu \in P^+(\pi)$ generates a crystal of highest weight $\lambda+\mu$. One says that $\mathcal{D}$ is closed (for tensor product) if this crystal is isomorphic to $D(\lambda+\mu)$. In particular by S_6^2 of 5.4.1 this holds when $D(\lambda) = B(\lambda)$ for all λ.

Proposition. *Let $\mathcal{D} = \{D(\lambda) : \lambda \in P^+(\pi)\}$ be a closed family of highest weight normal crystals. Then $D(\lambda) = B(\lambda)$ for all $\lambda \in P^+(\pi)$.*

First construct a crystal $D(\infty)$ of highest weight 0 with the property for each $\lambda \in P^+(\pi)$ there exists a strict embedding $D(\lambda) \hookrightarrow D(\infty) \otimes S_\lambda$ (notation 5.3.13).

Take $\lambda, \mu \in P^+(\pi)$. Then by normality $\varepsilon_i(d_\mu) = 0$ and $\varphi_i(d) \geq 0$ for all $d \in D(\lambda)$. Hence $\tilde{e}_i(d \otimes d_\mu) = (\tilde{e}_i d \otimes d_\mu)$ and $\tilde{f}_i(d \otimes d_\mu) = \tilde{f}_i d \otimes d_\mu$ unless $\varphi_i(d) = 0$. It follows that the map $d \mapsto d \otimes d_\mu$ is a full embedding of $D(\lambda)$ into $D(\lambda + \mu)$. Write $\lambda \succeq \mu$ if $\lambda - \mu \in P^+(\pi)$. It follows that for each $\nu \in \mathbb{N}\pi$ the $\#D(\lambda)_{\lambda-\nu}$ are increasing in λ for this order relation. On the other hand they are uniformly bound by the number of monomials in the $\tilde{f}_i$ of weight $-\nu$. Consequently there exists $\lambda \in P^+(\pi)$ such that $\#D(\lambda)_{\lambda-\nu}$ reaches its maximal value. As a set $D(\infty)$ is defined to be the disjoint union of these $D(\lambda)_{\lambda-\nu} : \nu \in \mathbb{N}\pi$. Define $wtd = -\nu$ for all $d \in D(\lambda)_{\lambda-\nu}$. Then for any $\nu_0 \in \mathbb{N}\pi$ it follows from the above $\bigcup_{\nu \leq \nu_0} D(\infty)_{-\nu}$ identifies with a full subcrystal of $D(\mu)$ (with weights translated by μ) for all $\mu \in P^+(\pi)$ sufficiently large. This defines $D(\infty)$ as an upper normal crystal. Finally observe as in 5.3.13 that for each $\lambda \in P^+(\pi)$ the map $d \mapsto d \otimes s_\lambda$ is a strict embedding of $D(\lambda)$ into $D(\infty) \otimes S_\lambda$. Let d_∞ be the unique element of $D(\infty)$ annihilated by all the $\tilde{e}_i$.

Now fix a monomial $\tilde{f}$ in the $\tilde{f}_j : j = 1, 2, \ldots, \ell$. Fix $i \in \{1, 2, \ldots, \ell\}$ and choose $\lambda, \mu \in P^+(\pi)$ with $(\alpha_i^\vee, \lambda) = 0$, $(\alpha_j^\vee, \mu) = 0 : j \neq i$ and $(\alpha_j^\vee, \lambda) \gg 0 : j \neq i$, $(\alpha_i^\vee, \mu) \gg 0$ with respect to $\tilde{f}$. Call $\tilde{f}'$ a submonomial of $\tilde{f}$ if it is obtained by deleting some of the powers of $\tilde{f}_i$. By normality the tensor product rule and the choices of λ, μ it follows that $\tilde{f} d_{\lambda+\mu} = \tilde{f}(d_\lambda \otimes d_\mu) = \tilde{f}' d_\lambda \otimes \tilde{f}_i^m d_\mu$ for some submonomial $\tilde{f}'$ of $\tilde{f}$ and some $m \in \mathbb{N}$. Moreover $\tilde{f}'$ and m are independent of the choices of λ, μ. Indeed this will follow from the tensor product rule if one can show for each submonomial $\tilde{f}''$ of $\tilde{f}$ that $\varphi_i(\tilde{f}'' d_\lambda)$ is independent of λ. In fact $\varphi_i(f'' d_\lambda) = (\alpha_i^\vee, wt \tilde{f}'' d_\lambda) + \varepsilon_i(\tilde{f}'' d_\lambda) = (\alpha_i^\vee, wt \tilde{f}'' d_\infty) - (\alpha_i^\vee, \lambda) + \varepsilon_i(\tilde{f}'' d_\infty) = \varphi_i(f'' d_\infty)$ since $(\alpha_i^\vee, \lambda) = 0$. This also shows that $\tilde{f}(d_\infty \otimes d_\mu) = \tilde{f}' d_\infty \otimes \tilde{f}_i^m d_\mu$. Since $\varepsilon_i(\tilde{f}_i^m d_\mu) = m = \varepsilon_i(c_i(-m))$ it follows from the tensor product rule that $\tilde{f}(d_\infty \otimes c_i(0)) = \tilde{f}' d_\infty \otimes c_i(-m)$. Yet $\tilde{f} d_{\lambda+\mu} \mapsto \tilde{f}(d_\lambda \otimes d_\mu)$ is a strict embedding of $D(\lambda + \mu)$ into $D(\lambda) \otimes D(\mu)$ and so one concludes that $\tilde{f} d_\infty \mapsto \tilde{f}' d_\infty \otimes c_i(-m)$ defines a strict embedding of $D(\infty)$ into $D(\infty) \otimes C_i$.

Finally as in 6.1.15 the above embedding results in a purely combinatorial description of $D(\infty)$ which is hence isomorphic to $B(\infty)$. Thus $D(\lambda)$ identifies with the subcrystal of $B(\infty) \otimes S_\lambda$ generated by $b_\infty \otimes s_\lambda$ and hence by 5.3.13 and 5.4.28 is isomorphic to $B(\lambda)$.

Remark. One sees that Kashiwara's embedding theorem (6.1.13) also follows from these considerations and in particular does not require the $\star$ operation; but only S_6^2.

6.4.22 To prove that $\mathcal{C} = \{C(\lambda) : \lambda \in P^+(\pi)\}$ is closed for tensor product one introduces a number of operations aimed at deforming $b_\lambda \otimes b_\mu$ into $b_{\lambda+\mu}$ without changing the crystals they define. The first operation is rather trivial consisting of reparametrizing the interval $[0, 1]$ through an increasing function and possibly omitting subintervals on which a given path function is constant. From the definitions of $\tilde{e}_i$, $\tilde{f}_i$ it follows that two path functions which are equal up to reparametrization generate the same crystal. This observation already settles the (admittedly trivial) case when $\ell = 1$.

Given $b, b' \in \mathbb{P}$ and $0 \leq s \leq s' \leq 1$ define their s, s' join $b_* \theta_s^{s'} * b' \in \mathbb{P}$ by

$$(b_* \theta_s^{s'} * b')(t) = \begin{cases} b(t) & : 0 \leq t \leq s \,, \\ b(s) & : s \leq t \leq s' \,, \\ b(s) + b'(t) - b'(s') & : s' \leq t \leq 1 \,. \end{cases}$$

This can be viewed as a concatenation of truncated paths. In particular if r is a strictly positive integer and $\delta, \delta' \in \frac{1}{r} P^+(\pi)$, then up to a reparametrization

$$b_\delta \otimes b_{\delta'} = b_{r\delta} * \theta_r^{1-1/r} * b_{r\delta'} \,.$$

Finally take $\lambda, \mu \in P^+(\pi)$ and set $\nu = \lambda + \mu$. For each $x \in [0,1] \cap \mathbb{Q}$ set $b^x = (1-x)(b_\lambda \otimes b_\mu) + xb_\nu$. This deforms $b_\lambda \otimes b_\mu$ into b_ν. For a given x one has $b^x = b_\delta \otimes b_{\delta'}$ where $\delta = (1-x)\lambda + \frac{1}{2}x\nu$, $\delta' = (1-x)\mu + \frac{1}{2}x\nu$. Then $\delta, \delta' \in \frac{1}{r}P^+(\pi)$ for some $r \in \mathbb{N}^+$. One should therefore examine how joined paths behave under $\tilde{e}_i, \tilde{f}_i$.

6.4.23 Fix $\lambda, \mu \in P^+(\pi)$ and $0 \leq s \leq s' \leq 1$. Recall the notation of 6.4.2 and the equivalence of paths and partitions described in 6.4.3. Then $C(\lambda) * \theta_s^{s'} * C(\mu)$ is defined to be the set of all paths $b = b'_* \theta_s^{s'} * b''$, where b' (resp. b'') is a path in $[0, s]$ (resp. $[s', 1]$) which expressed as a partition takes the form

$$b' = \sum_{j=u+1}^{r+1} (a_{j-1} - a_j) w_j \lambda + (s - a_u) w_u \lambda$$

$$\left(\text{resp. } b'' = (a_{u-1} - s') w_u \mu + \sum_{j=1}^{u-1} (a_{j-1} - a_j) w_j \mu\right),$$

satisfying the integrality conditions $(b(a_k), \beta_k^\vee) \in \mathbb{Z}$ for all k. As before $\{w_j\}$ is a Bruhat sequence, the $\{a_j\}$ form a decreasing sequence of rational numbers with $a_0 = 1$, $a_{r+1} = 0$, and with the further condition $a_{u-1} \geq s' \geq s \geq a_u$. It is convenient to view the sum $b' + b''$ as the partition defining b. When $\lambda = \mu$ and $s = s'$ the middle term just becomes $(a_{u-1} - a_u) w_u \lambda$. Furthermore in this case 6.4.6(*) shows that the integrality conditions are equivalent to the previous ones given in 6.4.2. In other words $b \in C(\lambda)$ in that case. More generally 6.4.6(*) holds for all $k \geq u$ with the $n_j \in \mathbb{Z}$ for $j \geq k$.

Observe that $b_\lambda * \theta_s^{s'} * b_\mu \in C(\lambda)_* \theta_s^{s'} * C(\mu)$. Indeed as a partition it takes the form $(1-s')\mu + s\lambda$, that is $r = 0$ and so there are no integrality conditions to check.

Call $b \in \mathbb{P}$ integral if for all i the maximum of $h_b^i(t)$ in $[0,1]$ is integer and hence equal to m_b^i. In this case b is called monotone if for all i and $t \in [0,1]$ for which $h_b^i(t) \geq m_b^i - 1$ and $dh^i(t)/dt > 0$ the function h_b^i increases monotonically beyond t till it first reaches the value m_b^i.

Let $\mathcal{A}$ denote the monoid generated by the $\tilde{e}_i, \tilde{f}_i : i = 1, 2, \ldots, \ell$.

Proposition

(i) *Each $b \in C(\lambda)_* \theta_s^{s'} * C(\mu)$ is integral and monotone.*

(ii) *For each i there exists k' (resp. k'') such that $e'_+(b) = a_{k'}$ (resp. $f_+^i(b) = a_{k''}$).*

(iii) *$C(\lambda) * \theta_s^{s'} * C(\mu)$ is $\mathcal{A}$ stable.*

(i) If h_b^i takes its maximal value at some $t < s$ or $t > s'$ then as in 6.4.6 one must have $t = a_k$ and $\beta_k = \alpha_i$. Then the assertion is an immediate consequence of the integrality conditions. Thus suppose the maximal value just occurs in the integral $[s, s']$ (where $h_b^i(t)$ is constant). Then $a_u < s$ implies that $(\alpha_i^\vee, w_u\lambda) < 0$; whilst $a_u = s$ and the condition $(\alpha_i^\vee, w_u\lambda) = 0$ (resp. $(\alpha_i^\vee, w_u\lambda) > 0$) substituted into 6.4.6(*) taking $k = u$ gives $h_b^i(s) \in \mathbb{Z}$. (In the second case if $s \neq 0$ one can choose $v \geq u$ minimal so that $a_v > a_{v+1}$. Then $(\alpha_i^\vee, w_v\lambda) < 0$ so one must have $\alpha_i = \beta_k$ for some $k \in \{v, v-1, \dots, u\}$ and consequently $h_b^i(s) = h_b^i(a_k) \in \mathbb{Z}$).

The above shows that one can assume $(\alpha_i^\vee, w_u\lambda) < 0$. In particular $s_i w_u < w_u$. Since $w_1 = e$ there exists $v < u$ maximal such that $s_i w_v > w_v$. Then $(\alpha_i^\vee, w_j\mu) \leq 0$ for $j \in \{v+1, \dots, u\}$. Moreover $(a_{j-1} - a_j)(\alpha_i^\vee, w_j u) = 0$ by the first supposition for the above choices of j (here a_u is taken to be s'). Consequently $h_b^i(s') = h_b^i(a_v) = -(\alpha_v^\vee, b(a_v)) \in \mathbb{Z}$ as required. This proves the first part of (i) and furthermore that the maximal value of h_b^i is reached on some a_k. Then the last part of (i) and (ii) follows as in 6.4.7 and 6.4.8, or directly by calculations similar to the above.

For (iii) first suppose $\tilde{e}_i b \neq 0$. The first part of the reasoning in 6.4.7 involves no integrality conditions and so applies to the present slightly more general situation. Thus the new Bruhat sequence and the expression for $\tilde{e}_i b$ as a partition take the form given in 6.4.7(*) which is as required. It remains to check the integrability conditions. The only non-trivial cases occur when the (new) $a_k \in [e_-, e_+]$. For e_- the corresponding new β is just α_i and then $(b(e_-), \alpha_i^\vee) \in \mathbb{Z}$ by definition of e_-. In the remaining cases one must show that $(s_i\beta_k^\vee, (\tilde{e}_i b)(a_k)) \in \mathbb{Z}$. Yet $(\tilde{e}_i b)(a_k) = b(e_-) + s_i(b(a_k) - b(e_-))$ and so the above expression equals

$$(\beta_k^\vee, s_i b(e_-) - b(e_-) + b(a_k)) = -(\beta_k^\vee, \alpha_i)(\alpha_i^\vee, b(e_-)) + (\beta_k^\vee, b(a_k)) \in \mathbb{Z}$$

by the definition of $b(e_-)$ and the integrability conditions on b. The case $\tilde{f}_i b \neq 0$ is similar.

6.4.24 Recall the notation of 6.4.22.

Corollary. *Each $b \in \mathcal{A}b^x$ is integral and monotone.*

6.4.25 Set $c = \max\{|(\alpha, \beta^\vee)| : \alpha, \beta \in \pi\}$. Given $b, b' \in \mathbb{P}$ define

$$d(b, b') = \max_{t \in [0,1], \alpha \in \pi} \{|(b(t) - b'(t), \alpha^\vee|\} .$$

Lemma. *Take $b, b' \in \mathbb{P}$ integral and monotone. Assume $d(b, b') < \varepsilon < 1$. Then*

(i) *$m_b^i = m_{b'}^i$ for all i.*
(ii) *If $\tilde{e}_i b \neq 0$, then $\tilde{e}_i b' \neq 0$ and $d(\tilde{e}_i b, \tilde{e}_i b') < 2c\varepsilon$.*
(iii) *If $\tilde{f}_i b \neq 0$, then $\tilde{f}_i b' \neq 0$ and $d(\tilde{f}_i b, \tilde{f}_i b') < 2c\varepsilon$.*

Fix i and define $M = \min\{(\alpha_i^\vee, b(t)) \mid t \in [0,1]\}$ and M' replacing b by b'. Clearly $|M - M'| < \varepsilon$ whilst $M, M' \in \mathbb{Z}$ by integrality. This forces $m_b := -M = -M' =: m_{b'}$. This proves (i) and by 6.4.5 the first parts of (ii) and (iii).

Consider (iii). Suppose $\tilde{f}_i b \neq 0$. The condition that b is monotone and the definition of $f_+(b)$ implies that the function $h_b(t)$ decreases monotonically in $[f_+^i(b), f_-^i(b)]$ and satisfies $h_b(t) \leq m_b - 1$ for all $t \in [f_-(b), 1]$. In particular $(\tilde{f}_i b)(t) = b(t) - \varphi(t)\alpha_i$ where $\varphi(t)$ increases monotonically in $[f_+(b), f_-(b)]$ from 0 to 1 and is constant outside this region. Writing $(\tilde{f}_i b')(t) = b(t) - \varphi'(t)\alpha_i$, a similar assertion applies to $\varphi'(t)$. Moreover the hypothesis $d(b, b') < 1$ and (i) implies that $f_+(b') < f_-(b)$ and $f_+(b) < f_-(b')$. One may assume $f_-(b') \leq f_-(b)$. Now $\varphi(t) = (b(t), \alpha_i^\vee) + m_b^i : t \in [f_+(b), f_-(b)]$ (resp. $\varphi'(t) = (b'(t), \alpha_i) + m_{b'}^i : t \in [f_+(b'), f_-(b')]$). Since these regions have a non-empty intersection and in them $\varphi(t), \varphi'(t)$ are monotone increasing it follows by (i) that $d(\varphi, \varphi') := \max_{t \in [0,1]} |\varphi(t) - \varphi'(t)| \leq \max_{t \in [0,1]} \mid (b(t) - b'(t), \alpha_i^\vee)| \leq d(b, b') < \varepsilon$. On the other hand

$$d(\tilde{f}_i b, \tilde{f}_i b') = d(b - \varphi\alpha_i, b' - \varphi'\alpha_i) \leq d(b, b') + cd(\varphi, \varphi') < \varepsilon + c\varepsilon \leq 2c\varepsilon$$

which proves (iii). The proof of (ii) is similar.

6.4.26 Recall 6.4.21, 6.2.44.

Theorem. *The set C is closed under tensor products.*

Recall the notation of 6.4.22, 6.4.23 and for each positive integer n, let $\mathcal{A}^n$ denote the subset of $\mathcal{A}$ of products of length n. Take $x, y \in [0,1] \cap \mathbb{Q}$. Then $d(b^x, b^y)$ is a multiple of $|x - y|$ depending only on λ, μ. Thus there exists an integer $m > 0$ such that $d(b^x, b^y) < (\frac{1}{2c})^n$ given $|x - y| < \frac{1}{m}$. By 6.4.24, 6.4.25 it follows that the subcrystals $\mathcal{A}^n b^x$, $\mathcal{A}^n b^y$ are isomorphic given $|x - y| < \frac{1}{m}$. Consequently $\mathcal{A}^n(b_\lambda \otimes b_\mu)$ and $\mathcal{A}^n b_{\lambda+\mu}$ are isomorphic. Since this holds for all n it proves the assertion.

6.4.27 By the uniqueness result of 6.4.21 one concludes that

Corollary. *For all $\lambda \in P^+(\pi)$ the Kashiwara crystal $B(\lambda)$ is isomorphic to the Littelmann path crystal $C(\lambda)$. In particular for any pair $\mu, \lambda \in P^+(\pi)$ and $b \in C(\mu)^\lambda$ the crystal generated by $b_\lambda \otimes b$ is isomorphic to $C(\lambda + wt b)$.*

6.5 Comments and Complements

6.5.1 Apart from 6.2.17–6.2.20, sections 6.1–6.3 are taken from work of M. Kashiwara [K2-K5]. The former is extracted from [JL2, 6.1] where it was originally derived using Lusztig's canonical basis which is now known [GrL1] to be the same as Kashiwara's; but whose construction is less elementary [L7–L12]. Since $G(b)$ is a basis for $U_q^{\mathbb{Z}}(\mathfrak{n}^-)$ over $\mathbb{Z}[q, q^{-1}]$ one can write

$$G(b)G(b') = \sum_{b'' \in B(\infty)} c_{b,b'}^{b''} G(b'')$$

where the structure constants $c_{b,b'}^{b''}$ lie in $\mathbb{Z}[q, q^{-1}]$. From Lusztig's canonical basis these structure constants have positive coefficients in the simply-laced case. I learnt the proof of 6.3.20 from O. Mathieu. It plays a key role in [M3].

6.5.2 The string property (6.3.14) of the Demazure modules was first noted by M. Demazure [D1] and was the basic clue to his character formula. However not only was his original proof faulty; but it is only in the crystal language that the string property really holds. (Recall the remarks in 4.4.4 and 6.3.7). The second part of 6.3.16 is due to Demazure [D2].

6.5.3 Section 6.4 is taken from a paper of P. Littelmann [Li2]. Note that 6.4.14 refines the following result of V. G. Kac and D. H. Peterson [Ka1, 11.3], namely that the set of weights of $V(\mu)$ coincides with the intersection of $\mu + \mathbb{Z}\pi$ with the convex hull of $W\mu$. The assertion in 6.4.16(∗) is an old result of R. Brauer and A. U. Klimyk. Littelmann informs me that for him the first inspiration for crystals comes from the Plaxique Monoide of Lascoux-Schützenberger [LaS1].

6.5.4 Littelmann ([Li2, 8.4] has conjectured that if $b \in \mathbb{P}$ is such that $b(t)$ is dominant for all $t \in [0, 1]$ then the crystal it generates is isomorphic to $C(wtb)$. Of this he recently gave a proof [Li3] which involves in particular new definitions of $\tilde{e}_i, \tilde{f}_i$ which nevertheless coincide with the old ones for integral, monotone paths. The weaker statement 6.4.26 needed here is inspired by this work but is shorter. In more detail 6.4.22 results from the analysis following [Li3, Lemma 5.8] and 6.4.25 is essentially [Li3, Proposition 3.1]. However 6.4.23 though inspired by [Li3, Proposition 5.6] is different in detail. The conclusion 6.4.27 has been reported by M. Kashiwara [K6], V. Lakshmibai [La3] and in some unpublished notes of S. Zelikson; but these proofs are different to the present one. A main interest of 6.4.27 is that the elements of $C(\lambda)$ can be easily tabulated (unlike those of $B(\lambda)$) whilst $B(\lambda)$ is directly related to the quantum group. In Lakshmibai's work [La3] the elements of $C(\lambda)$ are more directly related to the quantum group. For further results on bases see [BZ1, FS1].

6.5.5 The result in 6.4.18 was first proved independently by V. Kumar [Ku2] and O. Mathieu [M1]. It is known as the Parthasarathy-Ranga Rao-Varadarajan, or simply PRV, conjecture. Indeed these authors first established [PRV2] the result when $|W| < \infty$ and $w = w_0$. In this case the proof is rather easy and indeed the corresponding simple module can be interpreted as a unique "minimal $\mathfrak{k}$ type". This has some importance for unitary representations of $U(\mathfrak{g})$ especially those of highest weight [EJ1]. The result in 6.3.18 is not explicitly stated by Kashiwara in [K2–K5] although it was certainly known to him though I don't know at what instant in time. An equivalent version in the finite simply-laced case goes back to Lusztig [L8]. By 6.4.27 it is equivalent to Littelmann's (later) result 6.4.16. Of course this result is not so useful unless one has a good combinatorial description of crystals as say provided by the Littelmann theory.

6.5.6 The presentation (6.1.15) of $B(\infty)$ as an infinite product of the "one-dimensional" crystals C_i indexed by the simple roots is reminiscent of the Bott-Samelson resolution of the flag variety. Coincidently R. Bott recently suggested that one could use this resolution to introduce a large torus (being a product of the corresponding sequence of one-dimensional tori indexed by the same simple roots) acting on $V(\lambda)$ and whose weight spaces are at most one-dimensional. Kashiwara's construction achieves this, albeit in a combinatorial rather than geometric fashion. Recently M. Grossberg and Y. Karshon [GK1] have realized purely geometrically Bott's suggestion obtaining formulae, apparently similar to those of Kashiwara and Littelmann, for the extended character on this large torus.

6.5.7 In the finite simply-laced case Lusztig [L8, Sect. 7; L12, 4.2.1] gave a combinatorial description of his canonical basis for U^-, now known to be indexed by Kashiwara's crystal $B(\infty)$, in terms of an equivalence relation on the set $\mathbf{H}\times\mathbb{N}^n : n = \ell(w_0)$ where $\mathbf{H}$ denotes the set of reduced decompositions of w_0, whilst in Kashiwara's description only one (and any one) reduced decomposition is used. In contrast to Kashiwara (cf. 6.1.5) Lusztig attaches to each $\mathbf{c} \in \mathbb{N}^n$ the natural element of the PBW basis (10.5.2; [L12, 42.1.4]) for $U_q(\mathfrak{n}^-)$ corresponding to the linearly ordered system of positive roots defined by $h \in \mathbf{H}$ via say A.1.1. His equivalence relation just expresses the relation of two such bases in the crystal limit [L12, 42.1.5(a)]. Through its use he is able L8, Sect. 7 to say which elements of the canonical basis of $U_q(\mathfrak{n}^-)$ remain non-zero in $V(\lambda) : \lambda \in P^+(\pi)$ under the map $a \mapsto au_\lambda$, in terms of inequalities involving the $(\alpha_i^\vee, \lambda)$, and hence determine $B(\lambda)$. Though superficially similar to Littelmann's description of $B(\lambda)$ it appears to be quite different in detail. It requires a "canonical" decomposition of the Dynkin diagram which is only possible for C of finite type. The set of all possible Bruhat sequences used by Littelmann is much richer then $\mathbf{H}$ and its presence is not evident in Lusztig's analysis. By contrast Littelmann's description is poorly adapted to relating $B(\infty)$ to $B(\lambda)$.

Chapter 7. Structure Theorems for $U_q(\mathfrak{g})$

The base field will once again be denoted by k and assumed to be of characteristic zero. Apart from this the conventions of Chapter 5 will be retained and in particular $U_q(\mathfrak{g})$ is as defined as in 5.1.1. Notice that this means that the Cartan matrix is assumed integral.

7.1 Local Finiteness for the Adjoint Action

The aim of this section is to analyse the structure of $F(U_q(\mathfrak{g}))$ as defined in 1.3.1. This gives a mock version of the Peter-Weyl theorem (7.1.6) for $U_q(\mathfrak{g})$. Of course really the Peter-Weyl theorem applies to $R_q[G]$ (cf. 1.4.13) and so the result should be considered as a remnant of Drinfeld duality (cf. 7.1.23(iii)). It leads in a simple and elegant fashion to a description (7.1.17 (iii)) of the centre $Z(U_q(\mathfrak{g}))$. This is a quantum phenomenon in the best sense as it leads to further insight into the structure of $U(\mathfrak{g})$.

7.1.1 Recall the ad-invariant filtration $\mathcal{F}$ of $U = U_q(\mathfrak{g})$ defined in 5.3.1 and that in this the elements of $U^- = U_q(\mathfrak{n}^-)$ have degree 0. This led to an adjoint action of U which relative to each $\lambda \in P(\pi)$ could be twisted (5.3.10) making U^- isomorphic to $\delta M(-\frac{1}{2}\lambda)$ as a $U_q(\mathfrak{g})$ module. Recall 3.4.10, 4.1.4 and let $K(-\frac{1}{2}\lambda)^-$ denote the subspace of U^- corresponding to the unique simple submodule $V(-\frac{1}{2}\lambda)$ of $\delta M(-\frac{1}{2}\lambda)$. Similarly set $g_i = e_i t_i : i = 1, 2, \dots, \ell$ and let G^+ denote the subalgebra of U they generate. These elements have degree zero and G^+ viewed as a subspace of $gr_{\mathcal{F}} U$ is $\operatorname{ad} U$ invariant. The twisted action of U on U^- (resp. on G^+) is defined by setting

$$(\operatorname{ad}_\lambda e_i)1 = 0\,, \quad (\operatorname{ad}_\lambda t_i)1 = q^{-\frac{1}{2}(\lambda,\alpha_i)}1\,, \quad (\operatorname{ad}_\lambda f_i)1 = (1 - q^{-(\lambda,\alpha_i)})f_i\,,$$

$$(\text{resp. } (\operatorname{ad}_\lambda f_i)1 = 0\,, \quad (\operatorname{ad}_\lambda t_i)1 = q^{\frac{1}{2}(\lambda,\alpha_i)}1\,, \quad (\operatorname{ad}_\lambda e_i)1 = (q^{-(\lambda,\alpha_i)} - 1)g_i)\,,$$

and writing $(\operatorname{ad}_\lambda a)b = ((\operatorname{ad} a_1)b)(\operatorname{ad}_\lambda a_2)1$, $(\operatorname{ad}_\lambda a)c = ((\operatorname{ad}_\lambda a_1)1)(\operatorname{ad} a_2)c$ for all $a \in U$, $b \in U^-$, $c \in G^+$. Whilst this makes U^- isomorphic to $\delta M(-\frac{1}{2}\lambda)$ it makes G^+ isomorphic to the $\mathcal{O}$ dual of the universal lowest weight module of lowest weight $\frac{1}{2}\lambda$. The latter may be identified with $\delta M(-\frac{1}{2}\lambda)$ transported under the automorphism ι of U defined in 3.2.6. Let $K(\frac{1}{2}\lambda)^+$ denote the subspace of G^+ corresponding to the unique simple submodule $V(-\frac{1}{2}\lambda)^\iota$ of $\delta M(-\frac{1}{2}\lambda)^\iota$. Recall 4.3.5 and for each U module M set $I(M) = \{m \in M \mid \dim U_\alpha m < \infty, \forall \alpha \in \pi\}$.

Lemma. *Fix $\lambda \in Q(\pi)$. Then*

(i) *$U^- gr_\mathcal{F}\tau(\lambda)G^+$ is an $\operatorname{ad} U$ submodule of $gr_\mathcal{F}U$.*

(ii) *The map $\theta : u \otimes g \mapsto u \otimes gr_\mathcal{F}\tau(\lambda) \otimes g$ is a U module isomorphism of $\delta M(-\frac{1}{2}\lambda) \otimes \delta M(-\frac{1}{2}\lambda)^\iota$ onto $U^- gr_\mathcal{F}\tau(\lambda)G^+$.*

(iii) $I(U^- gr_\mathcal{F}\tau(\lambda)G^+) = \begin{cases} K(-\frac{1}{2}\lambda)^- gr_\mathcal{F}\tau(\lambda)K(\frac{1}{2}\lambda)^+ & : \lambda \in -2P^+(\pi) \\ 0 & : \textit{otherwise.} \end{cases}$

It is convenient to identify $\tau(\lambda)$ with $gr_\mathcal{F}\tau(\lambda)$. For (i) it remains to note that $(\operatorname{ad} f_i)\tau(\lambda) = (1 - q^{-(\lambda,\alpha_i)})f_i\tau(\lambda) \in U^-\tau(\lambda)$, whilst $(\operatorname{ad} e_i)\tau(\lambda) = (q^{-(\lambda,\alpha_i)} - 1)\tau(\lambda)g_i \in \tau(\lambda)G^+$. For (ii) it is enough to show that the twists cancel and for this it is enough to consider the action of generators on $1 \otimes 1 \in U^- \otimes G^+$. For example one has

$$\begin{aligned} e_i(1 \otimes 1) &= (\operatorname{ad}_\lambda e_i)1 \otimes (\operatorname{ad}_\lambda t_i^{-1})1 + 1 \otimes (\operatorname{ad}_\lambda e_i)1 \\ &= 1 \otimes (q^{-(\lambda,\alpha_i)} - 1)g_i , \end{aligned}$$

so

$$\theta(e_i(1 \otimes 1)) = (\operatorname{ad} e_i)\tau(\lambda) = (\operatorname{ad} e_i)\theta(1 \otimes 1) .$$

For (iii), let E be a $U_q(\mathfrak{b}^-)$ submodule of $I(U^- gr_\mathcal{F}\tau(\lambda)G^+)$. Since the action of $U_q(\mathfrak{b}^-)$ on G^+ is locally finite, there exists a finite dimensional $U_q(\mathfrak{b}^-)$ submodule F of $\delta M(-\frac{1}{2}\lambda)^\iota$ and a $U_q(\mathfrak{b}^-)$ module embedding $\psi : E \hookrightarrow \delta M(-\frac{1}{2}\lambda) \otimes F$. Now give F^* a left $(U_q(\mathfrak{b}^-)$ module structure through the antipode (1.4.8), that is write $(a\xi)(f) = \xi(\sigma(a)f)$ and consider $E \otimes F^*$, $\delta M(-\frac{1}{2}\lambda) \otimes F$ as $U_q(\mathfrak{b}^-)$ modules for the diagonal action. Then Frobenius reciprocity (A.2.15 (ii)) gives an isomorphism $T : \operatorname{Hom}_{U_q(\mathfrak{b}^-)}(E, \delta M(-\frac{1}{2}\lambda) \otimes F) \xrightarrow{\sim} \operatorname{Hom}_{U_q(\mathfrak{b}^-)}(E \otimes F^*, \delta M(-\frac{1}{2}\lambda))$ by $T_\psi(e \otimes \xi) = \xi(\psi(e))$. In particular $\operatorname{Im}\psi \subset \operatorname{Im} T_\psi \otimes F$. Yet $\operatorname{Im} T_\psi$ belongs to a $U_q(\mathfrak{b}^-)$ submodule of $\delta M(-\frac{1}{2}\lambda)$ for which the action of each $f_i : i = 1, 2, \dots, \ell$ is locally finite. Now each f_i generates an *Ore* subset (4.4.12) of U and so $\operatorname{Im} T_\psi$ can be assumed to belong to a U submodule of $\delta M(-\frac{1}{2}\lambda)$ with this property. Such a submodule is integrable and hence by 4.3.6(i) coincides with $V(-\frac{1}{2}\lambda)$ if $\lambda \in -2P^+(\pi)$ and is zero otherwise. Consequently $E \subset V(-\frac{1}{2}\lambda) \otimes F$ in the first case and is zero otherwise. A similar argument with $+$ and $-$ interchanged then gives (iii).

7.1.2 The $U^- gr_\mathcal{F}\tau(\lambda)G^+ : \lambda \in Q(\pi)$ form a direct sum decomposition of $gr_\mathcal{F}U$ as an $\operatorname{ad} U$ module. Set $R^+(\pi) = 2P^+(\pi) \cap Q(\pi)$.

Corollary

(i) $(\operatorname{ad} U)gr_\mathcal{F}\tau(\lambda) = K(-\frac{1}{2}\lambda)^- gr_\mathcal{F}\tau(\lambda)K(\frac{1}{2}\lambda)^+$

(ii) $I(gr_\mathcal{F}U) = \bigoplus_{\lambda \in -R^+(\pi)} (\operatorname{ad} U)gr_\mathcal{F}\tau(\lambda)$.

For (i) apply successively $\operatorname{ad} U^+$ and $\operatorname{ad} U^-$ to $gr_{\mathcal{F}}\tau(\lambda)$. Then (ii) follows from 7.1.1 (iii).

7.1.3 Now consider the adjoint action of U on itself, so in particular view $(\operatorname{ad} U)\tau(\lambda)$ as a submodule of U.

Lemma. $(\operatorname{ad} U)\tau(\lambda) \subset I(U) \iff \lambda \in -R^+(\pi)$.

The implication $\Longrightarrow$ follows from 7.1.1 (iii). Take $\lambda \in -R^+(\pi)$. Since $(\operatorname{ad} f_i)\tau(\lambda) = (1 - q^{-(\lambda,\alpha_i)})f_i\tau(\lambda) = (\operatorname{ad}_\lambda f_i)1$, the action of $\operatorname{ad} f_i$ on $\tau(\lambda)$ is locally nilpotent. Since f_i generates an *Ore* subset (4.4.12) of U it follows that the action of $\operatorname{ad} f_i$ on $(\operatorname{ad} U)\tau(\lambda)$ is locally nilpotent. A similar assertion holds for $\operatorname{ad} e_i$. Through triangular decomposition it follows that the action of $\operatorname{ad} U_{\alpha_i}$ on $(\operatorname{ad} U)\tau(\lambda)$ is locally finite. Since i is arbitrary this gives $\Longleftarrow$.

7.1.4 Since the quantum Serre relations are homogeneous one may consider U^+ (resp. U^-) as a graded algebra by defining each generator e_i (resp. f_i) to have degree 1. Let U_m^+ (resp. U_m^-) denote the subspace of U^+ (resp. U^-) of homogeneous elements of degree m and set

$$U_+^m = \bigoplus_{n=0}^{m} U_n^+ \,, \quad U_-^m = \bigoplus_{n=0}^{m} U_n^- \,.$$

Take $x \in U_r^+$, $y \in U_s^-$. By 5.1.1 it is immediate that

$$(*) \qquad xy - yx \in U_-^{s-1}U^0U_+^{r-1} \,.$$

Set $J^+ = \operatorname{Ann}_{\operatorname{ad} U^+} \tau(\lambda)$ which is independent of whether $\tau(\lambda)$ is considered as an element of U or of $gr_{\mathcal{F}}U$. Moreover it follows from weight space decomposition that J^+ is a homogeneous ideal, that is $J^+ = \oplus J_m^+$ where $J_m^+ = J^+ \cap U_m^+$. Similar considerations apply with $+$ and $-$ interchanged.

For each $m \in \mathbb{N}$ set

$$M_\lambda^m = \operatorname{ad}(U_-^m U^0 U_+^m)\tau(\lambda) \quad N_\lambda^m = \operatorname{ad}(U_-^m U^0 U_+^m) gr_{\mathcal{F}}\tau(\lambda) \,.$$

One has $gr_{\mathcal{F}}M_\lambda^m \supset N_\lambda^m$ since $\mathcal{F}$ is $\operatorname{ad} U$ stable. The λ subscript will often be omitted.

Lemma. *For all $\lambda \in Q(\pi)$, $m \in \mathbb{N}$ one has $gr_{\mathcal{F}}M_\lambda^m = N_\lambda^m$. In particular $gr_{\mathcal{F}}((\operatorname{ad} U)\tau(\lambda)) = (\operatorname{ad} U)(gr_{\mathcal{F}}\tau(\lambda))$ and $(\operatorname{ad} U)\tau(\lambda)$ is isomorphic to $(\operatorname{ad} U)gr_{\mathcal{F}}\tau(\lambda)$ as a U module.*

Since $N^m = (\operatorname{ad}_\lambda U_-^m)1 \otimes k(q)gr_{\mathcal{F}}\tau(\lambda) \otimes (\operatorname{ad}_\lambda U_+^m)1$ it is enough to show that

$$\dim M^m \le \dim(\operatorname{ad} U_-^m)\tau(\lambda) \dim(\operatorname{ad} U_+^m)\tau(\lambda) \,.$$

Set $M^{r,s} = (\operatorname{ad}(U_-^s U^0 U_+^r)\tau(\lambda)$. By $(*)$ above it follows that $J_s^-(\operatorname{ad} U_r^+)\tau(\lambda) \subset M^{r-1,s-1}$ and so $\dim M^{r,s}/(M^{r,s-1} + M^{r-1,s}) \le$

$\dim(U_s^-/J_s^-)\dim(\operatorname{ad} U_r^+)\tau(\lambda) = \dim(\operatorname{ad} U_s^-)\tau(\lambda)\dim(\operatorname{ad} U_r^+)\tau(\lambda)$. Consequently

$$\begin{aligned}\dim M^m &= \sum_{r,s=0}^{m} \dim M^{r,s}/(M^{r,s-1}+M^{r-1,s}) \\ &\leq \sum_{r,s=0}^{m} \dim(\operatorname{ad} U_s^-)\tau(\lambda)\dim(\operatorname{ad} U_r^+)\tau(\lambda) \\ &= \dim(\operatorname{ad} U_-^m)\tau(\lambda)\dim(\operatorname{ad} U_+^m)\tau(\lambda)\,,\end{aligned}$$

since $(\operatorname{ad} U_\pm^m)\tau(\lambda)$ is a direct sum of the $(\operatorname{ad} U_n^\pm)\tau(\lambda) : 0 \leq n \leq m$ by weight space decomposition.

7.1.5 Corollary. *The* $(\operatorname{ad} U)\tau(\lambda) : \lambda \in Q(\pi)$ *form a direct sum in* U.

If not there exists a finite subset $F \subset Q(\pi)$ and $m \in \mathbb{N}$ such that the $M_\lambda^m : \lambda \in F$ do not form a direct sum. Then

$$\sum_{\lambda\in F} \dim N_\lambda^m = \sum_{\lambda\in F} \dim M_\lambda^m > \dim \sum_{\lambda\in F} M_\lambda^m \geq \dim \sum_{\lambda\in F} N_\lambda^m = \sum_{\lambda\in F} \dim N_\lambda^m$$

where the last equality follows from the fact that the $(\operatorname{ad} U)gr_{\mathcal{F}}\tau(\lambda) : \lambda \in Q(\pi)$ form a direct sum in $gr_{\mathcal{F}}U$. This contradiction proves the assertion.

7.1.6 Consider U as a U module for the adjoint action.

Theorem

$$I(U) = \bigoplus_{\lambda\in R^+(\pi)} (\operatorname{ad} U)\tau(\lambda)\,.$$

In particular $I(U)$ *is a left coideal of* U.

In view of 7.1.3 and 7.1.5 it remains to show that $I(U) \subset \sum(\operatorname{ad} U)\tau(\lambda)$. Obviously $gr_{\mathcal{F}}I(U) \subset I(gr_{\mathcal{F}}U)$. Again $\lambda \in -R^+(\pi)$ implies that $gr_{\mathcal{F}}\tau(\lambda)$ has positive gradation degree. Then by 7.1.2 each homogeneous $a \in I(gr_{\mathcal{F}}U)$ has positive gradation degree and so each $a \in I(U)$ has positive filtration degree (with respect to $\mathcal{F}$). Then the assertion follows from 7.1.2 by induction on filtration degree. The last part follows from 1.3.5.

7.1.7 It is convenient to write $F(\lambda) = (\operatorname{ad} U)\tau(-2\lambda)$ and $G(\lambda) = (\operatorname{ad} U)gr_{\mathcal{F}}\tau(-2\lambda)$.

Lemma. *For all* $\lambda, \mu \in \frac{1}{2}Q(\pi)$ *one has with respect to multiplication in* $gr_{\mathcal{F}}U$ *that*

(i) $K(\lambda)^-K(\mu)^- = K(\lambda+\mu)^-$, $K(\lambda)^+K(\mu)^+ = K(\lambda+\mu)^+$
(ii) $G^+K(\lambda)^- = K(\lambda)^-G^+$, $U^-K(\lambda)^+ = K(\lambda)^+U^-$
(iii) $K(\lambda)^-K(\mu)^+ = K(\mu)^+K(\lambda)^-$
(iv) $G(\lambda)G(\mu) = G(\lambda+\mu)$.

For the first part of (i) observe that $K(\lambda)^-\tau(-2\lambda)$ is just the $\operatorname{ad} U^-$ submodule of $gr_{\mathcal{F}}U$ generated by $\tau(-2\lambda)$. Since $K(\lambda)^-\tau(-2\lambda)K(\mu)^-\tau(-2\mu) = K(\lambda)^-K(\mu)^-\tau(-2(\lambda+\mu))$ the assertion follows from the Leibnitz rule. The second part is similar.

Recall 5.1.1 (vii) and the definition of $\operatorname{ad}_\lambda$. Then for all $a \in K(\lambda)^-$ one has $(\operatorname{ad} e_i)a = q^{(\lambda,\alpha_i)}(\operatorname{ad}_{-2\lambda} e_i)a \in K(\lambda)^-$. On the other hand $(\operatorname{ad} e_i)a = e_i a t_i - a e_i t_i = q^{-(\xi,\alpha_i)} g_i a - a g_i$ for a of weight ξ. Thus $g_i K(\lambda)^- \subset K(\lambda)^- g_i + K(\lambda)^-$ for all i, giving $G^+K(\lambda)^- \subset K(\lambda)^-G^+$. The reverse inclusion is similar. This gives the first part of (ii). The second part obtains on interchanging + and − in the first part.

Since the multiplication map induces an isomorphism $U^-\otimes G^+ \xrightarrow{\sim} U^-G^+$ of vector spaces, it follows that $K(\lambda)^-G^+ \cap U^-K(\mu)^+ = K(\lambda)^-K(\mu)^+$. Applying (ii) to the left hand side then gives (iii).

Finally

$$\begin{aligned} G(\lambda)G(\mu) &= K(\lambda)^-\tau(-2\lambda)K(-\lambda)^+K(\mu)^-\tau(-2\mu)K(-\mu)^+ , \\ &= K(\lambda)^-K(\mu)^-\tau(-2\lambda)\tau(-2\mu)K(-\lambda)^+K(-\mu)^+ , \quad \text{by (iii)} \\ &= G(\lambda+\mu) , \quad \text{by (i).} \end{aligned}$$

Remark. It is *false* that $(\operatorname{ad}_{-2(\lambda+\mu)} a)(bc) = ((\operatorname{ad}_{-2\lambda} a_1)b(\operatorname{ad}_{-2\mu} a_2)c$ for $a \in U$; $b \in K(\lambda)^-$, $c \in K(\mu)^-$. This failure will be understood in 9.1.7, where a further modification is introduced to recover this property.

7.1.8 The above results have a significant interpretation for the Rosso form R. Apply the automorphism $\zeta_\varkappa$ of U (as in the beginning of chapter 5) to 3.3.3. Adjoin $q^{\frac{1}{2}}$ to $k(q)$. Then R is the $\operatorname{ad} U$ invariant bilinear form on U satisfying notably

(i) $$R(u\tau(\lambda)g, u'\tau(\lambda')g') = R(u,g')R(g,u')q^{-\frac{1}{2}(\lambda,\lambda')}$$

for all $u \in U^-_{-\mu}, u' \in U^-$; $g, g' \in G^+$; $\lambda, \lambda' \in Q(\pi)$. Here R restricts to a non-degenerate bilinear form on $G^+ \times U^-$ given by the adjoint action of U^- on G^+ viewed as a submodule of $gr_{\mathcal{F}}U$. Though this can be recovered from 3.3.3 it is useful to verify it directly. Take $g \in G^+_\mu$. Then $(\operatorname{ad} f_i)g$ computed in U has two terms, the first identifies with $(\operatorname{ad} f_i)(gr_{\mathcal{F}}g)$ whilst the second lies in $t_i^2G^+$. Again $(\operatorname{ad}\sigma(f_i))u\tau(\lambda) : u \in U^-_{-\mu+\alpha_i}$ has two terms. The first is $-q^{(\alpha_i,\mu)}f_iu\tau(\lambda)$ and the second is $q^{(\alpha_i,\alpha_i)}u\tau(\lambda)f_i = q^{(\alpha_i,\alpha_i-\lambda)}uf_i\tau(\lambda)$. Equating coefficients of $q^{-(\alpha_i,\lambda)}$ in the identity

$$R((\operatorname{ad} f_i)g, u\tau(\lambda)) = R(g, \operatorname{ad}\sigma(f_i)u\tau(\lambda))$$

then gives

$$R((\operatorname{ad} f_i)(gr_{\mathcal{F}}g), u) = -q^{(\alpha_i,u)}R(g, f_iu) .$$

Let $\varepsilon : U_q(\mathfrak{b}^+) \longrightarrow k(q)$ be the augmentation and $\star$ the antiautomorphism of U given in 6.1.5. Then for all $g \in G^+_\mu$, $u \in U^-_{-\mu}$ one obtains imposing the normalization condition $R(1,1) = 1$ that

(ii) $$R(g,u) = (-1)^{|\mu|}q^{-\frac{1}{2}(\mu,\mu)+(\rho,\mu)}\varepsilon((\operatorname{ad} u^\star)(gr_{\mathcal{F}}g)) .$$

A similar analysis starting from the ad-invariance of$R(u\tau(\lambda), g)$ gives $R(f_i u, g) = -q^{-(\alpha_i, \mu-\alpha_i)} R(u, (\operatorname{ad} f_i) g r_{\mathcal{F}} g)$ for all $u \in U^-_{-\mu+\alpha_i}$, $g \in G^+_\mu$ and hence that

(iii) $$R(u, g) = q^{2(\rho,\mu)} R(g, u) \ , \quad \text{for all} \ \ u \in U^-_{-\mu} \ , g \in G^+_\mu \ .$$

It is clear that (i)–(iii) determine R completely. Of course this analysis is insufficient to prove that R is ad-invariant for which one must go back to 3.3.3. If the Cartan form is non-degenerate then so is R. Otherwise one must augment T or correspondingly $Q(\pi)$, as discussed in 3.2.10, to include say $P(\pi)$. This process is trivial as far as the formulae are involved; but may make a significant difference to certain conclusions. (See for example 7.3.7).

7.1.9 The crucial and most striking property of R, besides its ad-invariance, is that it is compatible (7.1.8 (i)) with the triangular decomposition of U. This is particularly suitable for the study of highest weight modules as shown below. First a technical result is needed.

Lemma. *Let $\Omega \subset P(\pi)$ be Zariski dense in $\mathfrak{h}^*_{\mathbb{Q}}$. Then*

$$\bigcap_{\lambda \in \Omega} \operatorname{Ann}_U V(\lambda) = 0 \ .$$

Recall the definitions and notations of 3.4.10. The non-degeneracy of the Shapovalev form $\mathcal{S}^\Lambda$ on $V(\Lambda)$ implies that $a \in \operatorname{Ann}_U V(\Lambda) \iff \mathcal{S}^\Lambda(x, ay) = 0, \forall x, y \in U$. This translates to give

(*) $$\operatorname{Ann}_U V(\Lambda) = \{a \in U \mid \Lambda(\mathcal{P}(xay)) = 0, \ \forall x, y \in U\} \ .$$

In particular if $\Omega \subset P(\pi)$ is Zariski dense, then

(**) $$\bigcap_{\lambda \in \Omega} \operatorname{Ann}_U V(\lambda) = \{a \in U \mid \mathcal{P}(xay) = 0, \ \forall x, y \in U\} \ .$$

Thus one may assume $\Omega = P^+(\pi)$ without loss of generality.

Choose bases $\{y^i_{-\mu}\}$, $\{x^j_\nu\}$ of $U^-_{-\mu}$ and U^+_ν respectively. For each $0 \neq a \in U$, there exist $F \subset \mathbb{N}\pi$ finite, $t^{\mu,\nu}_{i,j} \in U^0$ such that

$$a = \sum_{\mu,\nu \in F} \sum_{i,j} y^i_{-\mu} t^{\mu,\nu}_{i,j} x^j_\nu \ .$$

Assume $\eta \in F$ minimal with property that some $t^{\mu,\eta}_{i,j} \neq 0$. By 3.4.10 and 4.3.6 (ii), $\det \mathcal{S}^\lambda_\eta = \det \mathcal{P}(x^j_\eta \varkappa(x^r_\eta))(\lambda)$ is non-zero for all $\lambda \in P^+(\pi)$ sufficiently large. Now suppose $a \in \operatorname{Ann} V(\lambda)$. Then

$$\sum_{\mu,\nu \in F} \sum_{i,j} y^i_{-\mu} t^{\mu,\nu}_{i,j} [x^j_\nu \varkappa(x^r_\eta) u_\lambda] = 0 \ .$$

Yet the square-bracketed term for $\eta < \nu$ vanishes and equals $\mathcal{P}(x^j_\eta \varkappa(x^r_\eta))(\lambda)$ if $\eta = \nu$. Consequently

$$\sum_{\mu \in F} \sum_i y^i_{-\mu} t^{\mu,\eta}_{i,j}(\lambda) u_\lambda = 0$$

for all j. The assumption on λ further implies that the $y^i_{-\mu} u_\lambda$ are linearly independent and so $t^{\mu,\eta}_{i,j}(\lambda) = 0$ for all i, j, μ. Yet this must hold for all $\lambda \in P^+(\pi)$ sufficiently large which is a Zariski dense subset of $\mathfrak{h}^*_{\mathbb{Q}}$ and so $t^{\mu,\eta}_{i,j} = 0$ for all i, j, μ, contradicting the choice of η.

7.1.10 By 7.1.9(*) one has a formula for $\operatorname{Ann}_U V(\Lambda)$ whilst 7.1.9(**) shows that $\mathcal{P}(I) \neq 0$ for any non-zero two-sided ideal of U. Actually both are improved by the

Lemma. *Let M be an $\operatorname{ad} U$ submodule of U. Then*

(i) $\Lambda(\mathcal{P}(M)) = 0 \iff M \subset \operatorname{Ann} V(\Lambda)$
(ii) $\mathcal{P}(M) = 0 \iff M = 0$.

Let $V(\Lambda)^-$ denote the T stable complement to $k(q)u_\Lambda$ in $V(\Lambda)$. Then

$$(*) \qquad \Lambda(\mathcal{P}(M)) = 0 \iff M u_\Lambda \subset V(\Lambda)^- \; .$$

Assume that $\Lambda(\mathcal{P}(M)) = 0$ and let $m \in M$ be a weight vector. Then $(t_i m t_i^{-1}) f_i = f_i m - (\operatorname{ad} f_i)m$ and so $MU^- \subset U^- M$ which gives $MV(\Lambda) = MU^- u_\Lambda \subset U^- M u_\Lambda \subset V(\Lambda)^-$ by (*). Since M is $\operatorname{ad} U$ stable one similarly obtains $UM = MU$ and then that $MV(\Lambda)$ is a U submodule of $V(\Lambda)$, hence zero by the above, that is $M \subset \operatorname{Ann}_U V(\Lambda)$. This proves $\Longrightarrow$ in (i), whilst $\Longleftarrow$ in (i) is immediate from $\Longrightarrow$ in (*). Finally (ii) results from (i) and 7.1.9.

7.1.11 The previous two lemmas are of course also valid for $U(\mathfrak{g})$. However using the Rosso form one obtains the following purely quantum result.

Given a subset $S \subset U$ set

$$S^\perp = \{u \in U \mid R(u, s) = 0 \, , \quad \text{for all} \quad s \in S\} \; .$$

Lemma. *For all $\lambda \in \frac{1}{2}Q(\pi)$ one has*

$$\operatorname{Ann}_U V(\lambda) = ((\operatorname{ad} \; U)\tau(-2\lambda))^\perp \; .$$

Given $M \subset U$, $\lambda \in Q(\pi)$ triangularity (7.1.8 (ii)) implies $R(M, \tau(\lambda)) = R(\mathcal{P}(M), \tau(\lambda)) = \mathcal{P}(M)(-\frac{1}{2}\lambda)$. Now assume M to be $\operatorname{ad} U$ invariant. Then

$$\begin{aligned} M \subset ((\operatorname{ad} U)\tau(-2\lambda))^\perp &\iff R(M, \tau(-2\lambda)) = 0 \, , \quad \text{by ad-invariance,} \\ &\iff \mathcal{P}(M)(\lambda) = 0 \, , \quad \text{by the above,} \\ &\iff M \subset \operatorname{Ann}_U V(\lambda) \, , \quad \text{by 7.1.10(i),} \end{aligned}$$

as required.

7.1.12 Corollary. *Suppose that $R^+(\pi)$ is Zariski dense in $\mathfrak{h}^*_{\mathbb{Q}}$. Then the restriction of R to $I(U)$ is non-degenerate.*

This follows from 7.1.6, 7.1.9, 7.1.11.

Remark. The hypothesis is equivalent to the Cartan matrix being non-degenerate. Of course defining $\check{U}$ as in 3.2.10, then R is non-degenerate on $I(\check{U})$.

7.1.13 The above result expresses the fact that $I(U)$ is rather large inside U. Another way of expressing this is as follows. Set $T_< = \tau(-R^+(\pi))$. By 7.1.6, $T_< \subset I(U)$ and is obviously an *Ore* subset. Set $J(U) = T_<^{-1}I(U)$ which identifies with a subalgebra of U.

Lemma. *One has $J(U) = U^-\tau(R(\pi))G^+$. In particular any $V(\Lambda)$ is simple as an $I(U)$ module and U is freely generated as a left (or right) $J(U)$ module by representatives of $\tau(Q(\pi)/R(\pi))$.*

One has $(\operatorname{ad} f_i)\tau(\lambda) = f_i\tau(\lambda) - \tau(\lambda)f_i = \tau(\lambda)(q^{(\alpha_i,\lambda)} - 1)f_i$. Taking $\lambda \in -R^+(\pi)$ with $(\alpha_i, \lambda) \neq 0$ it follows from 7.1.6 that $f_i \in J(U)$. Similarly $g_i \in J(U)$ and so $J(U) \supset U^-\tau(R(\pi))G^+$. On the other hand $U^-\tau(R(\pi))G^+$ is $\operatorname{ad} U$ stable via 5.1.1 (v) noting that $2Q(\pi) \subset R(\pi)$. Thus the opposite inclusion follows from 7.1.6.

7.1.14 Whereas U has negative components with respect to its filtration $\mathcal{F}$, that is $\mathcal{F}^r U \neq 0$ for $r < 0$, one has by 7.1.4 and 7.1.6 the

Lemma. *$\mathcal{F}^r U \cap I(U) = 0$ for $r < 0$ and reduces to scalars for $r = 0$.*

7.1.15 Assume that C is indecomposable and not of finite type. By A.1.2 the second condition is equivalent to $|W(C)| = \infty$.

Lemma

(i) *$|W\omega_i| = \infty$ for each fundamental weight ω_i.*
(ii) *$\dim V(\lambda) = \infty$ for all $\lambda \in P^+(\pi) \setminus \{0\}$.*
(iii) *$F(U)$, and hence $Z(U)$, reduces to scalars.*

Assume (i) fails and recall that $\rho = \sum \omega_i$. By A.1.1 (v), (vi) one has $\operatorname{Stab}_W \rho = \{e\}$ and so $|W\rho| = \infty$. Then $\mathcal{I} := \{i \in \{1, 2, \ldots, \ell\} \,\big|\, |W\omega_i| < \infty\}$ is a proper subset. Since C is indecomposable there exists $j \in \{1, 2, \ldots, \ell\} \setminus \mathcal{I}$ having a neighbour $i \in \mathcal{I}$, that is $(\alpha_j, \alpha_i) \neq 0$. For all $w \in W$ one has $w(2\omega_i - \alpha_i) = w\omega_i + ws_i\omega_i \in W\omega_i + W\omega_i$ which is a finite set by the definition of $\mathcal{I}$. Thus $|W(2\omega_i - \alpha_i)| < \infty$. Yet $2\omega_i - \alpha_i = \sum_s -(\alpha_s^\vee, \alpha_i)\omega_s \in P^+(\pi)$. Then by A.1.1 (vii) one has $\operatorname{Stab}_W(2\omega_i - \alpha_i) \subset \operatorname{Stab}_W \omega_j$ which forces $|W\omega_j| < \infty$ in contradiction to $j \notin \mathcal{I}$.

Take $\lambda \in P^+(\pi) \setminus \{0\}$. Then by A.1.1 (vii) and (i) one has $|W\lambda| = \infty$. By 4.3.6 (iii) this gives (ii).

Consider $(\operatorname{ad} U)\tau(\lambda) : \lambda \in -R^+(\pi)$. By 7.1.1 (iii) and 7.1.4 it is isomorphic to $V(-\frac{1}{2}\lambda) \otimes V(-\frac{1}{2}\lambda)^\iota$. Any finite dimensional submodule is automatically integrable and hence by 4.3.9 and (ii) must be isomorphic to the trivial module. Given $z \in V(-\frac{1}{2}\lambda) \otimes V(-\frac{1}{2}\lambda)^\iota$ one may write z as a finite sum of weight vectors $y_{-\mu} \otimes x_\mu$. Assume μ maximal with the property that such a term appears in z. Then if z is U invariant, $y_{-\mu}$ is U^- invariant and this forces $V(-\frac{1}{2}\lambda) = U^+ y_{-\mu}$ which is finite dimensional. By (ii) the latter implies that $\lambda = 0$. Hence (iii).

7.1.16 The above results take on a much more interesting form when C is of finite type, equivalently when $|W(C)| < \infty$. *This will be assumed in the remainder of section* 7.1. As in 1.4.8 give $V(\lambda)^*$ a left U module structure through $(u, \xi) \mapsto \xi\sigma(u)$. Then $V(\lambda)^*$ has lowest weight $-\lambda$. Set $U_+ = \ker \varepsilon$.

Lemma. *Assume C of finite type. Then for all $\lambda \in P^+(\pi)$ one has*

(i) $\dim V(\lambda) < \infty$. *In particular* $I(U) = F(U)$.
(ii) $V(\lambda)^\iota$ *and* $V(\lambda)^*$ *are both isomorphic to* $V(-w_0\lambda)$.
(iii) $V(\lambda) \otimes V(\lambda)^\iota \xrightarrow{\sim} \operatorname{End} V(\lambda)$ *as* $U \otimes U$ *modules.*
(iv) $(\operatorname{End} V(\lambda))^{\Delta(U)} = k(q)1d_{V(\lambda)}$.
(v) $\operatorname{End} V(\lambda) \xrightarrow{\sim} \Delta(U_+)(u_\lambda \otimes u_{-\lambda}) \oplus k(q)\operatorname{Id}_{V(\lambda)}$ *as* $\Delta(U)$ *modules.*

By A.1.4, $w_0\lambda$ is the lowest weight of $V(\lambda)$ whose weights must therefore lie in the finite set $(\lambda - \mathbb{N}\pi) \cap (w_0\lambda + \mathbb{N}\pi)$. Hence (i). Both $V(\lambda)^\iota$ and $V(\lambda)^*$ have lowest weight $-\lambda$, hence highest weight $-w_0\lambda$. Recalling 4.3.9 this gives (ii). The isomorphism of (iii) is given by $v \otimes \xi \mapsto (v' \mapsto \xi(v')v)$. Given $\theta \in (\operatorname{End} V(\lambda))$ it follows from weight space considerations that $\theta(u_\lambda)$ is a multiple of u_λ and hence θ is a multiple of $\operatorname{Id}_{V(\lambda)}$. Finally $k(q)\operatorname{Id}_{V(\lambda)}$ is complemented in $\operatorname{End} V(\lambda) \xleftarrow{\sim} V(\lambda) \otimes V(\lambda)^*$, by the kernel K of the $\Delta(U)$ module map $v \otimes \xi \mapsto \xi(v)$. Applying successively U^+ and U^- it follows that $u_\lambda \otimes u_{-\lambda}$ is a cyclic vector for $V(\lambda) \otimes V(\lambda)^*$ as a $\Delta(U)$ module. Thus $\Delta(U_+)(u_\lambda \otimes u_{-\lambda})$ is a submodule of codim ≤ 1. Yet $\Delta(U_+)K \subset \Delta(U)K \subset K$ whilst$\Delta(U_+)1d_{V(\lambda)} = 0$. Applying $\Delta(U_+)$ to $\Delta(U)(u_\lambda \otimes u_{-\lambda}) = K \oplus k(q)1d_{V(\lambda)}$ gives $\Delta(U_+)(u_\lambda \otimes u_{-\lambda}) = K$ and hence (v).

7.1.17 Take $\lambda \in -R^+(\pi)$ and let $y_{-\frac{1}{2}\lambda}$ (resp. $z_{-\frac{1}{2}\lambda}$) denote an element of $(\operatorname{ad} U)gr_{\mathcal{F}}\tau(\lambda)$ (resp. $(\operatorname{ad} U)\tau(\lambda)$) mapping to a non-zero multiple of $1d_{V(-\lambda/2)}$ under the isomorphisms of 7.1.1, 7.1.2, 7.1.4, 7.1.16. Let $Y(U)$ denote the subspace of $gr_{\mathcal{F}}U$ spanned by the $y_{-\frac{1}{2}\lambda} : \lambda \in -R^+(\pi)$. Let ψ denote the restriction of the projection $\mathcal{P} : U \longrightarrow U^0$ (3.4.10) to $Z(U)$. Define $w\tau(\lambda)$ (resp. $w.\tau(\lambda)$) by $(w\tau(\lambda))(\mu) = \tau(\lambda)(w^{-1}\mu)$ (resp. $(w.\tau(\lambda))(\mu) = \tau(\lambda)(w^{-1}.\mu)$ and extend to U^0 by linearity. One has

$$(w^{-1}.\tau(\lambda))q^{(\rho,\lambda)} = \tau(w^{-1}\lambda)q^{(\rho,w^{-1}\lambda)} \tag{$*$}$$

and

$$\hat{\tau}(\lambda) := \sum_{w\in W} \tau(w\lambda)q^{(\rho,w\lambda)} = q^{(\rho,\lambda)} \sum_{w\in W} w.\tau(\lambda) \ , \quad \text{for all} \quad \lambda \in P(\pi) \ .$$

Lemma. *For all* $\lambda, \mu \in \frac{1}{2}R^+(\pi)$

(i) $y_\lambda\, y_\mu = y_{\lambda+\mu}$ *up to a non-zero scalar. In particular* $Y(U)$ *is isomorphic to the semigroup algebra* $k(q)R^+(\pi)$.

(ii) $gr\ z_\lambda = y_\lambda$, *up to a non-zero scalar.*

(iii) $Z(U) = \bigoplus_{\lambda\in\frac{1}{2}R^+(\pi)} z_\lambda$.

(iv) $(k(q)\tau(R(\pi))^{W.} = \bigoplus_{\lambda\in R^+(\pi)} k(q)\hat{\tau}(-\lambda)$.

(v) ψ *is an isomorphism of* $Z(U)$ *onto* $(k(q)\tau(R(\pi))^{W.}$.

(i) is immediate from 7.1.7 (iv) and (ii) is immediate from 7.1.4. Assertion (iii) follows from 1.3.3, 7.1.6 and 7.1.16. Obviously $Z(U)$ is commutative and hence so is $Y(U)$.

Take $\lambda, \lambda' \in R^+(\pi)$ and suppose $\hat{\tau}(-\lambda)$, $\hat{\tau}(-\lambda')$ are proportional. Obviously $\lambda \in W\lambda'$. Then $\lambda = \lambda'$ by A.1.12(i). The proof of (iv) is then completed by A.1.12 (ii).

(v) From $U_+^- U \cap U^T = UU_+^+ \cap U^T$ it follows that the restriction of $\mathcal{P}$ to U^T is an algebra homomorphism. In particular ψ is an algebra homomorphism. It follows from definitions that $z \in Z(U)$ acts on $M(\lambda) : \lambda \in P(\pi)$ by the scalar $\psi(z)(\lambda)$. Now take $\mu \in P^+(\pi)$ and $w \in W$. Then the embedding $M(w.\mu) \hookrightarrow M(\mu)$ of 4.4.7gives $\psi(z)(\mu) = \psi(z)(w.\mu) = (w^{-1}.\psi(z))(u)$. Since $P^+(\pi)$ is Zariski dense in $h^*_{\mathbb{Q}}$ and (cf. 7.1.13) one has $Z(U) \subset U^-\tau(R(\pi))G^+$ it follows that $\operatorname{Im}\psi \subset (k(q)\tau(R(\pi)))^{W.}$.

Suppose $\psi(z) = 0$ for some $z \in Z(U)$. Then in particular $\psi(z)(\mu) = 0$ and so z annihilates $V(\mu)$ for all $\mu \in P^+(\pi)$. By 7.1.9 this implies $z = 0$.

Notice that $\mathcal{F}$ may be defined using the triangular decomposition $U^- \otimes U^0 \otimes G^+ \xrightarrow{\sim} U$, taking the elements of U^- and G^+ to have degree 0 and U^0 to be graded by taking each t_i to have degree 1. Again $\mathcal{P}$ may be identified with the projection onto U^0 defined by the decomposition $U = U^0 \oplus (U_+^- U + UG_+^+)$. From this it is immediate that

$$(**) \qquad gr_{\mathcal{F}}\mathcal{P}(u) = \mathcal{P}(gr_{\mathcal{F}}u) \ , \quad \text{for all} \quad u \in U \ .$$

Taking $z = z_\lambda$ it follows from (ii) and (**) that $gr_{\mathcal{F}}\psi(z_\lambda) = \tau(-2\lambda)$ for all $\lambda \in \frac{1}{2}R^+(\pi)$. Combined with (iii), (iv) and the observations of the previous two paragraphs, this gives (v).

Remark. In general $Y(U)$ and $Z(U)$ are not isomorphic. For example, take $|\pi| = 2$ and $C = \begin{pmatrix} 2 & -1 \\ -1 & 2 \end{pmatrix}$, that is of type A_2. Writing $\pi = \{\alpha_1, \alpha_2\}$ one has $R^+(\pi) = \mathbb{N}\{4\alpha_1 + 2\alpha_2,\ 2\alpha_1 + 2\alpha_2,\ 2\alpha_1 + 4\alpha_2\}$. Let $\{y_1, y_2, y_3\}$ (resp. $\{z_1, z_2, z_3\}$) be the corresponding set of generators of $Y(U)$ (resp. $Z(U)$). Then $y_1y_3 = y_2^3$ defines a surface which is singular at the origin, whilst $z_1z_3 = z_2^3 + 3$ defines a non-singular surface.

7.1.18 One may compute $\psi(z_\lambda)$ explicitly using the quantum trace. First recall that for any Hopf algebra H which is cocommutative one has $tr((\operatorname{ad} a)b, M) = \varepsilon(a)tr(b, M)$ for all $a, b \in H$ and any finite dimensional H module M. This does not quite hold here. However set $t_0 = \tau(-2\rho)$ and

$$tr_q(a, M) := tr(at_0, M)$$

for all $a \in U$ and any finite dimensional U module M.

Lemma. *For all $a, b \in U$ one has*

$$tr_q((\operatorname{ad} a)b, M) = \varepsilon(a)tr_q(b, M)\ .$$

Since $a \mapsto \operatorname{ad} a$ is an algebra homomorphism (1.3.1) it is enough to check this assertion with a being a generator. It is obvious if $a = t_i$. For $a = e_i$ one has

$$\begin{aligned} tr_q((\operatorname{ad} e_i)b, M) &= tr((e_i b t_i - b e_i t_i)t_0, M) \\ &= (q^{2(\rho,\alpha_i)-(\alpha_i,\alpha_i)} - 1)tr(b e_i t_i t_0, M) \\ &= 0\ . \end{aligned}$$

The case $a = f_i$ is similar.

7.1.19 Lemma. *For all $\mu \in \frac{1}{2}R^+(\pi)$ one has*

$$\psi(z_\mu) = \sum_{\nu \in P^+(\pi)} \hat{\tau}(-2\nu) \dim V(\mu)_\nu\ ,$$

up to a non-zero scalar.

In the isomorphism of 7.1.1 (ii) it is immediate that $gr_{\mathcal{F}}\tau(-2\mu)$ identifies with $u_\mu \otimes u_{-\mu} \in V(\mu) \otimes V(-\mu)$. Thus by 7.1.4, 7.1.16 (v) and the definition (7.1.17) of z_μ one obtains

$$\tau(-2\mu) = z_\mu \operatorname{mod}(\operatorname{ad} U_+)\tau(-2\mu)$$

up to a non-zero scalar. Then for all $\lambda \in P^+(\pi)$ it follows from 7.1.18 that

$$tr_q(\tau(-2\mu), V(\lambda)) = tr_q(z_\mu, V(\lambda))$$

and so

$$(*) \qquad \psi(z_\mu)(\lambda) = \frac{tr(\tau(-2(\mu+\rho), V(\lambda)))}{tr(\tau(-2\rho), V(\lambda))}\ .$$

The numerator on the right hand side above is exactly $\operatorname{ch} V(\lambda)$ at the evaluation $e^\nu \mapsto q^{-2(\mu+\rho,\nu)}$ and so by 4.3.6 (iv) recalling 4.2.16 one obtains

$$tr(\tau(-2(\mu+\rho)), V(\lambda)) = \frac{\sum_{w\in W}(-1)^{\ell(w)}q^{-2(w.\lambda,\mu+\rho)}}{\sum_{w\in W}(-1)^{\ell(w)}q^{-2(w.0,\mu+\rho)}}\ .$$

Setting $\mu = 0$ gives the denominator on the right hand side of $(*)$.

Observe that $(w.\lambda, \mu+\rho) = (\lambda+\rho, w^{-1}.\mu) + (\lambda-\mu, \rho)$. Substitution in the expression for $\psi(z_\mu)(\lambda)$ which results from the above and cancellation of common functors in q gives

$$\psi(z_\mu)(\lambda) = \frac{\sum_{w\in W}(-1)^{\ell(w)}q^{-2(w.\mu,\lambda+\rho)}}{\sum_{w\in W}(-1)^{\ell(w)}q^{-2(w.0,\lambda+\rho)}}$$

divided by the corresponding expression which results on setting $\lambda = 0$. The latter is a non-zero scalar and can be ignored. The above expression is just $\operatorname{ch} V(\mu)$ at the evaluation $e^\nu \mapsto q^{-2(\lambda+\rho,\nu)}$ and so

$$\begin{aligned}\psi(z_\mu)(\lambda) &= \sum_{\nu\in P(\pi)} q^{-2(\lambda+\rho,\nu)} \dim V(\mu)_\nu \\ &= \sum_{\nu\in P^+(\pi)} \sum_{w\in W} \tau(-2w\nu)(\lambda+\rho) \dim V(\mu) \,.\end{aligned}$$

This gives

$$\psi(z_\mu) = \sum_{\nu\in P^+(\pi)} \hat{\tau}(-2\nu) \dim V(\mu)_\nu$$

as required.

7.1.20 Not only does the above viewpoint give an elegant description of $Z(U)$; but it also results in a description of $Z(U^+)$. First the isomorphism $V(\lambda)^\iota \xrightarrow{\sim} K(\lambda)^+$ defined in 7.1.1 takes the highest weight vector $u_{-w_0\lambda}$ of $V(\lambda)^\iota$ to a vector $g_{\lambda-w_0\lambda} \in G^+$ of weight $\lambda - w_0\lambda$ which is $\operatorname{ad}_{-2\lambda} e_i$ invariant for all i. Moreover extending U to $\check{U} = U^-\tau(P(\pi))G^+$, this construction is valid for all $\lambda \in P^+(\pi)$. One can write $g_{\lambda-w_0\lambda} = y_{\lambda-w_0\lambda}\tau(\lambda-w_0\lambda) : y_{\lambda-w_0\lambda} \in U^+$. Set $z^+_{\lambda-w_0\lambda} := y_{\lambda-w_0\lambda}\tau(-\lambda-w_0\lambda)$ and $z_i^+ = z^+_{\omega_i-w_0\omega_i} : i = 1,2,\dots,\ell$. Note that $z^+_{\lambda-w_0\lambda} = z^+_{\lambda'-w_0\lambda'}$ implies both $\lambda \pm w_0\lambda = \lambda' \pm w_0\lambda'$ and so $\lambda = \lambda'$. Moreover $z^+_{\lambda-w_0\lambda} \in U_q(\mathfrak{b}^+)$ and $z^+_{\lambda-w_0\lambda} \in U_q(\mathfrak{n}^+)$ if and only if $\lambda \in S^+(\pi) := \{\lambda \in P^+(\pi) \mid \lambda + w_0\lambda = 0\}$. There is an element $\theta \in \operatorname{Aut}\pi$ of order ≤ 2 such that $w_0\omega_i = -\omega_{\theta(i)}$. Set $\mathcal{I} = \{i \in \{1,2,\dots,\ell\} \mid \theta(i) = i\}$. Then $S^+(\pi)$ is freely generated by the $\omega_i : i \in \mathcal{I}$ and the $\omega_i + \omega_{\theta(i)} : i \in \{1,2,\dots,\ell\} \setminus \mathcal{I}$. Finally one may remark that for all $a \in U$ one has $(\operatorname{ad} e_i)a = (e_i a - a e_i)t_i$ and so $(\operatorname{ad} e_i)a = 0 \iff e_i a = a e_i$. As in 3.1.12 write $U_q(\mathfrak{b}^+)$ (resp. $U_q(\mathfrak{n}^+)$) simply as V^+ (resp. U^+).

Lemma

(i) *The* $z^+_{\nu-w_0\nu} : \nu \in P^+(\pi)$ *are linearly independent and form a basis of* $(V^+)^{U^+}$.

(ii) *Up to a non-zero scalar* $z^+_{\nu-w_0\nu} = (z_1^+)^{r_1}\dots(z_\ell^+)^{r_\ell}$ *where* $\nu = \sum r_i\alpha_i$.

(iii) $Z(U^+)$ *is the polynomial algebra on generators* $z_i^+ : i \in \mathcal{I}$, $z_i^+ z^+_{\theta(i)} : i \in \{1,2,\dots,\ell\} \setminus \mathcal{I}$.

Since $(V^+)^{U^+}$ is $\operatorname{ad} T$ stable, it is the linear span of weight vectors $tx_\lambda : x_\lambda \in U_\lambda^+, t \in T$. Write $t = \sum c_\mu \tau(\mu) : c_\mu \in k(q)$. Yet by definition $e_i tx = txe_i$ which in view of the triangular decomposition of U implies that $\sum c_\mu \tau(\mu) q^{-(\mu,\alpha_i)} \in k(q)T$ for all i. Since C is of finite type, it is non-degenerate and then by a standard linear algebra argument the above condition implies that t is proportional to some $\tau(\mu)$. One may write $\tau(\mu)x_\lambda = g'_\lambda \tau(\mu-\lambda)$ for some $g'_\lambda \in G^+$. Then the $\operatorname{ad} e_i$ invariance of $g'_\lambda \tau(\mu-\lambda)$ is equivalent to the $\operatorname{ad}_{\lambda-\mu} e_i$ invariance of g'_λ. Under the isomorphism of 7.1.1 this means that $g'_\lambda \in (\delta M(\frac{1}{2}(\lambda-\mu))^\iota)^{U^+}$ and so generates a finite dimensional U submodule isomorphic to $V(\frac{1}{2}(\lambda-\mu))^\iota$. By 4.3.9 this implies that $\nu := \frac{1}{2}(\lambda-\mu) \in P^+(\pi)$ and $g'_\lambda = g_{\nu - w_0\nu}$ which forces $\lambda = \nu - w_0\nu$ and so $\mu = -\nu - w_0\nu$. Thus $\tau(\mu)x_\lambda = z^+_{\nu-w_0\nu}$, hence (i).

(ii) is immediate from 7.1.7 (i), whilst (iii) follows from (i).

7.1.21 Take $\lambda \in -R^+(\pi)$. Then 7.1.1, 7.1.4 and 7.1.6 (iii) give a U module isomorphism $(\operatorname{ad} U)\tau(\lambda) \xrightarrow{\sim} \operatorname{End} V(-\lambda/2)$. This may be obtained directly from the Rosso form as follows.

Lemma. *For all $\lambda \in Q(\pi)$ the restriction of R to $I(U)\times(\operatorname{ad} U)\tau(\lambda)$ factors to a non-degenerate $\operatorname{ad} U$ invariant bilinear form on $I(U)/\operatorname{Ann}_{I(U)} V(-\lambda/2) \times (\operatorname{ad} U)\tau(\lambda)$. Furthermore if $\lambda \in -R^+(\pi)$, then $F(U)/\operatorname{Ann} V(-\lambda/2) \xrightarrow{\sim} \operatorname{End} V(-\lambda/2)$.*

The first part (which holds for any integrable Cartan matrix) is immediate from 7.1.11. The second follows from the Jacobson density theorem using 7.1.13 and 7.1.16 (i), (iii).

Remark. Note that $\operatorname{End} V(-\lambda/2) : \lambda \in -R^+(\pi)$ is self-dual.

7.1.22 The Rosso form R on U defines a map $\mathcal{R} : U \longrightarrow U^*$ by $\mathcal{R}(u)(v) = R(u,v)$, for all $u, v \in U$. Define an adjoint action of U on U^* by $(a,\xi) \mapsto (\operatorname{ad}\sigma(a))^*\xi$. Since by definition $((\operatorname{ad}\sigma(a))^*\xi)(u) = \xi((\operatorname{ad}\sigma(a))u) = \xi(\sigma(a)_1 u \sigma(\sigma(a)_2)) = \xi(\sigma(a_2)u\sigma^2(a_1)) = (\sigma^2(a_1).\xi.\sigma(a_2))(u)$, one has $(\operatorname{ad}\sigma(a)^*)\xi = \sigma^2(a_1).\xi.\sigma(a_2)$ in terms of the U bimodule structure (1.1.5) of U^*. Moreover $((\operatorname{ad}\sigma(a))^*\mathcal{R}(u))(v) = \mathcal{R}(u)(\operatorname{ad}\sigma(a)v) = R(u, \operatorname{ad}\sigma(a)v) = R((\operatorname{ad} a)u, v) = \mathcal{R}((\operatorname{ad} a)u)(v)$ by the $\operatorname{ad} U$ invariance of R. Thus $(\operatorname{ad}\sigma(a))^*\mathcal{R}(u) = \mathcal{R}((\operatorname{ad} a)u)$, that is $\mathcal{R}$ is homomorphism of $\operatorname{ad} U$ modules. It is injective by the non-degeneracy of R itself a consequence of 3.3.3 (vi) and the non-degeneracy of C. Recall the notation of 1.4.4.

Lemma. *For all $\lambda, \mu \in Q(\pi)$ one has*

(i) *$e_i.\mathcal{R}(\tau(\lambda)) = \mathcal{R}(\tau(\lambda)).f_i = 0$ for all $i \in \{1, 2, \dots, \ell\}$.*
(ii) *$\tau(\mu).\mathcal{R}(\tau(\lambda)) = \mathcal{R}(\tau(\lambda)).\tau(\mu) = q^{-\frac{1}{2}(\lambda,\mu)}\mathcal{R}(\tau(\lambda))$.*
(iii) *$\mathcal{R}(\tau(\lambda)) \in U^\star \iff \lambda \in -R^+(\pi)$ and if either hold $\mathcal{R}(\tau(-2\lambda)) = c_{\lambda,\lambda}$.*
(iv) *If $\lambda \in -R^+(\pi)$, then $\mathcal{R}(F(\lambda)) = C^{V(-\lambda/2)}$.*

Consider $v = utg : u \in U^-$, $t \in U^0$, $g \in G^+$. Then $(\mathcal{R}(\tau(\lambda)).f_i)(v) = R(\tau(\lambda), f_i utg) = 0$ by 7.1.8 (i), (ii). Hence $\mathcal{R}(\tau(\lambda)).f_i = 0$. The first part of (i) is similar.

For all $\mu, \nu \in Q(\pi)$ one has $(\tau(\mu).\mathcal{R}(\tau(\lambda))(\tau(\nu)) = \mathcal{R}(\tau(\lambda)(\tau(\mu+\nu)) = R(\tau(\lambda), \tau(\mu+\nu)) = q^{-\frac{1}{2}(\lambda,\mu+\nu)} = q^{-\frac{1}{2}(\lambda,\mu)}\mathcal{R}(\tau(\lambda))(\tau(\nu))$ by 7.1.8 (i). Similarly one checks that both $\tau(\mu).\mathcal{R}(\tau(\lambda))$ and $\mathcal{R}(\tau(\lambda))$ vanish on $U_+^- U + UG_+^+$ which complements U^0 in U. This gives the assertion in (ii). The second is similar.

For (iii) consider $U.\mathcal{R}(\tau(\lambda))$ which by (i) and (ii) is a quotient of $M(-\lambda/2)$. Thus $\mathcal{R}(\tau(\lambda)) \in U^*$ implies that $-\lambda/2 \in P^+(\pi)$ and so $-\lambda \in 2P^+(\pi) \cap Q(\pi) = R^+(\pi)$. Conversely suppose $\lambda \in -R^+(\pi)$ so in particular $-\lambda/2 \in P^+(\pi)$. Then $\dim(\operatorname{ad} U)\tau(\lambda) < \infty$ by say 7.1.21. Consequently $\dim(\operatorname{ad}\sigma(U))^*\mathcal{R}(\tau(\lambda)) < \infty$ and so by the second part of (i) one has $\dim U^-.\mathcal{R}(\tau(\lambda)) < \infty$. Then $U.\mathcal{R}(\tau(\lambda)) = U^-.\mathcal{R}(\tau(\lambda))$ is finite dimensional as required.

Take $\lambda \in -R^+(\pi)$ and let $u \in V(-\lambda/2)$ (resp. $\xi \in V(-\lambda/2)^*$ considered as a right U module) be a non-zero vector of weight $-\lambda/2$. Then by (i)–(iii) it follows that $\mathcal{R}(\tau(\lambda)) = c_{\xi,u}$ up to a non-zero scalar. Then $\mathcal{R}((\operatorname{ad} U)\tau(\lambda)) = (\operatorname{ad} U)^* c_{\xi,u} = U^+.c_{\xi,u}.U^- = C^{V(-\lambda/2)}$, as required.

7.1.23 One may further use $\mathcal{R}$ to recover 7.1.6. This follows from the result below.

Let $R_q[G]$ denote the direct sum of the $C^{V(\lambda)} : \lambda \in P^+(\pi)$. It is a Hopf subalgebra of U^* (recall 1.4.9.).

Proposition. *For all $u \in U$ one has*

(i) $\dim U_q(\mathfrak{b}^+).\mathcal{R}(u) < \infty$, $\dim \mathcal{R}(u).U_q(\mathfrak{b}^-) < \infty$
(ii) $\mathcal{R}(u) \in U^* \iff u \in F(U)$.
(iii) *$\mathcal{R}$ induces an $\operatorname{ad} U$ module bijection of $F(\check{U})$ onto $R_q[G]$.*

(i) follows exactly as in 7.1.22 (i), (ii). Now take $a \in U_q(\mathfrak{b}^-)$ which one recalls is a Hopf subalgebra of U. Then (cf. 1.1.8, 2.1.1 (iv)) $\sigma^3(a_1).((\operatorname{ad}\sigma(a_2))^*\mathcal{R}(u)) = (\sigma^3(a_1)\sigma^2(a_2)).\mathcal{R}(u).\sigma(a_3) = \sigma^2(\sigma(a_1)a_2).\mathcal{R}(u).\sigma(a_3) = \mathcal{R}(u).\varepsilon(a_1)\sigma(a_2) = \mathcal{R}(u).\sigma(a)$. Since σ is bijective, it follows that $\mathcal{R}(u).U_q(\mathfrak{b}^+) \subset U_q(\mathfrak{b}^+).\mathcal{R}(\operatorname{ad} U_q(\mathfrak{b}^+))$. If $u \in F(U)$ the right hand side is finite dimensional by (i). Then by (i) again it follows that $\mathcal{R}(u).U$ is finite dimensional. Then $\mathcal{R}(u) \in U^*$ by 1.4.5 (ii). This establishes $\Longleftarrow$ in (ii). The converse is trivial. Similarly $\mathcal{R}(u) \in U^*$ for all $u \in F(\check{U})$. Finally surjectivity in (iii) follows from 7.1.22 (iv) extended in the obvious way.

7.1.24 Recall again that C is assumed of finite type. Consider U as an $\operatorname{ad} U$ module. One has the following characterization of $F(U)$.

Lemma. $F(U) = \operatorname{Soc} U$.

Every finite dimensional submodule of U belongs to $F(U)$ by definition, so it is enough to show that U admits no simple infinite dimensional submodule M. Such an M must satisfy $R(M, F(U)) = 0$ for otherwise it would be non-degenerately paired to some simple submodule of $F(U)$ and hence be finite dimensional. Then $M = 0$ by 7.1.12 and 7.1.16 (iii).

7.1.25 The results of this section are modified in an obvious fashion if U is replaced by $\check{U}$ replacing in particular $R(\pi)$ by $2P(\pi)$. This will often be assumed in the sequel. In general there are many ways to extend $\mathcal{F}$ to $\check{U}$. However if C is of finite type every $\lambda \in P(\pi)$ can be expressed uniquely in the form $\sum k_i\alpha_i$ with $k_i \in \mathbb{Q}$ and one has $\deg \tau(\lambda) = -\sum k_i$ in the definition of $\mathcal{F}$. The filtration $\mathcal{F}$ on $\check{U}$ is still indexed by $\mathbb{Z}$ by the following.

Lemma (C of finite type). *Take* $\lambda \in P(\pi)$ *and write* $\lambda = \sum k_i\alpha_i : k_i \in \mathbb{Q}$. *Then* $\deg \lambda := 2\sum k_i \in \mathbb{Z}$.

By 4.2.8 (ii) it follows that the half-sum ρ of the positive roots satisfies $s_\alpha\rho = \rho - \alpha$ and hence $(\rho, \alpha^\vee) = 1$, that is it coincides with ρ defined in 3.3.3. Similarly the half-sum $\rho^\vee$ of the positive coroots satisfies $(\rho^\vee, \alpha_i) = 1$. Then $2(\sum k_i) = 2(\rho^\vee, \lambda) \in \mathbb{Z}$ since $(\alpha^\vee, \lambda) \in \mathbb{Z}$ for every coroot $\alpha^\vee$.

7.2 Positivity of the Rosso Form

Assume C of finite type. The conclusion of 7.1.21 gives an isomorphism $(\operatorname{ad} U)\tau(\lambda) \xrightarrow{\sim} F(U)/\operatorname{Ann} V(-\lambda/2) : \lambda \in -R^+(\pi)$, of $\operatorname{ad} U$ modules. However it is not obvious that this map is the composition of the canonical maps $(\operatorname{ad} U)\tau(\lambda) \hookrightarrow F(U) \twoheadrightarrow F(U)/\operatorname{Ann} V(-\lambda/2)$. Indeed this would imply $(\operatorname{ad} U)\tau(\lambda) \cap \operatorname{Ann} V(-\lambda/2) = 0$ or equivalently by 7.1.11 that $(\operatorname{ad} U)\tau(\lambda) \cap ((\operatorname{ad} U)\tau(\lambda))^\perp = 0$. This last identity will be established for all $\lambda \in P(\pi)$. The proof is based on a positivity property of the Rosso form, which holds for C integrable, and the positive definiteness of C, which means that C is of finite type. In the first part of this section C is only assumed integrable. Since the character formula for an integrable highest weight module $V(\lambda)$ is independent of the choice of the base field k (except that $\operatorname{char} k$ is required to be zero) it follows that $V(\lambda)$ admits a $\mathbb{Q}$ basis. Consequently there will be no loss of generality in assuming $k = \mathbb{Q}$.

7.2.1 Recall 3.3.3. There is an antiautomorphism $\varkappa'$ of U interchanging f_i and $g_i = e_it_i$ which is the identity on T. It is an automorphism of coalgebras. Moreover for all $a \in U$ one has

$$(*) \qquad (\operatorname{ad} f_i)\varkappa'(a) = -\varkappa'((\operatorname{ad} e_i)a)\ , \quad (\operatorname{ad} e_i)\varkappa'(a) = -\varkappa'((\operatorname{ad} f_i)a)\ .$$

Recall (5.3.9) the bilinear form $(\,,)$ defined on $U_q(\mathfrak{n}^-)$. The following result may be anticipated from 6.2.17 and 7.1.8 (ii).

Lemma. *Take* $\mu = \sum k_i\alpha_i : k_i \in \mathbb{N}$. *Then for all* $a, b \in U^-_{-\mu}$ *one has*

$$R(\varkappa'(a), b) = \frac{q^{-(\rho,\mu)}}{\Delta(\mu)}(a,b), \quad \textit{where} \quad \Delta(\mu) := \prod_i (1-q_i^2)^{k_i} .$$

As in 6.2.17 one may write $R(\varkappa'(a), b) = q(\mu)(a, b)$ for some $q(\mu) \in \mathbb{Q}(q)$. Set $b = f_ic$. Then

$$\begin{aligned} R(\varkappa'(a), f_ic) &= -q^{-(\alpha_i,\mu)}R((\operatorname{ad} f_i)\varkappa'(a), c) , \quad \text{by 7.1.8} , \\ &= q_i^{-(\alpha_i^\vee,\mu)}R(\varkappa'((\operatorname{ad} e_i)a), c) , \quad \text{by } (*) , \\ &= -\frac{q(\mu-\alpha_i)}{(q_i-q_i^{-1})}q_i^{-2}(e_i'(a), c) , \quad \text{by 5.3.1 (or 6.2.17)} , \\ &= -\frac{q(\mu-\alpha_i)}{(q_i-q_i^{-1})}q_i^{-2}(a, f_ic) , \quad \text{by 5.3.9} . \end{aligned}$$

Thus $q(\mu)/q(\mu-\alpha_i) = q_i^{-1}/(1-q_i^2)$. Since $q(0) = 1$ the assertion results.

7.2.2 Define A as in 5.1.3. Let $\{G(b)\}_{b\in B(\infty)}$ be the basis of U^- defined by the conclusion of 6.2.16. Recall that by 6.2.8 (∞) one has $b = G(b) \bmod qL(\infty)$. Moreover $G(b) \in U_{-\mu}$ if $b \in B(\infty)_{-\mu}$. Then by 6.1.3 one has $(G(b), G(b')) \in A$ and $(G(b), G(b'))\, |_{q=0} = \delta_{b,b'}$. *In the remainder of this section one simply writes* b *for* $G(b)$ *and* B *for* $B(\infty)$. Adjoin $q^{1/2}$ to $\mathbb{Q}(q)$ and to A.

Take $b_i \in U_{-\xi_i} : i = 1, 2, 3, 4$ and $\lambda, \mu \in \mathbb{Q}(\pi)$. From 7.1.8 it is immediate that

$$(*) \qquad R(b_1\tau(\lambda)\varkappa'(b_2), \varkappa'(b_3\tau(\mu)\varkappa'(b_4))) = 0 , \quad \text{unless } \xi_1 = \xi_3 , \ \xi_2 = \xi_4 .$$

Moreover when the latter holds, 7.1.8 and 7.2.1 give

$$R(b_1\tau(\lambda)\varkappa'(b_2), \varkappa'(b_3\tau(\mu)\varkappa'(b_4))) = \frac{q^{-\frac{1}{2}(\lambda,\mu)}q^{(\xi_1-\xi_2,\rho)}}{\Delta(\xi_1)\Delta(\xi_2)}(b_3, b_1)(b_2, b_4) .$$

It follows that

$$(**) \qquad q^{\frac{1}{2}(\lambda,\mu)-(\xi_1-\xi_2,\rho)}R(b_1\tau(\lambda)\varkappa'(b_2), \varkappa'(b_3\tau(\mu)\varkappa'(b_4)))\Big|_{q=0} = \delta_{b_1,b_3}\delta_{b_2,b_4} .$$

7.2.3 Let ν be the $q^{1/2}$-adic valuation on $\mathbb{Q}(q^{1/2})$. That is for all $a \in \mathbb{Q}(q^{1/2})$, set $\nu(a) = \infty$ if $a = 0$, and otherwise let $m = \nu(a)$ be the largest half-integer such that $aq^{-m} \in A$.

Theorem. *Assume* C *of finite type. Then for all* $a \in U \setminus \{0\}$ *one has* $R(a, \varkappa'(a)) \neq 0$. *Moreover if* $m = \nu(R(a, \varkappa'(a)))$ *then* $q^{-m}R(a, \varkappa'(a))|_{q=0} > 0$.

For each $\xi = (\xi_1, \xi_2) \in Q^+(\pi) \times Q^+(\pi)$ let a_ξ denote a sum of elements of the form $c_\lambda b_{\xi_1} \tau(\lambda) \varkappa'(b_{\xi_2}) : c_\lambda \in \mathbb{Q}(q)$. One may write $a = \sum a_\xi$. By (*) of 7.2.2 one has $R(a_\xi, \varkappa'(a_{\xi'})) = 0$ if $\xi \neq \xi'$. This reduces the assertion to the case when only one ξ occurs in the sum.

With ξ fixed, write $a = a^\lambda$ if it takes the form $c_\lambda b_{\xi_1} \tau(\lambda) \varkappa'(b_{\xi_2}) : c_\lambda \in \mathbb{Q}(q)$. There may be several such terms. One may write $a = \sum a_i^\lambda$ where the a_i^λ can be assumed linearly independent. Given $a^\lambda = c_\lambda b_{\xi_1} \tau(\lambda) \varkappa'(b_{\xi_2})$, $a^\mu = c_\mu b'_{\xi_1} \tau(\mu) \varkappa'(b'_{\xi_2})$ write $a^\lambda \sim a^\mu$ if $b_{\xi_1} = b'_{\xi_1}$, $b_{\xi_2} = b'_{\xi_2}$. Then by (**) of 7.2.2 one has

$$(*) \qquad \nu(R(a^\lambda, \varkappa'(a^\mu))) \geq -\frac{1}{2}(\lambda, \mu) + (\xi_1 - \xi_2, \rho) + \nu(a^\lambda) + \nu(a^\mu)$$

with equality if and only if $a^\lambda \sim a^\mu$.

Since C is assumed of finite type and so positive definite one has

$$(**) \qquad (\lambda, \lambda) + (\mu, \mu) \geq 2(\lambda, \mu)$$

with a strict inequality if $\lambda \neq \mu$. Since $a_i^\lambda = a_j^\mu$ if and only if $\lambda = \mu$ and $a_i^\lambda \sim a_j^\mu$ it follows from (*) and (**) that

$$\nu(R(a_i^\lambda, \varkappa'(a_j^\mu))) \geq m := \min_{(s,\lambda)} \{\nu(R(a_s^\lambda, \varkappa'(a_s^\lambda)))\}$$

with equality if and only if $a_i^\lambda = a_j^\mu$. It follows that $m = \nu(R(a, \varkappa'(a)))$ and that

$$q^{-m} R(a, \varkappa'(a)) \Big|_{q=0} = \sum_{s,\lambda} q^{-m} R(a_s^\lambda, \varkappa'(a_s^\lambda)) \Big|_{q=0} > 0$$

by 7.2.2 (**) again.

7.2.4 Notice that the conclusion of 7.2.3 applies also to $\check{U}$. For the proof it is sufficient to adjoin further fractional powers of q. Again one may intepret q itself as a rational number. Then the results in 7.2.2, 7.2.3 can be expressed in the form

Corollary. *Assume C integral (resp. of finite type). Then for all $a \in U^-$ (resp. $a \in U$) non-zero there exists $q > 0$ such that $R(a, \varkappa'(a)) > 0$.*

Remark. Take $A = k[q, q^{-1}]$ in 3.4.5, 3.4.11. One would expect for U_A^- that the conclusion would hold for all q in the open interval $]1, 0[$. For this it is sufficient that R should be non-degenerate in this interval. This is a delicate point which only seems to have been established for C of finite type(10.5.2).

7.2.5 Assume C of finite type. Let V be a subspace of U such that $\varkappa'(V) \subseteq V$. (Since $\varkappa'$ is of order 2 this is equivalent to $\varkappa'(V) = V$). If $a \in V \setminus \{0\}$ then $R(a, \varkappa'(a)) \neq 0$. This gives the

Corollary *(C of finite type). For all $\lambda \in P(\pi)$ the restriction of R to $(\operatorname{ad} U)\tau(\lambda)$ is non-degenerate. In particular if $\lambda \in -2P^+(\pi)$ the composition of the canonical maps $(\operatorname{ad} U)\tau(\lambda) \hookrightarrow F(\check{U}) \twoheadrightarrow F(\check{U})/\operatorname{Ann}_{F(\check{U})} V(-1/2\lambda)$ is bijective.*

By 7.2.1 $(*)$, $\varkappa'(\operatorname{ad} U)\tau(\lambda)) \subset (\operatorname{ad} U)\varkappa'(\tau(\lambda)) = (\operatorname{ad} U)\tau(\lambda)$. The last part follows from the finite dimensionality of $(\operatorname{ad} U)\tau(\lambda) : \lambda \in -2P^+(\pi)$.

7.2.6 Take $\lambda, \mu \in -2P^+(\pi)$ and consider the image of $\tau(\lambda)$ in $\operatorname{End} V(-\frac{1}{2}\mu)$ given by the action of $\tau(\lambda)$ on $V(-\frac{1}{2}\mu)$. Identify $\operatorname{End} V(-\frac{1}{2}\mu)$ with $V(-\frac{1}{2}\mu) \otimes V(\frac{1}{2}\mu)$. Let $\{v_\nu^i\}$ be a basis for $V(-\frac{1}{2}\mu)_\nu$ and $\{\xi_{-\nu}^i\}$ be the corresponding dual basis for $V(\frac{1}{2}\mu)_{-\nu}$. Then the image of $\tau(\lambda)$ takes the form

$$\sum_{i,\nu} q^{(\nu,\lambda)}(v_\nu^i \otimes \xi_{-\nu}^i) \,.$$

This cannot be a cyclic vector for the diagonal action of U in $V(-\frac{1}{2}\mu) \otimes V(\frac{1}{2}\mu)$ if $\mu - \lambda \in P^+(\pi) \setminus \{0\}$ since $\dim(\operatorname{ad} U)\tau(\lambda) = (\dim V(-\frac{1}{2}\lambda))^2 < (\dim V(-\frac{1}{2}\mu))^2$. On the other hand it is easy to see that this does hold if $(\lambda - \mu, \alpha_i) \gg 0$ for all i. The conclusion of 7.2.5 expresses the remarkable fact that it already holds for $\lambda = \mu$.

Notice also that 7.2.5 means that $(\operatorname{ad} U)\tau(\lambda) : \lambda \in -2P^+(\pi)$ is isomorphic *as an algebra* to $\operatorname{End} V(-\frac{1}{2}\lambda)$; but only in the quotient space $F(\check{U})/\operatorname{Ann} V(-\frac{1}{2}\lambda)$. Notice that by 7.1.19 the central element of $(\operatorname{ad} U)\tau(\lambda)$ does not annihilate any finite dimensional module.

7.3 The Separation Theorem

It is assumed that C is of finite type in this section. Then $U_q(\mathfrak{g})$ specializes at $q = 1$ to $U(\mathfrak{g})$ which is the enveloping algebra of a semisimple Lie algebra $\mathfrak{g}$. An important structure theorem of B. Kostant asserts [Ko1] that $U(\mathfrak{g})$ is a free module over its centre $Z(\mathfrak{g})$, It is called a separation theorem because for $S0(3)$ it corresponds to separation of variables in polar co-ordinates. It is natural to ask if this is true also for $U_q(\mathfrak{g})$, after all by a theorem [Dr3] of V.G. Drinfeld these algebras have isomorphic completions. Actually it turns out to be *false* for $U := U_q(\mathfrak{g})$ but it does hold for $\check{U}$. This is less surprising as one already knows by virtue of 7.1.17 that $Z(\check{U})$ is the polynomial algebra on generators $z_{\omega_i} : i = 1, 2, \dots, \ell$, whilst $Z(U)$ need not be a polynomial algebra. For the separation theorem the precise difficulty is the following. Consider $P^+(\pi)$ as an ordered set for the order relation $\lambda \succeq \mu$ if $\lambda - \mu \in P^+(\pi)$.

Then any pair $\lambda, \mu \in P^+(\pi)$ admits a unique maximal element $\lambda \cap \mu \in P^+(\pi)$ satisfying $\lambda, \mu \succeq \lambda \cap \mu$. However it can happen that $\lambda, \mu \in R^+(\pi)$, yet $\lambda \cap \mu \notin R^+(\pi)$. Thus in the example of the remark in 7.1.17 one has $6\omega_1, 2(\omega_1 + \omega_2) \in R^+(\pi)$, yet $2\omega_1 = 6\omega_1 \cap 2(\omega_1 + \omega_2) \notin R^+(\pi)$.

7.3.1 The significance of the above order relation comes from the following result. Recall 7.1.1 (i) and set $K(\lambda) = K(\lambda)^- \otimes K(-\lambda)^+$.

Lemma. *For all* $\lambda, \mu \in P^+(\pi)$ *one has*

(i) $K(\mp\lambda)^\pm \cap K(\mp\mu)^\pm = K(\mp(\lambda \cap \mu))^\pm$
(ii) $K(\lambda) \cap K(\mu) = K(\lambda \cap \mu)$.

Suppose $\lambda \succeq \nu$, that is $\lambda - \nu \in P^+(\pi)$. Since $1 \in K(\lambda - \nu)^-$ it follows from 7.1.7 that $K(\nu)^- \subset K(\lambda)^-$. Taking $\nu = \lambda \cap \mu$ it follows that $K(\lambda \cap \mu)^- \subset K(\lambda)^- \cap K(\mu)^-$. For the converse fix $i \in \{1, 2, \ldots, \ell\}$. Then $(\alpha_i, \mu \cap \lambda) = \min\{(\alpha_i, \mu), (\alpha_i, \lambda)\}$. Suppose $(\alpha_i, \mu) \leq (\alpha_i, \lambda)$. Observe that $\mathrm{ad}_\nu f_i$ depends only on (α_i, ν). Thus if $\xi \in K(\mu)^-$, it follows that $\mathrm{ad}_{-2\mu} f_i$ and hence $\mathrm{ad}_{-2(\mu\cap\lambda)} f_i$ acts nilpotently on ξ. One concludes that $\mathrm{ad}_{-2(\mu\cap\lambda)} f_i$ acts nilpotently on $K(\mu)^- \cap K(\lambda)^-$. Since i is arbitrary the assertion follows. Interchanging $+$ and $-$ completes the proof of (i). Obviously (i) implies (ii).

7.3.2 Further to the above one has the following much deeper fact which is immediate from 6.2.20.

Theorem. *For all* $\lambda_0, \lambda_1, \ldots, \lambda_n \in P^+(\pi)$

(i) $K(\mp\lambda_0)^\pm \cap \sum_{j=1}^n K(\mp\lambda_i)^\pm = \sum_{j=1}^n K(\mp\lambda_0)^\pm \cap K(\mp\lambda_i)^\pm$
(ii) $K(\lambda_0) \cap \sum_{j=1}^n K(\lambda_i) = \sum_{j=1}^n K(\lambda_0) \cap K(\lambda_i)$.

7.3.3 Define a grading on $U^- \otimes G^+$ by taking the generators $y_i, g_i : i = 1, 2, \ldots, \ell$ to be of degree 1 and let $(U^- \otimes G^+)_+$ denote the graded complement to $k(q)$. Take $\lambda \in P^+(\pi)$. Observe that $K(\mp\lambda)^\pm$ and $K(\lambda)$ are graded subspaces of $U^- \otimes G^+$. Moreover the identity of U serves both as the unique up to scalars highest weight vector of $K(\lambda)^-$ and lowest weight vector of $K(-\lambda)^+$. It follows that the invariant element $y_\lambda \in K(\lambda)$ can be assumed to be of the form $y_\lambda = 1 \bmod (U^- \otimes G^+)_+$. Given any $a \in U^- \otimes G^+$ non-zero let $g(a)$ denote its *lowest* degree component and define $\ell \deg a$ to be the degree of $g(a)$. Thus $g(y_\lambda) = 1$ and $\ell \deg y_\lambda = 0$.

For all $\nu \in P^+(\pi)$, set $P^+_\nu(\pi) = \{\mu \in P^+(\pi) \mid \mu \preceq \nu\}$. Given $\mu \in P^+_\nu(\pi)$ set $K^\nu(\mu) := y_{\nu-\mu} K(\mu)$ which by 7.1.7 (iv) lies in $K(\nu)$.

Corollary. *For all* $\nu \in P^+(\pi)$ *and* $\lambda_0, \lambda_1, \ldots, \lambda_n \in P^+_\nu(\pi)$

$$K^\nu(\lambda_0) \cap \sum_{j=1}^n K^\nu(\lambda_j) = \sum_{j=1}^n K^\nu(\lambda_0 \cap \lambda_j) \, .$$

Since $\nu_j := \lambda_j - (\lambda_0 \cap \lambda_j) \in P^+(\pi)$ it follows by 7.1.17 (i) that $K^\nu(\lambda_0 \cap \lambda_j) = y_{\nu-(\lambda_0\cap\lambda_j)}K(\lambda_0 \cap \lambda_j) = y_{\nu-\lambda_j}y_{\lambda_j-(\lambda_0\cap\lambda_j)}K(\lambda_0 \cap \lambda_j) \subset y_{\nu-\lambda_j}K(\lambda_j) = K^\nu(\lambda_j)$ and similarly $K^\nu(\lambda_0 \cap \lambda_j) \subset K^\nu(\lambda_0)$. Thus it remains to establish the inclusion $\subset$ in the above.

Consider $a \in K^\nu(\lambda_0) \cap \sum_{j=1}^n K^\nu(\lambda_j)$ non-zero and write $a = \sum_{j=1}^n a_j : a_j \in K^\nu(\lambda_j)$ and let $\mathbf{a}$ denote the n tuple $\{a_j\}_{j=1}^n$. Then possibly dropping terms (of higher ℓ-degree) there exists a non-empty $\mathcal{I} \subset \{1, 2, \ldots, n\}$ for which $g(a_j) : j \in \mathcal{I}$ are non-zero, and either

(i) $g(a) = \sum_{j\in\mathcal{I}} g(a_j) \neq 0$

or

(ii) $\sum_{j\in\mathcal{I}} g(a_j) = 0$.

The inclusion $\subset$ will be established by decreasing induction on $\ell(\mathbf{a}) := \min_{j\in\{1,2,\ldots,n\}}\{\ell \deg a_j\}$ and by increasing induction on n. Since both sides lie in the finite dimensional subspace $K(\nu)$ the assertion is obvious for $\ell(\mathbf{a})$ sufficiently large. It is trivial when $n = 0$.

In case (i) it is clear that $g(a) \in K(\lambda_0) \cap \sum_{j\in\mathcal{I}} K(\lambda_j)$ and so by 7.3.1 (i), 7.3.2 (ii) one can write $g(a) = \sum_{j\in\mathcal{I}} b_j$ for some $b_j \in K(\lambda_0 \cap \lambda_j)$. Set $c_j = y_{\nu-(\lambda_0\cap\lambda_j)}b_j \in K^\nu(\lambda_0 \cap \lambda_j) \subset K^\nu(\lambda_0) \cap K^\nu(\lambda_j)$, $a'_j = a_j - c_j$ and $c = \sum c_j$. Then $a - c \in K^\nu(\lambda_0) \cap \sum_{j\in\mathcal{I}} K^\nu(\lambda_j)$. Since $g(c_j) = b_j$ it follows that $\ell(\mathbf{a} - \mathbf{c}) > \ell(\mathbf{a})$ and so the induction hypothesis applies.

Case (ii) is just case (i) with n decreased by 1. This is also true if $a = 0$. Thus by the induction hypothesis there exist $c_j \in K^\nu(\lambda_j)$ with $\sum c_j = 0$ and $g(a_j) = c_j$ for all $j \in \mathcal{I}$. Then $\ell \deg(a_j - c_j) > \ell \deg a_j$ for all $j \in \mathcal{I}$ whilst $a = \sum a'_j$, where $a'_j = a_j - c_j$. Then $\ell(\mathbf{a}') > \ell(\mathbf{a})$ and the induction hypothesis applies.

7.3.4 The conclusion of 7.3.3 is very strong and to convert it into a proof of the separation theory only requires some technical results of a fairly standard nature.

Lemma. *The algebras $U^+, U^-, G^+, U^0, gr_\mathcal{F}U, U$ are domains. In particular $gr_\mathcal{F}ab = (gr_\mathcal{F}a)(gr_\mathcal{F}b)$ for all $a, b \in U$.*

Suppose $a, b \in U^-\backslash\{0\}$ satisfy $ab = 0$. Write a, b as a sum of their T weight vectors $a_\mu, b_\nu : \mu, \nu \in \mathbb{Q}(\pi)$. Take μ (resp. ν) maximal such that a_μ (resp. b_ν) is non-zero. Then $a_\mu b_\nu = 0$ so one may assume a, b to be weight vectors. Now recall the action of U^+ on U^- via the skew derivations $e'_i : i = 1, 2, \ldots, \ell$ defined in 5.3.1. Since e'_i is locally nilpotent on weight vectors one may choose m (resp. n) maximal with the property that $e'^m_i(a) \neq 0$ (resp. $e'^m_i(b)$). By the Leibnitz rule it follows that $e'^{(m+n)}_i(ab)$ is a non-zero multiple of $e'^m_i(a)e'^n_i(b)$. In this manner and taking account of the fact that the weights of U^- belong to $-Q^+(\pi)$ it follows that there exist non-zero elements $a, b \in \cap \ker e'_i$ satisfying

$ab = 0$. This contradicts 5.3.5 and proves that U^- is a domain and hence so are the isomorphic algebras U^+, G^+ (cf. 3.2.6, 3.2.8).

Since U^0 is a Laurent polynomial algebra, it is a domain. Define a filtration $\mathcal{F}'$ on $gr_{\mathcal{F}}U = U^- \otimes U^0 \otimes G^+$ by taking the f_i, g_i to have degree 1 and the t_i, t_i^{-1} to have degree 0. Then $gr_{\mathcal{F}'}gr_{\mathcal{F}}U$ is just the smash product $(U^- \otimes G^+)\#U^0$. (A little more detail of this is given in 7.4.7). In particular $ab = ba$ for all $a \in U^-$, $b \in G^+$ and so this algebra is bigraded by the weight spaces of U^- and of G^+. Then $a, b \in U \setminus \{0\}$ satisfying $ab = 0$ can be assumed to be biweight vectors and using successively the action of U^+ on U^- and of U^- on G^+ can be assumed to lie in U^0 contradicting the first part. Hence $gr_{\mathcal{F}'}gr_{\mathcal{F}}U$ is a domain and so are $gr_{\mathcal{F}}U$ and U.

Remarks. In the above one may use 3.1.6 instead of 5.3.5 and then one only needs to assume that the Cartan matrix is symmetric. Combining 7.1.14 with the last part of 7.3.4 shows that only the scalars are invertible in $I(U)$. Actually one can do better.Define $\mathcal{F}'$ directly on $U = U^- \otimes U^0 \otimes G^+$. Then $gr_{\mathcal{F}'}U \cong (U^- \otimes G^+)\#U^0$ is a domain. In particular $(gr_{\mathcal{F}'}a)(gr_{\mathcal{F}'}b) = gr_{\mathcal{F}'}(ab)$ for all $a, b \in U$. Yet $\mathcal{F}'^{-1}(U) = 0$ so if a is invertible one has $a \in \mathcal{F}'^0(U) = U^0$. This proves that the set of invertible (and of group-like elements) of U is T.

7.3.5 Recall the filtration $\mathcal{F}$ of U defined in 7.1.1. Let Y_+ denote the ideal of $Y(U)$ generated by the $y_\lambda : \lambda \in R^+(\pi) \setminus \{0\}$ and set $J_+ = (gr_{\mathcal{F}}F(U))Y_+$. (One may recall here that by 7.1.2, 7.1.4, 7.1.6, 7.1.16 (i) one has $gr_{\mathcal{F}}F(U) = F(gr_{\mathcal{F}}U)$). Obviously J_+ is an $\operatorname{ad} U$ invariant graded subspace of $gr_{\mathcal{F}}F(U)$ which by 7.1.2 and 7.1.7 (iv) is generated by its intersection with the $G(\lambda) : \lambda \in R^+(\pi)$. Hence by complete reducibility (4.3.10) of the $G(\lambda)$, it admits a (not necessarily unique) graded $\operatorname{ad} U$ invariant complement H generated by its intersection $H(\lambda)$ with the $G(\lambda) : \lambda \in R^+(\pi)$. By 7.1.4 one may choose an $\operatorname{ad} U$ stable subspace $\mathbb{H}(\lambda)$ of $F(\lambda)$ such that $gr_{\mathcal{F}}\mathbb{H}(\lambda) = H(\lambda)$. Set $\mathbb{H} = \sum_{\lambda \in R^+(\pi)} \mathbb{H}(\lambda)$ which is, of course, a direct sum. Let $\mu_0 : H \otimes Y(U) \longrightarrow gr_{\mathcal{F}}F(U)$ (resp. $\mu : \mathbb{H} \otimes Z(U) \longrightarrow F(U)$) denote the multiplication map.

It is clear that the above construction applies to $\check{U}$ with $R^+(\pi)$ replaced by $2P^+(\pi)$. Denote the resulting quantities by $\check{Y}_+$, $\check{J}_+$, $\check{H}$, $\check{\mathbb{H}}$, $\mu_0^\vee$, $\mu^\vee$.

Lemma

(i) *The maps $\mu_0, \mu, \mu_0^\vee, \mu^\vee$ are surjective.*
(ii) *μ (resp. $\mu^\vee$) is injective if μ_0 (resp. $\mu_0^\vee$) is injective.*

Recalling 7.1.14 set $gr_r U = \mathcal{F}^r(U)/\mathcal{F}^{r-1}(U) : r \in \mathbb{Z}$. Then $gr_r F(U) = 0$ unless $r \geq 0$ and $gr_0 F(U)$ reduces to scalars. From the inclusion $gr_r F(U) \subset H + Y_+(gr_0 F(U) + \ldots + gr_{r-1}F(U))$ the surjectivity of μ_0 follows by induction on r. The surjectivity of $\mu_0^\vee$ is similar.

By 7.1.4–7.1.6, $gr_{\mathcal{F}}$ restricts to an isomorphism of $F(U)$ onto $gr_{\mathcal{F}}F(U)$ which by 7.1.17 (iii) further restricts to an isomorphism of $Z(U)$ onto $Y(U)$

and, by definition, an isomorphism of $\mathbb{H}$ onto H. By 7.3.4 multiplication is compatible with $gr_{\mathcal{F}}$ and so one has a commutative diagram

$$\begin{array}{ccc} \mathbb{H}\otimes Z(U) & \xrightarrow[\sim]{gr} & H\otimes Y(U) \\ \Big\downarrow{\scriptstyle \mu} & & \Big\downarrow{\scriptstyle \mu_0} \\ F(U) & \xrightarrow{\sim} & gr_{\mathcal{F}}F(U) \end{array}$$

from which the remaining parts of the lemma follow.

7.3.6 Recall the notation of 7.1.13 and set $\check{T}_< = \tau(-2P^+(\pi))$. Let $Y(U)^\wedge$ (resp. $Y(\check{U})^\wedge$) denote the set of linear characters (1.3.4) on $Y(U)$ (resp. $Y(\check{U})$). The following results stated for $\check{Y} = Y(\check{U})$ apply equally well to $Y(U)$. View $\mathcal{F}$ as defining a filtration of $gr_{\mathcal{F}}\check{U}$ in the obvious manner.

Lemma

(i) *The multiplication map $k(q)\check{T}_< \otimes \check{Y} \longrightarrow \check{T}_<\check{Y}$ is an isomorphism*

(ii) *For each $\chi \in \check{Y}^\wedge$ there is a linear isomorphism $gr_{\mathcal{F}}(\check{T}_< \ker\chi) \xrightarrow{\sim} \check{T}_<\check{Y}_+$ of graded vector spaces.*

Since $\varepsilon(f_i) = 0$ it follows that $\operatorname{ad} f_i$ acts by zero on $\check{Y}$ and hence f_i commutes with elements of $\check{Y}$. On the other hand for each polynomial p in the $\tau(-2\omega_j) : j = 1, 2, \ldots, \ell$ one has $(\operatorname{ad} f_i)p = f_i p - p f_i = f_i(p - \theta_i(p))$ where $\theta_i(p)$ is obtained from p by replacing $\tau(-2\omega_i)$ by $q_i^2\tau(-2\omega_i)$. Applied to a relation of the form $\sum y_j p_j = 0 : y_j \in \check{Y}, p_j \in \check{T}_<$ one obtains $\sum y_j \theta_i^n(p_j) = 0$ for all $n \in \mathbb{N}$, from which it follows that the p_j can be assumed independent of $\tau(-2\omega_i)$. This eventually gives (i).

Now $k(q)\check{T}_<$ is already a graded vector space, $gr \ker\chi = \check{Y}_+$ and $gr_{\mathcal{F}}$ is compatible with multiplication. Hence (ii) follows from (i).

7.3.7 The following result though superficially similar to 7.3.6 is much more subtle and holds only for $\check{U}$. Given $\chi \in Y(\check{U})^\wedge$ set $\check{J}_\chi = (gr_{\mathcal{F}}F(\check{U}))\ker\chi$.

Proposition. *For all $\chi \in Y(\check{U})^\wedge$ there is an isomorphism $gr_{\mathcal{F}}\check{J}_\chi \xrightarrow{\sim} \check{J}_+$ of graded vector spaces. In particular $\check{H} \cap \check{J}_\chi = 0$.*

One has $\check{J}_\chi = ((\operatorname{ad} U)\check{T}_<)\ker\chi = (\operatorname{ad} U)(\check{T}_< \ker\chi)$. Moreover since $\mathcal{F}$ is $\operatorname{ad} U$ invariant

$$(*) \qquad (\operatorname{ad} U)\mathcal{F}^m(\check{T}_< \ker\chi) \subset \mathcal{F}^m(\check{J}_\chi)$$

for all $m \in \mathbb{N}$.

Suppose equality fails in $(*)$. Then there exists $\nu \in P^+(\pi)$, $s \in \mathbb{N}^+$, $\mu_i \in P_\nu^+(\pi)$, $a_i \in (\operatorname{ad} U)\tau(-2\mu_i) : i = 1, 2, \ldots, s$ such that

$$\sum_{i=1}^{s}(y_{\nu-\mu_i}-\chi(y_{\nu-\mu_i}))a_i \in \mathcal{F}^m(\check{J}_\chi)$$

and yet $\deg_{\mathcal{F}} \tau(-2\nu) = n > m$. Cancellation of the degree n terms gives $\sum y_{\nu-\mu_i}a_i = 0$. Successive application of 7.3.3 to this identity either gives $\xi \in P^+_{\cap\mu_i}(\pi)$ such that $a_i = y_{\mu_i-\xi}b_i$ with $b_i \in (\operatorname{ad} U)\tau(-2\xi)$ satisfying $\sum b_i = 0$ or cancellations with s decreased by one which are dispensed with in a manner similar to the one described below. In the first case set $c_i = \sum_{j=1}^{i} b_j$. Then

$$\begin{aligned}\sum_{i=1}^{s}(y_{\nu-\mu_i}-\chi(y_{\nu-\mu_i}))a_i &= -\sum_{i=1}^{s}\chi(y_{\nu-\mu_i})y_{\mu_i-\xi}(c_i-c_{i-1})\\ &= -\sum_{i=1}^{s-1}(\chi(y_{\nu-\mu_i})y_{\mu_i-\xi}-\chi(y_{\nu-\mu_{i+1}})y_{\mu_{i+1}-\xi})c_i\,.\end{aligned}$$

Since $y-\chi(y) = 0$ on scalars one may assume $\deg_{\mathcal{F}} y_{\nu-\mu_i} > 0$ in the left hand side, that is $\mu_i \neq \nu$ all i. Then the right hand side belongs to $(\operatorname{ad} U)\mathcal{F}^{n-1}(\check{T}_< \ker\chi)$. Successive reduction gives the required contradiction. Finally by 7.3.6 (ii) equality in (*) gives the proposition.

7.3.8 From 7.3.5 and 7.3.7 the following separation theorem follows.

Theorem

(i) $\check{\mu}_0$ *is an isomorphism of* $\check{H}\otimes Y(\check{U})$ *onto* $gr_{\mathcal{F}}F(\check{U})$.

(ii) $\check{\mu}$ *is an isomorphism of* $\check{\mathbb{H}}\otimes Z(\check{U})$ *onto* $F(\check{U})$.

It remains to show that $\check{\mu}_0$ is injective. Consider a sum of the form $\sum h_\gamma\otimes y_\gamma : h_\gamma \in \check{H}, y_\gamma \in \check{Y}$ with the h_γ linearly independent over $k(q)$ and $\sum h_\gamma y_\gamma = 0$. One may assume $y_\gamma \neq 0$ for some γ and choose $\chi \in \check{Y}^\wedge$ such that $x_\gamma := \chi(y_\gamma) \neq 0$. Yet by 7.3.7 one has $\check{H}\cap \check{J}_\chi = 0$ and so the images of the h_γ are still linearly independent in $(gr_{\mathcal{F}}U)/\check{J}_\chi$. Then $x_\gamma = 0$ for all γ which is the required contradiction.

7.4 Noetherianity

Here noetherian means left and right noetherian. Since $U(\mathfrak{g})$ is noetherian when $\mathfrak{g}$ is a finite dimensional Lie algebra it is natural to suppose that $U = U_q(\mathfrak{g}_C)$ is noetherian when C is of finite type.Actually a more significant question is whether $F(U)$ is noetherian. The aim of the section is to develop techniques to establish such a result for $F(U)$ and for some related rings.

7.4.1 It is somewhat sobering to note that all proofs of noetherianity ultimately reduce to the Hilbert basis theorem now nearly a century old. The particular form needed here is the following. Let A be a k-algebra equipped with an automorphism θ and a θ-leftskew derivation ∂. Then (3.4.1) one may form the skew-polynomial extension $A\#_\partial k[x]$ defined to be $A \otimes k[x]$ as a vector space with the multiplication $(a \otimes x)(b \otimes x^n) = a\partial(b) \otimes x^n + a\theta(b) \otimes x^{n+1}$. The argument of the Hilbert basis theorem can be applied to show [MR1, 2.9 (iv)] that A noetherian implies $A \neq_\partial k[x]$ noetherian. Actually taking the filtration on the latter defined by the exponent of x one has $\partial = 0$ in the associated graded algebra and therefore the only really new feature is the automorphism θ.

A k-algebra A is said to be (finitely) quasi-commutative if it has (a finite number of) generators which commute up to elements of k^*. If the base field is replaced by $k(q)$ and the scalars are powers of q then A is called (finitely) q-commutative which here is the most relevant case. By the Hilbert basis theorem of the previous paragraph any finitely quasi-commutative algebra is noetherian.

An increasing filtration $\{\mathcal{F}^n A\}_{n\in\mathbb{Z}}$ of an algebra A is said to be bounded if $\mathcal{F}^n A = 0$ for some $n \in \mathbb{Z}$. Induction on filtration degree [Jac1, Chap.V, Thm. 4] shows that A is noetherian if $gr_\mathcal{F} A$ is noetherian for graded ideals. All this gives the

Lemma. *Let A be a k-algebra with an increasing bounded filtration $\mathcal{F}$. If $gr_\mathcal{F} A$ is finitely quasi-commutative, then A is noetherian.*

7.4.2 To 7.4.1 one may add the following remarks. In cases of present interest one has $\mathcal{F}^{-1}A = 0$ so then A is filtered by the natural numbers $\mathbb{N}$. If A is already graded by $\mathbb{N}$ and $\mathcal{F}$ is the filtration of A obtained taking $\mathcal{F}^s A$ to be the sum of the elements of degree $\leq s$ then trivially A is isomorphic to $gr_\mathcal{F} A$. However then the statement preceeding the lemma in 7.4.1 implies that A is noetherian if it is noetherian for just graded ideals.

In the above one may further replace $\mathbb{N}$ by any finitely generated free additive semigroup P^+ with identity 0.

Write $\lambda \succeq \mu$ (resp. $\lambda \succ \mu$) if there exists $\nu \in P^+$ (resp. and $\nu \neq 0$) such that $\lambda + \nu = \mu$. Let A be an algebra with identity 1. A P^+ filtration $\mathcal{F}$ of A is a family of subspaces $\mathcal{F}^\mu A : \mu \in P^+$ satisfying

(i) $1 \in \mathcal{F}^\mu A \subset \mathcal{F}^\lambda A$ if $\mu \prec \lambda$

(ii) $\bigcup_{\mu\in P^+} \mathcal{F}^\mu A = A$

(iii) $(\mathcal{F}^\mu A)(\mathcal{F}^\nu A) = \mathcal{F}^{\mu+\nu} A$.

In this case A is often said to be multi-filtered.

7.4.3 For each $\lambda \in P^+(\pi)$, recall that $K(\lambda)^- = (\mathrm{ad}_{-2\lambda} U^-)1$. As noted in 7.1.1, $K(\lambda)^-$ has the structure of a simple integrable module of highest weight λ and hence by 7.1.16 (i) is finite dimensional. Moreover the $\{K(\lambda)^- :$

$\lambda \in P^+(\pi)\}$ form an increasing bounded $P^+(\pi)$ filtration $\mathcal{F}_p$ of U^-. Indeed (i), (iii) of 7.4.2 obtain from 7.1.7 (i) and (ii) from 4.3.6 (ii).

For each $n \in \mathbb{N}$ set

$$P_n^+(\pi) := \{\mu \in P^+(\pi) \mid \mu = \sum n_i\omega_i \text{ with } \sum n_i \leq n\} .$$

Then

$$\mathcal{F}_1^n(U^-) := \sum_{\lambda \in P_n^+(\pi)} K(\lambda)^-$$

defines an increasing bounded filtration of U^-.

Proposition. *The graded algebra $gr_{\mathcal{F}_1}U^-$ is noetherian.*

By 7.4.1, it is enough to show that $gr_{\mathcal{F}_1}U^-$ admits an increasing bounded filtration $\mathcal{F}_2$ such that $gr_{\mathcal{F}_2}(gr_{\mathcal{F}_1}U^-)$ is finitely q-commutative. Observe that $\mathcal{F}_1^0(U^-) = k(q)$ and that $gr_{\mathcal{F}_1}U^-$ is generated by the image $k(q) \oplus gr^1_{\mathcal{F}_1}(U^-)$ of $\mathcal{F}_1^1(U^-)$ in $gr_{\mathcal{F}_1}(U^-)$. For each i, let $K(\omega_i)_+^-$ denote the unique T stable complement of $k(q)$ in $K(\omega_i)^-$. Identify $K(\omega_i)_+^-$ with its image in $gr^1_{\mathcal{F}_1}U^-$. If $i \neq j$, 7.3.1 (i) gives $K(\omega_i)_+^- \cap K(\omega_j)_+^- = 0$ and so by 7.3.2 (i) the $K(\omega_i)_+^- : i = 1, 2, \ldots, \ell$ form a *direct sum* in $gr_{\mathcal{F}_1}(U^-)$ generating $gr_{\mathcal{F}_1}U^-$ as an algebra with identity.

Fix i and set $\omega = \omega_i$. Let $K(\omega)^-_{-\mu}$ be a non-zero weight space of $K(\omega)^-$. Then $\mu \in Q^+(\pi)$ and define $|\mu|$ as in 3.1.2. In particular $-(\omega - w_0\omega)$ is the unique lowest weight of $K(\omega)^-$. Set $n(\omega) = |\omega - w_0\omega|$ and $n = \max n(\omega_i)$.

Let $f : [0, n] \longrightarrow [0, \frac{1}{2}]$ be a smooth function with a strictly decreasing second derivative taking rational values at integer points. Choose $m \in \mathbb{N}^+$ such that $mf(j) \in \mathbb{Z}$ for $j \in \{0, 1, 2, \ldots, n\}$. Given integers $0 \leq r' < r \leq s < s'$ such that $r + s = r' + s'$ one has

$$f(r) + f(s) > f(r') + f(s') . \tag{1}$$

For all $\mu \in Q^+(\pi) \setminus \{0\}$, set $d(\omega; 0) = 0$ and $d(\omega; \mu) = 1 + f(n(\omega) - |\mu|)$ if $\mu \neq 0$. Up to relabelling the fundamental weights, $gr_{\mathcal{F}_1}U^-$ is spanned by monomials of the form $a = a_1a_2 \ldots a_r : a_i \in K(\omega_i)^-_{-\mu_i}$. Set $d(a) = \sum d(\omega_i; \mu_i)$ and define $\mathcal{F}_2$ through

$$\mathcal{F}_2^t(gr_{\mathcal{F}_1}U^-) = \sum_{d(a) \leq t} a \, , \quad \text{for all } t \in \frac{1}{m}\mathbb{Z} . \tag{2}$$

By construction $gr_{\mathcal{F}_2}(gr_{\mathcal{F}_1}U^-)$ is generated by the images of the $K(\omega_i)^-_{-\mu_j}$. Again one may view these spaces as forming a direct sum in U^- generating the latter as an algebra with identity, so (2) also defines a filtration $\mathcal{F}_2$ of U^-. Since $d(\omega; \mu) + d(\omega'; \mu') \geq 2 > d(w''; \mu'') \geq 1$ for all $\mu, \mu', \mu'' \neq 0$ it follows that $gr_{\mathcal{F}_2}(gr_{\mathcal{F}_1}U^-)$ identifies with$gr_{\mathcal{F}_2}U^-$.

It remains to show that for any two weight vectors $a_{-\mu} \in K(\omega)^-_{-\mu}$, $b_{-\nu} \in K(\omega')^-_{-\nu}$ their images $\bar{a}_{-\mu}$, $\bar{b}_{-\nu}$ in $gr_{\mathcal{F}_2}(U^-)$ commute up to a power of q. For each $r \in \mathbb{Z}$ define the q-commutator

$$\{a_{-\mu}, b_{-\nu}\}_r = a_{-\mu}b_{-\nu} - q^r b_{-\nu}a_{-\mu} \,.$$

Then one requires $\{\bar{a}_{-\mu}, \bar{b}_{-\nu}\}_r = 0$ for some $r \in \mathbb{Z}$.

Take $\mu_0 = \omega - w_0\omega$. Then $(\mathrm{ad}_{-2\omega} f_i)a_{-\mu_0} = 0$ for all i, which translates to give $f_i a_{-\mu_0} - q^{(\alpha_i, \omega + w_0\omega)} a_{-\mu_0} f_i = 0$. Thus the required assertion already holds in U^- for the pair $a_{-\mu_0}$, $b_{-\nu}$. Recall further that $\mathrm{ad}_{-2\omega}\, e_i$ is independent of ω and this defines an action of U^+ on U^-. For all $a_{-\mu}$, $b_{-\nu}$ as above one has

(3) $$e_i\{a_{-\mu},\, b_{-\nu}\}_r = q^{(\alpha_i,\nu)}\{e_i a_{-\mu},\, b_{-\nu}\}_{r-(\alpha_i,\nu)} + \{a_{-\mu},\, e_i b_{-\nu}\}_{r+(\alpha_i,\mu)} \,.$$

This does not quite mean that the action of U^+ takes q-commutators into q-commutators. However suppose that $|\mu_0 - \mu| \leq |\nu_0 - \nu|$ with μ_0 as above and $\nu_0 = \omega' - w_0\omega'$. One can assume that $a_{-\mu}$ is obtained from $a_{-\mu_0}$ by applying some monomial in the $e_i : i = 1, 2, \ldots, \ell$. Then by repeated use of (3) and the first observation applied to pair $a_{-\mu_0}, b_{-\nu}$ it follows that for some $r, r_j \in \mathbb{Z}$ one has

(4) $$\{a_{-\mu},\, b_{-\nu}\}_r + \sum_j \{a_{-\mu_j},\, b_{-\nu_j}\}_{r_j} = 0$$

where $\mu_0 \geq \mu_j > \mu$ and $\mu_j + \nu_j = \mu + \nu$. If $\nu_j \neq 0$ then (1) gives

$$\begin{aligned} d(\omega;\ \mu) + d(\omega';\ \nu) &= 2 + f(|\mu_0 - \mu|) + f((\nu_0 - \nu|) \\ &> 2 + f(|\mu_0 - \mu_j|) + f(|\nu_0 - \nu_j|) \\ &= d(\omega;\ \mu_j) + d(\omega;\ \nu_j) \end{aligned}$$

and furthermore this trivially holds if $\nu_j = 0$. This inequality combined with (4) means that $\{\bar{a}_{-\mu}, \bar{b}_{-\nu}\}_r = 0$. Starting from $b_{-\nu_0}$ a similar result is obtained if $|\nu_0 - \nu| \leq |\mu_0 - \mu|$. This proves the proposition.

7.4.4 Corollary. *The isomorphic algebras U^-, U^+, G^+ are noetherian.*

7.4.5 Corollary. *The graded algebra $gr_{\mathcal{F}_p}U^-$ is noetherian.*

Indeed if $\mu \preceq \nu$ and $\nu \in P_n^+(\pi)$, then $\mu \neq \nu$ if and only if $\mu \in P_{n-1}^+(\pi)$. Thus $gr_{\mathcal{F}_1}U^-$ and $gr_{\mathcal{F}_p}U^-$ are isomorphic as algebras.

7.4.6 It is not hard to use 7.4.4 and 7.4.1 to establish the noetherianity of $gr_{\mathcal{F}}U$ and hence of U. However 7.4.5 will give the much more interesting result, namely the noetherianity of $F(U)$. By 7.1.13, 7.1.16 (i) this is stronger since it implies the noetherianity of $J(U)$ over which U is a free finitely generated left (or right) module.

The above claim and a number of similar assertions will follow by modifying slightly the following classical result.

Let A be a k-algebra with filtration $\mathcal{F}$. Then the Rees algebra $R(A, \mathcal{F})$, or simply R, of the pair $(A, \mathcal{F})$ is the graded vector space

$$R = \bigoplus_{n \in \mathbb{Z}} \mathcal{F}^n A$$

with multiplication $(a, b) \mapsto ab$ of $\mathcal{F}^m A \times \mathcal{F}^n A \longrightarrow \mathcal{F}^{m+n} A$ coming from A. Then if $\mathcal{F}$ is an increasing bounded filtration, the noetherianity of $R(A, \mathcal{F})$ is equivalent to that of $gr_{\mathcal{F}} A$.

First consider the multi-filtered version of the above. Let P^+ be a free finitely generated additive subgroup with identity 0. For all $\mu \in P^+$, let $\tau(\mu)$ denote the image of $1 \in \mathcal{F}^\mu A$ *considered as a subspace of* R. Then $\tau(\mu)\tau(\nu) = \tau(\mu+\nu)$ for all $\mu, \nu \in P^+$, and $\tau(0)$ is the identity of R. Moreover the $\tau(\mu) : \mu \in P^+$ form a basis for a polynomial subalgebra S of R. Then R is just the subalgebra of the polynomial extension $A \otimes S$ of A given by

$$R = \sum_{\mu \in P^+} (\mathcal{F}^\mu A)\tau(\mu) .$$

In particular $\tau(\mu)$ is central in R. In the twisted Rees algebra $R(A, \mathcal{F}, \chi)$ it is further assumed that the filtered algebra $\{A, \mathcal{F}\}$ is also graded by a second free abelian semigroup Q^+ (so that $\mathcal{F}^\mu A = \oplus_{\lambda \in Q^+} (\mathcal{F}^\mu A \cap A_\lambda)$) and that the multiplication is modified through a homomorphism $\chi : \nu \mapsto \chi_\nu$ of P^+ into the set characters on Q^+ with values in k^*. Precisely, $R(A, \mathcal{F}, \chi)$ is $\bigoplus_{\mu \in P^+} \mathcal{F}^\mu A \otimes \tau(\mu)$ as a vector space with multiplication given by $(a \otimes \tau(\mu))(b \otimes \tau(\nu)) = \chi_\mu(\lambda)(ab \otimes \tau(\mu + \nu))$ for all $b \in A_\lambda$.

Observe that S is a bialgebra for the coproduct $\Delta(\tau(\mu)) = \tau(\mu) \otimes \tau(\mu)$. Set $gr_\mu A = \mathcal{F}^\mu A / \sum_{\nu \prec \mu} \mathcal{F}^\nu A$. Then $gr_{\mathcal{F}} A = \bigoplus_{\mu \in P^+} gr_\mu A$. Define an action of S on $gr_{\mathcal{F}} A$ by $\tau(\mu).a = \chi_\mu(\lambda) a$ for all $a \in A_\lambda$. This makes $gr_{\mathcal{F}} A$ an S-algebra so one may form (1.1.8) the smash product $gr_{\mathcal{F}} A \# S$. Let J_+ be the left ideal of R generated by the $\tau(\mu) : \mu \in P_+ \setminus \{0\}$. Obviously J_+ is also a two-sided ideal of R and

$$R/J_+ = \bigoplus_{\mu \in P^+} (gr_\mu A)\tau(\mu)$$

which identifies as a subalgebra of $gr_{\mathcal{F}} A \# S$. In the case when χ is trivial one even has $R/J_+ = gr_{\mathcal{F}} A$.

Proposition. *The following are equivalent*

(i) $gr_{\mathcal{F}} A$ *is noetherian.*
(ii) R/J_+ *is noetherian.*
(iii) $R(A, \mathcal{F}, \chi)$ *is noetherian.*

Observe that $\bigoplus L_{\mu,\lambda}\tau(\mu) \mapsto \bigoplus L_{\mu,\lambda}$ is an order preserving bijection from the set of bigraded left (resp. right) ideals of R/J^+ onto the set of bigraded left (resp right) ideals of $gr_{\mathcal{F}}A$. By 7.4.2 this establishes the equivalence of (i), (ii). Clearly (iii) implies (ii). Finally assume (ii). Then by 7.4.1 the smash product $R/J_+\#S$ being an interated skew polynomial extension is noetherian. Consider the graded algebra

$$B = \bigoplus_{\mu\in P^+} B_\mu\,, \quad B_\mu := (R\tau(\mu)/\sum_{\nu\succ\mu} R\tau(\nu))\,.$$

The map $r\otimes\tau(\mu) \mapsto r\tau(\mu) \bmod \sum_{\nu\succ\mu} R\tau(\nu)$ factors to a surjection of $(R/J_+)\tau(\mu)$ onto B_μ and extends linearly to an algebra epimorphism of $R/J_+\#S$ onto B. Hence B is noetherian.

Let I be a graded left ideal of R. Then

$$J = \bigoplus_{\mu\in P_+} J_\mu\,, \quad J_\mu := (I\cap R\tau(\mu))/(I\cap\sum_{\nu\succ\mu} R\tau(\nu))$$

is a graded left ideal of B. Let $\{a_{i,\mu}\} : a_{i,\mu} \in J_\mu$ be a finite set of generators for J. Choose $b_{i,\mu} \in I\cap R\tau(\mu)$ homogeneous, such that $b_{i,\mu} - a_{i,\mu} \in I\cap\sum_{\nu\succ\mu} R\tau(\nu)$ and let I' be the left ideal of R they generate. Take $b\in I$ homogeneous, say $b \in (\mathcal{F}^\lambda A)\tau(\lambda)$. Then for all $\nu \in P^+$ the above construction gives $b(\nu) \in I'\cap(\mathcal{F}^\lambda A)\tau(\lambda)$ such that $b - b(\nu) \in R\tau(\nu)$. Yet $R\tau(\nu)\cap(\mathcal{F}^\lambda A)\tau(\lambda) = 0$ for $\nu\succ\lambda$ and so $b = b(\nu) \in I'$ for such a choice of ν. This shows that R is left noetherian for graded ideals and hence by 7.4.2 is left noetherian. A similar argument applies on the right and completes the proof.

7.4.7 The $P^+(\pi)$ filtration $\mathcal{F}_p$ of U^- defined in 7.4.3 can be equally well applied to the isomorphic algebra G^+ and together these form a $P^+(\pi)$ filtration $\mathcal{F}_p$ of the "diagonal" $D := \sum_{\mu\in P^+(\pi)} K(\mu)^-\otimes K(-\mu)^+$ in $U^-\otimes G^+$ (where weight vectors in U^- may only q-commute with those in G^+). By 7.1.7 (i), D is a subalgebra and as in 7.4.5 it follows that $gr_{\mathcal{F}_p}D$ is noetherian. For each $\mu\in P^+(\pi)$ let $\chi_\mu : \mu \in P^+(\pi)$ be the character on $Q^-(\pi)\times Q^+(\pi)$ given by $\chi_\mu(\lambda,\nu) = q^{-2(\mu,\lambda+\nu)}$, for all $\nu\in P^+(\pi)$.

Now recall (7.1.1) the filtration $\mathcal{F}$ of U. Then $gr_{\mathcal{F}}F(U)$ is a subalgebra of $gr_{\mathcal{F}}U = U^-\otimes U^0\otimes G^+$. In this recall that U^- (resp. G^+) can be given the structure of a graded algebra with graded generators f_i (resp. g_j) $: i,j = 1,2,\ldots,\ell$. Moreover $f_ig_j - q^{(\alpha_i,\alpha_j)}f_jf_i = \delta_{ij}/(q_i - q_i^{-1})$. Thus defining f_i, g_j to have degree 1 and the elements of U^0 to have degree 0 defines a filtration $\mathcal{F}'$ on $gr_{\mathcal{F}}U^-$. It is easily checked that $gr_{\mathcal{F}'}(gr_{\mathcal{F}}U)$ is isomorphic to the smash product $(U^-\otimes G^+)\#U^0$ where the group algebra U^0 of T acts by $\tau(\mu).f_i = q^{-(\mu,\alpha_i)}f_i, \tau(\mu).g_j = q^{(\mu,\alpha_j)}g_j$ on generators and where weight vectors in U^- may only q-commute with those in G^+. A similar construction applies to $F(\check{U})$. Recalling 7.1.2 one easily checks the

Lemma. *The graded algebra $gr_{\mathcal{F}'}(gr_{\mathcal{F}}F(\check{U}))$ is isomorphic to the twisted Rees algebra $R(D, \mathcal{F}_p, \chi)$. The subalgebra $gr_{\mathcal{F}'}(gr_{\mathcal{F}}F(U))$ is isomorphic to the corresponding Rees algebra obtained by replacing $P^+(\pi)$ by $\frac{1}{2}R^+(\pi)$.*

7.4.8 Obviously $\mathcal{F}'$ is a bounded filtration of $gr_{\mathcal{F}}U$. By 7.1.14 the restriction of $\mathcal{F}$ to $F(U)$ is also bounded. Corresponding assertions hold for $\check{U}$. In view of 7.4.6 one obtains the

Theorem. *$F(U)$ and $F(\check{U})$ are noetherian.*

7.4.9 As noted in 7.4.6 one deduces the

Corollary. *U and $\check{U}$ are noetherian.*

7.5 Comments and Complements

7.5.1 The results in section 7.1 were developed from [JL2] except that 7.1.21 (iii) and 7.1.23 are due to P. Caldero [C1-3] and 7.1.24 to C. Liu [Liu1]. See also [Ra1]. The quantum trace (7.1.18) has been noted by several authors in fact it is really a built-in aspect of Drinfeld's construction of $U_q(\mathfrak{g})$. It has its origins in the work of V. F. R. Jones [Jo1]. A notable application [A3] occurs when q is the root of unity as then the quantum trace of the identity (quantum dimension!) can vanish.

7.5.2 As noted in 5.3.3 the twisted action of U on U^- has a classical analogue. This is obtained by identifying $S(\mathfrak{n}^-)$ with the space of sections of $\mathcal{L}_\lambda$ on the open Bruhat cell. In particular 7.1.7 (i) is an analogue of what is often referred to as Cartan multiplication of simple highest weight modules.

7.5.3 The classical version of 7.1.9 (*) is due to M. Duflo [Du1]. It can be useful for studying primitive ideals.

7.5.4 The classical version of 7.1.17 (v) is a well-known and important result of Harish-Chandra. The difficulty in the proof resides mainly in establishing the surjectivity and it is notable that the quantum case appears easier and in particular does not require the preliminary step establishing complete reducibility (4.3.10). Actually the result in 7.1.19 which gives an explicit inverse to the Harish-Chandra map ψ was developed from an idea of G. Letzter which appears to be new even in the classical case. Instead of constructing invariants in $S(\mathfrak{h})$ and laboriously translating them to $U(\mathfrak{g})$ as in [Di2, 7.3.3, 7.4.5] one simply applies $\operatorname{ad} U(\mathfrak{g})$ to each $a \in S(\mathfrak{h})$. Then $(\operatorname{ad} U(\mathfrak{g}))a$ is a finite dimensional $U(\mathfrak{g})$ module with $\operatorname{ad}(U(\mathfrak{g})\mathfrak{g})a$ a submodule of codimension ≤ 1. Admitting complete reducibility this gives a central element z (possibly zero) whose trace on any finite dimensional module coincides with that of a.

From this the Harish-Chandra map is easily seen to be surjective. Independent proofs in the quantum case were obtained by V.G. Drinfeld [Dr3], by M. Rosso [R2] (who obtained a slightly weaker result) and by C. De Concini and V. Kac [DK1] who expressed this result introducing an extended Weyl group (8.4.2). Together with C. Procesi [DKP1] they further determined the centre of $U_q(\mathfrak{g})$ when q is a root of unity.

7.5.5 It is clear that the $z^+_{\nu-w_0\nu} : \nu \in P^+(\pi)$ specialize in the sense of 3.4.5 to central elements of $U(\mathfrak{n}^+)$. One can ask if the $z^+_{\omega_i-w_0\omega_i}$ become the generators of the centre $Z(\mathfrak{n}^+)$ which is also a polynomial algebra. The latter were determined by B. Kostant (unpublished) and in [J2]. At first this seemed to be so obviously true that it didn't seem worth checking. Actually it is false; but only just - occasionally $z^+_{\omega_i-w_0\omega_i}$ specializes to the square of generator. At least compared to the treatment in [J2] the construction of these central elements is much easier in the quantum case. Moreover the existence of the above mentioned square roots only needs [J2, 3.6] and not the more technical [J2, 4.12]. In type $\mathfrak{sl}(n)$ the generators of $Z(\mathfrak{n}^+)$ were first constructed by J. Dixmier [Di1] and appear as the determinants of upper right hand minors. Similarly in the quantum case they are quantum determinants [C2]. Again the quantum case is more natural since determinants really lie in the function algebra or Hopf dual of $U(\mathfrak{n}^+)$. In the quantum case, by Drinfeld duality (3.1.11, 3.2.1) this is essentially the same object. In some sense Dixmier's construction was fortutious since the entries in the upper right hand minors *commute* and so also serve as functions on $SL(n)$. However this is false for the entries of a general minor in $\mathfrak{sl}(n)$; yet there are also circumstances in which these are important, particularly in the quantization of orbital varieties [J9, J12, Ben1] arising in the orbit method of representation theory.

7.5.6 The results in section 7.2 are taken from [JL4]. The separation theorem of section 7.3 is taken from [JL2], the classical version being due to B. Kostant [Ko1]. In Kostant's theory the key point is the primeness of the ideal I_+ of $S(\mathfrak{g})$ generated by the positive degree invariant homogeneous elements. In some sense J_+ is the analogue of Kostant's ideal. However neither J_+ nor $\check{J}_+$ is prime in general. Moreover these ideals have a more precise classical analogue in $S(\mathfrak{g})$ which can be described as follows.

Recall the twisted action of $U(\mathfrak{g})$ on $S(\mathfrak{n}^-)$ described in 7.5.2. For each $\mu \in P^+(\pi)$ one obtains a subspace of $S(\mathfrak{n}^-)$ isomorphic to $\mathcal{V}(\mu)$ as a $U(\mathfrak{g})$ module and the corresponding dual module in $S(\mathfrak{n}^+)$. Just as in 7.1.17 (i) this gives rise to an invariant element $y^0_\mu \in S(\mathfrak{n}^-) \otimes S(\mathfrak{n}^+)$. One may also obtain y^0_μ from y_μ by specialization in the sense of 3.4.6. The ideal in $S(\mathfrak{g})$ generated by the $y^0_\mu : \mu \in P^+(\pi) \setminus \{0\}$ is a more precise analogue of J_+. It is not prime in general. These questions were studied extensively in [JL2,JL3].

7.5.7 Section 7.4 is taken mainly from [J10, Sect. 6]. The proof of 7.4.6 is a slight modification of an analysis in the classical (untwisted) case. I learnt the latter from O. Gabber. The noetherianity of U was first established by C. De Concini and V. Kac [DK1] using a PBW type basis constructed independently by G. Lusztig [L6] and S. Z. Levendorskii and Ya. S. Soibelman [LeS1] from the quantum Weyl group (10.2). This leads to a filtration of U (and of $U^{\pm}$) such that the associated graded ring is q-commutative [DK1, Proposition 1.7] but not commutative. This argument fails to give the noetherianity of $F(U)$.

7.5.8 Recently J. Bernstein and V. Lunts [BL1] found a very simple proof that $U(\mathfrak{g})$ (resp. $S(\mathfrak{g})$) is free over $Z(\mathfrak{g})$ (resp. $S(\mathfrak{g})^G$). This goes as follows. The triangular decomposition $S(\mathfrak{h}) \otimes S(\mathfrak{n}^+) \otimes S(\mathfrak{n}^-) \xrightarrow{\sim} S(\mathfrak{g})$ makes $S(\mathfrak{g})$ free over $S(\mathfrak{h})$. Let H denote the subspace of W harmonic polynomials (A.1.21) of $S(\mathfrak{h})$. The multiplication map is an isomorphism $S(\mathfrak{h})^W \otimes H \xrightarrow{\sim} S(\mathfrak{h})$ of graded vector spaces [Bb1, Chap. V, 5.5, Theorem 4]. Moreover the projection $S(\mathfrak{g}) \longrightarrow S(\mathfrak{h})$ defined by restriction of functions restricts [Di2, 7.3.5] to an isomorphism ψ of $S(\mathfrak{g})^G$ onto $S(\mathfrak{h})^W$ of graded algebras. Now take the filtration $\mathcal{F}$ on $S(\mathfrak{g})$ defined by degree on $S(\mathfrak{h})$; but extended by the trivial filtration on $A := S(\mathfrak{n}^+) \otimes S(\mathfrak{n}^-)$. Then given $a \in S(\mathfrak{h})^W$ homogeneous of degree d one obtains $\psi^{-1}(a) - a \in \mathcal{F}^{d-1}(S(\mathfrak{g}))$. Thus $\psi^{-1} \otimes 1d : S(\mathfrak{g})^W \otimes (H \otimes A) \longrightarrow S(\mathfrak{g})^G \otimes (H \otimes A)$ induces an isomorphism of the free $S(\mathfrak{h})^W$ module $S(\mathfrak{g})$ onto the $S(\mathfrak{g})^G$ module $gr_{\mathcal{F}} S(\mathfrak{g})$. Thus $S(\mathfrak{g})$ is free over $S(\mathfrak{g})^G$. Similarly $U(\mathfrak{g})$ is free over $Z(\mathfrak{g})$. This gives Kostant's separation theorem, but still needs [Di2, 8.1.1] for the primeness of I^+ and the description of regular functions of the nilpotent cone, crucial to the representation theory of $U(\mathfrak{g})$.

The above degree argument fails for $\check{U}$. However set $\check{U}_2^0 = k(q)\tau(2P(\pi))$. Then $\check{U}$ is free over $\check{U}_2 := U^- \otimes \check{U}_2^0 \otimes G^+$ and $Z(\check{U}_2) = Z(\check{U})$. By A.1.23(i) the $\tau(2\delta_w) : w \in W$ are free generators for $\check{U}_2^0$ over $(\check{U}_2^0)^W$ and furthermore these elements q-commute with any weight vector. Then the isomorphism

$$(*) \qquad \check{U}_2 \xrightarrow{\sim} Z(\check{U}_2) \otimes \sum_{w \in W} U^- \otimes k(q)\tau(2\delta_w) \otimes G^+$$

of $Z(\check{U}_2)$ modules obtains as follows.

Take $\lambda \in P^+(\pi)$. As in 8.2.9 every element of $(\operatorname{ad} U)\tau(-2\lambda)$ is a sum of the form $e_\nu''(y_{-\mu})\tau(2(\nu - \lambda))\, g_{\nu'}$ for some non-zero $y_{-\mu} \in K(\lambda)^-_{-\mu}, g_{\nu'} \in G^+_{\nu'}$. One may assume $\mu - \nu \geq 0$ and then applying $e'_{\mu-\nu}$ to the first factor, using 5.3.4, 5.3.5, gives as in 8.2.9 that $K(\lambda)^-_{-\nu} \neq 0$. Thus $\lambda - \mu, \lambda - \nu \in \Omega(V(\lambda))$ and so $\|\lambda - \mu + \lambda - \nu\| = \|2\lambda\|$ implies $\mu = \nu = \lambda - w\lambda$ for some $w \in W$. In particular the central element $z_\lambda \in (\operatorname{ad} U)\tau(-2\lambda)$ has a non-zero component in $U^-_{-\gamma}\tau(-2\delta)G^+_\gamma$ for $\gamma > 0$ and $-\delta$ dominant only if $(\delta, \delta) < (\lambda, \lambda)$.

Given $\delta \in P(\pi)$ choose $w \in W$ such that $w\delta \in P^+(\pi)$. By A.1.23(i) the coefficient $z_y \in (\mathbb{Z}e^{P(\pi)})^W$ of e^{δ_y} in the expansion of e^δ obtains as the cofactor $\det(e^{x\delta_u})$, in which δ replaces δ_y, divided by $\det(e^{x\delta_u})$. It follows as in A.1.22 that a dominant term e^ν in z_y satisfies $w\delta - \rho_y - \nu \geq 0$. Thus $(\nu, \nu) \leq (w\delta - \rho_y, w\delta - \rho_y) \leq (\delta, \delta)$ with equality if and only if $y = 1$. Then $\nu = w\delta$ forces $w\delta = \delta$ via A.1.20(i). Translated to the present situation one concludes in the notation of 7.1.17 that

$$\tau(-2x\lambda) \in k(q)\hat{\tau}(-2\lambda) + \sum_{w\in W} \sum_{\nu\in P^+(\pi)|(\nu,\nu)<(\lambda,\lambda)} k(q)\hat{\tau}(-2\nu)\tau(2\delta_w) ,$$

for all $x \in W$, where the first term occurs only if $-x\lambda$ is dominant.

Finally recall 7.1.19. It is then clear that $(*)$ obtains by induction on $(\mu,\mu) : \mu \in P^+(\pi)$. (The above proof was obtained following some discussions with P. Caldero.)

Notice that $(*)$ does not immediately provide 7.3.8 which concerns $F(\check{U})$ nor what will be needed for Chapter 8.

Chapter 8. The Primitive Spectrum of $U_q(\mathfrak{g})$

In this chapter it is assumed throughout that the Cartan matrix C is of finite type. The base field k is assumed of characteristic zero and $\overline{k(q)}$ denotes the algebraic closure of $k(q)$. The goal is to determine $\text{Prim}(U_q(\mathfrak{g}) \otimes_{k(q)} \overline{k(q)})$.

8.1 The Poincaré Series of the Harmonic Space

It is convenient to work mainly with $\check{U}$ as the bad properties of U become increasingly serious as the analysis becomes finer.

8.1.1 Recall that the filtration $\mathcal{F}$ on U extends to $\check{U}$ and then by 7.1.14, 7.1.25 restricts to a filtration on $F(\check{U})$ indexed by $\mathbb{N}$. For each $m \in \mathbb{N}$ set $F_m(\check{U}) = \mathcal{F}^m(F(\check{U}))/\mathcal{F}^{m-1}(F(\check{U}))$. By 7.1.2, 7.1.4, 7.1.6 there is an isomorphism of U modules

$$F_m(\check{U}) \xrightarrow{\sim} \bigoplus_{\lambda \in 2P^+(\pi) | \deg \lambda = m} (\text{ad}\, U)(gr_{\mathcal{F}} \tau(-\lambda)) .$$

Since $F_m(\check{U})$ is already finite dimensional, it is of finite length as a U module. Thus in the notation of 4.1.5 one may define for each $\mu \in P^+(\pi)$ the Poincaré series

$$R_\mu(z) = \sum_{m=0}^{\infty} [F_m(\check{U}) : V(\mu)] z^m .$$

8.1.2 It follows from 4.3.10 that $F(\check{U})$ is semisimple. For each simple finite-dimensional U module V let $F(\check{U})_V$ denote the sum of all submodules of $F(\check{U})$ isomorphic to V.

Lemma. *$F(\check{U})_V$ is a finitely generated $Z(\check{U})$ module.*

Let φ denote the restriction of $\mathcal{P}$ to $F(\check{U})^T$. Since $U_+^- U \cap U^T = UU_+^+ \cap U^T$, then by 7.1.13 it follows that φ is a homomorphism of $F(\check{U})^T$ into $k(q)\tau(2P(\pi))$. Set $r = \dim V$ and fix a basis $\{v_s\}$ of the zero weight space V_0 of V. For any homomorphism $\theta : V \to F(\check{U})$ set $\Phi_\theta = \varphi(\theta(v_s))_{s=1}^r \in$

$(k(q)\tau(2P(\pi))^r$. Suppose $\Phi_\theta = 0$. Then for all $\lambda \in P^+(\pi)$ one has $\theta(V_0)u_\lambda = 0$. Then $\theta(V)u_\lambda \in V(\lambda)^-$ and hence $\mathcal{P}(\theta(V))(\lambda) = 0$ by 7.1.10 (*). Since $P^+(\pi)$ is Zariski dense, $\mathcal{P}(\theta(V)) = 0$ and so $\theta(V) = 0$ by 7.1.10 (ii). This proves that $\theta \mapsto \Phi_\theta$ is injective.

Each $z \in Z(\check{U})$ acts on $F(\check{U})$ by left multiplication and the map $\theta \mapsto z\theta$ gives $\mathrm{Hom}_U(V, F(\check{U}))$ a $Z(\check{U})$ module structure. Since φ is a homomorphism it follows that $\Phi_{z\theta} = \varphi(z)\Phi_\theta$.

By 7.1.17 (v) one has $\varphi(Z(\check{U})) = (k(q)\tau(2P(\pi)))^{W\cdot}$ which further identifies with $(k(q)\tau(2P(\pi))^W$ via the map $\tau(\lambda) \mapsto \tau(\lambda)q^{(\rho,\lambda)}$. Then by A.1.19 it follows that $k(q)\tau(2P(\pi))$ and hence $\mathrm{Im}\,\Phi$ is a finitely generated $\varphi(Z(\check{U}))$ module. Combined with the conclusions of the first two paragraphs this proves the lemma.

8.1.3 The above result can be expressed via the separation theorem (7.3.8) as saying that $[\mathbb{H} : V(\mu)] = [\check{H} : V(\mu)] < \infty$ for all $\mu \in P^+(\pi)$. Of course $\check{H}$ is also a graded subspace of $gr_{\mathcal{F}}F(\check{U})$ so writing $\check{H}_m = F_m(\check{U}) \cap \check{H}$ there is a polynomial $P_\mu(z)$ defined by

$$P_\mu(z) := \sum_{m=0}^{\infty} [\check{H}_m : V(\mu)]z^m .$$

Moreover it is clear from 7.3.8 that

$$(*) \qquad R_\mu(z) = P_\mu(z)R_0(z) \quad \text{for all} \quad \mu \in P^+(\pi) .$$

Of course $R_0(z)$ is just the Poincaré series for the trivial module, that is for the graded algebra $Y(\check{U})$. In view of 7.1.17 (i) this is a polynomial algebra on generators $y_i := y_{\omega_i}$. Since $y_i \in F_{\deg \omega_i}(\check{U})$ one has

$$(**) \qquad R_0(z) = \prod_{i=1}^{\ell}(1 - z^{\deg \omega_i})^{-1} .$$

8.1.4 It is immediate from 8.1.2 and 7.3.8 that $gr_{\mathcal{F}}F(\check{U})_V$ is a finitely generated $Y(\check{U})$ module. Yet it is worth obtaining this directly as it leads to a second proof of 7.3.8 (but still using 7.3.3).

Define $\mathrm{End}\, V(\lambda)$ as a U module for the diagonal action (cf. 7.1.16). Recall the notation of 7.1.7.

Lemma. *For all* $\lambda, \mu, \nu \in P^+(\pi)$.

(i) *There is an injection* $\mathrm{End}\, V(\lambda) \hookrightarrow \mathrm{End}\, V(\lambda + \nu)$ *of* U *modules.*
(ii) $[\mathrm{End}\, V(\lambda) : V(\mu)] \leq \dim V(\mu)_0$ *with equality for* $\lambda \gg \mu$.
(iii) *For each* $i \in \{1, 2, \dots, \ell\}, \mu \in P^+(\pi)$ *there exists* $n \in \mathbb{N}^+$ *such that* $[y_i G(\nu + (m-1)\omega_i) : V(\mu)] = [G(\nu + m\omega_i) : V(\mu)]$, *for all* $m \geq n$.

Via 7.1.16 (iii) and 7.1.2 one may identify $\operatorname{End} V(\lambda)$ with $G(\lambda)$ as a U module. Multiplication by y_ν gives via 7.1.7 (iv) a homomorphism of $G(\lambda)$ into $G(\lambda+\nu)$ which is injective by 7.3.4. Hence (i).

Identifying $\operatorname{End} V(\lambda)$ with $V(\lambda)\otimes V(\lambda)^*$ it follows from 4.3.10 and Frobenius reciprocity that

$$[\operatorname{End} V(\lambda) : V(\mu)] = \dim \operatorname{Hom}_U(V(\lambda)\otimes V(\lambda)^*, V(\mu)) = N^{\mu}_{\lambda,\lambda_0}$$

in the notation of 6.3.18, 6.3.20 (Remark). Now $N^{\mu}_{\lambda,\lambda_0} = N^{\lambda}_{\lambda,\mu_0}$ which by 6.3.20 equals $\dim V(\mu_0)_0 = \dim V(\mu)_0$ for $\lambda \gg \mu$. Combined with (i) this proves (ii). Now 7.3.4 and 7.1.7 (iv) give $[G(\nu) : V(\mu)] = [y_i G(\nu) : V(\mu)] \leq [G(\nu+\omega_i) : V(\mu)] \leq \dim V(\mu)_0$ by (ii). This forces (iii).

8.1.5 To show that 8.1.4 (iii) implies $gr_{\mathcal{F}} F(\check{U})_V$ to be finitely generated over $Y(\check{U})$, define for each $j \in \{0,1,2,\dots,\ell\}$ the $Y^j := k(q)[y_1,\dots,y_j]$ submodule

$$G^j := \sum\{G(\nu) \,\Big|\, \nu \in \sum_{i=1}^{j} \mathbb{N}\omega_i\} \,.$$

Then $G^0 = k(q) = Y^0$. Assume that G^j_V is generated over Y^j by finitely many $G(\nu)_V$. Then by 8.1.4 (iii) it follows that G^{j+1}_V is generated over Y^{j+1} by finitely many $G(\nu)_V$. Hence the assertion. One may add that this result combined with A.3.9 implies that $R_\mu(z)/R_0(z)$ is a polynomial $P_\mu(z)$, satisfying $P_\mu(1) = \dim V(\mu)_0$ by 8.1.4 (ii). Recombined with 7.3.8 this gives in particular the

Lemma. *For all $\mu \in P^+(\pi)$ one has* $[\check{\mathbb{H}} : V(\mu)] = \dim V(\mu)_0$.

8.1.6 The computation of $R_\mu(z)$ requires a precise expression for the left hand side of 8.1.4 (ii). The first step obtains from the following general result for inducing modules over Hopf algebras.

Let A be a Hopf algebra and B a subalgebra which is a right coideal. Let M be a B module and $\varphi : m \mapsto 1\otimes m$ the canonical map of M into $A\otimes_B M$ which in turn identifies with $A\varphi(M)$. Then the coproduct Δ on A defines an A (resp. B) module structure on $(A\otimes_B M)\otimes N$ (resp. $\varphi(M)\otimes N$). Similarly if B is a left coideal $N\otimes\varphi(M)$ acquires a B module structure.

Lemma

(i) *The map $\theta : a\otimes m\otimes n \mapsto a_1\otimes(m\otimes\sigma(a_2)n)$ factors to an A module isomorphism of $(A\otimes_B M)\otimes N$ onto $A\otimes_B(\varphi(M)\otimes N)$.*

(ii) *If σ is bijective, the map $\theta' : n\otimes\sigma(a)\otimes m \mapsto \sigma(a_1)\otimes a_2 n\otimes m$ factors to an A module isomorphism of $N\otimes(A\otimes_B M)$ onto $A\otimes_B(N\otimes\varphi(M))$.*

(i) Since $\varphi(M) \otimes N$ is a B submodule of the A module $A\varphi(M) \otimes N$ it follows that $\varphi(M) \otimes N$ identifies with its image in $A \otimes_B (\varphi(M) \otimes N)$. Universality gives a surjection ψ of the latter onto $A(\varphi(M) \otimes N)$ viewed as an A submodule of $A\varphi(M) \otimes N$. Yet for all $a \in A$, $m \in \varphi(M)$, $n \in N$ one has $am \otimes n = a_1 m \otimes \varepsilon(a_2)n = a_1 m \otimes a_2 \sigma(a_3)m = a_1(m \otimes \sigma(a_2)n) \in A(\varphi(M) \otimes N)$ so ψ is a surjection onto $(A \otimes_B M) \otimes N$ and satisfies $\psi(a_1 \otimes m \otimes \sigma(a_2)n) = (a \otimes m) \otimes n$. Again $\theta(a \otimes bm \otimes n) = a_1 \otimes (bm \otimes \sigma(a_2)n) = a_1 \otimes b_1(m \otimes \sigma(b_2)\sigma(a_2)n) = (ab)_1 \otimes (m \otimes \sigma((ab)_2)n) = \theta(ab \otimes m \otimes n)$ and $\theta(a_1 \otimes m \otimes a_2 n) = a \otimes m \otimes n$, so θ factors to a surjection of $(A \otimes_B M) \otimes N$ onto $A \otimes_B (\varphi(M) \otimes N)$. Furthermore $\psi\theta = 1d_{(A \otimes_B M) \otimes N}$ and so θ is injective.

(ii) As in (i) universality gives a surjection ψ' of $A \otimes_B (N \otimes \varphi(M))$ onto $A(N \otimes \varphi(M))$ viewed as an A submodule of $N \otimes A\varphi(M)$. Yet $n \otimes \sigma(a)m = n \otimes \sigma(a_1 \varepsilon(a_2))m = \varepsilon(a_2)n \otimes \sigma(a_1)m = \sigma(a_2)a_3 n \otimes \sigma(a_1)m = \sigma(a_1)(a_2 n \otimes m) \in A(N \otimes \varphi(M))$. Thus if σ is surjective, ψ' has image $N \otimes (A \otimes_B M)$ and satisfies $\psi'(\sigma(a_1) \otimes a_2 \otimes m) = n \otimes (\sigma(a) \otimes m)$. Then (ii) obtains as in (i).

8.1.7 The following holds for any pair $U = U_q(\mathfrak{g}_C)$, $V^+ = U_q(\mathfrak{b}_C^+)$ with C an arbitrary symmetric Cartan matrix. Take $A = U$, $B = V_+$ in 8.1.6. By triangular decomposition A is a free right B module and so φ of 8.1.6 is injective. Also σ is bijective. Take $\Lambda \in T^\wedge$, $\mu \in P(\pi)$, recalling the notation of 3.4.9, 3.4.10. The V^+ module $1_\Lambda \otimes V(\mu)$, or $V(\mu) \otimes 1_\Lambda$, admits a decreasing filtration $V_1 \supseteq V_2 \supseteq \ldots \supseteq V_{n+1}$ with one-dimensional quotients $1_{\Lambda q^{\nu_i}}$ where ν_i runs over the set $\Omega(V(\mu))$ of weights of $V(\mu)$ counted with multiplicities and ordered so that $\nu_i \not> \nu_{i+1}$. By 8.1.6 and 3.4.9 one obtains the

Corollary. *The U modules $M(\Lambda) \otimes V(\mu)$ and $V(\mu) \otimes M(\Lambda)$ admit decreasing filtrations $M_1 \supseteq M_2 \supseteq \ldots \supseteq M_{n+1}$ with quotients isomorphic to $M(\Lambda q^{\nu_i})$: $\nu_i \in \Omega(V(\mu))$ and ordered as above. In particular in the Grothendieck group*

$$[V(\mu) \otimes M(\Lambda)] = [M(\Lambda) \otimes V(\mu)] = \sum_{\nu \in \Omega(V(\mu))} [M(\Lambda q^\nu)] \ .$$

Remark. Such a filtration is often called a Verma flag.

8.1.8 Recall the notation of 3.4.8.

Lemma. *For all $\lambda, \mu \in P^+(\pi)$ one has*

$$[\operatorname{End} V(\lambda) : V(\mu)] = \sum_{x,y \in W} (-1)^{\ell(x)+\ell(y)} P(x.\lambda - y.\lambda + \mu) \ .$$

It is convenient to let $V(-\lambda)$ denote the integrable module with lowest weight $-\lambda$. One has $V(\lambda)^* \cong V(-\lambda)$. By 4.3.6 (iv) and 8.1.7

$$[\operatorname{End} V(\lambda)] = [V(\lambda) \otimes V(-\lambda)] = \sum_{y \in W} \sum_{\nu \in \Omega(V(-\lambda))} (-1)^{\ell(y)} [M(y.\lambda + \nu)] \ .$$

Hence

$$\begin{aligned}[\operatorname{End} V(\lambda) : V(\mu)] &= \sum_{y\in W} (-1)^{\ell(y)} \dim V(-\lambda)_{\mu - y.\lambda} \\ &= \sum_{y\in W} (-1)^{\ell(y)} \dim V(\lambda)_{y.\lambda-\mu} \\ &= \sum_{x,y\in W} (-1)^{\ell(x)+\ell(y)} P(x.\lambda - y.\lambda + \mu) \, .\end{aligned}$$

8.1.9 Define the formal power series

$$S(z) = \sum_{\lambda\in P^+(\pi)} \sum_{x\in W} (-1)^{\ell(x)} e^{x.\lambda-\lambda} z^{\deg \lambda} = \sum_{x\in W} \frac{(-1)^{\ell(x)} e^{x.0}}{\prod_{i=1}^{\ell}(1 - z^{\deg \omega_i} e^{x\omega_i - \omega_i})}$$

and

$$Q(z) = \sum_{x\in W} (-1)^{\ell(x)} e^{x.0} \prod_{i=1}^{\ell} \left[\frac{1 - z^{\deg \omega_i}}{1 - z^{\deg \omega_i} e^{x\omega_i - \omega_i}}\right] .$$

By 8.1.8, $V(\mu) : \mu \in P^+(\pi)$ occurs in $\operatorname{End} V(\lambda)$ only if $\mu \in Q(\pi)$. Recall the notation of 3.4.8 and write D_- simply as D. Set $P(\mu) = 0$ if $\mu \in Q(\pi)\backslash Q^+(\pi)$.

Proposition. *For all* $\mu \in Q^+(\pi)$

(i) $R_\mu(z)$ *is the coefficient of* e^0 *in* $(\operatorname{ch} V(\mu))S(z)$.
(ii) $P_\mu(z)$ *is the coefficient of* e^0 *in* $(\operatorname{ch} V(\mu))Q(z)$.

By 7.1.2, 8.1.1 and 8.1.8, $R_\mu(z)$ is the coefficient of $e^{-\mu}$ in

$$\begin{aligned} R(z) :&= \sum_{\mu\in Q(\pi)} \sum_{\lambda\in P^+(\pi)} [\operatorname{End} V(\lambda) : V(\mu)] z^{\deg \lambda} e^{-\mu} \\ &= \sum_{\mu\in Q(\pi)} \sum_{\lambda\in P^+(\pi)} \sum_{x,y\in W} (-1)^{\ell(x)+\ell(y)} P(\mu) e^{-\mu - y.\lambda + x.\lambda} z^{\deg \lambda} \\ &= D \sum_{\lambda\in P^+(\pi)} \Big(\sum_{x,y\in W} (-1)^{\ell(x)+\ell(y)} e^{-y(\lambda+\rho)+x(\lambda+\rho)} \Big) z^{\deg \lambda} \\ &= D \sum_{\lambda\in P^+(\pi)} \Big\{ \sum_{y\in W} y \Big(\sum_{x\in W} (-1)^{\ell(x)} e^{x.\lambda - \lambda} \Big) \Big\} z^{\deg \lambda} \\ &= D \sum_{y\in W} yS(z) \, . \end{aligned}$$

Set $y_*\lambda = y(\lambda - \rho) + \rho$. Since $y_*(De^\xi) = (-1)^{\ell(y)} De^{y\xi}$ for all $\xi \in P(\pi)$ one obtains

$$(*) \qquad R(z) = \sum_{y\in W} (-1)^{\ell(y)} y_*(DS(z)) \, .$$

Yet by 4.2.16 and 4.3.6 (iv) the coefficient of e^0 in $(\operatorname{ch} V(\mu))S(z)$ is just the sum of the coefficients of $e^{-y.\mu}$ in $(-1)^{\ell(y)}DS(z)$. Since $-y.\mu = y_*(-\mu)$, the latter is just the coefficient of $e^{-\mu}$ in $(-1)^{\ell(y)}y_*^{-1}(DS(z))$. Thus (i) results from (*). Then (ii) results from (i) and 8.1.3 (*), (**).

8.1.10 Set $V = V(\mu)$. Then for all $n \in \mathbb{N}$ one has

$$(*) \qquad \sum_{i=1}^{\ell} (\deg \omega_i) \dim V_{n\alpha_i} = \frac{1}{2} \sum_{\alpha \in \Delta^+} (\deg \alpha) \dim V_{n\alpha} \,.$$

Indeed it is sufficient to take C indecomposable and to show that all partial sums over a fixed root length satisfy (*). Since C is assumed positive definite it follows from A.1.13 that roots of a given length are W conjugate. Then by 4.3.6 (iii), $\dim V_{n\alpha}$ is constant on roots of a given length, so it suffices to show that the half-sum of the positive roots of a given length is the sum of the fundamental weights corresponding to the simple roots of that length. This follows from 4.2.8 (ii) as in 7.1.25 noting that the set of roots of a given length is W stable.

8.1.11 For each rational function $F(z)$ let $F'(1)$ denote the value of its derivative at $z = 1$. In the product occurring in the expression for $Q(z)$, suppose $\ell(x) > 1$. Then by A.1.1 (vii) the denominator has at least two terms with no zeros at $z = 1$. It follows that

$$Q'(1) = \sum_{i=1}^{\ell} \left(\frac{e^{-\alpha_i}}{1 - e^{-\alpha_i}} \right) \deg \omega_i \,.$$

Then by 8.1.9 (ii) and 8.1.10 (*) one obtains the

Corollary. *For all* $\mu \in Q^+(\pi)$

$$P'_\mu(1) = \frac{1}{2} \sum_{n \in \mathbb{N}^+} \sum_{\alpha \in \Delta^+} \dim V(\mu)_{n\alpha} \deg \alpha \,.$$

8.1.12 By A.1.1 (vii) it follows that $Q(1) = 1$. Thus by 8.1.9 (ii) one has $P_\mu(1) = \dim V(\mu)_0 =: m$, a result which was already noted in 8.1.5. Recalling 8.1.3 this means that one can choose $a_{ij} \in \mathbb{H} : i, j = 1, 2, \ldots, m$ such that $\{a_{ij}\}_{i=1}^m$ is a basis for the jth copy of V in $\mathbb{H}$. In the notation of 3.4.10, let $\mathcal{P}_V$ denote the matrix with entries $\{\mathcal{P}(a_{ij})\}_{i,j=1}^m$. It is called a quantum Parthasarathy-Ranga Rao-Varadarajan, or simply PRV, form. Remarkably $\det \mathcal{P}_V$ admits a factorization similar to that of a quantum Shapovalev determinant (4.1.16). This is shown below.

8.2 Factorization of the Quantum PRV Determinants

8.2.1 Since C is assumed of finite type, hence positive definite it follows that $(\alpha,\alpha) > 0$ for all $\alpha \in \Delta$. Set $d_\alpha = (\alpha,\alpha)/2$ and $\alpha^\vee = 2\alpha/(\alpha,\alpha)$. For each $\alpha \in \Delta^+$ and each $n \in \mathbb{N}^+$ set $\Lambda_{n,\alpha} = \{\lambda \in \mathfrak{h}^*_{\mathbb{Q}} \mid (\lambda+\rho,\alpha^\vee) = n\}$ and $\Lambda^0_{n,\alpha} = \{\lambda \in \Lambda_{n,\alpha} \mid (\lambda+\rho,\beta^\vee) \notin \mathbb{Z}, \forall \beta \in \Delta^+, \beta \neq \alpha\}$. Given $\lambda \in \mathfrak{h}^*_{\mathbb{Q}}$, view $q^{(\lambda,\alpha_i)}$ as an element of $\overline{k(q)}^*$ and $q^\lambda : t_i \mapsto q^{(\lambda,\alpha_i)}$ as a character on T. Let $M(\lambda)$ denote the Verma module with highest weight λ.

Lemma. *For all* $\lambda \in \Lambda_{n,\alpha}$

(i) $M(s_\alpha.\lambda)$ *is a submodule of* $M(\lambda)$.
(ii) *If* $\lambda \in \Lambda^0_{n,\alpha}$, *then* $M(s_\alpha.\lambda)$ *and* $M(\lambda)/M(s_\alpha.\lambda)$ *are simple.*

Recall that $(\beta,\beta) > 0$ for all $\beta \in Q^+(\pi)$ and that $\Delta = \Delta_{re} = \Delta_{\text{irr}}$ by A.1.3 (ii). Then (i) follows from 4.4.9 whilst 4.1.7, 4.1.8 give $M^1(s_\alpha.\lambda) = 0$ hence the simplicity of $M(s_\alpha.\lambda)$ and $M^1(\lambda) = M(s_\alpha.\lambda)$ hence the simplicity of $M(\lambda)/M(s_\alpha.\lambda)$.

8.2.2 For the moment take $U = U_q(\mathfrak{g}_C)$ with C an arbitrary symmetric Cartan matrix. Define an order relation on $T^\wedge$ through $\Lambda \leq \Lambda'$ if $\Lambda = \Lambda' q^{-\beta}$ for some $\beta \in Q^+(\pi)$.

Proposition. *Take* $\Lambda, \Lambda' \in T^\wedge$ *with* $\Lambda \leq \Lambda'$. *Consider the following assertions*

(i) $M(\Lambda)$ *is projective in* $\mathcal{O}$.
(ii) $\mathrm{Hom}_U(M(\Lambda), M(\Lambda')) \neq 0 \Longrightarrow \Lambda = \Lambda'$.
(iii) $\Lambda(\tau(2\beta))q^{2(\rho,\beta)+(\beta,\beta)} \neq 1$ *for all* $\beta = m\gamma : \gamma \in \Delta^+,\ m \in \mathbb{N}^+$.
(iv) $[M(\Lambda') : V(\Lambda)] \neq 0 \Longrightarrow \Lambda = \Lambda'$.

Then (iii) $\Longrightarrow$ (iv) $\Longrightarrow$ (i) $\Longrightarrow$ (ii) *and* (ii) $\Longrightarrow$ (iii) *if* C *is integrable.*

Not (i) $\Longrightarrow$ *Not* (iv). The short exact sequence

$$0 \longrightarrow K \longrightarrow M \xrightarrow{\varphi} M(\Lambda) \longrightarrow 0$$

splits if $\varphi^{-1}(u_\Lambda)$ is a primitive vector. Consider the composed map $M \xrightarrow{\varphi} M(\Lambda) \xrightarrow{\pi} V(\Lambda)$. Then the inverse image of $\pi(u_\Lambda)$ in M is also not a primitive vector. Applying the $\mathcal{O}$ dual functor δ gives a non-split embedding $V(\Lambda) \hookrightarrow \delta M$. Then δM admits a weight $\Lambda' > \Lambda$. Moreover since $\delta M \in \mathcal{O}$ there can only be finitely many such weights, even counted with multiplicity, and one can choose Λ' maximal. Then a weight vector $u_{\Lambda'} \in \delta M$ of weight Λ' is primitive. If $V(\Lambda) \cap Uu_{\Lambda'} = 0$, then the composed map $V(\Lambda) \longrightarrow \delta M \longrightarrow \delta M/Uu_{\Lambda'}$ is still a non-split embedding. Consequently one may assume that $V(\Lambda) \hookrightarrow Uu_{\Lambda'}$ and by universality a subquotient of $M(\Lambda')$.

Not (iv) $\Longrightarrow$ *Not* (iii). Suppose $[M(\Lambda') : V(\Lambda)] \neq 0$ for some $\Lambda' > \Lambda$. Then 4.1.7 (i) gives $[M(\Lambda' q^{-\beta}) : V(\Lambda)] \neq 0$ for some $\beta \in Q^+(\pi)$ satisfying $\Lambda'(\tau(2\beta))q^{2(\rho,\beta)-(\beta,\beta)} = 1$. Moreover by 4.1.8 one has $\beta = m\gamma$ for some $\gamma \in \Delta^+$, $m \in \mathbb{N}^+$. Then either $\Lambda = \Lambda' q^{-\beta}$ giving *Not* (iii) or $\Lambda < \Lambda' q^{-\beta}$ and the process can be repeated to eventually give the first conclusion.

Not (ii) $\Longrightarrow$ *Not* (i). Given $M(\Lambda) \xrightarrow{\varphi} M(\Lambda')$ set $M = M(\Lambda')/\varphi(\ker \mathcal{S}^\Lambda)$. Then $V(\Lambda)$ is a proper submodule of the indecomposable module M. Applying δ gives a surjection $\delta M \longrightarrow V(\Lambda)$. Then $M(\Lambda)$ is not projective for otherwise the inverse image of $\bar{u}_\Lambda \in V(\Lambda)$ would be a primitive vector contradicting the indecomposability of δM.

Not (iii) $\Longrightarrow$ *Not* (ii) for C integrable. Suppose $\Lambda(\tau(2\beta)q^{2(\rho,\beta)+(\beta,\beta)} = 1$ for some $\beta \in Q^+(\pi) \setminus \{0\}$. Then Λq^β satisfies (ii) of 4.4.10 and so $M(\Lambda)$ is a submodule of $M(\Lambda q^\beta)$.

Remark. Suppose $\Lambda = q^\lambda : \lambda \in \mathfrak{h}^*_{\mathbb{Q}}$ viewed as a homomorphism of T into $\overline{k(q)}^{*\ell}$. Then the condition $\Lambda(\tau(2\gamma))q^{2(\rho,\gamma)+m(\gamma,\gamma)} \neq 1$ for all $\gamma \in \Delta^+, m \in \mathbb{N}^+$ is equivalent to $2(\lambda+\rho,\gamma)+m(\gamma,\gamma) \neq 0$. For $\gamma \in \Delta^+_{re}$ this can be expressed as $(\lambda+\rho,\gamma^\vee) \notin -\mathbb{N}^+$. For C of finite type all roots are real and such λ are called dominant. In this case $\Lambda \in T^\wedge$ is called dominant if $\Lambda(\tau(2\gamma))q^{2(\rho,\gamma)+m(\gamma,\gamma)} \neq 1$ for all $\gamma \in \Delta^+$, $m \in \mathbb{N}^+$. This terminology is less natural in the general case.

8.2.3 Assume again that C is of finite type. Given U modules M, N the space of homomorphisms $\operatorname{Hom}(M,N)$ has a natural bimodule structure given by $(a.\psi.b)(m) = a(\psi(bm))$ for all $a, b \in U$, $m \in M$, $\psi \in \operatorname{Hom}(M,N)$. This is generally too big to be of great interest. However U admits a natural "adjoint action" on $\operatorname{Hom}(M,N)$ defined by setting $(\operatorname{ad} a)\psi = a_1.\psi.\sigma(a_2)$. Define

$$F(M,N) = \{\psi \in \operatorname{Hom}(M,N) \mid \dim(\operatorname{ad} U)\psi < \infty\} ,$$

that is $F(M,N) = F(\operatorname{Hom}(M,N))$ for the adjoint action of U. Observe in particular that $F(M) := F(M,M)$ is a subalgebra of $\operatorname{End} M$. By complete reducibility $F(M,N)$ is a direct sum of simple finite dimensional U modules (4.3.10).

Whereas $F(M,N)$ is no longer a U bimodule, it is an $F(U)$ bimodule and of course an $\operatorname{ad} U$ module. Moreover these two actions are compatible in the following sense. By 1.3.5 and 7.1.6 it follows that $F(U)$ is a left coideal. Define an $F(U)$ bimodule H to have a compatible $\operatorname{ad} U$ action if

$$(*) \qquad ((\operatorname{ad} a_1)h).a_2 = a.h, \ \forall\, a \in F(U), h \in H .$$

Viewing $F(M,N)$ as a submodule $\operatorname{Hom}(M,N)$ one easily checks that it satisfies $(*)$.

Lemma. *Let H be an $F(U)$ bimodule with a compatible* $\operatorname{ad} U$ *action. Then for any* $\operatorname{ad} U$ *stable subspace V of H one has $F(U)V = VF(U)$.*

The inclusion $F(U)V \subset VF(U)$ is immediate from $(*)$. Take $\lambda \in -R^+(\pi)$. Then $\tau(\lambda) \in F(U)$ and for all $v \in V$ one has $\tau(\lambda)(\operatorname{ad}\tau(-\lambda)v) = v\tau(\lambda)$, so $V\tau(\lambda) \subset \tau(\lambda)V$. Since U is generated by quasi-primitive elements (4.3.12) it easily follows that $V((\operatorname{ad}U)\tau(\lambda)) \subset (\operatorname{ad}U)(\tau(\lambda)V) \subset F(U)V$ and so $VF(U) \subset F(U)V$ by 7.1.6 as required.

8.2.4 For each $\Lambda \in T^\wedge$ set $\Lambda q^{P(\pi)} = \{\Lambda q^\mu : \mu \in P(\pi)\}$.

Proposition. *For all $\Lambda, \Lambda' \in T^\wedge$, $\mu \in P(\pi)$, $\nu \in P^+(\pi)$ one has*

(i) $[F(M(\Lambda), M(\Lambda')) : V(\mu)] = 0$ *unless* $\Lambda' \in \Lambda q^{P(\pi)}$.
(ii) *If Λ is dominant* $[F(M(\Lambda), M(\Lambda q^\mu)) : V(\nu)] = \dim V(\nu)_\mu$.

By Frobenius reciprocity and the finite dimensionality of $V(\nu)$

$$\begin{aligned}
[F(M(\Lambda), M(\Lambda')) : V(\nu)] &= \dim \operatorname{Hom}_U(V(\nu), \operatorname{Hom}(M(\Lambda), M(\Lambda'))) \\
&= \dim \operatorname{Hom}_U(V(\nu) \otimes M(\Lambda), M(\Lambda')) \\
&= \dim \operatorname{Hom}_U(M(\Lambda), V(\nu)^* \otimes M(\Lambda')) \\
&\leq \sum_{\xi \in \Omega(V(-\nu))} \dim \operatorname{Hom}_U(M(\Lambda), M(\Lambda' q^\xi))
\end{aligned}$$

where the last step obtains from 8.1.7. Clearly $\operatorname{Hom}_U(M(\Lambda), M(\Lambda' q^\xi)) = 0$ unless $\Lambda = \Lambda' q^{\xi-\mu}$ for some $\mu \in Q^+(\pi)$. This gives (i). If Λ is dominant, then $M(\Lambda)$ is projective in $\mathcal{O}$ by 8.2.2. Since $V(\nu)^* \otimes M(\Lambda') \in Ob\mathcal{O}$ by 4.1.4, equality holds above. Finally by 8.2.2, the above sum with $\Lambda' = \Lambda q^\mu$ equals $\dim V(-\nu)_{-\mu} = \dim V(\nu)_\mu$.

Remark. The conclusion of (i) does not quite imply that $F(M(\Lambda), M(\Lambda')) = 0$ because of the presence (4.3.9) of non-trivial one-dimensional U modules.

8.2.5 Let M be a finitely generated U modules and let $d(M)$ denote its Gelfand-Kirillov dimension (A.3.6).

Lemma. *Let M, M' be finitely generated U modules. Then*

(i) $F(M, M') = 0$ *if $d(M') < d(N)$ for every simple quotient N of M.*
(ii) $F(M, M') = 0$ *if $d(M) < d(N')$ for every simple submodule N' of M'.*

Consider (i). By Frobenius reciprocity it is enough to show that $\operatorname{Hom}_U(M, V^* \otimes M') = 0$ for every simple finite dimensional U module V. Given $\varphi \in \operatorname{Hom}_U(M, V^* \otimes M')$ non-zero, one has $d(\varphi(M)) > d(M')$ by the hypotheses of (i). Yet by A.3.6 (iv), since V^* is finite dimensional $d(V^* \otimes M') = d(M') \geq d(\varphi(M))$, contradicting the previous inequality. For (ii) it is enough to show that $\operatorname{Hom}_U(V \otimes M, M') = 0$ and the proof is similar to the first part.

8.2.6 Recall the notation of 8.2.1 and its conclusion.

Lemma. *For all* $\alpha \in \Delta^+$, $n \in \mathbb{N}^+$, $\lambda \in \Lambda^0_{n,\alpha}$, $\mu \in P^+(\pi)$ *one has* $[F(V(\lambda), V(\lambda)) : V(\mu)] = \dim V(\mu)_0 - \dim V(\mu)_{n\alpha}$.

One has $(\lambda + \rho, \beta^\vee) \notin -\mathbb{N}^+$ for all $\beta \in \Delta^+$ by the hypotheses and so $M(\lambda)$ is projective by 8.2.2. Now $V(\lambda) = M(\lambda)/M(s_\alpha.\lambda)$ is a proper quotient of the Verma module $M(\lambda)$ and hence by A.3.7 (iii) satisfies $d(V(\lambda)) < d(M(\lambda)) = d(M(s_\alpha.\lambda))$. By the simplicity (8.2.1 (i)) of $M(s_\alpha.\lambda)$ it follows that $F(M(s_\alpha.\lambda), N(\lambda)) = 0$ by 8.2.5 (i) and so the natural injection $F(V(\lambda), V(\lambda)) \hookrightarrow F(M(\lambda), V(\lambda))$ is surjective. Since $M(\lambda)$ is projective one has a short exact sequence

$$0 \longrightarrow F(M(\lambda), M(s_\alpha.\lambda)) \longrightarrow F(M(\lambda), M(\lambda)) \longrightarrow F(M(\lambda), N(\lambda)) \longrightarrow 0$$

and from 8.2.4 (ii) that

$$\begin{aligned}&[F(M(\lambda), M(\lambda)) : V(\mu)] - [F(M(\lambda), M(s_\alpha\lambda)) : V(\mu)] \\ &\quad = \dim V(\mu)_0 - \dim V(\mu)_{s_\alpha.\lambda-\lambda}\,.\end{aligned}$$

Yet $\dim V(\mu)_{s_\alpha.\lambda-\lambda} = \dim V(\mu)_{-n\alpha} = \dim V(\mu)_{n\alpha}$ by 4.3.6 (iii). Hence the assertion.

8.2.7 Fix $V = V(\mu) : \mu \in P^+(\pi)$ and recall the notation of 8.1.12.

Lemma. *For each character* Λ *of* $\check{T}$ *with values in* $\overline{k(q)}^{*\ell}$ *one has*

(i) $rk\,\mathcal{P}_V(\Lambda) = [F(\check{U})/\operatorname{Ann}_{F(\check{U})} V(\Lambda) : V]$.
(ii) $\det \mathcal{P}_V \neq 0$.
(iii) Λ *is a zero of* $\det \mathcal{P}_V$ *of order* $\geq [\operatorname{Ann}_{\check{\mathbb{H}}} V(\Lambda) : V]$.
(iv) *If* $\lambda \in \Lambda^0_{n,\alpha}$, *then* λ *is a zero of* $\det \mathcal{P}_V$ *of order* $\geq \dim V_{n\alpha}$.

Let M be an $\operatorname{ad} U$ submodule of $F(\check{U})$. Then $M \subset \operatorname{Ann} V(\Lambda) \iff \Lambda(\mathcal{P}(M)) = 0$ by 7.1.10 (i). Since $Z(\check{U})$ acts by scalars on $V(\Lambda)$ it follows from 7.3.5 (i) that $\check{\mathbb{H}} + \operatorname{Ann}_{F(\check{U})} V(\Lambda) = F(\check{U})$ which gives an isomorphism $\check{\mathbb{H}}/\operatorname{Ann}_{\check{\mathbb{H}}} V(\Lambda) \xrightarrow{\sim} F(\check{U})/\operatorname{Ann}_{F(\check{U})} V(\Lambda)$ of $\operatorname{ad} U$ modules. All this gives (i).

If $\det \mathcal{P}_V = 0$ then there exist scalars c_j not all zero such that $\mathcal{P}(\sum_j a_{ij}c_j) = 0$ with a_{ij} defined as in 8.1.12. This means that $\mathcal{P}(V) = 0$ for the corresponding copy of V in $\check{\mathbb{H}}$. However this conclusion contradicts 7.1.10 (ii). Hence (ii).

For each copy of V in $\operatorname{Ann}_{\check{\mathbb{H}}} V(\Lambda)$ one has $\Lambda(\mathcal{P}(V)) = 0$ and so Λ is a common zero for the row in $\mathcal{P}_V$ corresponding to this copy. Hence (iii).

It is clear that the action of $F(\check{U})$ on $V(\lambda) = M(\lambda)/M(s_\alpha.\lambda)$ defines an injection of $F(\check{U})/\operatorname{Ann}_{F(\check{U})} V(\lambda) \hookrightarrow F(V(\lambda), V(\lambda))$ and so $[\check{\mathbb{H}}/\operatorname{Ann}_{\check{\mathbb{H}}} V(\lambda) : V] \leq \dim V_0 - \dim V_{n\alpha} = [\check{\mathbb{H}} : V] - \dim V_{n\alpha}$. Hence (iii) implies (iv).

8.2.8 Define U_2^0 as in 3.3.4. For any $V = V(\mu) : \mu \in P^+(\pi)$ one can assume in computing $\det \mathcal{P}_V$ that the jth copy of V in $\check{\mathbb{H}}$ lies in $(\operatorname{ad} U)\tau(-2\lambda_j)$ for some $\lambda_j \in P^+(\pi)$. From the relations in 5.1.1 it is immediate that $\mathcal{P}((\operatorname{ad} U)\tau(-2\lambda_j)) \subset U_2^0\tau(-2\lambda_j)$. One concludes that there exist $\lambda_1, \lambda_2, \dots, \lambda_m \in P^+(\pi)$ such that

$$(*) \qquad \det \mathcal{P}_V \in U_2^0(\prod_{j=1}^{m} \tau(-2\lambda_j)) .$$

Recall the notation of 8.2.1.

Lemma. *For each $\alpha \in \Delta^+$, $n \in \mathbb{N}^+$ the polynomial*

$$(\tau(\alpha)^2 - q^{2d_\alpha(n-(\rho,\alpha^\vee))})^{\dim V_{n\alpha}}$$

divides $\mathcal{P}_V$.

By $(*)$, $\det \mathcal{P}_V \in U_2^0$ up to units. The polynomial $p(\alpha) := \tau(a)^2 - q^{2(\alpha,\lambda)} = \tau(\alpha)^2 - q^{2d_\alpha(n-(\rho.\alpha^\vee))}$ is irreducible in U_2^0 and has a zero of order one at $\lambda \in \Lambda_{n,\alpha}^0$. Set $\mathfrak{h}_\alpha^\perp = \{\beta \in \mathfrak{h}_\mathbb{Q}^* \mid (\beta, \alpha) = 0\}$ and extend τ to $\mathfrak{h}_\alpha^\perp + 2\mathbb{Z}\alpha$. Each $\beta \in Q(\pi)$ can be expressed uniquely as $\beta = (\beta, \alpha^\vee)\alpha + \gamma$ for some $\gamma \in \mathfrak{h}_\alpha^\perp$ and so up to units $\det \mathcal{P}_V$ can be expressed uniquely as a finite sum of the form

$$\sum_{i \geq i_0} p(\alpha)^i P_i \quad \text{for some} \quad P_i \in k[q]\tau(\mathfrak{h}_\alpha^\perp) \quad \text{with} \quad P_{i_0} \neq 0 .$$

Since $\Lambda_{n,\alpha}^0 - \frac{1}{2}(\lambda, \alpha^\vee)\alpha$ is dense in $\mathfrak{h}_\alpha^\perp$ and C is positive definite there exists $\lambda \in \Lambda_{n,\alpha}$ which is not a zero of P_{i_0}. Then by 8.2.7 (iv) it follows that $i_0 \geq \dim V_{n\alpha}$, which gives the required assertion.

8.2.9 To proceed further it is necessary to refine 8.2.8 $(*)$. Recall the notation of 7.1.7 and 8.1.7.

Lemma. *For all $\lambda \in \frac{1}{2}P(\pi)$ one has*

$$\mathcal{P}(F(\lambda)) = \sum k(q)\{\tau(-2\mu) \mid \mu \in \Omega(V(\lambda))\} .$$

By definition and triangular decomposition $F(\lambda) = (\operatorname{ad} U^+)(\operatorname{ad} U^-)\tau(-2\lambda) = (\operatorname{ad} U^+)(K(\lambda)^-\tau(-2\lambda))$. Now $\Omega(K(\lambda)^-) = \Omega(V(\lambda)) - \lambda$. So $K(\lambda)^-_{-\mu} \neq 0 \iff \lambda - \mu \in \Omega(V(\lambda))$. Since $\mathcal{P}(\operatorname{ad} U^+)(K(\lambda)^-\tau(-2\lambda))) = \mathcal{P}((\operatorname{ad} U^+)K(\lambda)^-)\tau(-2\lambda)$ it suffices to show that

$$(*) \qquad \mathcal{P}((\operatorname{ad} U^+)K(\lambda)^-) = \sum k(q)\{\tau(2\mu) \mid K(\lambda)^-_{-\mu} \neq 0\} .$$

Take $y_{-\mu} \in K(\lambda)^-_{-\mu}$. Then by 5.1.1 and 5.3.1

$$\begin{aligned}(\operatorname{ad} e_i)y_{-\mu} &= e_i y_{-\mu} t_i - y_{-\mu} e_i t_i \\ &= \frac{1}{q_i - q_i^{-1}} \left[q_i^{2-(\alpha_i^\vee,\mu)} e_i''(y_{-\mu}) t_i^2 - q_i^{(\alpha_i^\vee,\mu)-2} e_i'(y_{-\mu}) \right] .\end{aligned}$$

Recall (5.3.4) that the e_i'', e_j' commute up to a factor of q. Then for each monomial e_μ in the $e_i : i = 1, 2, \ldots, n$ of weight μ it follows that $\mathcal{P}((\operatorname{ad} e_\mu)y_{-\mu})$ is a sum of the form

$$(e_\nu''(e_{\mu-\nu}'(y_{-\mu})))\tau(2\nu)$$

where e_ν'' (resp. $e_{\mu-\nu}'$) is a monomial in the e_i'' (resp. e_i') of weight ν (resp. $\mu-\nu$). Since $e_i' = (\operatorname{ad} t_i)(\operatorname{ad} e_i)$ up to a scalar and from the formulae of 7.1.1 and the definition of $K(\lambda)^-$ both $\operatorname{ad} t_i$ and $\operatorname{ad} e_i$ leave $K(\lambda)^-$ stable (for any λ) it follows that $e_{\mu-\nu}'(y_{-\mu}) \in K(\lambda)_{-\nu}^-$. Thus $\mathcal{P}(e_\nu''(e_{\mu-\nu}'(y_{-\mu}))) \in k(q)$ if $K(\lambda)_{-\nu}^- \neq 0$ and is zero otherwise. In particular the inclusion $\subset$ holds in $(*)$.

To obtain equality in $(*)$ it suffices to show that $e_\mu''(y_{-\mu}) \neq 0$ for some monomial e_μ''. This holds for any $y_{-\mu} \in U_{-\mu}^-$ because $\bigcap_{i=1}^\ell \ker e_i''$ is reduced to scalars. (The latter is proved exactly as in 5.3.5).

8.2.10 The factorization of the PRV determinants can now be described.

Theorem. *For all $\mu \in P^+(\pi)$ one has*

$$\det \mathcal{P}_{V(\mu)} = \prod_{n\in\mathbb{N}^+} \prod_{\alpha\in\Delta^+} (\tau(\alpha) - q^{2d_\alpha(n-(\rho,\alpha^\vee))}\tau(\alpha)^{-1})^{\dim V(\mu)_{n\alpha}}$$

up to a non-zero element of $k(q)$.

Set $V = V(\mu)$ and following 8.2.8 assume that the jth copy of V lies in $F(\lambda_j)$. Set $\lambda = \sum_{j=1}^m \lambda_j$. By the definition of $P_\mu(z)$ one obtains $P_\mu'(1) = \sum_j \deg \lambda_j = \deg \lambda$, so 8.1.11 gives

$$(*) \qquad \deg \lambda = \frac{1}{2} \sum_{n\in\mathbb{N}^+} \sum_{\alpha\in\Delta^+} (\dim V_{n\alpha}) \deg \alpha\ .$$

Now 8.2.9 gives $\mathcal{P}(F(\lambda_j))\tau(2\lambda_j) \subset k(q)\{\tau(2\mu) \mid \mu \in \lambda_j + \Omega(V(-\lambda_j))\}$. In particular, viewed as a polynomial in the $t_i : i = 1, 2, \ldots, \ell$ each $a \in \mathcal{P}(F(\lambda_j))\tau(2\lambda_j)$ satisfies $\deg a \leq \deg \tau(2\lambda_j) + \deg \tau(-2w_0\lambda_j) \leq 4 \deg \tau(\lambda_j)$. Thus $(\det \mathcal{P}_V)\tau(2\lambda) \in k(q)[t_1^2, t_2^2, \ldots, t_\ell^2]$ and has degree $\leq 4 \sum \deg \tau(\lambda_j) = 2 \deg \lambda$. Yet by 8.2.8

$$(**) \qquad (\det \mathcal{P}_V)\tau(2\lambda) = p \prod_{n\in\mathbb{N}^+} \prod_{\alpha\in\Delta^+} (\tau(\alpha)^2 - q^{2d_\alpha(n-(\rho,\alpha^\vee))})^{\dim V_{n\alpha}}$$

for some non-zero polynomial p. Then $2\deg\lambda \geq \deg(\det \mathcal{P}_V)\tau(2\lambda) = \deg p + 2 \deg \lambda$ by $(*)$ and so $\deg p \leq 0$. Thus p is a scalar. Moreover $2 \deg \lambda = \deg(\det \mathcal{P}_V)\tau(2\lambda)$ and so the unique highest degree term, namely $\tau(-2w_0\lambda)$ must occur in $\det \mathcal{P}_V$. Resubstitution in $(**)$ then gives

$$\lambda - w_0\lambda = \sum_{n\in\mathbb{N}^+} \sum_{\alpha\in\Delta^+} \alpha \dim V_{n\alpha}\ .$$

It remains to show that $-w_0\lambda = \lambda$ for then resubstitution in $(*)$ gives the conclusion of the theorem.

Recall that $\check{\mathbb{H}}$ is generated by its intersections $\check{\mathbb{H}}(\mu)$ with the $F(\mu) : \mu \in P^+(\pi)$. Then the required identity will result given $[\check{\mathbb{H}}(\mu) : V] = [\check{\mathbb{H}}(-w_0\mu) : V]$ or equivalently from

$$[\check{H}(\mu) : V] = [\check{H}(-w_0\mu) : V] \,, \quad \text{for all} \quad \mu \in P^+(\pi)$$

where $\check{H}(\mu) := G(\mu) \cap \check{H}$.

As a U module $G(\mu)$ is isomorphic to $V(\mu) \otimes V(-\mu)$ which in turn is isomorphic to $G(-w_0\mu)$. By 7.1.7 and 7.1.17 (i), the U module isomorphism θ sending each $G(\mu)$ to $G(-w_0\mu)$ is an algebra involution and takes y_μ to $y_{-w_0\mu}$. In particular J_+ is θ stable and so the same may be assumed of $\check{H}$. This proves the above equality and hence the theorem.

8.3 Verma Module Annihilators

A key fact in the structure theory of $U(\mathfrak{g})$ is that the annihilator of any Verma module is generated by its intersection with the centre $Z(\mathfrak{g})$. A natural question is whether this holds for any Verma module $M(\Lambda)$ of $U = U_q(\mathfrak{g})$ or at least if $\operatorname{Ann}_{F(U)} M(\Lambda)$ is generated by $\operatorname{Ann}_{Z(U)} M(\Lambda)$. This turns out to be *false*, a difficulty which can be traced to the failure of the separation theorem. However it does hold for $F(\check{U})$ as shown below. This leads to an important equivalence of categories theorem which eventually determines $\operatorname{Prim} U_q(\mathfrak{g})$.

8.3.1 Let Λ be a character on $\check{T}$.

Lemma. *The following two conditions are equivalent*

(i) $\operatorname{Ann}_{\check{\mathbb{H}}} V(\Lambda) = 0$,
(ii) $M(\Lambda) \xrightarrow{\sim} V(\Lambda)$.

A comparison of 4.1.16 with 8.2.10 shows that Λ is a zero of some Shapovalev determinant $\det \mathcal{S}_\nu$ if and only if Λ is a zero of some PRV determinant $\det \mathcal{P}_{V(\mu)}$. Since the former means that $M(\Lambda) \twoheadrightarrow V(\Lambda)$ has a non-zero kernel and the latter that $\operatorname{Ann}_{\check{\mathbb{H}}} V(\Lambda) \neq 0$ the claim follows.

8.3.2 Recall (7.1.17) that the isomorphism $\tau : P(\pi) \xrightarrow{\sim} \check{T}$ defines an action of W on $\check{T}$. Define an action on the set $\check{T}^\wedge$ of characters on $\check{T}$ by transport of structure, that is $(w\Lambda)(\tau(\beta)) := \Lambda(w^{-1}\tau(\beta)) = \Lambda(\tau(w^{-1}\beta))$ for all $w \in W$, $\beta \in P(\pi)$, $\Lambda \in \check{T}^\wedge$.

Take $\Lambda \in \check{T}^\wedge$ and set $\Lambda_i = \Lambda(\tau(\omega_i)) : i = 1, 2, \ldots, \ell$. Define an action of $\mathbb{Z}_2^\ell$ on $\check{T}^\wedge$ by letting its ith generator multiply Λ_i by -1 leaving the remaining Λ_j fixed. For each $\gamma \in Q(\pi^\vee)$ define a character $(-1)^\gamma$ on $\check{T}^\wedge$ by $(-1)^\gamma(\tau(\omega)) = (-1)^{(\gamma,\omega)}$ for all $\omega \in P(\pi)$. Then the above action of $\mathbb{Z}_2^\ell \xleftarrow{\sim} Q(\pi^\vee)/Q(2\pi^\vee)$ may be identified with the map $(\gamma, \Lambda) \mapsto (-1)^\gamma \Lambda$. Moreover from the action of W on $Q(\pi^\vee)$ one obtains an action of W on $\mathbb{Z}_2^\ell$ and one may form the semidirect product $\check{W} := \mathbb{Z}_2^\ell \rtimes W$. Clearly Λ, Λ' are conjugate under $\check{W}$ if and only if Λ^2, Λ'^2 are conjugate under W. One calls $\check{W}$ the extended Weyl group.

Take $z \in Z(\check{U})$. From the definition (7.1.17) of the Harish-Chandra map ψ it is immediate that $zu_\Lambda = \Lambda(\psi(z))u_\Lambda$ and so z acts on $M(\Lambda)$ by the scalar $\Lambda(\psi(z))$. Thus $\operatorname{Ann}_{Z(\check{U})} M(\Lambda) = \{z \in Z(\check{U}) \mid \Lambda(\psi(z)) = 0\}$ is a maximal ideal. Let χ denote the homomorphism of $\check{T}^\wedge$ into $\operatorname{Max} Z(\check{U})$ which results.

Lemma

(i) $\chi(\Lambda_1) = \chi(\Lambda_2) \iff \Lambda q^\rho,\ \Lambda' q^\rho$ *are conjugate under* $\check{W}$.

(ii) *Over* $\overline{k(q)}$, *the map* χ *is surjective.*

By 7.1.19 there exist linear combinations z_i' of the z_{ω_i} such that $\psi(z_i') = \hat{\tau}(-2\omega_i) : i = 1, 2, \ldots, \ell$. Consequently

$$\Lambda(\psi(z_i')) = \sum_{y \in W} (\Lambda q^\rho)(\tau(-2y\omega_i)) \ .$$

View Λ as an indeterminate and set ${}^t\Lambda_i = (\Lambda q^\rho)(\tau(-\omega_i))$. Let A be the Laurent polynomial algebra generated by the ${}^t\Lambda_i^2 : i = 1, 2, \ldots, \ell$ and B the subalgebra generated by the invariants

$$\Omega_i := \sum_{y \in W} y({}^t\Lambda_i^2) \ .$$

Identify A with the group algebra $k(q)P(\pi)$ of $P(\pi)$ via the map $\omega_i \mapsto (\Lambda q^\rho)(\tau(-2\omega_i))$. Then the subalgebra B identifies with $(k(q)P(\pi))^W$ and (A.1.23) is a free A module on $|W|$ generators, so one may write $A = B \otimes F$.

Assertion (i) results from the fact that any two maximal ideals of A over a maximal ideal of B are conjugate under W. This is a general property for any finite group G acting on any commutative ring A with $B = A^G$. For assertion (ii) let I be a maximal ideal of B, then I generates a proper ideal $I \otimes F$ of A. Let J be a maximal ideal of A containing $I \otimes F$. Over $\overline{k(q)}$ one has codim $J = 1$ and so it defines a character Λ on $\tau(P(2\pi))$ which may be lifted to a character on $\check{T}$. By construction $\chi(\Lambda) = I$, hence (ii).

8.3.3 For the moment just assume C symmetric. Take $M \in Ob\mathcal{O}$. It follows from 4.1.4 (i), (iii) that $\Omega(M)$ admits a maximal element Λ and then any vector $u_\Lambda \in M$ of weight Λ is primitive. Then the submodule $N = Uu_\Lambda$ generated by u_Λ is a quotient of $M(\Lambda)$ and $M/N \in Ob\mathcal{O}$. This procedure may be repeated to give a possibly infinite increasing filtration $N_1 \subsetneq N_2 \subsetneq \ldots$ whose quotients N_{i+1}/N_i are images of Verma modules $M(\Lambda_i)$. Then by 4.1.4 (ii), (iii) any weight vector belongs to N_i for i sufficiently large and consequently through 4.1.4 (i) one concludes that $M = \bigcup N_i$. Since a central element of $U_q(\mathfrak{g})$ or more generally of $U_q^{\mathcal{O}}(\mathfrak{g})$, acts by scalars on a Verma module one obtains the

Lemma. *Take $M \in Ob\mathcal{O}$. The centre $Z(U)$ acts locally finitely on M, that is* $\dim Z(U)m < \infty$ *for all $m \in M$. More generally any $z \in Z(U_q^{\mathcal{O}}(\mathfrak{g}))$ acts locally finitely on M, that is $k(q)[z]m$ is finite dimensional.*

Remark. If C is not of finite type, $Z(U)$ is reduced to scalars by 7.1.15 (iii) so the first part is of little interest in that case. On the other hand if C is of finite type $Z(U)$ is already adequately large as shown in the next result.

8.3.4 Return to the assumption that C is of finite type. By 8.3.3 each $M \in Ob\mathcal{O}$ may be written as a direct sum of its $Z(U)$ primary components. Given $J \in \operatorname{Max} Z(U)$, let M^J be the primary component of M corresponding to J.

Lemma. *For all $M \in Ob\mathcal{O}$, $J \in \operatorname{Max} Z(U)$, M^J has finite length.*

One may pass to the algebraic closure $\overline{k(q)}$ since this can at most increase length and of course one may assume $M = M^J$. Recall 4.1.5 and suppose $[M : V(\Lambda)] \neq 0$. Since $V(\Lambda)$ can be viewed as a subquotient of M, it follows that $JV(\Lambda) = 0$. Since $|\check{W}| < \infty$ it follows from 8.3.2 (i) that there can be only finitely many simples with this property. Then $[M]$ is a finite sum of simples in the Grothendieck group and hence M has finite length.

8.3.5 It follows from 8.3.4 that any Verma module $M(\Lambda)$ has finite length, in particular $\operatorname{Soc} M(\Lambda) \neq 0$.

Lemma. *For each $\Lambda \in \check{T}^\wedge$, there exists $\Lambda' \in \check{T}^\wedge$ with $\Lambda' \leq \Lambda$ such that* $\operatorname{Soc} M(\Lambda) = M(\Lambda')$.

Any simple submodule of a Verma module $M(\Lambda)$ is a highest weight module $L(\Lambda')$ and since $M(\Lambda)$ is a free U^- module so is $L(\Lambda')$ which in turn implies that $L(\Lambda')$ is also a Verma module. Finally $\operatorname{Soc} M(\Lambda)$ is simple by A.3.7.

8.3.6 There is an important connection between spaces of Verma module maps and so-called principal series modules.

Take $M \in Ob\mathcal{O}$. It is convenient to consider M^* as a left U module via the antiautomorphism $\varkappa'$ defined in 7.2.1 and to take $\delta' M$ to be the largest submodule of M^* on which T acts locally finitely. Since $\varkappa, \varkappa'$ coincide on T one has $\delta M = \delta' M$. However $\varkappa'$ is also an automorphism of coalgebras whilst $\varkappa$ is not. Notice one may also define a projection $\mathcal{P}'$ (which actually coincides with $\mathcal{P}$) of U onto U^0 with respect to $U = U^0 \oplus (U_+^- U + U G_+^+)$, where G^+ is the subalgebra generated by the $y_i : i = 1, 2, \dots \ell$ and a Shapovalev form by

$\mathcal{S}'(a,b) = \mathcal{P}'(\varkappa'(a)b)$. This is contravariant with respect to $\varkappa'$, so in particular for all $y, y' \in U_{-\nu}$ one has $\mathcal{S}'(y,y') = \tau(\nu)q^{-\frac{1}{2}(\nu,\nu)-(\rho,\nu)}\mathcal{S}(y,y')$. This means that $\mathcal{S}$ and $\mathcal{S}'$ are interchangeable except for studying the limit $q \longrightarrow 0$.

Let M, N be left U modules. Given $\theta \in \operatorname{Hom}(M,N)$ define $\delta'\theta \in \operatorname{Hom}(N^*, M^*)$ by transport of structure, that is $(\delta'\theta)\xi(m) = \xi(\theta m)$ for all $m \in M$, $\xi \in N^*$. With M^*, N^* viewed as left U modules via δ' as above one obtains $\delta'(a.\theta.b) = \varkappa'(b).\delta'\theta.\varkappa'(a)$ for all $a, b \in U$. Define a $U \otimes U$ module structure on $\operatorname{Hom}(M,N)$ by $(a \otimes b).\theta = a.\theta.\varkappa'(b)$ and let $\tau \in \operatorname{Aut} U \otimes U$ be the transposition $a \otimes b \mapsto b \otimes a$. Then δ' is a $U \otimes U$ module monomorphism of $\operatorname{Hom}(M,N)$ into $(\operatorname{Hom}(N^*, M^*))^\tau$. Take $M,\ N \in Ob\mathcal{O}$ and $\theta \in F(M,N)$. Then $(\delta'\theta)(\delta'N) \subset \delta'M$ and furthermore one may show that δ' restricts to an $F(U) \otimes F(U)$ module isomorphism of $F(M,N)$ onto $F(\delta'N, \delta'M)^\tau$. For the last part note that $\iota := \varkappa'\sigma = \sigma^{-1}\varkappa'$ and is the involution defined in 3.2.6 of U. It is an automorphism of algebras and an antiautomorphism of coalgebras. Then $\delta'((\operatorname{ad} a)\theta) = \delta'(a_1\theta\sigma(a_2)) = \varkappa'\sigma(a_2)\delta'\theta\varkappa'(a_1) = \iota(a)_1\delta'\theta\sigma(\iota(a)_2)$, that is

$$(*) \qquad \delta'((\operatorname{ad} \iota(a))\theta) = (\operatorname{ad} a)\delta'\theta$$

from which the assertion results.

Now consider $(M \otimes N)^*$ as a left $U \otimes U$ module via $\varkappa' \times \varkappa'$. Then Frobenius reciprocity (A.2.15) gives a $U \otimes U$ module isomorphism T : $\operatorname{Hom}(N, M^*) \xrightarrow{\sim} (M \otimes N)^*$ by $T_\theta(m \otimes n) = \theta(n)(m)$. In particular $T_{(\operatorname{ad} a)\theta} = T_{a_1\theta\sigma(a_2)} = \Delta^\iota(a)T_\theta$ where Δ^ι is the algebra homomorphism $a \mapsto (1 \otimes \iota)\Delta(a)$ of U into $U \otimes U$. Suppose now that $M, N \in Ob\mathcal{O}$. Viewing $(M \otimes N)^*$ as a U module with respect to Δ^ι it follows that T restricts to an $F(U) \otimes F(U)$ module isomorphism of $F(N, \delta'M)$ onto $F(M \otimes N)^*$. When $M = M(\Lambda)$, $N = M(\Lambda')$ one writes $F(M \otimes N)^*$ simply as $F(\Lambda, \Lambda')$. It is called the principal series modules with respect to (Λ, Λ'). The above conclusions may be summarized in the

Lemma. *For all $M, N \in Ob\mathcal{O}$ there exist $F(U) \otimes F(U)$ module isomorphisms*

(i) $F(M,N) \xrightarrow{\sim} F(\delta N, \delta M)^\tau$,
(ii) $F(N, \delta M) \xrightarrow{\sim} F(M \otimes N)^*$.

8.3.7 Retain the notation of 8.3.6 and write $V^+ = U_q(\mathfrak{b}^+)$. By triangular decomposition of $U \otimes U$ it follows that $(V^+ \otimes V^+) \cap \Delta^\iota(U) = \Delta^\iota(U^0)$ and that the multiplication map induces a linear isomorphism of $(V^+ \otimes V^+) \otimes_{\Delta^\iota(U^0)} \Delta^\iota(U)$ onto $U \otimes U$. This and Frobenius reciprocity gives isomorphisms

$$\begin{aligned}(M(\Lambda) \otimes M(\Lambda'))^* &\xrightarrow{\sim} \operatorname{Hom}_{V^+ \otimes V^+}(U \otimes U, \operatorname{Hom}(\Lambda \otimes \Lambda', k(q))) \\ &\xrightarrow{\sim} \operatorname{Hom}_{\Delta^\iota(U^0)}(\Delta^\iota(U), \operatorname{Hom}(\Lambda \otimes \Lambda', k(q))) \\ &\xrightarrow{\sim} (\Delta^\iota(U) \otimes_{\Delta^\iota(U^0)} (\Lambda \otimes \Lambda'))^*\end{aligned}$$

of $U \otimes U$ and $\Delta^\iota(U)$ modules respectively. In this $\Lambda \otimes \Lambda'$ is just $\Lambda\Lambda'^{-1}$ as a U^0 module and by Frobenius reciprocity

$$\mathrm{Hom}_U((U \otimes_{U^0} \Lambda\Lambda'^{-1}), V) \xrightarrow{\sim} \mathrm{Hom}_{U^0}(\Lambda\Lambda'^{-1}, V) .$$

Combined the above gives the

Proposition. *For all $\Lambda, \Lambda' \in T^\wedge$, $\mu \in P(\pi)$, $\nu \in P^+(\pi)$*

(i) $[F(M(\Lambda) \otimes M(\Lambda'))^* : V(\nu)] = 0$ *unless* $\Lambda' \in \Lambda q^{P(\pi)}$,
(ii) $[F(M(\Lambda) \otimes M(\Lambda q^\mu))^* : V(\nu)] = \dim V(\nu)_\mu$.

8.3.8 Combining 8.3.6 (ii) and 8.3.7 gives the

Corollary. *Suppose $M(\Lambda')$ is simple. Then for all $\nu \in P^+(\pi)$ one has* $[F(M(\Lambda')) : V(\nu)] = \dim V(\nu)_0$.

Remark. By 8.2.4 (ii) this also holds if $M(\Lambda')$ is projective. The next result shows that it holds in general.

8.3.9 Take Λ as above. Let $\varphi_\Lambda : F(\check{U}) \longrightarrow F(M(\Lambda))$ denote the map defined by the action of $\check{U}$ on $M(\Lambda)$.

Theorem

(i) $\mathrm{Ann}_{F(\check{U})} M(\Lambda) = F(\check{U})\, \mathrm{Ann}_{Z(\check{U})} M(\Lambda) = \ker \varphi_\Lambda$,
(ii) φ_Λ *is surjective and factors to an* $\mathrm{ad}\, U$ *module isomorphism of* $\check{\mathbb{H}}$ *onto* $F(M(\Lambda))$.

Let $M(\Lambda') \hookrightarrow M(\Lambda)$ be as in the conclusion of 8.3.5. Then $\mathrm{Ann}_{\check{\mathbb{H}}} M(\Lambda) \subset \mathrm{Ann}_{\check{\mathbb{H}}} M(\Lambda') = 0$ by 8.3.1. Yet $\check{\mathbb{H}} + F(\check{U})\, \mathrm{Ann}_{Z(\check{U})} M(\Lambda) = F(\check{U})$ and so this gives (i).

For (ii) observe that $M(\Lambda)/M(\Lambda')$ is a proper quotient of $M(\Lambda)$ and so $d(M(\Lambda)/M(\Lambda')) < d(M(\Lambda)) = d(M(\Lambda'))$ by A.3.7. Then by 8.2.5 (ii) and 8.3.5 one obtains $F(M(\Lambda)/M(\Lambda'), M(\Lambda)) = 0$. Consequently the natural map $F(M(\Lambda), M(\Lambda)) \longrightarrow F(M(\Lambda'), M(\Lambda))$ is injective. Similarly 8.2.5 (i) gives $F(M(\Lambda'), M(\Lambda)/M(\Lambda')) = 0$ and hence an isomorphism $F(M(\Lambda'), M(\Lambda)) \xrightarrow{\sim} F(M(\Lambda'), M(\Lambda'))$. Combined with (i) and 7.3.8 (ii) this gives $\mathrm{ad}\, U$ module embeddings

$$\check{\mathbb{H}} \hookrightarrow F(M(\Lambda)) \hookrightarrow F(M(\Lambda'))$$

which are isomorphisms by 8.1.5 and 8.3.8.

8.3.10 Both 7.3.8 and 8.3.9 fail for $F(U)$. Indeed take $\mathfrak{g} = \mathfrak{sl}(3)$. Set $y_i = y_{\omega_i} : i = 1, 2$. Then $G(\omega_i)$ is a direct sum of $k(q)y_i$ and a simple module with highest weight $\alpha_1 + \alpha_2$. Let $a_i \in G(\omega_i)$ be its unique up to scalars highest weight vector. Since $2\omega_i \notin Q(\pi)$ it follows from 7.1.4 that $a_i \notin gr_{\mathcal{F}} F(U)$. Yet by 7.1.7 (iv) and 7.1.4, $a_1 y_2 \in G(\omega_1)G(\omega_2) = G(\omega_1 + \omega_2) \subset gr_{\mathcal{F}} F(U)$. Similarly $a_1 y_1^2 \in gr_{\mathcal{F}} F(U)$. The relation $(a_1 y_2) y_1^3 = (a_1 y_1^2) y_1 y_2$ shows that $gr_{\mathcal{F}} F(U)$ is not free over $Y(U)$.

Now set $z_i = z_{\omega_i}$ and let $b_i \in F(\omega_i)$ be a vector of weight $\alpha_1 + \alpha_2$. Then $b_1 z_2 \in F(\omega_1)F(\omega_2) \subset F(\omega_1+\omega_2)+F(0) \subset F(U)$. Similarly $b_2 z_1$, $b_1 z_1^2$, $b_2 z_2^2 \in F(U)$. Moreover since $F(\check{U})_{V(\alpha_1+\alpha_2)}$ is generated over $Z(\check{U}) = k(q)[z_1, z_2]$ by b_1, b_2 it follows that $F(U)_{V(\alpha_1+\alpha_2)}$ is generated over $Z(U) = k(q)[z_1^3, z_2^3, z_1 z_2]$ by $b_1 z_2$, $b_2 z_1$, $b_1 z_1^2$, $b_2 z_2^2$.

Now over $\overline{k(q)}$ it follows from 8.3.2 (ii) that there exists a Verma module $M(\Lambda)$ such that $z_1, z_2 \in \operatorname{Ann}_{Z(\check{U})} M(\Lambda)$. (An explicit calculation shows that $\Lambda = \gamma q^{-\rho}$ where γ is a primitive cube root of unity). Then $F(U)_{V(\alpha_1+\alpha_2)} \subset \operatorname{Ann}_{F(U)} M(\Lambda)$ whilst $b_1 z_2 \notin F(U) \operatorname{Ann}_{Z(U)} M(\Lambda)$. Moreover $[F(U)/\operatorname{Ann}_{F(U)} M(\Lambda) : V(\alpha_1 + \alpha_2)] = 0$ and so $F(U)/\operatorname{Ann}_{F(U)} M(\Lambda)$ is a proper submodule of $F(M(\Lambda))$.

8.3.11 It is clear from 7.1.13 that any weight module (i.e. satisfying 4.1.4 (i)) which is simple as a $\check{U}$ module, is simple both as a $F(\check{U})$ module and as an $F(U)$ module. Then 8.3.5 implies that $\operatorname{Ann}_{F(\check{U})} M(\Lambda)$ and $\operatorname{Ann}_{F(U)} M(\Lambda)$ are primitive ideals.

Lemma. *For all $\Lambda \in T^\wedge$ the quotient algebra $F(U)/\operatorname{Ann}_{F(U)} M(\Lambda)$ is a domain.*

Consider the adjoint action of V^- on $F(U)/\operatorname{Ann}_{F(U)} M(\Lambda)$. By the argument of 7.3.4 it suffices to show that $(F(U)/\operatorname{Ann}_{F(U)} M(\Lambda))^{U^-}$ is a domain. Through the action of $F(U)$ on $M(\Lambda)$ the latter embeds in $(\operatorname{End} M(\Lambda))^{U^-}$.

Since $M(\Lambda)$ is freely generated by U^- over u_Λ there exists for each $a \in (\operatorname{End} M(\Lambda))^{U^-}$ a unique element $\gamma(a) \in U^-$ such that $\gamma(a)u_\Lambda = au_\Lambda$. Obviously γ is a linear map of $(\operatorname{End} M(\Lambda))^{U^-}$ into U^-.

It is clear that $\ker\gamma$ is $\operatorname{ad} T$ stable. Consider a weight vector $a_\mu \in (\operatorname{End} M(\Lambda))^{U^-}$ of weight μ. One has

$$0 = (\operatorname{ad} f_i)a_\mu = (f_i a_\mu - t_i a_\mu t_i^{-1} f_i) = f_i a_\mu - q^{(\alpha_i,\mu)} a_\mu f_i \ .$$

Hence $f_\nu a_\mu = q^{(\nu,\mu)} a_\mu f_\nu$ for every weight vector $f_\nu \in U^-$ of weight ν. It follows in particular that γ is injective. Moreover $\gamma(a_\mu a_\nu)u_\Lambda = a_\mu a_\nu u_\Lambda = a_\mu \gamma(a_\nu)u_\Lambda = q^{-(\mu,\nu)}\gamma(a_\nu)a_\mu u_\Lambda = q^{-(\mu,\nu)}\gamma(a_\nu)\gamma(a_\mu)u_\Lambda$ and so $\gamma(a_\mu a_\nu) = q^{-(\mu,\nu)}\gamma(a_\nu)\gamma(a_\mu)$.

To show that $(\operatorname{End} M(\Lambda))^{U^-}$ is a domain, it suffices as in 7.3.4 to show that for any non-zero weight vectors a_μ, a_ν one has $a_\mu a_\nu \neq 0$. If not $\gamma(a_\nu)\gamma(a_\mu) = 0$. Since U^- is a domain by 7.3.4 and γ is injective this is a contradiction proving the assertion and hence the lemma.

Remark. Similarly $F(\check{U})/\operatorname{Ann}_{F(\check{U})} M(\Lambda)$ is a domain.

8.4 Equivalence of Categories

It is assumed throughout this section that the Cartan matrix is of finite type. Let $\mathcal{V}$ denote the category of finite dimensional modules with weights of the form $q^\nu : \nu \in Q(\pi)$.

8.4.1 Recall 8.2.3. An $F(\check{U})$ bimodule H with a compatible $\operatorname{ad}\check{U}$ action is said to belong to the Harish-Chandra category $\mathcal{H}$ if

(i) *As an* $\operatorname{ad}\check{U}$ *module* H *is a direct sum of simple finite dimensional modules in* $\mathcal{V}$ *each simple occurring with finite multiplicity.*

(ii) *As both a left and a right* $F(\check{U})$ *module,* $\operatorname{Ann}_{Z(\check{U})} H$ *has finite codimension in* $Z(\check{U})$.

Recall 8.3.2. For each $\Lambda \in \check{T}^\wedge$ let $\mathcal{H}_{\chi(\Lambda)}$ denote the full subcategory of $\mathcal{H}$ of all right $F(\check{U})$ modules whose annihilators contain $\chi(\Lambda)$. Recalling 8.3.3 and 8.3.4, let $\mathcal{O}_f$ denote the full subcategory of $\mathcal{O}$ of all $N \in Ob\mathcal{O}$ for which $\operatorname{Ann}_{Z(\check{U})} N$ has finite codimension. Set $\mathcal{O}_f^\Lambda = \mathcal{O}_f \cap \mathcal{O}^\Lambda$. By 8.3.4 the objects of $\mathcal{O}_f$ have finite length. Observe that a weight of the form $\Lambda q^\mu : \mu \in Q(\pi)$ is completely determined by its restriction to $-2P^+(\pi)$. Indeed only μ has to be determined and this is possible because $-2P^+(\pi)$ generates $\mathfrak{h}^*_{\mathbb{Q}}$ over $\mathbb{Q}$.

Lemma. *For all* $N \in Ob\mathcal{O}_f^\Lambda$ *one has* $F(M(\Lambda), N) \in Ob\mathcal{H}_{\chi(\Lambda)}$.

As a left (resp. right) $F(\check{U})$ module the annihilator of $F(M(\Lambda), N)$ contains $\operatorname{Ann}_{Z(\check{U})} N$ (resp. $\chi(\Lambda)$) so (ii) holds. Since N has finite length it is enough to check (i) for N simple, say $N = V(\Lambda')$ with $\Lambda' \in \Lambda q^{Q(\pi)}$. Then 8.3.6 (ii) gives maps

$$F(M(\Lambda), V(\Lambda')) \hookrightarrow F(M(\Lambda), \delta M(\Lambda')) \xrightarrow{\sim} F(M(\Lambda') \otimes M(\Lambda))^* .$$

Then the required assertion follows from 8.3.7 (ii).

8.4.2 Take $H \in Ob\mathcal{H}$ and let H' be an $F(\check{U})$ bisubmodule generated by a finite dimensional subspace V'. One may write $H' = F(\check{U})V'F(\check{U})$. By 8.4.1 (i), V' is contained in a finite dimensional $\operatorname{ad}\check{U}$ stable subspace V of H and then 8.2.3 gives $H'' := F(\check{U})V = VF(\check{U}) \in Ob\mathcal{H}$. Recall (7.4.8) that $F(\check{U})$ is noetherian and let $\mathcal{H}^f$ denote the full subcategory of $\mathcal{H}$ of finitely generated left (or right) $F(\check{U})$ modules. For each $\Lambda \in \check{T}^\wedge$, set $\mathcal{H}^f_{\chi(\Lambda)} = \mathcal{H}^f \cap \mathcal{H}_{\chi(\Lambda)}$.

Now let H be an $F(\check{U})$ bimodule with a compatible $\operatorname{ad}\check{U}$ action. For example $F(\check{U})$ itself. Given $V \in \mathcal{V}$ one may define $V \otimes H$ to have the obvious right $F(\check{U})$ module structure and to be a left $F(\check{U})$ module through $a(v \otimes h) = (\operatorname{ad} a_1)v \otimes a_2 h$ for all $a \in F(\check{U})$ recalling here that $F(\check{U})$ is a left coideal (7.1.6). Finally define $V \otimes H$ to be an $\operatorname{ad}\check{U}$ module through the coproduct. Then for all $a \in F(\check{U})$

$$\begin{aligned}((\operatorname{ad} a_1)(v\otimes h))a_2 &= (\operatorname{ad} a_1)v \otimes ((\operatorname{ad} a_2)h)a_3 \\ &= (\operatorname{ad} a_1)v \otimes a_2 h\ , \quad \text{by compatibility in } H \\ &= a(v\otimes h)\ .\end{aligned}$$

So $V\otimes H$ has a compatible $\operatorname{ad}\check{U}$ action. Observe that if $H \in Ob\mathcal{H}$ (resp. $\mathcal{H}_{\chi(\Lambda)}$, $\mathcal{H}^f$, $\mathcal{H}^f_{\chi(\Lambda)}$) then $V\otimes H \in Ob\mathcal{H}$ (resp. $\mathcal{H}_{\chi(\Lambda)}$, H^f, $H^f_{\chi(\Lambda)}$).

Now suppose $H \in Ob\mathcal{H}$ is finitely generated as a right $F(\check{U})$ module by some $V \in \mathcal{V}$. Compatibility in H gives $(\operatorname{ad} a_1)va_2 f = avf$ for all $v \in V$, $f \in F(\check{U})$ and so by universality one may view H as an image of $V\otimes F(\check{U})$. Finally if $H \in Ob\mathcal{H}_{\chi(\Lambda)}$ then by 8.3.9 one may view H as an image of $V\otimes F(M(\Lambda))$. These conclusions may be summarized as follows.

Lemma. *Take $\Lambda \in \check{T}^\wedge$.*

(i) *Every $H \in Ob\mathcal{H}$ (resp. $\mathcal{H}_{\chi(\Lambda)}$) is a union of objects in $\mathcal{H}^f$ (resp. $\mathcal{H}^f_{\chi(\Lambda)}$).*

(ii) *For each $V \in \mathcal{V}$ there is an exact functor $V\otimes -$ on $\mathcal{H}$ (resp. on $\mathcal{H}_{\chi(\Lambda)}$) defined by $H \mapsto V\otimes H$ which restricts to $\mathcal{H}^f$ (resp. $\mathcal{H}^f_{\chi(\Lambda)}$).*

(iii) *Every $H \in Ob\mathcal{H}^f_{\chi(\Lambda)}$ is an image of $V\otimes F(M(\Lambda))$ for some $V \in \mathcal{V}$.*

8.4.3 From now on fix Λ to be dominant (8.2.2). Then $M(\Lambda)$ is projective in $\mathcal{O}^\Lambda_f$ and so the functor $T : N \longrightarrow F(M(\Lambda), N)$ from $\mathcal{O}^\Lambda_f$ to $\mathcal{H}_{\chi(\Lambda)}$ is exact.

Lemma. *For all $H \in Ob\mathcal{H}^f_{\chi(\Lambda)}$ one has $H\otimes_{F(\check{U})} M(\Lambda) \in Ob\mathcal{O}^\Lambda_f$.*

By (i) of 8.4.1, H is a direct sum of its $\operatorname{ad}\check{T}$ weight spaces and its weights take the form $q^\mu : \mu \in Q(\pi)$. Given $\lambda \in -2P^+(\pi)$ one has $\tau(\lambda) \in F(\check{U})$ and then by compatibility $\tau(\lambda)h_\mu \otimes m = (\operatorname{ad}\tau(\lambda)h_\mu)\tau(\lambda)\otimes m = q^{(\mu,\lambda)}h_\mu \otimes \tau(\lambda)m$. Thus the weights of $H\otimes_{F(\check{U})} M(\Lambda)$ lie in $\Lambda q^{Q(\pi)}$ and so uniquely extend to $\tau(P(\pi))$ weights in $\Lambda q^{Q(\pi)}$. Again compatibility gives $H\otimes_{F(\check{U})} M(\Lambda)$ a $F(\check{U})$ module structure and hence a $\tau(P(\pi))F(\check{U})$ module structure. Applying 7.1.13 this gives $H\otimes_{F(\check{U})} M(\Lambda)$ a $\check{U}$ module structure and the resulting module satisfies 4.1.4 (i). Finally by 8.4.2 (ii) there exists $V \in \mathcal{V}$ such that $H\otimes_{F(\check{U})} M(\Lambda) \twoheadleftarrow V\otimes F(M(\Lambda))\otimes_{F(\check{U})} M(\Lambda) \cong V\otimes M(\Lambda)$ which by 8.1.7 belongs to $\mathcal{O}^\Lambda_f$.

8.4.4 Let $T' : H \mapsto H\otimes_{F(\check{U})} M(\Lambda)$ denote the functor from $\mathcal{H}^f_{\chi(\Lambda)}$ to $\mathcal{O}^\Lambda_f$ defined by the conclusion of 8.4.3. For all $H \in Ob\mathcal{H}_{\chi(\Lambda)}$, $N \in Ob\mathcal{O}^\Lambda$ Frobenius reciprocity gives an isomorphism

$$(*)\quad \operatorname{Hom}_{F(\check{U})-F(\check{U})}(H, \operatorname{Hom}(M(\Lambda), N)) \overset{F}{\underset{\sim}{\longrightarrow}} \operatorname{Hom}_{F(\check{U})}(H\otimes_{F(\check{U})} M(\Lambda), N)$$

via $F_\psi(h\otimes m) = \psi(h)(m)$. In particular if H is finitely generated one obtains an isomorphism $\operatorname{Hom}_{F(\check{U})-F(\check{U})}(H, TT'H) \xrightarrow{\sim} \operatorname{Hom}_{F(\check{U})}(T'H, T'H)$. Let θ_H be an inverse image of the identity on $T'H$ under the above isomorphism.

Take $V \in \mathcal{V}$ and recall 8.4.2 (i). It is obvious $V \otimes -$ commutes with $\mathcal{T}'$. From the canonical isomorphism $V \otimes F(M(\Lambda), N) \xrightarrow{\sim} F(M(\Lambda), V \otimes N)$ it also commutes with $\mathcal{T}$.

Set $\check{T}_< = \tau(-2P^+(\pi))$ which is obviously an *Ore* subset of $F(\check{U})$. Given $H \in Ob\mathcal{H}_{\chi(\Lambda)}$ let H^0 denote its right $\check{T}_<$ torsion submodule. Obviously H^0 is an $F(\check{U})$ bisubmodule of H. By compatibility it follows that H^0 is $\operatorname{ad}\tau(\lambda) : \lambda \in \tau(-2P^+(\pi))$ stable, hence a direct sum of its weight subspaces (in particular H^0 is $\operatorname{ad}\check{T}$ stable). Now let $h_\nu \in H^0$ be a weight vector and suppose $h_\nu \tau(\mu) = 0$ for some $\mu \in -2P^+(\pi)$. As shown in 7.1.13 given $\lambda \in -2P^+(\pi)$ such that $(\lambda, \alpha_i) \neq 0$ one has $f_i\tau(\lambda) \in F(\check{U})$. Then compatibility implies that

$$(**) \qquad [(\operatorname{ad} f_i)(\operatorname{ad}\tau(\lambda))h_\nu]\tau(\lambda) + [(\operatorname{ad}\tau(\lambda)t_i)h_\nu]f_i\tau(\lambda) = f_i\tau(\lambda)h_\nu \, .$$

Hence $[(\operatorname{ad} f_i)h_\nu]\tau(\lambda+\mu) = 0$. Similarly one shows that $[(\operatorname{ad} e_i)h_\nu]\tau(\lambda+\mu) = 0$. Thus H^0 is also $\operatorname{ad}\check{U}$ stable. Again H^0 is also the left $\check{T}_<$ torsion submodule of H^0 by compatibility.

Proposition. *For all $H \in \mathcal{H}^f_{\chi(\Lambda)}$*

(i) *θ_H is an isomorphism of $F(\check{U})$ bimodules.*
(ii) $H^0 = 0$.

Take $H = F(M(\Lambda))$ which by 8.3.9 is isomorphic to $F(U)/\operatorname{Ann} M(\Lambda)$. Hence $\mathcal{T}'H = M(\Lambda)$ and so $\mathcal{T}\mathcal{T}'H = H = F(M(\Lambda))$. This forces θ_H to be an isomorphism. Take $V \in \mathcal{V}$. Since $V \otimes -$ commutes with $\mathcal{T}$ and $\mathcal{T}'$ it follows that θ_H is also an isomorphism when $H = V \otimes F(M(\Lambda))$.

Now take $H \in \mathcal{H}^f_{\chi(\Lambda)}$ arbitrary. By 8.2.2 (i) there exists $V_1 \in \mathcal{V}$ such that H is an image of $V_1 \otimes F(M(\Lambda))$. Yet $\ker(V_1 \otimes F(M(\Lambda)) \longrightarrow H) \in \mathcal{H}^f_{\chi(\Lambda)}$ by 7.4.8 and so there exists $V_2 \in \mathcal{V}$ giving an exact sequence

$$V_2 \otimes F(M(\Lambda)) \longrightarrow V_1 \otimes F(M(\Lambda)) \longrightarrow H \longrightarrow 0 \, .$$

Set $\theta_i = \theta_{V_i \otimes F(M(\Lambda))}$. Since $\mathcal{T}'$ (resp. $\mathcal{T}$) is right exact (resp. exact) and commutes with $V_i \otimes -$ one obtains the commuting diagram

$$\begin{array}{ccccc}
V_2 \otimes F(M(\Lambda)) & \longrightarrow & V_1 \otimes F(M(\Lambda)) & \longrightarrow & H \longrightarrow 0 \\
\theta_2 \wr \downarrow & & \theta_1 \wr \downarrow & & \theta_H \downarrow \\
V_2 \otimes F(M(\Lambda)) & \longrightarrow & V_1 \otimes F(M(\Lambda)) & \longrightarrow & \mathcal{T}\mathcal{T}'H \longrightarrow 0
\end{array}$$

This gives (i). Moreover $\mathcal{T}\mathcal{T}'H$ being a submodule of the $\check{U}$ bimodule $\operatorname{Hom}(M(\Lambda), \mathcal{T}'H)$ has no right $\check{T}_<$ torsion and so $H^0 = 0$.

8.4.5 Let H be an $F(\check{U})$ bimodule with a compatible $\mathrm{ad}\,\check{U}$ action. Suppose H is a direct sum of weight spaces $H_\nu : \nu \in Q(\pi)$. Then the weight space decomposition of H is determined by the adjoint action of just $\check{T}_<$. Suppose further that H has no right $\check{T}_<$ torsion. By compatibility $[(\mathrm{ad}\,\tau(\lambda))h]\tau(\lambda) = \tau(\lambda)h$ for all $h \in H$, $\lambda \in -2P^+(\pi)$ and so $\mathrm{ad}\,\tau(\lambda)h$ is determined by the $\check{T}_<$ bimodule structure of H. Again from 8.4.4 (**) it follows that $(\mathrm{ad}\,f_i)h_\nu$ is determined by the $F(\check{U})$ bimodule structure of H. Similarly so is $(\mathrm{ad}\,f_i)h_\nu$. Consequently any $F(\check{U})$ bimodule isomorphism between such modules extends uniquely to an $\mathrm{ad}\,\check{U}$ module isomorphism.

As already noted in 8.4.3 any $F(\check{U})$ module isomorphism between objects in $\mathcal{O}^\Lambda$ extends uniquely to a $\check{U}$ module isomorphism of objects in $\mathcal{O}^\Lambda$ via 7.1.13.

Combining the above two observations with 8.4.4 (ii) one concludes from (*) of 8.4.4 the

Corollary. *For all $H \in \mathcal{H}^f_{\chi(\Lambda)}$, $N \in Ob\mathcal{O}^f_\Lambda$ Frobenius reciprocity gives an isomorphism*

$$\mathcal{H}om(H, TN') \xrightarrow[\sim]{\alpha} \mathcal{H}om(T'H, N)$$

where $\mathcal{H}om$ means homomorphism in the category.

8.4.6 Let $\mathcal{S}_{\chi(\Lambda)}$ denote the set of simple objects in $\mathcal{H}_{\chi(\Lambda)}$. By 8.4.2 (i), $\mathcal{S}_{\chi(\Lambda)} \subset \mathcal{H}^f_{\chi(\Lambda)}$.

Lemma. *If $V \in Ob\mathcal{O}^\Lambda$ is simple then $TV = 0$ or $TV \in Ob\mathcal{S}_{\chi(\Lambda)}$. Moreover every $H \in Ob\mathcal{S}_{\chi(\Lambda)}$ is so obtained.*

Suppose $TV \neq 0$ and not simple. By 8.4.2 (i), TV admits a proper submodule $H \in Ob\mathcal{H}^f_{\chi(\Lambda)}$. Let $i : H \hookrightarrow TV$ denote the embedding that ensues. By 8.4.5, $\alpha(i)$ is a non-zero homomorphism from $T'H$ to V, hence surjective. Since T is exact $T(\alpha(i))$ is a surjection of $TT'H$ onto TV. By 8.4.4 the map $\theta_H : h \mapsto (m \to (h \otimes m))$ of H into $TT'H$ is an isomorphism. The composed map $H \longrightarrow TT'H \twoheadrightarrow TV$ is given by $h \mapsto (m \mapsto i(h)m)$ which is just the original embedding i. This contradiction proves that TV is simple.

Conversely suppose $H \in Ob\mathcal{S}_{\chi(\Lambda)}$. Then $H \xrightarrow{\sim} TT'H$ by 8.4.4 so in particular $T'H \in Ob\mathcal{O}^\Lambda$ and is non-zero. By finiteness of length in $\mathcal{O}^\Lambda$ there exists a simple quotient V of $T'H$. By 8.4.5, $\mathcal{H}om(H, TV) \neq 0$ and so $H \xrightarrow{\sim} TV$ by the first part.

8.4.7 By 4.1.4 every simple quotient in $Ob\mathcal{O}^\Lambda$ is some $V(\Lambda q^\mu) : \mu \in Q(\pi)$ and then by 8.3.2 (i) there are at most $|\check{W}|$ simples with the same action of the centre. Then by 8.4.6 the same holds true in $\mathcal{H}_{\chi(\Lambda)}$ and hence in $\mathcal{H}$. By the finite multiplicity hypothesis of 8.4.1 (i) only finitely many simples of a given type can occur in any $H \in Ob\mathcal{H}$. Taking account of primary decomposition with respect to the centre (8.4.2 (i)) one obtains the

Corollary. *Every $H \in Ob\mathcal{H}$ has finite length. In particular $\mathcal{H} = \mathcal{H}^f$.*

8.4.8 Recall 8.3.2. Define a translated action of the extended Weyl group $\check{W}$ on $\check{T}^\wedge$ by $\check{w}.\Lambda = \check{w}(\Lambda q^\rho)q^{-\rho}$ for all $\check{w} \in \check{W}$. Recall (8.2.2) the order relation $\leq$ on $\check{T}^\wedge$. Call Λ strongly dominant if Λ is a maximal element of $\check{W}.\Lambda$. Recall 8.2.2 and A.1.16.

Lemma. *A strongly dominant character Λ is dominant. The converse is true if Λ^2 is torsion-free.*

By definition Λ is not dominant if $(\Lambda q^\rho)^2(\tau(\alpha)) = q_\alpha^{-2m}$ for some $\alpha \in \Delta^+$, $m \in \mathbb{N}^+$. This by A.1.15 the above equality becomes $s_\alpha.\Lambda^2 = (\Lambda q^{m\alpha})^2$ and so Λ is not maximal in its $\check{W}$. orbit. The converse obtains from A.1.16 (ii).

Remark. Using primary decomposition with respect to $Z(\check{U})$ it follows from 8.3.2 (i) that $M(\Lambda)$ is projective in $\mathcal{O}$ when Λ is strongly dominant. The example in A.1.16 shows that dominant need not imply strongly dominant and so by 8.2.2 not all the projective Verma modules are recovered in the above fashion.

8.4.9 For all $\Lambda \in \check{T}^\wedge$ set $\check{W}_\Lambda = \{\check{w} \in \check{W} \mid \check{w}\Lambda \in \Lambda q^{Q(\pi)}\}$. Call Λ regular if $\mathrm{Stab}_{\check{W}_\lambda}(\Lambda q^\rho) = \{e\}$.

Lemma. *Take $\mu \in P^+(\pi)$ and suppose $\Lambda \in \check{T}^\wedge$ is strongly dominant and regular. Then $\check{W}.(\Lambda q^\mu) \cap \Lambda q^{\Omega(V(\mu))} = \Lambda q^\mu$.*

Take $\check{w} \in \check{W}$. If $\check{w}.(\Lambda q^\mu) \cap \Lambda q^{\Omega(V(\mu))} \neq 0$, then $\check{w} \in \check{W}_\Lambda$. In this case $\check{w}^{-1}.\Lambda = \Lambda q^{-\gamma}$ for some $\gamma \in Q^+(\pi)$ since Λ is strongly dominant. Let w denote the image of $\check{w}$ under the projection $\mathbb{Z}_2^\ell \rtimes W \longrightarrow W$. Then $\check{w}.(\Lambda q^\mu) = (\check{w}.\Lambda)q^{w\mu} = \Lambda q^{w(\mu+\gamma)}$. Yet $(w(\mu+\gamma), w(\mu+\gamma)) = (\mu,\mu)+2(\mu,\gamma)+(\gamma,\gamma) \geq 0$ with a strict inequality if $\gamma \neq 0$. Yet $(\nu,\nu) \leq (\mu,\mu)$ for all $\nu \in \Omega(V(\mu))$ by 4.4.3. Hence $\check{w}^{-1}.\Lambda = \Lambda$ and so $\check{w} = e$ by the regularity of Λ, as required.

8.4.10 The following result is the basis of the so-called translation principle.

Corollary. *Take $\mu \in P^+(\pi)$ and assume $\Lambda \in \check{T}^\wedge$ strongly dominant and regular. Then for all $\check{w} \in \check{W}$ the Verma module $M(\check{w}.(\Lambda q^\mu))$ is a direct summand of $V(\mu) \otimes M(\check{w}.\Lambda)$.*

Recall 8.1.7 and take the $Z(\check{U})$ primary component of $M := V(\mu) \otimes M(\check{w}.\Lambda)$ corresponding to the maximal ideal $\chi(\Lambda q^\mu)$. By 8.1.7 and 8.3.2 (i) the Verma modules occurring in M counted with multiplicities take the form $M(\Lambda')$ with $\Lambda' \in (\check{W}\check{w}).(\Lambda q^\mu) \cap (\check{w}.\lambda)q^{\Omega(V(\mu))}$. Since $\Omega(V(\mu))$ is W stable this equals $\check{w}.(\check{W}.\Lambda q^\mu \cap \Lambda q^{\Omega(V(\mu))}) = \check{w}.(\Lambda q^\mu)$ by 8.4.9, as required.

Remarks. Of course the conclusion also holds for $M(\check{w}.\Lambda)\otimes V(\mu)$. Probably it also holds with Λ dominant. It fails if Λ is not regular.

8.4.11 Take $\Lambda \in \check{T}^\wedge$ dominant. Then by 8.4.4, 8.4.6 the exact functor $\mathcal{T} : Ob\mathcal{O}^f_\Lambda \longrightarrow Ob\mathcal{H}_{\chi(\Lambda)}$ is an equivalence of categories exactly when $\mathcal{T}V \neq 0$ for every V simple. This may fail if Λ is not regular.

Theorem. *Assume Λ strongly dominant and regular. Then $\mathcal{T} : M \mapsto F(M(\Lambda), M)$ is an equivalence of categories.*

It remains to show that $F(M(\Lambda), V(\Lambda')) \neq 0$ for all $\Lambda' \in \Lambda q^{Q(\pi)}$. Take $\mu \in P^+(\pi)$. The canonical isomorphism $\operatorname{Hom}(V(\mu), k(q)) \otimes \operatorname{Hom}(M(\Lambda), M) \xrightarrow{\sim} \operatorname{Hom}(V(\mu)\otimes M(\Lambda), M)$ of $\check{U}$ bimodules restricts (cf. 7.1.6) to an isomorphism $V(\mu)^* \otimes F(M(\Lambda), M) \xrightarrow{\sim} F(V(\mu) \otimes M(\Lambda), M)$ of $F(\check{U})$ bimodules. Considered as a right $Z(\check{U})$ module, the primary component of $F(V(\mu)\otimes M(\Lambda), M)$ corresponding to $\chi(\Lambda q^\mu)$ is $F(M(\Lambda q^\mu), M)$ by 8.4.10 and the hypothesis on Λ. Hence it suffices to show that $F(M(\Lambda q^\mu), V(\Lambda')) \neq 0$ for some $\mu \in P^+(\pi)$.

Write $\Lambda' = \Lambda q^\beta : \beta \in Q(\pi)$ and choose $\mu \in P^+(\pi)$ such that $\nu := \mu - \beta \in P^+(\pi)$. Then by 8.3.6 (i) and 8.3.7 (ii) one has

$$[F(M(\Lambda q^\mu), \delta M(\Lambda')) : V(\nu)) = \dim V(\nu)_{\mu-\beta} = 1 \ ,$$

whilst

$$[F(M(\Lambda q^\mu), \delta M(\Lambda' q^{-\gamma})) : V(\nu)] = \dim V(\nu)_{\mu-\beta+\gamma} = 0$$

whenever $\gamma \in Q^+(\pi) \setminus \{0\}$. Yet $V(\Lambda')$ is a submodule of $\delta M(\Lambda')$ and the second equality shows that the quotient M, being such that $[M]$ is a sum of the $[\delta M(\Lambda' q^{-\gamma})] : \gamma \in Q^+(\pi) \setminus \{0\}$, satisfies $[F(M(\Lambda q^\mu), M) : V(\nu)] = 0$. Hence $F(M(\Lambda q^\mu), V(\Lambda')) \neq 0$ by the first equality.

8.4.12 Take $\Lambda \in \check{T}^\wedge$ dominant. Let $(\operatorname{Spec} F(\Lambda))^U$ denote the set of $\operatorname{ad} U$ invariant prime ideals of $F(\Lambda)$. Note that an ideal of $F(\check{U})$ is $\operatorname{ad} U$ invariant if and only if it is $\operatorname{ad} \check{U}$ invariant.

Theorem

(i) *Let J be an* $\operatorname{ad} U$ *invariant ideal of $F(\Lambda)$. Then*

$$J = \operatorname{Ann}_{F(\check{U})}(M(\Lambda)/JM(\Lambda)) \ .$$

(ii) $(\operatorname{Spec} F(\Lambda))^U = \{\operatorname{Ann}_{F(\check{U})} V(\Lambda') : \Lambda' \in W_\Lambda.\Lambda\}$.

Set $M = M(\Lambda)/JM(\Lambda)$. Clearly $J \operatorname{Ann}_{F(\check{U})} M$. On the other hand the image of $J \otimes_{F(\check{U})} M(\Lambda)$ in $F(\Lambda) \otimes_{F(\check{U})} M(\Lambda) \xrightarrow{\sim} M(\Lambda)$ is exactly $JM(\Lambda)$ and so $M = \mathcal{T}'(F(\Lambda)/J)$ by right exactness of $\mathcal{T}'$. Then 8.4.4 (i) gives $F(\Lambda)/J \xrightarrow{\sim} \mathcal{T}M = F(M(\Lambda), M)$ from which the opposite inclusion results.

Take $P \in (\operatorname{Spec} F(\Lambda))^U$. By (i), $P = \operatorname{Ann}_{F(\check{U})} M$ for some quotient M of $M(\Lambda)$. Yet $M \in Ob\mathcal{O}_f^\Lambda$ so has finite length (8.3.4). Thus the minimal primes over P take the form $\operatorname{Ann}_{F(\check{U})} L(\Lambda')$ where $L(\Lambda') \in \operatorname{Soc} M$ and P being prime must be one of them. Finally $\Lambda' \in \check{W}.\Lambda \cap \Lambda q^{Q(\pi)} = \check{W}_\Lambda.\Lambda$ by 8.3.2 (i). The opposite inclusion in (ii) is obvious.

8.4.13 Recall 7.1.13. Set $\check{T}_< = \tau(-2P^+(\pi))$ which is an Ore subset of $F(\check{U})$. Then $J(\check{U}) := \check{T}_<^{-1}F(\check{U}) = U^-\tau(2P(\pi))G^+$ and is $\operatorname{ad} U$ stable. Let $(\operatorname{Spec} F(\check{U}))^U$ denote the set of $\operatorname{ad} U$ invariant prime ideals of $F(\check{U})$.

Proposition

(i) *Every ideal of* $J(\check{U})$ *is* $\operatorname{ad} U$ *invariant.*

(ii) *The map* $P \mapsto P \cap F(\check{U})$ *is a bijection of* $\operatorname{Spec} J(\check{U})$ *onto* $(\operatorname{Spec} F(\check{U}))^U$ *with inverse* $Q \mapsto \check{T}_<^{-1}Q$.

An ideal J of $J(\check{U})$ is clearly $\operatorname{ad} t : t \in \check{T}_<$ invariant. Yet since the weights $\check{U}$ have no torsion (in the sense of A.1.16) this forces J to be $\operatorname{ad} t : t \in \check{T}$ invariant. The proof of (ii) is then completed using 8.4.4 (**) and its analogue for $\operatorname{ad} e_i$ as in 8.4.5.

Since $F(\check{U})$ is noetherian (7.4.8) then by A.2.8, the map $P \mapsto P \cap F(\check{U})$ is an embedding of $\operatorname{Spec} J(\check{U})$ into $\operatorname{Spec} F(\check{U})$. By (i) its image lies in $(\operatorname{Spec} F(\check{U}))^U$.

It remains to establish surjectivity in (ii). Consider $Q \in (\operatorname{Spec} F(\check{U}))^U$. By A.2.8, it is enough to show that $Q \cap \check{T}_< = \emptyset$. Since $F(\check{U})$ is noetherian one may assume Q to be a maximal invariant ideal and the base field to be replaced by its algebraic closure $\overline{k(q)}$.

Recall 7.3.8 and let $\check{\mathbb{H}}_+$ denote the $\operatorname{ad} \check{U}$ invariant complement in $\check{\mathbb{H}}$ of the trivial $\operatorname{ad} \check{U}$ module (which occurs exactly once in $\check{\mathbb{H}}$). Then $F(\check{U}) = (\check{\mathbb{H}}_+ \otimes Z(\check{U})) \oplus Z(\check{U})$ is the decomposition of $F(\check{U})$ into the isotypical component, namely $Z(\check{U})$, corresponding to the trivial module and the sum of the remaining isotypical components. Then $Q \cap Z(\check{U})$ is the trivial component of Q and is contained in a maximal ideal of $Z(\check{U})$ which by 8.3.2 and 8.4.8 can be assumed to be $\operatorname{Ann}_{Z(\check{U})} M(\Lambda)$ with $\Lambda \in \check{T}^\wedge$ dominant. If $1 \in Q + \operatorname{Ann}_{F(\check{U})} M(\Lambda)$ then $1 \in Q \cap Z(\check{U}) + \operatorname{Ann}_{Z(\check{U})} M(\Lambda)$ which contradicts the choice of Λ. Consequently $Q \supset \operatorname{Ann}_{F(\check{U})} M(\Lambda)$ and it remains to apply 8.4.12 (ii).

8.4.14 Set $\Gamma = P(\pi)/2P(\pi)$. Choose a set $\{\gamma\}_{\gamma \in \Gamma}$ of representatives in $P(\pi)$. Then $\check{U}_\gamma := U^-\tau(2P(\pi) + \gamma)G^+ = \tau(\gamma)J(\check{U}) : \gamma \in \Gamma$ defines a Γ grading of $\check{U}$. Obviously $1 \in \check{U}_\gamma\check{U}_{-\gamma}$, each $\check{U}_\gamma$ is $\operatorname{ad} \check{U}$ invariant and is a left coideal in virtue of 7.1.6. In other words the conditions (i)–(iii) of 1.3.8 are satisfied. Hence the conclusion of 1.3.13 can be applied. Here the Γ orbits in $\check{U}_0 = J(\check{U})$ are singletons by 8.4.13 (i) so one obtains the

Theorem. *The map taking P to the minimal primes over $P\check{U}$ is an isomorphism of* $\operatorname{Spec} J(\check{U})$ *onto the space of Γ orbits in* $\operatorname{Spec} \check{U}$.

8.4.15 There is a generalization [MR1, Chap. 1] of the nullstellensatz to non-commutative k-algebra A. When A is noetherian it says that

(i) *Let A be a simple module. Then* $\operatorname{End}_A M$ *is algebraic over k.*
(ii) *Every $P \in$* $\operatorname{Spec} A$ *is an intersection of primitive ideals of A.*

8.4.16 By A.3.4 an algebra A satisfies the nullstellensatz if it admits a bounded filtration $\mathcal{F}$ such that $gr_{\mathcal{F}}A$ is finitely quasi-commutative (7.4.1).

Proposition. *The algebras U, $\check{U}$, $F(U)$, $F(\check{U})$ satisfy the nullstellensatz.*

Recall the filtration $\mathcal{F}_2$ on U^- defined in 7.4.3 and that $U^- \cong G^+$. Let $\mathcal{F}_0$ denote the trivial filtration on U^0. Then $\mathcal{F}_3 := \mathcal{F}_2 \times \mathcal{F}_0 \times \mathcal{F}_2$ defines a filtration on $U = U^- \otimes U^0 \otimes G^+$ and as in 7.4.7 it follows that $gr_{\mathcal{F}_3}U$ is isomorphic to the smash product $(gr_{\mathcal{F}_2}U^- \otimes gr_{\mathcal{F}_2}G^+)\#U^0$. Since $gr_{\mathcal{F}_2}U^-$ and $gr_{\mathcal{F}_2}G^+$ are finitely q-commutative (7.4.3) so is $gr_{\mathcal{F}_3}U$. This proves the assertion for U and similarly for $\check{U}$.

For $F(\check{U})$ (or $F(U)$) let $\mathcal{F}$, $\mathcal{F}'$ be the filtrations defined in 5.3.1 and 7.4.7. Clearly $gr_{\mathcal{F}_3}F(\check{U}) = gr_{\mathcal{F}_3}(gr_{\mathcal{F}'}(gr_{\mathcal{F}}F(\check{U})))$.

By 7.1.2, 7.1.4, 7.1.6 one has

$$gr_{\mathcal{F}'}gr_{\mathcal{F}}F(\check{U}) = \bigoplus_{\lambda \in P^+(\pi)} (K(\lambda)^- \otimes K(-\lambda)^+)\#k(q)\tau(-2\lambda)\,,$$

and so

$$gr_{\mathcal{F}_3}F(\check{U}) = \bigoplus_{\lambda \in P^+(\pi)} (gr_{\mathcal{F}_2}K(\lambda)^- \otimes gr_{\mathcal{F}_2}K(-\lambda)^+)\#k(q)\tau(-2\lambda)\,.$$

By construction (see after 7.4.3 (2)) one has $gr_{\mathcal{F}_2}K(\lambda)^- gr_{\mathcal{F}_2}K(\mu)^- = gr_{\mathcal{F}_2}K(\lambda)^-K(\mu)^- = gr_{\mathcal{F}_2}K(\lambda+\mu)^-$ by 7.1.7 (i), with a similar assertion for $+$ and $-$ interchanged. Again by construction weight vectors in $gr_{\mathcal{F}_2}K(\lambda)^-$ (resp. in $gr_{\mathcal{F}_2}K(-\lambda)^+$) q-commute. Since $P^+(\pi)$ is finitely generated as a semigroup this proves that $gr_{\mathcal{F}_3}F(\check{U})$ is finitely q-commutative. Replacing $2P^+(\pi)$ by $R^+(\pi)$ give the corresponding assertion for $F(U)$.

8.4.17 Theorem. *Over $\overline{k(q)}$ one has*

(i) $\operatorname{Prim} U_q(\mathfrak{g}) = \{\operatorname{Ann} V(\Lambda) : \Lambda \in T^\wedge\}$.
(ii) $\operatorname{Prim} \check{U}_q(\mathfrak{g}) = \{\operatorname{Ann} V(\Lambda) : \Lambda \in \check{T}^\wedge\}$.

Consider first (ii). Take $P \in \operatorname{Prim}\check{U}$. Then 8.4.16 and 8.4.15 (i) imply $P \cap Z(\check{U}) \in \operatorname{Max} Z(\check{U})$, so by 8.3.2 and 8.4.8 there exists $\Lambda \in \check{T}^\wedge$ dominant such that $P \supset \operatorname{Ann} M(\Lambda)$. Then $P \cap F(\Lambda) \in (\operatorname{Spec} F(\Lambda))^U$ by 8.4.13 and 1.3.13. Conclude using 8.4.12 (i) and 1.3.13 again. Finally (i) obtains from (ii) and 1.3.13 using the gradation of $\check{U}$ induced by $\Gamma := P(\pi)/Q(\pi)$.

8.4.18 Of course a result similar to 8.4.17 applies to $(\operatorname{Prim} F(U))^U$. Again not only does the conclusion of 8.4.15 apply to U, $\check{U}$, $F(U)$, $F(\check{U})$; but also combining this result for say $\check{U}$ with the correspondence given by 8.4.13 and 8.4.14 gives the

Corollary. *Every $P \in (\operatorname{Spec} F(U))^U$ is an intersection of* $\operatorname{ad} U$ *invariant primitive ideals.*

8.5 Comments and Complements

8.5.1 The main results of this chapter are taken from [JL3]. Technical improvements were made possible by the noetherianity of $F(\check{U})$ and the systematic use of 1.3.13. In the generality given 8.2.2 is new and so is 8.4.13 (ii). For U (or $\check{U}$) the conclusion of 8.4.16 is due to J. C. McConnell [Mc1] whilst for $F(U)$ (or $F(\check{U})$ it is new. The translation principle is due to J.-C. Jantzen, 8.4.10 being a quantum version of [Ja1, Chap. 2].

8.5.2 In the enveloping algebra case the conclusion of 8.4.17 is due to M. Duflo [Du1]. This itself was developed from work of D. P. Zelobenko on principal series modules in the so-called complex case. The use of the latter in studying $\operatorname{Prim} U(\mathfrak{g})$ was first introduced by J. Dixmier [Di2]. Extensive use of many results in this theory were applied to the study of $\operatorname{Prim} U(\mathfrak{g})$ as the reader may discern from [BG1, Ja2, J3, J5, J8]. Notable recent developments occur in [McG1]. Much of this theory carries over to $U_q(\mathfrak{g})$.

8.5.3 For $U(\mathfrak{g})$ the conclusion of 8.4.12 (i) was a positive answer to a problem of Dixmier resolved in [J3]. Shortly after J. Bernstein and S. Gelfand independently obtained their equivalence of categories theorem [BG1] of which 8.4.11 is the quantum version. The category $\tilde{\mathcal{O}}$ obtained by adjoining all extensions of modules in $\mathcal{O}$ essentially corresponds to the Harish-Chandra category $\mathcal{H}$. This was developed in the enveloping algebra case by W. Soergel [So1] and recently in the quantum case by C. Liu [Liu2].

8.5.4 The ideals occurring in the right hand sides of 8.4.17 need not be distinct. Deciding this question is reduced by 8.3.2 to just $\check{W}$. orbits and can be anticipated to be a combinatorial aspect of the extended Weyl group. In the enveloping algebra case a solution was proposed in [J4] in which left cells were first defined and completely settled shortly afterwards by D. Vogan [V1]. The parameters needed to compute cells were the Jordan-Holder multiplicities of Verma modules whose precise form was shortly after conjectured by D. A. Kazhdan and G. Lusztig [KL1]. This was soon established independently by J. L. Brylinski and M. Kashiwara [BK1] and A. A. Beilinson and I. N. Bernstein [BB1]. At first it was not possible to independently prove all properties of cells implied by their primitive ideal origin coming from GK dimension and Goldie rank [J17]. Yet their study was developed mainly by G. Lusztig [L1, L13, L14, J18] into a remarkably rich theory whose relation to Weyl group representations and the geometry of orbits predicted by primitive ideal theory came to have an independent life. The ordering in Prim $U(\mathfrak{g})$ it implies carries over to Prim $U_q(\mathfrak{g})$ by virtue of the result noted in 4.5.2 in the case of characters of the form $q^\lambda : \lambda \in \mathfrak{h}^*_{\mathbb{Q}}$. The general question remains open. Even for $U(\mathfrak{g})$ the description of Spec $U(\mathfrak{g})$ as an ordered set or as a topological space remains open [So2].

8.5.5 The enveloping algebra version of 8.3.11 is an old result of N. Conze [Coz1]. The general question of when $U(\mathfrak{g})/P : P \in \text{Prim}\, U(\mathfrak{g})$ is a domain was settled by C. Moeglin [Mo2] in type A_n. For arbitrary $\mathfrak{g}$ it is related to the unitarity of the corresponding principal series modules after D. A. Vogan [V3]. These questions have received extensive study respectively by W. M. McGovern [McG1] and D. Barbasch and D. A. Vogan [BV2]; but have not yet been completely settled. The more general question of computing the Goldie ranks of primitive quotients was extensively studied in [J17] of which [Ja2] gives amongst other things an exposition. Its relations to geometry referred to in 8.4.5 was also studied by D. Barbasch and D. A. Vogan [BV1], R. Hotta [Ho1], W. Rossmann [Ro1], M. Vergne [Ve1], in [J13] and by W. Borho, J.-L. Brylinski and R. MacPherson [BBM1]. Most of the theory (but not yet all) carries over intact to $U_q(\mathfrak{g})$.

8.5.6 The enveloping algebra version of 8.2.10 is an old result of K. P. Parthasarathy, R. Ranga Rao and V. S. Varadarajan [PRV1] which notably preceded the factorization of the Shapovalev determinants. The prohibitive length of the original proof caused it to be omitted from [Di2]. The analysis in 8.2 is developed from the much shorter proof in [J12]. In the latter a sum rule, analogous to Jantzen's sum rule 4.1.7, is given for the still unresolved problem of determining the multiplicities of simple $U(\mathfrak{g})$ bimodule subquotients in $U(\mathfrak{g})/P : P \in \text{Prim}\, U(\mathfrak{g})$.

8.5.7 Let M be a simple $U(\mathfrak{g})$ (resp. $U_q(\mathfrak{g})$ module). In general the embedding $U(\mathfrak{g})/\text{Ann}\, M$ (resp. $F(U)/\text{Ann}\, M$) in $F(M)$ is strict. The precise description of $F(M)$ is a very interesting question which has particular importance in the study of Goldie ranks (8.5.5) and relations to nilpotent orbits in $\mathfrak{g}^*$. More generally D. A. Vogan [V4] and W. M. McGovern [McG1,2] have intensively studied Dixmier algebras which are in particular over-algebras of primitive quotients admitting an action of the algebraic group. See also [J9]. One may define a quantum Dixmier algebra to be an overalgebra of $F(\check{U})/P : P \in (\text{Prim}\, F(\check{U}))^U$ which as a bimodule is an object in $\mathcal{H}$.

8.5.8 The inclusion $(\text{Spec}\, F(\check{U}))^U \hookrightarrow \text{Spec}\, F(\check{U})$ is strict [JL3, 5.15]. Of course by 8.4.13 (ii) the image is exactly the set of primes for which the quotient algebra has no $\check{T}_<$ torsion.

8.5.9 The result in 8.2.4 has the following generalization to $U_q(\mathfrak{g}_C)$ with C integrable. Given U modules M, M' consider $\mathrm{Hom}(M, M')$ as an $\mathrm{ad}\, U$ module (8.2.3) and set $I(M, M') := I(\mathrm{Hom}(M, M'))$ using the notation of 7.1.1. Then by 8.3.6 $(*)$ one has for all $\nu \in P^+(\pi)$, $\Lambda, \Lambda' \in T^\wedge$ that $\mathrm{Hom}_U(V(\nu)^\iota, \mathrm{Hom}(M(\Lambda), M(\Lambda'))) \cong \mathrm{Hom}_U(V(\nu), I(M(\Lambda), M(\Lambda'))^\tau) \cong \mathrm{Hom}_U(V(\nu), I(\delta' M(\Lambda'), \delta' M(\Lambda))) \cong \mathrm{Hom}_U(V(\nu), \mathrm{Hom}(\delta' M(\Lambda'), \delta' M(\Lambda))) \cong \mathrm{Hom}_U(V(\nu) \otimes \delta' M(\Lambda'), \delta' M(\Lambda)) \cong \mathrm{Hom}_U(M(\Lambda), V(\nu) \otimes M(\Lambda')) \cong V(\nu)_{\Lambda\Lambda'^{-1}}$ if Λ is dominant. Thus under this last hypothesis $I(M(\Lambda), M(\Lambda'))$ as an $\mathrm{ad}\, U$ module contains the direct sum

$$\bigoplus_{\nu \in P^+(\pi)} (V(\nu)^\iota)^{\dim V(\nu)_{\Lambda\Lambda'^{-1}}} .$$

When C is of finite type, it is exactly this direct sum and moreover $V(\nu)^\iota \cong V(\nu)^*$.

8.5.10 In the enveloping algebra case 8.3.9 (i) is an easy consequence of Kostant's result on the primeness of the ideal in $S(\mathfrak{g})$ described in 7.5.6 and furthermore for simple Verma modules 8.3.9 (ii) follows from Kostant's separation theorem and Frobenius reciprocity (8.3.8) as noted in N. Conze-Berline and M. Duflo [CD1]. For an arbitrary Verma module the use of Gelfand-Kirillov dimension (viz. 8.2.5) was introduced in [J16, Sect. 6]. It is still an indispensable tool. A more general result relating an arbitrary $F(M(\lambda), M(\mu))$ to a principal series modules is given in [GJ1].

8.5.11 The Poincaré series for the harmonics in the classical case, that is for $S(\mathfrak{g})$, was first considered by B. Kostant [Ko1]. He showed that the isotypical component of the adjoint represention occurs as the $\left\{\frac{\partial y}{\partial x} : x \in \mathfrak{g},\ y \in S(\mathfrak{g})^G\right\}$ and that the last occurrence of any $\mathcal{V}(\mu)$ was in degree $|\mu|$. These degrees are called generalized exponents and their generating function, the analogue of $Q(z)$, was found by W. H. Hesselink [He1] to be given by the particularly simple expression

$$\prod_{\alpha \in \Delta^+} \left(\frac{1 - e^{-\alpha}}{1 - z e^{-\alpha}}\right)$$

of which R. K. Brylinski [Bry1] found another interesting interpretation, and further discussed in the quantum context in [J12, Sect. 4].

8.5.12 In [J12, Sect. 3] it is shown that $\mathrm{Ann}_{\check{U}}\, M(\Lambda)$ is generated by its intersection with $Z(\check{U})$.

Chapter 9. Structure Theorems for $R_q[G]$

The conventions of Chapter 8 are retained.

9.1 Commutativity Relations

9.1.1 A finite dimensional module over $U = U_q(\mathfrak{g}_C)$ has a filtration whose quotients are simple integrable highest weight modules and hence by 4.3.10 is semisimple. Thus the Hopf dual U^* satisfies the direct sum decomposition of 1.4.13. By 7.1.15 (ii) this is only of interest when C is of finite type as will be henceforth assumed.

It is convenient to define $R_q[G]$ (or simply, R) to be the subspace of U^* spanned by the $C^{V(\mu)} : \mu \in P^+(\pi)$. By the remark in 1.4.9 this is a Hopf subalgebra of U^* and by 4.3.9 differs from the latter only by tensoring over the group algebra of $\mathbb{Z}_2^{|\pi|}$. On the other hand $R_q[G]$ has the advantage of being an integral domain (9.1.9 (i)). Here G is just a symbol; but if $k = \mathbb{C}$ it may also be viewed as the unique connected, simply connected algebraic group over k with Lie algebra $\mathfrak{g}$. Then $R_q[G]$ specializes to $R[G]$ at $q = 1$.

It is convenient to continue to consider $V(\mu)^*$ as a right U module though this may cause some confusion. Thus $V(\mu)^*$ has highest weight μ and the corresponding highest weight vector is annihilated by the $f_i : i = 1, 2, \dots, \ell$ rather than by the e_i. By 1.4.4 this convention and possible confusion carries over to the $c_{\xi,v}^{V(\mu)} : \xi \in V(\mu)^*$, $v \in V(\mu)$ which are besides simply written as $c_{\xi,v}^{\mu}$ or $c_{\xi,v}$. Let $\Omega(V(\mu))$ denote the set of weights of $V(\mu)$ counted with their multiplicities.

Given weight vectors $\xi \in V(\mu)^*_\lambda$, $v \in V(\mu)_\eta$ it is convenient to write $c_{\xi,v}^{\mu}$ as $c_{\lambda,\eta}^{\mu}$. If $\dim V(\mu)^*_\lambda = \dim V(\mu)_\eta = 1$ for example (4.3.6 (iii)) if $\lambda, \eta \in W\mu$ then the ambiguity this causes is very minor. However in all cases for which this convention is adopted it will mean that any formula involving these quantities hold irrespective of the choice of the weight vectors.

9.1.2 Set $V_+^{\pm} = \ker \varepsilon|_{V^{\pm}}$. Then 5.1.1 (vii) gives the

Lemma. *For all $\mu \in Q^+(\pi)$, $a_\mu \in U_\mu^+$, $b_{-\mu} \in U_{-\mu}^+$*

(i) $\Delta(a_\mu) = a_\mu \otimes \tau(-\mu) + 1 \otimes a_\mu \mod V_+^+ \otimes V_+^+$,
(ii) $\Delta(b_{-\mu}) = b_{-\mu} \otimes 1 + \tau(\mu) \otimes b_{-\mu} \mod V_+^- \otimes V_+^-$.

9.1.3 Recall the conclusion of A.1.4.

Lemma. *For all* $\mu, \nu \in P^+(\pi)$ *one has*

$$(*) \qquad c^{\mu}_{\mu,w_0\mu} c^{\nu}_{w_0\nu,\nu} = c^{\nu}_{w_0\nu,\nu} c^{\mu}_{\mu,w_0\mu} \,.$$

Take $a_\lambda \in U^+_\lambda$, $b_{-\eta} \in U^-_{-\eta}$. Then 9.1.2 gives

$$a_\lambda(u_{w_0\mu} \otimes u_\nu) = a_\lambda u_{w_0\mu} \otimes \tau(-\lambda)u_\nu = q^{-(\lambda,\nu)}(a_\lambda u_{w_0\mu} \otimes u_\nu)$$

and

$$(\xi_\mu \otimes \xi_{w_0\nu})b_{-\eta} = \xi_\mu \tau(\eta) \otimes \xi_{w_0\nu} b_{-\eta} = q^{(\eta,\mu)}(\xi_\mu \otimes \xi_{w_0\nu} b_{-\eta}) \,.$$

Hence (1.4.6)

$$\begin{aligned}(c^{\mu}_{\mu,w_0\mu} c^{\nu}_{w_0\nu,\nu})(b_{-\eta}a_\lambda) &= (\xi_\mu \otimes \xi_{w_0\nu})(b_{-\eta}a_\lambda(u_{w_0\mu} \otimes u_\nu)) \\ &= q^{-(\lambda,\nu)+(\eta,\mu)} \xi_\mu(a_\lambda u_{w_0\mu}) \xi_{w_0\nu}(b_{-\eta}u_\nu) \,.\end{aligned}$$

The above expression vanishes unless $\lambda = \mu - w_0\mu$, $\eta = \nu - w_0\nu$, in which case the exponent in q is zero. A similar calculation shows that the right hand side of $(*)$ takes the same value on $b_{-\eta}a_\lambda$. Recalling the triangular decomposition of U completes the proof.

9.1.4 Act on both sides of 9.1.3 $(*)$ by U^+ (resp. U^-) interchanging μ, ν in the second case. Then 9.1.2 gives the

Corollary. *For all* $\eta, \lambda \in \Omega(V(\mu))$ *one has*

(i) $c^{\mu}_{\eta,\lambda} c^{\nu}_{w_0\nu,\nu} = q^{(\lambda,\nu)-(\eta,w_0\nu)} c^{\nu}_{w_0\nu,\nu} c^{\mu}_{\eta,\lambda}$.
(ii) $c^{\nu}_{\nu,w_0\nu} c^{\mu}_{\eta,\lambda} = q^{(\lambda,w_0\nu)-(\eta,\nu)} c^{\mu}_{\eta,\lambda} c^{\nu}_{\nu,w_0\nu}$.

9.1.5 Define an order relation $\leq$ on $P(\pi)$ through $\mu \geq \nu$ if $\mu - \nu \in Q^+(\pi)$. Fix $\eta, \lambda \in \Omega(V(\nu))$ and let $J^+_\nu(\eta', \lambda')$ (resp. $J^-_\nu(\eta', \lambda')$) denote the left ideal of $R_q[G]$ generated by the $c^{\nu}_{\eta'',\lambda''}$ with either $\lambda'' > \lambda'$ and $\eta'' \leq \eta'$ or $\lambda'' \geq \lambda'$ and $\eta'' < \eta'$ (resp. $\lambda'' < \lambda'$, $\eta'' \geq \eta'$ or $\lambda'' \leq \lambda'$ and $\eta'' > \eta'$). Act by U^- (resp. U^+) on both sides of 9.1.4 (i) (resp. 9.1.4 (ii)). Applying 9.1.2 and induction on $\leq$ gives the

Proposition. *For all* $\mu, \nu \in P^+(\pi)$; $\lambda, \eta \in \Omega(V(\mu))$, $\lambda', \eta' \in \Omega(V(\nu))$ *one has*

(i) $c^{\mu}_{\eta,\lambda} c^{\nu}_{\eta',\lambda'} = q^{(\lambda,\lambda')-(\eta,\eta')} c^{\nu}_{\eta',\lambda'} c^{\mu}_{\eta,\lambda} \mod J^+_\nu(\eta', \lambda')$.

Moreover $J^+_\nu(\eta', \lambda')$ *is a two-sided ideal.*

(ii) $c^{\nu}_{\eta',\lambda'} c^{\mu}_{\eta,\lambda} = q^{(\lambda',\lambda)-(\eta',\eta)} c^{\mu}_{\eta,\lambda} c^{\nu}_{\eta',\lambda'} \mod J^-_\nu(\eta', \lambda')$.

Moreover $J^-_\nu(\eta', \lambda')$ *is a two-sided ideal.*

Remarks. Comparison of (i), (ii) might lead to think that the exponent in q has the wrong symmetry. Actually this "wrong symmetry" leads to a significant simplification in the structure of $\operatorname{Prim} R_q[G]$. In the definition of $J^{\pm}$ one can obviously replace left by right.

9.1.6 Let $R_q[G/N^+]$ (resp. $R_q[G/N^-]$) denote the subspace of $R_q[G]$ spanned by the $c^{\mu}_{\xi,\mu}$ (resp. $c^{\mu}_{\xi,w_0\mu}$): $\mu \in P^+(\pi)$, $\xi \in V(\mu)^*$. It is clear that $R_q[G/N^{\pm}]$ is just the set of invariants of $R_q[G]$ with respect to the left action of $U^{\pm}$ and so (2.4.6) is the quantum analogue of the algebra of functions on $G/N^{\pm}$ where $N^{\pm}$ denotes the subgroup of G corresponding (2.4.2) to $\mathfrak{n}^{\pm}$. It is in particular a subalgebra of $R_q[G]$ and by 1.4.15 a left coideal as may also be verified directly from 1.4.7 (1). It will be denoted simply by $R^{\pm}$. It is clear that as a right U module R^+ (resp. R^-) is a direct sum of the $V^+(\mu)^* := \{c^{\mu}_{\xi,\mu} : \xi \in V(\mu)^*\}$ (resp. $V^-(\mu)^* := \{c^{\mu}_{\xi,w_0\mu} : \xi \in V(\mu)^*\}$) which is isomorphic to $V(\mu)^*$. Moreover by 1.4.6 (ii) these modules satisfy the Cartan multiplication rule, for example

$$(*) \qquad V^+(\mu)^*\, V^+(\nu)^* = V^+(\mu+\nu)^* \,, \quad \forall\, \mu,\nu \in P^+(\pi) \,.$$

On the other hand by 9.1.4 (i) taking $\lambda = \mu$, $\eta = w_0\mu$

$$(**) \qquad c^{\mu}_{w_0\mu,\mu} c^{\nu}_{w_0\nu,\nu} = c^{\nu}_{w_0\nu,\nu} c^{\mu}_{w_0\mu,\mu} \,, \quad \forall\, \mu,\nu \in P^+(\pi) \,.$$

The algebra R^+ (or R^-) may be described through generators and relations as follows. For each $i \in \{1,2,\ldots,\ell\}$ fix a vector $u_i \in V(\omega_i)^*$ of weight $w_0\omega_i$ and set

$$V = \bigoplus_{i=1}^{\ell} V(\omega_i)^* \,.$$

Consider the tensor algebra $\mathcal{T}(V)$ as a right U module through the coproduct. By 5.3.20 there exists for all $n \in \mathbb{N}^+$ and all $i_1, i_2, \ldots, i_n \in \{1,2,\ldots,\ell\}$ a surjection of right U modules

$$V(\omega_{i_1})^* \otimes V(\omega_{i_2})^* \otimes \ldots \otimes V(\omega_{i_n})^* \longrightarrow V(\omega_{i_1} + \omega_{i_2} + \ldots + \omega_{i_n})^* \,.$$

Let $K_{i_1,i_2,\ldots,i_n}$ denote its kernel. For each $i,j \in \{1,2,\ldots,\ell\}$ let $E_{i,j}$ denote the right U module generated by $u_i \otimes u_j - u_j \otimes u_i$. Let $\mathcal{K}$ denote the two-sided ideal of $\mathcal{T}(V)$ generated by the all $K_{i_1,i_2,\ldots,i_n}$ and all the $E_{i,j}$. It is clear from $(*)$, $(**)$ above that there is a surjection $\mathcal{T}(V)/\mathcal{K} \twoheadrightarrow R^+$ of U algebras. Since $u_{i_1} \otimes u_{i_2} \otimes \ldots \otimes u_{i_n}$ is independent of ordering in the quotient and is a generator for $V(\omega_{i_1}+\omega_{i_2}+\ldots+\omega_{i_n})^*$, the multiplicity of $V(\mu)^* : \mu \in P^+(\pi)$ in $\mathcal{T}(V)/\mathcal{K}$ is at most one. This proves injectivity. A similar description applies to R^- which is hence isomorphic to R^+ as a U algebra.

9.1.7 Recall 7.1.1. Comparison of 7.1.7 and 9.1.6 leads to a further description of the U algebras $R^{\pm}$. Indeed by 7.1.7 (i)

$$\bigoplus_{\lambda \in P^+(\pi)} K(\lambda)^- \tau(-\lambda)$$

is a subalgebra of $U^-\check{T}$ and may be given a U module structure by identifying $K(\lambda)^-\tau(-\lambda)$ with $K(\lambda)^-$ given the action $(u,a) \mapsto (\text{ad}_{-2\lambda}\, u)a$. (Notice the factor of two!) To show that it is a U-algebra it suffices to establish the Leibnitz rules

$$(*) \quad a.\tau(-\mu-\nu) = (a_1.\tau(-\mu))(a_2.\tau(-\nu)) \ , \quad \forall\, a \in U \ , \quad \forall\, \mu,\nu \in P^+(\pi) \ ,$$

since the action of U on $K(\lambda)^-$ is defined on 1 and extended by the Leibnitz rule to U^-. In $(*)$ it is enough to let a run over the generators. The assertion is obvious for the e_i, t_i, whilst for the f_i the relations in 5.1.1 and 7.1.1 give

$$\begin{aligned}
&(f_i.\tau(-\mu))\tau(-\nu) + (t_i.\tau(-\mu))(f_i.\tau(-\nu)) \\
&\quad = ((\text{ad}_{-2\mu}\, f_i)1)\tau(-\mu-\nu) + ((\text{ad}_{-2\mu}\, t_i)1)\tau(-\mu)((\text{ad}_{-2\nu}\, f_i)1)\tau(-\nu) \\
&\quad = (1-q^{2(\mu,\alpha_i)}) f_i\tau(-\mu-\nu) + q^{(\mu,\alpha_i)}\tau(-\mu)(1-q^{2(\nu,\alpha_i)}) f_i\tau(-\nu) \\
&\quad = (1-q^{2((\mu+\nu),\alpha_i)}) f_i\tau(-\mu-\nu) = f_i.\tau(-\mu-\nu)
\end{aligned}$$

as required.

Recall 1.4.8. The left U module $(V^+(-w_0\lambda)^*)^\sigma$ is isomorphic to $V(\lambda)$, so may be identified with $K(\lambda)^-\tau(-\lambda)$ and then $c^{-w_0\lambda}_{-\lambda,-w_0\lambda}$ becomes $\tau(-\lambda)$. In this it follows from 7.1.7 (i) and the relations in $\check{T}$ that 9.1.6 $(*)$, $(**)$ hold. Similarly the subspace

$$\bigoplus_{\lambda \in P^+(\pi)} \tau(-\lambda)K(-\lambda)^+$$

of $\check{T}G^+$ admits a U algebra structure with action $(u,a) \mapsto (\text{ad}_{-2\lambda}\, u)a$. As above $(V^-(\lambda)^*)^\sigma$ may be identified with $\tau(-\lambda)K(-\lambda)^+$ and then $c^\lambda_{\lambda,w_0\lambda}$ becomes $\tau(-\lambda)$. Summarizing

Lemma. *There are U-algebra isomorphisms*

(i) $R^+ \xrightarrow{\sim} \bigoplus_{\lambda\in P^+(\pi)} K(\lambda)^-\tau(-\lambda)$

(ii) $R^- \xrightarrow{\sim} \bigoplus_{\lambda\in P^+(\pi)} \tau(-\lambda)K(-\lambda)^+$

In particular R^+, R^- are noetherian.

The last part follows by identifying the right hand side as the twisted Rees algebra $R(U^-,\mathcal{F}_p,\chi)$ (resp. $R(G^+,\mathcal{F}_p,\chi)$) where $\chi_\mu(\nu) = q^{-(\mu,\nu)}$ and applying 7.4.5, 7.4.6.

Remarks. The isomorphism of R^+, R^- as U-algebras now also follows from the involution $\varkappa'$ of 7.2.1. Notice that $(V^+(\lambda)^*)^\sigma$ (resp. $(V^-(\lambda)^*)^\sigma$) can also be identified with $\tau(-\lambda)K(-\lambda)^+$ (resp. $K(-w_0\lambda)^-\tau(w_0\lambda)$) and then $c^\lambda_{\lambda,\lambda}$ (resp. $c^\lambda_{w_0\lambda,w_0\lambda}$) becomes $\tau(-\lambda)$ (resp. $\tau(w_0\lambda)$).

9.1.8 Let $R^+ \circledast R^-$, or simply, $\hat{R}$ denote the $k(q)$-algebra generated by the elements of R^+ and R^- satisfying the relations of these algebras together with additional relations which derive from 9.1.3(*) by allowing U to act on the right (as in 9.1.4, 9.1.5). It is immediate that the multiplication map $R^+ \otimes R^- \longrightarrow R$ factors through $R^+ \circledast R^-$ to a surjection $R^+ \otimes R^- \longrightarrow R^+ \circledast R^-$ of right $T \otimes T$ and right $\Delta(U)$ modules. Thus the kernel is generated by its intersection with the $V^+(\mu)^* \otimes V^-(\nu)^* : \mu, \nu \in P^+(\pi)$. Recall that multiplication in R is given by the transpose Δ^* of the coproduct Δ on U. View $R^\pm$ as subspaces of R.

Lemma. *For all $\mu, \nu \in P^+(\pi)$ the restriction of Δ^* to $V^+(\mu)^* \otimes V^-(\nu)^*$ is injective. In particular $R^+ \circledast R^-$ is isomorphic to $R^+ \otimes R^-$ as a vector space.*

Let $\{\xi_i\}$ (resp. $\{\xi'_j\}$) be a basis for $V(\mu)^*$ (resp. $V(\nu)^*$). One must show that the $c_{\xi_i, u_\mu} c_{\xi'_j, u'_{w_0\nu}} = c_{\xi_i \otimes \xi'_j, u_\mu \otimes u'_{w_0\nu}}$ are linearly independent. Now $U^- U^+ (u_\mu \otimes u'_{w_0\nu}) = U^-(u_\mu \otimes V(\lambda)) = V(\mu) \otimes V(\nu)$ and so $c_{\xi_i \otimes \xi'_j, u_\lambda \otimes u'_{w_0\nu}}(U)$ $= (\xi_i \otimes \xi'_j)(U(u_\mu \otimes u'_{w_0\nu})) = (\xi_i \otimes \xi'_j)(V(\mu) \otimes V(\nu))$ which establishes the required assertion.

9.1.9 Recall the analysis of 7.3.4.

Lemma

(i) *R is a domain.*
(ii) *$R^+ \circledast R^-$ is a domain.*

(i) Suppose $cc' = 0$ for some $c, c' \in R$ non-zero. Via the $T - T$ action one may assume c, c' to be weight vectors. Since the action of $U^+ - U^-$ on $T - T$ weight vectors in R is by locally nilpotent skew derivations one may assume that c, c' are $U_q(\mathfrak{b}^+) - U_q(\mathfrak{b}^-)$ invariant. Then there exists $\mu, \nu \in P^+(\pi)$ such that $c = c^\mu_{\mu,\mu}, c' = c^\nu_{\nu,\nu}$. Yet $(cc')(1) = c(1)c'(1) \neq 0$ giving the required contradiction.

(ii) Suppose $cc' = 0$ for some $c, c' \in R^+ \circledast R^-$ non-zero. Similar to the above one may assume them to be $T \otimes T$ weight vectors. Then the assertion follows from (i) and 9.1.8.

9.1.10 For all $\mu \in P^+(\pi)$ assume in the definition of $c_\mu := c^\mu_{\mu,\mu} := c^\mu_{\xi_\mu, u_\mu}$ that $\xi_\mu(u_\mu) = 1$. Take $\mu, \nu \in P^+(\pi)$. Then $(\xi_\mu \otimes \xi_\nu)(u_\mu \otimes u_\nu) = \xi_\mu(u_\mu)\xi_\nu(u_\nu) = 1$ and since $P^+(\pi)$ is a free abelian semigroup one may assume $\xi_\mu \otimes \xi_\nu = \xi_{\mu+\nu}$ and $u_\mu \otimes u_\nu = u_{\mu+\nu}$. With these conventions $c_\mu c_\nu = c_{\mu+\nu} = c_\nu c_\mu$. Similarly one may assume that the $d_\mu := c^\mu_{w_0\mu, w_0\mu}$ satisfy $d_\mu d_\nu = d_{\mu+\nu} = d_\nu d_\mu$.

Now for each $w \in W$ set $c_{w\mu} := c^\mu_{w\mu,\mu}$, $d_{w\mu} := c^\mu_{ww_0\mu, w_0\mu}$ which are thus defined up to scalars. It will be shown that these scalars can be chosen so that

$$(*) \qquad c_{w\mu} c_{w\nu} = c_{w(\mu+\nu)}\ , \quad d_{w\mu} d_{w\nu} = d_{w(\mu+\nu)}\ , \quad \forall \mu, \nu \in P^+(\pi)\ .$$

Fix $i \in \{1, 2, \dots, \ell\}$. By 7.1.1 and 1.2.13

$$\Delta(e_i^m) = \sum_{n=0}^{m} q_i^{-n(m-n)} \begin{bmatrix} m \\ n \end{bmatrix}_{q_i} e_i^n \otimes e_i^{m-n} t_i^{-n} .$$

Now suppose c_{-r} (resp. c_{-s}) is a lowest weight vector for the $r+1$ (resp. $s+1$) dimensional U_i module of lowest weight $q_i^{-\frac{1}{2}r\alpha_i}$ (resp. $q_i^{-\frac{1}{2}s\alpha_i}$). Then $t_i^{-1}c_{-s} = q_i^s c_{-s}$ and so the above formula for $\Delta(e_i^{r+s})$ gives

$$e_i^{r+s}(c_{-r} \otimes c_{-s}) = \begin{bmatrix} r+s \\ r \end{bmatrix}_{q_i} e_i^r c_{-r} \otimes e_i^s c_{-s}$$

or in terms of the divided powers $e_i^{(r)} := e_i^r/[r!]_{q_i}$ that

(**) $$e_i^{(r+s)}(c_{-r} \otimes c_{-s}) = e_i^{(r)} c_{-r} \otimes e_i^{(s)} c_{-s} .$$

Now observe that $c_\mu \in (V^+(\mu)^*)^\sigma \cong V(-\mu)$ is a lowest weight vector for the left U module $V(-\mu)$. Then the remaining $c_{w\mu} : w \in W$ can be obtained by applying appropriate divided powers of the $e_i : i = 1, 2, \dots, \ell$. Thus the first assertion in (*) follows from (**). The second assertion is similar.

The above analysis shows that $c_w := \{c_{w\mu} : \mu \in P^+(\pi)\}$ (resp. $d_w := \{d_{w\mu} : \mu \in P^+(\pi)\}$) is a multiplicatively closed subset of R^+ (resp. R^-) of commuting elements.

Lemma. *For all $w \in W$*

(i) *c_w is Ore in R^+.*
(ii) *d_w is Ore in R^-.*
(iii) *Both c_w and d_w are Ore in R.*

The assertions for $w = w_0$ follow from 9.1.4. Then the general case follows from A.2.9.

9.1.11 Set $c = c_{w_0}$, $d = d_{w_0}$. The localized algebra $c^{-1}R^+$ contains

$$S^+ := \sum_{\mu \in P^+(\pi)} c_{w_0\mu}^{-1} V^+(\mu)^*$$

which by 9.1.7 (i) is isomorphic to U^-. Set $c_i = c_{w_0\omega_i}, r_i = c_{w_0 s_i \omega_i, \omega_i}^{\omega_i}$. Then the $x_i := c_i^{-1} r_i : i = 1, 2, \dots, \ell$ form a set of generators of S^+. Similarly the localized algebra $d^{-1}R^-$ contains

$$S^- := \sum_{\mu \in P^+(\pi)} d_{w_0\mu}^{-1} V^-(\mu)^*$$

which by 9.1.7 (ii) is isomorphic to G^+ being in turn generated by the $y_i := d_i^{-1} s_i$ where $d_i = d_{-\omega_i}$, $s_i = c_{-w_0 s_i \omega_i, -\omega_i}^{-w_0\omega_i} : i = 1, 2, \dots, \ell$.

From 9.1.4 one obtains

$$s_j c_i = q^{-(\omega_i,\alpha_j)} c_i s_j \ , \quad r_j d_i = q^{(\omega_i,\alpha_j)} d_i r_j \ .$$

Let δ_{ij} denote the Kronecker delta. Define i_0 by $\alpha_{i_0} = -w_0\alpha_i$. Applying f_{i_0} on the right to the first of these relations gives

$$\delta_{ij} d_j c_i + q^{(\alpha_i, s_j\omega_j)} s_j r_i = q^{-(\omega_i,\alpha_j)} (\delta_{ij} q^{-(\alpha_i,\omega_i)} c_i d_j + r_i s_j) \ .$$

By 9.1.3 (*) one has $d_j c_i = c_i d_j$ and so

$$\begin{aligned}\delta_{ij}(1-q_i^{-2}) &= q^{-(\omega_i,\alpha_j)}(c_i d_j)^{-1} r_i s_j - q^{(\alpha_i,s_j\omega_j)}(c_i d_j)^{-1} s_j r_i \\ &= x_i y_j - q^{-(\alpha_i,\alpha_j)} y_j x_i \ .\end{aligned}$$

Apart from the factor $(1-q_i^{-2})$, which can be absorbed in x_i, this is just the relation of 3.1.3 (*). Since the x_i (resp. y_j) already satisfy the quantum Serre relations it follows that the subalgebra of the localization $c^{-1}d^{-1}(R^+ \circledast R^-)$ they generate is just the quantum Weyl algebra $A_q(\mathfrak{g}_C)$ which in particular is $U^- \otimes G^+$ as a vector space. Now take the filtration $\mathcal{F}_p$ on U^- (and on G^+) defined in 7.4.3. Then $gr_{\mathcal{F}_p \times \mathcal{F}_p}(A_q(\mathfrak{g}_C)) \cong gr_{\mathcal{F}_p} U^- \otimes gr_{\mathcal{F}_p} G^+$ (where now weight vectors in $gr_{\mathcal{F}_2} U^-$ must q-commute with those in $gr_{\mathcal{F}_2} G^+$) is noetherian by (the argument of) 7.4.5. Moreover it is clear from 9.1.7 and the above formulae that $R^+ \circledast R^-$ is the twisted Rees algebra

$$R(A_q(\mathfrak{g}_C),\, \mathcal{F}_p \times \mathcal{F}_p,\, \chi \times \chi) \quad \text{where} \quad \chi_\mu(\nu) = q^{-(\mu,\nu)} \ .$$

Thus 7.4.6 gives the

Proposition. *The algebra $R^+ \circledast R^-$ is noetherian.*

9.1.12 Take $\mu \in P^+(\pi)$. By 7.1.16 (ii) there exists a unique up to scalars right U invariant element $x_\mu \in V^+(\mu)^* \otimes V^-(-w_0\mu)^*$. Let X denote the subspace of $R^+ \circledast R^-$ generated by the $x_\mu : \mu \in P^+(\pi)$. Every right U invariant element belongs to X.

Lemma

(i) $X \subset Z(R^+ \circledast R^-)$.

(ii) *X is a polynomial algebra on generators $x_{\omega_i} : i = 1, 2, \ldots, \ell$.*

It is clear that x_μ has zero weight as a $\Delta(T) - \Delta(T)$ weight vector. Then by 9.1.4 (i) (resp. 9.1.4 (ii)) it follows that x_μ commutes with $c^\nu_{w_0\nu,\nu}$ (resp. $c^\nu_{\nu,w_0\nu}$). The latter generates $V^+(\nu)^*$ (resp. $V^-(\nu)^*$) as a right U module. Since x_μ is invariant (i) follows.

It follows from the multiplication rules in $R^+ \circledast R^-$ (in particular 9.1.3 (*)) that $V^+(\nu)^* V^-(-w_0\mu)^* = V^-(-w_0\mu)^* V^+(\nu)^*$. Combined with the Cartan multiplication rule (for example 9.1.6) it follows that $x_\mu x_\nu = x_{\mu+\nu}$ up to a non-zero scalar (4.1.9 (ii)) which by (i) is independent of ordering and hence can be assumed to be equal to 1. Since the x_μ are obviously linearly independent (ii) follows.

9.1.13 Recall 9.1.9, 9.1.10 and the notation of 9.1.11. From 9.1.4 one obtains the

Lemma. *For all* $\mu \in P^+(\pi)$

(i) $c_{w_0\mu}^{-1} d_{-\mu}$ *is central in* $c^{-1}R$.
(ii) $d_{w_0\mu}^{-1} c_{-\mu}$ *is central in* $d^{-1}R$.

9.1.14 An argument closely paralleling 9.1.9 gives

Lemma. *The set of invertible elements of* R *reduces to scalars.*

9.1.15 It is clear from the calculations in 9.1.11 that the localized algebra $c^{-1}d^{-1}(R^+ \circledast R^-)$ is isomorphic to the smash product $A_q(\mathfrak{g}_C)\#(\check{U}^0 \otimes \check{U}^0)$. In particular it is isomorphic to $U^- \otimes U^+ \otimes \check{U}^0 \otimes \check{U}^0$ as a vector space. Let $\mathcal{F}_0$ denote the trivial filtration on U^0 and $\mathcal{F}_2$ the filtration on U^- and on G^+ defined in 7.4.3. Consider the filtration induced on $R^+ \circledast R^-$ by $\mathcal{F}_4 := \mathcal{F}_2 \otimes \mathcal{F}_2 \otimes \mathcal{F}_0 \otimes \mathcal{F}_0$. Then by 9.1.7 one has

$$gr_{\mathcal{F}_4}(R^+ \circledast R^-) = \bigoplus_{\lambda,\mu \in P^+(\pi)} (gr_{\mathcal{F}_2} K(\lambda)^- \otimes_{gr_{\mathcal{F}_2}} K(-\mu)^+)\#(k(q)\tau(-\lambda) \otimes k(q)\tau(-\mu)) .$$

(Notice this doesn't quite have the symmetry one might naively expect). As in 8.4.16 one checks that this algebra is finitely q-commutative. Then by A.3.4 one obtains the

Proposition. *The algebra* $R^+ \circledast R^-$ *satisfies the nullstellensatz.*

9.2 Surjectivity and Injectivity Theorems

9.2.1 Recall (4.3.10) the semisimplicity of finite dimensional U modules.

Lemma. *For all* $\lambda, \mu, \nu \in P^+(\pi)$ *one has*

(i) $V(\mu) \otimes V(\lambda) \hookrightarrow \bigoplus\{V(\mu+\nu) \mid \nu \in \Omega(V(\lambda)),\ \mu+\nu \in P^+(\pi)\}$.
(ii) $\mathrm{Hom}_U(V(\lambda), V(-w_0\mu) \otimes V(\nu)) \hookrightarrow V(\lambda)_{\nu-\mu}$ *as* T *modules.*
(iii) *Both maps are isomorphisms if* $\mu + \Omega(V(\lambda)) \subset P^+(\pi)$.

(i) is a special case of 6.4.16. It may also be obtained in a more elementary fashion from 8.1.7. Then (i), Frobenius reciprocity and the isomorphism $V(-w_0\mu)^* \cong V(\mu)$ of left U modules give $\mathrm{Hom}_U(V(\lambda), V(-w_0\mu) \otimes V(\nu)) \xrightarrow{\sim} \mathrm{Hom}_U(V(\mu) \otimes V(\lambda), V(\nu)) \hookrightarrow V(\lambda)_{\nu-\mu}$. Again 6.4.16 gives (iii).

Remark. The hypothesis of (iii) holds if $\mu \gg \lambda$, that is if $(\mu - \lambda, \alpha_i^\vee)$ is sufficiently large for all i.

9.2.2 Set $P^{++}(\pi) = \rho + P^+(\pi)$. By A.1.1 (vii) one has $P^{++}(\pi) = \{\lambda \in P^+(\pi) \mid \mathrm{Stab}_W \lambda = \{e\}\}$ and such weights are generally called regular. However this does not quite coincide with the notion of regular given in 8.4.9 since the translation by ρ had already been introduced there.

Set

$$R^{++} = \bigoplus_{\mu \in P^{++}(\pi)} V^+(\mu)^* , \quad R^{--} = \bigoplus_{\mu \in P^{++}(\pi)} V^-(\mu)^*$$

which are subalgebras of R^+, R^-. For what follows ρ may be replaced by any positive integer multiple of itself.

Proposition. *The multiplication map $\Delta^* : R^+ \circledast R^- \longrightarrow R$ is surjective even when restricted to $R^{++} \circledast R^{--}$. In particular R is noetherian and satisfies the nullstellensatz.*

Take $\lambda \in P^+(\pi)$. View $V(\lambda)^*$ and R as right U modules. Then by 4.3.10 the conclusion of 1.4.13 applies and $V(\lambda)^*$ occurs exactly $\dim V(\lambda)^*$ times in R. Through the left $\Delta(T)$ action it follows that the restriction of Δ^* to $V^+(-w_0\mu)^* \otimes R^-$ is generated by its intersection with the $V^+(-w_0\mu)^* \otimes V^-(\nu)^*$ and hence by 9.1.8 is injective. Then by 9.2.1 the multiplicity of $V(\lambda)^*$ in the image of $V^+(-w_0\mu)^* \otimes R^{--}$ is $\dim V(\lambda)^*$ for all $\mu \gg \lambda$. In view of the definition (9.1.1) of R this proves the required surjectivity.

9.2.3 Given $\mu \in P^+(\pi)$ write $\mu = \sum k_i \varpi_i$ and set $\|\mu\| = \sum k_i$. Through the Cartan multiplication rule 9.1.6 (∗) one may define a gradation $\mathcal{G}$ on R^+ by taking the elements of $V^+(\mu)^*$ to have degree $\|\mu\|$. A similar result holds for R^- and then $\mathcal{G} \times \mathcal{G}$ defines a gradation of $R^+ \circledast R^-$ with values in $\mathbb{N}$. Obviously each gradation subspace $(\mathcal{G} \times \mathcal{G})_m(R^+ \circledast R^-)$, or simply $(R^+ \circledast R^-)_m$, is a finite dimensional right U module. One may therefore define for each $\mu \in P^+(\pi)$ the Poincaré series

$$\mathcal{R}_\mu(z) := \sum_{m=0}^{\infty} [(R^+ * R^-)_m : V(\mu)^*] z^m .$$

Recalling 9.1.12 one has

$$\mathcal{R}_0(z) = \sum_{m=0}^{\infty} (\dim X_m) z^m = \frac{1}{(1 - z^2)^\ell} ,$$

where $\ell = |\pi|$. Recall the notation of 8.1.2.

Proposition. *For all* $\mu \in P^+(\pi)$

(i) $(R^+ \circledast R^-)_{V(\mu^*)}$ *is a finitely generated* X *module.*
(ii) $\mathcal{P}_\mu(z) := \mathcal{R}_\mu(z)/\mathcal{R}_0(z)$ *is a polynomial in* z.
(iii) $\mathcal{P}_\mu(1) = \dim V(\mu)^*$.

As in 9.1.12 the multiplication rules in $R^+ \circledast R^-$ give

$$(*) \qquad x_\lambda(V^+(\nu)^* \otimes V^-(-w_0\nu')) \subset V^+(\nu+\lambda)^* \otimes V^-(-w_0(\nu'+\lambda))^*$$

for all $\lambda, \nu, \nu' \in P^+(\pi)$. Set $V^* = V(\mu)^*$. It follows from $(*)$ and 9.1.9 (ii) that left multiplication by the invariant element x_λ defines an injection of $(V^+(\nu)^* \otimes V^-(-w_0\nu')^*)_{V^*}$ into $(V^+(\nu+\lambda)^* \otimes V^-(-w_0(\nu'+\lambda)^*))_{V^*}$. Now by 9.2.1 (ii) the multiplicity of $V(\mu)^*$ in $V^+(\nu)^* \otimes V^-(\nu')^*$ is uniformly bounded by $\dim V(\mu)^*_{\nu-\nu'}$. Use the convention that $V^\pm(\xi)^* = 0$ if $\xi \in P(\pi) \setminus P^+(\pi)$. It follows exactly as in 8.1.5 that any fixed "diagonal" namely

$$\left(\bigoplus_{\lambda \in P(\pi)} V^+(\nu+\lambda)^* \otimes V^-(-w_0(\nu'+\lambda))^* \right)_{V^*}$$

is finitely generated over X. Since only finitely many such diagonals are non-zero (i.e. those for which $\nu - \nu' \in \Omega(V(\mu)^*)$) the assertion of (i) follows.

(ii) follows from (i) by A.3.9(2). Finally (iii) is clear from 9.2.1 and the calculation in (i).

9.2.4 Since the trivial right U module occurs with multiplicity one in R it follows that the invariants $x_\mu : \mu \in P^+(\pi)$ have scalar images in R. These elements may be described canonically as follows. Let $\{v_i\}$ be a basis for $V(\mu)$ formed from weight vectors with $v_1 = u_\mu$ and $\{\xi_i\}$ the dual basis for $V(\mu)$ (resp. $V(\mu)^*$). Then $\xi_1 = \xi_\mu$. One checks that $x_\mu := \sum_i c^\mu_{\xi_i,u_\mu} \otimes \sigma^{*-1}(c^\mu_{\xi_u,v_i})$ satisfies $(\Delta^* x_\mu)(a) = \varepsilon(a)$ for all $a \in U$. Hence $\Delta^* x_\mu$ is a bi-invariant element of R, so equal to 1. Now take $V = V(-w_0\mu)$ in 1.4.8 and identify $V(-w_0\mu)^\sigma$ with $V(\mu)^*$ through 7.1.16. Then $\varphi^{-1}(\xi_\mu)$ is a lowest weight vector $u_{-\mu} \in V(-w_0\mu)$ and $\sigma^*(c^{-w_0\mu}_{\varphi^*(v_i),u_{-\mu}}) = c^\mu_{\xi_\mu,v_i}$ by 1.4.8. Then $x_\mu \in V^+(\mu)^* \otimes V^-(\mu)^*$ and so by 9.1.8 is the required invariant element. Setting $\eta_i = \varphi^*(v_i)$, then $\{\eta_i\}$ is the dual basis in $V(-w_0\mu)^*$ corresponding to $\{\xi_i\}$ and one may write

$$(*) \qquad x_\mu = \sum_i c^\mu_{\xi_i,u_\mu} \otimes c^{-w_0\mu}_{\eta_i,u_{-\mu}} \;.$$

Corollary. *The kernel of the multiplication map* $\Delta^* : R^+ \circledast R^- \longrightarrow R$ *is generated by its intersection with* X *namely* $\sum_i (x_{\omega_i} - 1)X$.

Recall the notation of 9.2.4. It is enough to show the restriction of Δ^* to each X module $(R^+ \circledast R^-)_{V^*}$ has the above property.

For each $\mathbf{a} = (a_1, a_2, \ldots, a_\ell) \in k(q)^\ell$ let $X(\mathbf{a}) \in \operatorname{Max} X$ denote the ideal generated by the $x_{\omega_i} - a_i : i = 1, 2, \ldots, \ell$. Consider the left action of $\tau(\lambda)$ on the first factor R^+ in $R^+ \circledast R^-$. This sends x_μ to $q^{(\lambda,\mu)} x_\mu$ and hence $X(\mathbf{a})$ to $X(\tau(-\lambda)\mathbf{a})$ where $(\tau(-\lambda)\mathbf{a})_i := q^{-(\lambda,\omega_i)} a_i$. In particular the T orbit Ω of $\mathbf{1} := (1, 1, \ldots, 1)$ is Zariski dense in $\operatorname{Max} X$. Moreover it is immediate that

$$[R^+ \circledast R^- / (R^+ \circledast R^-) X(\mathbf{a}) : V^*]$$

is independent of $\mathbf{a} \in \Omega$ and it remains to show that this common value is $[R : V^*]$ which by 1.4.13 equals $\dim V^*$.

Since $M := (R^+ \circledast R^-)_{V^*}$ is finitely generated over the polynomial ring X the generic flatness lemma (A.3.5) gives $f \in X$ non-zero such that $M_f := X_f \otimes_X M$ is free over the localized ring X_f. Choose $\mathbf{a} \in \Omega$ such that $f \notin X(\mathbf{a})$. Then $X(\mathbf{a})_f$ is a maximal ideal of X_f and the dimension of $M_f / M_f X(\mathbf{a})_f = M / M X(\mathbf{a})$ is just the rank of M_f over X_f which by 9.2.3 (iii) equals $(\dim V^*)^2$ as required.

9.2.5 Take $\lambda \in P^+(\pi)$ and set $V^* = V(\lambda)^*$, $M = (R^+ \circledast R^-)_{V^*}$. As shown in 9.2.2 there exists $\mu \in P^+(\pi)$ such that $(V^+(\mu)^* \otimes R^-)_{V^*}$ surjects to R_{V^*}. By semisimplicity there exists a right U module $\mathcal{H}_{V^*}$ complementing $\ker \Delta^*$ in $(V^+(\mu)^* \otimes R^-)_{V^*}$. Then 9.1.4 gives the direct sum decomposition

$$\mathcal{H}_{V^*} \oplus M X(\mathbf{1}) = M \ .$$

Using the left action of T on the first factor, $\mathbf{1}$ can be replaced in the above by any $\mathbf{a} \in \Omega$. Moreover T can be enlarged in an obvious way to make $\Omega = k(q)^{*\ell}$ or even $\overline{k(q)}^{*\ell}$ when $\overline{k(q)}$ replaces $k(q)$.

Let $\mathcal{O}$ be the multiplicatively closed subset of X generated by the x_{ω_i} : $i = 1, 2, \ldots, \ell$. Over $\overline{k(q)}$, $\operatorname{Max} \mathcal{O}^{-1} X$ identifies with $\overline{k(q)}^{*\ell}$. It follows from the above and A.2.12 that M is freely generated by $\mathcal{O}^{-1} X$ over $\mathcal{H}_{V^*}$. Finally setting $\mathcal{H} = \oplus_{\mu \in P^+(\pi)} \mathcal{H}_{V(\mu)^*}$ one obtains the

Corollary. $\mathcal{O}^{-1}(R^+ \circledast R^-) = \mathcal{H} \otimes \mathcal{O}^{-1} X$.

9.2.6 One may ask if $R^+ \circledast R^-$ is free over X which would be is a much deeper result than 9.2.5. It turns out that this can be proved by an analysis exactly following 7.3. As the result does not seem to be of great interest (partly because $R^+ \circledast R^-$ is of lesser importance than R) the proof is only briefly indicated.

By 9.1.7 and 9.1.11 one obtains the vector space isomorphism

$$(*) \qquad R^+ \circledast R^- = \bigoplus_{\lambda,\mu \in P^+(\pi)} K(\lambda)^- \otimes K(-\mu)^+ \otimes k(q)\tau(-\lambda) \otimes k(q)\tau(-\mu) \ .$$

Moreover as an algebra the factor $K(\lambda)^-$ (resp. $K(-\mu)^+$) is just that defined in 7.1.1 being in particular a subspace of U^- (resp. G^+). Moreover the canonical generators x_i (resp. y_j) of U^- (resp. G^+) satisfy (9.1.11) the relation $x_i y_j - q^{-(\alpha_i,\alpha_j)} y_j x_i = \delta_{ij}(1-q_i^{-2})$. This only differs trivially from the relation (cf. 7.4.7) found previously namely $f_i g_j - q^{(\alpha_i,\alpha_j)} g_j f_i = \delta_{ij}/(q_i - q_i^{-1})$. This already gives a somewhat surprising result.

Lemma. *There is an algebra isomorphism of $gr_{\mathcal{F}} F(\check{U})$ onto the diagonal component*

$$\bigoplus_{\lambda \in P^+(\pi)} K(\lambda)^- \otimes K(-\lambda)^+ \otimes k(q)\tau(-2\lambda)$$

of $R^+ \circledast R^-$.

9.2.7 Recall the notation of 7.3.3. For all $\mu, \nu \in P^+(\pi)$, set $K(\mu,\nu) = K(\mu)^- \otimes K(-\nu)^+$ and for all $\lambda \in P^+_\mu(\pi) \cap P^+_\nu(\pi)$ set $K^{\mu,\nu}(\mu-\lambda, \nu-\lambda) := x_\lambda K(\mu-\lambda, \nu-\lambda)$ which by 9.2.3 $(*)$ lies in $K(\mu,\nu)$. Then the common basis theorem 6.2.19 gives the following result which is nothing more than a generalization of 7.3.3 to the off-diagonal case.

Corollary. *For all $\mu, \nu \in P^+(\pi)$ and $\lambda_0, \lambda_1, \ldots, \lambda_n \in P^+_\mu(\pi) \cap P^+_\nu(\pi)$*

$$\begin{aligned} K^{\mu,\nu}(\mu-\lambda_0, \nu-\lambda_0) \cap \sum_{j=1}^n K^{\mu,\nu}(\mu-\lambda_j, \nu-\lambda_j) \\ = \sum_{j=1}^n K^{\mu,\nu}(\mu-\lambda_0 \cap \mu-\lambda_j, \nu-\lambda_0 \cap \nu-\lambda_j)\ . \end{aligned}$$

9.2.8 By 9.1.9 (ii) it follows that $K^{\mu,\nu}(\mu-\lambda, \nu-\lambda)$ is isomorphic to $K(\mu-\lambda, \nu-\lambda)$ as a right U module. It is clear that 9.2.7 and 9.2.1 permit one to determine $(R^+ \circledast R^-)/(R^+ \circledast R^-)X(\mathbf{0})$ as a right U module. The result is the following

Proposition. *For all $\mu \in P^+(\pi)$*

$$[(R^+ \circledast R^-)/(R^+ \circledast R^-)X(\mathbf{0}) : V(\mu)^*] = \dim V(\mu)^*\ .$$

9.2.9 Through the multiplication rules in $R^+ \circledast R^-$ it follows that the decomposition $(*)$ of 9.2.6 is a $P^+(\pi) \times P^+(\pi)$ grading which is moreover U equivariant. Thus by semisimplicity (4.3.10) there exists a graded U stable complement $\mathcal{H}$ of $(R^+ \circledast R^-)X(\mathbf{0})$ in $R^+ \circledast R^-$. Through the grading it follows as in 7.3.5(i) that $\mathcal{H}X = R^+ \circledast R^-$. By A.3.9 one has $[R^+ \circledast R^- / (R^+ \circledast R^-)X(\mathbf{a}) : V(\mu)^*] \geq \mathcal{P}_\mu(1)$. On the other hand $gr((R^+ \circledast R^-)X(\mathbf{a})) \supset (R^+ \circledast R^-)X(\mathbf{0})$ and so the above estimate coupled with 9.2.8 and 9.2.3 (iii) gives equality. Consequently $\mathcal{H} \cap (R^+ \circledast R^-)X(\mathbf{a}) = 0$ for all $\mathbf{a} \in k(q)^\ell$. As in 7.2.8 this proves the

Theorem. *The multiplication map defines an isomorphism of $\mathcal{H} \otimes X$ onto $R^+ \circledast R^-$.*

Remark. Of course 9.2.5 which was not used here is a consequence of this result; but the proof in 9.2.5 is much more elementary.

9.2.10 Recall the notation $V^+ = U_q(\mathfrak{b}^+) = \tau(Q(\pi))U^+$ and set $\check{V}^+ = \check{U}_q(\mathfrak{b}^+) = \tau(P(\pi))U^+$ with similar formulae for $V^-, \check{V}^-$. The skew-Hopf pairing (3.1.8) from which U was built (3.2.5) becomes in the conventions of Chapter 5 (sending x_α to $f_{-\alpha}$, $y_{-\alpha}$ to e_α and q to q^{-1}) a skew-Hopf pairing φ of V^+, V^-. Adjoining appropriate fractional powers of q to $k(q)$ this extends to a skew-Hopf pairing φ of $\check{V}^+, \check{V}^-$ through

(1) $\varphi(\tau(\gamma)x,\ \tau(\delta)y) = q^{-(\gamma,\delta)}\varphi(x,y)$, for all $x \in U^+, y \in U^-, \gamma, \delta \in P(\pi)$.

Recall further that φ is non-degenerate, $\varphi(x,1) = \varepsilon(x)$ and satisfies (3.1.10 (v), (vii)) the rules (making it a skew-Hopf pairing)

(2) $$\varphi(x, yy') = (\varphi \times \varphi)(\Delta(x), y \otimes y')$$

(3) $$\varphi(xx', y) = (\varphi \times \varphi)(x' \otimes x, \Delta(y))$$

for all $x, x' \in \check{V}^+$; $y, y' \in \check{V}^-$.

In particular the map $\Phi : a \mapsto (b \mapsto \varphi(a,b))$ of $\check{V}^+$ into $(\check{V}^-)^*$ is injective, an antihomomorphism of algebras and a homomorphism of coalgebras. Similarly $\Phi' : b \mapsto (a \mapsto \varphi(a,b))$ of $\check{V}^-$ into $(\check{V}^+)^*$ is injective, a homomorphism of algebras and an antihomomorphism of coalgebras.

9.2.11 The Hopf algebra embedding $V^\pm \hookrightarrow U$ gives a Hopf algebra homomorphism $U^* \longrightarrow (V^\pm)^*$ which is *not* of course surjective. Let $R_q[B^\pm]$ denote the image of its restriction $\Psi^\pm$ to $R_q[G]$.

Recall 9.1.5 and set

$$J^\pm = \sum_{\mu \in P^+(\pi)} J^\pm_\mu(\mu,\mu) .$$

Proposition

(i) $\ker \Psi^+ = J^+$, $\ker \Psi^- = J^-$.
Take $\lambda \in P^+(\pi)$ *and* $\xi \in V(\lambda)^*_\mu$.

(ii) *There exists a unique* $g \in G^+_{\lambda-\mu}$ *such that* $\Psi^-(c^\lambda_{\xi,\lambda}) = \Phi(\tau(-\lambda)g)$.

(iii) *There exists a unique* $f \in U^-_{w_0\lambda-\mu}$ *such that* $\Psi^+(c^\lambda_{\xi,w_0\lambda}) = \Phi'(\tau(-\mu)f)$.

(i) By definition of Ψ^+ one has

(*) $$\ker \Psi^+ = \{c \in R \mid c(V^+) = 0\} .$$

By definition $J^+_\mu(\mu,\mu)$ is generated by the $c^\mu_{\eta,\mu}$ with $\eta < \mu$. Take $a \in V^+$. Since $c^\mu_{\eta,\mu}(a) = 0$ if $\eta < \mu$ then $\ker \Psi^+ \supset J^+$. On the other hand it is immediate from (*) that the restriction of Ψ^+ to R^- is injective. Again the presentation of 9.2.4 (*) gives

(**) $$1 = \Psi^+(1) = \Psi^+(\Delta^* x_\mu) = c^\mu_{\mu,\mu} c^{-w_0\mu}_{-\mu,-\mu} .$$

Then by 9.2.2 and recalling 9.1.10 (ii) it follows that $R_q[B^+] = \operatorname{Im}\Psi^+ = d_e^{-1}R^-$. In particular Ψ^+ and the canonical projection $R \longrightarrow R/J^+$ have the same image on each finite sum $\sum_{\mu<\lambda} V^+(\mu)^*V^-(\mu)^*$ in R^+R^- whose union is R by 9.2.2. Thus the inclusion $\ker\Psi^+ \supset J^+$ is an equality. Similarly $\ker\Psi^- = J^-$.

(ii) Take $b = \tau(\nu)f : f \in U^-_{-\eta}$. Then $c^\lambda_{\xi,\lambda}(b) = \xi(bu_\lambda) = q^{(\nu,\mu)}\xi(fu_\lambda)$ if $\eta = \lambda - \mu$ and is zero otherwise. Since φ restricts to a non-degenerate pairing on $U^+_\eta \times U^-_{-\eta}$ there exists a unique $e \in U^+_{\lambda-\mu}$ such that $\varphi(e,f) = c^\lambda_{\xi,\lambda}(f)$. Then 9.2.10 (1) gives $\varphi(\tau(-\mu)e, \tau(\nu)f) = q^{(\mu,\nu)}c^\lambda_{\xi,\lambda}(f) = c^\lambda_{\xi,\lambda}(b)$. Consequently $\Phi(\tau(-\lambda)g) = \Psi^-(c^\lambda_{\xi,\lambda})$ where $g = \tau(\lambda-\mu)e \in G^+_{\lambda-\mu}$, as required. The proof of (iii) is similar.

9.2.12 One concludes from 9.2.11 (i) that J^+ (resp. J^-) is rather larger than at first apparent and indeed for all $\lambda \in P^+(\pi)$; $\mu, \nu \in \Omega(V(\lambda))$ one has $c^\lambda_{\mu,\nu} \in J^+$ (resp. $c^\lambda_{\mu,\nu} \in J^-$) unless $\mu \geq \nu$ (resp. $\mu \leq \nu$). Thus the matrices that these coefficients form can be regarded as being lower (resp. upper) triangular in the respective quotient algebras.

Define $\varkappa'$ as in 7.2.1, and $\iota := \varkappa'\sigma$ as in 8.3.6.

Corollary

(i) $\Phi^- := \Phi\varkappa'$ *is a Hopf algebra isomorphism of* $\check{V}^-$ *onto* $R_q[B^-]$.

(ii) $\Phi^+ := \Phi'\iota$ *is a Hopf algebra isomorphism of* $\check{V}^+$ *onto* $R_q[B^+]$.

9.2.13 It is clear that Ψ^+ (resp. Ψ^-) is a map of $U_q(\mathfrak{b}^+)$ (resp. $U_q(\mathfrak{b}^-)$) bimodules. Recall that the multiplication map μ in $U_q(\mathfrak{g})$ transposes to the coproduct μ^* on $R_q[G]$.

Lemma. *The composed map* $\Psi = (\Psi^- \otimes \Psi^+)\mu^*$ *of* R *into* $R/J^- \otimes R/J^+$ *is injective.*

From 1.4.14 (i) and the above, Ψ is a map of $U_q(\mathfrak{b}^+) - U_q(\mathfrak{b}^-)$ bimodules. Thus as in say 9.1.9 if $\ker\Psi$ is non-zero it must contain a multiple of some $c^\lambda_{\lambda,\lambda}$. Yet up to a scalar which is non-zero by 9.2.11 (**) one has $\Psi(c^\lambda_{\lambda,\lambda}) = c^\lambda_{\lambda,\lambda} \otimes c^\lambda_{\lambda,\lambda}$. This contradiction proves the lemma.

9.2.14 Whereas Ψ is an algebra homomorphism it fails to be a Hopf algebra map. One may also calculate $\operatorname{Im}\Psi$ up to localization at the *Ore* subset c_e defined in 9.1.10. This is based on the formulae established below.

$$(*) \quad \Psi(c^\lambda_{\lambda,\mu}) = c^\lambda_{\lambda,\lambda} \otimes c^\lambda_{\lambda,\mu}\,, \quad \Psi(c^\lambda_{\mu,\lambda}) = c^\lambda_{\mu,\lambda} \otimes c^\lambda_{\lambda,\lambda} \quad \text{for all} \quad \mu \in \Omega(V(\lambda))\,.$$

By 9.2.12 there exists $f_{-(\lambda-\mu)} \in U^-_{-(\lambda-\mu)}$ such that $\Phi^-(f_{-(\lambda-\mu)}\tau(-\lambda)) = \Psi^-(c^\lambda_{\mu,\lambda})$. In particular if $\lambda = \varpi_i$, $\mu = s_i\varpi_i$ then $f_{-(\lambda-\mu)} = f_i$ which is one of the defining generators of U^-. Again $\Psi^-(c^\lambda_{\lambda,\lambda}) = \Phi^-(\tau(-\lambda))$. This settles the left hand equality in $(*)$.

The right hand equality is more subtle. First by 1.4.8 one can write $c^{\lambda}_{\lambda,\mu} = \sigma^{*-1}(c^{-w_0\lambda}_{-\mu,-\lambda})$ up to non-zero scalars. Since $\Phi^+\sigma^{-1} = \sigma^{*-1}\Phi^+$ it follows from 9.2.12(ii) that there exists $e_{\lambda-\mu} \in U^+_{\lambda-\mu}$ such that $\Phi^+(\tau(\lambda)e_{\lambda-\mu}) = \Psi^+(\sigma^{*-1}(c^{-w_0\lambda}_{-\mu,-\lambda}))$. In particular if $\lambda = \omega_i$ and $\mu = s_i\omega_i$, then $e_{\lambda-\mu} = e_i$ which is one of the defining generators of $\check{V}^+$. Again $\Psi^+(c^{\lambda}_{\lambda,\lambda}) = \Phi^+(\tau(\lambda))$. In view of 9.2.11, 9.2.12 all this gives the

Proposition. *The map $\Psi : R \longrightarrow R/J^- \otimes R/J^+$ extends to an injection of $c_e^{-1}R$ into $\check{V}^- \otimes \check{V}^+$. Moreover $\Psi(c_e^{-1}R)$ (resp. $\Psi(c_e^{-1}R^+)$, $\Psi(c_e^{-1}R^-)$) is generated by the $\tau(-\lambda) \otimes \tau(\lambda) : \lambda \in P(\pi)$ over $U^- \otimes U^+$ (resp. $U^- \otimes 1$, $1 \otimes U^+$).*

Remark. Notice that it is the antidiagonal copy of the torus that occurs in the image and not the diagonal copy.

9.3 The Adjoint Action

9.3.1. Recall the subalgebras (9.1.11)

$$S^+ = \sum_{\mu\in P^+(\pi)} c^{-1}_{w_0\mu}V^+(\mu)^* \,, \quad S^- = \sum_{\mu\in P^+(\pi)} d^{-1}_{w_0\mu}V^-(\mu)^*$$

of $c^{-1}R^+$ and $d^{-1}R^-$ respectively.

Proposition

(i) $c^{-1}_{w_0\mu}V^+(\mu)^*$ *is an* $\operatorname{ad} R$ *subalgebra of* S^+.

(ii) $d^{-1}_{w_0\mu}V^-(\mu)^*$ *is an* $\operatorname{ad} R$ *subalgebra of* S^-.

By 9.1.13 (i) the element $d^{-1}_{-\mu}c_{w_0\mu}$ is central in $d^{-1}R$ and hence transforms under $\operatorname{ad} R$ via the coidentity. Substitute the formula $\Delta^2(f_i) = f_i \otimes 1 \otimes 1 + t_i \otimes f_i \otimes 1 + t_i \otimes t_i \otimes f_i$ into 1.4.14 (iii) taking $\xi \in R$, $\eta \in d^{-1}_{-\mu}V^+(\mu)^*$ to be $T-T$ weight vectors and $a = f_i$. Then the central term on the right hand side is proportional to $(\operatorname{ad}\xi)(\eta.f_i)$ whilst the first and third terms lie in $(\operatorname{ad} R)\eta$. Now $d_{-\mu} = c^{-w_0\mu}_{-w_0\mu,-\mu}$ transforms as a right U^- eigenvector via the coidentity, whilst $c_{w_0\mu} = c^{\mu}_{w_0\mu,\mu}$ generates $V^+(\mu)^*$. Induction on weights thus gives $(\operatorname{ad} R)\eta \subset d^{-1}_{-\mu}V^+(\mu)^*$ for all $\eta \in d^{-1}_{-\mu}V^+(\mu)^*$. Multiplication by the invariant element $c^{-1}_{w_0\mu}d_{-\mu}$ then gives (i). The proof of (ii) is similar.

9.3.2 It follows from 1.3.1 that S^+ and S^- are locally finite R modules for the adjoint action.

Lemma

(i) $\operatorname{Ann}_R S^+ \supset J^-$.

(ii) $\operatorname{Ann}_R S^- \supset J^+$.

Clearly J^- is a V^- bimodule contained in the augmentation ideal of R. It therefore annihilates the invariant element $d_{-\mu}^{-1} c_{w_0\mu}$. Applying 1.4.14 (iii) as in 9.3.1 (i) then gives (i). The proof of (ii) is similar.

9.3.3 The conclusion of 9.3.2 (i) combined with 9.2.11 (ii) and 9.2.12(i) makes S^+ a $\check{V}^-$-algebra. Moreover (9.1.11) as an algebra S^+ is isomorphic to U^-. It is convenient to replace here U^- by the isomorphic algebra U^+ since the latter has a $\check{V}^-$-algebra structure through the filtration introduced in 5.3.1 and the adjoint action. Similar considerations apply to S^-.

Lemma

(i) *As a $\check{V}^-$-algebra S^+ is isomorphic to U^+. In particular equality holds in 9.3.2(i) and the U^- invariant elements of S^+ are the scalars.*

(ii) *As a $\check{V}^+$-algebra S^- is isomorphic to U^-. In particular equality holds in 9.3.2(ii) and the U^+ invariant elements of S^- are the scalars.*

(i) It is enough to check that the generator x_i (given in 9.1.11) has weight α_i and satisfies $(\operatorname{ad}\Phi^-(f_j))x_i = \delta_{ij}$ up to a non-zero scalar and a permutation of indices preserving the Cartan matrix.

Take $\xi = c_i := c_{\omega_i,\omega_i}^{\omega_i}$, $a = f_j$ in 1.4.14 (iii). Since $c_i.f_j = 0$ and $\sigma(f_j).c_i \in J^-$ it follows from 9.3.2 (i) that

$$(*) \qquad [(\operatorname{ad} c_i)\eta].f_j = q^{(\omega_i,\alpha_j)}(\operatorname{ad} c_i)(\eta.f_j)\ , \quad \text{for all} \quad \eta \in S^+\ .$$

Let η be the $\operatorname{ad} R$ invariant element $d_{-\mu}^{-1} c_{w_0\mu}$. In the notation of 9.1.11 one has $\eta.f_j = \eta x_{j_0}$. Substitution in $(*)$ above gives $(\operatorname{ad} c_i)x_{j_0} = q^{-(\omega_i,\alpha_j)}x_{j_0}$. Yet (9.2.11, 9.2.12) one has $\Phi(\tau(-\omega_i)) = c_i$. Consequently x_j has weight a_{j_0}.

It follows from the above that the weights of S^+ lie in $\mathbb{N}\pi$. Thus from weight space considerations $(\operatorname{ad}\Phi^-(f_{i_0}))x_j = \gamma_j\delta_{ij}$ for some $\gamma_j \in k(q)$. If $\gamma_j = 0$ for some j, then x_j would be an $\operatorname{ad} R$ eigenvector and hence satisfy $x_j a \in Rx_j$ for all $a \in R$. However this is incompatible with its commutation rule (9.1.11) with y_j. This completes the proof (i). The proof of (ii) is similar.

9.3.4 One may view 9.1.7 as a consequence of 7.1.22(iv). However $\mathcal{R}$ even when restricted to $(\operatorname{ad} U^{\pm})\tau(-2\lambda) : \lambda \in P^+(\pi)$ is not an algebra map owing to the $\frac{1}{2}$ factor in the Rosso form as compared to the expression for the skew-Hopf pairing φ. This explains why 2λ had to be replaced by λ in 9.1.7.

Recall that S^+ (resp. S^-) is a right $\check{U}$ module and for the adjoint action is a left $\check{V}^-$ (resp. $\check{V}^+$) module through the isomorphism of 9.3.3.

Proposition

(i) $(\operatorname{ad}\varkappa'(e))(s) = s.e$ *for all* $s \in S^+$, $e \in \check{V}^+$.

(ii) $(\operatorname{ad}\varkappa'(f))(s) = s.f$ *for all* $s \in S^-$, $f \in \check{V}^-$.

Since $\varkappa'$ is an antiautomorphism of algebras and an automorphism of coalgebras, it follows that both sides of (i) define a right $\check{V}^+$ algebra structure on S^+. Thus it suffices to verify the assertion on the generators x_i (given in 9.1.11) of S^+. In view of 9.3.3(i) it suffices to observe that $x_{i_0}.e_j = \delta_{ij}$ up to non-zero scalars which may be viewed as fixing the isomorphism $S^+ \xrightarrow{\sim} U^+$ of $\check{V}^-$-algebras. Similarly (ii) follows from 9.3.3(ii).

9.3.5 The adjoint action is particularly useful in studying prime ideals of R. However here it is inconvenient to have to localize. This difficulty may be avoided by twisting the adjoint action in a manner analogous to 7.1.1.

By 9.1.4 (ii), $d_{w_0\nu} : \nu \in P^+(\pi)$ is a normal element of R and therefore defines an automorphism $\tau_\nu : a \mapsto d_{w_0\nu} a d_{w_0\nu}^{-1}$. Let $\mathcal{T}$ denote the commutative semigroup $\{\tau_\nu\}_{\nu\in P^+(\pi)}$. Set $\tau_{-\nu} = \tau_\nu^{-1}$. Observe that $\tau_\nu(c^\mu_{\eta,\lambda}) = q^{(\lambda,w_0\nu)-(\eta,\nu)} c^\mu_{\eta,\lambda}$. By the multiplication rules in 1.4.6 it follows that $c^\mu_{\eta,\lambda} \mapsto q^{(\lambda-\eta,\nu)} c^\mu_{\eta,\lambda}$ is also an automorphism $\hat{\tau}_\nu$ of R. Recall 1.3.6 (2).

Lemma. *With respect to the coproduct μ^* and antipode σ^* of R one has $(\tau_\nu \otimes \tau_\nu)(1\otimes\sigma^*)\mu^* = (1\otimes\sigma^*)\mu^*\hat{\tau}_\nu$ for all $\nu \in P^+(\pi)$.*

Recall (1.4.8) that $\sigma^*(c^\mu_{\eta,\lambda}) = c^{-w_0\mu}_{-\lambda,-\eta}$ up to a non-zero scalar. Then by 1.4.7 (1)

$$
\begin{aligned}
(\tau_\nu \otimes \tau_\nu)(1\otimes\sigma^*)\mu^*(c^\mu_{\eta,\lambda}) &= \sum_{\varsigma\in\Omega(V(\mu)^*)} (\tau_\nu \otimes \tau_\nu)(1\otimes\sigma^*)\left(c^\mu_{\eta,\varsigma}\otimes c^\mu_{\varsigma,\lambda}\right)\\
&= \sum_{\varsigma\in\Omega(V(\mu)^*)} \tau_\nu(c^\mu_{\eta,\varsigma})\otimes\tau_\nu(c^{-w_0\mu}_{-\lambda,-\varsigma})\\
&= q^{(\lambda-\eta,\nu)}(1\otimes\sigma^*)\mu^*(c^\mu_{\eta,\lambda})\\
&= (1\otimes\sigma^*)\mu^*\hat{\tau}_\nu(c^\mu_{\eta,\lambda})\ ,
\end{aligned}
$$

as required.

9.3.6 In $b^{-1}R$ or $d^{-1}R$, τ_ν is just an inner automorphism and $(\mathrm{ad}_{\tau_\nu} a)b = d_{w_0\nu}(\mathrm{ad}\, a)(d^{-1}_{w_0\nu} b)$ for all $a, b \in R$. Again set $\nu_0 = -w_0\nu$ then by 9.1.4 (i) the map $a \mapsto c_{w_0\nu} a c^{-1}_{w_0\nu}$ is the automorphism τ_{ν_0} and becomes inner in $b^{-1}R$ or $c^{-1}R$. In this 9.3.4 becomes on writing $\Theta^\pm = (\Phi^\pm)^{-1}\Psi^+$ the

Corollary. *For all $a \in R$, $\nu \in P^+(\pi)$,*

(i) $(\mathrm{ad}_{\tau_{\nu_0}} a)s = s.\varkappa'^{-1}(\Theta^-(a))$, *for all $s \in V^+(\nu)^*$.*
(ii) $(\mathrm{ad}_{\tau_\nu} a)s = s.\varkappa'^{-1}(\Theta^+(a))$, *for all $s \in V^-(\nu)^*$.*

9.3.7 For each $w \in W$ set

$$
J^+_w = \sum_{\nu\in P^+(\pi)} J^+_\nu(w\nu,\nu)\ , \quad J^-_w = \sum_{\nu\in P^+(\pi)} J^-_\nu(ww_0\nu, w_0\nu)\ .
$$

Given $a \in R$, let $\bar{a}$ denote its image in R/J_w^+ (resp. R/J_w^-). Then by 9.1.5 (i) (resp. 9.1.5 (ii)) $\bar{c}_w = \{\bar{c}_{w\mu}\}_{\mu \in P^+(\pi)}$ (resp. $\bar{d}_w = \{\bar{d}_{w\mu}\}_{\mu \in P^+(\pi)}$) is Ore in R/J_w^+ (resp. R/J_w^-).

Lemma. *For all $\bar{a} \in R/J_w^+$, $\bar{b} \in R/J_w^-$ one has*

(i) $(\mathrm{ad}_{\tau_{\nu_0}} c^\mu_{\eta,\lambda})\bar{c}_{w\nu} = q^{(\eta, w_0\nu - w\nu)}\delta_{\eta,\lambda}\bar{c}_{w\nu}$

(ii) $\bar{c}^{-1}_{w\nu}(\mathrm{ad}_{\tau_{\nu_0}} c^\mu_{\eta,\lambda})\bar{a} = q^{(\eta, w_0\nu - w\nu)}(\mathrm{ad}\, c^\mu_{\eta,\lambda})\bar{c}^{-1}_{w\nu}\bar{a}$

(iii) $(\mathrm{ad}_{\tau_\nu} c^\mu_{\eta,\lambda})\bar{d}_{w\nu} = q^{(\eta, ww_0\nu - \nu)}\delta_{\eta,\lambda}\bar{d}_{w\nu}$

(iv) $\bar{d}^{-1}_{w\nu}(\mathrm{ad}_{\tau_\nu} c^\mu_{\eta,\lambda})\bar{b} = q^{(\eta, ww_0\nu - \nu)}(\mathrm{ad}\, c^\mu_{\eta,\nu})\bar{d}^{-1}_{w\nu}\bar{b}$.

For (i) use of 9.1.5 gives

$$\begin{aligned}(\mathrm{ad}_{\tau_{\nu_0}} c^\mu_{\eta,\lambda})\bar{c}_{w\nu} &= \sum_{\zeta \in \Omega(V(\mu)^*)} \tau_{\nu_0}(c^\mu_{\eta,\zeta})\bar{c}_{w\nu}\sigma^*(c^\mu_{\zeta,\lambda}) \\ &= \sum_{\zeta \in \Omega(V(\mu)^*)} q^{(\eta, w_0\nu) - (\zeta,\nu)} q^{(\zeta,\nu) - (\eta, w\nu)} \bar{c}_{w\nu} c^\mu_{\eta,\zeta}\sigma^*(c^\nu_{\zeta,\lambda}) \\ &= q^{(\eta, w_0\nu - w\nu)}\delta_{\eta,\lambda}\bar{c}_{w\nu}\end{aligned}$$

where one may notice that twisting eliminates dependence of the q-exponent on the summation variable. Then (ii) follows from (i) and (iii), (iv) are similar.

9.3.8 Fix $P \in \operatorname{Spec} R$. For each $\mu \in P^+(\pi)$ set

$$C_P^+(\mu) = \{\eta \in \Omega(V(\mu)) \mid c^\mu_{\eta,\mu} \notin P\}\,, \quad C_P^-(\mu) = \{\eta \in \Omega(V(\mu)) \mid c^\mu_{\eta,w_0\mu} \notin P\}\,.$$

If $C_P^+(\mu) = \emptyset$ set $D_P^+(\mu) = \emptyset$. Otherwise let $D_P^+(\mu)$ denote the set of minimal elements of $C_P^+(\mu)$. Define $D_P^-(\mu)$ similarly; but with maximal replacing minimal.

Proposition

(i) *There exists $w \in W$ such that $D_P^+(\mu) = \{w\mu\}$ for all $\mu \in P^+(\pi)$.*

(ii) *There exists $w' \in W$ such that $D_P^-(\mu) = \{w'w_0\mu\}$ for all $\mu \in P^+(\pi)$.*

Suppose $D_P^+(\mu_i) \neq \emptyset$ and take $\eta_i \in D_P^+(\mu_i) : i = 1, 2$. Then by definition $P \supset J^+_{\mu_i}(\eta_i, \mu_i)$. Let $\bar{c}_i$ denote the image of $c^{\mu_i}_{\eta_i,\mu_i}$ in R/P. By 9.1.5 (i) it follows that $\bar{c}_i$ is normal in R/P. Yet $\bar{c}_i \neq 0$ by definition of $D_P^+(u_i)$ and then P being prime implies that $\bar{c}_i$ is a non-zero divisor. Taking $\lambda = \mu_1, \lambda' = \mu_2$, $\eta = \eta_1, \eta' = \eta_2$ in 9.1.5 (i) gives

$$\bar{c}_2\bar{c}_1 = q^{(\mu_1,\mu_2) - (\eta_1,\eta_2)}\bar{c}_2\bar{c}_1\,.$$

Interchanging the roles of $\bar{c}_1$, $\bar{c}_2$ gives the same (!) exponent in q which must therefore vanish. That is $(\mu_1, \mu_2) = (\eta_1, \eta_2)$.

By 9.2.2 one must have $C_P^+(\mu) \neq 0$ for some $\mu \in P^{++}(\pi)$. For such an element it follows from the above and A.1.17 that $D_P^+(\mu) = \{w\mu\}$ for some $w \in W$. Then 9.1.10 (*) forces $w\omega_i \in C_P^+(\omega_i)$ for all i.

Assume that $w\omega_i \notin D_P^+(\omega_i)$ for some i. Then by definition there exists $\xi \in \mathbb{N}\pi \setminus \{0\}$ such that $w\omega_i - \xi \in D_P^+(\omega_i)$. Set $c = c^{\omega_i}_{w\omega_i-\xi,\omega_i}$. Then $c \notin P$ and its image $\bar{c}$ in R/P is normal, hence a non-zero divisor. Consequently $cc_{w\mu} \notin P$. By 1.4.6 this gives $w(\omega_i + \mu) - \xi \in D_P^+(\omega_i + \mu)$. Yet $\omega_i + \mu$ is regular so this contradicts the previous result.

Assume that there exists $\nu \in P^+(\pi)$ such that $D_P^+(w\nu) \supsetneq \{w\nu\}$. Then by A.1.17 and the above there exists $y \in W$ such that $y\nu \in D_P^+(w\nu) \setminus \{w\nu\}$. As in the previous paragraph this forces $w\mu + y\nu \in C_P^+(\mu + \nu)$. Since $\mu + \nu$ is regular the previous result gives $w\mu + y\nu \geq w(\mu + \nu)$ and so $y\nu \geq w\nu$. This contradicts $y\nu$ and $w\nu$ being distinct minimal elements of $C_P^+(\nu)$ and completes the proof of (i). The proof of (ii) is similar.

9.3.9 Fix $\mathbf{w} = (w_+, w_-) \in W \times W$ and let $\mathbf{X}(\mathbf{w})$ denote the set of all $P \in \operatorname{Spec} R$ such that $D_P^+(\mu) = \{w_+\mu\}$ and $D_P^-(\mu) = \{w_-w_0\mu\}$ for all $\mu \in P^+(\pi)$.

Corollary

$$\operatorname{Spec} R_q[G] = \coprod_{\mathbf{w} \in W \times W} \mathbf{X}(\mathbf{w}) .$$

9.3.10 The above result has the following consequence which is far from being a priori obvious.

Corollary. *For all $P \in \operatorname{Spec} R$ one has $P \cap c_e = \emptyset$. In particular the map $P \mapsto c_e^{-1}P$ is an isomorphism of* $\operatorname{Spec} R$ *onto* $\operatorname{Spec} c_e^{-1}R$.

By 9.3.9 there exists $\mathbf{w} = (w_+, w_-) \in W \times W$ such that $c_{w_+} \cap P = \emptyset$. Suppose $c_\mu \in P$ for some $\mu \in P^+(\pi)$. Then by 9.3.6(i), 9.3.7(ii) and 9.3.8 (ii) one obtains $1 = \bar{c}^{-1}_{w_+\mu}\bar{c}_{w_+\mu} \in (\operatorname{ad} R)\bar{c}^{-1}_{w_+\mu}\bar{c}_\mu = 0$. This contradiction proves the first part. The second part follows from A.2.8 and the noetherianity (9.2.2) of R.

9.3.11 A further non-trivial consequence of the above is the following. Set $J = J^+ + J^-$.

Lemma. *Let M be a finite dimensional R module*

(i) *M has no c_e torsion.*
(ii) *There exists $n \in \mathbb{N}^+$ such that $J^n M = 0$.*
(iii) *Over $\overline{k(q)}$ every simple finite dimensional R module is one dimensional.*

For (i) one may assume M simple. Then *Not* (i) means that M is c_e torsion and the finite dimensionality of M implies that $\operatorname{Ann} M \cap c_e \neq \emptyset$ contradicting 9.3.10.

For (ii) it is enough to show that a finite dimensional R module N satisfying $JN = N$ is the zero module. Set $\tilde{R} = \Psi(c_e^{-1}R)$ which contains the subalgebra $\check{U}_a^0$ generated by the $\tau(-\lambda) \otimes \tau(\lambda) : \lambda \in P(\pi)$. By (i) and 9.2.13 one may assume that N is an $\tilde{R}$ module. Under adjoint action the $\check{U}_a^0$ eigenspaces define a $Q^+(\pi) \times Q^+(\pi)$ gradation of $\check{V}^- \otimes \check{V}^+$ and of $\tilde{R}$. Moreover $\tilde{J} := \Psi(c_e^{-1}J) = \check{U}_a^0(U_+^- \otimes U^+ + U^- \otimes U_+^+)$ which complements the zero weight space in $\tilde{R}$. Write $\bar{N} := N \otimes_{k(q)} \overline{k(q)}$ as a direct sum of its generalized $\check{U}_a^0$ eigenspaces. If $\bar{N} \neq 0$, then at least one such eigenspace must be absent from $\tilde{J}\bar{N}$ proving (ii).

Finally by (i) and (ii) any simple finite dimensional R module is a simple module over the commutative subalgebra generated by $c_\mu : \mu \in P^+(\pi)$. Hence (iii).

Warning. It is false that c_μ is a normalizing element of R even though this is true in $c_e^{-1}R$.

9.3.12 Let M be a finite dimensional R module. By 9.3.11(i) the canonical homomorphism $m \mapsto 1 \otimes m$ of M into $c_e^{-1}R \otimes_R M$ is an embedding. Identify $c_e^{-1}R$ with its image under Ψ.

Proposition. *Let M be a finite dimensional $c_e^{-1}R$ module. There exists a finite dimensional $\check{V}^- \otimes \check{V}^+$ module N containing M as a submodule.*

Let $\check{U}_a^0$ (resp. $\check{U}_d^0$) denote the group algebra of the antidiagonal (resp. diagonal) copy $\tau(-\lambda) \otimes \tau(\lambda)$ (resp. $\tau(\lambda) \otimes \tau(\lambda)) : \lambda \in P(\pi)$ of the torus $\check{T}$ in $\check{V}^- \otimes \check{V}^-$. Then $A := \Psi(c_e^{-1}R)$ is just the smash product $(U^- \otimes U^+)\#\check{U}_a^0$ and $B := \check{V}^- \otimes \check{V}^+$ the smash product $A\#\check{U}_d^0$.

Set $I = (U^- \otimes U^+)_+$ which is stable under the adjoint action of $\check{T} \otimes \check{T}$. By 9.3.11 (ii) there exists $n \in \mathbb{N}^+$ such that $J^n M = 0$. Thus M is an $\bar{A} := (U^- \otimes U^+)/I^n\#\check{U}_a^0$ module and it suffices to construct a $\bar{B} := \bar{A}\#\check{U}_d^0$ module N containing M.

Let $K \subset \check{U}_d^0$ be an ideal of codimension 1. Since $\bar{A}$ is a direct sum of finitely many $\check{U}_d^0$ eigenspaces there exists an ideal $L \subset \check{U}_d^0$ of finite codimension such that $\bar{A}L \subset K\bar{A}$, that is $AL \subset KA \bmod \check{U}_d^0 AI^n$.

The induced module $B \otimes_A M$ is isomorphic to $\check{U}_d^0 \otimes M$ as a left $\check{U}_d^0$ module and hence contains $L \otimes M$ as a subspace. Yet $B(L \otimes M) \subset K \otimes M$ and so the composed map

$$M \overset{m \mapsto 1\otimes m}{\longrightarrow} B \otimes_A M \twoheadrightarrow (B \otimes_A M)/B(L \otimes M) \twoheadrightarrow B \otimes_A M/K \otimes M \overset{\sim}{\longrightarrow} M$$

of A modules is the identity map. Taking $N = (B \otimes_A M)/B(L \otimes M)$ which as a vector space is an image of $(B \otimes_A M)/(L \otimes M) \cong (\check{U}_d^0/L) \otimes M$ and hence finite dimensional, proves the required assertion.

9.3.13 Combined with 1.4.17 the above result already gives some information on the second Hopf dual of $U_q(\mathfrak{g})$. Recall 9.2.14 and 1.4.17.

Corollary. *By restriction Ψ^* gives a surjection* $\check{U}_q(\mathfrak{b}^-)^* \otimes \check{U}_q(\mathfrak{b}^+)^* \xrightarrow{\sim} (\check{U}_q(\mathfrak{b}^-) \otimes \check{U}_q(\mathfrak{b}^+))^* \twoheadrightarrow R_q(G)^*$ *of coalgebras.*

9.3.14 Set $\mathbf{w}_0 = (w_0, w_0)$. The following leads to a description of $\mathbf{X}(\mathbf{w}_0)$. By 9.1.10 $(*)$ the $c_{w_0\mu}$ (resp. $d_{-\mu}$) : $\mu \in P^+(\pi)$ form a commuting family. Then by 9.1.3 the $c_{w_0\mu}, d_{-\nu} : \mu, \nu \in P^+(\pi)$ form a commuting family. Set $b_{w_0\mu} = c_{w_0\mu}d_{-\mu}$. Then by 9.1.10 (iii) the set $\mathcal{b} := \{b_{w_0\mu} : \mu \in P^+(\pi)\}$ is Ore in R. Let A^+ (resp. A^-) denote the subalgebra of the centre of $c^{-1}R$ (resp. $d^{-1}R$) generated by the $c_{w_0\mu}^{-1}d_{-\mu}$ (resp. $d_{w_0\mu}^{-1}c_{-\mu}$) : $\mu \in P^+(\pi)$ and set $A = A^+A^-$. Then A^+ (resp. A^-) is a polynomial algebra on generators $a_{\omega_i} := c_{w_0\omega_i}^{-1}d_{-\omega_i}$ (resp. $a_{\omega_i}^{-1}$). Let S (resp. B) be the subalgebra of $\mathcal{b}^{-1}R$ generated by S^+, S^- (resp. and A). Obviously S and B are $\operatorname{ad} R$ invariant. As in 8.4.12 let $(\operatorname{Spec} B)^R$ denote the set of $\operatorname{ad} R$ invariant prime ideals of B.

Proposition

(i) *The multiplication map $S^+ \otimes S^- \longrightarrow S$ (resp. $S^+ \otimes S^- \otimes A \longrightarrow B$) is an isomorphism of vector spaces.*
(ii) *A is the set of* $\operatorname{ad} R$ *invariant elements of B.*
(iii) *The map $P \mapsto P \cap A$ is an isomorphism of* $(\operatorname{Spec} B)^R$ *onto* $\operatorname{Spec} A$.
(iv) *For all $P \in (\operatorname{Spec} B)^R$ the quotient algebra B/P is a domain.*

Surjectivity in (i) is clear from the relations in 9.1.11. Both injectivity in (i) and the assertion in (ii) will follow from the proof of (iii), (iv) corresponding to the special case $P = 0$.

Take $P \in (\operatorname{Spec} B)^R$ and set $Q = B(P \cap A)$. Since A belongs to the centre of B it follows that Q is an $\operatorname{ad} R$ stable ideal of B contained in P. Since P is prime, $P \cap A \in \operatorname{Spec} A$.

Suppose $b, b' \in B$ with $bb' \in Q$. By surjectivity in (i) one may write $b = \sum s_i \otimes a_i$, $b' = \sum s'_j \otimes a'_j$ with the $\{s_i\}$ (resp. $\{s'_j\}$) linearly independent elements of S and $a_i, a'_j \in A$. If the a_i (resp. a'_j) are not all zero, there exists a point m in the irreducible affine variety $\operatorname{Max} A/P \cap A$ such that the $a_i(m)$ (resp. $a_j(m)$) are not all zero. Then $b(m) := \sum s_i \otimes a_i(m) \neq 0$ and similarly $b'(m) \neq 0$. Yet $b(m)b'(m) = 0$ contradicting that S is a subalgebra of $d^{-1}R$ which by 9.1.9 (i) is a domain over any field extension. Hence B/Q is a domain. Of course the special case when $P = 0$ is immediate from 9.1.9 (i).

Set $\mathcal{S} := A \setminus P \cap A$ which is an Ore subset of A and hence of B. Since Q is prime it follows that P/Q has no $\mathcal{S}$ torsion. Set $F = \mathcal{S}^{-1}(A/P \cap A)$. To show that $P = Q$ it is enough to show that $\mathcal{S}^{-1}(B/Q) = S \otimes F$ has no proper $\operatorname{ad} R$ stable ideals.

Through 9.3.2 one has $(\operatorname{ad} a)s_+s_- = (\operatorname{ad} \Psi^-(a_1)s_+)(\operatorname{ad} \Psi^+(a_2)s_-)$ for all $s_\pm \in S^\pm$, $a \in R$. In particular by 9.2.14 $(*)$ the $c_{\lambda,\lambda}^\lambda : \lambda \in P^+(\pi)$ act like group-like elements. Consequently the action of R can be extended to an action of $c_e^{-1}R$. It is then immediate from 9.2.14 and 9.3.3 that every proper ideal of $S \otimes F$ contains a non-zero $\operatorname{ad} c_e^{-1}R$ invariant element and such an element is a scalar. This contradiction completes the proof of the proposition.

9.3.15 The extent to which B fails to be the whole of $b^{-1}R$ is examined below.

Lemma. *For all $\lambda \in P^+(\pi)$ one has $b_{w_0\lambda}^{-1} \in S \subset B$.*

Recall (9.2.4) that

$$\sum_{\nu \in \Omega(V(\lambda))} c_{\nu,\lambda}^{\lambda} c_{-\nu,-\lambda}^{-w_0\lambda} = 1 \,.$$

By 9.1.4 (ii) this can be written (changing possibly scalars in the sum) as

$$\sum_{\nu \in \Omega(V(\lambda))} (c_{w_0\lambda}^{-1} c_{\nu,\lambda}^{\lambda})(d_{-\lambda}^{-1} c_{-\nu,-\lambda}^{-w_0\lambda}) = b_{w_0\lambda}^{-1} \,.$$

Since each term in the left hand side belongs to S the assertion follows.

9.3.16 Identify $P(\pi)/2P(\pi) = P^+(\pi)/2P^+(\pi)$ with $\mathbb{Z}_2^\ell$. For each $\mu \in P^+(\pi)$ let $d(\mu)$ denote its image in $\mathbb{Z}_2^\ell$. The Cartan multiplication rule (9.1.6 (∗)) implies that R^+ is $P^+(\pi)$ graded and thus admits a $\mathbb{Z}_2^\ell$ gradation by taking $V^+(\mu)^*$ to have degree equal to $d(\mu)$. Similarly R^- is $\mathbb{Z}_2^\ell$ graded by taking $V^-(\mu)^*$ to have degree equal to $d(-w_0\mu) = d(w_0\mu)$. Recalling 9.1.8 take the $\mathbb{Z}_2^\ell$ gradation on $R^+ \otimes R^-$ obtained by adding degrees of the factors. Then (notation 9.2.4) $x_\mu : \mu \in P^+(\pi)$ has degree 0, so by 9.2.4 this $\mathbb{Z}_2^\ell$ gradation passes to the quotient R compatible with the right (but not left) action of U. A simple verification using 1.4.7 (1), 1.4.8 and 9.2.4 gives the

Lemma. *The above $\mathbb{Z}_2^\ell$ gradation of R satisfies the conditions* (i)–(iii) *of* 1.3.8.

9.3.17 Recall the notation of 1.3.1. Observe that b^{-1} is Ore in B. Moreover extending the above $\mathbb{Z}_2$ gradation of R to $b^{-1}R$ one has $bB = (b^{-1}R)_0$. (Here and below bB means the localization of B at its Ore set b^{-1}).

Proposition. *One has $B = F(b^{-1}R)$. In particular $A = Z(b^{-1}R)$.*

By 9.1.13 and 9.3.3 it remains to show that $F(b^{-1}R) \subset B$. By 9.3.16, $F(b^{-1}R)$ is a sum of its $\mathbb{Z}_2^\ell$ graded components. Take $a \in F(b^{-1}R)$ non-zero and homogeneous. Multiplying by the central element $d_{w_0\mu}^{-1}c_{-\mu}$ one may assume $a = d_{w_0\mu}^{-1}c : c \in R$, $\mu \in 2P^+(\pi)$. Now $d_{w_0\mu} = c_{\mu,w_0\mu}^{\mu}$ is invariant for the right action of V^-, whilst the right action of V^- on c generates a finite dimensional subspace of R. The compatibility of this action with $\operatorname{ad} R$ (1.4.13 (iii)) means that one may first assume that c is a T-eigenvector and then (as in the proof of 9.3.1) that it is V^- invariant. Then $c = c^{\nu}_{\nu,v}$ for some $v \in V(\nu)$. Since $V^-U^+ = U$ it follows that $\Psi^+(c) \neq 0$. Moreover a calculation similar to that in 9.2.11 gives $g \in G^+$ such that $\Psi^+(c) = \Phi^+(\tau(-\nu)g)$.

On the other hand $c^{\mu}_{w_0\mu,w_0\mu}c^{-w_0\mu}_{-w_0\mu,-w_0\mu} = 1 \bmod J^+$ and so it follows from 9.1.4 (ii) and 9.1.5 (i) that

$$(\operatorname{ad} c^{\mu}_{\eta,\lambda})(d_{w_0\mu}d_{\mu}) = \delta_{\eta,\lambda}q^{(\eta,\mu-w_0\mu)}d_{w_0\mu}d_{\mu} \bmod J^+ .$$

Consequently the hypothesis on $a = d^{-1}_{w_0\mu}c$ implies that $\Psi^+(d_\mu c) = \Phi^+(\tau(w_0\mu - \nu)g)$ is $\operatorname{ad}\check{V}^+$ locally finite. By 7.1.1 (iii) this implies that $w_0\mu - \nu \in -2P^+(\pi)$. Since $\mu \in 2P^+(\pi)$ one concludes that $\nu \in 2P^+(\pi)$ or equivalently that $a \in bB$.

It remains to show that $F(bB) = B$. By 9.3.15 one has $b_{w_0\mu} \in \operatorname{Fract} S$ and so by 9.3.14 (i) that $bB \cap (\operatorname{Fract} A)B \subset (\operatorname{Fract} S)A \cap S(\operatorname{Fract} A) = S \otimes A = B$. Thus it suffices to show that any finite dimensional $\operatorname{ad} R$ invariant subspace M of bB lies in $S(\operatorname{Fract} A)$.

The action of $\operatorname{ad} R$ on A is trivial (9.3.14) and so the action of $\operatorname{ad} R$ on $B = SA$ extends to an action on $S(\operatorname{Fract} A)$. The action of $\operatorname{ad} R$ on $S = S^+ \otimes S^-$ is described by 9.3.2 and 9.3.3. In particular the $U^- \otimes 1$ (resp. $1 \otimes U^+$) factor in $\Psi(c_e^{-1}R)$ acts on just S^+ (resp. S^-) and this action is its adjoint action on U^+ (resp. U^-) described in 5.3.1. In particular this action gives as in 3.1.7 a non-degenerate pairing on $U^+ \times S^-$ (resp. $U^- \times S^+$). Given $\mu = \sum r_i\alpha_i \in Q^+(\pi)$ set $|\mu| = \sum r_i$. Then for all $\mu, \nu \in Q^+(\pi)$ there exist by the above pairing, bases $\{u^i_{-\mu}\}, \{u^j_{\nu}\}$ of U^-, U^+ and bases $\{s^m_{\mu}\}, \{s^n_{-\nu}\}$ of S^+, S^- such that for all $r, r' \in \mathbb{N}$, $\mu, \nu, \mu', \nu' \in Q^+(\pi)$ satisfying $r = |\mu| \geq |\nu|$, $r' = |\mu'| \geq |\nu'|$ one has

$$(*) \qquad \operatorname{ad}(u^i_{-\mu} \otimes u^j_{\mu'})(s^m_{\nu} \otimes s^n_{\nu'}) = \delta_{i,m}\delta_{\mu,\nu}\delta_{j,n}\delta_{\mu',\nu'} .$$

By 9.2.13 and 9.3.11 (i), it follows that M has the structure of an $\tilde{R} = \Psi(c_e^{-1}R)$ module. Moreover setting $\tilde{J} = \Psi(c_e^{-1}J)$ there exists (9.3.11 (ii)) an $n \in \mathbb{N}^+$ such that $\tilde{J}^n M = 0$. Then by $(*)$ the assertion reduces to the case $n = 1$.

Set $I = \{a \in B \mid aM \subset B\}$ which is a non-zero left ideal of B. Since $JM = 0$ by hypothesis, one has

$$(\operatorname{ad} c^{\mu}_{\eta,\lambda})(a\, M) = ((\operatorname{ad} c^{\mu}_{\eta,\lambda})a)(\operatorname{ad} c^{\mu}_{\lambda,\lambda})M = ((\operatorname{ad} c^{\mu}_{\eta,\lambda})a)M .$$

Consequently I is $\operatorname{ad} R$ invariant and so by 9.3.3, as in the last part of the proof of 9.3.14, I contains a non-zero element $a \in A$. Then $aM \subset A$ by 9.3.14 (ii) and so $M \subset \operatorname{Fract} A$ as required.

9.3.18 Let R_0 denote the zero component of R corresponding to the $\mathbb{Z}_2^{\ell}$ gradation defined in 9.3.16. Recall the twisted adjoint action (9.3.5) and in particular the definition of the semigroup T. Then in the notation of 1.3.6 one has the

Corollary. $TF(R) = R_0$.

9.3.19 As noted in 9.1.11, S is a quantum Weyl algebra and in particular noetherian. Hence B is noetherian. By 9.3.15 one has $\mathfrak{b}^{-1} \subset S$ and so 9.3.3 gives $P \cap \mathfrak{b}^{-1} = \emptyset$ for all $P \in (\operatorname{Spec} B)^R$. Then by A.2.8 the map $P \mapsto \mathfrak{b}P$ of $(\operatorname{Spec} B)^R$ into $(\operatorname{Spec} \mathfrak{b}B)^R$ is bijective. (Here $\mathfrak{b}B$ means $(\mathfrak{b}^{-1})^{-1}B$, that is the localization of B at its Ore set $\mathfrak{b}^{-1}$.) Recall the notation of 9.3.5. Then $\tau_\nu : \nu \in P^+(\pi)$ is an automorphism of $\mathfrak{b}B$ which by 1.3.15 is *inner* if $\nu \in 2P^+(\pi)$. Yet its eigenvalues are powers of q and so have no torsion. Consequently these automorphisms leave ideals of $\mathfrak{b}B$ invariant. Since the $d_{w_0\nu} : \nu \in P^+(\pi)$ generate $\mathfrak{b}^{-1}R$ over $\mathfrak{b}B$ it follows that every ideal of $\mathfrak{b}B$ is $\operatorname{ad} R$ invariant. In particular $(\operatorname{Spec} \mathfrak{b}B)^R = \operatorname{Spec} \mathfrak{b}B$. Extend the $\Gamma := \mathbb{Z}_2^\ell$ gradation on R to $\mathfrak{b}^{-1}R$. Then the zero component $(\mathfrak{b}^{-1}R)_0$ equals $\mathfrak{b}B$. By 9.2.2, R is noetherian, hence so is $\mathfrak{b}^{-1}R$. Then by 9.3.16 the conclusion of 1.3.13 applies. (Moreover by the above and 1.3.14, the $\Gamma = P^+(\pi)/2P^+(\pi)$ orbits on $\operatorname{Spec} \mathfrak{b}B$ are trivial). Finally by the map $P \mapsto \mathfrak{b}^{-1}R$ is a bijection of $\mathbf{X}(\mathbf{w}_0)$ onto $\operatorname{Spec} \mathfrak{b}^{-1}R$. Thus 9.3.14 gives the

Theorem. *The map $P \mapsto A \cap \mathfrak{b}^{-1}P$ is an isomorphism of the space of Γ orbits in $\mathbf{X}(\mathbf{w}_0)$ onto* $\operatorname{Spec} A$.

9.3.20 Recall (9.1.12) the definition of the subalgebra $X \subset R^+ \circledast R^-$.

Theorem

(i) *$Z(R)$ reduces to scalars.*

(ii) *$Z(R^+ \circledast R^-) = X$.*

By 9.3.17 one has $Z(R) = R \cap A$. The right hand side is stable for the left action of T. Yet A is just the Laurent polynomial algebra on the a_{ω_i} which have T weight $-2\omega_i$. Thus if $R \cap A$ is not reduced to scalars it contains a non-trivial monomial in the $a_{\omega_i}, a_{\omega_j}^{-1} : i, j = 1, 2, \ldots, \ell$. Through 9.1.10 (*) such an element takes the form $c_{w_0\mu}^{-1} d_{-\mu}$ or $d_{-\mu}^{-1} c_{w_0\mu}$ for some $\mu \in P^+(\pi)$. Suppose $c_{w_0\mu}^{-1} d_{-\mu} = a \in R$. Then $d_{-\mu} = c_{w_0\mu} a$. Yet $d_{-\mu}$ is invariant for the left action of U^-. Since this action is locally finite on R, it follows as in 9.1.9 that $c_{w_0\mu}$ must be U^- invariant and so $\mu = 0$. A similar argument applies to $d_{-\mu}^{-1} c_{w_0\mu}$ and completes the proof of (i).

By 9.1.12 (i) it suffices to show that $z \in Z(R^+ \circledast R^-)$ belongs to X. By 9.2.5 one may write z in the form $z = \sum h_i \otimes x_i : h_i \in U$, $x_i \in \mathcal{O}^{-1}X$ with the h_i linearly independent. By 9.2.4 the surjection $(R^+ \circledast R^-) \longrightarrow R$ may be extended to a surjection of $\mathcal{O}^{-1}(R^+ \circledast R^-) \longrightarrow R$ in which the x_i become scalars and the h_i remain linearly independent. Yet by (i) the image of z is central hence scalar. Thus only one term can occur in the sum and this belongs to X, as required.

9.4 The R-Matrix

Let A be a bialgebra. As noted in 1.1.4 the category of A modules is closed under tensor product; but the map $\tau_{M,N} : m \otimes n \mapsto n \otimes m$ of $M \otimes N$ onto $N \otimes M$ may fail to be an A module map unless A is cocommutative. The R-matrix (when it exists) redresses this difficulty. The theory is briefly discussed here.

9.4.1 A tensor product on a category $\mathcal{C}$ is a functor $\mathcal{T}(M,N) \mapsto M \otimes N$ of $Ob\mathcal{C} \times Ob\mathcal{C} \longrightarrow Ob\mathcal{C}$. In particular given $M_1, M_2, N_1, N_2 \in Ob\mathcal{O}$, $f, f' \in \mathrm{Hom}(M_1, M_2)$, $g, g' \in \mathrm{Hom}(N_1, N_2)$ then $f \otimes f', g \otimes g' \in \mathrm{Hom}(M_1 \otimes M_2, N_1 \otimes N_2)$ are defined and satisfy

$$(f \otimes f')(g \otimes g') = fg \otimes f'g' .$$

Let $\mathcal{T}'$ denote the functor $(M,N) \mapsto N \otimes M$.

A category $\mathcal{C}$ with a tensor product is called monoidal if there is a functorial isomorphism $\mathcal{T}(\mathcal{T} \times \mathrm{Id}) \longrightarrow \mathcal{T}(\mathrm{Id} \times \mathcal{T})$ such that $\varphi_{M_1,M_2,M_3} : \mathcal{T}(\mathcal{T}(M_1, M_2), M_3) = (M_1 \otimes M_2) \otimes M_3 \longrightarrow M_1 \otimes (M_2 \otimes M_3)$ which is an isomorphism for all $M_1, M_2, M_3 \in Ob\mathcal{C}$, satisfies certain natural compatibility conditions (the pentagon identities). Here φ_{M_1,M_2,M_3} is just assumed to be the identity map.

A monoidal category $\mathcal{C}$ is said to be braided if there exists a functorial isomorphism $\mathcal{T} \longrightarrow \mathcal{T}'$ such that $R_{M,N} : \mathcal{T}(M,N) = M \otimes N \longrightarrow N \otimes M$, which is an isomorphism for all $M, N \in Ob\mathcal{O}$, satisfies certain natural compatibility conditions (the hexagon identities). In the simplest case $R_{M,N} = \tau_{M,N}$; but this will not be assumed here.

The hexagon identities can be expressed as

$$(H_1) \qquad R_{L,M\otimes N} = (\mathrm{Id}_M \otimes R_{L,N})(R_{L,M} \otimes \mathrm{Id}_N)$$

$$(H_2) \qquad R_{L\otimes M,N} = (R_{L,N} \otimes \mathrm{Id}_M)(\mathrm{Id}_L \otimes R_{M,N})$$

for all $M, N, L \in Ob\mathcal{C}$.

9.4.2 Let A be a Hopf algebra over k and suppose for each $M \in \mathrm{Mod}_f A$ (notation 1.4.5) that there exists $S_M \in \mathrm{End}_k M$ such that the diagram

$$\begin{array}{ccc} M & \xrightarrow{\ S_M\ } & M \\ {\scriptstyle f}\downarrow & & \downarrow{\scriptstyle f} \\ M' & \xrightarrow{\ S_{M'}\ } & M' \end{array}$$

commutes for all $f \in \mathrm{Hom}_A(M, M')$. Then for all $m \in M$, $\xi \in M'^*$ one has $f^*(\xi)(S_M m) = \xi(f S_M m) = \xi(S_{M'} f m)$ and so by 1.4.18 there exists $\mathbf{S} \in A^{**}$ such that $\mathbf{S}(c^M_{\xi,m}) = \xi(S_M m)$.

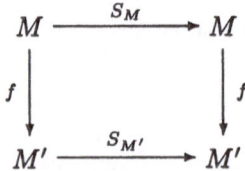

The tensor product on $\mathcal{C} = \mathrm{Mod}_f A$ makes $\mathcal{C}$ a monoidal category through the co-associativity (1.1.2) of Δ and 1.1.4. A braiding on $\mathcal{C}$ assigns to each pair $M, N \in \mathrm{Mod}_f A$ a linear map $R_{M\otimes N} := \tau_{N,M} R_{M,N} \in \mathrm{End}_k(M \otimes N)$ and the functoriality of $\mathcal{T} \longrightarrow \mathcal{T}'$ gives a commuting diagram as above with M replaced by $M \otimes N$ and S_M by $R_{M\otimes N}$. It follows that there exists $\mathbf{R} \in (A \otimes A)^{**} = (A^* \otimes A^*)^*$ such that

$$\mathbf{R}(c^{M\otimes N}_{\xi\otimes\eta, m\otimes n}) = (\xi \otimes \eta)(R_{M\otimes N}(m \otimes n)) = (\xi \otimes \eta)(\tau_{N,M} R_{M,N}(m \otimes n))$$

for all $M, N \in \mathrm{Mod}_f A$, $m \in M$, $n \in N$, $\xi \in M^*$, $\eta \in N^*$. Then the condition that $R_{M,N}$ is an A module isomorphism is equivalent to $\mathbf{R}$ possessing an inverse in $(A^* \otimes A^*)^*$ and to

$$(H_0') \qquad \mathbf{R}\Delta(x) = \Delta^{\mathrm{opp}}(x)\mathbf{R} \quad \text{for all} \quad x \in A$$

where $\Delta(x), \Delta^{\mathrm{opp}}(x)$ are viewed as elements of $(A \otimes A)^{**}$ under the canonical embedding.

Let $p_{12} : A \otimes A \longrightarrow A \otimes A \otimes A$ be the embedding $p_{12}(a \otimes b) = a \otimes b \otimes 1$ and similarly define p_{13}, p_{23}. Then by 1.4.17, p_{ij}^* restricts to a homomorphism p_{ij}^* of $A^* \otimes A^* \otimes A^* \longrightarrow A^* \otimes A^*$ and so defines a homomorphism $p_{ij}^{**} : (A^* \otimes A^*)^* \longrightarrow (A^* \otimes A^* \otimes A^*)^*$ which will simply be denoted by p_{ij}. Set $\mathbf{R}_{ij} = p_{ij}(\mathbf{R})$. Again $\mathrm{Id}_A \otimes \Delta : A \otimes A \longrightarrow A \otimes A \otimes A$ defines by twofold transposition a homomorphism $(A^* \otimes A^*)^* \longrightarrow (A^* \otimes A^* \otimes A^*)^*$. Consider

$$(H_1') \qquad (\mathrm{Id}_A \otimes \Delta)(\mathbf{R}) = \mathbf{R}_{13}\mathbf{R}_{12}$$

$$(H_2') \qquad (\Delta \otimes \mathrm{Id}_A)(\mathbf{R}) = \mathbf{R}_{13}\mathbf{R}_{23}$$

Lemma. *Under the correspondence $R_{M,N} \longrightarrow \mathbf{R}$ one has*

(i) $(H_1) \iff (H_1')$.
(ii) $(H_2) \iff (H_2')$.

Take $L, M, N \in \mathrm{Mod}_f A$. Then the action of $\mathbf{R}$ on $M \otimes N$ prescribed by the conclusion of 1.4.11 is just that of $\tau_{N,M} R_{M,N}$. Then $\mathbf{R}_{12}$ acts on $L \otimes M \otimes N$ by $\tau_{M,L} R_{L,M} \otimes \mathrm{Id}_N$ and $\mathbf{R}_{13}$ acts by $(\tau_{M,L} \otimes 1d_N)(1d_M \otimes \tau_{N,L} R_{L,N})(\tau_{L,M} \otimes 1d_N)$. Hence $\mathbf{R}_{13}\mathbf{R}_{12}$ acts by $(\tau_{M,L} \otimes 1d_N)(1d_M \otimes \tau_{N,L})(\mathrm{Id}_M \otimes R_{L,N})(R_{L,M} \otimes \mathrm{Id}_N)$ which is just the right hand side of (H_1) followed by the cyclic permutation $m \otimes n \otimes \ell \mapsto \ell \otimes m \otimes n$.

For all $a \in C^L$, $b \in C^M$, $c \in C^N$ one has

$$((1d_A \otimes \Delta)(\mathbf{R}))(a \otimes b \otimes c) = \mathbf{R}((1d_A \otimes \Delta)^*(a \otimes b \otimes c)) = \mathbf{R}(a \otimes bc) \ .$$

Moreover $bc \in C^{M\otimes N}$ where $M \otimes N$ is viewed as an A module via Δ. It follows that $(1d_A \otimes \Delta)(\mathbf{R})$ acts on $L \otimes M \otimes N$ by $\tau_{M\otimes N, L} R_{L, M\otimes N}$ which is the left hand side of (H_1) followed by the exchange of L and $M \otimes N$. The latter coincides with the cyclic permutation above. Hence (i). The proof of (ii) is similar.

9.4.3 An element $\mathbf{R} \in (A^\star \otimes A^\star)^*$ defines a pairing $\varphi : A^\star \times A^\star \longrightarrow k$ by $\varphi(b,a) = \mathbf{R}(a \otimes b)$ for all $a, b \in A^\star$.

Recall that μ^* is the coproduct on $A^\star$. Recall 3.2.1, 3.2.2.

Lemma

(i) $(H_1') \iff \varphi(ab,c) = (\varphi \times \varphi)(a \otimes b, \mu^{*\,\mathrm{opp}}(c)),\ \forall\, a,b,c \in A^\star$

(ii) $(H_2') \iff \varphi(a,bc) = (\varphi \times \varphi)(\mu^*(a), b \otimes c),\ \forall\, a,b,c \in A^\star$

(iii) *If* $(H_1'), (H_2')$ *hold, then* $\mathbf{R}$ *is invertible* $\iff \varphi(a,1) = \varepsilon(a) = \varphi(1,a)$, *for all* $a \in A^\star$. *In particular if* $\sigma_{A^\star}$ *is invertible,* φ *is a skew-Hopf pairing.*

Take $a,b,c \in A^\star$. Then for all $x, y \in A$ one has $p_{12}^*(a \otimes b \otimes c)(x \otimes y) = (a \otimes b \otimes c)(x \otimes y \otimes 1) = (a \otimes b)(x \otimes y)c(1)$ and so $p_{12}^*(a \otimes b \otimes c) = \varepsilon(c)(a \otimes b)$. Consequently $\mathbf{R}_{12}(a \otimes b \otimes c) = \varepsilon(c)\mathbf{R}(a \otimes b)$. Similarly $\mathbf{R}_{13}(a \otimes b \otimes c) = \varepsilon(b)\mathbf{R}(a,c)$. Then

$$\begin{aligned}(\mathbf{R}_{13}\mathbf{R}_{12})(a \otimes b \otimes c) &= (\mathbf{R}_{13} \otimes \mathbf{R}_{12})\mu^*(a \otimes b \otimes c)\\ &= \mathbf{R}_{13}(a_1 \otimes b_1 \otimes c_1)\mathbf{R}_{12}(a_2 \otimes b_2 \otimes c_2)\\ &= \varepsilon(b_1)\varepsilon(c_2)\varphi(c_1,a_1)\varphi(b_2,a_2)\\ &= \varphi(c,a_1)\varphi(b,a_2)\end{aligned}$$

whilst

$$(1d_A \otimes \Delta)(\mathbf{R})(a \otimes b \otimes c) = \mathbf{R}(a,bc) = \varphi(bc,a)\ .$$

Hence (i). The proof of (ii) is similar.

Assume $\mathbf{R}$ has inverse $\mathbf{R}^{-1}$ and set $\gamma(a) = \mathbf{R}(1 \otimes a),\quad \delta(b) = \mathbf{R}^{-1}(1 \otimes b)$. Then $\gamma(a_1)\delta(a_2) = \varepsilon(a) = \delta(a_1)\gamma(a_2)$, and so $\varphi(a,1) = \gamma(a) = \gamma(a_1)\varepsilon(a_2) = \gamma(a_1)\gamma(a_2)\delta(a_3) \overset{(H_2')}{=} \gamma(a_1)\delta(a_2) = \varepsilon(a)$. A similar argument using (H_1') gives $\varphi(1,a) = \varepsilon(a)$.

Conversely suppose $\varphi(b,1) = \varepsilon(b)$, and define $\tilde{\mathbf{R}} \in (A^\star \otimes A^\star)^*$ by $\tilde{\mathbf{R}}(a \otimes b) = \mathbf{R}(\sigma(a) \otimes b)$. Then $(\mathbf{R}\tilde{\mathbf{R}})(a \otimes b) = \varphi(b_1,a_1)\varphi(b_2,\sigma(a_2)) \overset{(H_1')}{=} \varphi(b, a_1\sigma(a_2)) = \varepsilon(a)\varphi(b,1) = \varepsilon(a)\varepsilon(b)$, that is $\mathbf{R}\tilde{\mathbf{R}} = 1$. Similarly $\tilde{\mathbf{R}}\mathbf{R} = 1$. (One does not need $\varphi(1,a) = \varepsilon(a)$ because this follows from (H_1'), (H_2'), $\varphi(a,1) = \varepsilon(a)$ and the existence of an antipode).

9.4.4 Under the hypothesis that φ is a skew-Hopf pairing, one may form the Drinfeld double $\mathcal{D}(A^\star, A^\star)$ defined by 3.2.2 - 3.2.4. As a vector space this is just $A^\star \otimes A^\star$. Recall that Δ^* is multiplication in $A^\star$. Then by 3.2.3 the map $\Delta^* : a \otimes b \mapsto ab$ is a morphism of coalgebras.

Proposition. $(H_0') \iff \Delta^*$ *is a morphism of algebras.*

Evaluating the identity given in Remark 3.2.2 on $\Delta(x)$ with $x \in A$ gives

$$\begin{aligned}&\varphi(a_1,b_1)(b_2 \otimes a_2)(\Delta(x)) - \varphi(a_2,b_2)(a_1 \otimes b_1)(\Delta(x))\\ &= (\mathbf{R}(b_1 \otimes a_1))(b_2 \otimes a_2)(\Delta(x)) - (\mathbf{R}(b_2 \otimes a_2))(b_1 \otimes a_1)(\Delta^{\mathrm{opp}}(x))\\ &= (\mathbf{R}\Delta(x))(b \otimes a) - (\Delta^{\mathrm{opp}}(x)\mathbf{R})(b \otimes a)\end{aligned}$$

and hence the assertion.

9.4.5 Retain the hypothesis that φ is a skew-Hopf pairing. For any Hopf algebra A let A^{opp} (resp. A_{opp}) denote the same algebra as A but with the opposed product (resp. coproduct). Then the map $\Phi' : b \mapsto (a \mapsto \varphi(a,b))$ of A^{*}_{opp} into A^{**} is a morphism of Hopf algebras. Moreover the duality between A^{**} and A^{*}_{opp} is a skew-Hopf pairing and so one may form the Drinfeld double $\mathcal{D}(A^{**}, A^{*}_{\mathrm{opp}})$. The composed map $\mu(\mathrm{Id} \otimes \Phi') : \mathcal{D}(A^{**}, A^{*}_{\mathrm{opp}}) = A^{**} \otimes A^{*}_{\mathrm{opp}} \longrightarrow A^{**} \otimes A^{**} \xrightarrow{\mu} A^{**}$ is a morphism of coalgebras.

Proposition. $(H_0') \Longleftrightarrow \mu(\mathrm{Id} \otimes \Phi')$ *is a morphism of algebras.*

Let $M, N \in \mathrm{Mod}_f A$. Extend $M, N, M \otimes N$ to A^{**} modules as prescribed by 1.4.11. Then by 1.4.11 one has $R_{M,N} \in \mathrm{End}_A(M \otimes N) \Longleftrightarrow R_{M,N} \in \mathrm{End}_{A^{**}}(M \otimes N)$. The latter condition can be expressed as follows. Take $m \in M$, $n \in N$, $x \in A^{**}$, $c \in A^{*}$. Let $m \mapsto m_1 \otimes c_2$ (resp. $n \mapsto n_1 \otimes d_2$) be defined by the right A^{*} comodule structure of M (resp. N) given in 1.4.11. Then

$$xR_{M,N}(m \otimes n) = x(n_1 \otimes m_1)\mathbf{R}(c_2 \otimes d_2) = (x_1 n_1 \otimes m_1) x_2(c_2)\Phi'(c_3)(d_2) \ ,$$

whilst (writing $x_2 n \mapsto n_1' \otimes d_2'$ for the right A^{*} action)

$$R_{M,N}x(m \otimes n) = x_1(c_2)R_{M,N}(m_1 \otimes x_2 n) = x_1(c_3)\Phi'(c_2)(d_2')(n_1' \otimes m_1) \ .$$

Now $m_1 \otimes c_2 = m_i \otimes \sum_i c^M_{\xi_i, m}$, so if $m \in M$ and $M \in \mathrm{Mod}_f A$ are arbitrary the equality that results from the above is equivalent to

$$x_2(c_1)\Phi'(c_2)(d_2)x_1 n_1 = x_1(c_2)\Phi'(c_1)(d_2')n_1'$$

which is just

$$x_2(c_1)(\mu(\mathrm{Id} \otimes \Phi')(x_1 \otimes c_2))n = x_1(c_2)(\mu(\mathrm{Id} \otimes \Phi')(1 \otimes c_1)(x_2 \otimes 1)n \ .$$

Since $n \in N$ and $N \in \mathrm{Mod}_f A$ are arbitrary this is equivalent to

$$\mu(\mathrm{Id} \otimes \Phi')(x_2(c_1)(x_1 \otimes c_2) - x_1(c_2)(1 \otimes c_1)(x_2 \otimes 1)) = 0 \ .$$

Finally the expression in brackets is just the identity which defines (Remark 3.2.2) the multiplication rule in $\mathcal{D}(A^{**}, A^{*}_{\mathrm{opp}})$.

9.4.6 Given a Hopf algebra A an element $\mathbf{R} \in (A^{*} \otimes A^{*})^{*}$ satisfying $(H_i') : i = 0, 1, 2$ is called an R-matrix for A. As one can see its existence is intimately connected to the Drinfeld double construction. However it does not seem quite correct to say that a Drinfeld double possesses an R-matrix. Indeed let A, B be Hopf algebras with a skew-Hopf pairing φ and consider the following "natural" construction of an R-matrix on $\mathcal{D}(A, B)$.

The Hopf algebra embedding $A \hookrightarrow \mathcal{D}(A,B)$ (resp. $B \hookrightarrow \mathcal{D}(A,B)$) transposes to a Hopf algebra map $\mathcal{D}(A,B)^{*} \longrightarrow A^{*}$ (resp. $\mathcal{D}(A,B)^{*} \longrightarrow B^{*}$) whose image will be denoted by $\mathcal{B}$ (resp. $\mathcal{A}$). On the other hand $\Phi : a \mapsto (b \mapsto \varphi(a,b))$ (resp. $\Phi' : b \mapsto (a \mapsto \varphi(a,b))$) is a Hopf algebra map of A^{opp} into B^{*}

(resp. $B_{\rm opp}$ into A^*). Consider the fortuitous situation when $\mathcal{A} = A^{\rm opp}$ and $\mathcal{B} = B_{\rm opp}$. Then φ defines a skew-Hopf pairing $(a,b) \mapsto \varphi(a,b)$ (resp. $(b,a) \mapsto \varphi(a,b)$) sending $\mathcal{A}\times\mathcal{B}$ (resp. $\mathcal{B}\times\mathcal{A}$) to k. Let φ_ℓ (resp. φ_r) denote the skew-Hopf pairing on $\mathcal{D}(A,B)^*$ defined by the composition with the canonical projection namely $\mathcal{D}(A,B)^* \times \mathcal{D}(A,B)^* \longrightarrow \mathcal{A}\times\mathcal{B}$ (resp. $\mathcal{B}\times\mathcal{A}$) $\longrightarrow k$.

Theorem. *The element $a\otimes b \mapsto \varphi_\ell(a,b)$ (resp. $\varphi_r(a,b)$) of $(\mathcal{D}(A,B)^* \otimes \mathcal{D}(A,B)^*)^*$ is an R-matrix for $\mathcal{D}(A,B)$.*

By 9.4.3 it remains to verify (H_0'). By 9.4.5 it is enough to show that $\mu(\mathrm{Id}\otimes\Phi_\ell') : \mathcal{D}(\mathcal{D}(A,B)^{**}, \mathcal{D}(A,B)^*_{\rm opp}) \longrightarrow \mathcal{D}(A,B)^{**}$, (resp. $\mu(\mathrm{Id}\otimes\Phi_r') : \mathcal{D}(\mathcal{D}(A,B)^{**}, \mathcal{D}(A,B)^{*\,\rm opp}) \longrightarrow \mathcal{D}(A,B)^{**}$) is an algebra map. Moreover by 1.4.11, as one sees from the proof of 9.4.5, one may replace $\mathcal{D}(A,B)^{**}$ in the left hand factor by $\mathcal{D}(A,B)$ and then since the latter is generated by A and B to just take these two subalgebras there.

The algebra structure on $\mathcal{D}(A,\mathcal{D}(A,B)^*_{\rm opp})$ is defined by restricting (the duality) evaluation to A. Thus $\mathcal{D}(A,B)^*_{\rm opp}$ may be replaced by $\mathrm{Im}(\mathcal{D}(A,B)^*_{\rm opp} \longrightarrow A^*_{\rm opp}) = \mathcal{B}_{\rm opp} = B$ where to take account of the last identification, evaluation is replaced by the pairing $(a,b) \mapsto \varphi(a,b)$. It follows that this algebra structure is exactly that of the original Drinfeld double $\mathcal{D}(A,B)$. Similarly for the algebra structure on $\mathcal{D}(B,\mathcal{D}(A,B)^*)$ one may replace $\mathcal{D}(A,B)^*$ by $\mathrm{Im}(\mathcal{D}(A,B)^* \longrightarrow B^*) = \mathcal{A}^{\rm opp} = A$ with evaluation replaced by the pairing $(b,a) \mapsto \varphi(a,b)$, so again the algebra structure is that of the original Drinfeld double.

On the other hand $\Phi_\ell' : \mathcal{D}(A,B)^*_{\rm opp} \longrightarrow \mathcal{D}(A,B)^{**}$ (resp. $\Phi_r' : \mathcal{D}(A,B)^{*\,\rm opp} \longrightarrow \mathcal{D}(A,B)^{**}$) factors to a map of $\mathcal{B}_{\rm opp}$ (resp. $\mathcal{A}^{\rm opp}$) into $\mathcal{D}(A,B)^{**}$ which is just the canonical embedding of B (resp. A) in $\mathcal{D}(A,B)$. Combined with the previous observation this gives the assertion of the theorem.

9.4.7 Recall (9.2.11, 9.2.12) the Hopf algebra isomorphism $\Phi : \check{U}_q(\mathfrak{b}^+) \xrightarrow{\sim} \mathrm{Im}(R_q[G] \longrightarrow \check{U}_q(\mathfrak{b}^-)^*)^{\rm opp}$ (resp. $\Phi' : \check{U}_q(\mathfrak{b}^-) \xrightarrow{\sim} \mathrm{Im}(R_q[G] \longrightarrow \check{U}_q(\mathfrak{b}^+)^*)_{\rm opp}$) defined by the skew-Hopf pairing $\varphi : \check{U}_q(\mathfrak{b}^+) \times \check{U}_q(\mathfrak{b}^-) \longrightarrow \overline{k(q)}$. Thus the above theorem applies to this situation with the following rather trivial modification. Instead of taking $\mathcal{C}$ to be the category of all finite dimensional $\mathcal{D}(\check{U}_q(\mathfrak{b}^+), \check{U}_q(\mathfrak{b}^-))$ modules, one takes it to be the category of finite dimensional modules which factor to $\check{U}_q(\mathfrak{g})$ modules having moreover weights in $P(\pi)$, that is to say exactly those modules used in constructing $R_q[G]$. Then the construction of 9.4.6 gives an R matrix (in fact two!) with respect to this smaller category. These can of course be viewed as R-matrices for $\check{U}_q(\mathfrak{g})$.

9.4.8 Each $\Lambda \in \check{T}^\wedge$ defines an element of $\mathrm{Aut}\, U_q(\mathfrak{n}^+)$ (resp. $\mathrm{Aut}\, U_q(\mathfrak{n}^-)$) by $\Lambda(e_\alpha) = \Lambda(t_\alpha)e_\alpha$ (resp. $\Lambda(f_{-\alpha}) = \Lambda(t_\alpha^{-1})f_{-\alpha}$) and hence an action of $\check{T}^\wedge$ on $U_q(\mathfrak{n}^+)$ (resp. $U_q(\mathfrak{n}^-)$). Since the subalgebra $\check{U}^0$ of $(k(q)\check{T}^\wedge)^*$ is separating it follows from 1.4.11 that this action may be extended to an action of $(\check{U}^0)^*$ from which one may define the smash product $U(\mathfrak{n}^\mp)\#(\check{U}^0)^*$ which is a Hopf algebra in an obvious fashion. Extend φ to a (non-degenerate) skew-Hopf pairing $U_q(\mathfrak{n}^+)\#(\check{U}^0)^\star \times \check{U}_q(\mathfrak{b}^-)$ (resp. $\check{U}_q(\mathfrak{b}^-) \times U_q(\mathfrak{n}^-)\#(\check{U}^0)^\star) \longrightarrow \overline{k(q)}$ by replacing the pair $\tau(\gamma), q^{-(\gamma,\delta)}$ by $\Lambda, \Lambda(\tau(-\delta))$ (resp. $\tau(\delta), q^{-(\gamma,\delta)}$ by $\Lambda, \Lambda\tau(-\gamma))$ in 9.2.10 (1).

Lemma. *Over $\overline{k(q)}$ one has*

(i) $(\check{U}^0)^\star \xleftarrow{\sim} U(\text{Lie}\,\check{T})\#\overline{k(q)}\check{T}^\wedge$

(ii) $(\check{U}_q(\mathfrak{b}^-))^\star \overset{\Phi}{\underset{\sim}{\longleftarrow}} (U_q(\mathfrak{n}^+)\#(\check{U}^0)^\star)^{\text{opp}}$

(iii) $(\check{U}_q(\mathfrak{b}^+))^\star \overset{\Phi'}{\underset{\sim}{\longleftarrow}} (U_q(\mathfrak{n}^-)\#(\check{U}^0)^\star)_{\text{opp}}$.

(i) is immediate from 2.1.8. For surjectivity in (iii) consider $(U_q(\mathfrak{b}^+))^\star$. Set $J_+ = U_q(\mathfrak{n}^+)_+$. It follows as in 9.3.11 (ii) that any finite dimensional $\check{U}_q(\mathfrak{b}^+)$ module M is annihilated by some power of J_+ . Yet J_+ is stable for the adjoint action of $\check{T}$ and so the smash product $U_n := U_q(\mathfrak{n}^+)/J_+^n\#\check{U}^0$ is defined and is a quotient of $\check{U}_q(\mathfrak{b}^+)$. Fix $n \in \mathbb{N}^+$. For each ideal I of $\check{U}^0$ of finite codimension one may form the induced module $M_n := U_n \otimes_{\check{U}^0} (\check{U}^0/I)$ whose annihilator in $U_q(\mathfrak{n}^+)$ (resp. $\check{U}^0$) is J_+^n (resp. an ideal I_n of finite codimension in $\check{U}^0$ contained in I). In particular $(\check{U}^0)^\star = \lim_{\rightarrow}(\check{U}^0/I)^* = \lim_{\rightarrow}(\check{U}^0/I_n)^*$. It follows that $\lim_{\rightarrow}(U_n/\operatorname{Ann} M_n)^* = \lim_{\rightarrow}(U_q(\mathfrak{n}^+)/J_+^n)^*\#(\check{U}^0/I_n)^* = (U_q(\mathfrak{n}^+)/J_+^n)^*\#(\check{U}^0)^\star$. Finally $\lim_{\rightarrow}(U_q(\mathfrak{n}^+)/J_+^n)^*$ identifies with the graded dual of $U_q(\mathfrak{n}^+)$ and hence with $U_q(\mathfrak{n}^-)$ via the pairing φ so surjectivity results from the first observation. Surjectivity in (ii) is similar.

9.4.9 One may now derive from 9.3.13 a description of $R_q[G]^\star$. By transposing the commuting square

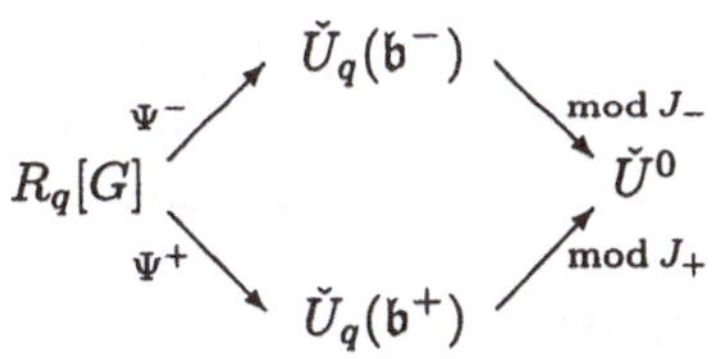

it follows that the subalgebras $(\check{U}^0)^\star$ of $\check{U}_q(\mathfrak{b}^-)^\star$ and $\check{U}_q(\mathfrak{b}^+)^\star$ have a common image in $R_q[G]^\star$ under $\Psi^\star$. Thus by 9.3.13 and 9.4.8, the restriction of $\Psi^\star$ to $\check{U}_q(\mathfrak{b}^-)^\star \otimes U_q(\mathfrak{b}^-)_{\text{opp}}$ has image $R_q[G]^\star$. Yet Ψ^+ is surjective so $\check{U}_q(\mathfrak{b}^-)^\star$ identifies with a subalgebra of $R_q[G]^\star$. Form the Drinfeld double $\mathcal{D}(\check{U}_q(\mathfrak{b}^-)^\star, \check{U}_q(\mathfrak{b}^-)_{\text{opp}})$ through the natural pairing. Using the R-matrix for $\check{U}_q(\mathfrak{g})$ corresponding to φ_r gives by 9.4.5 a Hopf algebra map $\mathcal{D}(R_q[G]^\star, R_q[G]_{\text{opp}}) \longrightarrow R_q[G]^\star$. Since Φ'_r factors to an injection of $\check{U}_q(\mathfrak{b}^-)_{\text{opp}}$ into $\check{U}_q(\mathfrak{b}^+)^\star$ restricting the first factor gives a Hopf algebra surjection $\mathcal{D}(\check{U}_q(\mathfrak{b}^-)^\star, \check{U}_q(\mathfrak{b}^-)_{\text{opp}}) \longrightarrow R_q[G]^\star$. It is clear that the kernel of this map is generated by $u \otimes 1 - 1 \otimes u : u \in \check{U}^0$. In particular $R_q[G]^\star$ is generated over $U_q(\mathfrak{g})$ by $(\check{U}^0)^\star$ with the relations coming from the adjoint action. In particular one has the

Proposition. *The multiplication map gives an isomorphism $U_q(\mathfrak{n}^-)\otimes(\check{U}^0)^\star\otimes U_q(\mathfrak{n}^+) \xrightarrow{\sim} R_q[G]^\star$ of vector spaces.*

9.5 Comments and Complements

9.5.1 A slightly weaker version of 9.1.5 is due to Y. Soibelman [S1] who derived it using the R-matrix. (Indeed multiplication in $R_q[G]$ is defined by the R-matrix for U by 9.4.4). He also stated 9.2.2 without proof referring only to a paper of N. Y. Reshetikhin and M. A. Semenov-Tian-Shansky who do not explicitly claim this result and whose analysis is somewhat heuristic [RS1]. In any case the proof of 9.2.2 applies equally well to $R[G]$ which is of course a classical object; but surprisingly this result does not seem to have been known previously. Moreover the most obvious geometric argument (using $G(u_\mu \otimes u_{w_0\nu})$) fails. For $G = SL(n)$ one can easily give a quite explicit proof of 9.2.2 by calculations involving appropriate (quantum) minors. (See for example [NYM1].) The remaining parts of 9.1–9.3 are mainly developed from [J11] with the notable technical improvement described in [HLT1, Proposition 4.6] (the use of φ in 9.2.11). In this 9.2.9 and 9.3.18 are new. C. De Concini drew my attention to injectivity in 9.2.14.

9.5.2 By $(H_0'), (H_2')$ of 9.4.2 one obtains $\mathbf{R}_{12}\mathbf{R}_{13}\mathbf{R}_{23} = \mathbf{R}_{12}(\Delta\otimes 1d_A)(\mathbf{R}) = (\Delta^{\text{opp}} \otimes 1d_A)(\mathbf{R})\mathbf{R}_{12} = \mathbf{R}_{23}\mathbf{R}_{13}\mathbf{R}_{12}$. The resulting identity is known as the quantum Yang-Baxter equation. It arises in various problems involving n particle problems. For example each $\mathbf{R}_{ij}$ may be viewed as a matrix describing the crossing of the i, j^{th} pair. A basic problem is to find matrices satisfying these identities and to a large extent Quantum Groups were invented (by V. G. Drinfeld [Dr1] and M. Jimbo [Ji1,2]) for that purpose. The formalism in 9.4.1 goes back to early work of S. Maclane [Mac1, 2] and of J. D. Stasheff [Sta1], whilst the extraction (9.4.2) of an $\mathbf{R}$ matrix from the Drinfeld double is due to V. G. Drinfeld. However he (and his followers!) usually take $\mathbf{R}$ to be an element of $A \otimes A$ and the resulting formalism, which requires the existence of dual bases, often only really makes sense if A is finite dimensional or infinite sums (with appropriate convergence conditions which are usually not specified) are permitted. This is perfectly valid for certain models at least if some care is exercised. In any case the formalism developed in 9.4.2 and 9.4.6 follows work of D. Gaitsgory [Ga1] who has shown that if $\mathfrak{g}$ is simple then the two solutions given in 9.4.6 essentially exhaust all R-matrices for $U_q(\mathfrak{g})$. For a related result see [KT1]. It would appear that in a purely algebraic context a Drinfeld double should not always admit an R-matrix. The formalism of braided monoidal categories and its relationship with knot invariants and conformal field theory has undergone an explosive development which can be partially ascertained from [Dr4, Ks1, RT1, ShS1, T1] and references therein. The Yang-Baxter equation has a classical version which is briefly discussed in A.4.8.

9.5.3 The description of $R_q[G]^*$ given in 9.3.13 and 9.4.8 is new. It shows that $R_q[G]^*$ is very nearly just $U_q(\mathfrak{g})$ in sharp contrast to the classical situation. Indeed by 2.1.8 one has $R[G]^* = U(\mathfrak{g})\#kG$ which is significantly bigger than $U(\mathfrak{g})$. It essentially resolves the problem of how to recover $U_q(\mathfrak{g})$ from $R_q[G]$. This is important in the sense that Y. Manin [Man1] has shown how to directly derive $R_q[SL(n)]$ as functions on a "group of symmetries" of quantum n-space. This theory is discussed briefly in A.4.9.

9.5.4 One can ask if the ideal $\mathcal{K}$ defined in 9.1.6 is generated by just the $K_{i,j}$ and the $E_{i,j}$. Indeed the corresponding classical object $R[G/N^+]$ is of course a commutative algebra and has a similar module structure. A result of B. Kostant (reproduced in [LT1]) shows that as an algebra it is generated by the fundamental modules $\mathcal{V}(\omega_i) : i = 1, 2, \ldots, \ell$ (this part is clear) satisfying the pair relations $\mathcal{K}_{i,j}$ defined by the short exact sequence

$$0 \longrightarrow \mathcal{K}_{i,j} \longrightarrow \mathcal{V}(\omega_i) \otimes \mathcal{V}(\omega_j) \longrightarrow \mathcal{V}(\omega_i + \omega_j) \longrightarrow 0 .$$

Actually his argument is disarmingly simple (and elegant). The Casimir element $J \in U(\mathfrak{g})$ acts by the scalar $(\mu + \rho, \mu + \rho)$ on $\mathcal{V}(\mu)$. Thus (4.2.8 (i)) the value that this scalar takes on $\mathcal{V}(\omega_{i_1} + \ldots \omega_{i_n})$ is distinct from all those values taken on simple submodules in $\mathcal{K}_{i_1, i_2, \ldots, i_n}$. On the other hand J is quadratic in the elements of $\mathfrak{g}$ which are primitive elements. Thus in computing J on an n-fold tensor product at worst only *pairs* of modules are involved and so only the pair relations are needed for J to act as a scalar. Combined with the first observation this proves the assertion. In the quantum case central elements are not just quadratic in quasi-primitive elements so this proof fails. Yet the assertion has been verified by a different method for $R_q[SL(n)]$ in unpublished work of A. Braverman.

9.5.5 Take $\lambda, \mu \in P^+(\pi)$. One can show that the Cartan multiplication rule (c.f. 9.1.6(∗)) implies that

$$(*) \qquad \sum_{w \in W} V(\lambda) u_{w\mu} = V(\lambda + \mu)$$

if $\lambda \gg \mu$. Indeed this is obtained [J12] through the specialization $R_q[G/N^+] \longrightarrow R[G/N^+]$ at $q = 1$ and by interpreting $R[G/N^+]$ as the algebra of homogeneous functions on the flag variety G/B^+. In this latter situation the assertion is a trivial consequence of Hilbert's nullstellensatz [JPP1]. The result can be used to give a really elementary proof [JPP1] of the Beilinson-Bernstein equivalence of categories theorem [BB1]. This has also a quantum version [J10] obtained from (∗). However there does not seem to be any representation theoretic proof of (∗).

Chapter 10. The Prime Spectrum of $R_q[G]$

10.1 Highest Weight Modules

A highest weight module for R is a module generated by a one dimensional R^+ module. Their theory is superficially similar to highest weight modules for U. Thus there are universal highest weight modules (analogous to Verma modules) and these admit unique simple quotients (10.1.5). However not all characters on R^+ can give rise to such a module (10.1.3) and $\operatorname{Prim} R$ is not exhausted (10.1.7) by their annihilators. On the other hand they satisfy a remarkable tensor product theorem (10.1.18) which gives rise to a braid group action (10.2) on U.

10.1.1 Let Λ be a non-zero character on R^+. In a manner analogous to 9.3.8 define for each $\mu \in P^+(\pi)$ the set

$$C_\Lambda(\mu) = \{\eta \in \Omega(V(\mu)) \mid \Lambda(c^\mu_{\eta,\mu}) \neq 0\} .$$

If $C_\Lambda(\mu) = \emptyset$, set $D_\Lambda(\mu) = \emptyset$. Otherwise let $D_\Lambda(\mu)$ denote the set of minimal elements of $C_\Lambda(\mu)$.

Suppose $D_\Lambda(\mu_1) \neq \emptyset$, $C_\Lambda(\mu_2) \neq \emptyset$ and take $\eta_1 \in D_\Lambda(\mu_1), \eta_2 \in C_\Lambda(\mu_2)$. Set $c_i = c^{\mu_i}_{\eta_i,\mu_i} : i = 1, 2$. Then by 9.1.5 (i) one has

$$\Lambda(c_2)\Lambda(c_1) = q^{(\mu_1,\mu_2)-(\eta_1,\eta_2)}\Lambda(c_2)\Lambda(c_1) .$$

Yet $\Lambda(c_i)$ is a non-zero scalar, so this forces $(\mu_1, \mu_2) = (\eta_1, \eta_2)$. Then by A.1.17 there exists $w \in W$ such that $\eta_i = w\mu_i$. Again if μ is regular $D_\Lambda(\mu) \neq \emptyset$ implies $C_\Lambda(\omega_i) \neq \emptyset$ for all i by the Cartan multiplication rule.

On the other hand it is clear from 9.1.6 that one may have $D_\Lambda(\mu) = \emptyset$. To avoid this bad situation first recall (9.2.2) the definition of R^{++}. The above analysis gives

Proposition. *Suppose Λ does not vanish on R^{++}. Then there exists $w \in W$ such that $C_\Lambda(\mu) = \{w\mu\}$ for all $\mu \in P^+(\pi)$.*

10.1.2 Fix $w \in W$ and recall that $c_{w\mu}$ spans the $w\mu$ weight subspace of $V^+(\mu)^*$ considered as a right T module. Let $V^+(\mu)^{*w\mu}$ denote its unique T stable complement in $V^+(\mu)^*$. Using also the left T action it easily follows that $V^+(\mu)^{*w\mu}V^+(\nu)^{*w\nu} \subset V^+(\mu+\nu)^{*w(\mu+\nu)}$ for all $\mu, \nu \in P^+(\pi)$ and so the sum

$$I_w^+ := \bigoplus_{\mu \in P^+(\pi)} V^+(\mu)^{*w\mu}$$

is a two-sided ideal of R^+. Through the multiplication rules of 9.1.10 (∗) it follows that each ℓ-tuple $\Lambda_w^+ = \{\Lambda_{w,i}^+\}_{i=1}^{\ell} \in k(q)^\ell$ can be viewed as the unique character on R^+ satisfying $\ker \Lambda_w^+ \supset I_w^+$ and $\Lambda_w^+(c_{w\omega_i}) = \Lambda_{w,i}^+$ for all i. Moreover Λ_w^+ does not vanish on R^{++} if and only if $\Lambda_{w,i}^+ \in k(q)^*$ for all i. By 10.1.1 all characters with this non-vanishing property are so obtained. A similar construction gives a unique character Λ_w^- on R^- for each ℓ-tuple $\{\Lambda_w^-(d_{w\omega_i})\}_{i=1}^{\ell} \in k(q)$.

10.1.3 To each character Λ on R^+ let $k(q)v_\Lambda$ denote the corresponding one dimensional R^+ module and $N(\Lambda) := R \otimes_{R^+} k(q)v_\Lambda$ the induced module. Recall $\hat{R} = R^+ \circledast R^-$. As in 9.1.8 it easily follows that the multiplication map $R^- \otimes R^+ \longrightarrow \hat{R}$ is bijective. Moreover it is also convenient to rewrite the invariant elements $x_\mu : \mu \in P^+(\pi)$ of 9.2.4 in the form

$$(*) \qquad x_\mu = \sum_i \sigma^*(c^\mu_{\xi_\mu, u_i}) \otimes c^\mu_{\xi_i, \mu} \, .$$

Lemma. *The following are equivalent*

(i) $N(\Lambda) \neq 0$.
(ii) $\Lambda(R^{++}) \neq 0$.

Set $\hat{N}(\Lambda) = \hat{R} \otimes_{R^+} k(q)v_\Lambda$. By the above remark $\hat{N}(\Lambda)$ is isomorphic to R^- as a left R^- module. By 9.2.2 one has $\hat{N}(\Lambda)/(\ker \Delta^*)v_\Lambda \xrightarrow{\sim} N(\Lambda)$ where v_Λ is viewed as an element of $\hat{N}(\Lambda)$. *Not* (ii) implies that $\Lambda(V^+(\omega_i)^*) = 0$ for some i and then $x_{\omega_i} v_\Lambda = 0$. By 9.2.4 this forces $N(\Lambda) = 0$.

Conversely suppose $\Lambda(R^{++}) \neq 0$. By 10.1.2 one has $\Lambda = \Lambda_w^+$ for some $w \in W$. Define $\mu_0 = -w_0\mu$ and $\omega_{i_0} = -w_0\omega_i$ and fix scalars by setting $d_{w\mu_0} = \sigma^*(c^\mu_{\mu,w\mu})$ in the expression (∗) for x_μ. Then $x_{\omega_i} v_\Lambda = d_{w\omega_{i_0}} \Lambda_w^+(c_{w\omega_i}) v_\Lambda$. Let L_w^- denote the left ideal of R^- generated by the $d_{w\omega_{i_0}} - \Lambda_w^+(c_{w\omega_i})^{-1}$. Since $\Delta^* x_\mu = 1$ for all $\mu \in P^+(\pi)$ it follows that $N(\Lambda)$ is isomorphic to R^-/L_w^- as a left R^- module. It is non-zero since $L_w^- \subset \ker \Lambda_w^-$, where Λ_w^- is the non-zero character on R^- defined through 10.1.2 by $\Lambda_w^-(d_{w\omega_{i_0}}) = \Lambda_w^+(c_{w\omega_i})^{-1}$.

10.1.4 Define the subspaces $\{\mathcal{F}^i R^- : i \in \mathbb{N}\}$ inductively through

$$\mathcal{F}^0 R^- = \bigoplus_{\mu \in P^+(\pi)} k(q) d_{w_0\mu} \, , \quad \mathcal{F}^i R^- = \mathcal{F}^{i-1} R^- + \sum_{j=1}^{\ell} (\mathcal{F}^{i-1} R^-) \cdot e_j \, .$$

It is clear that each $\mathcal{F}^i R^-$ is a $T-T$ stable subspace of R^-. Since the e_j act like T-skew derivations and $d_{w_0\mu} \cdot U^+ = V^-(\mu)^*$, it follows that $\{\mathcal{F}^i R^-\}_{i\in\mathbb{N}}$ is a filtration for R^-.

Fix $\mu, \nu \in P^+(\pi)$. Then 9.1.4 gives $d_{w_0\nu} c^\mu_{\xi,\mu} = q^{(\mu,w_0\nu)-(\xi,\nu)} c^\mu_{\xi,\mu} d_{w_0\nu}$ and so through the right action of the e_j it follows by induction on i using 9.1.2 (ii) that if $c^\nu_{\eta,w_0\nu} \in \mathcal{F}^i R^-$ then

$$(*) \qquad c^\nu_{\eta,w_0\nu} c^\mu_{\xi,\mu} = q^{(\mu,w_0\nu)-(\xi,\eta)} c^\mu_{\xi,\mu} c^\nu_{\eta,w_0\nu} \operatorname{mod}(\mathcal{F}^{i-1}R^-)R^+ .$$

In particular $(\mathcal{F}^i R^-)R^+ = R^+(\mathcal{F}^i R^-) \operatorname{mod}(\mathcal{F}^{i-1}R^-)R^+$ and so by 9.2.2 these subspaces form a filtration of R.

Now fix $w \in W$ and let Λ^+_w be the character on R^+ defined in 10.1.2. Set $\mathcal{F}^i(N(\Lambda^+_w)) = (\mathcal{F}^i R^-)v_{\Lambda^+_w}$ (notation 10.1.3) which makes $N(\Lambda^+_w)$ a filtered module over the filtered ring $(R, \mathcal{F})$. Then $(*)$ above gives the

Lemma. *The R^+ module $gr_{\mathcal{F}} N(\Lambda^+_w)$ is a direct sum of one dimensional modules whose characters are amongst those given by the ℓ-tuples $q^{(\omega_i, w^{-1}\eta - w_0\nu)}\Lambda^+_{w,i}$ where $\nu \in P^+(\pi)$ and $\eta \in \Omega(V(\nu))$. Moreover the module with character $\Lambda^+_{w,i}$ occurs with multiplicity one.*

For the last part, note that $(\omega_i, w^{-1}\eta - w_0\nu) = 0$ for all i implies $\eta = ww_0\nu$. Yet (10.1.3) the corresponding element $c^\nu_{ww_0\nu,w_0\nu} = d_{w\nu}$ acts on $v_{\Lambda^+_w}$ by a scalar and so the assertion follows.

Remark. Observe that the exponent of q above is positive.

10.1.5 Retain the above notation. By 10.1.4 it follows that no proper R submodule of $N(\Lambda^+_w)$ contains a copy of the one-dimensional R^+ module with character Λ^+_w. Consequently the sum of any two proper R submodules is again proper. Since R is noetherian (9.2.2) and $N(\Lambda^+_w)$ is cyclic, this establishes the

Corollary. *$N(\Lambda^+_w)$ admits a unique simple quotient $\mathcal{S}(\Lambda^+_w)$. It is the unique simple R module generated by a one dimensional R^+ module with character Λ^+_w.*

10.1.6 Recall (9.2.11, 9.3.11) the definition of the Hopf ideal J of R. It is clear that as a Hopf algebra R/J is spanned by the $c_\mu : \mu \in P^+(\pi)$ and their inverses, satisfying the relations $c_\mu c_\nu = c_{\mu+\nu}$, $\mu^*(c_\nu) = c_\nu \otimes c_\nu$, $\sigma^*(c_\nu) = c_\nu^{-1}$ and so is isomorphic to $\check{U}^0$. In particular each ℓ-tuple $\Lambda = \{\Lambda_i\}_{i=1}^\ell \in k(q)^{*\ell}$ can be viewed as the character on R satisfying $\ker \Lambda \supset J$ and $\Lambda(c_{\omega_i}) = \Lambda_i$. It is clear that all linear characters (1.3.4) on R are so obtained. Let $\mathcal{S}(\Lambda)$ denote the corresponding one dimensional R module. It is immediate from 10.1.5 that $\mathcal{S}(\Lambda) \cong \mathcal{S}(\Lambda^+_e)$ given $\Lambda_i = \Lambda^+_{e,i}$ for all i. More generally for each $w \in W$ define $\mathcal{S}(w)$ to be $\mathcal{S}(\Lambda^+_w)$ when $\Lambda^+_{w,i} = 1$ for all i, and let v_w denote the image of $v_{\Lambda^+_w}$ in $\mathcal{S}(w)$. The following is a trivial consequence of 10.1.5 and 1.4.7 (1).

Lemma. *Fix $w \in W$. Given $\Lambda_i = \Lambda^+_{w,i}$ for all i, one has $\mathcal{S}(\Lambda^+_w) \cong \mathcal{S}(w) \otimes \mathcal{S}(\Lambda) \cong \mathcal{S}(\Lambda) \otimes \mathcal{S}(w)$.*

Remark. This result can be expressed by saying that up to the action of $R^\wedge$ there exists for each $w \in W$ exactly one simple highest weight module for R, namely $\mathcal{S}(w)$.

10.1.7 Set $J(w) = \operatorname{Ann} \mathcal{S}(w)$. Obviously $J(w) \in \operatorname{Prim} R \subset \operatorname{Spec} R$. Recall 9.3.9.

Lemma. *For all $w \in W$ one has $J(w) \in \mathbf{X}(\mathbf{w})$, where $\mathbf{w} = (w, w)$.*

Suppose $J(w) \in \mathbf{X}(\mathbf{w})$ with $\mathbf{w} = (w_+, w_-)$. By 9.1.5 (i) it follows that for all $\mu \in P^+(\pi)$ the image $\bar{c}_{w_+\mu}$ of $c_{w_+\mu}$ in $R/J(w)$ is normal and non-zero. Then $0 \neq \bar{c}_{w_+\mu}(R/J(w))v_w = (R/J(w))c_{w_+\mu}v_w$ which forces $w_+ = w$. The second case is more subtle. By definition of w_- one has $J(w) \supset J^-_{w_-w_0\nu}(w_-w_0\nu, w_0\nu)$ for all $\nu \in P^+(\pi)$. Then 9.1.5 (ii) gives $J^-_w(d_{w_-\nu}v_w) = 0$ and

$$c_{w\mu}d_{w_-\nu}v_w = q^{(\mu, w^{-1}w_-w_0\nu - w_0\nu)}d_{w_-\nu}v_w \ .$$

Now $(\mu, w^{-1}w_-w_0\nu - w_0\nu) \geq 0$ with equality for μ, ν regular implying $w = w_-$ by A.1.17. Suppose a strict inequality holds. By 10.1.4, it follows that $Rd_{w_-\nu}v_w = R^-d_{w_-\nu}v_w$ admits no copy of the R^+ module $k(q)v_w$ and hence is zero in $\mathcal{S}(w)$, that is $d_{w_-\nu}v_w = 0$. Yet by 9.1.5 (ii) again, the image of $d_{w_-\nu}$ in $R/J(w)$ is normal and non-zero so this gives a contradiction as in the first case. Hence $w_- = w$.

10.1.8 A more precise description of the annihilators of the $S(w)$ involves the Demazure modules $V_w(\lambda)$ described in section 4.4. Only the elementary theory given there is required and not the deeper results of 6.3.

Fix $w \in W$. For each $\lambda \in P^+(\pi)$, let $V^+_w(\lambda)^\perp$ (resp. $V^-_w(\lambda)^\perp$) denote the orthogonal of $V^+_w(\lambda) = U_q(\mathfrak{b}^+)u_{w\lambda}$ (resp. $V^-_w(\lambda) := U_q(\mathfrak{b}^-)u_{ww_0\lambda}$) in $V(\lambda)^*$ identified with $V^+(\lambda)^*$ (resp. $V^-(\lambda)^*$). Set

$$Q^+_w = \sum_{\lambda \in P^+(\pi)} V^+_w(\lambda)^\perp \ , \quad Q^-_w = \sum_{\lambda \in P^+(\pi)} V^-_w(\lambda)^\perp \ .$$

Proposition. *For each $w \in W$ the subspace $Q^\pm_w$ is a completely prime ideal of $R^\pm$ stable for the right action of $U_q(\mathfrak{b}^\pm)$.*

Both parts are similar and so only + is considered. By transport of structure it is immediate that $V^+_w(\lambda)$ is $V^+ := U_q(\mathfrak{b}^+)$ stable. Then for Q^+_w to be a right ideal, it is enough to show that

$$(*) \qquad V^+_w(\lambda)^\perp c_\mu \subset V^+_w(\lambda + \mu)^\perp$$

for all $\lambda, \mu \in P^+(\pi)$. Yet $c \in V_w^+(\lambda)^\perp \iff (c.V^+)u_{w\lambda} = 0 \iff c_{w\lambda} \notin c.V^+$. Thus (*) is equivalent to

$$(**) \qquad c_{w(\lambda+\mu)} \notin (V_w^+(\lambda)^\perp c_\mu).V^+ .$$

By 9.1.10 (*) one has $c_{w\lambda}c_{w\mu} = c_{w(\lambda+\mu)}$. Conversely suppose that there exist weight vectors $a_\xi \in V_w^+(\lambda)^\perp$, $b_\eta \in V^+(\mu)^*$ such that $a_\xi b_\eta = c_{w(\lambda+\mu)}$. By 4.4.3 (v) the weights of $V_w^+(\lambda)$ lie in $w\lambda + \mathbb{N}(\Delta^+ \cap w\Delta^-)$ and so $\xi = w\lambda + \delta$ with $\delta \in \mathbb{N}w^{-1}\Delta^-$. Then $\eta = w\mu - \delta$ and so by 4.4.3 (i) one has $(\mu, \mu) \geq (\eta, \eta) = (\mu, \mu) + (\delta, \delta) - 2(w\mu, \delta) \geq (\mu, \mu) + (\delta, \delta)$ which forces $\delta = 0$. Consequently $a_\xi = c_{w\lambda}$, $b_\eta = c_{w\mu}$ up to scalars. Thus $Not(**)$ implies $c_{w\lambda} \in V_w^+(\lambda)^\perp$ contradicting $\xi_{w\lambda}(u_{w\lambda}) \neq 0$. This proves that Q_w^+ is a right ideal. A similar argument shows that it is a left ideal. Note that $c_{w\lambda}$ has been shown to have a non-zero image $\bar{c}_{w\lambda}$ in R^+/Q_w^+.

Since $V_w^+(\lambda) = k(q)u_{w\lambda} \oplus U_+^+.V_w^+(\lambda)$ it follows from the isomorphism $V(\lambda)^*/V_w^+(\lambda)^\perp \xrightarrow{\sim} V_w^+(\lambda)^*$ that every V^+ invariant vector of weight $w\lambda$ in R^+/Q_w^+ is proportional to $\bar{c}_{w\lambda}$. It follows that R^+/Q_w^+ is a domain just as in 9.1.9 (i) using 9.1.10 (*).

10.1.9 Recall (9.3.5) the automorphisms $\tau_\nu : \nu \in P^+(\pi)$ of R, 9.3.1 and the notation of 9.3.6 and 10.1.3.

Lemma. *Take $\lambda \in P^+(\pi)$. Let $F(\lambda)$ be an $\mathrm{ad}_{\tau_{\lambda_0}} R$ (resp. $\mathrm{ad}_{\tau_\lambda} R$) invariant subspace of $V^+(\lambda)^*$ (resp. $V^-(\lambda)^*$). Then $R^-F(\lambda) = F(\lambda)R^-$ (resp. $R^+F(\lambda) = F(\lambda)R^+$) holds both in R and in $\hat{R}$.*

Take $F(\lambda) \subset V^-(\lambda)^*$. One has in R that

$$(*) \qquad \tau_\lambda(a)b = [(\mathrm{ad}_{\tau_\lambda} a_1)b]a_2$$

for all $a \in R^+$, $b \in F(\lambda)$. Moreover since R^+ is a left coideal one can assume $a_2 \in R^+$ in (*). This proves that $R^+F(\lambda) \subset F(\lambda)R^+$.

For the reverse inclusion recall that even as an $\mathrm{ad}_{\tau_\lambda} R$ module $F(\lambda)$ admits a filtration $F(\lambda) = F_1 \supsetneq F_2 \supsetneq \dots \supsetneq F_{n+1} = 0$ by submodules such that the quotients are one-dimensional, annihilated by J (notation 9.3.11) and on which the $c^\mu_{\mu,\mu} : \mu \in P^+(\pi)$ act by non-zero scalars. (This is for example obvious from 9.3.6 (ii)). An easy induction argument on filtration length then proves the assertion. The case $F(\lambda) \subseteq V^+(\lambda)^*$ is similar.

The last part follows from the first part and 9.1.8 which implies that the restrictions of μ^* to the subspaces $R^- \otimes F(\lambda)$ and $F(\lambda) \otimes R^-$ (resp. $R^+ \otimes F(\lambda)$ and $F(\lambda) \otimes R^+$) are injective.

10.1.10 Take $\mathbf{w} = (w_+, w_-) \in W \times W$ and set $\hat{Q}_{\mathbf{w}} := Q^+_{w_+} \otimes R^- + R^+ \otimes Q^-_{w_-}$ viewed as a subspace of $\hat{R}$. Let $Q_{\mathbf{w}}$ denote its image in R.

Corollary. *In either R or $\hat{R}$ one has $R^{\pm}Q^{\mp}_{w_{\mp}} = Q^{\mp}_{w_{\mp}}R^{\pm}$. In particular $\hat{Q}_{\mathbf{w}}$ (resp. $Q_{\mathbf{w}}$) is a completely prime ideal of $\hat{R}$ (resp. ideal of R).*

Since $V^-_w(\lambda) \subset V^-(\lambda)^*$ is V^- stable under right action, it is $\operatorname{ad}_{\tau_\lambda} R$ stable by 9.3.6 (ii). Then $R^+V^-_w(\lambda) = V^-_w(\lambda)R^+$ by 10.1.9. A similar assertion holds by 9.3.6 (i) with $+$ and $-$ interchanged. This proves the first part. For complete primeness in the second part use the right $V^+ \otimes V^-$ action on $R^+/Q^+_{w_+} \otimes R^-/Q^-_{w_-}$ as in the proof of 10.1.8.

Remarks. Recall the notation of 9.3.7. Clearly $Q^+_{w_+}R^-$ (resp. $Q^-_{w_-}R^+$) is a two-sided ideal of R containing $J^+_{w_+}$ (resp. $J^-_{w_-}$). Again $Q_{\mathbf{w}}$ is their sum. It will eventually be shown that the images of the $c_{w_+\lambda}$ and $d_{w_-\lambda} : \lambda \in P^+(\pi)$ in $R/Q_{\mathbf{w}}$ are non-zero divisors and the latter is even a domain. This will prove that $Q_{\mathbf{w}} \in \mathbf{X}(\mathbf{w})$.

10.1.11 Recall (9.2.11) that $J^+ \cap R^- = 0$ and $J^- \cap R^+ = 0$. Combined with 9.1.5 this leads to the following commutation relations.

Lemma. *For all $\mu, \nu \in P^+(\pi)$, $\eta \in \Omega(V(\mu))$ one has*

(i) $c^\nu_{\nu,\nu}c^\mu_{\eta,\mu} = q^{-(\nu,\eta)+(\nu,\mu)}c^\mu_{\eta,\mu}c^\nu_{\nu,\nu}$

(ii) $c^\nu_{w_0\nu,w_0\nu}c^\mu_{\eta,w_0\mu} = q^{(w_0\nu,\eta)-(\nu,\mu)}c^\mu_{\eta,w_0\mu}c^\nu_{w_0\nu,w_0\nu}$

Consider (ii). In 9.1.5 (i) replace ν, η', λ' by $w_0\nu$ and λ by $w_0\mu$. Since all the terms in (ii) lie in R^- and $J^+ \cap R^- = 0$ this gives (ii).

Consider (i). In 9.1.5 (ii) replace ν, η', λ' by ν and λ by μ. As above this gives (i).

10.1.12 Recall the decomposition of Spec R given in 9.3.9.

Proposition. *Take $\mathbf{w} = (w_+, w_-)$ and $P \in \mathbf{X}(\mathbf{w})$.*

(i) *Given $\mu \in P^+(\pi), \xi \in V(\mu)^*_\eta$ such that $c^\mu_{\xi,\mu} \notin P$ and $c^\mu_{\xi.a,\mu} \in P$, $\forall a \in U^+_+$ then $\eta = w_+\mu$.*

(ii) *Given $\mu \in P^+(\pi)$, $\xi \in V(\mu)^*_\eta$ such that $c^\mu_{\xi,w_0\mu} \notin P$ and $c^\mu_{\xi.a,w_0\mu} \in P$, $\forall a \in U^-_+$ then $\eta = w_-w_0\mu$.*

Consider (i). If $\nu \neq w_+\nu$ then there exists $a_\zeta \in U^+_+$ of weight $\zeta = \nu - w_+\nu$ such that $\xi_\nu.a_\zeta = \xi_{w_+\nu}$ (recall the remarks in 9.1.1). Applying a_ζ on the right to 10.1.11 (i) with $c^\mu_{\eta,\mu} = c^\mu_{\xi,\mu}$ gives through the hypothesis on P

$$c^\nu_{w_+\nu,\nu}c^\mu_{\eta,\nu} = q^{-(w_+\nu,\eta)+(\nu,\mu)}c^\mu_{\eta,\mu}c^\nu_{w_+\nu,\nu} \bmod P \,.$$

Yet $P \supset J^+_{w_+}$ so 9.1.5 (i) applies with $\lambda = \mu$, $\eta' = w_+\nu$, $\lambda' = \nu$ giving a similar expression; but with the exponent in q having the opposite sign!

Since $c^{\nu}_{w_+\nu,\nu} \notin P$ it has a normal image which must be a non-zero divisor. This forces $(w_+\nu,\eta) = (\nu,\mu)$. Taking ν regular implies $\eta = w_+\mu$, as required. The proof of (ii) follows in a similar fashion from 10.1.11 (ii) and 9.1.5 (ii).

10.1.13 Recall the notation of 10.1.10.

Corollary. *Every $P \in \mathbf{X}(\mathbf{w})$ contains $Q_{\mathbf{w}}$.*

Suppose $P \not\supset Q_{\mathbf{w}}$. Recall that $Q_{\mathbf{w}}$ is generated by certain elements of the form $c^{\mu}_{\xi,\mu}$ (resp. $c^{\mu}_{\xi,w_0\mu}$) lying in $Q^+_{w_+}$ (resp $Q^-_{w_-}$).

Suppose $c^{\mu}_{\xi,\mu} \notin P$ and yet $c^{\mu}_{\xi,\mu} \in Q^+_{w_+}$ with $\xi \in V(\mu)^*_{\eta}$. Assume η minimal with this property. Since $Q^+_{w_+}$ is stable (10.1.8) for the right action of U^+ it follows that $c^{\mu}_{\xi.a,\mu} \in P$ for all $a \in U^+_+$. By 10.1.12 (i) this forces $\eta = w_+\mu$ contradicting that $c^{\mu}_{w_+\mu,\mu} \notin Q^+_{w_+}$. A similar argument applies to the second case.

10.1.14 Return to the simple modules $\mathcal{S}(w) : w \in W$. By 10.1.7 and 10.1.13 it follows in particular that $\operatorname{Ann}\mathcal{S}(w) \supset Q^+_w$ whilst $c_{w\lambda} \notin \operatorname{Ann}\mathcal{S}(w)$ for all $\lambda \in P^+(\pi)$. If $w \neq w_0$ there exists $i \in \{1,2,\dots,\ell\}$ such that $s_iw > w$. Write $\mathcal{S}(s_i)$ simply as $\mathcal{S}(i)$. Let φ_i denote the canonical embedding of $U_i := U_{\alpha_i}$ (4.3.5) in U and let $\varphi_i^\star$ denote the restriction of φ_i^* to $R_q[G]$.

Since every finite dimensional U_i module occurs (4.3.1, 4.3.4, 4.3.6) in some finite dimensional U module it follows that $\operatorname{Im}\varphi_i^\star$ is just the index 2 subalgebra of U_i^* described in 9.1.1. It may be conveniently denoted by $R_q[SL(2)_i]$. This gives a second construction of $\mathcal{S}(i)$. Namely apply the construction of 10.1.5 to $R_q[SL(2)_i]$ to obtain a module $\mathcal{S}'(i)$ associated to the unique non-identity element s_i of the Weyl group of $SL(2)$. Then view $\mathcal{S}'(i)$ as a $R_q[G]$ module via $\varphi_i^\star$. A fortiori, $\mathcal{S}'(i)$ is simple and using 10.1.5 one easily checks that $\mathcal{S}'(i)$ is isomorphic to $\mathcal{S}(i)$. It is moreover easy to write down generators and relations for $R_q[SL(2)]$ and therefore give a rather explicit description of $\mathcal{S}'(i)$.

Lemma. *Assume $s_iw > w$. Then*

(i) $\operatorname{Ann}(\mathcal{S}(i) \otimes \mathcal{S}(w)) \supset Q^+_{s_iw}$.
(ii) $c_{s_iw\lambda}(a \otimes b) = c^{\lambda}_{s_iw\lambda,w\lambda}a \otimes c_{w\lambda}b$, *for all $a \in \mathcal{S}(i)$, $b \in \mathcal{S}(w)$; $\lambda \in P^+(\pi)$, given that $\xi \in V(\lambda)^*_{w\lambda}$ and $v \in V(\lambda)_{w\lambda}$ arising in the right hand side, satisfy $\xi(v) = 1$.*

By 4.4.3 (v) the hypothesis implies that $V_{s_iw}(\lambda)$ is U_i stable. Through the Shapovalev form (3.4.10) on $V(\lambda)$ one may identify $V_{s_iw}(\lambda)^\perp \subset V(\lambda)^*$ with a U_i stable subspace of $V(\lambda)$ complementing $V_{s_iw}(\lambda)$. Fix an orthonormal basis $\{v_j\}_{j=1}^n$ of $V(\lambda)$ compatible with the inclusion $V_w(\lambda) \subset V_{s_iw}(\lambda)$ and this decomposition, that is $v_j \in V_w(\lambda)$ for $j \le s := \dim V_w(\lambda)$ (resp. $v_j \in V_{s_iw}(\lambda)$, for $j \le r := \dim V_{s_iw}(\lambda)$) and $v_j \in V_{s_iw}(\lambda)^\perp$ for $j > r$. This also serves as the dual basis for $V(\lambda)^*$. Then for all $a \in \mathcal{S}(i), b \in \mathcal{S}(w)$; $\xi \in V(\lambda)^*$ one has

$$(*) \qquad c^\lambda_{\xi,\lambda}(a \otimes b) = \sum_{j=1}^n c^\lambda_{\xi,v_j} a \otimes c^\lambda_{v_j,\lambda} b \,.$$

Since $\operatorname{Ann} \mathcal{S}(w) \subset Q^+_w$ one has $c^\lambda_{v_j,\lambda} b = 0$ for $j > s$. On the other hand since $a \in \mathcal{S}(i)$ one can write $c^\mu_{\xi,v_j} a = \varphi^\star_i(c^\mu_{\xi,v_j})a$. Now for all $v \in V_{s_i w}(\lambda)$, $\xi \in V_{s_i w}(\lambda)^\perp$, $x \in U_i$ one has $c^\mu_{\xi,v}(x) = \xi(xv) = 0$. In particular $\varphi^\star_i(c^\mu_{\xi,v_j}) = 0$ if $j \le r$ and $\xi \in V_{s_i w}(\lambda)^\perp$. This proves (i).

Now take $\xi \in V(\lambda)^*$ of weight $s_i w\lambda$. The only contributions to the right hand side of $(*)$ occur when $j \le s$. These are non-zero contributions only if $c^\mu_{\xi,v_j}(x) \neq 0$ for some $x \in U_i$. One can assume the v_j to be weight vectors and $v_s = u_{w\lambda}$. Since the U_i module generated by ξ has weights $\{s_i w\lambda, s_i w\lambda - \alpha_i, \dots, w\lambda\}$, it follows from the definition of $V_w(\lambda)$ that the only non-zero term corresponds to $j = s$. Hence (ii).

10.1.15 Since $\operatorname{Ann} \mathcal{S}(w) \supset Q_{\mathbf{w}} \supset J^+_w + J^-_w$ it follows that the $c_{w\lambda}, d_{w\lambda} : \lambda \in P^+(\pi)$ have normal images $\bar{c}_{w\lambda}, \bar{d}_{w\lambda}$ in $R / \operatorname{Ann} \mathcal{S}(w)$. More precisely 9.1.5 (i), 9.1.5 (ii) give

$$(*) \qquad \begin{aligned} \bar{c}^\mu_{\eta,w_0\mu} \bar{c}_{w\lambda} &= q^{(\lambda,w_0\mu)-(\eta,w\lambda)} \bar{c}_{w\lambda} \bar{c}^\mu_{\eta,w_0\mu} \\ \bar{c}^\mu_{\eta,w_0\mu} \bar{d}_{w\lambda} &= q^{(ww_0\lambda,\eta)-(\lambda,\mu)} \bar{d}_{w\lambda} \bar{c}^\mu_{\eta,w_0\mu} \end{aligned}$$

for all $\eta \in \Omega(V(\mu))$. In particular the $\bar{d}_{w\mu}$ commute with the $\bar{c}_{w\lambda}$. Set $\lambda_0 = -w_0\lambda$. Recalling 10.1.3 it follows that both $c_{w\lambda}$ and $d_{w\lambda_0}$ act on $c^\mu_{\eta,w_0\mu} v_w$ by the same scalar $q^{(w^{-1}\eta - w_0\mu,\lambda)}$. Set

$$\Omega(w) = \bigcup_{\mu \in P^+(\pi)} (w^{-1}\Omega(V^-_w(\mu)) - w_0\mu) \subset Q^+(\pi)$$

and for each $\gamma \in Q(\pi)$, let $\mathcal{S}(w)_\gamma$ denote the subspace of $\mathcal{S}(w)$ in which the commuting elements $\bar{c}_{w\lambda}, \bar{d}_{w\lambda_0} : \lambda \in P^+(\pi)$ act by the scalar $q^{(\gamma,\lambda)}$.

Lemma

(i) $\mathcal{S}(w) = \bigoplus_{\gamma \in \Omega(w)} \mathcal{S}(w)_\gamma$.
(ii) $\mathcal{S}(w)_0 = k(q) v_w$.
(iii) $\dim \mathcal{S}(w)_\gamma < \infty$ *for all* $\gamma \in \Omega(w)$.

Recall that $\mathcal{S}(w) = R^- v_w$. Yet by 10.1.7 and 10.1.13 one has $\operatorname{Ann} \mathcal{S}(w) \supset Q^-_w$ so R^- can be replaced by R^-/Q^-_w. The latter can be identified with the direct sum of the $V^-_w(\mu)^*$ which are in turn spanned by the $c^\mu_{\xi,w_0\mu} : \xi \in V^-_w(\mu)^*$. Since $\Omega(V^-_w(\mu)^*) = \Omega(V^-_w(\mu))$ the assertion of (i) follows from $(*)$.

For (ii) observe that $\gamma := w^{-1}\eta - w_0\mu \ge 0$ for $\eta \in \Omega(V(\mu))$ and moreover this is a strict inequality unless $\gamma = ww_0\mu$. However in this case the corresponding element is just $d_{w\mu}$. For any other $d = c^\mu_{\xi,w_0\mu} : \xi \in V^-_w(\mu)^*$ one has

(**) $\quad d\mathcal{S}(w)_\gamma \subset \mathcal{S}(w)_{\gamma'}$ for some $\gamma' \in \Omega(w)$ with $\gamma' > \gamma$.

This gives (ii) and since R^- is finitely generated, (iii) also follows.

10.1.16 By 10.1.15 it is natural to define a formal character for $\mathcal{S}(w)$. However here one should note that the eigenspaces are defined with respect to the $c_{w\lambda} : \lambda \in P^+(\pi)$ which also depends on w. Thus given an R module M which is a direct sum of finite dimensional weight spaces $M_\gamma : \gamma \in Q^+(\pi)$ in which the $c_{w\lambda} : \lambda \in P^+(\pi)$ act by the scalars $q^{(\gamma,\lambda)}$ define

$$\operatorname{ch}_w M := \sum_{\gamma \in Q^+(\pi)} (\dim M_\gamma) e^\gamma .$$

Then $\operatorname{ch}_w \mathcal{S}(w)$ is defined; but it is not assumed that $\operatorname{ch}_y \mathcal{S}(w)$ is defined for $y \neq w$.

Lemma. *Assume $s_i w > w$. Then*

$$\operatorname{ch}_{s_i w}(\mathcal{S}(i) \otimes \mathcal{S}(w)) = w^{-1}(\operatorname{ch}_{s_i} \mathcal{S}(i)) \operatorname{ch}_w \mathcal{S}(w) .$$

This follows from 10.1.14 (ii) noting that $\varphi_i^*(c^\lambda_{s_i w\lambda, w\lambda})$ is just the $\mathfrak{sl}(2)$ matrix element $c^\nu_{s_i\nu,\nu}$ with respect to the $\mathfrak{sl}(2)$ module of highest weight $\nu := w\lambda$. Thus it acts on $\mathcal{S}(i)_\gamma$ by the scalar $q^{(\gamma,\nu)} = q^{(w^{-1}\gamma,\lambda)}$.

Remark. Recall 10.1.14. Then by 10.1.15 (i) one has $\mathcal{S}(i)_\gamma = \mathcal{S}'(i)_\gamma \neq 0$ implies $\gamma \in \mathbb{N}\alpha_i$. Since $w^{-1}\alpha_i \in \Delta^+$ it follows by 10.1.15 (i) and the above that $(\mathcal{S}(i) \otimes \mathcal{S}(w))_\gamma \neq 0$ implies $\gamma \in Q^+(\pi)$. One may also check that $\operatorname{ch}_i \mathcal{S}(i) = (1 - e^{\alpha_i})^{-1}$.

10.1.17 Recall 3.2.6, 8.3.6. The involution $\iota = \varkappa'\sigma$ of U transposes to an involution ι^* of R which is an antiautomorphism of algebras and an automorphism of coalgebras. For each finite dimensional simple U module V the Shapovalev form $\mathcal{S}'$ defines an isomorphism $\varphi : v \mapsto (v' \mapsto \mathcal{S}'(v', v))$ of V onto V^*. Then for all $v, v' \in V$, $a \in U$ one has $\varkappa'^*(c^V_{\varphi(v),v'})(a) = \varphi(v)(\varkappa'(a)v') = \mathcal{S}'(v, \varkappa'(a)v') = \mathcal{S}'(av, v') = c^V_{\varphi(v'),v}(a)$, that is $\varkappa'^*(c^V_{\varphi(v),v'}) = c^V_{\varphi(v'),v}$. Since $\mathcal{S}'$ respects weight space decomposition one can choose an orthonormal basis $\{v_i\}$ for V consisting of weight vectors. Then $\{\xi_i := \varphi(v_i)\}$ is a dual basis for V^*. Using this pair in 10.1.3(*) and recalling the definition given there of $d_{w\mu_0}$ one obtains $\iota^*(c_{w\mu}) = \sigma^*\varkappa'^*(c^\mu_{w\mu,\mu}) = \sigma^*(c^\mu_{\mu,w\mu}) = d_{w\mu_0}$. Thus for all $\mu \in P^+(\pi)$, $c_{w\mu}$ and $\iota^*(c_{w\mu})$ act by the *same* scalar on weight subspaces of $\mathcal{S}(w)$.

Fix $w \in W$ and give $\mathcal{S}(w)^*$ a left R module structure through ι^*. Obviously $\operatorname{Ann} \mathcal{S}(w)^* = \iota^*(\operatorname{Ann} \mathcal{S}(w))$. Thus by 10.1.7 and the above, the images of $c_{w\lambda}, d_{w\lambda_0}$ in $R/\operatorname{Ann} \mathcal{S}(w)^*$ are normal. Thus the direct sum $\mathcal{S}(w)^\succ$ of the common eigenspaces of the $c_{w\lambda}, d_{w_0\lambda} : \lambda \in P^+(\pi)$ in $\mathcal{S}(w)^*$ is an R submodule. Moreover $\mathcal{S}(w)^\succ$ is exactly the graded dual of $\mathcal{S}(w)$ with the gradation defined by the weight space decomposition (10.1.15). Again by the result of the previous paragraph $\mathrm{ch}_w \mathcal{S}(w)^{\succ} = \mathrm{ch}_w \mathcal{S}(w)$. Finally by 10.1.15 (**) the zero weight subspace of $\mathcal{S}(w)^{\succ}$ is an R^+ submodule and so by 10.1.5 it generates an R submodule of $\mathcal{S}(w)^{\succ}$ surjecting to $\mathcal{S}(w)$. Comparison of formal characters forces an isomorphism $\mathcal{S}(w)^{\succ} \xrightarrow{\sim} \mathcal{S}(w)$. This conclusion may be expressed as follows.

Lemma. *Fix $w \in W$. Then $\mathcal{S}(w)$ admits a non-degenerate invariant form giving an orthogonal decomposition into the joint $c_{w\lambda} : \lambda \in P^+(\pi)$ eigenspaces.*

Remark. A similar argument shows that $\mathcal{S}(\Lambda_w^+)$ and $\mathcal{S}((\Lambda_w^+)^{-1})$ are paired.

10.1.18 Take the product form an $\mathcal{S}(i) \otimes \mathcal{S}(w)$ which is of course non-degenerate and invariant. Moreover if $s_i w > w$ then by 10.1.14 (ii) it gives an orthogonal decomposition of $\mathcal{S}(i) \otimes \mathcal{S}(w)$ into the joint $c_{s_i w\lambda} : \lambda \in P^+(\pi)$ eigenspaces which are furthermore finite dimensional. (Notice that this means that the $d_{s_i w\lambda_0}$ act by the same scalars on a given eigenspace).

Theorem. *Suppose $s_i w > w$. Then $\mathcal{S}(i) \otimes \mathcal{S}(w)$ is simple and isomorphic to $\mathcal{S}(s_i w)$.*

Let M be a proper submodule of $\mathcal{S}(i) \otimes \mathcal{S}(w)$. Then its orthogonal $M^\perp$ for the form is a submodule satisfying $M \cap M^\perp = 0$. By the above property of eigenspaces one also has $M + M^\perp = \mathcal{S}(i) \otimes \mathcal{S}(w)$.

Since the "highest weight subspace" $(\mathcal{S}(i) \otimes \mathcal{S}(w))_0$ on which the $c_{s_i w\lambda} : \lambda \in P^+(\pi)$ act by 1 is one-dimensional by 10.1.15 (ii) and 10.1.16, it can be assumed not to belong to M. Then $M_\gamma \neq 0$ for some $\gamma \in Q^+(\pi) \setminus \{0\}$ and take γ minimal with this property. By 10.1.14 (i) one has $\operatorname{Ann} M \supset Q^+_{s_i w} R^- \supset J^+_{s_i w}$ and so 9.1.5 (i) gives

$$c^\mu_{\eta,\mu} c_{s_i w\lambda} = q^{(\mu,\lambda)-(\eta, s_i w\lambda)} c_{s_i w\lambda} c^\mu_{\eta,\mu} \bmod \operatorname{Ann} M$$

for all $\eta \in \Omega(V(\mu))$. Thus $c^\mu_{\eta,\mu} M_\gamma \subset M_{\gamma'}$ where $\gamma' = \gamma - (\mu - w^{-1} s_i \eta) \leq \gamma$ with equality if and only if $\eta = s_i w\mu$ (that is if $c^\mu_{\eta,\mu} = c_{s_i w\mu}$). By the minimality of γ one obtains $c^\lambda_{\eta,\lambda} M_\gamma = 0$ unless $\eta = s_i w\lambda$. Consequently for all $m \in M_\gamma$ one has $m = x_\lambda m = d_{s_i w\lambda_0} c_{s_i w\lambda} m = q^{2(\lambda,\gamma)} m$ which is clearly absurd. Hence $\mathcal{S}(i) \otimes \mathcal{S}(w)$ is simple and so isomorphic to $\mathcal{S}(s_i w)$ via 10.1.5.

Remarks. Strictly speaking the form was not needed and the same argument goes through with quotients. However one does need to also control the eigenvalues of the $d_{s_i w\lambda_0}$ and so the argument would not have been significantly shorter. Replacing k by $\mathbb{R}$ and taking $q \in \mathbb{R}^*$ in general position, the form can be assumed to be positive definite. Then one could have first proved that $\mathcal{S}(i) \otimes \mathcal{S}(w)$ is semisimple. However note that say $\mathcal{S}(i) \otimes \mathcal{S}(i)$

which also admits a positive definite form is not semisimple as may checked by an explicit calculation. The trouble is that it does not admit a convenient decomposition into finite dimensional eigenspaces.

10.1.19 The importance of 10.1.18 is that one can construct $\mathcal{S}(w)$ inductively on $\ell(w)$ from the modules $\mathcal{S}(i) : i = 1, 2, \ldots, \ell$ which are easy to write down. Let S_i^- denote the subalgebra of $d^{-1}R^-$ constructed in 9.3.1 relative to $R_i := R[SL(2)_i]$. It is rather obvious that $\mathcal{S}(i)$ can be identified with S_i^- as an R_i^+ module. In particular $\mathrm{ch}_i\, \mathcal{S}(i) = (1 - e^{\alpha_i})^{-1}$. A corresponding interpretation will be given for $\mathcal{S}(w)$ in the general case (10.3.10). For the moment, observe that this, 10.1.16 and 10.1.18 give the

Corollary. *(Notation* A.1.1*)*

$$\mathrm{ch}_w\, \mathcal{S}(w) = \prod_{\alpha \in S(w)} (1 - e^{\alpha})^{-1} .$$

10.2 The Quantum Weyl Group

A main interest in highest weight modules is that they give rise to a braid group action on $U_q(\mathfrak{g})$. This becomes the usual Weyl group action when $q \longrightarrow 1$. For this reason it is called the quantum Weyl group.

10.2.1 For each $w \in W$, define $N_w \in R^*$ through $N_w(a) = \langle v_w, av_w \rangle$, for all $a \in R$, where $\langle , \rangle$ denotes the form in the conclusion of 10.1.17. Since $\langle v_w, av_w \rangle = \langle \iota^*(a)v_w, v_w \rangle = \langle v_w, \iota^*(a)v_w \rangle$, one has $\iota(N_w) = N_w$.

Lemma. *For all* $w, w' \in W$ *satisfying* $\ell(w) + \ell(w') = \ell(ww')$ *one has* $N_w N_{w'} = N_{ww'}$.

By 10.1.18 and A.1.1 the hypothesis implies that $\mathcal{S}(w) \otimes \mathcal{S}(w') \cong \mathcal{S}(ww')$ where moreover $v_{ww'}$ identifies with $v_w \otimes v_{w'}$ and the form on $\mathcal{S}(ww')$ coincides with the product form on $\mathcal{S}(w) \otimes \mathcal{S}(w')$. Then for all $c \in R$

$$\begin{aligned}(N_w N_{w'})(c) &= N_w(c_1) N_{w'}(c_2) \\ &= \langle v_w, c_1 v_w \rangle \langle v_{w'}, c_2 v_{w'} \rangle \\ &= \langle v_{ww'}, c v_{ww'} \rangle = N_{ww'}(c) ,\end{aligned}$$

as required.

10.2.2 Let V be a finite dimensional U module with weights in $P(\pi)$. By 1.4.11 it follows that N_w induces an element N_w^V of $\mathrm{End}\, V$. To compute this it suffices by 10.2.1 to take $w = s_i$ and in this case one writes N_{s_i} simply as N_i.

Lemma. *For each* $\lambda \in \Omega(V)$, $v_\lambda \in V_\lambda \setminus \{0\}$, *one has* $N_i v_\lambda \in V_{s_i\lambda} \setminus \{0\}$. *In particular* N_w^V *is invertible in* End V *and hence* N_w *is invertible in* R^*.

Recall the notation of 10.1.14. Then $N_i(c) = \langle v_{s_i}, cv_{s_i} \rangle = \langle v_{s_i}, \varphi_i^*(c) v_{s_i} \rangle$ and thus N_i factors through the surjection $\varphi_i^* : R \longrightarrow R_i$. In particular N_i leaves each U_i submodule of V invariant. By 4.3.4 it suffices to consider simple submodules and by 4.3.1 these are determined by their lowest weights. Indeed (cf. 9.1.10) let u_{-r} denote the lowest weight vector for the $r+1$ dimensional U_i module $V^{(r)}$ of lowest weight $q_i^{-\frac{1}{2}r\alpha_i}$. Applying $\varkappa$ to 4.3.1 (*) one obtains

$$\varkappa'(e_i) e_i^s u_{-r} = t_i^{-1}[f_i, e_i^s] u_{-r} = q_i^{r+2-2s}[s]_{q_i}[r-s+1]_{q_i} e_i^{s-1} u_{-r} .$$

Thus one may choose a basis of $V^{(r)}$ orthonormal for the Shapovalev form S' defined in 8.3.6 by setting

$$(1) \qquad u_{2s-r}^{(r)} = q_i^{-\frac{1}{2}(r-s+1)s} \begin{bmatrix} r \\ s \end{bmatrix}_{q_i}^{-\frac{1}{2}} e_i^{(s)} u_{-r}^{(r)} : s = r, r-1, \ldots, 0 .$$

This serves as a basis for $V^{(r)*}$ and one defines $c_{m,n}^{(r)} \in R_i$ or simply $c_{m,n}$ by $c_{m,n}^{(r)}(a) = S'(u_m^{(r)}, a u_n^{(r)})$, $\forall a \in U$. One must show that $N_i(c_{m,n}) \neq 0 \Longleftrightarrow m+n=0$.

From 9.1.4 (i) one obtains

$$c_{-1,1} c_{m,n} = q_i^{-(m+n)/2} c_{m,n} c_{-1,1} .$$

Recalling that v_{s_i} is a $c_{-1,1}$ eigenvector $\Longrightarrow$ follows.

In the remainder of the calculation, drop the i subscript take $r=1$ and set $a = c_{1,1}$, $b = c_{1,-1}$, $c = c_{-1,1}$, $d = d_{-1,-1}$. From 9.1.4 one obtains the relations

$$(2) \qquad \begin{aligned} &ba = q^{-1}ab , \quad ca = q^{-1}ac , \quad bc = cb \\ &db = q^{-1}bd , \quad dc = q^{-1}cd , \quad ad - da = (q - q^{-1})bc \end{aligned}$$

which also hold in $R^- \otimes R^+$ and are incidentally also the defining relations of the co-ordinate ring of the so-called quantum 2×2 matrices. By say 9.1.6 and 9.1.8 the ordered monomials in the a, b, c, d form a PBW-type basis for this algebra. The additional relation (9.2.4) holding in R is just

$$(3) \qquad da - q^{-1}bc = 1$$

where the left hand-side (or a multiple of it) is the so-called 2×2 quantum determinant.

Set $u = u_1$, $v = u_{-1}$. Then $ev = q^{\frac{1}{2}}u$ and $fu = q^{-\frac{1}{2}}v$. Consequently $e.b = q^{\frac{1}{2}}a$, $t.b = q^{-1}b$, $b.t = qb$, $b.e = q^{\frac{1}{2}}d$, $a.e = q^{\frac{1}{2}}c$. Using 5.1.1 (vii) this gives

$$(4) \qquad e.b^m = q^{\frac{1}{2}} q^{m-1} [m]_q b^{m-1} a , \quad b^m.e = q^{\frac{1}{2}} q^{-(m-1)} [m]_q b^{m-1} d .$$

which through the identity $[m]_q = q^{m-1} + q^{-1}[m-1]_q$ combine to give

$$(5) \qquad q^{-1}e.b^m.e = [m]_q^2 b^{m-1}c + [m]_q[m-1]_q b^{m-2} .$$

Now $av_s = 0$, whilst $cv_s = v_s$ by definition and $bv_s = -qv_s$ by (3). Define $D \in \operatorname{End} k(q)[b]$ by

$$(6) \qquad Db^m = [m]_q^2 b^{m-1} + [m]_q[m-1]_q b^{m-2} .$$

Since $e.b^m c^n.e = (e.b^m.e)c^n$ for all $m, n \in \mathbb{N}$, it follows from (4), (5) and A.3.2 that

$$(7) \quad q^{-n}N(e^n.b^m.e^n) = (D^n b^m)\big|_{b=-q} = (-1)^{m-n}q^{(n+1)(m-n)} \frac{[m!]_q[n!]_q}{[(m-n)!]_q}$$

for all $n = 0, 1, 2, \ldots, m$. Since $b^m = c^{(m)}_{m,-m}$. One concludes from (1) and (7) that

$$(8) \qquad N(c^{(m)}_{m-2n,-(m-2n)}) = (-1)^{m-n}q^{(n+1)(m-n)}$$

as required.

Equivalently (recovering the ith subscript) one has

$$(9) \qquad N_i u^{(m)}_{-(m-2n)} = (-1)^{m-n}q_i^{(n+1)(m-n)}u^{(m)}_{(m-2n)} .$$

10.2.3 In view of 10.2.2 one may define for each $w \in W$ an inner automorphism $r_w : a \mapsto N_w a N_w^{-1}$ of R^* commuting with ι. The following result is immediate from 10.2.2.

Lemma. *For all $w \in W$, $\lambda \in P(\pi)$ one has $r_w(\tau(\lambda)) = \tau(w\lambda)$, that is r_w restricts to the automorphism of T given by w.*

10.2.4 One may anticipate that r_w should restrict to an automorphism of $R^\star$ and in view of 9.4.9 and 10.2.3 to even restrict to an automorphism of U. Though this approach does not seem to be successful one may prove by explicit computation that each r_i restricts to an automorphism of U. In this it is convenient to define a slightly larger algebra than U as the formulae becomes more symmetric.

Set $K_i^{\pm 1} = \tau(\pm\alpha_i/2)$, $E_i = e_i K_i$, $F_i = K_i^{-1}f_i : i = 1, 2, \ldots, \ell$. For each i these generate a Hopf subalgebra U_i' of R^* which together generate a Hopf subalgebra $U_q'(\mathfrak{g})$, or simply U', containing U. Moreover $\varkappa'(E_i) = F_i$, $\varkappa'(F_i) = E_i$ and $\iota(E_i) = -q_i^{-1}F_i$, $\iota(F_i) = -q_iE_i$. Again $e_i^s = q^{-\frac{1}{2}s(s-1)}E_i^sK_i^{-s}$ and so 10.2.2 (1) gives in terms of divided powers

$$(*) \qquad u^{(r)}_{2s-r} = \begin{bmatrix} r \\ s \end{bmatrix}_{q_i}^{-\frac{1}{2}} E_i^{(s)}u^{(r)}_{-r} : s = r, r-1, \ldots, 0 .$$

Lemma. *For all $i = 1, 2, \ldots, \ell$ one has*

(i) $r_i(t_iE_i) = -F_i$

(ii) $r_i(F_it_i^{-1}) = -E_i$.

Omit the i subscript. Then $(*)$ gives $u_m = E^{(m)}u_{-m}$ and

$$E^{(n)}u_{m-2n} = \begin{bmatrix} m \\ n \end{bmatrix}_q^{\frac{1}{2}} E^{(n)}E^{(m-n)}u_{-m} = \begin{bmatrix} m \\ n \end{bmatrix}_q^{-\frac{1}{2}} u_m \, .$$

By comparison with $(*)$ one concludes that $\mathcal{S}'(u_{2n-m+2}, Eu_{2n-m}) = \mathcal{S}'(u_{m-2n}, Eu_{m-2n-2})$. Applying $\varkappa'$ one may write for a fixed pair m, n

$$(**) \qquad Eu_{2n-m} = \gamma u_{2n+2-m} \, , \qquad Fu_{m-2n} = \gamma u_{m-2n-2}$$

for some $\gamma \in \overline{k(q)}^*$. Then by 10.2.2 (9)

$$\begin{aligned} NEN^{-1}u_{m-2n} &= (-1)^{m-n}q^{-(n+1)(m-n)}NEu_{2n-m} \\ &= (-1)^{m-n}q^{-(n+1)(m-n)}(-1)^{m-n-1}q^{(m-n-1)(n+2)}\gamma u_{m-2n-2} \\ &= -K^2Fu_{m-2n} \end{aligned}$$

hence (i). Then $-E = q^{-1}\iota(F) = -q^{-1}\iota r(t\,E) = q^{-2}r(t^{-1}F) = r(Ft^{-1})$, as required.

10.2.5 Set $a_{ij} := -(\alpha_i^\vee, \alpha_j)$.

Proposition. *For all $i, j \in \{1, 2, \ldots, \ell\}$ distinct one has*

(i) $r_i(E_j) = \sum_{s=0}^{a_{ij}}(-1)^s q_i^s E_i^{(s)} E_j E_i^{(a_{ij}-s)}$

(ii) $r_i(F_j) = \sum_{s=0}^{a_{ij}}(-1)^s q_i^{-s} F_i^{(a_{ij}-s)} F_j\, F_i^{(s)}$.

Since ι commutes with r_i it suffices to establish (i). Drop the i subscript and write U_i' simply as U. Set $E_j = E'$, $F_j = F'$, $K_j = K'$, $a_{ij} = a$. By 10.2.4 (ii) one has $r(F) = -EK^{-2}$. Hence EK^{-2} commutes with $r(E')$ in view of 5.1.1 (v). On the other hand the Serre relation $y_q^{\alpha_i, \alpha_j} = 0$ of 4.3.5 rewritten in the conventions of 5.1.1 becomes $(\operatorname{ad} E)^{a+1}E'K' = 0$. Set $P(E, E') = ((\operatorname{ad} E)^a E'K')K^{-a}K'^{-1}$. Then this Serre relation becomes $[EK^{-2}, P(E, E')] = 0$. Now let $V^{(m)}$ be a U submodule of some finite dimensional $U_q(\mathfrak{g})$ module $V(\mu)$. Given $\gamma \in k(q)$ such that

$$(1) \qquad r(E')u_{-m}^{(m)} = \gamma P(E, E')u_{-m}^{(m)}, \quad \text{for all } m \in \mathbb{N}\, ,$$

it follows on applying EK^{-2} to both sides that $r(E') = \gamma P(E, E')$.

Set $V = V^{(m)}$ and let V' denote the $a+1$ dimensional $\operatorname{ad} U$ module with basis $\{(\operatorname{ad} E)^n E'K'\}_{n=0}^a$. The action of $U_q'(\mathfrak{g})$ on $V(\mu)$ gives a U module homomorphism $v' \otimes v \mapsto v'v$ of $V' \otimes V$ into $V(\mu)$. Through the isomorphism $V^{(a)} \otimes V^{(m)} \cong V^{(m+a)} \oplus V^{(m+a-2)} \oplus \ldots \oplus V^{|m-a|}$ one can write

$$(2) \qquad E'u_m^{(m)} = \sum_{j=0}^{a} \gamma_j u_{m-a}^{(m+a-2j)}$$

for some $\gamma_j \in k(q)$ with $\gamma_j = 0$ if $m + a - 2j < |m - a|$.

In what follows it is convenient to write $[s]_q$ simply as $[s]$ for all integer s. Then 10.2.4 (∗) gives

$$(3)\qquad E^i u_j^{(m)} = \left(\frac{[((m-j)/2)!][((m+j)/2+i)!]}{[((m-j)/2-i)!][((m+j)/2)!]}\right)^{\frac{1}{2}} u_{j+2i}^{(m)},$$

which on use of $\varkappa'$ implies

$$(4)\qquad F^i u_j^{(m)} = \left(\frac{[((m+j)/2)!][((m-j)/2+i)!]}{[((m+j)/2-i)!][((m-j)/2)!]}\right)^{\frac{1}{2}} u_{j-2i}^{(m)}.$$

Combined (3), (4) and 5.1.1 (v) give

$$(5)\qquad E'E^i u_{-m}^{(m)} = \frac{[i!]}{[(m-i)!]} E'F^{m-i}u_m^{(m)} = \frac{[i!]}{[(m-i)!]} F^{m-i}E'u_m^{(m)}.$$

Consequently

$$\begin{aligned} P(E,E')u_{-m}^{(m)} &= \sum_{i=0}^{a}(-1)^i q^{-i}\begin{bmatrix} a \\ i \end{bmatrix} E^{a-i}E'E^i u_{-m}^{(m)} \\ &= \sum_{j=0}^{a}\gamma_j \sum_{i=0}^{a}(-1)^i q^{-i}\begin{bmatrix} a \\ i \end{bmatrix} \frac{[i!]}{[(m-i)!]} E^{a-i}F^{m-i}u_{m-a}^{(m+a-2j)} \\ &= \sum_{j=0}^{a}\gamma_j \sum_{i=0}^{a}(-1)^i q^{-i}\begin{bmatrix} a \\ i \end{bmatrix} \frac{[i!][(m+a-i-j)!]}{[(m-i)!][(i-j)!]} u_{a-m}^{(m+a-2j)} \\ &= [a!]q^{-a}\sum_{j=0}^{a}\gamma_j(-1)^j q^{(a-j)(m-j+1)}u_{a-m}^{(m+a-2j)} \end{aligned}$$

by (2)–(5) and A.3.1.

Then by 10.2.2 (9)

$$\begin{aligned} NE'N^{-1}u_m^{(m)} &= \sum_{j=0}^{a}\gamma_j N u_{m-a}^{(m+a-2j)} \\ &= \sum_{j=0}^{a}\gamma_j(-1)^{a-j}q^{(a-j)(m-j+1)}u_{a-m}^{(m+a-2j)} \\ &= \frac{(-1)^a}{[a!]} q^a P(E,E')u_{-m}^{(m)}, \end{aligned}$$

which is (1) with $\gamma = (-1)^a q^a [a!]^{-1}$. Substitution in the above gives (i).

10.2.6 It is straightforward to compute $r_i(e_j)$ and $r_i(f_j)$ from 10.2.4 and 10.2.5. In particular one obtains the

Theorem. *For all $w \in W$, the inner automorphism r_w of $R_q[G]^*$ restricts to an automorphism of $U_q(\mathfrak{g})$.*

10.2.7 Let ζ denote the involution $q \longrightarrow q^{-1}$ of $k(q)$ extended to a ζ-linear antiautomorphism of U' as follows. First note that U' admits a triangular decomposition $U'^{-} \otimes \tau(\frac{1}{2}P(\pi)) \otimes U'^{+} \xrightarrow{\sim} U'$ where U'^{+} (resp. U'^{-}) denotes the subalgebra generated by the E_i (resp. F_i). Then one may define ζ on U'^{+} (resp. U'^{-}) by taking it to be the identity on monomials since the coefficients in the Serre relations are ζ-invariant. (This follows from 3.1.14 (iii) and 3.2.8 or by direct computation 4.3.13). Finally take $\zeta(\tau(\lambda)) = \tau(\lambda)$ for all $\lambda \in \frac{1}{2}P(\pi)$. One checks that $\omega := \zeta\varkappa'$ is a ζ-linear Hopf algebra involution of U'. Define the ζ-linear map ζ^* of R by $\zeta^*(c)(u) = \zeta(c(\zeta(u)))$. Transposing again defines ω as a ζ-linear involution of R^*.

Recall that N_i factors through the surjection $\varphi_i^\star : R^\star \longrightarrow R_i^\star$ to a map $\bar{N}_i : R_i^\star \longrightarrow k(q)$. Define $sg_i \in \operatorname{Aut} R_i$ to be $(-1)^m$ on the $e_{r,s}^{(m)}$ and set $N_i' = \bar{N}_i sg_i \varphi_i^\star$.

Lemma. *For all $i \in \{1, 2, \ldots, \ell\}$ one has $\omega(N_i') = N_i^{-1}$.*

Let V be a simple finite dimensional U' module. Give V a ζ-linear structure compatible with that of U' by fixing a ζ-invariant highest weight vector of norm 1 relative to the Shapovalev form $\mathcal{S}'$. Take a decomposition of V into a direct sum of simple U_i' modules which are ζ-invariant and pairwise orthogonal with respect to $\mathcal{S}'$. (This is possible by 4.3.4 and the fact that non-isomorphic U_i' modules are already pairwise orthogonal). On each simple U_i' module $V^{(m)}$ choose an orthonormal basis $\{u_{m-2n}^{(m)}\}_{n=0}^{m}$ as in 10.2.2. Then as noted in 10.1.17 it follows that $\varkappa'^*(c_{m-2r,m-2s}^{(m)}) = c_{m-2s,m-2r}^{(m)}$ in the notation of 10.2.2. For all $m \in \mathbb{N}$ one has $\zeta u_{-m}^{(m)} = \varepsilon u_{-m}^{m}$ with $\varepsilon = \pm 1$. Then 10.2.4 (*) gives $\zeta(u_{-(m-2n)}^{(m)}) = \varepsilon(-1)^n u_{-(m-2n)}^{(m)}$ and so $\zeta^*(c_{m-2r,m-2s}^{(m)}) = (-1)^{r+s} c_{m-2r,m-2s}^{(m)}$. Then by 10.2.2 (9) one obtains

$$\begin{aligned}
(\omega N_i')(c_{-(m-2n),m-2n}^{(m)}) &= \zeta(N_i'(\varkappa'^*\zeta^*(c_{-(m-2n),m-2n}^{(m)}))) \\
&= \zeta(N_i(c_{m-2n,-(m-2n)}^{(m)})) \\
&= (-1)^{m-n} q^{-(n+1)(m-n)} = N_i^{-1}(c_{-(m-2n),m-2n}^{(m)}) \,.
\end{aligned}$$

Since both expressions vanish on the remaining matrix coefficients this proves the lemma.

Remarks. The fact that N_i' must be used in place of N_i is apparent from 10.2.2 (9) which shows that $N_i^2(c_{r,s}^{(m)})|_{q=1} = (-1)^m$. This reflects a difficulty already present in G for which it is not always possible to choose a representative $n_i \in N(T)$ of $s_i \in W = N(T)/T$ of order 2. One may check that $r_i' : a \mapsto N_i' a N_i'^{-1}$ restricts to an automorphism of U which differs from r_i in a similar fashion. Namely 10.2.4 are unchanged; but in 10.2.5 a factor of $(-1)^{a_{ij}}$ must be introduced in the right hand sides.

10.2.8 Consider ΔN_w. One has $(\Delta N_w)(a\otimes b) = N_w(ab) = \langle v_w, abv_w\rangle$ which in general is unrelated to $(N_w\otimes N_w)(a\otimes b) = \langle v_w, av_w\rangle\langle v_w, bv_w\rangle$. Yet $N_w\otimes N_w$ is invertible in $(R\otimes R)^*$ so there exists a unique invertible element $\mathbf{R}_w \in (R\otimes R)^*$ such that $\Delta N_w = \mathbf{R}_w(N_w\otimes N_w)$. Then if $V\subset U$ is a coideal one obtains

$$\Delta(r_w(V)) \subset \mathbf{R}_w(r_w(V)\otimes U + U\otimes r_w(V))\mathbf{R}_w^{-1}$$

which indicates how $r_w(V)$ fails to be a coideal.

Set $\mathbf{R}_0 = \mathbf{R}_{w_0}$. Using 9.2.11 and 10.2.2 one checks that *ker* $\mathbf{R}_0 \supset J^+\otimes R + R\otimes J^-$. The resulting pairing on $\check{V}^-\times\check{V}^+$ is just the skew-Holf pairing φ of 3.1.10 except that it is trivial on torus elements. (One may use 10.3.10 to see that it comes from the action of U^- on U^+). Recalling 9.4.7 this recovers directly the remarkable observation of S.Z. Levendorskii and Ya. S. Soibelman ([LeS1], 3.4) that $\mathbf{R}_0$ is essentially (but not quite) Drinfeld's R-matrix.

10.2.9 Set $V = UTU_+^+$ which is a left ideal and a coideal of U.

Proposition. *For all $w\in W$ the left ideal $r_w(V)$ is a coideal of U.*

Consider the action of r_w on $\text{Mod}_f\, U$ defined by transport of structure. By 4.3.6(iii) and 10.2.3 it follows that $r_w(V(\lambda))\cong V(\lambda)$ and $r_w(u_\lambda) = u_{w\lambda}$ for all $\lambda\in P^+(\pi)$. Yet $V = \bigcap_{\lambda\in P^+(\pi)} \text{Ann}_U\, u_\lambda$ and $r_w(V) = \bigcap_{\lambda\in P^+(\pi)}\text{Ann}_U\, u_{w\lambda}$. Since $u_{w\lambda}\otimes u_{w\mu}$ in $V(\lambda)\otimes V(\mu)$ identifies with $u_{w(\lambda+\mu)}$ in $V(\lambda+\mu)$, the assertion follows.

10.3 Prime and Primitive Ideals of $R_q[G]$

Recall 9.3.9. The aim of this section is to describe each $\mathbf{X(w)}$. Apart from a few technical details which have themselves an intrinsic beauty this follows the special case described in 9.3.19.

10.3.1 Fix $\mathbf{w} = (w_+, w_-)\in W\times W$. Write w_+ simply as w. Recall (10.1.8) that $R^+/Q^+_{w_+}\hookrightarrow \hat{R}/Q_{\mathbf{w}}$ is a domain and may be identified as a direct sum of the $V^+_{w_+}(\lambda)^* : \lambda\in P^+(\pi)$. A similar description holds for $R^-/Q^-_{w_-}$. In particular $c_{w\lambda} : \lambda\in P^+(\pi)$ has a non-zero image $\bar{c}_{w\lambda}$ in the quotient. (In the sequel the bar will generally be omitted). Since (10.1.10) the two-sided ideal $Q^+_w R^-$ contains J^+_w these elements satisfy

(1) $$c_{w\lambda}V^\pm_{w_\pm}(\mu)^* = V^\pm_{w_\pm}(\mu)^* c_{w\lambda}\ , \quad \forall\mu\in P^+(\pi)$$

by 9.1.5 (i). Then the Cartan multiplication rule (9.1.6 (*)) gives

(2) $$c^{-1}_{w\lambda}V^+_w(\lambda)^* c^{-1}_{w\mu}V^+_w(\mu)^* = c^{-1}_{w(\lambda+\mu)}V^+_w(\lambda+\mu)^*\ ,$$

and so defines

(3)
$$S_w^+ := \varinjlim_{\lambda \in P^+(\pi)} c_{w\lambda}^{-1} V_w^+(\lambda)^* ,$$

which is a subalgebra of $c_w^{-1}(R^+/Q_w^+)$. By 9.1.10 (*), the subalgebra T_w^+ generated by the $c_{w\lambda}^{\pm 1} : \lambda \in P^+(\pi)$ is isomorphic to the group algebra of $P(\pi)$. With respect to the automorphism $s \mapsto c_{w\lambda} s c_{w\lambda}^{-1}$ of S_w^+ defined by (1) one may form the smash product $S_w^+ \# T_w^+$ and then the multiplication map gives an isomorphism

(4)
$$S_w^+ \# T_w^+ \xrightarrow{\sim} c_w^{-1}(R/Q_w^+)$$

of algebras.

By 10.1.8 the right $U_q(\mathfrak{b}^+)$ action on R^+ factors to R^+/Q_w^+ and by 9.3.6 (i) identifies with the left twisted adjoint action of R which by 9.3.7 (ii) further identifies with the adjoint action of R on S_w^+ (which is hence $\operatorname{ad} R$ stable, locally finite and with annihilator containing $\ker \Psi^- = J^-$). Since the $U_q(\mathfrak{b}^+)$ invariant elements of R/Q_w^+ are just the $c_{w\lambda} : \lambda \in P^+(\pi)$ it follows that

(5) *The* $\operatorname{ad} R$ *socle of* S_w^+ *reduces to scalars* .

Finally the right T action on S_w^+ gives the latter a formal character which by (2) is just the limiting value of $e^{-w\lambda} \operatorname{ch} V_w(\lambda)$. This can be calculated to be (notation A.1.1)

(6)
$$\operatorname{ch} S_w^+ = \prod_{\beta \in S(w^{-1})} (1 - e^{\beta})^{-1} .$$

Here one may use either the Demazure character formula 6.3.16 (i) or the following more elementary reasoning using induction on $\ell(w)$.

Choose $\alpha \in \pi$ such that $\ell(s_\alpha w) < \ell(w)$. Then $\operatorname{ch} V_w(\lambda) = s_\alpha \operatorname{ch} V_w(\lambda)$ by 4.4.3 (v) and 4.3.1. For fixed $\nu \in Q(\pi)$ one has $(\alpha^\vee, \lambda - \nu) > 0$ given $\lambda \gg 0$ and in this case $x_{-\alpha}$ acts injectively on $V_{s_\alpha w}(\lambda)_{\lambda - \nu}$. Thus by 4.4.3 (v) one has asymptotically (that is retaining only the coefficients of $e^{\lambda - \nu}$ for $\lambda \gg \nu$)

$$\begin{aligned} \operatorname{ch} V_w(\lambda) &= s_\alpha(\operatorname{ch} U_\alpha^- \operatorname{ch} V_{s_\alpha w}(\lambda)) \\ &= (1 - e^{\alpha})^{-1} s_\alpha(e^{s_\alpha w \lambda} \operatorname{ch} S_{s_\alpha w}^+) . \end{aligned}$$

Hence by A.1.1 (i), (ii) and the induction hypothesis

$$\operatorname{ch} S_w^+ = (1 - e^{\alpha})^{-1} s_\alpha(\operatorname{ch} S_{s_\alpha w}^+) = \prod_{\beta \in S(w^{-1})} (1 - e^{\beta})^{-1} .$$

Now take $w = w_-$ and define S_w^- (resp. T_w^-) to be the subalgebra of $d_w^{-1}(R^-/Q_w^-)$ obtained on replacing $+$ by $-$ and $c_{w\lambda}$ by $d_{w\lambda}$. Through this substitution, (1)–(5) are defined to become (1′)–(5′) and (6) is replaced by

(6′)
$$\operatorname{ch} S_{w_0 w w_0^{-1}}^- = \prod_{\beta \in S(w^{-1})} (1 - e^{-\beta})^{-1} .$$

10.3.2 Set $\lambda_0 = -w_0\lambda : \lambda \in P^+(\pi)$. In $\hat{R}/\hat{Q}_{\mathbf{w}}$ the $c_{w_+\lambda}$ and $d_{w_-\mu_0}$ commute up to scalars. Set $b_{\mathbf{w},\lambda} := c_{w_+\lambda}d_{w_-\lambda_0} : \lambda \in P^+(\pi)$. Recalling 9.1.10 (∗) and 10.3.1 (1), (1′) it follows that $b_{\mathbf{w}} := \{b_{\mathbf{w},\lambda}\}_{\lambda\in P^+(\pi)}$ is Ore in $\hat{R}/\hat{Q}_{\mathbf{w}}$. Set $\hat{R}_{\mathbf{w}} = b_{\mathbf{w}}^{-1}(\hat{R}/\hat{Q}_w)$. Since $V^+(\lambda)^*V^-(\mu)^* = V^-(\mu)^*V^+(\lambda)^*$ in $\hat{R}$, which will be called the exchange rule, it follows that $S^+_{w_+}S^-_{w_-} = S^-_{w_-}S^+_{w_+}$ which is therefore a subalgebra $S_{\mathbf{w}}$ of $\hat{R}_{\mathbf{w}}$. By 10.3.1 (1′) one may form the smash product $S_{\mathbf{w}}\#T^-_{w_-}$ which is a subalgebra of $\hat{R}_{\mathbf{w}}$. From (5), (5′) of 10.3.1 and 9.2.14 (cf. proof of 9.3.14) one has

(1) *The* $\operatorname{ad} R$ *socle of* $S_{\mathbf{w}}$ *reduces to scalars .*

Again $T^+_{w_+}T^-_{w_-} = T^-_{w_-}T^+_{w_+}$ is a subalgebra $T_{\mathbf{w}}$ of $\hat{R}_{\mathbf{w}}$ and by (4), (4′) of 10.3.1 the multiplication map gives an isomorphism

(2) $$S_{\mathbf{w}} \otimes T_{\mathbf{w}} \xrightarrow{\sim} \hat{R}_{\mathbf{w}}$$

of vector spaces. As in 9.3.15 one checks that

(3) $$b^{-1}_{\mathbf{w},\lambda}x_\lambda \in S_{\mathbf{w}} \quad \textit{for all} \quad \lambda \in P^+(\pi)\ .$$

Define the Ore subset $\mathcal{O}$ of X as in 9.2.5. Recall that $X = Z(\hat{R})$ by 9.3.20.

In $\mathcal{O}^{-1}\hat{R}_{\mathbf{w}}$ one can write $c^{-1}_{w_+\lambda} = x^{-1}_\lambda a_\lambda$ for some $a_\lambda \in S_{\mathbf{w}}\#T^-_{w_-}$. This gives an algebra isomorphism

(4) $$c_{w_+}(\mathcal{O}^{-1}X \otimes S_{\mathbf{w}}\#T^-_{w_-}) \xrightarrow{\sim} \mathcal{O}^{-1}\hat{R}_{\mathbf{w}}$$

where c_{w_+} on the left hand side means localization at the Ore subset $c^{-1}_{w_+} := \{c^{-1}_{w_+\lambda}\}_{\lambda\in P^+(\pi)}$ of the bracketed term.

Set $R_{\mathbf{w}} = b_{\mathbf{w}}^{-1}(R/Q_{\mathbf{w}})$. By 9.2.4 the evaluation map $x_\lambda \mapsto 1$ sends $\mathcal{O}^{-1}\hat{R}_{\mathbf{w}}$ onto $R_{\mathbf{w}}$. By (4) this gives an isomorphism

(5) $$c_{w_+}(S_{\mathbf{w}}\#T^-_{w_-}) \xrightarrow{\sim} R_{\mathbf{w}}$$

which (10.1.10) shows in particular that

(6) $R_{\mathbf{w}}$ *is a domain .*

Since R is noetherian (9.2.2), the definition of $\mathbf{X}(\mathbf{w})$, 10.1.13 and A.2.8 give the

Proposition. *The map* $P \mapsto b_{\mathbf{w}}^{-1}(P/Q_{\mathbf{w}})$ *is an isomorphism of* $\mathbf{X}(\mathbf{w})$ *onto* Spec $R_{\mathbf{w}}$.

10.3.3 Just as in 9.3.14–9.3.20 the algebra $R_{\mathbf{w}}$ should be replaced by a slightly smaller subalgebra $B_{\mathbf{w}}$. This is defined below.

For each $\mu \in P^+(\pi)$, set $a_{\mathbf{w},\mu} = c_{w_+\mu}^{-1} d_{w_-\mu_0} \in R_{\mathbf{w}}$. Then 1.3.6 (4), 1.4.7, 9.3.5 and 9.3.7 give (in $R/Q_{\mathbf{w}}$)

$$(\operatorname{ad} c_{\eta,\lambda}^{\nu}) a_{\mathbf{w},\mu} = \sum_{\zeta \in \Omega(V(\nu))} ((\operatorname{ad}_{\tau_{-\mu_0}} \hat{\tau}_{\mu_0}(c_{\eta,\zeta}^{\nu})) c_{w_+\mu}^{-1})(\operatorname{ad}_{\tau_{\mu_0}} c_{\zeta,\lambda}^{\nu}) d_{w_-\mu_0}$$
$$= \delta_{\eta,\lambda} q^{-(\lambda, w_0\mu - w_+\mu)} q^{(w_- w_0 \mu_0 - \mu_0, \lambda)} a_{\mathbf{w},\mu} \,,$$

so

(1) $$(\operatorname{ad} c_{\eta,\lambda}^{\nu}) a_{\mathbf{w},\mu} = \delta_{\eta,\lambda} q^{(\lambda, w_+\mu - w_-\mu)} a_{\mathbf{w},\mu} \,.$$

Since $c_{w_+\mu}, d_{w_-\nu_0}$ commute up to non-zero scalars (mod $Q_{\mathbf{w}}$) it follows from 9.1.10 (*) that $a_{\mathbf{w},\mu} a_{\mathbf{w},\nu} = a_{\mathbf{w},\mu+\nu}$, up to non-zero scalars. Thus one may define $a_{\mathbf{w},\mu} : \mu \in P(\pi)$ up to scalars as elements of $R_{\mathbf{w}}$ by taking appropriate ratios. Let $A_{\mathbf{w}}$ denote the subalgebra of $R_{\mathbf{w}}$ they span.

Let $B_{\mathbf{w}}$ denote the subalgebra of $R_{\mathbf{w}}$ generated by $S_{\mathbf{w}}$ and $A_{\mathbf{w}}$. By (1), (1′), (4), (4′) of 10.3.1 it follows that the multiplication map gives an isomorphism

(2) $$S_{\mathbf{w}} \otimes A_{\mathbf{w}} \xrightarrow{\sim} B_{\mathbf{w}}$$

of ad R modules.

Set $P_{\mathbf{w}}(\pi) = \{\mu \in P(\pi) | w_+\mu - w_-\mu = 0\}$ which is a (free) additive subgroup of $P(\pi)$. By (1) each $a_{\mathbf{w},\mu} : \mu \in P_{\mathbf{w}}(\pi)$ is ad R invariant hence central in $R_{\mathbf{w}}$. Let $Z_{\mathbf{w}}$ be the subalgebra of $A_{\mathbf{w}}$ they span. Obviously $Z_{\mathbf{w}}$ is a Laurent polynomial algebra on $rk\, P_{\mathbf{w}}(\pi)$ generators. Set $s(\mathbf{w}) = \ell - rk\, P_{\mathbf{w}}(\pi)$. Fix representatives of $P^{\mathbf{w}}(\pi) := P(\pi)/P_{\mathbf{w}}(\pi)$ in $P(\pi)$.

Let I be an ideal of $Z_{\mathbf{w}}$ and set $Z = Z_{\mathbf{w}}/I$. It is immediate that $Q := B_{\mathbf{w}} I$ is an ad R invariant ideal of $B_{\mathbf{w}}$. Furthermore by (2) it follows that as an ad R module

(3) $$B_{\mathbf{w}}/Q = \bigoplus_{\mu \in P^{\mathbf{w}}(\pi)} M_\mu \quad \textit{where} \quad M_\mu = S_{\mathbf{w}} a_{\mathbf{w},\mu} Z \,.$$

By (1) above and 10.3.2 (1) it follows that

(4) *The* ad R *socle of* M_μ *is* $a_{\mathbf{w},\mu} Z$.

By (1), (1′) of 10.3.1 and because $S_{\mathbf{w}}$ is a subalgebra one has

(5) $$M_\mu M_\nu \subset M_{\mu+\nu} \quad \textit{for all} \quad \mu, \nu \in P^{\mathbf{w}}(\pi) \,.$$

Proposition
(i) *For each* $P \in (\operatorname{Spec} B_{\mathbf{w}})^R$, *the quotient* $B_{\mathbf{w}}/B_{\mathbf{w}}(P \cap Z_{\mathbf{w}})$ *is a domain.*
(ii) *The map* $P \mapsto P \cap Z_{\mathbf{w}}$ *is an isomorphism of* $(\operatorname{Spec} B_{\mathbf{w}})^R$ *onto* $\operatorname{Spec} Z_{\mathbf{w}}$.

Take $P \in (\operatorname{Spec} B_{\mathbf{w}})^R$. Then $P \cap Z_{\mathbf{w}} \in \operatorname{Spec} Z_{\mathbf{w}}$. Now take any $I \in \operatorname{Spec} Z_{\mathbf{w}}$ and Q, M_μ as above.

To show that $B_{\mathbf{w}}/Q$ is a domain, observe first that $P^{\mathbf{w}}(\pi)$ is torsion-free and hence can be identified with a subgroup of $P(\pi)$. Then by (1), (5) the argument of 7.3.4 applies and shows that a pair $a, b \in B_{\mathbf{w}}/Q$ satisfying $ab = 0$ can be assumed to belong to some M_μ. Since $a_{\mathbf{w},\mu}$ is invertible in $B_{\mathbf{w}}$ one can assume $a, b \in S_{\mathbf{w}} \otimes Z$. Since $S_{\mathbf{w}}$ is a subalgebra of $\hat{R}_{\mathbf{w}}$ which is a domain (10.1.10) over any field extension this leads to a contradiction just as in the proof of 9.3.14. This establishes (i) and surjectivity in (ii).

Finally suppose $P/Q \neq 0$. Since the adjoint action on $B_{\mathbf{w}}/Q$ is locally finite one has $\operatorname{Soc} P/Q \neq 0$. Yet by (3), (4) above the socle of $B_{\mathbf{w}}/Q$ is a direct sum of the $a_{\mathbf{w},\mu}Z : \mu \in P^{\mathbf{w}}(\pi)$ which by (1) are pairwise non-isomorphic. Thus $a_{\mathbf{w},\mu}z \in P/Q$ for some $z \in Z$ non-zero. Since $a_{\mathbf{w},\mu}$ is invertible in $B_{\mathbf{w}}$ this forces $z \in P/Q$, contradicting the definition of Q.

10.3.4 It is clear from (1), (1′), (4), (4′) of 10.3.1 that one may form the iterated smash product $(S_{\mathbf{w}} \# T^-_{w_-}) \# T^+_{w_+}$ and this is isomorphic to $\hat{R}_{\mathbf{w}}$. Since the latter is noetherian by 9.1.11 and the $T^{\pm}_{w_\pm}$ are isomorphic to $\check{U}^0$ as Hopf algebras it follows that

(1) $$S_{\mathbf{w}} \textit{ is noetherian} .$$

Again from (1), (1′) of 10.3.1 and 10.3.2 (1) it follows that $B_{\mathbf{w}}$ is an iterated Ore extension of $S_{\mathbf{w}}$. Then by (1) and 7.4.1 one has

(2) $$B_{\mathbf{w}} \textit{ is noetherian} .$$

Take $\lambda \in P^+(\pi)$. By 10.3.2 (3) one has $b^{-1}_{\mathbf{w},\lambda} \in S_{\mathbf{w}}$ viewed as an element of $R_{\mathbf{w}}$. Let $b_{\mathbf{w}}B_{\mathbf{w}}$ denote the localization of $B_{\mathbf{w}}$ at its Ore subset $b^{-1}_{\mathbf{w}} := \{b^{-1}_{\mathbf{w},\lambda} : \lambda \in P^+(\pi)\}$. Let R_0 denote the zero component of R for the gradation defined in 9.3.16. Then

$$\begin{aligned}(R_0)_{\mathbf{w}} &:= b^{-1}_{\mathbf{w}}(R_0/R_0 \cap Q_{\mathbf{w}}) \\ &= b^{-1}_{\mathbf{w}} \left(\sum_{\lambda,\mu \mid \lambda+\mu \in 2P^+(\pi)} V^+_{w_+}(\lambda)^* V^-_{w_-}(\mu_0)^* \right)\end{aligned}$$

which is generated over $S_{\mathbf{w}}$ by the elements $c_{w_+\lambda}d_{w_-\mu_0}$, and their inverses, satisfying $\lambda + \mu \in 2P^+(\pi)$. It follows that

(3) $$(R_0)_{\mathbf{w}} = b_{\mathbf{w}}B_{\mathbf{w}} .$$

Given $P \in (\operatorname{Spec} B_{\mathbf{w}})^R$ then $P \cap b^{-1}_{\mathbf{w}} = \emptyset$. Otherwise $b^{-1}_{\mathbf{w},\lambda} \in P \cap S_{\mathbf{w}}$ for some $\lambda \in P^+(\pi)$ and by 10.3.2 (1) this implies that $1 \in P$. By (2) and A.2.8, this gives

The map $P \mapsto b_{\mathbf{w}}P$ *is an isomorphism of* $(\operatorname{Spec} B_{\mathbf{w}})^R$ *onto* $(\operatorname{Spec} b_{\mathbf{w}}B_{\mathbf{w}})^R$.
(4)

For each $\lambda \in P^+(\pi)$, $c_{w\lambda}^{-2}$ is a non-zero multiple of $b_{\mathbf{w},\lambda}^{-1} a_{\mathbf{w},\lambda}$ and so this element belongs to $B_{\mathbf{w}}$ and its inverse to $b_{\mathbf{w}} B_{\mathbf{w}}$. Thus $b \mapsto c_{w\lambda} b c_{w\lambda}^{-1}$ which is an automorphism of $b_{\mathbf{w}} B_{\mathbf{w}}$ is inner if $\lambda \in 2P^+(\pi)$. Yet its eigenvalues are powers of q and so have no torsion. Consequently these automorphisms leave ideals of $b_{\mathbf{w}} B_{\mathbf{w}}$ invariant. Since the $c_{w_+}\lambda : \lambda \in P^+(\pi)$ generate $b_{\mathbf{w}}^{-1} R_{\mathbf{w}}$ over $b_{\mathbf{w}} B_{\mathbf{w}}$ it follows that

$$(5) \qquad (\operatorname{Spec} b_{\mathbf{w}} B_{\mathbf{w}})^R = \operatorname{Spec} b_{\mathbf{w}} B_{\mathbf{w}} \ .$$

Set $\Gamma = P(\pi)/2P(\pi)$ and recall 1.3.13, 9.3.16.

Theorem. *The map $P \mapsto Z_{\mathbf{w}} \cap b_{\mathbf{w}}^{-1}(P/Q_{\mathbf{w}})$ is an isomorphism of the space of Γ orbits in $\mathbf{X}(\mathbf{w})$ onto* Spec $Z_{\mathbf{w}}$.

Take $P \in \mathbf{X}(\mathbf{w})$. Then $R_0 \cap P \supset R_0 \cap Q_{\mathbf{w}}$ by 10.1.13. Moreover $b_{\mathbf{w}} \cap (R_0 \cap P) = \emptyset$ by the definition of $\mathbf{X}(\mathbf{w})$. By 1.3.10, $R_0 \cap P$ is semiprime and so at least one of the primes P_0 minimal over it satisfies $b_{\mathbf{w}} \cap P_0 = \emptyset$. Set $\bar{P}_0 = P_0/R_0 \cap Q_{\mathbf{w}}$ and omit the bar. Then $b_{\mathbf{w}}^{-1} P_0 \in \operatorname{Spec}(R_0)_{\mathbf{w}}$. By (3) and (5) it follows that $b_{\mathbf{w}}^{-1} P_0$ is ad R stable and hence so is P_0. Applying 1.3.11 it follows that $R_0 \cap P$ is ad R stable and hence prime. The conclusion of the theorem then follows from 1.3.13, 1.3.14, 10.3.3 (ii), 10.3.2 and (3)–(5) above.

10.3.5 Proposition. *The algebra $R/Q_{\mathbf{w}}$ is a domain. In particular $Q_{\mathbf{w}}$ is the unique minimal element of $\mathbf{X}(\mathbf{w})$.*

Since $R_{\mathbf{w}}$ is domain (10.3.2 (6)) it is enough to show that the $c_{w_+\mu}, d_{w_-\mu} : \mu \in P^+(\pi)$ are non-zero divisors. Both cases are similar. By noetherianity and A.2.3, it is enough to check that if $a \in R/Q_{\mathbf{w}}$ satisfies $c_{w_+\mu} a = 0$, then $a = 0$. Since $Q_{\mathbf{w}}$ is left T invariant and $c_{w_+\mu}$ is a left T eigenvector, the same can be assumed of a, say of eigenvalue $\lambda \in P(\pi)$. By definition of $Q_{\mathbf{w}}$ one can take

$$(*) \qquad a \in \sum_{i=1}^{n} V_{w_+}^+(\lambda_i^+)^* V_{w_-}^-(\lambda_i^-)^* : \lambda_i^\pm \in P^+(\pi)$$

with $\lambda = \lambda_i^+ + w_0 \lambda_i^+$ for all i. Set $\nu_i = -w_0 \lambda_i^+ + \lambda_i^-$. By 1.3.6 (4) and 9.3.6 the ith summand in the right hand side is $\operatorname{ad}_{\tau_{\nu_i}} R$ locally finite. Since $\nu_i = \lambda (\operatorname{mod} 2P(\pi))$ one may choose $\zeta^i \in P^+(\pi)$ such that $2\zeta^i + \nu^i =: \nu$ for all i. Yet by 9.2.4 one has

$$1 \in V_{w_+}^+(-w_0\zeta_i)^* V_{w_-}^-(\zeta_i)^* \,(\operatorname{mod} Q_{\mathbf{w}})$$

whilst the right hand side is $\operatorname{ad}_{\tau_{2\zeta_i}} R$ locally finite. Then by Cartan multiplication (9.1.6 (*)) and the exchange rule (10.3.2) one may assume $\nu = \nu_i$ for all i without loss of generality. Let M denote the right hand side of (*) viewed as an $\operatorname{ad}_{\tau_\nu} R$ module.

On the other hand $k(q)c_{w_+\mu}$ is right $U_q(\mathfrak{b}^-)$ invariant (mod $Q^+_{w_+}$) and so by 9.3.6 (1) is $\operatorname{ad}_{\tau_{\mu_0}} R$ invariant. Applying $\operatorname{ad}_{\tau_{\mu_0+\nu}} R$ to $c_{w_+\mu}a$ and using 1.3.6 (4) one can assume that $a \in \operatorname{Soc} M$. Since M is finite dimensional the action of R can be extended (9.3.11) to a $c_e^{-1}R$ action. Then by 9.2.14 and 9.3.6, it follows that

$$\operatorname{Soc} M = \sum_{i=1}^{n} k(q) c_{w_+\lambda_i^+} d_{w_-\lambda_i^-} \ .$$

This belongs to $T_{\mathbf{w}}$ which by 10.3.2 (4) is a subalgebra of $R_{\mathbf{w}}$. Since $c_{w_+\mu} \in T_{\mathbf{w}}$ and $R_{\mathbf{w}}$ is a domain (10.3.2 (6)) this gives the required contradiction.

10.3.6 Obviously 9.3.9 and 10.3.4 determines $\operatorname{Spec} R$ as a set; but not quite as a topological space. For this one must know when $J \in \operatorname{Spec} R$ contains an intersection $I = \bigcap_{P \in \mathcal{I}} P$ for any $\mathcal{I} \subset \operatorname{Spec} R$. One may assume $J \in \mathbf{X}(\mathbf{w})$ and define $J_{\mathbf{w}'} = \cap\{P | P \in \mathcal{I} \cap \mathbf{X}(\mathbf{w}')\}$ for all $\mathbf{w}' \in W \times W$. Take the product Bruhat order on $W \times W$.

Lemma

(i) *$J \supset J_{\mathbf{w}''}$ for some $\mathbf{w}'' \in W \times W$.*
(ii) *$\mathbf{w} \leq \mathbf{w}''$.*
(iii) *If $J = I$, then $J = J_{\mathbf{w}''}$ and $\mathbf{w} = \mathbf{w}''$.*

Trivially $J \supset \cap J_{\mathbf{w}'}$ and since J is prime it must contain one of them, say $J \supset J_{\mathbf{w}''}$. Hence (i). Then 10.3.5 gives $J \supset Q_{\mathbf{w}''}$. By definition $c_{w_+\lambda} \notin J$ for any $\lambda \in P^+(\pi)$ and so $c_{w_+\lambda} \notin Q_{\mathbf{w}''}$, equivalently $c_{w_+\lambda}$ has a non-zero image in $R^+/Q^+_{w''_+}$. By definition (10.1.8) of the latter this means that $u_{w_+\lambda} \in V^+_{w''_+}(\lambda)$ which by 4.4.5 means that $w_+ \leq w''_+$. Similarly $w_- \leq w''_-$. Hence (ii). If $J = I$ then $J_{\mathbf{w}'} \supset J_{\mathbf{w}''}$ for all $\mathbf{w}'$ so $J = J_{\mathbf{w}''}$. Then by 10.3.5 one obtains $Q_{\mathbf{w}} \subset J_{\mathbf{w}''}$ and so $\mathbf{w}'' \leq \mathbf{w}$ as before. Combined with (ii) this gives (iii).

10.3.7 Let $\mathbf{X}^{\max}(\mathbf{w})$ denote the set of maximal ideals of $\mathbf{X}(\mathbf{w})$.

Theorem

$$\operatorname{Prim} R_q[G] = \coprod_{\mathbf{w} \in W \times W} \mathbf{X}^{\max}(\mathbf{w}) \ .$$

For the inclusion $\subset$ let M be a simple R module and set $P = \operatorname{Ann} M$. Then $P \in \operatorname{Spec} R$ and so $P \in \mathbf{X}(\mathbf{w})$ for some $\mathbf{w} \in W \times W$ by 9.3.9. Write $\mathbf{w} = (w_+, w_-)$. Suppose $c_{w_+\mu} m = 0$ for some $m \in M \setminus \{0\}$. Since $c_{w_+\mu}$ is normal in R/P this implies that $c_{w_+\mu} M = 0$ contradicting that $P \in \mathbf{X}(\mathbf{w})$. A similar result holds for $d_{w_-\mu}$. Consequently M has no $b_{\mathbf{w}}$ torsion and so $b_{\mathbf{w}}^{-1} M$ is a non-zero simple module for $R_{\mathbf{w}}$. Set $I = Z_{\mathbf{w}} \cap b_{\mathbf{w}}^{-1}(P/Q_{\mathbf{w}})$ and $Z = Z_{\mathbf{w}}/I$ which is contained in the centre of $R_{\mathbf{w}}/b_{\mathbf{w}}^{-1}(P/Q_{\mathbf{w}})$ and hence in $\operatorname{End}_{R_{\mathbf{w}}}(b_{\mathbf{w}}^{-1} M)$. Thus Z is field, I is maximal and by 10.3.4 so is P.

For the reverse inclusion consider $P \in \mathbf{X}^{\max}(\mathbf{w})$. By 9.2.2 it is an intersection of primitive ideals. By 10.3.7, P must be one of these. This completes the proof of the theorem.

10.3.8 The above result may be given a more precise form over $\overline{k(q)}$. Recall the notation of 1.3.4 and 9.3.11. It is clear from 9.3.11 and 9.2.14 that $R^\wedge$ identifies with $\check{T}^\wedge$. More precisely $\chi(J) = 0$ for all $\chi \in R^\wedge$ and furthermore by 9.2.4 (*) one has $\chi(c_\mu) = \chi(d_{\mu_0})^{-1}$. This gives

$$\theta_\chi(c_{w_+\mu}) = \chi(c_\mu)c_{w_+\mu}\ , \quad \theta_\chi(d_{w_-\mu_0}) = \chi(c_\mu)^{-1}d_{w_-\mu_0}$$

and so

$$\theta_\chi(a_{\mathbf{w},\mu}) = \chi(c_\mu)^{-2}a_{\mathbf{w},\mu}\ , \quad \text{for all} \quad \mu \in P^+(\pi)\ .$$

Thus over $\overline{k(q)}$ the action of $R^\wedge$ on $\operatorname{Max} Z_\mathbf{w}$ is transitive. Moreover the characters on $\Gamma := P(\pi)/2P(\pi)$, that is those which take value 1 on the $c_\mu : \mu \in 2P(\pi)$ act trivially on $\operatorname{Max} Z_\mathbf{w}$; but exactly implement the Γ orbits in the conclusion of 10.3.4. This proves the

Theorem. *Over* $\overline{k(q)}$ *each* $\mathbf{X}^{\max}(\mathbf{w}) : \mathbf{w} \in W \times W$ *is a single* $R^\wedge$ *orbit. In particular* Prim R *is a disjoint union of* $R^\wedge$ *orbits parametrized by* $W \times W$.

10.3.9 Given $P \in \operatorname{Spec} R$ it is easy to determine the Gelfand-Kirillov dimension $d(R/P)$ of the quotient algebra in terms of the parametrization given by 9.3.9 and 10.3.4. Take $\mathbf{w} = (w_+, w_-)$ and set $\ell(\mathbf{w}) = \ell(w_+) + \ell(w_-)$. Recall the definition of $s(\mathbf{w})$ given in 10.3.3. By A.1.18, it is just the least number of ways of writing $w_-^{-1}w_+$ as a product of reflections. In particular $\ell(\mathbf{w}) + s(\mathbf{w})$ is always even.

Proposition. *For all* $P \in \mathbf{X}(\mathbf{w})$ *one has*

(i) $d(R/P) = \ell(\mathbf{w}) + s(\mathbf{w}) + d(Z_\mathbf{w}/Z_\mathbf{w} \cap b_\mathbf{w}^{-1}(P/Q_\mathbf{w}))$
(ii) $d(R/P) = \ell(\mathbf{w}) + s(\mathbf{w})$ *if and only if* $P \in \operatorname{Prim} R$.

By (6) and (6′) of 10.3.1 one has via A.3.6

$$d(S_{w_+}^+) = \ell(w_+)\ , \quad d(S_{w_-}^-) = \ell(w_-)\ .$$

Consequently $d(S_\mathbf{w}) = \ell(\mathbf{w})$. Identifying $P^\mathbf{w}(\pi)$ with a subgroup of $P(\pi)$ defines a subalgebra $A^\mathbf{w}$ of $A_\mathbf{w}$ such that $A_\mathbf{w} = A^\mathbf{w} \otimes Z_\mathbf{w}$. Moreover $d(A^\mathbf{w}) = s(\mathbf{w})$. Since $R_\mathbf{w}$ is a finite extension of $B_\mathbf{w}$ one obtains writing $\bar{P} = P/Q_\mathbf{w}$, $I = Z_\mathbf{w} \cap b_\mathbf{w}^{-1}(\bar{P})$ that

$$d(R/\bar{P}) = d(B_\mathbf{w}/B_\mathbf{w} \cap b_\mathbf{w}^{-1}\bar{P}) = d(S_\mathbf{w}) + d(A^\mathbf{w}) + d(Z_\mathbf{w}/I)$$

giving (i). If $P \in \operatorname{Prim} R$, then $I \in \operatorname{Max} Z_\mathbf{w}$ so $d(Z_\mathbf{w}/I) = 0$ by the nullstellensatz. Hence (ii).

10.3.10 Consider the diagonal case $\mathbf{w} = (w, w) : w \in W$. In view of 10.3.1 (3) and recalling 4.4.2 it is natural to view S_w^+ as the quantized algebra of functions on the Bruhat cell defined by w. Then $S_{\mathbf{w}}$ should be identified with the algebra of differential operators on this cell. This leads one to expect S_w^+ to have the structure of a simple $S_{\mathbf{w}}$ module. Since the multiplication map gives an isomorphism

$$S_w^+ \otimes S_w^- \xrightarrow{\sim} S_{\mathbf{w}}$$

of vector spaces (by the corresponding result for $\hat{R}_{\mathbf{w}}$) one may define S_w^+ as an $S_{\mathbf{w}}$ module by identifying it with the module induced from the trivial module $\mathbb{1}$ for S_w^-. Moreover by 10.3.2 (5) one may further view S_w^+ as the $R_{\mathbf{w}}$ module induced from the one dimensional R_w^- module $\mathbb{1}$ in which the above S_w^- action is extended by requiring the generators $d_{w\mu} : \mu \in P^+(\pi)$ of T_w^- to act by 1, it being clear from the expression for x_μ that then the $c_{w\mu} : \mu \in P^+(\pi)$ also act by 1. Then the central elements $a_{\mathbf{w},\mu}$ act by 1 on $\mathbb{1}$ and hence by 1 on S_w^+. In particular the image of $B_{\mathbf{w}}$ (resp. R) in End S_w^+ coincides with image of $S_{\mathbf{w}}$ (resp. $R_{\mathbf{w}}$).

Proposition. *Under these identifications S_w^+ is a simple R module and a faithful simple $S_{\mathbf{w}}$ module.*

For the first part it is enough to show that every non-zero R submodule of S_w^+ contains the cyclic vector $\mathbb{1}$. Recall (1.3.1) the right adjoint action which in the present notation becomes $(\operatorname{Ad} a)s = \sigma^*(a_1)sa_2$ for all $a, s \in R$. One checks that

(1) $$(\operatorname{Ad} a)\iota^*(s) = \iota^*((\operatorname{ad} \varkappa'^*(a)s) .$$

Given $a \in R^+$, say $a = c^\mu_{\nu,\mu}$, and $s \in S_w^+$ one has

$$((\operatorname{Ad} a)s)\mathbb{1} = \sum \sigma^*(c^\mu_{\nu,\xi_i})sc^\mu_{\xi_i,\mu}\mathbb{1} = \sigma^*(c^\mu_{\nu,w\mu})s\mathbb{1}$$

and so

(2) $$((\operatorname{Ad} R^+)s)\mathbb{1} \subset Rs\mathbb{1} .$$

Now up to a non-zero scalar $\sigma^*(c^\mu_{\mu,w\mu}) = c^{\mu_0}_{ww_0\mu_0,w_0\mu_0} = d_{w\mu_0}$. Thus taking $\nu = \mu$ above, gives $((\operatorname{Ad} c_\mu)s)\mathbb{1} = d_{w\mu_0}s\mathbb{1}$. By say 10.3.1 (1′) it follows that S_w^+ is a direct sum of its $d_{w\mu_0}$ eigenspaces with the eigenvalues being powers of q and in particular non-zero. Thus if a subspace of S_w^+ is $d_{w\mu_0}$ stable, it is also $d_{w\mu_0}^{-1}$ stable.

View S_w^- as an R module under adjoint action. By 9.3.7 (iv) and 9.3.6 (ii) it follows that $\operatorname{Ann}_R S_w^- \supset \ker \Psi^+ = J^+$. Using 1.4.8 and 10.3.17 one checks that ι^* interchanges S_w^+ and S_w^- whilst $\varkappa'^*$ interchanges J^+ and J^-. Finally by 9.2.11, Ψ^- extends to a bijection of $c_e^{-1}R^+$ onto R/J^-. Substitution in (1) then give

(3) $$\iota^*(\operatorname{ad} \varkappa'^*(R)s) = \iota^*(\operatorname{ad} \varkappa'^*(R/J^-)s) = (\operatorname{Ad} c_e^{-1}R^+)\iota^*(s)$$

for all $s \in S_w^-$. Then (2) and the remark that follows give

$$R\, s\mathbb{1} \supset \iota^*(\operatorname{ad} R(\iota^*(s)))\mathbb{1} \tag{4}$$

for all $s \in S_w^+$. Finally suppose $s \neq 0$. Then by 10.3.1 (5′) the right hand side of (4) contains the scalars, proving the required assertion.

It is immediate from 10.3.2 (1) that $\operatorname{Ann}_{S_{\mathbf{w}}} S_w^+ = 0$. Finally let M be a non-zero $S_{\mathbf{w}}$ submodule of S_w^+ and recall 10.3.2 (3). Then M is a direct sum of its $b_{\mathbf{w},\lambda}^{-1} a_{\mathbf{w},\lambda} = c_{w\lambda}^{-2}$ eigenspaces and hence is also $c_{w\lambda}$ stable. This makes M an $R_{\mathbf{w}}$ module hence equal to S_w^+ by the first part.

Remarks. Unlike the Weyl algebra (1.2.10) $S_{\mathbf{w}}$ fails to a simple algebra. Of course by 10.3.2 (1) it has no proper ad R invariant ideals. One may easily recognize 10.3.10 as giving a second construction of the module $\mathcal{S}(w)$ described in 10.1.6. Moreover as shown above the $d_{w\mu_0}$ spectrum of $S_w^+\mathbb{1}$ coincides with the $\operatorname{Ad} c_\mu$ spectrum of S_w^+ which by (1) is just the $\operatorname{ad} c_\mu$ spectrum of $\iota^*(S_w^+) = S_w^-$. Since $\operatorname{Ann} S_w^- \supset J^+$ the latter coincides via 9.2.11 (iii) with the spectrum of S_w^- viewed as a right $\tau(-\mu)$ module. In this manner the character formulae of 10.3.1 (6′) and 10.1.19 become equivalent recalling (10.1.15) the manner in which the $d_{w\mu_0}$ (or $c_{w\mu_0}$) eigenvalues are defined.

10.3.11 Let $\operatorname{Rat} R$ denote the set of rational ideals of R, that is those which stay primitive under any field extension. It is immediate from 10.3.8 that for each $\mathbf{w} \in W \times W$ the subset $\operatorname{Rat} R \cap \mathbf{X}(\mathbf{w})$ consists of those ideals corresponding under 10.3.4 to ideals of codimension 1 in $Z(\mathbf{w})$. Through 9.2.2, 10.3.5, 10.3.6 and 10.3.8 it is immediate that

Lemma
(i) $\operatorname{Rat} R \cap \mathbf{X}(\mathbf{w})$ *is a single* $R^\wedge$ *orbit.*
(ii) $Q_{\mathbf{w}} = \cap\{P | P \in \operatorname{Rat} R \cap \mathbf{X}(\mathbf{w})\}$.

10.3.12 Take $\chi \in R^\wedge$. Then $\ker \chi = J$ in the notation of 9.3.11. Furthermore R/J identifies with $\check{T}$ through 9.2.14. Thus $R^\wedge = \check{T}^\wedge \hookrightarrow T^\wedge$. (Over $k(q)$ this last map becomes an isomorphism).

With respect to the left θ_χ^ℓ (resp. right θ_χ^r) winding automorphism one has $\theta_\chi^\ell(c_{\mu,\lambda}) = \chi(c_{\lambda,\lambda})c_{\mu,\lambda}$ (resp. $\theta_\chi^r(c_{\mu,\lambda}) = \chi(c_{\mu,\mu})c_{\mu,\lambda}$). Neither are Hopf algebra automorphisms in general. However one checks (for any Hopf algebra) that the composition $\theta_{\operatorname{ad}\chi} := \theta_\chi^\ell \theta_{\chi^{-1}}^r$ is a Hopf algebra automorphism. Notice that θ_χ^ℓ (resp. θ_χ^r) extends the left (resp. right) action of T.

10.4 Hopf Algebra Automorphisms

Here the automorphism group $\mathrm{AUT}(U_q(\mathfrak{g}))$ of $U_q(\mathfrak{g})$ as a Hopf algebra will be computed. The result is similar to that for $\mathrm{AUT}(U(\mathfrak{g}))$ except that G must be replaced by $T^\wedge$. Consequently $U_q(\mathfrak{g})$ has very few Hopf algebra automorphisms. Corresponding there can be very few Hopf algebra maps between the different $U_q(\mathfrak{g})$ in sharp contrast to the situation for $U(\mathfrak{g})$ as described briefly in 1.2.7. Again regarding $S^+_{w_0} = S^+ \cong U^-$ as being the quantum analogue of functions on the open Bruhat cell, one has no Weyl group action to obtain the corresponding algebras on its W-translates. Here a different construction will be used. Its importance lies in finding a quantum analogue of the Beilinson-Bernstein equivalence of categories [BB1].

10.4.1 Take $R = R_q[SL(2)]$ and recall the notation and formulae (2), (3) of 10.2.2. Here it may be convenient to replace q by a fractional multiple of itself - a process called rescaling. Then if R_i is a $k(q_i)$ algebra: $i = 1, 2$ with $q_1^r = q_2^s = q : r, s \in \mathbb{Z}$, a homomorphism $R_1 \to R_2$ is defined to be a $k(q)$-algebra homomorphism $R_1 \otimes_{k(q_1)} k(q) \longrightarrow R_2 \otimes_{k(q_2)} k(q)$. Write s for the unique simple reflection of the Weyl group. Recall (9.3.9, 10.3.7) the classification of $\operatorname{Prim} R$. Call a quotient $R/P : P \in \operatorname{Prim} R$, or the ideal P, of type $\mathbf{w}$ if $P \in \mathbf{X}(\mathbf{w})$.

Lemma. *Take $P, P' \in \operatorname{Rat} R$. Suppose that $R/P \cong R/P'$ possibly up to rescaling. Then either P and P' are of the same type or P is of type (s, e) and P' of type (e, s) (or vise versa).*

From 10.3.9 (ii) one checks that $d(R/P) = 0$ if P is of type (e, e) and equals 2 otherwise.

Let a_i, b_i, c_i, d_i denote the generators of $R_i := R_{q_i}[SL(2)] : i = 1, 2$ described in 10.2.2 (2). Let P_1 (resp. P_2) be a rational ideal of R_1 (resp. R_2) of type (e, s) (resp. (s, s)). Let $A_i = R_i/P_i : i = 1, 2$, To the relations (2), (3) of 10.2.2 one must add the additional relation $c_1 = 0$ (resp. $b_2 = \lambda c_2 : \lambda \in k(q)^*$) in R_1/P_1 (resp. R_2/P_2). Thus A_i is generated by just the images of a_i, b_i, d_i. Moreover in A_1 one has $d_1 = a_1^{-1}$.

Let $\varphi : A_2 \longrightarrow A_1$ be an isomorphism. The ideals of codimension 1 of A_i correspond to the rational ideals of R_i of type (e, e) and so have common intersection $\langle b_i \rangle$. Hence $\varphi(\langle b_2 \rangle) = \langle b_1 \rangle$. Thus $\varphi(b_2)$ which is a generator of $\langle b_1 \rangle$ takes the form $x b_1$ for some $x \in k(q)[a_1, a_1^{-1}]$ which through the commutation rules of 10.2.2 (2) must be a non-zero scalar. Through $\operatorname{Aut} R_1$ one can assume $\varphi(b_2) = b_1$.

By 10.1.19 the spectrum of b_2 on the faithful simple A_2 module $\mathcal{S}(s)$ takes the form $\{q_2^n : n \in \mathbb{N}\}$. Viewed as an A_1 module through φ such a spectrum is incompatible with the first relation of 10.2.2 (2) and the fact that $a_1^n \in A_1$ for all $n \in \mathbb{Z}$. This and a similar argument for type (s, e) proves the lemma.

10.4.2 Take $R = R_q[SL(2)]$ and identify $T^\wedge$ with $k(q)^*$. Then for all $\lambda, \mu \in k(q)^*$ the map $\varphi_{\lambda,\mu} : \begin{pmatrix} a & b \\ c & d \end{pmatrix} \mapsto \begin{pmatrix} \lambda a & \mu b \\ \mu^{-1}c & \lambda^{-1}d \end{pmatrix}$ extends to an automorphism of R. Then the map $(\lambda, \mu) \mapsto \varphi_{\lambda,\mu}$ defines an embedding of $T^\wedge \times T^\wedge$ into $\operatorname{Aut} R$ and hence an action of $T^\wedge \times T^\wedge$ on R. Recall that $\varkappa'^*$ is an automorphism of R fixing a, d and interchanging b, c.

Set $T_d^\wedge = \{\varphi_{1,\mu} : \mu \in k(q)^*\}$.

Lemma

(i) $\operatorname{Aut}(R) = (T^\wedge \times T^\wedge) \rtimes \{1, \varkappa'^*\}$.

(ii) $\operatorname{AUT}(R) = T_d^\wedge$.

(iii) *Given an algebra (resp. Hopf algebra) isomorphism* $\varphi : R_{q_1}[SL(2)] \xrightarrow{\sim} R_{q_2}[SL(2)]$ *then* $q_1 = q_2$ *or* $q_1 = q_2^{-1}$ *(resp.* $q_1 = q_2$*).*

Adopt the notation of 10.4.1. Let φ be an algebra isomorphism of R_1 onto R_2. By 10.4.1 either $\varphi(\langle b_1 \rangle) = \langle b_2 \rangle$ or $\varphi(\langle b_1 \rangle) = \langle c_2 \rangle$. Using $\varkappa'^*$ it can be assumed that the first holds. Then $\varphi(b_1) = xb_2$ for some $x \in R_2$. Similarly $\varphi^{-1}(b_2) = yb_1$ for some $y \in R_1$ and then $b_1 = \varphi^{-1}\varphi(b_1) = \varphi^{-1}(x)yb_1$. By 9.1.9 and 9.1.14 this forces x to be a scalar which through the $T^\wedge \times T^\wedge$ action can be assumed to be 1. Moreover $\varphi(c_1) = \alpha c_1$ for some $\alpha \in k(q)^*$. Through the automorphism $x \mapsto b_1 x b_1^{-1}$ of R_1 which fixes b_1, c_1 and sends a_1 to $q_1^{-1}a_1$ and d_1 to $q_1 d_1$ it follows using 10.2.2 (3) that there exist $i, j \in \mathbb{N}^+$, $x_1, x_2 \in k(q_2)[b_2, c_2]$ non-zero such that $\varphi(a_1) = x_1 a_2^i$ and $q_2^i = q_1$ or $\varphi(a_1) = x_2 d_2^j$ and $q_2^{-j} = q_1$. A similar result for φ^{-1} forces $i, j = 1$ and then that x_1, x_2 are non-zero scalars which using the $T^\wedge \times T^\wedge$ action can be assumed to be 1. Moreover $\varphi(d_1) = \beta d_2$ in the first case and $\varphi(d_1) = \beta' a_2$ in the second case with β, β' scalar. From the formula for the coproduct the second case is excluded for a Hopf algebra isomorphism. This proves (iii).

Now consider (i), (ii) dropping the subscripts. By the last relation of 10.2.2 (2) one obtains $\beta = \alpha$ and then from 10.2.2 (3) that $\alpha = 1$. Hence $\varphi = 1d$ proving (i). Using the formula for the coproduct then gives (ii).

Remark. Take $\chi \in R^\wedge$ and set $\chi(a) = \alpha$, $\chi(d) = \delta$. Then $\alpha\delta = 1$ and (notation 10.3.12) $\theta_{\operatorname{ad}\chi} : \begin{pmatrix} a & b \\ c & d \end{pmatrix} \mapsto \begin{pmatrix} a & \alpha^2 b \\ \alpha^{-2}c & d \end{pmatrix}$. Thus in general the map $\chi \mapsto \theta_{\operatorname{ad}\chi}$ on $R^\wedge$ into $\operatorname{AUT}(R)$ is not surjective; but it is so over $\overline{k(q)}$. Of course one can recover $T_d^\wedge$ by taking all characters $\chi : R \longrightarrow \overline{k(q)}^*$ such that $\theta_{\operatorname{ad}\chi}$ preserves R.

10.4.3 Take $R = R_q[G]$ arbitrary. Recall (10.1.14) the Hopf algebra surjections $\varphi_i^\star : R \longrightarrow R_{q_i}[SL(2)_i] =: R_i$ and set $J_i = \ker \varphi_i^\star$. Given $P_i \in \operatorname{Rat} R_i$ its inverse image $P = \varphi_i^{\star-1}(P_i)$ lies in $\operatorname{Rat} R$. Set $W_i = \{e, s_i\}$ which may be viewed as the Weyl group for R_i or as the subgroup of the Weyl group W for $R_q[G]$ generated by s_i. Then (9.3.9) there exists $\mathbf{w} \in W_i \times W_i$ such that P_i is of type $\mathbf{w}$ and it is clear from definitions that $P \in \mathbf{X}(\mathbf{w})$. Then by 10.3.11 one obtains $J_i = Q_{(s_i, s_i)}$.

Proposition. *Let $\varphi : R_q[G] \longrightarrow R_{q'}[SL(2)]$ be an algebra surjection. Then there exists $i \in \{1, 2, \ldots, \ell\}$ such that $\ker \varphi = J_i$. In particular $q' = q_i$ or q_i^{-1}.*

Set $R = R_q[G]$, $S = R_{q'}[SL(2)]$. By 10.3.11 (i) the rational ideals of S of type (s, s) form a single $S^\wedge$ orbit $\{P_\chi : \chi \in S^\wedge\}$. Fix χ and let $p_\chi : S \longrightarrow S/P_\chi$ denote the canonical projection. Then by 10.3.9 (ii) one has $d(R/\ker(p_\chi \varphi)) = 2$. Yet $\varphi^{-1}(P_\chi) = \ker p_\chi \varphi$ is a rational ideal of R and so by 10.3.9 (ii) again it must be of type (s_j, e), (e, s_j) or (s_j, s_j) for some simple reflection s_j. By 10.4.1 only the last case is possible. Set $S_j^\wedge = \{\chi \in S^\wedge \mid \varphi^{-1}(P_\chi)$ *is of type* $(s_j, s_j)\}$. Then the irreducible set $S^\wedge$ is a finite union of the $S_j^\wedge$ one of which must be dense, say $S_i^\wedge$. This gives

$$\ker \varphi = \varphi^{-1}(0) = \varphi^{-1}\Big(\bigcap_{\chi \in S_i^\wedge} P_\chi\Big) = \cap_{\chi \in S_i^\wedge} \varphi^{-1}(P_\chi) .$$

Now by 10.1.13 and the above, the right side contains J_i. The latter is a prime ideal satisfying $d(R/J_i) = 3 = d(R/\ker \varphi)$. By A.2.7 and A.3.6 (ii), this forces equality and proves the first part. The last part follows from 10.4.2 (iii).

10.4.4 Combining 10.4.2 and 10.4.3 one obtains the

Corollary. *Let $\varphi : R_q[G] \longrightarrow R_{q'}[SL(2)]$ be a surjection of Hopf algebras. Then there exists $i \in \{1, 2, \ldots, \ell\}$ such that $q' = q_i$ and $\varphi = \varphi_i^\star$ up to* $\mathrm{AUT}(R_{q'}[SL(2)])$.

10.4.5 For each $R_q[G]$ with torus T, let $T_d^\wedge$ denote the set of all $\theta_{\mathrm{ad}\,\chi}$ preserving $R_q[G]$ where χ is a linear character on $R_q[G]$ with values in $\overline{k(q)}$. By 10.3.12, $T_d^\wedge$ is a subgroup of $\mathrm{AUT}(R_q[G])$.

An embedding $\theta : \pi \hookrightarrow \pi'$ of simple systems is said to be a Dynkin diagram embedding if there exists a rational number $r_\theta > 0$ such that $(\theta(\alpha), \theta(\beta)) = r_\theta(\alpha, \beta)$ for all $\alpha, \beta \in \pi$. If θ is also surjective, it is called a Dynkin diagram automorphism. From the construction of $U_q(\mathfrak{g})$ it is clear that a Dynkin diagram embedding defines an embedding $\varphi_\theta : U_q(\mathfrak{g}) \hookrightarrow U_{q'}(\mathfrak{g}')$ of Hopf algebras given $q^{r_\theta} = q'$. From the classification (4.3.9, 4.3.10) of finite dimensional $U_q(\mathfrak{g})$ modules it follows (as in 10.1.14) that φ_θ transposes to a surjection $\varphi_\theta^\star : R_{q'}[G'] \twoheadrightarrow R_q[G]$ of Hopf algebras.

Theorem. *Let $\varphi : R_{q'}[G'] \longrightarrow R_q[G]$ be a surjection of Hopf algebras. Then up to $T_d^\wedge$ one has $\varphi = \varphi_\theta^\star$, $q^r = q'$ for some Dynkin diagram embedding θ.*

Let $\varphi_i^\star : R_q[G] \longrightarrow R_i : i = 1, 2, \ldots, rk\ G$ be defined as in 10.4.3. Then by 10.4.4, the composition $\varphi_i^\star \varphi$ equals some $\Phi_{\theta(i)}^\star : R_{q'}[G'] \longrightarrow R_{\theta(i)}$ up to $\mathrm{AUT}(R_i)$ with $q^{(\alpha_i, \alpha_i)} = q'^{(\alpha_{\theta(i)}, \alpha_{\theta(i)})}$. In particular there exists a rational number $r_\theta > 0$ such that $(\alpha_{\theta(i)}, \alpha_{\theta(i)}) = r_\theta(\alpha_i, \alpha_i), \forall i$. From the remark following 10.4.2 one may choose $\psi \in T_d^\wedge$ such that $\varphi_i^\star \psi \varphi = \Phi_{\theta(i)}^\star$ for all i. Replace φ by $\psi\varphi$. By transposition $\varphi^\star \varphi_i^{\star\star} = \Phi_{\theta(i)}^{\star\star}$ where $\varphi_i^{\star\star}, \Phi_{\theta(i)}^{\star\star}$ extend $\varphi_i, \Phi_{\theta(i)}$. Thus for all i the Hopf algebra map $\varphi^\star : R_q[G]^\star \longrightarrow R_{q'}[G']^\star$ restricts to an isomorphism of $\operatorname{Im}\varphi_i$ onto $\operatorname{Im}\Phi_{\theta(i)}$. The Serre relations (4.3.13) between the $e_i, e_j : i \neq j$ then force θ to be a Dynkin diagram embedding.

10.4.6 Let D denote the subgroup of $\mathrm{AUT}(R_q[G])$ defined by the group of Dynkin diagram automorphisms.

Corollary. $\mathrm{AUT}(R_q[G]) = T_d^\wedge \rtimes D$.

10.4.7 Since any Hopf algebra injection $\varphi : U_q(\mathfrak{g}) \hookrightarrow U_{q'}(\mathfrak{g}')$ gives by transposition a Hopf algebra surjection $\varphi^\star : R_{q'}[G'] \longrightarrow R_q[G]$, assertions analogous to 10.4.5, 10.4.6 hold for $U_q(\mathfrak{g})$.

10.4.8 Recall the notation of 9.1.11. One may view S^+ as the q-analogue of regular functions on the w_0 translate of the open Bruhat cell. In view of 9.1.10 one may define

$$S^w := \sum_{\mu \in P^+(\pi)} c_{w\mu}^{-1} V^+(\mu)^* \ .$$

Take $\mu, \nu \in P^+(\pi)$, $a \in V^+(\mu)^*$. Since c_w is Ore in R^+ there exist $\lambda, \lambda' \in P^+(\pi)$, $b \in V^+(\lambda')^*$ such that $c_{w\lambda} a = b c_{w\nu}$. Through the grading implied by the Cartan multiplication rule (9.1.6 (*)) and because R^+ is a domain (9.1.9) it follows that $\lambda + \mu = \lambda' + \nu$. Since $\dim V^+(\mu)^* < \infty$ one can further choose λ such that

$$c_{w\lambda} V^+(\mu)^* \subset V^+(\lambda + \mu - \nu) c_{w\nu} \ .$$

From this relation one concludes that S^w is a subalgebra of $\operatorname{Fract} R^+$. It is natural to regard S^w as the q-analogue of regular functions on the w-translate of the open Bruhat cell.

10.4.9 By 4.3.12 the right action of $U_q(\mathfrak{g})$ on R^+ extends to $c_w^{-1} R^+$ and one easily checks that it leaves S^w invariant. From 9.1.6 (*) and 9.1.10 (*) one has an embedding $c_{w\mu}^{-1} V^+(\mu)^* \hookrightarrow c_{w(\mu+\nu)}^{-1} V^+(\mu+\nu)^*$, so one may also regard S^w as a direct limit of the $c_{w\mu}^{-1} V^+(\mu)^*$. In particular S^w admits a formal character. By 4.3.6 one has $\operatorname{ch}(c_{w\mu}^{-1} V^+(\mu)^*) = e^{-w\mu} \operatorname{ch} V(\mu) = w(e^{-\mu} \operatorname{ch} V(\mu))$ whilst the limiting value of $e^{-\mu} \operatorname{ch} V(\mu)$ is just $\prod_{\beta \in \Delta^+} (1 - e^{-\beta})^{-1}$. This gives the

Lemma. *For all $w \in W$,*

$$\operatorname{ch} S^w = w\Big(\prod_{\beta \in \Delta^+} (1 - e^{-\beta})^{-1} \Big) \ .$$

10.4.10 Set $S = S^e$. Recall (10.2.6) the definition of $r_w \in \operatorname{Aut} U$. Consider S^{r_w} as an U module by transport of structure. Recall the notation of 4.1.

Lemma. *As U modules $S^w \cong S^{r_w}$.*

Set $N = (S^w)^{r_w^{-1}}$. It is enough to show that $N \cong \delta M(0)$ and as in 5.3.6 it suffices by 10.4.9 to show that N^{U^+} reduces to scalars.

Take $\nu \in P^+(\pi)$. One checks that $c_{w\nu} S^w$ is a U submodule of R^+. Set $N(\nu) = (c_{w\nu} S^w)^{r_w^{-1}}$. Then $\operatorname{ch} N(\nu) = w^{-1}(e^{w\nu} \operatorname{ch} S^w) = e^\nu(\operatorname{ch} \delta M(0)) = \operatorname{ch} \delta M(\nu)$. Consequently $N(\nu) \in Ob\mathcal{O}$ and has the same composition factors as $\delta M(\nu)$. In particular $V(\nu)$ is a subquotient N_1/N_2 of $N(\nu)$.

Take $b \in N^{U^+}$ which can be assumed to be a weight vector of weight $\mu \in -\mathbb{N}\pi$. Viewed as an element S^w it is an $r_w(U^+)$ invariant vector of weight $w\mu$. By 10.2.9, $c_{w\nu} b$ is again $r_w(U^+)$ invariant and of weight $w(\mu + \nu)$. Now choose ν sufficiently large so that the injection $V(\nu)_{\nu+\mu} \longrightarrow \delta M(\nu)_{\nu+\mu}$ is surjective (4.3.6). This forces $c_{w\nu} b$ to belong to N_1 and to be non-zero in the quotient $N_1/N_2 \cong V(\nu)$. Yet it is U^+ invariant and this forces $\mu = 0$. Then 10.4.9 implies that b is scalar.

10.5 Comments and Complements

10.5.1 The construction of the modules $\mathcal{S}(w) : w \in W$ and the quantum Weyl group is due to Y. Soibelman [S1]. However the present analysis is rather different in detail incorporating several results from [J11]. The classification of $\operatorname{Prim} R_q[G]$ is taken from [J11]. The result in 10.3.8 settled a conjecture of T. Hodges and T. Levasseur [HL1] which they had previously proved [HL2] for the special case $G = SL(n)$. The result in 10.2.9 is just [J10, 3.7]. A result of a similar nature can be found in [AJS1, c.6] but the proof is far more complicated and needs the details of Soibelman's construction. Sections 10.4.1–10.4.7 are due to A. Braverman [Br1]. Aut $U_q(\mathfrak{sl}(2))$ had been determined earlier by J. Alev and M. Chamarie [AC1] whilst 10.4.7 which is slightly weaker than 10.4.6 has been recently reported by W. Chin and I. M. Musson [CM1]. Finally 10.4.8–10.4.10 is taken from [J10].

10.5.2 The Soibelman automorphisms $r_w : w \in W$ coincide with those written down independently by G. Lusztig [L2, 6, 12]. They lead to a PBW type basis [L2, 6, LeS1] for $U_q(\mathfrak{g})$. Actually in the notation of 3.4.11 this leads [L6] to an A basis for U_A. Moreover in this basis the scalars a_ν occurring in 4.1.13 can be computed [DK1, Sect. 1] and shown to have zeros only when q is a root of unity. By 3.4.12 this is equivalent to showing that the non-degenerate pairing $\varphi : V^- \times V^+ \longrightarrow k(q)$ may degenerate at specialization only when q is a root of unity. I do not know if such a result holds in general for C integrable. Notice that 10.2.2(9) defines the action of N_i on any integrable $U_q(\mathfrak{g}_C)$ module without restriction on C. Moreover the calculations and conclusions of 10.2.3–10.2.9 (with R^* replaced by the algebra of matrix coefficients of integrable modules) go over to $U_q(\mathfrak{g}_C) : C$ integrable without change and the formulae only involve divided powers. (See also [L12, 41.2]). For further results consult [B1, Ch1, KR1, Maj1, MS1].

10.5.3 One may define a quantized version D^w of the algebra of the differential operators on each w-translate of the open Bruhat cell generalizing S_{w_0}. Just as in 10.3.10 one may show that S^w is a faithful simple D^w module. For each $\lambda \in P^+(\pi)$ regular one may define a faithfully flat embedding of the minimal primitive quotient $U/\operatorname{Ann} M(w_0.\lambda)$ into the direct sum of the $D^w : w \in W$. This gives a quantum analogue of the Beilinson-Bernstein equivalence of categories [BB1] as interpreted by T. Hodges and S.P. Smith [HS1]. These results together with the noetherianity of S^w, D^w are proved in [J10]. These are interesting algebras which have probably few isomorphisms amongst themselves though their skew-field of fractions $\operatorname{Fract} S^w : w \in W$ for example all coincide. In the simplest case when $\mathfrak{g}$ is of type A_2 an unpublished result of M. Gorelik shows that $S^{s_\alpha} : \alpha \in \pi$ has centre reduced to scalars, whereas by say 7.1.20 (iii) the centre of S^e is a polynomial algebra in one variable.

10.5.4 In [AD1], J. Alev and F. Dumas have shown that $\operatorname{Fract} U^-$ is, in type A_n, much more commutative than one might imagine and in fact the algebra of fractions on quantum m-space over an appropriate centre. One may ask if this holds for C of finite or of affine type (A.3.8).

10.5.5 Take $R = R_q[G]$. In [J11] it is shown that R/P is a domain for every $P \in \operatorname{Spec} R$. This generalized a result of K. Goodearl and E. Letzter [GL1] for the case $G = SL(n)$.

Appendix

The following gives with less details of proofs than the main text some background material.

A.1 The Root System and the Weyl Group of a Cartan Matrix C

Recall the notations and conventions of 3.1. When C is symmetric the construction of $U_q(\mathfrak{g}_C)$ gives rise to a root system $\Delta_{\text{mult}}(C, q)$ counted with multiplicities (3.4.7). When C is specializable the construction of the Kac-Moody Lie algebra gives rise to a root system $\Delta_{\text{mult}}(C)$ counted with multiplicities (4.2.1). When both are defined one has $\Delta_{\text{mult}}(C, q) \supset \Delta_{\text{mult}}(C)$; but little else can be said about either and they even fail to be stable under the Weyl group $W(C)$. The situation changes dramatically when C is integrable. Then they are equal (4.3.8) and stable under the Weyl group (4.2.8 (ii)). The following section examines resulting combinatorial properties of Δ and W in this case.

A.1.1 Take $\alpha \in \pi$ and define $s_\alpha \in \operatorname{Aut} \mathfrak{h}^*$ by $s_\alpha \lambda = \lambda - (\alpha^\vee, \lambda)\alpha$. One calls s_α the simple reflection defined by α. Set $s_i = s_{\alpha_i}$. Let W denote the subgroup of $\operatorname{Aut} \mathfrak{h}^*$ generated by the $s_\alpha : \alpha \in \pi$. It leaves the Cartan form and Δ invariant (4.2.5, 4.2.8). Thus for each $w \in W$ one may define

$$S(w) = \{\alpha \in \Delta^+ \mid w\alpha \in \Delta^-\} = \Delta^+ \cap w^{-1}\Delta^- .$$

In particular $S(e) = \emptyset$ where e denotes the identity of W.

If $s_{i_1} s_{i_2} \dots s_{i_r} = w$ and r takes its minimal value amongst all possible expressions for w, then it is called a reduced decomposition for w and r its reduced length $\ell(w)$. For such a reduced decomposition set $\beta_j = w_{j+1}^{-1}\alpha_{i_j} : j = 1, 2, \dots, r$ where $w_{r+1} = e$, $w_j = s_{i_j} \dots s_{i_r} : j \leq r$. Let $\omega_i : i = 1, 2, \dots, \ell$ be elements of $\mathfrak{h}^*_{\mathbb{Q}}$ chosen so that $(\alpha_i^\vee, \omega_j) = \delta_{ij}$ and let ρ denote their sum. They are called the fundamental weights and are unique if C is of finite type. Then $P(\pi) \supset \sum \mathbb{Z}\omega_i$, $P^+(\pi) \supset \sum \mathbb{N}\omega_i$ (notation 4.2.8).

Lemma

(i) $S(w^{-1}) = -wS(w)$, $\forall w \in W$.

(ii) *For all* $\alpha \in \pi$,

$$S(s_\alpha w) = \begin{cases} S(w) \cup \{w^{-1}\alpha\} : w^{-1}\alpha \in \Delta^+ \ , \\ S(w) \setminus \{-w^{-1}\alpha\} : w^{-1}\alpha \in \Delta^- \ . \end{cases}$$

(iii) $S(w) = \emptyset \iff w = e$.
(iv) $|S(w)| = \ell(w)$. *Moreover* $w^{-1}\alpha \in \Delta^- \iff \ell(s_\alpha w) \leq \ell(w)$ *and*

$$S(w) = \{\beta_j\}_{j=1}^r$$

defined as above with respect to a reduced decomposition of w.
(v) $\rho - w^{-1}\rho = \sum_{\beta \in S(w)} \beta$.
(vi) *The map* $w \mapsto S(w)$ *is injective.*
(vii) *The stabilizer of* $\lambda \in P^+(\pi)$ *in* W *is generated by the simple reflections it contains.*

(i) obtains from $wS(w) = w\Delta^+ \cap \Delta^- = -S(w^{-1})$. (ii) follows from $\Delta^+ \setminus \{\alpha\}$ being s_α stable (4.2.8 (ii)). It implies that $|S(w)| \leq \ell(w)$. For (iii) suppose $w \neq e$ satisfies $S(w) = \emptyset$. Take a reduced decomposition for w as above and set $\alpha_{i_r} = \alpha$. Choose $j \leq r$ maximal such that $w_j\alpha \in \Delta^+$. Then $j < r$ and setting $\beta = \alpha_{i_j}$ it follows from 4.2.8 (ii) that $w_{j+1}\alpha = -\beta$. Set $y = w_{j+1}s_\alpha$. Then $y\alpha = \beta$ and this implies that $s_\beta = ys_\alpha y^{-1}$. Then $w_j = y$ forcing $\ell(w) \leq j - 1 + \ell(y) \leq j - 1 + r - j - 1 \leq r - 2$ which is a contradiction.

(iv) obtains by induction on $|S(w)|$. By (iii) it holds when $|S(w)| = 0$. Set $S_r = \{w \in W$ *such that* $|S(w)| = r\}$. Assume (iv) established for S_r. Consider $y \in S_{r+1}$. By (ii) there exists $w \in S_r$, $\alpha \in \pi$ such that $y = s_\alpha w$ and $S(y) = S(w) \cup \{w^{-1}\alpha\}$. Moreover

$$1 + \ell(w) \geq \ell(y) \geq |S(y)| = 1 + |S(w)| = 1 + \ell(w) \ ,$$

as required.

Note that $\rho - s_\alpha\rho = \alpha$. Then (v) obtains from $\rho - (s_\alpha w)^{-1}\rho = \rho - w^{-1}\rho + w^{-1}\alpha$ and (ii). Then $\rho = w^{-1}\rho$ would imply $S(w) = \emptyset$ and so $w = e$ by (iii). Then $S(x) = S(y)$ implies $x^{-1}\rho = y^{-1}\rho$ and so $x = y$ which is (vi).

For (vii), set $\pi' = \{\alpha \in \pi | (\alpha^\vee, \lambda) = 0\}$ and let W' be the subgroup of W generated by the $s_\alpha : \alpha \in \pi'$. It is enough to show that $\lambda \geq w\lambda$ with equality if and only if $w \in W'$ and this will be achieved by induction on $\ell(w)$. It is trivial if $\ell(w) = 0$, that is if $w = e$. Otherwise by (i), (iv) there exists $\alpha \in \pi$, $y \in W$ such that $w = ys_\alpha$, $\ell(y) = \ell(w) - 1$ and $y\alpha \in \Delta^+$. Then $\lambda - w\lambda = \lambda - y\lambda + (\alpha^\vee, \lambda)y\alpha$ from which the assertion follows.

A.1.2 The conclusion of (vi) above implies that $|W| < \infty$ if $|\Delta^+| < \infty$. The converse is immediate from 4.2.16. Moreover it leads to a very interesting order relation $\overset{D}{\leq}$ (the weak Bruhat or Duflo order) defined by $x \overset{D}{\leq} y$ if $S(x) \subset S(y)$.

A.1.3 The subset $W\pi$ of Δ is called the set of real roots Δ_{re}. By 4.2.8 (ii) every $\beta \in \Delta_{\text{re}}$ occurs with multiplicity 1 and of course satisfies $(\beta, \beta) > 0$. Again if $r\beta \in \Delta_{\text{re}}$ then $r = \pm 1$. The remaining roots in $\Delta_{\text{im}} := \Delta \setminus \Delta_{\text{re}}$ are called imaginary for the following reason.

Lemma

(i) Δ^+_{im} *is* W *stable.*
(ii) $(\beta, \beta) \leq 0$ *for all* $\beta \in \Delta_{\text{im}}$. *In particular* $W\pi = \Delta$ *if* C *is of finite type.*
(iii) $2(\rho, \beta) - (\beta, \beta) > 0$ *for all* $\beta \in \Delta^+ \setminus \pi$.

(i) is immediate from 4.2.8 (ii) and the definition of Δ^+_{im}. For (ii) assume $\beta \in \Delta^+_{\text{im}}$, satisfies $(\beta, \beta) > 0$ and $|\beta|$ is minimal for this property. By (i) one has $s_\alpha\beta \in \Delta^+_{\text{im}}$ for all $\alpha \in \pi$, whilst $(s_\alpha\beta, s_\alpha\beta) = (\beta, \beta) > 0$. Consequently $|s_\alpha\beta| \geq |\beta|$ which implies that $(\alpha^\vee, \beta) \leq 0$. Since $\alpha \in \pi$ is arbitrary, this gives $(\beta, \beta) \leq 0$ which is a contradiction.

Consider (iii). Since $(\rho, \beta) > 0$ for all $\beta \in \Delta^+$ one may assume by (i) that $\beta = w\alpha$ for some $w \in W$, $\alpha \in \pi$. This may be written $\beta^\vee = w\alpha^\vee$ and so by 4.2.8 (ii) one has $\beta^\vee \in \Delta^\vee$. Thus $\beta^\vee$ takes the form $\sum k_i\alpha_i^\vee$ with $k_i \in \mathbb{N}$. Then $(\rho, \beta^\vee) = \sum k_i \geq 1$ and equality implies that $\beta^\vee \in \pi^\vee$, equivalently $\beta \in \pi$ contrary to the hypothesis. Hence $2(\rho, \beta) - (\beta, \beta) > 0$ as required.

A.1.4 One may describe the image of the map $S : w \mapsto S(w)$. Call a subset $T \subset \Delta^+$ closed if $\beta, \gamma \in T$ and $\beta + \gamma \in \Delta^+$ imply $\beta + \gamma \in T$ and coclosed if $\Delta^+ \setminus T$ has this property. Call $T \subset \Delta^+$ complete if it is both closed and coclosed. Obviously any $S(w)$ is complete and finite.

Lemma. $\operatorname{Im} S$ *consists of the complete finite subsets of* Δ^+. *In particular if* $|\Delta^+| < \infty$ *there exists a unique longest element* $w_0 \in W$ *and it satisfies* $w_0\Delta^+ = \Delta^-$.

The proof is by induction on cardinality. Suppose T is complete and finite. If $T \neq \emptyset$ then since it is coclosed there exists $\alpha \in T \cap \pi$. Then by 4.2.8 (ii) one has $s_\alpha T \cap \Delta^+ = s_\alpha(T \setminus \{\alpha\})$ which is obviously closed and has complement $s_\alpha(\Delta^+ \setminus T) \cup \{\alpha\}$ which closed because T is complete. By the induction hypothesis $s_\alpha(T \setminus \{\alpha\}) = S(w)$ and then by A.1.1 (i), (ii) one has $S(ws_\alpha) = s_\alpha S(w) \cup \{\alpha\} = T$.

A.1.5 Fix a reduced decomposition for $w \in W$ as in A.1.1 and set $w'_j = e$ if $j = 0$ and $w'_j = s_{i_1}s_{i_2}\ldots s_{i_j}$ if $j > 0$. By A.1.1 (i), (iv) one has $S(w^{-1}) = \{\gamma_j\}_{j=1}^r$ where $\gamma_j = w'_{j-1}\alpha_{i_j}$. Moreover $s_{\gamma_j}w = s_{i_1}\ldots\hat{s}_{i_j}\ldots s_{i_r}$ where $\hat{s}$ denotes omission of s.

Lemma. *For all* $w \in W$, $\gamma \in \Delta^+_{\text{re}}$ *one has* $\ell(s_\gamma w) \neq \ell(w)$ *and* $\ell(s_\gamma w) < \ell(w) \iff \gamma \in S(w^{-1})$.

Set $S = \{\beta \in S(w^{-1}) \mid s_\gamma\beta \in \Delta^+\}$, $S' = S(w^{-1}) \setminus S$. Then $\beta \in S'$ implies $s_\gamma\beta = \beta - n\gamma$ with $n \in \mathbb{N}^+$. Suppose $w^{-1}\gamma \in \Delta^+$. Then $S(w^{-1}s_\gamma) \supset s_\gamma S \amalg S' \amalg \{\gamma\}$. Consequently A.1.1 (iv) gives $\ell(s_\gamma w) = \ell(w^{-1}s_\gamma) \geq 1 + \ell(w)$. To conclude replace w by $s_\gamma w$.

A.1.6 Take $w \in W, \gamma \in S(w^{-1})$. It need not happen that $\ell(s_\gamma w) = \ell(w)-1$. However given $w, w' \in W$, $\gamma \in \Delta^+_{\text{re}}$ such that $w' = s_\gamma w$ and $\ell(w') = \ell(w) - 1$ one writes $w' \overset{\gamma}{\longleftarrow} w$, or simply $w' \longleftarrow w$. The Bruhat order $\leq$ on W is defined by setting $w' \leq w$ if $w' = w$ or if there exists a chain $w' = w_1 \longleftarrow w_2 \longleftarrow \ldots \longleftarrow w_n = w$.

Lemma. *Suppose $w' \overset{\gamma}{\longleftarrow} w$. Take $\alpha \in \pi$ and suppose $\gamma \neq \alpha \in \pi$. Then either $s_\alpha w' \overset{s_\alpha\gamma}{\longleftarrow} s_\alpha w$ or $s_\alpha w' \overset{\alpha}{\longleftarrow} w' \overset{\gamma}{\longleftarrow} w \overset{\alpha}{\longleftarrow} s_\alpha w$.*

If $\ell(s_\alpha w) < \ell(w)$, then one can take $\alpha_{i_1} = \alpha$. Then the hypothesis $\gamma \neq \alpha$ and the definition of $\longleftarrow$ forces $\ell(s_\alpha w') = \ell(w') - 1 = \ell(w) - 2 = \ell(s_\alpha w) - 1$ so the first conclusion holds. This is also true if $\ell(s_\alpha w) > \ell(w)$ and $\ell(s_\alpha w') > \ell(w')$. Otherwise the second conclusion holds.

Remark. It follows from A.1.1 (ii), (iv) that $\ell(w)$ is even if and only if w is a product of an even number of simple reflections. In particular if $w, w' \in W$, $\gamma \in \Delta'_{\text{re}}$ satisfy $w' = s_\gamma w$ with $\ell(w') < \ell(w) - 1$, then $\ell(w') \leq \ell(w) - 3$. Then for any $\alpha \in \pi$ one has $\ell(s_{s_\alpha\gamma}s_\alpha w) = \ell(s_\alpha s_\gamma w) \leq \ell(w) - 2 \leq \ell(s_\alpha w) - 1$. Consequently if $w' \overset{\gamma}{\twoheadleftarrow} w$ is defined to mean that $\ell(s_\gamma w) < \ell(w)$, then if $\gamma \neq \alpha \in \pi$ one has $s_\alpha w' \overset{s_\alpha\gamma}{\twoheadleftarrow} s_\alpha w$.

A.1.7 Proposition. *Take $w, w' \in W$ with $w \geq w'$.*

(i) $w \leq e \Longrightarrow w = e$.
(ii) *If $\alpha \in \pi$ and $\ell(s_\alpha w) < \ell(w)$ then $w \geq s_\alpha w$.*
(iii) *If $\alpha \in \pi$ either $s_\alpha w \geq s_\alpha w'$ or $w \geq s_\alpha w'$.*
(iv) *If $\alpha \in \pi$ either $s_\alpha w \geq s_\alpha w'$ or $s_\alpha w \geq w'$.*
(v) *The Bruhat order is the unique order relation on W satisfying* (i)–(iv).

(i) and (ii) are immediate. (iii) is trivial if $w = w'$, otherwise it is proved by induction on chain length. Suppose $w_1 \overset{\gamma}{\longleftarrow} w_2$. If $\gamma = \alpha$, then $s_\alpha w' = w_2 \leq w$. If $\gamma \neq \alpha$, then $s_\alpha w' \leq s_\alpha w_2$ by A.1.6, whilst by the induction hypothesis $s_\alpha w_2$ is either $\leq s_\alpha w$ or $\leq w$. The proof of (iv) is similar; but the induction is started from the other end of the chain.

For (v) let $\leq'$ be an order relation satisfying (i)–(iii) and $\leq''$ an order relation satisfying (i), (ii), (iv). Show that $w' \leq'' w$ implies $w' \leq' w$ by induction on $\ell(w)$. It holds for $\ell(w) = 0$ by (i). Otherwise there exists $\alpha \in \pi$ such that $\ell(s_\alpha w) < \ell(w)$. Applying (ii) gives $w \geq' s_\alpha w$ whilst by (iv) and the induction hypothesis $s_\alpha w$ is either $\geq' w'$ or $\geq' s_\alpha w'$. In the second case applying (iii) gives either $w \geq' w'$ or $w' \geq' s_\alpha w \geq' w$ as required.

Remark. If $|W| < \infty$ one may include the property $w \geq w_0 \Longrightarrow w = w_0$ in (i). Since $s_\alpha w \geq w \Longleftrightarrow s_\alpha w w_0 \leq w w_0$, it follows that $w' \leq'' w \Longleftrightarrow w' w_0 \geq' w w_0$. From this and the proof of (v) one concludes that either condition (iii) or (iv) becomes superfluous.

A.1.8 From the remark in A.1.6 it follows that replacing $w' \longleftarrow w$ by $w' \twoheadleftarrow w$ gives a second order relation on W satisfying (i)–(iv) of A.1.7 which by A.1.7 (v) is just the Bruhat order. This gives the following useful result.

Corollary. *Let $w = s_{i_1} s_{i_2} \ldots s_{i_r} \in W$ be a reduced decomposition. Then $w' \leq w$ if and only if it can be obtained by omitting some of the s_{α_i} in the expansion for w.*

A.1.9 For each distinct pair $\alpha, \beta \in \pi$ set $m_{\alpha,\beta} = (\alpha^\vee, \beta)(\beta^\vee, \alpha) = 4(\alpha,\beta)^2/(\alpha,\alpha)(\beta,\beta)$ which is a positive integer. One easily checks that $(s_\alpha s_\beta)^{m_{\alpha,\beta}} = (s_\beta s_\alpha)^{m_{\alpha,\beta}}$ if $m_{\alpha,\beta} = 0, 2, 3$ and $s_\alpha s_\beta s_\alpha = s_\beta s_\alpha s_\beta$ if $m_{\alpha,\beta} = 1$. These together with the empty relation for all other values of $m_{\alpha,\beta}$ are known as the Coxeter relations.

Lemma

(i) *Any two reduced decompositions of $w \in W$ can be transformed into one another by just using the Coxeter relations.*

(ii) *Consider W as the quotient of the free group on the set of generators $s_\alpha : \alpha \in \pi$. Then the relations of W are the $s_\alpha^2 = 1 : \alpha \in \pi$ together with the Coxeter relations.*

The proof of (i) is by induction on $\ell(w)$. Take reduced decompositions of w ending in s_α and in $s_\beta : \alpha, \beta \in \pi$ and write $w = w_1 s_\alpha = w_2 s_\beta : \ell(w_i) = \ell(w) - 1$. One may assume $\alpha \neq \beta$. Let $\Delta^+_{\alpha,\beta}$ be the subset of Δ generated by $\{\alpha, \beta\}$. By A.1.1 (i), (ii) and A.1.4 one has $\Delta^+_{\alpha,\beta} \subset S(w)$ and so $|\Delta^+_{\alpha,\beta}| \leq |S(w)| = \ell(w) < \infty$ by A.1.1 (iv). Now $s_\alpha s_\beta$ leaves $\mathbb{Q}\alpha + \mathbb{Q}\beta$ invariant and with respect to this subspace $\det(s_\alpha s_\beta - \lambda 1d) = \lambda^2 + (2 - m_{\alpha,\beta})\lambda + 1$. This has real roots $\neq \pm 1$ if $m_{\alpha,\beta} > 4$ and so $s_\alpha s_\beta$ has infinite order in this case and also when $m_{\alpha,\beta} = 4$. Again by A.1.4 this forces $\Delta^+_{\alpha,\beta}$ to be infinite. Consequently $m_{\alpha,\beta} = 0, 1, 2, 3$. Since $\Delta^+_{\alpha,\beta} \subset S(w)$ it follows that $\beta \in S(w_1)$, $\alpha \in S(w_2)$. Then w_1 (resp. w_2) can be assumed to have a reduced decomposition $w_3 s_\beta$ (resp. $w_4 s_\alpha$) by A.1.1 (iv) and the induction hypothesis. If $m_{\alpha,\beta} = 0$, then $w_3 = w_4$ and the assertion holds in this case. Otherwise the above process can be repeated (that is $w_3 = w_5 s_\alpha$, $w_4 = w_6 s_\beta$) and eventually gives the required assertion.

(ii) Write $w = s_{i_1} s_{i_2} \ldots s_{i_r}$ and choose t maximal so that $w_t := s_{i_1} s_{i_2} \ldots s_{i_t}$ is reduced. Suppose $t < r$. Then $\ell(w_t s_{i_{t+1}}) < \ell(w_t)$, so by (i) and A.1.1 (iv) using the Coxeter relations one can rewrite w_t as a product of t simple reflections ending in $s_{i_{t+1}}$. Using $s^2_{i_{t+1}} = 1$, one can then write w_{t+1} as a product of $t-1$ simple reflections. In this fashion it can be assumed that w is written in reduced form. Applying (i) completes the proof of (ii).

Remark. A pair (W, S) satisfying (ii) is called a Coxeter group.

A.1.10 The Hecke algebra $H(W)$ is defined to be the ring over $\mathbb{Q}[q, q^{-1}]$ with generators T_α and relations $(T_\alpha - q)(T_\alpha + 1) = 0 : \alpha \in \pi$ together with Coxeter relations. By A.1.9 (i), $T_w : w \in W$ may be defined by taking a reduced decomposition.

The singular Hecke algebra is defined by taking $q = 0$ and replacing T_α by $-T_\alpha$ so then $T_\alpha^2 = T_\alpha$. One may view it as defining a new associative multiplication $(w, w') \mapsto w_* w'$ in W defined by $s_{\alpha *} s_\alpha = s_\alpha : \alpha \in \pi$ and the Coxeter relations. It is immediate from 6.3.15 that the Demazure operators $\Delta_w : w \in W$ form a $\mathbb{Q}$ basis of the singular Hecke algebra.

Lemma. *Take $w, w' \in W$. Then $w_* w' \geq w, w'$.*

It suffices to take $w' = s_\alpha : \alpha \in \pi$, in which case the assertion is obvious.

A.1.11 Consider $\nu \in WP^+(\pi)$. One may write $\nu = w\lambda : w \in W$, $\lambda \in P^+(\pi)$. If $\lambda \in P^+(\pi) + \rho$, then by A.1.1 (vii), w is uniquely determined and one writes $sg(\nu) = (-1)^{\ell(w)}$ in this case. Otherwise $\operatorname{Stab}_W \nu = w^{-1}(\operatorname{Stab}_W \lambda)w$. Then A.1.1 (vii) gives $\sum_{y \in \operatorname{Stab}_W \nu} (-1)^{\ell(y)} = 0$ and one writes $sg(\nu) = 0$ in this case.

In general $WP^+(\pi) \subsetneq P(\pi)$. For example by A.1.3 (i) the set $\{\beta \in \Delta_{\text{im}}^+ \mid (\beta, \beta) < 0\}$ is W stable and of course has null intersection with $P^+(\pi)$.

Lemma. *Take $\nu \in P(\pi)$. Then $\nu \in WP^+(\pi) \iff \nu$ is the weight of an integrable highest weight module. In particular $WP^+(\pi)$ is additively closed.*

The implication $\Longrightarrow$ is immediate from $\operatorname{ch} V(\lambda) : \lambda \in P^+(\pi)$ being W stable (4.3.6 (iii)). Conversely suppose ν is a weight of $V(\lambda) : \lambda \in P^+(\pi)$. Then for each $w \in W$ one has $w\nu = \lambda - \xi$ for some $\xi \in \mathbb{N}\pi$. Choose w such that $|\xi|$ (notation 5.3.15) is minimal. Then $(\alpha^\vee, w\nu) \geq 0$ for all $\alpha \in \pi$, as required. The last part follows by taking the tensor product of two integrable highest weight modules which by 4.3.10 is a direct sum of integrable highest weight modules.

A.1.12 The following completes A.1.11.

Lemma

(i) *Take $\lambda, \lambda' \in P^+(\pi)$. If $\lambda \in W\lambda'$ then $\lambda = \lambda'$.*
(ii) *Assume C of finite type. Then $P(\pi) = WP^+(\pi)$.*

One may write $\lambda = w\lambda'$ with w of minimal length. Take $\alpha \in \pi$ and suppose $\ell(s_\alpha w) < \ell(w)$. Then $w^{-1}\alpha \in \Delta^-$ by A.1.1(iv) and so $0 \leq (\alpha, \lambda) = (w^{-1}\alpha, \lambda') \leq 0$ which gives $\lambda = s_\alpha w\lambda'$ contradicting the hypothesis on w. Thus $S(w^{-1}) \cap \pi = \emptyset$ which implies (A.1.1 (iii), A.1.4) that $w = e$. Hence (i). For (ii) recall (A.1.2) that $|W| < \infty$ under the hypothesis. Then take a maximal element in the finite set $W\lambda$ and conclude as in A.1.11.

A.1.13 Assume C indecomposable of finite type and recall [Kal, Chap. 4] the classification of such matrices by Dynkin diagrams. By A.1.3 (ii) every $\alpha \in \Delta$ is conjugate to some $\alpha \in \pi$ whilst the simple roots of a given length form a connected subset of the Dynkin diagram and are hence W conjugate. Moreover there can be at most two root lengths. Call a root short if (α, α) takes its minimal value and long otherwise. Thus the set Δ_s (resp. Δ_ℓ) of short (resp. long) roots form a single W orbit. In particular there is a unique (short) root $\alpha_0 \in \Delta_s \cap P^+(\pi)$ and if $\Delta_\ell \neq \emptyset$ a unique (long) root $\beta_0 \in \Delta_\ell \cap P^+(\pi)$.

Set $\pi_s = \pi \cap \Delta_s$, $\pi_\ell = \pi \cap \Delta_\ell$. If $\pi_\ell \neq \emptyset$ one may write $\pi_\ell = \{\alpha_1, \alpha_2, \dots, \alpha_i\}$, $\pi_s = \{\alpha_{i+1}, \dots, \alpha_\ell\}$ for some i, where $(\alpha_r, \alpha_s) \neq 0$ if and only if $r = s \pm 1$. Set $\alpha'_j = \alpha_j + \alpha_{j+1} + \dots + \alpha_{i+1}$ if $j \leq i$ and $\alpha'_j = \alpha_j$ otherwise. Then $\pi' = \{\alpha'_j : j = 1, 2, \dots, \ell\}$ consists of short roots and is even a simple system relative to Δ_s^+. Moreover every $\alpha \in \Delta$ can be written uniquely as a linear combination of the elements of π'.

A.1.14 Assume C of finite type. Define the affine Weyl group W_a as the subgroup of Aut $h_{\mathbb{R}}^*$ generated by the reflections $s_{\alpha,n} : \lambda \mapsto s_\alpha \lambda + n\alpha : \alpha \in \Delta, n \in \mathbb{Z}$. Then W_a contains the subgroup R generated by the translations $r_\alpha : \lambda \mapsto \lambda + \alpha : \alpha \in \Delta$ on which W acts by conjugation. Clearly W_a is an image of the semidirect product $R \rtimes W$. Moreover by A.1.13, R is generated by the $r_\alpha : \alpha \in \Delta_s$ and hence W_a is generated by $s_0 := s_{\alpha_0,1}$ and the $s_i := s_{\alpha_i} : i = 1, 2, \dots, \ell$. One has [Bb1, Chap. V, Sect. 3.3] the

Proposition. *Take $\lambda \in \mathfrak{h}_{\mathbb{R}}^*$. Then* $\text{Stab}_{W_a} \lambda$ *is generated by the $s_{\alpha,n} : \alpha \in \Delta, n \in \mathbb{Z}$ it contains.*

A proof is briefly indicated. For each $\lambda \in \mathfrak{h}_{\mathbb{R}}^* \setminus \{0\}$ set $\lambda^\perp = \{\mu \in \mathfrak{h}_{\mathbb{R}}^* | (\lambda, \mu) = 0\}$. Since $\mathfrak{h}_{\mathbb{R}}^* = \mathbb{R}P(\pi)$ it follows from A.1.12 that the set $D := \{\lambda \in \mathfrak{h}_{\mathbb{R}}^* | (\lambda, \alpha_i^\vee) \geq 0, \forall i\}$ is a fundamental domain for W. It is bounded by the hyperplanes $H_i := \alpha_i^\perp : i = 1, 2, \dots, \ell$ of reflection for the s_i. Set $D_a := \{\lambda \in D \mid (\lambda, \alpha_0^\vee) \leq 1\}$. Then D_a is bounded by the additional hyperplane $H_0 := \alpha_0^\perp + \frac{1}{2}\alpha_0$ of reflection for s_0. Let H denote any hyperplane obtained by translating the $H_i : i = 0, 1, \dots, \ell$ under W_a and let s_H be the reflection it defines. Let D_a^0 denote the interior of D_a.

Take $\lambda \in \mathfrak{h}_{\mathbb{R}}^*$. Choose $\mu \in D_a^0$ and a ball B of centre μ such that $B \cap W_a \lambda \neq \emptyset$. This intersection is finite since $|W| < \infty$ and $B \cap RF$ is finite for any finite set F. Thus there exists $\nu \in W_a \lambda$ such that $\|\mu - \nu\|$ is minimal. Take $H = H_i : i = 0, 1, 2, \dots, \ell$. If $\nu \notin H$, then the inequality $\|\nu - \mu\|^2 \leq \|s_H \nu - \mu\|^2$ shows that ν and μ lie on the same side of H. Consequently $\nu \in D_a$ and so $\mathfrak{h}_{\mathbb{R}}^* = W_a D_a$.

It remains to prove the assertion for $\lambda \in D_a$. For any $i = 0, 1, 2, \dots, \ell$ let P_i denote the set of all $w \in W_a$ such that D_a^0 and $w(D_a^0)$ lie on the same side of H_i. Clearly $e \in P_i$ and $s_i P_i \cap P_i = \emptyset$. Take $w \in P_i$ and suppose $w s_j \notin P_i$. Then D_a^0 and $s_j D_a^0$ are on opposite sides of $w^{-1} H_i$ and share H_j as a common boundary. This forces $w^{-1} H_i = H_j$ and hence $w s_j = s_i w$.

Take $w \in W$ and define its reduced length $\ell(w)$ with respect to the $s_i : i = 0,1,2,\ldots,\ell$ as in A.1.1. Take a reduced decomposition $w = s_1 s_2 \ldots s_r$ and set $w_j = s_1 s_2 \ldots s_j$. Assume $w \notin P_i$. Then by the above there exists $j : 1 \leq j \leq r$ such that $w_{j-1} \in P_i$; but $w_j = w_{j-1}s_j \notin P_i$. Hence $s_i w_{j-1} = w_{j-1}s_j$. Consequently $\ell(s_i w) < \ell(w)$. Replacing w by $s_i w$ proves that

$$(*) \qquad w \notin P_i \Longleftrightarrow \ell(s_i w) < \ell(w) .$$

Now suppose $\lambda, \mu \in D_a$ and that there exists $w \in W_a$ such that $w\lambda = \mu$. It will be shown by induction on $\ell(w)$ that $\lambda = \mu$ and that w is a product of the s_j stabilizing λ. Take s_i such that $\ell(s_i w) < \ell(w)$ Then $(*)$ implies that D_a^0 and $w(D_a^0)$ lie on opposite sides of H_i. Hence $D_a \cap wD_a \subset H_i$, so $\mu \in H_i$. Then $\mu = s_i w\lambda$ and the result is obtained by the induction hypothesis. This proves the proposition and that D_a is a fundamental domain for W_a. Furthermore $W_a \xleftarrow{\sim} R \rtimes W$.

Remarks. A fortiori the conclusion holds in $\mathfrak{h}^*_{\mathbb{Q}}$. Set $S_a = \{s_i : i = 0,1,\ldots,\ell\}$. By $(*)$ it follows as in A.1.9 that (W_a, S_a) is a Coxeter group. In particular $W_a \cong W(\tilde{C})$ where $\tilde{C}$ obtained from C by adding the column $\begin{pmatrix} -\alpha_0 \\ \alpha_i \end{pmatrix}$ and row $(-\alpha_0, \alpha_i)$.

A.1.15 Assume C of finite type and set $\check{T} = \tau(P(\pi))$. Define an action of W_a on $\check{T}$ through τ and transport this to $\check{T}^\wedge$ by 1.4.8. For each $\Lambda \in \check{T}^\wedge$ set $\Delta_\Lambda = \{\alpha \in \Delta \mid \Lambda(\tau(\alpha)) \in q^{\mathbb{Z}}\}$, $W_\Lambda = \{w \in W \mid w\Lambda \in \Lambda q^{\mathbb{Q}(\pi)}\}$.

Lemma. *For all $\alpha \in \Delta$, $n \in \mathbb{Z}$ one has*

$$(*) \qquad s_\alpha \Lambda = \Lambda q^{-n\alpha} \Longleftrightarrow \Lambda(\tau(\alpha)) = q_\alpha^n$$

where $q_\alpha = q^{(\alpha,\alpha)/2}$. In particular $s_\alpha \in W_\Lambda$ for all $\alpha \in \Delta_\Lambda$.

For all $\nu \in P(\pi)$, the right hand side of $(*)$ gives $(s_\alpha\Lambda)(\tau(\nu)) = \Lambda(\tau(\nu))\Lambda(\tau(\alpha))^{-(\alpha^\vee,\nu)} = \Lambda(\tau(\nu))q^{-n(\alpha,\nu)} = (\Lambda q^{-n\alpha})(\tau(\nu))$ and hence the left hand side. Conversely the left hand side gives $\Lambda(\tau(\nu))\Lambda(\tau(\alpha))^{-(\alpha^\vee,\nu)} = (s_\alpha\Lambda)(\tau(\nu)) = \Lambda(\tau(\nu)){q_\alpha}^{-n(\alpha^\vee,\nu)}$. Yet by A.1.3 (ii) each $\alpha \in \Delta$ can be conjugated into π and so there exists $\nu \in P(\pi)$ such that $(\alpha^\vee, \nu) = 1$. Substitution in the above gives the right hand side.

A.1.16 It is *false* in general that W_Λ is generated by the $s_\alpha : \alpha \in \Delta_\Lambda$. Indeed take C as in the Remark 7.1.17, that is of type A_2 and let γ be a primitive cube root of unity. Define $\Lambda \in T^\wedge$ by $\Lambda(t_i) = \gamma : i = 1,2$. Then $\Delta_\Lambda = \emptyset$; yet $W_\Lambda = \{1, s_1s_2, s_2s_1\}$. For the general case set $\Lambda_i = \Lambda(\tau(\varpi_i))$ and let Γ to be the subgroup of $k(q)^*$ they generate. Since Γ is a finitely generated abelian group it can be written as the direct product $\Gamma_0 \times F$ of its torsion subgroup Γ_0 which is finite and a free finite rank abelian group F. It is convenient to express the generators of F in the form q^{e_i} where the e_i, including $e_0 = 1$ if necessary, form a basis of a $\mathbb{Q}$ vector space $\mathbb{V}$. Then Λ can be written as $e^{2\pi i\nu}q^\lambda$ where $\lambda = \sum_{i=0}^m \lambda_i e_i : \lambda_i \in Q(\pi)$ and $\nu \in \frac{1}{n}Q(\pi^\vee)$ for some $n \in \mathbb{N}^+$. Call Λ torsion-free if $\nu = 0$.

Let $p : W_a \longrightarrow W$ be the projection defined by the isomorphism $W_a \overset{\sim}{\longleftarrow} R \rtimes W$. It is immediate that

(1) $W_\Lambda = p(\text{Stab}_{W_a} \Lambda)$

whilst

(2) $\text{Stab}_{W_a} \Lambda = \text{Stab}_W \, e^{2\pi i\nu} \cap \text{Stab}_{W_a} \lambda$

and

(3) $\text{Stab}_{W_a} \lambda = \text{Stab}_{W_a} \lambda_0 \cap \bigcap_{i=1}^m \text{Stab}_W \lambda_i$.

From A.1.14 one has the

Corollary

(i) $\text{Stab}_{W_a} \lambda$ *is generated by the* $s_{\alpha,n} : \alpha \in \Delta,\ n \in \mathbb{Z}$ *it contains.*

(ii) *If* Λ *is torsion-free, then* W_Λ *is generated by the* $s_\alpha : \alpha \in \Delta_\Lambda$.

It follows from A.1.1 (vii) and A.1.12 (ii) by induction on m that $\bigcap_{i=1}^m \text{Stab}_W \lambda_i$ is generated by the $s_\alpha : \alpha \in \pi_0$ for some subset π_0 of a W translate of π. Then (2) follows from (3) and A.1.14 with π replaced by π_0. Alternatively embed $\mathbb{V}$ in $\mathbb{R}$ viewed as a $\mathbb{Q}$ vector space. Finally (ii) follows from (i) and (1), (2).

A.1.17 For all $\lambda \in P^+(\pi)$ let $\Omega(V(\lambda))$ denote the set of weights of the simple highest weight module $V(\lambda)$.

Lemma. *Take* $\lambda_1, \lambda_2 \in P^+(\pi)$. *Then* $(\mu_1, \mu_2) \leq (\lambda_1, \lambda_2)$ *for all* $\mu_i \in V(\lambda_i) : i = 1, 2$. *Moreover if* λ_2 *is regular equality implies* $\mu_1 = w\lambda_1$ *for some* $w \in W$. *Then if* λ_1 *is also regular* $\mu_2 = w\lambda_2$.

By A.1.11 one may assume $\mu_1 \in P^+(\pi)$. Since $\mu_i \in V(\lambda_i)$ one may write $\mu_i = \lambda_i - \gamma_i$ for some $\gamma_i \in \mathbb{N}\pi$. Then $(\mu_1, \mu_2) = (\mu_1, \lambda_2 - \gamma_2) \leq (\mu_1, \lambda_2) = ((\lambda_1 - \gamma_1), \lambda_2) \leq (\lambda_1, \lambda_2)$. Equality forces $(\gamma_1, \lambda_2) = 0$ and $(\mu_1, \gamma_2) = 0$. Hence the assertions.

A.1.18 Assume C of finite type. For each $w \in W$ set $P_w(\pi) = \{\mu \in P(\pi) | w\mu = \mu\}$ which is a free subgroup of $P(\pi)$. Set $s(w) = \ell - rk\ P_w(\pi)$.

Lemma. *There exist* $s := s(w)$ *linearly independent positive roots* $\gamma_1, \gamma_2, \ldots, \gamma_s$ *such that* $w = s_{\gamma_1} s_{\gamma_2} \ldots s_{\gamma_s}$. *Moreover* w *cannot be written as product of fewer reflections.*

The proof is by induction on $|\pi|$, it being trivial for $|\pi| = 0$. Set $\mathfrak{h}_w^* = \{\lambda - w\lambda | \lambda \in \mathfrak{h}_\mathbb{Q}^*\}$. Then $\mathfrak{h}_\mathbb{Q}^*$ is an orthogonal direct sum of $\mathfrak{h}_w^*$ and $\mathbb{Q}P_w(\pi)$. In particular $s(w) = \dim \mathfrak{h}_w^*$. If $s(w) < \ell$ one can choose $\nu \in P_w(\pi)$ non-zero. By A.1.12 there exists $y \in W$ such that $y\nu \in P^+(\pi)$. Set $\pi' = \{\alpha \in \pi | s_\alpha y\nu = y\nu\}$ which is a strict subset of π. By A.1.1 (vii), ywy^{-1} belongs to the subgroup of W generated by the $s_\alpha : \alpha \in \pi'$. Thus $\mathfrak{h}_w^*$ can be assumed equal to $\mathfrak{h}_\mathbb{Q}^*$ by the induction hypothesis. In this case for any (positive) root γ there exists $\lambda \in \mathfrak{h}_\mathbb{Q}^*$ such that $\lambda - w\lambda = \gamma$, which forces $(\lambda, \gamma^\vee) = 1$ and so $s_\gamma w\lambda = s_\gamma(\lambda - \gamma) = \lambda$. Then as before the induction hypothesis applies and $s_\gamma w$ can be assumed to be a product $s_{\gamma_2}s_{\gamma_2}\ldots s_{\gamma_t}$ of linearly independent reflections satisfying $(\gamma_i, \lambda) = 0$. Set $\gamma = \gamma_1$. Since $(\gamma_1, \lambda) \neq 0$ and $\mathfrak{h}_w^*$ is contained in the linear span of the γ_i, the assertion results.

A.1.19 Set $I = \{1, 2, \ldots, \ell\}$. For each subset $J \subset I$ let W_J (resp. W^J) denote the subgroup (resp. subset) of W generated by the $s_i : i \in J$ (resp. of $w \in W$ satisfying $w\alpha_i > 0$ for all $i \in J$).

Lemma

(i) *For all $w_1, w_2 \in W$ one has $\ell(w_1w_2) = \ell(w_1) + \ell(w_2) \iff S(w_1) \cap S(w_2^{-1}) = \emptyset \iff w_1w_2 \overset{D}{\geq} w_2$.*

(ii) *Each $w \in W$ can be written uniquely in the form $w = w_1w_2$ with $w_1 \in W^J$, $w_2 \in W_J$ and then $\ell(w) = \ell(w_1) + \ell(w_2)$.*

(i) Suppose $S(w_1) \cap S(w_2^{-1}) = \emptyset$. One checks from A.1.1(i) that $S(w_1w_2) \supset w_2^{-1}S(w_1) \coprod S(w_2)$. In particular equality holds and $w_1w_2 \overset{D}{\geq} w_2$. The converses are similar.

(ii) By A.1.1(i), (iv) one has $\ell(ws_i) < \ell(w) \iff w\alpha_i < 0$ and so existence of the pair w_1, w_2 follows by induction on $\ell(w)$. Moreover the hypothesis of (i) is satisfied and so the last part is obtained. Finally if $w = w_1'w_2'$ is a second such presentation then applying (i) to the identity $w_1 = w_1'(w_2'w_2^{-1})$ gives $\ell(w_1) = \ell(w_1') + \ell(w_2'w_2^{-1})$ and conversely $\ell(w_1') = \ell(w_1) + \ell(w_2w_2'^{-1})$. This forces $\ell(w_2'w_2^{-1}) = 0$ so $w_2' = w_2$ as required.

A.1.20 For each $w \in W$ set $I(w) = I \cap S(w)$. By A.1.4 the set of positive roots generated by $J := I(w)$ is finite and hence so is W_J. Recalling A.1.4 let w_J denote the unique longest element of W_J. Then $ww_J \in W^J$. Define

$$\rho_w = \sum_{i \in I(w^{-1})} \varpi_i \,, \quad \delta_w = w^{-1}\rho_w \,, \quad \delta_w^* = -w^{-1}(\rho - \rho_w) \,.$$

Lemma. *For all $x, y, z \in W$ one has*

(i) *If $\ell(x) \leq \ell(y)$ then $\rho_y \geq x\delta_y$ with equality if and only if $x = y$.*

(ii) *If $\delta_x - \delta_y^* = z^{-1}\rho$ and $z\delta_x = \rho_x$, then $x \overset{D}{\leq} z \overset{D}{\leq} y$.*

(iii) *Take $w_1, w_2, \ldots, w_{n+1} = w_1 \in W$. Then the $\delta_{w_{i+1}} - \delta_{w_i}^* : i = 1, 2, \ldots, n$ cannot all be regular unless $w_1 = w_2 = \ldots = w_n$.*

(i) One has $x\delta_y = xy^{-1}\rho_y \le \rho_y$ by 4.3.6. Equality implies $xy^{-1} \in W_J$, where $J = \{i \mid (\alpha_i, \alpha_j) = 0, \forall j \in I(y^{-1})\}$, by A.1.1(vii). In particular $yx^{-1} \in W_{I\setminus I(y^{-1})}$ whilst $y^{-1} \in W^{I\setminus I(y^{-1})}$ by definition. Consequently $\ell(x^{-1}) \ge \ell(y^{-1}) + \ell(yx^{-1})$ by A.1.19(ii). Recalling (A.1.1) that $\ell(x) = \ell(x^{-1})$ the conclusion of (i) obtains from the hypothesis.

(ii) The hypothesis gives $zy^{-1}(\rho - \rho_y) = \rho - \rho_x$. Then by A.1.12(i) one has $\rho - \rho_y = \rho - \rho_x$, so $I(y^{-1}) = I(x^{-1}) =: J$ and $zx^{-1} \in W_{I\setminus J}$, $zy^{-1} \in W_J$ by A.1.1(vii). Now $y^{-1}w_J \in W^J$. Yet $z^{-1} = (y^{-1}w_J)(w_J yz^{-1})$ so A.1.19(ii) gives $\ell(z^{-1}) = \ell(y^{-1}w_J) + \ell(w_J yz^{-1}) = \ell(y^{-1}) - \ell(w_J) + \ell(w_J) - \ell(yz^{-1})$. Then $\ell(z) + \ell(yz^{-1}) = \ell(y)$ so $y \overset{D}{\geq} z$ by A.1.19(i). Again $x^{-1} \in W^{I\setminus J}$ by definition, so $\ell(z^{-1}) = \ell(x^{-1}) + \ell(xy^{-1})$ by A.1.19(ii) and then $z \overset{D}{\geq} x$ by A.1.19(i).

(iii) Set $w_{i+1} = x, w_i = y$. By A.1.11 there exists $z \in W$ and $\mu \in P^+(\pi)$ such that $\delta_x - \delta_y^* = x^{-1}\rho_x + y^{-1}(\rho - \rho_y) = z^{-1}\mu$ which being regular means that $\mu = \nu + \rho$ with $\nu \in P^+(\pi)$. Then

$$(\rho, \rho) \le (\rho, \mu) = (\rho, zx^{-1}\rho_x) + (\rho, zy^{-1}(\rho - \rho_y)) \le (\rho, \rho_x) + (\rho, \rho) - (\rho, \rho_y) ,$$

that is $(\rho, \rho_x) \ge (\rho, \rho_y)$. Thus regularity for all i forces equality throughout which gives in particular $\mu = \rho$ and $\rho_x = zx^{-1}\rho_x = z\delta_x$. Then $x \overset{D}{\leq} y$ by (ii). In view of A.1.1(vi) this forces the required conclusion.

A.1.21 Assume $|W| < \infty$. View $\tau(P(\pi))$ as the multiplicative group $e^{P(\pi)} := \{e^\lambda\}_{\lambda \in P(\pi)}$. Its group algebra is a Laurent polynomial algebra, hence a unique factorization domain. Consider the expression

$$D(\lambda) := \sum_{w \in W} (-1)^{\ell(w)} e^{w(\lambda+\rho)-\rho} : \lambda \in P(\pi) .$$

Take $\alpha \in \Delta^+$. Since $s_\alpha . D(\lambda) = -D(\lambda)$, it is divisable by $(1 - e^{-\alpha})$ and hence by

$$D := \prod_{\alpha \in \Delta^+} (1 - e^{-\alpha}) .$$

A similar argument shows that each term $\sum_{w \in W} \frac{(-1)^{\ell(w)}}{n!} w^{-1}\rho^n$ in the expansion of $D(\rho)$ is divisable by $\prod_{\alpha \in \Delta^+} \alpha$. Consequently this sum vanishes for $n < |\Delta^+|$ whilst by 4.2.16 it equals $\prod_{\alpha \in \Delta^+} \alpha$ when $n = |\Delta^+|$. It follows from 4.2.15 that for $\lambda \in P^+(\pi)$ one obtains the Weyl dimension formula

$$\mathcal{D}(\lambda + \rho) := \dim \mathcal{V}(\lambda) = \prod_{\alpha \in \Delta^+} \frac{(\lambda + \rho, \alpha^\vee)}{(\rho, \alpha^\vee)} .$$

Consequently $\mathcal{D}(\lambda)$ takes integer values for all $\lambda \in P(\pi)$ and is a W-harmonic polynomial, that is it satisfies

$$(*) \qquad \sum_{w \in W} \mathcal{D}(w\lambda + \mu) = |W| \mathcal{D}(\mu) , \ \forall \lambda, \mu \in P(\pi) .$$

Indeed taking $n = |\Delta^+|$ and up to a fixed scalar the left hand side equals

$$\sum_{w,y\in W}(-1)^{\ell(y)}\rho^n(yw\lambda+y\mu)=\sum_{w,y\in W}(-1)^{\ell(y)}\rho^n(w\lambda+y\mu)\,,$$
$$=\sum_{w,y\in W}(-1)^{\ell(y)}\rho^n(y\mu)\,,\quad\text{by the above vanishing}\,,$$
$$=|W|\mathcal{D}(\mu)\,.$$

A.1.22 Set $D_0=\prod_{\alpha\in\Delta^+}(e^{\alpha/2}-e^{-\alpha/2})=e^\rho D$ which belongs to $\tau(P(\pi))$ identified with $e^{P(\pi)}$. For each $y\in W$ define the polynomial $\mathcal{D}_y$ on $\mathfrak{h}^*_{\mathbb{Q}}$ through

$$\mathcal{D}_y(\lambda)=\mathcal{D}(\lambda-\delta^*_y)\,,\quad\forall\lambda\in\mathfrak{h}^*_{\mathbb{Q}}\,.$$

Theorem
(i) $\det(e^{x\delta_y})_{x,y\in W}=D_0^{\frac{1}{2}|W|}$.
(ii) $\det(\mathcal{D}_y(\delta_x))_{x,y\in W}=1$ *(up to a sign).*

The first part of the proof of (i) shows that the right hand side divides the left. It is valid for arbitrary $\delta_y\in P(\pi)$. Subtracting the rows $x\delta_y$, $s_\alpha x\delta_y$ shows (as in A.1.21) that $(1-e^{-\alpha})^{\frac{1}{2}|W|}$ divides $\det(e^{x\delta_y})$ and then by unique factorization so does $D_0^{\frac{1}{2}|W|}$. Ordering rows and columns by the length function of W and using A.1.20(i) shows that the highest degree term of the left hand side (seting $d=\frac{1}{2}|W|$) is

$$\prod_{w\in W}e^{\rho_w}=e^{d\rho}=(\prod_{\alpha\in\Delta^+}e^{\alpha/2})^d\,.$$

Invariance (up to a sign) of the left hand side under w_0 gives a similar expression for the lowest order term. Together with the first observation, this completes the proof of (i).

For (ii) first observe that $\mathcal{D}(\lambda)=0$ unless λ is regular. Then by A.1.20 (iii) any contribution to $\det\mathcal{D}_y(\delta_x)$ not coming from a product of diagonal terms is zero. Yet $\delta_x-\delta^*_x=x^{-1}\rho_x+x^{-1}(\rho-\rho_x)=x^{-1}\rho$. Now $\mathcal{D}(x^{-1}\rho)=(-1)^{\ell(x)}\mathcal{D}(\rho)=(-1)^{\ell(x)}$ through say the corresponding symmetry of the Weyl character formula. Hence (ii).

A.1.23 Let H (resp. $H_{\mathbb{Z}}$) denote the $\mathbb{Z}$ module of W harmonic polynomials on $\mathfrak{h}^*_{\mathbb{Q}}$ (resp. and taking integer values on $P(\pi)$).

Corollary
(i) *The* $e^{\delta_w}:w\in W$ *form a free basis of the* $(\mathbb{Z}e^{P(\pi)})^W$ *module* $\mathbb{Z}e^{P(\pi)}$.
(ii) *The* $\mathcal{D}_w:w\in W$ *form a free basis of the* $\mathbb{Z}$ *module* $H_{\mathbb{Z}}$.

(i) Take $e^\delta\in\mathbb{Z}e^{P(\pi)}$. The solution of the linear equations $\sum_{y\in W}z_y(e^{x\delta_y})=e^{x\delta}$ for the z_y is given uniquely as the determinant in which δ replaces δ_y in $\det(e^{x\delta_y})$ divided by the latter. By A.1.22(i) and the remark in the proof the latter belongs to $\mathbb{Z}e^{P(\pi)}$ and is W-invariant because both factors transform like the sign representation. Hence (i). Finally it is clear from (*) that $\dim H \leq |W|$ and so (ii) results from A.1.22(ii).

A.2 Excerpts from Ring Theory

Some basic results in ring theory needed in the text are reviewed. A few are quite recent. In what follows A denotes a ring and left can be replaced by right throughout.

A.2.1 Let M be an A module. For each subset V (resp. B) of M (resp. A) define

$$\operatorname{Ann}_A V = \{a \in A \mid aV = 0\} \quad (\text{resp. } \operatorname{Ann}_M B = \{m \in M | Bm = 0\}) .$$

Obviously $V \subset \operatorname{Ann}_M(\operatorname{Ann}_A V)$ and $B \subset \operatorname{Ann}_A(\operatorname{Ann}_M B)$. If M is a simple module, then $K := \operatorname{End}_A M$ is a skew field by Schur's lemma. An induction argument on dimension proves the

Lemma. *Let M be a simple A module and V a finite dimensional K-subspace of M. Then $V = \operatorname{Ann}_M(\operatorname{Ann}_A V)$.*

A.2.2 Let M be a simple A module, K and V as above. If $v \notin V$, then $(\operatorname{Ann}_A V)v = M$ by the lemma. Consequently for any $w \in M$ there exists $a \in A$ such that $aV = 0$ and $av = w$. This yields the Jacobson density theorem. If M is faithful and finite dimensional over K the latter implies the Artin-Wedderburn theorem which can be expressed as

Theorem. *Assume A admits a simple faithful module M finite dimensional over $K := \operatorname{End}_A M$. Then $A \xrightarrow{\sim} \operatorname{End}_K M$.*

A.2.3 A multiplicatively closed subset $S \subset A$ is said to be left Ore in A if for all $(s, a) \in S \times A$ there exists $(t, b) \in S \times A$ such that $ta = bs$. Let M be a left A module. Then the Ore condition implies that $T_S(M) := \{m \in M | sm = 0 \textit{ for some } s \in S\}$ is a submodule called the S-torsion submodule of M. Then M is said to be S-torsion (resp. S-torsion-free) if $T_S(M) = M$ (resp. $T_S(M) = 0$). Observe that $M/T_S(M)$ is S-torsion free.

By the above $T_S(A)$ is a two-sided ideal of A and so the quotient ring $\bar{A} = A/T_S(A)$ is defined. In $\bar{A}$ the image $\bar{S}$ of S is left Ore and further its elements are non-zero divisors on the left. Replacing A by $\bar{A}$ one can assume $T_S(A) = 0$.

To get the elements of S to be non-zero divisors on the right requires an additional hypothesis. For this define a left ideal of A having the form $\ell_A(T) = \{a \in A | at = 0, \forall t \in T\}$ for some subset T of A, to be a left annihilator. Then A is said to satisfy the ascending chain condition (acc) on left annihilators if every ascending chain of left annihilators is stationary. This is of course satisfied by a left noetherian ring.

Lemma. *Let A be a ring satisfying acc on left annihilators and S an Ore subset of A. If $T_S(A) = 0$, then the elements of S are non-zero divisors.*

Fix $s \in S$. For each integer $i > 0$ set $L_i = \{a \in L | as^i = 0\}$. If $L_1 \neq 0$ there exists $j \in \mathbb{N}^+$ such that $L_{j-1} \subsetneq L_j = L_{j+1}$ by acc. Choose $a \in L_j \setminus L_{j-1}$ and $b \in A$, $t \in S$ such that $ta = bs$. Then $0 = tas^j = bs^{j+1}$ so $b \in L_{j+1} = L_j$. Consequently $0 = bs^j = tas^{j-1}$ and so $as^{j-1} = 0$ since $T_S(A) = 0$. This contradiction gives $L_1 = 0$ as required.

A.2.4 Let S be an Ore subset of A consisting of non-zero divisors. Then one may construct an overring $S^{-1}A$ in which the elements of S are rendered invertible and whose inverses generate $S^{-1}A$ over A. For this set $(s, a) \sim (t, b)$ if there exist $u, v \in A$ such that $us = vt \in S$ and $ua = vb$. One checks that this defines an equivalence relation on $S \times A$ and one writes $S^{-1}A = S \times A / \sim$. Fix pairs $(s, a), (t, b) \in S \times A$. By the left Ore condition there exists $(u, v) \in S \times A$ such that $w := vs = ut \in S$. Then $(s, a) \sim (vs, va) = (w, va)$ and $(t, b) \sim (ut, ub) = (w, ub)$. One defines

$$\text{(sum)} \qquad (s, a) + (t, b) = (w, va + ub)\ .$$

Again there exists $(u, c) \in S \times A$ such that $ua = ct$ and one defines

$$\text{(product)} \qquad (s, a)(t, b) = (us, cb)\ .$$

One checks that these two relations give $S^{-1}A$ a ring structure. Furthermore $(s, 1) : s \in S$ is inverse to $(1, s)$ and one writes simply $(s, 1) = s^{-1}$ and $(1, a) = a$.

A.2.5 Recall A.2.3. Let A be a ring which is either commutative or left noetherian and S a left Ore subset. Define $S^{-1}A$ to be $\bar{S}^{-1}\bar{A}$. The canonical map $\bar{a} \mapsto (1, \bar{a})$ of $\bar{A}$ into $\bar{S}^{-1}\bar{A}$ defines by composition with the projection $A \longrightarrow A/T_S(A)$ a canonical map φ of A into $S^{-1}A$ with kernel $T_S(A)$. Then $S^{-1}A$ may be given a right A module structure through $(x, a) \mapsto x\varphi(a)$.

In the above framework one may define a functor F_S from the category of left A modules to the category of left $S^{-1}A$ modules through $F_S(M) := S^{-1}A \otimes_A M$. One checks that F_S is exact and that the canonical map $\varphi : m \mapsto (1, m)$ has kernel $T_S(M)$. Moreover given a homomorphism $f : N \longrightarrow M$ of S modules then the diagram

$$\begin{array}{ccc} N & \xrightarrow{\ f\ } & M \\ \downarrow{\scriptstyle \varphi} & & \downarrow{\scriptstyle \varphi} \\ F_S(N) & \xrightarrow{\ F_S(f)\ } & F_S(M) \end{array}$$

is commutative.

A.2.6 A prime ideal P of A is a two-sided ideal such that if I, J are two-sided ideals satisfying $IJ \subset P$ then either $I \subset P$ or $J \subset P$. An ideal P is said to be completely prime if A/P is a domain. This is a stronger condition. A (left) primitive ideal of A is the annihilator of a (left) simple A module. One checks that the set $\operatorname{Prim} A$ of (left) primitive ideals of A is contained in the set $\operatorname{Spec} A$ of prime ideals of A. If H is a Hopf algebra and A an H-algebra then $(\operatorname{Spec} A)^H := \{P \in \operatorname{Spec} A \mid H(P) \subset P\}$ and $(\operatorname{Prim} A)^H := \operatorname{Prim} A \cap (\operatorname{Spec} A)^H$.

One may endow $\operatorname{Spec} A$ with a topology (the Jacobson topology) which generalizes the Zariski topology of algebraic geometry, namely for each subset $\mathcal{I} \supset \operatorname{Spec} A$ define its closure $\bar{\mathcal{I}}$ to be the set of all prime ideals containing the intersection $\bigcap_{P \in \mathcal{I}} P$. Then for any two rings A and B, an order preserving map φ taking the set of ideals (resp. prime ideals) of A into the set of ideals (resp. prime ideals) of B is continuous. For example if S is Ore in A then one may take $B = S^{-1}A$ and $\varphi(I) = S^{-1}I$. Again if $\psi : B \longrightarrow A$ is a surjection of rings, then one may take φ to be ψ^{-1}. On the other hand for a ring embedding $A \hookrightarrow B$ intersection may fail to preserve prime ideals.

A.2.7 To an ideal I of A one may associate the set $\mathcal{P}(I)$ of minimal primes over I. If A is (left) noetherian then $\mathcal{P}(I)$ is always finite and moreover the intersection $\bigcap_{P \in \mathcal{P}(I)} P$ is a nilpotent ideal (see [Di2, 3.1.10] for example). An ideal is said to be semiprime if it is the intersection of the minimal primes over it. A ring A is said to be prime (resp. semiprime) if 0 is a prime (resp. semiprime) ideal.

A ring is said to be left Goldie if

1) *A satisfies acc on left annihilators.*
2) *A does not admit infinite direct sums of left ideals.*

For example any left noetherian ring is left Goldie.

A left ideal L of A is said to be essential if L has non-zero intersection with any non-zero left ideal of A. For example any non-zero two-sided of a prime ring has this property. A key observation is that if s is a non-zero divisor in A and L a non-zero left ideal satisfying $L \cap As = 0$ then $L + Ls + Ls^2 + \ldots$, is an infinite direct sum of left ideals. In other words As is an essential left ideal in a left Goldie ring. The converse also holds but this is more difficult. It leads to the following quite remarkable result of A.W. Goldie. A ring A is said to admit a (classical) left ring of quotients $S^{-1}A$ if the set of non-zero divisors S of A is Ore.

Theorem. [MR1, 2.3.6]. *The following are equivalent,*

(i) *A is semiprime (resp. prime) left Goldie.*
(ii) *A admits a left ring of quotients $S^{-1}A$ which is semisimple (resp. simple) artinian.*

A.2.8 Let A be a prime left Goldie ring and S the set of non-zero divisors of A. Then by A.2.2 and A.2.7 the ring of quotients $S^{-1}A$ is a matrix ring over a skew-field K say $M_n(K)$. One calls n the Goldie rank $rk\ A$ of A and K the Goldie skew-field of A. A difficulty with Goldie rings is that their defining properties do not pass to quotients. In most applications of A.2.7 one should know that A is left noetherian and then 2.2.7 applies to A/P for each $P \in \operatorname{Spec} A$. For example one has the following

Proposition. *Let A be a left noetherian ring and S a left Ore subset of A. Then the map $P \mapsto S^{-1}P$ is an order isomorphism (with inverse $Q \mapsto A \cap Q$) of $\{P \in \operatorname{Spec} A | P \cap S = \emptyset\}$ onto $\operatorname{Spec} S^{-1}A$.*

Take $P \in \operatorname{Spec} A$. Then $S^{-1}P \in \operatorname{Spec} S^{-1}A$. If $T_S(R/P) \neq 0$, then it contains a regular element of A/P which forces $P \cap S \neq 0$. Otherwise $A/P \hookrightarrow S^{-1}(A/P) \xrightarrow{\sim} S^{-1}A/S^{-1}P$ by exactness of localization. Hence $P = A \cap S^{-1}P$. Take $Q \in \operatorname{Spec} S^{-1}A$. To show that $A \cap Q \in \operatorname{Spec} A$ one reduces to the case $Q = 0$ by exactness of localization. Then the assertion follows from (ii) $\Longrightarrow$ (i) of A.2.7.

A.2.9 Let A be a k-algebra. Recall the definitions of 1.2.11. Take $\theta \in \operatorname{Aut} A$ and let ∂ be a locally nilpotent θ-leftskew (or rightskew) derivation of A such that $\theta\partial\theta^{-1}$ is a multiple of ∂. If $0 \neq a \in A$ let $\deg_\partial a$ denote the least integer $r \geq 0$ such that $\partial^{r+1}a = 0$. If $a = 0$ set $\deg_\partial a = -\infty$. Let $a, b \in A$ be θ-eigenvectors. Then ab is a θ-eigenvector and one checks that

$$(*) \qquad \deg_\partial(ab) \leq \deg_\partial a + \deg_\partial b\ .$$

More generally setting $r = \deg_\partial a$, $s = \deg_\partial b$ one has

$$\partial^t(ab) = \sum_{i=0}^{r+s-t} c_{t,i}(\partial^{r-i}a)(\partial^{t-r+i}b)$$

for some $c_{t,i} \in k$. Call ∂ left (resp. right) regular if $c_{t,0} \neq 0$ for all $t \in \{r, r+1, \ldots, r+s\}$ (resp. $c_{t,r+s-t} \neq 0$ for all $t \in \{s, s+1, \ldots, r+s\}$). It is immediate that either condition implies equality in $(*)$. Both conditions hold if $\theta = 1d$, $\operatorname{char} k = 0$ or for the skew derivations described in 3.1 used in the construction of $U_q(\mathfrak{g})$ assuming that $\operatorname{char} k = 0$ and q is not a root of unity.

Let $e \in A$ be a θ-eigenvector such that $\{e^n\}_{n \in \mathbb{N}}$ is Ore in A. Set $r = \deg_\partial e$, $f = \partial^r e$. Assume that A is a domain and a direct sum of its θ-eigenspaces.

Lemma. *If ∂ is left (resp. right) regular, then $\{f^m\}_{m \in \mathbb{N}}$ is left (resp. right) Ore in A.*

Take $a \in A$. It is enough to show that there exists $b \in A$, $m \in \mathbb{N}^+$ such that $f^m a = bf$. Here a may be assumed to be a non-zero θ-eigenvector and one sets $s = \deg_\partial a$. By hypothesis there exists $b_1 \in A$, $m_1 \in \mathbb{N}^+$ such that $e^{m_1}a = b_1 e$. From this relation b_1 can be assumed to be a non-zero θ-eigenvector and by equality in $(*)$ to satisfy $\deg_\partial b_1 = r(m_1-1)+s$. Applying $\partial^{m_1 r}$ to this relation gives $c_i, c_i' \in k$ such that

$$c_0 f^{m_1} a - c_0'(\partial^{r(m_1-1)} b_1) f + a' = 0 \tag{1}$$

where

$$a' = \sum_{i=1}^{s} c_i(\partial^{rm_1-i}e)(\partial^i a) - c_i'(\partial^{r(m_1-1)+i} b_1)(\partial^{r-i}e) \, .$$

One checks that

$$\partial^s a' = c f^{m_1}(\partial^s a) - c'(\partial^{r(m_1-1)+s} b_1) f \tag{2}$$

for some $c, c' \in k$. Now apply ∂^s to (1). This shows the terms on the right hand side of (2) to be proportional and furthermore to be multiples of $\partial^s([\partial^{r(m_1-1)} b_1] f)$. Consequently (1) can be rewritten in the form

$$c_0 f^{m_1} a - c_0''(\partial^{r(m_1-1)} b_1) f + a'' = 0 \tag{3}$$

for some θ-eigenvector a'' satisfying $\deg_\partial a'' < s$. Moreover by the left regularity of ∂ one has $c_0 \neq 0$.

If $s = 0$ then $a' = 0$ in (1) and the required assertion follows. Otherwise proceed by induction on $\deg_\partial a$. Applied to a'' the induction argument gives $m_2 \in \mathbb{N}^+$ such that $f^{m_2} a'' \in Af$. Then (3) gives $c_0 f^{m_1+m_2} a \in Af$ as required. Similar reasoning gives the assertion on the right.

A.2.10 Let A be a ring. The intersection of all its (left) primitive ideals $J(A)$ is called the Jacobson radical of A. One easily checks that $a \in J(A)$ if and only if $1 - a$ is invertible in A.

Lemma. *Let M be a finitely generated A module, satisfying $J(A)M = M$. Then $M = 0$.*

Assume $M \neq 0$. The hypothesis passes to quotients and so M may be replaced by a non-zero cyclic quotient Am. Then $m = am$ for some $a \in J(A)$. Inverting $(1 - a)$ gives the required contradiction.

A.2.11 Let A be a commutative ring. Given any $P \in \operatorname{Spec} A$ the subset $S := A \setminus P$ is multiplicatively closed and trivially Ore in A. One denotes $S^{-1}A$ by A_P. It is called the localization of A at P. If P is a maximal ideal then $S^{-1}P$ is the unique maximal ideal of A_P, that is to say A_P is a local ring and $J(A_P) = S^{-1}P$.

Corollary. *Let M be a finitely generated A module. If $\mathfrak{m}M = M$ for all $\mathfrak{m} \in \operatorname{Max} A$ then $M = 0$.*

Assume $M \neq 0$. Replacing A by $A/\operatorname{Ann}_A M$ one can assume M to be faithful. Choose any $\mathfrak{m} \in \operatorname{Max} A$ and set $S = A \setminus \mathfrak{m}$. If M is S-torsion then, since M is finitely generated and A is commutative, $S \cap \operatorname{Ann}\ M \neq \emptyset$ contradicting fidelity. Thus $S^{-1}M$ is a non-zero finitely generated module over the local ring $S^{-1}A$ satisfying the hypothesis of A.2.10. This contradiction proves the corollary.

A.2.12 The above result applies to the situation described in 9.2.5. Let A be a finitely generated commutative k-algebra and M a finitely generated A module admitting a k-subspace H for which $H \oplus IM = M$ is a direct sum for all $I \in \operatorname{Max} A$. Then the multiplication map $A \otimes H \longrightarrow M$ is injective. Its cokernel M/AH satisfies the hypothesis of A.2.11 and is hence zero. This gives an isomorphism $A \otimes H \xrightarrow{\sim} M$.

A.2.13 Let F denote the free k-algebra on generators $x_1, x_2, \ldots, x_n$. Give the set X of all monomials in $x_1, x_2, \ldots, x_n$ the lexicographic ordering $\leq$. This is a partial ordering compatible with left and right multiplication satisfying the descending chain condition dcc. Any order on X with these properties will do. Let S denote a subset of the pairs of the form $\sigma = (w_\sigma, f_\sigma)$ with $w_\sigma \in X$ and $f_\sigma \in F$ being a linear combination of monomials $< w_\sigma$. For all $\sigma \in S$ and $x, y \in X$ let $r_{x\sigma y}$ denote the linear map sending $xw_\sigma y$ to $xf_\sigma y$ and fixing all other monomials. Let R denote the semigroup generated by the $r_{x\sigma y} : \sigma \in S;\ x, y \in X$. Call $x \in X$ reduced if $r(x) = x$ for all $r \in R$. An ambiguity of S is a 5-tuple $\{\sigma, \tau;\ x, y, z\} \in S^2 \times X^3$ for which $w_\sigma = xy$, $w_\tau = yz$. Such an ambiguity is said to be resolvable if there exists $r, r' \in R$ such that $r(f_\sigma z) = r(xf_\tau)$. The Bergman diamond lemma [Be1] asserts.

Lemma. *If every ambiguity of S is resolvable then the reduced elements form a k-basis for the quotient algebra $F/ < w_\sigma - f_\sigma : \sigma \in S >$.*

A.2.14 The above result applies to the situation described in 3.1.3. View $\tilde{A}_C$ as a quotient of the free algebra F on generators ordered as $x_{\alpha_1}, x_{\alpha_2}, \ldots, x_{\alpha_\ell}, y_{-\alpha_1}, \ldots, y_{-\alpha_\ell}$. Let S denote the sets $\sigma_{ij} = (y_{-\alpha_j}x_{\alpha_i}, q^{(\alpha_i,\alpha_j)}(x_{\alpha_i}y_{-\alpha_j} - \delta_{i,j})) : i, j = 1, 2, \ldots, \ell$. Then $\tilde{A}_C = F/\langle w_\sigma - f_\sigma : \sigma \in S\rangle$. In this case S does not admit any ambiguities. Further $z \in X$ is reduced if and only if it is a product of the form xy with x a monomial in the x_{α_i} and y a monomial in the $y_{-\alpha_j}$. This proves the assertion in 3.1.3, namely that the multiplication map $\tilde{U}^+ \otimes \tilde{U}^- \longrightarrow \tilde{A}_C$ is a linear bijection.

Let $\mathfrak{g}$ be a finite dimensional Lie algebra with basis $\{x_i\}_{i=1}^n$. Recall 1.2.6 and view $U(\mathfrak{g})$ as a quotient of the free algebra $F = \mathcal{T}(\mathfrak{g})$ on generators ordered as $x_1, x_2, \ldots, x_n$. Order the monomials by degree and within any fixed degree lexicographically. This order relation is still linear and satisfies dcc. Let S denote the sets $\sigma_{ij} = (x_j x_i, x_i x_j - [x_i, x_j]) : i, j = 1, 2, \ldots,$ with $i < j$. Then $U(\mathfrak{g}) = F/\langle w_\sigma - f_\sigma : \sigma \in S\rangle$. Since each w_σ has degree 2, ambiguities can only occur through monomials of degree 3 and take the form $\{\sigma_{\ell,j}, \sigma_{j,i}, x_\ell, x_j, x_i\}$ with $w_{\sigma_{\ell,j}} = x_\ell x_j, w_{\sigma_{j,i}} = x_j x_i$. Through the Jacobi identity such an ambiguity can be easily shown to be resolvable. Finally it is clear that $x \in X$ is reduced if and only if x takes the form $x_1^{i_1} x_2^{i_2} \ldots x_n^{i_n}$:

$i_j \in \mathbb{N}$. Such a monomial is called standard. Applying A.2.13 proves the Poincaré-Birkhoff-Witt (PBW) theorem namely

Theorem. *The standard monomials form a basis for* $U(\mathfrak{g})$.

A.2.15 Let A be a Hopf algebra. Frobenius reciprocity with respect to A takes several forms and needs a little extra care if A is not cocommutative. Let M, M' be a A modules and view $\mathrm{Hom}(M, M')$ as an A module through

$$(a.\psi)(m) = a_1\psi(\sigma(a_2)m), \quad \text{for all } m \in M, \ a \in A .$$

Let N be an A module. For each $\varphi \in \mathrm{Hom}_A(N, \mathrm{Hom}(M, M'))$ define $T_\varphi \in \mathrm{Hom}(N \otimes M, M')$ through $T_\varphi(n \otimes m) = \varphi(n)(m)$. Then for all $a \in A$ one has

$$\begin{aligned} T_\varphi(a(n \otimes m)) =& \varphi(a_1 n)(a_2 m) = (a_1\varphi(n))(a_2 m) \\ &= a_1\varphi(n)(\sigma(a_2)a_3 m) = a\varphi(n)(m) = aT_\varphi(n \otimes m) \end{aligned}$$

that is T_φ is a homomorphism of A modules. This gives the

Lemma. *The map* $\varphi \mapsto T_\varphi$ *is an isomorphism of* $\mathrm{Hom}_A(N, \mathrm{Hom}(M, M'))$ *onto* $\mathrm{Hom}_A(N \otimes M, M')$ *with inverse the map* $\psi \mapsto S_\psi$ *where* $S_\psi(n)(m) = \psi(n \otimes m)$ *for all* $n \in N$, $m \in M$.

A.2.16 Let A, M, N be as in A.2.15 and let E, F be finite dimensional A modules. Give E^* (resp. F^*) a left A module structure as in 1.4.8, that is $a\xi = \xi\sigma(a)$ for all $a \in A$, $\xi \in E^*$ (resp. F^*).

For each $\psi \in \mathrm{Hom}_A(N, F \otimes M)$ define $S_\psi \in \mathrm{Hom}(F^* \otimes N, M)$ through $S_\psi(\xi \otimes n) = \xi(\psi(n))$ for all $\xi \in F^*$, $n \in N$. Then $S_\psi(a(\xi \otimes n)) = (a_1\xi)\psi(a_2 n) = (a_1\xi)(a_2\psi(n))$. Now write $\psi(n) = f' \otimes m''$ where the primes designate a sum. Since $(a_1\xi)(a_2 f') = \xi(\sigma(a_1)a_2 f') = \varepsilon(a_1)\xi(f')$. One obtains $S_\psi(a(\xi \otimes n)) = \xi(f')am'' = a\xi(\psi(n)) = aS_\psi(\xi \otimes n)$, that is S_ψ is a homomorphism of A modules. Conversely if S_ψ is an A module homomorphism, so is ψ.

For each $\psi \in \mathrm{Hom}(M \otimes E, N)$ define $R_\psi \in \mathrm{Hom}(M, N \otimes E^*)$ through $R_\psi(m)(e) = \psi(m \otimes e)$. Assume R_ψ is an A module homomorphism. Writing $R_\psi(m) = n' \otimes \xi''$, one obtains $\psi(a(m \otimes e)) = R_\psi(a_1 m)(a_2 e) = (a_1 R_\psi(m))(a_2 e) = a_1(n' \otimes \xi'')(a_2 e) = a_1 n'(a_2\xi'')(a_3 e) = an'\xi''(e) = a(n' \otimes \xi'')(e) = aR_\psi(m)(e) = a\psi(m \otimes e)$. The converse is similar. One concludes

Lemma. *Let* M, N, E, F *be* A *modules with* E, F *finite dimensional.*

(i) *The map* $\psi \mapsto S_\psi$ *is an isomorphism of* $\mathrm{Hom}_A(N, F \otimes M)$ *onto* $\mathrm{Hom}_A(F^* \otimes N, M)$.
(ii) *The map* $\psi \mapsto R_\psi$ *is an isomorphism of* $\mathrm{Hom}_A(M \otimes E, N)$ *onto* $\mathrm{Hom}_A(M, N \otimes E^*)$.
(iii) *The map* $n \otimes \xi \mapsto (e \mapsto \xi(e)n)$ *is an isomorphism* $N \otimes E^* \xrightarrow{\sim} \mathrm{Hom}(E, N)$ *of* A *modules.*

A.2.17 The following result is due to Ado. A proof may be found in [Di2, 2.5.5]. It was generalized to positive characteristic by Iwasawa [Jac1, Chap. VI].

Theorem. *Every finite dimensional Lie algebra admits a faithful finite dimensional module.*

A.2.18 Recall the notation of 1.4.9. The following result is due to Harish-Chandra. A proof may be found in [Di2, 2.5.7].

Theorem. *Let $\mathfrak{g}$ be a finite dimensional Lie algebra. Then*

$$\bigcap_{V \in \mathrm{Mod}_f\, U(\mathfrak{g})} \mathrm{Ann}_{U(\mathfrak{g})} V = 0 \ .$$

Remarks. Ado's theorem is trivial for $\mathfrak{g}$ semisimple. The proof of Harish-Chandra's theorem in the semisimple case can be accomplished as in 7.1.9.

A.3 Combinatorial Identities and Dimension Theory

The first part of the section illustrates a technique for obtaining q-binomial identities. The second part, quite unrelated to the first, describes some dimension theory.

A.3.1 Let k be a field of characteristic zero and recall the q-binomial coefficients defined in 1.2.12.

Lemma. *For all $m, n \in \mathbb{N}$ and $j \in \{0, 1, \dots, n\}$ one has*

$$\sum_{i=0}^{n} (-1)^i \frac{[(m+n-i-j)!]}{[(m-i)!][(i-j)!][(n-j)!]} q^{n-1} = (-1)^j q^{(n-j)(m-j+1)} \ .$$

Define $\theta \in \operatorname{Aut} k(q)[a]$ and the θ-leftskew derivation ∂ as in 1.2.12. Combining 1.2.12(1) and 1.2.14(3) gives

$$\partial^s \Bigl(\prod_{i=1}^{n} (a + q^{2(i-1)})\Bigr) = \partial^s (exp_q \partial) a^n = (exp_q \partial) \partial^s a^n$$

$$= q^{sn - \frac{1}{2}s(s+1)} \frac{[n!]}{[(n-s)!]} \prod_{i=1}^{n-s} (a + q^{2(i-1)}) \ .$$

Let σ (resp. τ) be the automorphism of $k(q)[a, b]$ defined by $\sigma(a) = q^2 a$, $\sigma(b) = b$ (resp. $\tau(a) = a$, $\tau(b) = q^{-2} b$) and ∂_a (resp. ∂_b) the σ-leftskew (resp. τ-leftskew) derivation of $k(q)[a, b]$ defined by $\partial_a a = 1$, $\partial_a b = 0$ (resp. $\partial_b a = 0$, $\partial_b b = 1$). Replacing a by ab^{-1} in the above one obtains

$$(1)\quad \partial_a^s\Big(\prod_{i=1}^{n}(a+q^{2(i-1)}b)\Big)=q^{sn-\frac{1}{2}s(s+1)}\,\frac{[n!]}{[(n-s)!]}\,\prod_{i=1}^{n-s}(a+q^{2(i-1)}b)\,.$$

Interchanging a, b and q, q^{-1} gives

$$(2)\quad \partial_b^r\Big(\prod_{i=1}^{n}(a+q^{2(i-1)}b)\Big)=q^{rn-\frac{1}{2}r(r+1)}\,\frac{[n!]}{[(n-r)!]}\,\prod_{i=1}^{n-r}(a+q^{2(i-1)}b)\,.$$

Combined these imply

$$(3)\quad \Big\{\partial_a^{n-j}\partial_b^{j}a^{m-j}\prod_{i=1}^{n}(a+q^{2(i-1)}b)\Big\}_{a=1,b=-1}=q^{2(m-j)(n-j)}q^{\frac{1}{2}n(n-1)}[n!]\,.$$

On the other hand using 1.2.13(2) to replace the product, the left hand side becomes

$$\sum_{i=0}^{n}\begin{bmatrix} n\\ i\end{bmatrix}q^{(n-1)i}(\partial_a^{n-j}a^{m+n-i-j})_{a=1}(\partial_b^{j}b^{i})_{b=-1}$$

$$=\sum_{i=0}^{n}\begin{bmatrix} n\\ i\end{bmatrix}q^{(n-1)i}q^{(n-j)(m+n-i-j)-\frac{1}{2}(n-j)(n-j+1)}q^{-(ij-\frac{1}{2}j(j+1))}$$

$$\times\frac{[m+n-i-j)!][i!]}{[(m-i)!][(i-j)!]}\,(-1)^{i-j}$$

$$=(-1)^{j}[n!]q^{\frac{1}{2}n(n-1)}q^{(m-j-1)(n-j)}\sum_{i=0}^{n}\frac{(-1)^{i}[(m+n-i-j)!]q^{n-i}}{[(n-i)!][(m-i)!][(i-j)!]}\,.$$

Comparison with (3) gives the required result.

A.3.2 Following 10.2.2 (6) define $D\in \operatorname{End} k(q)[b]$ through

$$D_a^m=[m]^2a^{m-1}+[m][m-1]a^{m-2}\,.$$

Lemma. *For all* $m\in\mathbb{N}$, $n\in\{0,1,2,\dots,m\}$

$$(1)\qquad (D^na^m)\Big|_{a=-q^{-1}}=(-1)^{m-n}q^{-(m-n)(n+1)}\,\frac{[m!][n!]}{[(m-n)!]}\,.$$

Let $\sigma\in \operatorname{Aut} k(q)[a]$ be defined by $\sigma(a)=qa$ and $\partial\in\operatorname{End} k(q)[a]$ by $\partial b=1$ and $\partial(bc)=\sigma(b)\partial c+(\partial b)\sigma^{-1}(c)$, for all $b,c\in k(q)[a]$. Then as in 1.2.12 one obtains $\partial a^n=[n]a^{n-1}$ and so $D=\partial(a+1)\partial$.

Consequently

$$D^na^m=D^{n-1}\partial(a+1)\partial a^m=[m]^2D^{n-1}a^{m-1}+[m][m-1]D^{n-1}a^{m-2}\,.$$

Writing $a_{m,n}=(D^na^m)_{a=-q^{-1}}$ it follows that

$$(2)\qquad a_{m,n}=[m]^2a_{m-1,n-1}+[m][m-1]a_{m-2,n-1}\,.$$

Assertion (1) is proved by induction on $m+n$. it is clear for $m+n=0$. Then by the induction hypothesis and (2)

$$a_{m,n} = [m]\left\{[m](-1)^{m-n}q^{-(m-n)n}\,\frac{[(m-1)!][(n-1)!]}{[(m-n)!]}\right.$$
$$\left. + [m-1](-1)^{m-n-1}q^{(-(m-n-1)n}\,\frac{[(m-2)!][(n-1)!]}{[(m-n-1)!])}\right\}$$
$$= (-1)^{m-n}q^{-(m-n)n}\,\frac{[m!][(n-1)!]}{[(m-n)!]}\,\{[m]-q^n[m-n]\}\ .$$

Since $[m]-q^n[m-n] = q^{-(m-n)}[n]$, the required identity results.

A.3.3 Let K be a commutative ring. Let A be a K-algebra, that is to say a ring which is also a K module such that $ma = am$ for all $a \in A$, $m \in K$. A K-filtration $\mathcal{F}$ (resp. K-gradation $\mathcal{G}$) on A is a family of K modules $\mathcal{F}^n(A)$ (resp. $\mathcal{G}_n(A)$) : $n \in \mathbb{Z}$ satisfying $\mathcal{F}^m(A) \subset \mathcal{F}^n(A)$ if $m \leq n$ and

(i) $$\bigcup_{n\in\mathbb{Z}} \mathcal{F}^n(A) = A \qquad (\text{resp. } \bigoplus_{n\in\mathbb{Z}} \mathcal{G}_n(A) = A)$$

(ii) $$\mathcal{F}^m(A)\mathcal{F}^n(A) \subset \mathcal{F}^{m+n}(A) \quad (\text{resp. } \mathcal{G}_m(A)\mathcal{G}_n(A) \subset \mathcal{G}_{m+n}(A))\ ,$$ $$\forall m,n \in \mathbb{Z}\ .$$

If $\mathcal{G}$ is K-gradation on A, then trivially $\mathcal{F}^n(A) = \oplus_{m\leq n}\mathcal{G}_m(A)$ is a K-filtration on A. Conversely if $\mathcal{F}$ is a K-filtration on A, then

$$gr_m(A) = \mathcal{F}^m(A)/\mathcal{F}^{m-1}(A) : m \in \mathbb{Z}$$

defines a K-gradation on the associated graded algebra $gr_{\mathcal{F}}A$ which by definition is

$$gr_{\mathcal{F}}(A) := \bigoplus_{m\in\mathbb{Z}} gr_m(A)\ .$$

If A admits an identity (as will be assumed henceforth) then it is assumed that $1 \in \mathcal{F}^\circ A$. In some sense $gr_{\mathcal{F}}A$ is an approximation to A and to some extent properties of $gr_{\mathcal{F}}A$ can be lifted to A. Often this is possible only when $\mathcal{F}$ is bounded (7.4.1) that is when $\mathcal{F}^{-n}(A) = 0$ for some $n \in \mathbb{Z}$ necessarily > 0. Observe that for a bounded filtration $(\mathcal{F}^{-1}(A))^n = 0$. In particular if $gr_{\mathcal{F}}A$ is quasi-commutative, then the image of $\mathcal{F}^{-1}(A)$ in $gr_{\mathcal{F}}A$ generates a nilpotent ideal which can often be assumed zero without loss of generality (for example in discussing the nullstellensatz). In the following it is assumed that $\mathcal{F}^{-1}(A) = 0$.

A.3.4 Let K be a commutative domain and A a K-algebra. If $0 \neq f \in K$, denote by K_f the localization of K at the Ore subset $\{f^n\}_{n\in\mathbb{N}}$ and set $A_f = K_f \otimes_K A$. Let M be a left A module and set $M_f = A_f \otimes_A M$. One calls A generically flat over K if for each finitely generated left A module M there exists $0 \neq f \in K$ such that M_f is free over K_f. This is mainly of interest for K finitely generated over a field k and then it can be assumed to be a polynomial ring. For the nullstellensatz and during certain proofs (for example, (iii) below) one also needs the above property to hold for the polynomial extension $A[x]$. Define A to be (n, m)-generically flat over K if $A[x_1, x_2, \ldots, x_n, y_1, y_2, \ldots, y_m]$ is generically flat over $K[y_1, y_2, \ldots, y_m]$ and $(\mathbb{N}, m)$-generically flat if this holds for all n.

Generic flatness is established in the present cases of interest through a number of steps indicated below. For the proofs the reader is referred to [MR1].

(i) By [MR1, 9.3.5] it is enough to consider M above cyclic. In particular K is generally flat over itself since a cyclic K-module M is either free or $M_f = 0$ (hence free!) for any $0 \neq f \in K$ annihilating the cyclic vector.
(ii) By [MR1, 9.4.6] if A is a left noetherian K-algebra generically flat over K then A is $(\mathbb{N}, 0)$-generically flat over K.
(iii) Suppose $A \subset B$ are K-algebras and B is generated over A by some $b \in B$ satisfying $bA = Ab$. Then by [MR1, 9.4.10] if A is $(\mathbb{N}, m)$-generically flat over K so is B.
(iv) Let A be a K-algebra with a bounded K-filtration $\mathcal{F}$ (that is $\mathcal{F}^{-1}A = 0$). Then by [MR1, 9.4.9] if $g_{\mathcal{F}}A$ is (n, m)-generically flat over K so is A.

Recall 7.4.1 and 8.4.15.

Proposition. *Let A be a k-algebra with a bounded filtration $\mathcal{F}$ such that $gr_{\mathcal{F}}A$ is finitely quasi-commutative. Then A is (n, m)-generically flat over k for all $n, m \in \mathbb{N}$. In particular A satisfies the nullstellensatz* (8.4.15(i), (ii)).

Take $K = k[y_1, \ldots, y_m]$ in (i)–(iii) and $K = k$ in (iv). Then by (i) and (ii), K is $(\mathbb{N}, 0)$-generally flat over itself. Hence by (iii) and the hypothesis, $gr_{\mathcal{F}}A$ is $(\mathbb{N}, 0)$-generally flat over K. Then $gr_{\mathcal{F}}A$ is $(\mathbb{N}, m)$-generically flat over k and by (iv) so is A. For the last part, note that A is noetherian by 7.4.1 and apply [MR1, 9.3.13 with $K = k$] recalling [MR1, 9.1.2, 9.1.4].

A.3.5 One may extract from A.3.4 the following

Lemma. *Let K be a finitely generated commutative integral k-algebra. Let A be a K-algebra with a bounded K-filtration such that $gr_{\mathcal{F}}A$ is finitely quasi-commutative. Let M be a finitely generated A module. Then there exists $0 \neq f \in K$ such that M_f is a free K_f module.*

Remarks. This of course applies when $A = K$ given the trivial filtration. A further application is to the nth Weyl algebra (1.2.10) taking $K = k[a_1, a_2, \ldots, a_n]$ and giving $A = A_n$ the K-filtration defined by the degree of differential operators. Now let M be a simple A_n module and M_f be as in the conclusion of A.3.5. If $M_f \neq 0$ then as a K-module it admits a one-dimensional quotient K_λ. By Frobenius reciprocity $\mathrm{Hom}_A(M, \mathrm{Hom}_K(A, K_\lambda)) \xrightarrow{\sim} \mathrm{Hom}_K(A \otimes_A M, K_\lambda) = \mathrm{Hom}_K(M, K_\lambda) \neq 0$. This gives an embedding of

M into the co-induced module $\mathrm{Hom}_K(A, K_\lambda)$. If $M_f = 0$ then M is a quotient of A/Af. This gives a partial description of simple A_n modules which becomes a little more satisfactory when $n = 1$. Unfortunately this analysis does not immediately apply to the quantum Weyl algebra A_C owing to the non-commutativity of U^+.

A.3.6 Let A be a k-algebra generated by some finite dimensional subspace X. It is convenient to assume that $1 \in X$ so then the $X^m : m \in \mathbb{N}$ form an increasing family of subspaces whose union is A. Let M be an A module generated by some finite dimensional subspace M^0 and set $M^m = X^m M^0$. For example one can take A itself. Define the Gelfand-Kirillov dimension $d_A(M)$ of M over A by

$$d_A(M) = \overline{\lim}_{m\to\infty} \left(\frac{\log \dim M^m}{\log m}\right)$$

which one can easily show is independent of the choice of generating subspaces. If A is fixed one writes d_A simply as d. Note that for an algebra $d(A)$ means $d_A(A)$.

If A has an $\mathbb{N}$ gradation and M is a graded A module (where both have finite dimensional graded subspaces) then one may replace M^m in the above by $\sum_{n\leq m} \mathcal{G}_n(M)$. This applies in particular to algebras with a formal character.

Some elementary properties of A are the following. In this note that if A is a Hopf algebra and V (resp. M) is a finite dimensional (resp. finitely generated) module then (cf. 8.1.6) one has $AM^0 \otimes V \subset A(M^0 \otimes V)$ and $V \otimes \sigma(A)M^0 \subset A(V \otimes M^0)$. Thus $M \otimes V$ (resp. $V \otimes M$) is finitely generated (resp. is finitely generated if σ is surjective).

Lemma. *Let $N \subset M$ be finitely generated A modules and V a finite dimensional A module. Then*

(i) $d(M) \geq \max\{d(N), d(M/N)\}$.
(ii) $d(A/Aa) \leq d(A) - 1$ *if $a \in A$ is a non-zero divisor.*
(iii) $d(M) = \max_i d(Am_i)$, *for any finite set $m_1, m_2, \dots, m_r$ of generators of M.*
(iv) *If A is a Hopf algebra (resp. and σ is surjective), then $d(M \otimes V) = d(M)$ (resp. $d(V \otimes M) = d(M)$).*
(v) *If A is a pointed Hopf algebra and B an A-algebra generated by V, then $d(B\#A) = d(A) + d(B)$.*

Remarks. Taking the supremum of all finite systems of generators, $d(A)$ can be defined when A is not finitely generated. Yet then (v) may fail. One need not have equality in (i). Finally $d(M)$ need not be finite nor integer. By pointed in (v) one may assume $\Delta X \subset X \otimes X$ and so $XV \subset VX$ in $B\#A$.

A.3.7 A useful application of the Gelfand-Kirillov dimension to Verma modules derives from A.3.6 (ii) when the Cartan matrix is of finite type. Under this assumption one has

Lemma
(i) $d_U(M) = d_{U^-}(M)$ *for all* $M \in \mathrm{Ob}\mathcal{O}$.
(ii) $d(M) = \ell(w_0)$, *for any Verma module* M.
(iii) $d(M) < \ell(w_0)$ *for any proper quotient of a Verma module.*
(iv) *Non-zero submodules of a Verma module have a non-zero intersection.*

(i) is clear. For (ii) recall that a Verma module is isomorphic to U^- as a U^- module and use the character formula of 3.4.7 together with A.1.1 and A.1.4. For (iii) recall that U^- is a domain (7.3.4) and apply A.3.6(ii). Finally (iii) follows from A.3.6 (i) and (iii).

A.3.8 Let A be an integral k-algebra of finite Gelfand-Kirillov dimension. Then by A.3.6 (i), (ii) any two non-zero left ideals have a non-zero intersection. Thus $S = A \setminus \{0\}$ is an Ore subset of A and $\mathrm{Fract}\, A := S^{-1}A$ is a skew-field in which case A is said to admit a classical skew-field of fractions. Given $a, b \in A$ non-zero, then $Aa \cap Ab \neq 0$ unless a, b generate a free subalgebra of A. Thus if A is a finitely generated domain of subexponential growth (that is if the X^m defined as in A.3.6 satisfy $\overline{\lim}_{m\to\infty} \log\log\dim X^m / \log m < 1$) then A admits a classical skew-field of fractions.

An integral indecomposable Cartan matrix C is said to be of affine type if $(\alpha, \alpha) \geq 0$ for all $\alpha \in \Delta(C)$ with equality on some line through $\Delta(C)$, equivalently if C is positive semidefinite of corank 1.

Lemma. *Assume* C *indecomposable. The* k*-algebra* $U_q(\mathfrak{n}_C^-)$ *admits a classical field of fractions if and only if* C *is of finite type or affine.*

Since $U_q(\mathfrak{n}_C^-)$ is a domain (7.3.4) and has the same growth rate as $U(\mathfrak{n}_C^-)$ by 3.4.7 and 4.3.8, it is enough to determine when the latter has subexponential growth. By [Kal, 9.14], the Lie algebra $\mathfrak{n}_C^-$ has exponential growth unless C is of affine (or finite) type in which case the growth rate is polynomial of degree 1 (or is eventually zero). It then suffices to apply the following fact. If $\mathfrak{n} = \bigoplus_{i \in \mathbb{N}} \mathfrak{n}_i$ is a graded Lie algebra and $U(\mathfrak{n}) = \bigoplus_{i\in\mathbb{N}} U(\mathfrak{n})_i$ the induced gradation, then $G(i) := \dim U(\mathfrak{n})_i$ has subexponential growth if and only if this is true of $g(i) := \dim \mathfrak{n}_i$. Indeed $g(i)$ has subexponential growth if and only if $\sum g(i)x^i$ has radius of convergence ≥ 1 and of course a similar assertion holds for $G(i)$. Yet

$$\sum_{i=0}^{\infty} G(i)x^i = \prod_{i=1}^{\infty} \left(\frac{1}{1-x^i}\right)^{g(i)}$$

and so by taking log of the right hand side this series has the same radius of convergence as the first. Hence the assertion and the lemma.

Remark. If C is of affine type, then $\mathfrak{n}_C^- = \mathfrak{n}_{C_0}^- \otimes k[t,t^{-1}]$ for some Cartan matrix C_0 of finite type. Quite possibly $\mathrm{Fract}(U_q(\mathfrak{n}_C^-))$ is just an infinite transcendental extension of $\mathrm{Fract}(U_q(\mathfrak{n}_{C_0}^-))$.

A.3.9 Let A be a graded polynomial algebra on generators $x_1, x_2, \ldots, x_n$ of degrees $1 \leq d_2, \ldots, d_n < \infty$. Let M be a graded A module generated by finitely many homogeneous elements of degree ≥ 0. Let M_m denote the subspace of M of homogeneous elements of degree m and define the Poincaré series $R_M(q)$ of M through

$$R_M(q) = \sum_{m=0}^{\infty} (\dim M_m) q^m .$$

Let K (resp. C) denote the kernel (resp. cokernel) of the endomorphism $a \mapsto x_1 a$ of M. Then

$$(1) \qquad R_M(q)(1-q^{d_1}) = R_C(q) - q^{d_1} R_K(q) .$$

Noting that K and C are finitely generated over $k[x_2, \ldots, x_n]$ an easy induction on (1) gives

$$(2) \qquad R_M(q) = \frac{P_M(q)}{\prod_{i=1}^n (1-q^{d_i})}$$

for some polynomial P_M, called the Hilbert-Samuel polynomial.

If M is a free A module, it is immediate that $P_M(1) = rk\ M$. If $P_M(q)$ is not divisible by $(1-q)$, equivalently if $P_M(1) \neq 0$, then $d(M) = n$. If M is torsion with respect to any non-empty Ore subset S of A it follows from A.3.6 (i)-(iii) that $d(M) < n$. Thus

$$(3) \qquad \textit{If } M \textit{ is } S\textit{-torsion } (S \neq \emptyset) \textit{ then } P_M(1) = 0 .$$

Lemma. *Let P be an ideal of A of codimension 1. Then*

$$P_M(1) \leq \dim(M/PM) .$$

Let V complement PM in M and set $N = AV$. Then $M = N + PM$ and V complements PN in N. Hence the composed map $N \hookrightarrow M \longrightarrow M/PM$ factors to an isomorphism $N/PN \xrightarrow{\sim} M/PM$ and on the other hand $P(M/N) = M/N$. Since M/N is finitely generated it follows by Nakayama's lemma (A.2.10) that M/N is torsion with respect to $S := A \setminus P$. Hence by (3) one has $P_{M/N}(1) = 0$. This gives

$$P_M(1) = P_N(1) \leq \dim V = \dim(M/PM)$$

as required.

Remark. By generic flatness (A.3.5) there exists $0 \neq f \in A$ such that M_f is free over A_f. Then equality holds in the lemma if $P_f \neq A_f$, that is if $f \notin P$. If equality holds for the "special fibre" namely when P is the augmentation ideal, then M is a free A module. Of course such an equality is usually difficult to establish.

A.3.10 For an algebra A with a "good" filtration the Hilbert-Samuel polynomial may be used to derive properties of GK dimension, for example equality in A.3.6 (i) and that $d_A(M)$ is integer-valued. Of course the Hilbert-Samuel polynomial earlier played a key role in algebraic geometry and in particular in dimension theory. For example a result used here is the following. A proof may be found in [Sh1, Chap. 1].

Lemma. *Let* $\varphi : X \longrightarrow Y$ *be a morphism of algebraic varieties such that* $\varphi^* : R[Y] \longrightarrow R(X)$ *is injective. Then* $\dim \varphi^{-1}(y) \geq \dim X - \dim Y$ *with equality on some non-empty open set contained in* $\operatorname{Im} \varphi$.

A.4 Remarks on Constructions of Quantum Groups

Several important ideas in the construction of quantum groups have been ignored in the main text. This is partly because it is quite unnecessary to understand the origins of $U_q(\mathfrak{g})$ to study the representation theory or primitive spectrum of it or its Hopf dual. It is also because we do not wish to give a logical and comprehensive analysis of such constructions. The present section partially attenuates this default. However it is not intended to be a historical presentation. Already Drinfeld's ICM address [Dr2] reviews the early work in the field and it is also striking how much it has influenced more recent papers. Again deformation theory which has its origins much further back, particularly in the pioneering work of M. Gerstenhaber [Ge1], is not considered here even though it is often viewed as a central component to quantization. Here again for representation theory it is not so important and there is nearly always a level of refinement where the quantized object is significantly different and paradoxically may be *easier* to study.

A.4.1 Let A be a commutative k-algebra. A Poisson bracket $\mathcal{P}$ on A is a linear map $(a, b) \mapsto \{a, b\}$ of $A \times A$ into A making A a Lie algebra and such that for each $a \in A$ the map $b \mapsto \{a, b\}$ is a derivation of A. Call an ideal I of A a Poisson ideal if $\{A, I\} \subset I$. If I is a Poisson ideal then $\mathcal{P}$ defines a Poisson bracket on A/I. Let $(\operatorname{Spec} A)^{\mathcal{P}}$ denote the set of all prime Poisson ideals of A. Given $P \in (\operatorname{Spec} A)^{\mathcal{P}}$ set $K = \operatorname{Fract}(A/P)$. Then $\mathcal{P}$ defines a Poisson bracket on K and define P to be symplectic if $\{b \in K | \{a, b\} = 0, \forall b \in K\}$ reduces to scalars. Let $(\mathit{Sp\ell}\ A)^{\mathcal{P}}$ denote the set of symplectic ideals of A.

The problem of determining $(\mathit{Sp\ell}\ A)^{\mathcal{P}}$ may be viewed as a close analogue of the problem discussed in 10.3. However in the commutative set-up it also has an important geometric interpretation. For simplicity assume k algebraically closed and of characteristic zero (an assumption retained throughout A.4.). Assume further that A is finitely generated with no nilpotent elements. Then A is just the algebra of regular functions on the affine algebraic variety $X := \operatorname{Max} A$. Given $x \in X$, let $\mathfrak{m}_x \in \operatorname{Max} A$ be the ideal it defines. Then for all $a, b \in \mathfrak{m}_x$ the map $a \otimes b \mapsto \{a, b\}(x)$ is antisymmetric with kernel containing $\mathfrak{m}_x^2 \otimes \mathfrak{m}_x + \mathfrak{m}_x \otimes \mathfrak{m}_x^2$ and hence defines an element $\Pi(x) \in T_{x,X} \wedge T_{x,X}$. For each $a \in A$ define $da : X \to T^*(X)$ through $x \mapsto (d_x a) := a - a(x) \bmod \mathfrak{m}_x^2$. Then a map $\Pi : X \longrightarrow T_*(X) \wedge T_*(X)$ is called a Poisson tensor on X if $x \mapsto (\Pi, da \otimes db)(x) := (\Pi(x), d_x a \otimes d_x b)$ is a regular function on X and the map $a \otimes b \mapsto \{a, b\} := (\Pi, da \otimes db)$ satisfies the Jacobi identity and hence is a Poisson bracket on A. In terms of generators $x_1, x_2, \ldots, x_n$ of A, and writing $c_{r,s} = \{x_r, x_s\}$, the Jacobi identity becomes

$$\sum_{i=1}^{n} c_{i,r} \frac{\partial c_{s,t}}{\partial x_i} + c_{i,s} \frac{\partial c_{t,r}}{\partial x_i} + c_{i,t} \frac{\partial c_{r,s}}{\partial x_i} = 0 \, .$$

A smooth affine variety X equipped with a Poisson tensor Π is said to be symplectic if $\Pi(x)$, viewed as an antisymmetric bilinear form on $\mathfrak{m}_x/\mathfrak{m}_x^2$, is nondegenerate at each $x \in X$.

Assume X is smooth and for each $d \in \mathbb{N}$ set $X_d = \{x \in X \mid \operatorname{codim} \ker \Pi(x) = d\}$. Then $\partial X_d := \bigcup_{i<d} X_i$ is just the zero set of the ideal I_d generated by all $d \times d$ minors of the matrix $\{c_{i,j}\}_{i,j=1}^n$. Using (*) one may check that I_d is a Poisson ideal of A, as are also the minimal primes over I_d. This decomposes each ∂X_d into irreducible components whose intersection with X_{d-1} are locally closed irreducible subsets which for the various d form a disjoint union of X. They are called the Poisson sheets of X. Each such sheet S is a disjoint union of symplectic subvarieties corresponding to the maximal Poisson ideals of $k[S]$. Thus X itself is a disjoint union of symplectic subvarieties called the symplectic leaves of X. Note that one may write $k[S] = (A/P)_Q$ where P (resp. Q) is the element of $(\operatorname{Spec} A)^{\mathcal{P}}$ defining $\bar{S}$ (resp. $\partial S := \bar{S} \setminus S$). If L is a maximal Poisson ideal of $k[S]$ then the inverse image of $L \cap A/P$ in A belongs to $(\mathcal{S}p\ell\ A)^{\mathcal{P}}$.

A smooth algebraic variety X is said to be Poisson variety if it admits a finite open affine covering $\{U_\alpha\}$ together with a Poisson tensor Π^α on each U_α such that $\Pi^\alpha(x) = \Pi^\beta(x)$ for all $x \in U_\alpha \cap U_\beta$. As in the affine case, X may be decomposed into a disjoint union of its symplectic leaves. Again if X, Y are Poisson varieties, then the product $X \times Y$ has a natural Poisson structure defined locally by $\{f_1, f_2\}(x, y) = \{f_1^x, f_2^x\}(y) + \{f_1^y, f_2^y\}(x)$ where f^x (resp. f^y) denotes the function on Y (resp. X) defined by $f^x(y) = f(x, y)$ (resp. $f^y(x) = f(x, y)$).

A.4.2 A Poisson algebraic group G is an algebraic group which is a Poisson variety such that the multiplication map $G \times G \longrightarrow G$ is a morphism of Poisson varieties. If ℓ_x (resp. r_y) denotes the map $G \longrightarrow G$ defined by left (resp. right) multiplication by $x \in G$ (resp. $y \in G$) then this condition can be expressed as

(1) $\{f,g\}(xy) = \{\ell_x^* f, \ell_x^* g\}(y) + \{r_y^* f, r_y^* g\}(x)\ , \quad \forall f,g \in R[G]\ ; \quad x,y \in G\ .$

Recall (2.1.7) that ℓ_x (resp. r_y) induces a linear isomorphism $\ell_x : T_{z,G} \longrightarrow T_{xz,G}$ (resp. $r_y : T_{z,G} \longrightarrow T_{zy,G}$). Set $\mathfrak{g} = \operatorname{Lie} G = T_{e,G}$. Then Π gives rise to a map $\Pi_\ell : G \longrightarrow \mathfrak{g} \times \mathfrak{g}$ defined by $\Pi_\ell(x) = \ell_{x^{-1}} \Pi(x)$ which may simply be written as $x^{-1}.\Pi(x)$. Similarly $\Pi_r : G \longrightarrow \mathfrak{g} \times \mathfrak{g}$ is defined by $\Pi_r(x) = r_x^{-1} \Pi(x) = \Pi(x).x^{-1}$. It follows from (1) that

(2) $$\Pi(xy) = x.\Pi(y) + \Pi(x).y$$

or equivalently

(3) $$\Pi_\ell(xy) = \Pi_\ell(y) + \operatorname{ad}_{y^{-1}}(\Pi_\ell(x))$$

where $\operatorname{ad}_{y^{-1}}$ denotes the linear isomorphism $a \mapsto y^{-1}.a.y$ of $\mathfrak{g}$.

Observe in particular that (2) implies $\Pi(e) = 0$. Recall 2.2.2 and identify $T_{0,\mathfrak{g}\times\mathfrak{g}}$ with $\mathfrak{g} \times \mathfrak{g}$. Then $\varphi := d_e \Pi_\ell$ is a linear map of $\mathfrak{g}$ into $\mathfrak{g} \times \mathfrak{g}$. Given $\xi, \eta \in \mathfrak{g}^* = \mathfrak{m}_e/\mathfrak{m}_e^2$ choose $a, b \in R[G]$ such that $\xi = d_e a, \eta = d_e b$. Then

(4) $\varphi^*(\xi,\eta) = \varphi^*(d_e a, d_e b) = d_e(\Pi_\ell^*(d_e a, d_e b)) = d_e(\Pi, da \otimes db) = d_e\{a,b\}\ .$

It is clear that $(\xi,\eta) \mapsto \varphi^*(\xi,\eta)$ is bilinear and antisymmetric. Since $\varphi^*\{\varphi^*(d_e a, d_e b), d_e c\} = d_e\{\{a,b\},c\}$ it follows from (4) that

(5) $$\varphi^* : \mathfrak{g}^* \times \mathfrak{g}^* \quad \textit{is a Lie product on} \quad \mathfrak{g}^*\ .$$

On the other hand take $\xi, \eta \in \mathfrak{g}^*$. Then $\Psi := \Pi_\ell^*(\xi,\eta)$ and $\Phi_y := \Pi_\ell^*((\operatorname{ad}_y)^*(\xi,\eta))$ are elements of $R[G]$ and by (3) satisfy

(6) $\Psi(xy) - \Psi(yx) = (\Phi_{y^{-1}}(x) - \Psi(x)) - (\Phi_{x^{-1}}(y) - \Psi(y))\ , \quad \forall x,y \in G\ .$

Differentiated at the identity of G this relation implies

(7) $$\varphi[x,y] = (\operatorname{ad} \Delta(x))\varphi(y) - (\operatorname{ad} \Delta(y))\varphi(x)\ , \quad \forall x,y \in \mathfrak{g}$$

where $\Delta : \mathfrak{g} \longrightarrow \mathfrak{g} \times \mathfrak{g}$ is the coproduct given by 1.2.5 and 1.2.6.

A.4.3 Let $\mathfrak{g}$ be a Lie algebra. Then pair $(\mathfrak{g}, \varphi)$, where $\varphi : \mathfrak{g} \longrightarrow \mathfrak{g} \times \mathfrak{g}$ satisfies (5), (7) above, is said to be a Lie bialgebra. In this case $\mathfrak{p} := \mathfrak{g} \oplus \mathfrak{g}^*$ has a natural Lie algebra structure. Indeed take the scalar product on $\mathfrak{p}$ in which $\mathfrak{g}, \mathfrak{g}^*$ are isotropic and which restricted to $\mathfrak{g} \times \mathfrak{g}^*$ and $\mathfrak{g}^* \times \mathfrak{g}$ is the natural pairing. The Lie product on $\mathfrak{p}$ extends that defined on $\mathfrak{g}$ and on $\mathfrak{g}^*$ by taking elements of the form $[x,\xi] : x \in \mathfrak{g},\ \xi \in \mathfrak{g}^*$ to be determined by

$$([x,\xi],y) = -(\xi,[x,y]) \quad \text{and} \quad ([x,\xi],\eta) = (x,[\xi,\eta])\ , \quad x,y \in \mathfrak{g}\ ;\ \xi,\eta \in \mathfrak{g}^*\ .$$

The relation (7) is equivalent to the Jacobi identity for $\mathfrak{g}$, $\mathfrak{g}^*$ extending to $\mathfrak{p}$. Conversely a Manin triple consists of a Lie algebra $\mathfrak{p}$ admitting a non-degenerate invariant scalar product together with isotropic Lie subalgebras $\mathfrak{p}_1, \mathfrak{p}_2$ of which $\mathfrak{p}$ is the direct sum. Taking $\mathfrak{p}_1 = \mathfrak{g}$, $\mathfrak{p}_2 = \mathfrak{g}^*$ in the above establishes an equivalence between Manin triples and Lie bialgebras. Then for any Kac-Moody Lie algebra $\mathfrak{g} = \mathfrak{n}^- \oplus \mathfrak{h} \oplus \mathfrak{n}^+$ one can give $\mathfrak{k} := \{(x, y) \in \mathfrak{g} \times \mathfrak{g} \mid x = y\}$ the structure of Lie bialgebra. Indeed take the scalar product on $\mathfrak{p} = \mathfrak{g} \times \mathfrak{g}$ to be $((x_1, y_1), (x_2, y_2)) \mapsto K(x_1, x_2) - K(y_1, y_2)$ where K is the invariant scalar product on $\mathfrak{g}$ defined in 4.2.2. In this example $\mathfrak{k}$ is isomorphic to $\mathfrak{g}$ whilst $\mathfrak{k}^* = (\mathfrak{n}^+, 0) + (0, \mathfrak{n}^-) + \mathfrak{h}_a$, where $\mathfrak{h}_a = \{(h, -h) \mid h \in \mathfrak{h}\}$, is a degenerate version of $\mathfrak{k}$ much as $R[G]$ is via 9.2.14 a degenerate version of $U_q(\mathfrak{g})$. Again one may view $\mathfrak{p} = \mathfrak{g} \times \mathfrak{h} \hookrightarrow \mathfrak{g} \times \mathfrak{g}$ as a part of the Manin triple $(\mathfrak{p}, \mathfrak{p}_1, \mathfrak{p}_2)$ completed by $\mathfrak{p}_1 = (\mathfrak{n}^+, 0) \oplus \mathfrak{h}_d$, $\mathfrak{p}_2 = (\mathfrak{n}^-, 0) \oplus \mathfrak{h}_a$ where $\mathfrak{h}_d = \{(h, h) \mid h \in \mathfrak{h}\}$. In this case $\mathfrak{p}_1, \mathfrak{p}_2$ are isomorphic to Borel subalgebras and $\mathfrak{g}$ is obtained from $\mathfrak{p}$ by factoring out the ideal $(0, \mathfrak{h})$ much as $U_q(\mathfrak{g})$ is obtained as the quotient (3.2.9) of the Drinfeld double $\mathcal{D}(V^-, V^+)$.

A.4.4 Let $(\mathfrak{p}, \mathfrak{p}_1, \mathfrak{p}_2)$ be a Manin triple. If $\mathfrak{p}$ is a subalgebra of $\mathfrak{gl}(V)$ for some V finite dimensional so that $\mathfrak{p}_1, \mathfrak{p}_2$ are algebraic subalgebras, then $\mathfrak{p}$ is also algebraic (2.4.4). Consequently (2.4.2) there exist unique irreducible algebraic subgroups P, P_1, P_2 of $\mathrm{GL}(V)$ with Lie algebras $\mathfrak{p}, \mathfrak{p}_1, \mathfrak{p}_2$. The Poisson bracket on $R[P_1]$ is given by the Lie bracket on $\mathfrak{p}_2$. Indeed a basis $\{x_i\}_{i=1}^n$ for $\mathfrak{m}_e/\mathfrak{m}_e^2 = \mathfrak{p}_2$ forms a system of free generators for the local ring $R[P_1]_{\mathfrak{m}_e}$ into which $R[P_1]$ embeds and one defines

$$(8) \qquad \{a, b\} = \sum_{i,j=1}^{n} \frac{\partial a}{\partial x_i} \frac{\partial b}{\partial x_j} [x_i, x_j], \quad \forall a, b \in R[P_i]_{\mathfrak{m}_e}.$$

Via (7) this is compatible with (1) and the original Lie bracket on $\mathfrak{p}_2$ is recovered from (4).

As noted in 2.4.1 the subset P_1P_2 of P is open in its closure. Moreover since $\mathfrak{p}_1 + \mathfrak{p}_2 = \mathfrak{p}$ it follows from 2.4.1 (iii) that $\overline{P_1P_2} = P$. Since $P_1 \cap P_2$ is an algebraic subgroup (2.4.1) with Lie algebra $\mathfrak{p}_1 \cap \mathfrak{p}_2 = 0$ it is in fact a finite group Γ. One calls P a double algebraic group if $P_1P_2 = P$ and $P_1 \cap P_2 = \{e\}$. In this case each $g \in P$ can be written uniquely in the form $g_1g_2 : g_i \in P_i$ and this defines a projection $p : g \mapsto g_1$ of P onto P_1. Then [Sem1] the symplectic leaves of P_1 are just the images of the left cosets P_2g under p. More generally [HL1] one has

Theorem. *Suppose $P_1P_2 = P$. Then the symplectic leaves of P_1 are just the connected components of the inverse images in P_1 of the P_2 orbits in $P/P_2 \xrightarrow{\sim} P_1/\Gamma$.*

A.4.5 In the example of A.4.4 take $\mathfrak{g}$ a finite dimensional semisimple Lie algebra $\mathfrak{g} \hookrightarrow \mathfrak{gl}(V)$ where V is chosen so that every simple highest weight module $V(\lambda) : \lambda \in P(\pi)$ occurs in $\mathcal{T}(V)$. Let G (resp. H) be the unique irreducible algebraic subgroup of $\mathrm{GL}(V)$ with Lie algebra $\mathfrak{g}$ (resp. $\mathfrak{h}$). Then P identifies with $G \times G$ and P_1 with the diagonal copy $\{(g,g) : g \in G\}$ of G in $G \times G$. Set $H_d = \{(h,h) \mid h \in H\}, H_a = \{(h,h^{-1}) \mid h \in H\}$. Then $H_a P_2$ is just the Borel subalgebra $B \times B$ of $G \times G$. Since $G \times G/B \times B$ is a complete variety [Sp1] its P_1 orbits are closed. Consequently P_1P_2/P_2 is closed and so $P_1P_2 = P$. Clearly $\Gamma = H_d \cap H_a = \{(h,h) \mid h \in H$ with $h^2 = 1\}$ and so identifies with $P(\pi)/2P(\pi)$ by the choice of V. Through the Bruhat decomposition of $G \times G$ into double $B \times B$ cosets parametrized by $W \times W$ it is not difficult to compute the symplectic leaves of G using A.4.4. In particular [HL1] for each $\mathbf{w} \in W \times W$ the symplectic leaves of type $\mathbf{w}$ form a single H_d orbit $\mathcal{O}_{\mathbf{w}}$ and have all constant dimension $\ell(\mathbf{w}) + s(\mathbf{w})$. Moreover the $\mathcal{O}_{\mathbf{w}}$ are necessarily irreducible and together form a finite disjoint union of P_1. Hence they are the Poisson sheets of P_1. It is clear that $\bar{\mathcal{O}}_{\mathbf{w}}$ corresponds to the prime ideal $Q_{\mathbf{w}}$ of $R_q[G]$ described in 10.3.5 by specialization at $q = 1$.

A.4.6 The 1-cocycle condition A.4.2(7) is satisfied by taking $\varphi : \mathfrak{g} \longrightarrow \mathfrak{g} \otimes \mathfrak{g}$ to be a coboundary, that is $\varphi(x) = [\Delta(x), r]$ for some $r \in \mathfrak{g} \otimes \mathfrak{g}$. Set $\tau(x \otimes y) = y \otimes x$, $\forall x, y \in \mathfrak{g}$ and $s = r + \tau(r)$. Define $p_{12} : \mathfrak{g} \otimes \mathfrak{g} \longrightarrow \mathfrak{g} \otimes \mathfrak{g} \otimes \mathfrak{g}$ by $p_{12}(x \otimes y) = x \otimes y \otimes 1$ and $p_{ij} : i \leq i < j \leq 3$ similarly, putting $p_{ji} = p_{ij}\tau$. Set $r_{ij} = p_{ij}(r)$, $s_{ij} = p_{ij}(s)$ and define elements of $\mathfrak{g} \otimes \mathfrak{g} \otimes \mathfrak{g}$ by

$$[\![r,r]\!] = [r_{12}, r_{23}] + [r_{13}, r_{23}] + [r_{12}, r_{13}]$$
$$[\![r,r]\!]_\tau = [r_{12}, r_{23}] + [r_{23}, r_{31}] + [r_{31}, r_{12}]$$

where the right hand sides are defined by the Lie product on $\mathfrak{g}$.

Lemma. *Suppose* $[\Delta(x), s] = 0$ *for all* $x \in \mathfrak{g}$. *Then*

$$[\Delta^2(x), [\![r,r]\!]] = [\Delta^2(x), [\![r,r]\!]_\tau]$$

for all $x \in \mathfrak{g}$.

Write $r = r_1 \otimes r_2$, where as usual this notation means an appropriate sum. Then $[r_{12} + r_{21}, r_{13} + r_{23}] = [s \otimes 1, (r_1 \otimes 1 + 1 \otimes r_1) \otimes r_2] = 0$, by the hypothesis. A cyclic permutation of indices gives $[r_{31} + r_{13}, r_{32} + r_{12}] = 0$. Hence

$$\begin{aligned} [\![r,r]\!] - [\![r,r]\!]_\tau &= [r_{12} - r_{23}, r_{13} + r_{31}] \\ &= -[r_{23} + r_{32}, r_{13} + r_{31}] \\ &= -[s_{23}, s_{13}] \end{aligned}$$

which commutes with $\Delta^2(x)$ by the hypothesis.

A.4.7 Retain the above notation.

Proposition. *The following two assertions are equivalent.*

(i) $\varphi^* : \mathfrak{g}^* \otimes \mathfrak{g}^* \longrightarrow \mathfrak{g}^*$ *is a Lie product.*
(ii) $[\Delta(x), s] = 0$ *and* $[\Delta^2(x), [\![r,r]\!]] = 0$, $\forall x \in \mathfrak{g}$.

Clearly $\varphi(x) + \tau\varphi(x) = [\Delta(x), r + \tau(r)] = [\Delta(x), s]$ and so antisymmetry is equivalent to the first part of (ii). The Jacobi identity on $\mathfrak{g}$ is equivalent to

$$\tau_{123}(1d \otimes \varphi)\varphi(x) = 0\,, \quad \forall x \in \mathfrak{g}$$

where τ_{123} denotes the endomorphism $x \otimes y \otimes z \mapsto x \otimes y \otimes z + y \otimes z \otimes x + z \otimes x \otimes y$ of $\mathfrak{g}^{\otimes 3}$.

Write as before $r = r_1 \otimes r_2 = r_1' \otimes r_2'$ (where two expressions are needed to avoid confusion). Then

$$\begin{aligned}(1 \otimes \varphi)\varphi(x) &= (1 \otimes \varphi)([x, r_1] \otimes r_2 + r_1 \otimes [x, r_2]) \\ &= [x, r_1] \otimes \{[r_2, r_1'] \otimes r_2' + r_1' \otimes [r_2, r_2']\} \\ &\quad + r_1 \otimes \{[x, r_2] \otimes r_1'] \otimes r_2' + r_1' \otimes [[x, r_2], r_2']\}\,.\end{aligned}$$

Then writing $x_1 = x \otimes 1 \otimes 1,\ x_2 = 1 \otimes x \otimes 1,\ x_3 = 1 \otimes 1 \otimes x$ and applying τ_{123} this gives

$$\begin{aligned}\tau_{123}(1 \otimes \varphi)\varphi(x) &= \tau_{123}\{[r_2, r_1'] \otimes r_2' \otimes [x, r_1] + r_1' \otimes [r_2, r_2'] \otimes [x, r_1] \\ &\quad + [[x, r_2], r_1'] \otimes r_2' \otimes r_1 + r_1' \otimes [[x, r_2], r_2'] \otimes r_1\} \\ &= \tau_{123}\{[x_3, [r_{31}, r_{12}]] + [[x_3, r_{32}], r_{12}] + [[x_1, r_{31}], r_{12}] \\ &\quad + [[x_2, r_{32}], r_{12}]\}\,.\end{aligned}$$

Yet

$$\begin{aligned}[[x_3, r_{32}], r_{12}] + [[x_2, r_{32}], r_{12}] &= [[x_2 + x_3, r_{32}], r_{12}] \\ &= -[[x_2 + x_3, r_{23}], r_{12}]\,, \quad \text{by the first part of (ii)}\,, \\ &= -[[x_2, r_{12}], r_{23}] + [x_2, [r_{12}, r_{23}]] + [x_3, [r_{12}, r_{23}]]\,.\end{aligned}$$

Resubstitution then gives

$$\begin{aligned}\tau_{123}(1 \otimes \varphi)\varphi(x) &= \tau_{123}\{[x_3, [r_{31}, r_{12}]] + [[x_1, r_{31}], r_{12}] - [[x_2, r_{12}], r_{23}] \\ &\quad + [x_2, [r_{12}, r_{23}]] + [x_3, [r_{12}, r_{23}]]\}\,.\end{aligned}$$

The cyclic sum forces the cancellation of the second and third terms and replacing the first term by a cyclic permutation of itself gives

$$\begin{aligned}\tau_{123}(1 \otimes \varphi)\varphi(x) &= \tau_{123}[\Delta^2 x, [r_{12}, r_{23}]] \\ &= [\Delta^2 x, [\![r, r]\!]_\tau] \\ &= [\Delta^2 x, [\![r, r]\!]], \quad \text{by A.4.6.}\end{aligned}$$

Hence the assertion.

A.4.8 The equation $[\![r, r]\!] = 0$ is known as the classical Yang-Baxter equation. The weaker statement $[\Delta^2 x, [\![r, r]\!]] = 0$, $\forall x \in \mathfrak{g}$ expresses the $\mathfrak{g}$ invariance of the expression $[\![r, r]\!]$. It is called the generalized classical Yang-Baxter equation. If r is antisymmetric, that is if $r = -\tau(r)$ then the above result can be expressed in the form.

Theorem. *The pair* $(\mathfrak{g}, \varphi)$ *where* $\varphi : x \mapsto [\Delta(x), r]$, *defines a Lie bialgebra structure if and only if* r *satisfies the generalized classical Yang-Baxter equation.*

A.4.9 Yu. I. Manin has given a simple recipe for constructing a significant number of Hopf algebras including $R_q[SL(n)]$. The theory is reviewed briefly below. No conditions are needed on the base field k.

Let V be an n dimensional k vector space. Let W be a subspace of $V \otimes V$ and $\langle W \rangle$ the ideal generated by W in $T := \mathcal{T}(V)$. A quadratic algebra A is one of the form $\mathcal{T}(V)/\langle W \rangle$. Its quadratic dual $A^!$ is defined to be $\mathcal{T}(V^*)/\langle W^\perp \rangle$ where $W^\perp$ is the orthogonal of W in $V^* \otimes V^*$. Let e be the image of $1d_{V \otimes V}$ under the isomorphism $\mathrm{End}(V \otimes V) \xrightarrow{\sim} V \otimes V \otimes V^* \otimes V^*$. One easily verifies the following simple but crucial fact

$$e \in W \otimes V^* \otimes V^* + V \otimes V \otimes W^\perp . \tag{1}$$

Let $\{v_i\}_{i=1}^n$ be a basis for V and $\{\xi_i\}_{i=1}^n$ the dual basis for V^*. The $z_{i,j} := \xi_i \otimes v_j$ form a basis for $V^* \otimes V$. Define a coproduct Δ on $F := \mathcal{T}(V^* \otimes V)$ through

$$\Delta(z_{i,j}) = \sum_{m=1}^{n} z_{i,m} \otimes z_{m,j} . \tag{2}$$

Let $\tau_{23} : V^* \otimes V^* \otimes V \otimes V \longrightarrow V^* \otimes V \otimes V^* \otimes V$ be the interchange of the elements in the second and third factors. One checks that (2) implies the rather surprising fact

$$\Delta \tau_{23}(\zeta \otimes u) = (\tau_{23} \otimes \tau_{23})(\zeta \otimes e \otimes u) , \quad \forall \zeta \in V^* \otimes V^* , \quad u \in V \otimes V . \tag{3}$$

Substitution from (1) then implies that $I := \tau_{23}(W^\perp \otimes W)$ satisfies

$$\Delta(I) \subset I \otimes F + F \otimes I . \tag{4}$$

Again T (resp. $T^{\succ}$) has the structure of a right (resp. left) F comodule $\nu : T \longrightarrow T \otimes F$ (resp. $\lambda : T^{\succ} \longrightarrow F \otimes T^{\succ}$) defined through

$$\nu(v_i) = \sum_{m=1}^{n} v_m \otimes z_{m,i} , \quad \lambda(\xi_i) = \sum_{m=1}^{n} z_{i,m} \otimes \xi_m . \tag{5}$$

One checks that

$$\nu(u) = (1d \otimes \tau_{23})(e \otimes u) , \quad \lambda(\zeta) = (\tau_{23} \otimes 1d)(\zeta \otimes e) \tag{6}$$

for all $u \subset V \otimes V$; $\zeta \in V^* \otimes V^*$.

Substitution from (1) then implies that

$$\nu(W) \subset T \otimes I + W \otimes F , \quad \lambda(W^\perp) \subset I \otimes T^{\succ} + F \otimes W^\perp . \tag{7}$$

It is clear that (4) implies that $\langle I \rangle$ is a coideal of F and then (7) implies that ν (resp. λ) induces a right (resp. left) $F/\langle I \rangle$ comodule structure on A (resp. $A^!$). Of course in a similar fashion $\mathcal{T}(V \otimes V^*)/\langle \tau_{23}(W \otimes W^\perp) \rangle$ admits $A^!$ (resp. A) as a right (resp. left) comodule. Finally one may identity V with V^* through the linear isomorphism taking v_i to ξ_i. Equivalently one may assume V given a scalar product. Then the quotient algebra $M_W(V) = \mathcal{T}(V \otimes V)/J$, where $J := I +{}^t I$ with ${}^t z_{i,j} = z_{j,i}$, is defined. The coidentity on $M_W(V)$ is defined by $\varepsilon(z_{i,j}) = \delta_{ij}$.

Theorem. *The natural coproduct gives $M_W(V)$ the structure of a bialgebra admitting A and $A^!$ as left and right comodules.*

Remarks. If $W = k\{v_i \otimes v_j - v_j \otimes v_i \mid i, j = 1, 2, \ldots, n\}$, then A is just the symmetric algebra $S(V)$ over V and $A^!$ the exterior algebra $\wedge^* V$. Moreover $M_W(V)$ is the algebra of regular functions on $\operatorname{End} V$. When

$$(*) \qquad W_q := k\{v_i \otimes v_j - q v_j \otimes v_i \mid 1 \le i < j \le n\}$$

then A is called the algebra of regular functions on quantum n-space and $M_{W_q}(V)$, or simply $M_q(V)$, the algebra of regular functions on the "space of quantum matrices".

A.4.10 One may give a second description of $M_q(V)$ and with it an intrinsic description of the quantum determinant $\det_q(V)$ as follows.

Take $\mathfrak{g} = \mathfrak{sl}(n)$, $q = q_i$, $\forall i$ and set $V = V(\omega_1)$. Let $\{v_i\}_{i=1}^n$ be a basis for V chosen so that $e_i v_{i+1} = v_i$ and v_{i+1} has weight $\omega_1 - \alpha_1 - \alpha_2 - \ldots - \alpha_i$ for $1 \le i < n$. One checks using 5.1.1 (vii) that W_q in A.4.9($*$) is a $U := U_q(\mathfrak{sl}(n))$ submodule of $V \otimes V$. From say the Weyl character formula one may check that $V(\omega_1) \otimes V(\omega_i) \cong V(\omega_1 + \omega_i) + V(\omega_{i+1}) : i = 1, 2, \ldots, n-1$ identifying $V(\omega_n)$ with the trivial module. In particular $W_q \cong V(\omega_2)$ and $W_q^\perp = V(2\omega_1)$. Then A (resp. $A^!$) identifies with the quantum symmetric algebra $S_q(V)$ (resp. quantum exterior algebra $\wedge_q^*(V)$) which as a $U_q(\mathfrak{sl}(n))$ module is given by

$$S_q(V) = \bigoplus_{n \in \mathbb{N}} V(n\omega_1)\,, \quad \wedge_q^*(V) = \bigoplus_{i=1}^{n} V(\omega_i)\,.$$

Moreover the above isomorphism gives an explicit description of $\wedge_q^*(V)$ in terms of the given basis for V. A further use of the Weyl character formula shows that the trivial module occurs in $V^{\otimes n}$ with multiplicity 1 and then may be identified with $V(\omega_n)$. Moreover the latter may be described explicitly as follows.

Identify W with the symmetric group S_n. One checks that $\ell(s_i w) > \ell(w) \iff r := w^{-1}(i+1) > w^{-1}(i) =: s$. Then $e_i v_{w(r)} = v_{w(s)}$ whilst $e_i v_{w(j)} = 0$ for $j \ne r$. Again $t_i v_i = q v_i$, $t_i v_{i+1} = q^{-1} v_{i+1}$, $t_i v_j = v_j$ for $j \notin \{i, i+1\}$. Since

$$\Delta^{n-1}(e_i) = \sum_{j=1}^{n-1} 1d^{\otimes j} \otimes e_i \otimes (t_i^{-1})^{\otimes n-j-1}$$

it follows that

$$\begin{aligned} &e_i(v_{w(1)} \otimes v_{w(2)} \otimes \ldots \otimes v_{w(n)}) \\ &= \begin{cases} v_{w(1)} \otimes \ldots \otimes v_{w(r-1)} \otimes v_i \otimes \ldots \otimes v_{w(n)} & : \ell(s_i w) > \ell(w) \\ q^{-1} v_{w(1)} \otimes \ldots \otimes v_{w(r-1)} \otimes v_i \otimes \ldots \otimes v_{w(n)} & : \ell(s_i w) < \ell(w) \, . \end{cases} \end{aligned}$$

Consequently

$$u := \sum_{w \in W} (-q)^{\ell(w)} v_{w(1)} \otimes v_{w(2)} \otimes \ldots \otimes v_{w(n)}$$

satisfies $e_i u = 0$ for all i. Since it has zero weight, it must also satisfy $f_i u = 0$ for all i. It is therefore the required generator of $V(\omega_n)$.

Now consider the n-fold tensor product of $C^V \cong V^* \otimes V$. By the above it admits a unique up to scalars U bi-invariant element $\mathrm{Det}_q(V)$ which may be written as

$$\mathrm{Det}_q(V) = \frac{1}{n!} \sum_{x,y \in W} (-q)^{\ell(x)+\ell(y)} c_{x(1),y(1)} \otimes c_{x(2),y(2)} \otimes \ldots \otimes c_{x(n),y(n)} \, .$$

However this is *not* the quantum determinant in the usually accepted sense. Rather the latter is its image $\det_q(V)$ in $M_q(V)$. This algebra may be identified with the quotient of $\mathcal{T}(C^V)$ by the quadratic relations K obtained from 9.1.3 and U bimodule action. Indeed observe in the construction of A.4.9 that J is trivially a U bimodule for the above identification of $W, W^\perp$. One checks that J contains the relation of 9.1.3, so $J \supset K$. On the other hand the simplicity of $W, W^\perp$ as U modules forces the simplicity of I and its transpose ${}^tI$ as U bimodules. Since the relation of 9.1.3 is t-invariant one has ${}^tK = K$ and hence $K = J$ as required. As in 9.1.12(i) it follows from 9.1.4 and the invariance of $\det_q(V)$ that it is central in $M_q(V)$.

In the classical case (that is when $q = 1$) K implies exactly the commutativity of the $c_{i,j}$. In the quantum case it follows from 9.1.5 that up to an appropriate ordering $M_q(V)$ is spanned by the standard monomials in the $c_{i,j}$. Moreover the relations between the $c_{i,j}$ only involve powers of q and so specialization at $q = 1$ (defined by the $k[q, q^{-1}]$ ring they generate) recovers the above polynomial algebra. In particular the standard monomials must form a basis for $M_q(V)$ as may also be verified using the diamond lemma (A.2.13) and 9.1.5.

It is clear from the above that $M_q(V)$ is an integral domain of GK dimension n^2. Again one may adjoin a central group-like element to $U_q(\mathfrak{sl}(n))$ acting on $V(\omega_m)$ as m times the identity. The resulting Hopf algebra is denoted by $U_q(\mathfrak{gl}(n))$ and its Hopf dual (relative to modules with integral weights) by $R := R_q[\mathrm{GL}(n)]$. It has Gelfand-Kirillov dimension equal to $\dim \mathrm{GL}(n) = n^2$

by the growth estimate which follows from the Peter-Weyl theorem, that is 1.4.13(∗). On the other hand the subalgebra of R generated by C^V satisfies the relations K and hence is an image of $M_q(V)$. Then the inverse of $\det_q(V)$, which is central, generates R over this subalgebra. The above GK dimension estimate combined with A.3.6 (ii) implies that

$$(*) \qquad M_q(V)[\det_q(V)^{-1}] \xrightarrow{\sim} R_q[\mathrm{GL}(n)] .$$

Finally by the Peter-Weyl theorem any left $U_q(\mathfrak{gl}(n))$ invariant element is also right invariant. Hence $\det_q(V)$ can be simply written as

$$\det_q(V) = \sum_{w \in W} (-q)^{\ell(w)} c_{1,w(1)} c_{2,w(2)} \cdots c_{n,w(n)} .$$

A.5 Comments and Complements

A.5.1 The results in A.1.1–A.1.19 are classical excepting A.1.16 which is a slight variation on [Ja1, 1.3]. A more leisurely treatment of A.1.1 can be found in [Ca1] where also the Bruhat decomposition is described. Several further results of A.1 can be found in [Bb1].

A.5.2 Assume C of finite type. Fix $\lambda \in P^+(\pi)$ and set $J(w.\lambda) = \mathrm{Ann}_{U(\mathfrak{g})} \mathcal{V}(w.\lambda) : w \in W$. An early result [Du1] of M. Duflo asserted that $J(x.\lambda) \supset J(y.\lambda)$ if $x \overset{D}{\leq} y$. This induces an order relation on left cells which for example in $\mathfrak{sl}(n)$ are in bijection with the set $St(n)$ of Standard Tableaux attached to the partitions of n. A. Melnikov [Me1] has described this order relation on $St(n)$ in terms of the Robinson-Schensted algorithm and has shown that it is strictly weaker than that defined by inclusion of ideals.

A.5.3 The notion of real and imaginary roots is due to V. Kac who obtained their principal properties and also classified Cartan matrices of integrable type [Ka1].

A.5.4 The Hecke algebra played a fundamental role in the representation theory of finite Chevalley groups and was notably a starting point of the Kazhdan-Lusztig conjecture [KL1,2]. The singular Hecke algebra can be realized through products of almost primitive ideals [J14] and the monoid generated by the Enright functors [J15]. In the Kostant-Kumar theory [KoK1] it is used to describe the singularities of Schubert varieties.

A.5.5 The analysis in A.1.20–A.1.23 is a combination of the work of S. G. Hulsurkar [Hu1] and R. Steinberg [St1]. It is unclear to me if A.1.23(i) and A.1.23 (ii) are equivalent. D. A. Vogan [V2] remarked that A.1.23 (i), which is due to R. Steinberg, implies that the $\mathcal{D}_w : w \in W$ form a basis of H; but this is weaker than S. G. Hulsurkar's little known and earlier result A.1.23 (ii) conjectured by D. N. Verma. J. Bernstein pointed out to me

[J13, Sect. 6] the fact that the $\mathcal{D}_\lambda : \mu \mapsto \mathcal{D}(\mu + \lambda) : \lambda \in P(\pi)$ generate $H_{\mathbb{Z}}$ is a consequence of one being able to resolve an arbitrary coherent sheaf on the flag variety G/B in terms of the invertible sheaves $\mathcal{L}_\lambda$, the latter assertion being a classical result of A. Borel.

A.5.6 The theorem in A.2.7 is due to A. W. Goldie being one of the most remarkable results in noetherian ring theory for generality and usefulness. The result in A.2.8 is taken from [BGR1] and A.2.9 from [J10]. Of course A.2.10 is (the non-commutative version of) Nakayama's lemma, whilst A.2.13 is G. M. Bergman's Diamond Lemma for Ring Theory [Be1].

A.5.7 The q-binomial identity and many such results can be found in [An1]. Generic flatness first arose in algebraic geometry (A. Grothendieck). Its application to the non-commutative nullstellensatz arise from work of D. Quillen, P. Gabriel and M. Duflo. For more details the reader may consult [MR1, Chap.9]. The classification of simple A_1 modules alluded to in A.3.5 (Remark) is due to R. Block [Bl1]. More details on GK dimension may be obtained from [MR1] and [KrL1]. The trick using radius of convergence in A.3.8 was explained to me by O. Mathieu. Of course $G(i)$ is a partition function and much more precise results on its growth rate are known from classical number theory.

A.5.8 J. C. McConnell and J. T. Stafford [MS1, 1.7] showed that a bounded filtration $\mathcal{F}$ on a ring A for which $gr_{\mathcal{F}}A$ is commutative and finitely generated can be modified without losing these properties so that the graded subspaces are finite dimensional. Here commutative can be replaced by quasi-commutative [Mc1, 4.7].

A.5.9 The results in A.4 have their origins in the work of V. G. Drinfeld. In particular A.4.8 is his often quoted interpretation of the Yang-Baxter equation of which the details are seldom given. Use of not necessarily antisymmetric r matrices is due to N. Yu. Reshetikhin and M. A. Semenov-Tian-Shansky [RS1, 1.1] though they assume r to satisfy the ordinary Yang-Baxter equation. The result in A.4.7 is a common generalization of these works as taken from the dissertation of M. Kebe whom I would also like to thank for showing me how to do this terrible calculation. The treatment in A.4.2 is developed from J.-H. Lu and A. Weinstein [LW1]. The description of symplectic leaves in the case of a double Lie group is due to M. A. Semenov-Tian-Shansky [Sem1, 1.1]. The more detailed conclusion when $\Gamma \neq \{e\}$ is taken from T. Hodges and T. Levasseur [HL1, appendix] as is also the precise description of the symplectic leaves in A.4.5.

A.5.10 When $\mathfrak{g}$ is simple the classical Yang-Baxter equation has the solution $r(x,y) = \frac{1}{(x-y)}\Sigma(e_i \otimes e_i)$, where $\{e_i\}$ is an orthonormal basis for $\mathfrak{g}$, due to P. P. Kulish and E. K. Skylanin [KS1]. For a discussion of the various types of solutions and a historical presentation see V. G. Drinfeld [Dr2, Sect. 4]. The more precise nature of these various solutions was studied by A. A. Belavin and V. G. Drinfeld [BD1] and by A. Stolin [Sto1].

A.5.11 A basic problem is to pass from the classical Yang-Baxter equation to the quantum one described in 9.4 and 9.5.2. Significant results have been noted by V. G. Drinfeld [Dr2], M. Gerstenhaber and S. D. Schack [GS1], N. Yu. Reshetikhin [Re2], V. Turaev [T2] in particular. A detailed analysis of the deformation theory involved and the precise information on proofs can be found in the book of S. Shnider and S. Steinberg [ShS1]. For more recent results, see [GS1, GGS1]. The construction of A.4.9 is due to Y. I. Manin [Man1, 2].

A.5.12 The description in A.4.10 of $R_q[SL(n)]$ and of $\det_q(V)$ is taken from [S2] which is in turn attributed to V. G. Drinfeld and to N. Yu. Reshetikhin, L. A. Takhtajan and L. D. Fadeev [FRT1]. For an application to $M_q(V)$ see [LevS1]. For a discussion of quantum groups in the context of ring theory, see [Sm1].

A.5.13 D. R. Farkas and G. Letzter [FL1] have observed that for any non-commutative algebra A with a Poisson bracket, the following identity $[a,c]\{b,d\} = \{a,c\}[b,d]$, $\forall a,b,c,d \in A$ holds where $[a,b]$ is the usual commutator. This indicates that the Poisson bracket is less interesting in the non-commutative case.

A.5.14 The term Poisson sheets was motivated by a paper of J. Dixmier [Di3]. In the coadjoint case $X = \mathfrak{g}^*$ with $\mathfrak{g}$ semisimple, these sheets have a very nice classification [Di3], [Bo2].

A.5.15 A special and important case of a quadratic algebra (A.4.9) is a Koszul algebra. For example both $S_q(V)$ and $\wedge_q^*(V)$ are Koszul and it is possible that these should be the only ones worth considering in Manin's construction. Koszul algebras have been intensively studied for a number of reasons, see for example [AV1] and [BGS1].

Bibliography

Abe, E.

[Ab1] Hopf Algebras. Cambridge Tracts in Mathematics 74. Cambridge University Press, Cambridge 1977

Alev, J., Chamarie, M.

[AC1] Dérivations et Automorphismes de quelques algèbres quantiques. Comm. Algebra **20** (1992) 1787–1802

Alev, J., Dumas, F.

[AD1] Sur le corps des fractions de certaines algèbres quantiques. J. Algebra (in press)

Andersen, H.H.

[A1] Schubert varieties and Demazure's character formula. Invent. Math. **5** (1985) 611–618

[A2] Jantzen's filtrations of Weyl modules. Math. Z. **194** (1987) 127–142

[A3] Tensor products of quantized tilting modules. Comm. Math. Phys. **149** (1992) 149–159

Andersen, H.H., Jantzen, J.C., Soergel, W.

[AJS1] Representations of quantum groups at a pth root of unity and of semisimple groups in characteristic p: independence of p. Astérisque **220** (1994)

Andersen, H.H., Paradowski, J.

[AP1] Fusion categories arising from semisimple Lie algebras. Comm. Math. Phys. (in press)

Andersen, H.H., Polo, P., Kexin, W.

[APK1] Representations of quantum algebras. Invent. Math. **104** (1991) 1–59

[APK2] Injective modules for quantum groups. Amer. J. Math. **114** (1992) 571–604

Andrews, G.E.

[An1] The theory of partitions. In: Encyclopedia of Mathematics and its Applications, vol. 2. Addison-Wesley, London 1976

Artin, M., Van Den Bergh, M.

[AV1] Twisted homogeneous coordinate rings. J. Algebra **133** (1990) 249–271

Barbasch, D., Vogan, D.A.

[BV1] Primitive ideals and orbital integrals in complex classical groups. Math. Ann. **259** (1982) 153–199

[BV2] Unipotent representations of complex semisimple groups. Ann. Math. **121** (1985) 41–110

Beck, J.

[B1] Braid group action and quantum affine algebras. Comm. Math. Phys. (in press)

Beilinson, A., Bernstein, J.

[BB1] Localisation de $\mathfrak{g}$ modules. C.R. Acad. Sci. Paris (A) **292** (1981) 15–18

Beilinson, A., Ginzburg, V., Soergel, W.

[BGS1] Koszul duality patterns in representation theory. J. Amer. Math. Soc. (in press)

Belavin, A.A., Drinfeld, V.G.

[BD1] On the solutions of the classical Yang-Baxter equations. Funct. Anal. Appl. **16** (1982) 159–180

Benlolo, E.

[Ben1] Sur la quantification de certaines variétés orbitales. Bull. Sci. Math. **3** (1994) 225–241

Berenstein, A., Zelevinsky, A.

[BZ1] String bases for quantum groups of type A_r. I.M. Gelfand Seminar. Adv. Sov. Math. **16**, Part 1 (1993) 51–89

Bergman, G.M.

[Be1] The diamond lemma for ring theory. Adv. Math. **29** (1978) 178–218

Bernstein, I.N., Gelfand, S.I.

[BG1] Tensor products of finite and infinite dimensional representations of semisimple Lie algebras. Compositio Math. **41** (1980) 245–285

Bernstein, I.N., Gelfand, I.M., Gelfand, S.I.

[BGG1] Schubert cells and cohomology of the spaces G/P. Russian Math. Surveys **28** (1973) 1–26

[BGG2] Differential operators on the base affine space and a study of $\mathfrak{g}$ modules. In: I.M. Gelfand (ed.), Lie Groups and their representations, A. Hilger, London 1975, pp. 22–64

[BGG3] Category of $\mathfrak{g}$ modules. Funct. Anal. Appl. **10** (1976) 87–92

Bernstein, J., Lunts, V.

[BL1] A simple proof of Kostant's theorem that $U(\mathfrak{g})$ is free over its center. Preprint 1994

Björk, J.-E.

[Bj1] Rings of Differential Operators. North Holland, Amsterdam 1979

Block, R.

[Bl1] The irreducible representations of the Lie Algebra $\mathfrak{sl}(2)$ and of the Weyl algebra. Adv. Math. **39** (1981) 69–110

Borel, A.

[Bor 1] Linear algebraic groups. Benjamin, New York 1969

Borho, W.

[Bo1] Nilpotent Orbits, Primitive Ideals and Characteristic Classes (A Survey). In: ICM proceedings, Berkeley 1986, pp. 350–359

[Bo2] Über Schichten halbeinfacher Lie-Algebren. Invent. Math. **65** (1981) 283–317

Borho, W., Brylinski, J.-L. MacPherson, R.

[BBM1] Nilpotent Orbits, Primitive Ideals, and Characteristic Classes. Progress in Mathematics **78**, Birkhäuser, Boston 1989

Borho, W., Gabriel, P., Rentschler, R.

[BGR1] Primideale in einhüllenden auflösbarer Lie-Algebren. Lecture Notes in Mathematics, vol. 357. Springer, Berlin Heidelberg New York 1973

Bourbaki, N.

[Bb1] Groupes et algèbres de Lie. Hermann, Paris (Ch. I: 1971, Ch. II/III: 1972, Ch. IV–VI: 1968, Ch. VII/VIII: 1975)

Braverman, A.

[Br1] On embeddings of quantum groups. C.R. Acad. Sci. Paris (I) **319** (1994) 111–115

Brown, K.A.

[Bw1] The Representation Theory of Noetherian Rings. In: S. Montgomery, L. Small (eds.), Noncommutative Rings, MSRI Publ. **24**. Springer, Berlin 1992, pp. 1–25

Brylinski, J.-L., Kashiwara, M.

[BK1] Kazhdan-Lusztig conjecture and holonomic systems. Invent. Math. **64** (1981) 387–410

Brylinski, R.K.

[Bry1] Limits of weight spaces, Lusztig's q-analogs, and fiberings of adjoint orbits. J. Amer. Math. Soc. **2** (1989) 517–533

Caldero, P.

[C1] Éléments ad-finis de certains groupes quantiques. C.R. Acad. Sci. Paris (I) **316** (1993) 327–329

[C2] Générateurs du centre de $U_q(\mathfrak{sl}(N+1))$. Bull. Sci. Math. **118** (1994) 177–208

[C3] Sur le centre de $U_q(\mathfrak{n}^+)$. Beiträge zur Algebra und Geometrie **35** (1994) 13–23

[C4] Etude des q-commutations dans l'algèbre $U_q(\mathfrak{n}^+)$. Preprint 1994

Carter, R.W.

[Ca1] Simple Groups of Lie Type. Interscience, John Wiley, London 1972

Cherednik, I.

[Ch1] Quantum Knizhnik-Zamolodchikov Equations and Affine Root Systems. Comm. Math. Phys. **150** (1992) 109–136

Chin, W., Musson, I.M.

[CM1] The coradical filtration for quantized enveloping algebras. Preprint 1994

Cigler, J.

[Ci1] Operatormethoden für q-Identitäten. Mh. Math. **88** (1979) 87–105

Cohen, M.

[Co1] Smash products, inner actions and quotient rings. Pac. J. Math. **125** (1986) 45–65

Conze, N.

[Coz1] Algèbres d'opérateurs différentiels et quotients des algèbres enveloppantes. Bull. Soc. Math. France **102** (1974) 379–415

Conze-Berline, N., Duflo, M.

[CD1] Sur les représentations induites des groupes semi-simples complexes. Compositio Math. **34** (1977) 307–336

De Concini, C.
[Del] Poisson Algebraic groups and representations of Quantum groups at roots of 1. In: Progress in Mathematics **120**, Birkhäuser, Boston 1994, pp. 93–119

De Concini, C., Kac, V.G.
[DK1] Representations of quantum groups at roots of 1. In: Progress in Mathematics **92**, Birkhäuser, Boston 1990, pp. 471–506

De Concini, C., Kac, V.G., Procesi, C.
[DKP1] Quantum coadjoint action. J. Amer. Math. Soc. **5** (1992) 151–890

Demazure, M.
[D1] Désingularisation des variétés de Schubert généralisées. Ann. Ec. Norm. Sup. **7** (1974) 53–88
[D2] Une nouvelle formule des caractères. Bull. Sc. Math. **98** (1974) 163–172

Deodhar, V.V.
[Del] On a construction of representations and a problem of Enright. Invent. Math. **57** (1980) 101–118

Dixmier, J.
[Di1] Sur les représentations unitaires des groupes de Lie nilpotents IV. Can. J. Math. **11** (1959) 321–344
[Di2] Algèbres enveloppantes. Cahiers scientifiques, vol. 37. Gauthier-Villars, Paris 1974
[Di3] Polarisations dans les algèbres de Lie semi-simples complexes. Bull. Sci. Math. **99** (1975) 45–63

Drinfeld, V.G.
[Dr1] Hopf algebras and the quantum Yang-Baxter equation. Sov. Math. Dokl. **32** (1985) 254–258
[Dr2] Quantum Groups. In: ICM proceedings, Berkeley 1986, pp. 798–820
[Dr3] On almost cocommutative Hopf algebras. Leningrad Math. J. **1** (1990) 321–342
[Dr4] Quasi-Hopf algebras. Leningrad Math. J. **1** (1990) 1419–1457
[Dr5] On some unsolved problems in quantum group theory. In: P.P. Kulish (ed.), Quantum Groups. Lecture Notes in Mathematics, vol. 1510, Springer, Berlin 1992, pp. 1–8

Duflo, M.
[Du1] Sur la classification des idéaux primitifs dans l'algèbre enveloppante d'une algèbre de Lie semi-simple. Ann. Math. **105** (1977) 107–120

Enright, T.J, Joseph, A.
[EJ1] An intrinsic analysis of unitarizable highest weight modules. Math. Ann. **288** (1990) 571–594

Faddeev, L.D., Reshetikhin, N.Y., Takhtajan, L.A.
[FRT1] Quantization of Lie groups and Lie algebras. Leningrad Math. J. **1** (1990) 193–225

Farkas, D.R., Letzter, G.
[FL1] Ring theory from symplectic geometry. Israel J. Math. (in press)

Frenkel, E., Szenes, A.
[ES1] Crystal bases, dilogarithm identities and torsion in algebraic K-groups. J. Amer. Math. Soc. (in press)

Gabber, O.
[G1] The integrability of the characteristic variety. Amer. J. Math. **103** (1981) 445–468

Gabber, O., Joseph, A.
[GJ1] On the Bernstein-Gelfand-Gelfand resolution and the Duflo sum formula. Compositio math **43** (1981) 107–131
[GJ2] Towards the Kazhdan-Lusztig conjecture. Ann. Ec. Norm. Sup. **14** (1981) 261–302

Gabriel, P.
[Gab1] Unzerlegbare Darstellungen I. Manuscripta Math. **6** (1972) 71–103

Gaitsgory, D.
[Ga1] On the existence and uniqueness of the R-matrix. J. Algebra (in press)

Gekhtman, M.I., Stolin, A.
[GS1] Orbits of coadjoint representation and Yang-Baxter equation. Preprint 1994

Gerstenhaber, M.
[Ge1] On the deformations of rings and algebras. Ann. Math. **79** (1964) 59–103

Gerstenhaber, M., Giaquinto, A., Schack, S.D.
[GGS1] Construction of quantum groups from Belavin-Drinfeld infinitesimals. In: A. Joseph and S. Shnider (eds.), Quantum Deformations of Algebras and their Representations, Israel Math. Conf. Proc. **7** (1983) 25–64

Gerstenhaber, M., Schack, S.D.
[GS1] Quantum groups as deformations of Hopf algebras. Proc. Nat. Acad. Sci. **87** (1990) 478–481

Goodearl, K.R., Letzter, E.
[GL1] Prime factor algebras of the coordinate ring of quantum matrices. Proc. Amer. Math. Soc. (in press)

Grojnowski, I., Lusztig, G.
[GrL1] A comparison of bases of quantized enveloping algebras. Contemp. Math. **153** (1993) 11–19

Grossberg, M., Karshon, Y.
[GK1] Bott towers, complete integrability and the extended character of representations. Duke Math. J. (in press)

Gurevich, D.I.
[Gu1] Algebraic aspects of the quantum Yang-Baxter equation. Leningrad Math. J. **2** (1991) 803–828

Hadžiev, Dž,
[H1] Some questions in the theory of vector invariants. Math. USSR Sbornik **1** (1967) 383–396

Hesselink, W.H.
[He1] Characters of the Nullcone. Math. Ann. **252** (1980) 179–182

Hochschild, G., Serre, J.-P.
[HSe1] Cohomology of Lie algebras. Ann. Math. **57** (1953) 591–605

Hodges, T.J., Levasseur, T.
[HL1] Primitive ideals of $\mathbb{C}_q[SL(3)]$. Comm. Math. Phys. **156** (1993) 581–605
[HL2] Primitive ideals of $\mathbb{C}_q[SL(n)]$. J. Algebra **156** (1993) 581–605

Hodges, T.J., Levasseur, T., Tors, M.
[HLT1] Algebraic Structure of Multiparameter Quantum Groups. Preprint 1994

Hodges, T.J., Smith, S.P.
[HS1] Sheaves of noncommutative algebras and the Beilinson-Bernstein equivalence of categories. Proc. Amer. Math. Soc. **93** (1985) 379–386

Hotta, R.
[Ho1] On Joseph's construction of Weyl group representations. Tohoku Math. J. **36** (1984) 49–74

Hulsurkar, S.G.
[Hu1] Proof of Verma's conjecture on Weyl's dimension polynomial. Invent. Math. **27** (1974) 45–52

Irving, R.S.
[I1] The Socle Filtration of a Verma Module. Ann. Ec. Norm. Sup. **21** (1988) 47–65
[I2] BGG Algebras and the BGG Reciprocity Principle. J. Algebra **135** (1990) 363–380

Jacobson, N.
[Jac1] Lie Algebras. Interscience, John Wiley, New York 1962

Jantzen, J.C.
[Ja1] Moduln mit einem höchsten Gewicht. Lecture Notes in Mathematics, vol. 750, Springer, Berlin Heidelberg New York 1979
[Ja2] Einhüllende Algebren halbeinfacher Lie-Algebren. Springer, Berlin 1983
[Ja3] Representations of Algebraic Groups. Pure and Applied Mathematics **131**, Academic Press

Jimbo, M.
[Ji1] A q-difference analogue of $U(\mathfrak{g})$ and the Yang-Baxter equation. Lett. Math. Phys. **10** (1985) 63–69
[Ji2] Quantum R matrix for the generalized Toda system. Comm. Math. Phys. **102** (1986) 537–547

Jones, V.F.R.
[Jo1] Hecke algebra representations of braid groups, and link polynomials. Ann. Math. **126** (1987) 335–388

Joseph, A.
[J1] A generalization of the Gelfand-Kirillov conjecture. Amer. J. Math. **99** (1977) 1151–1165
[J2] A preparation theorem for the prime spectrum of a semisimple Lie algebra. J. Algebra **48** (1977) 241–289
[J3] Dixmier's problem for Verma and principal series submodules. J. London Math. Soc. **20** (1979) 193–204
[J4] W-module structure in the primitive spectrum of the enveloping algebra of a semisimple Lie algebra. In: Non-Commutative Harmonic Analysis, Lecture Notes in Mathematics, vol. 728, Springer, Berlin Heidelberg New York 1979, pp. 116–135
[J5] Primitive Ideals in Enveloping Algebras. ICM proceedings, Warsaw 1983
[J6] On the Demazure character formula. Ann. Ec. Norm. Sup. **18** (1985) 381–419
[J7] Rings which are modules in the BGG $\mathcal{O}$ category. J. Algebra **113** (1988) 110–126

[J8] The primitive spectrum of an enveloping algebra. Astérisque **173-174** (1989) 13–53

[J9] Some Ring Theoretic Techniques and Open Problems in Enveloping Algebras. In: S. Montgomery and L. Small (eds.), Noncommutative Rings, MSRI Publ. **24**, Springer, Berlin Heidelberg New York 1992, pp. 27–67

[J10] Faithfully flat embeddings for minimal primitive quotients of quantized enveloping algebras. In: A. Joseph and S. Shnider (eds.), Quantum Deformations of Algebras and their representations, Israel Math. Conf. Proc. **7** (1993), pp. 79–106

[J11] On the prime and primitive spectra of the algebra of functions on a quantum group. J. Algebra (in press)

[J12] Enveloping Algebras: Problems old and new. In: Progress in Mathematics **123**, Birkhäuser, Boston 1994, pp. 385–413

[J13] Characteristic polynomials for orbital varieties. Ann. Ec. Norm. Sup. **22** (1989) 569–603

[J14] On the annihilators of the simple subquotients of the principal series **10** (1977) 419–440

[J15] The Enright functor on the Bernstein-Gelfand-Gelfand category $\mathcal{O}$. Invent. Math. **67** (1982) 423–445

[J16] Kostant's problem, Goldie rank, and the Gelfand-Kirillov conjecture. Invent. Math. **56** (1980) 193–204

[J17] Goldie rank in the enveloping algebra of a semisimple Lie algebra, I–III. J. Algebra **65** (1980) 269–283, 284–306; **73** (1981) 295–326

[J18] On the cyclicity of vectors associated to Duflo involutions. In: Lecture Notes in Mathematics, vol. 1243, Springer, Berlin Heidelberg New York 1987, pp. 144–188

Joseph, A., Letzter, G.

[JL1] Local finiteness for the adjoint action for quantized enveloping algebras. J. Algebra **153** (1992) 289–318

[JL2] Separation of variables for quantized enveloping algebras. Amer. J. Math. **116** (1994) 127–177

[JL3] Verma module annihilators for quantized enveloping algebras. Ann. Ec. Norm. Sup. (in press)

[JL4] Rosso's form and quantized Kac-Moody algebras. Preprint 1994

Joseph, A., Perets, G.S., Polo, P.

[JPP1] Sur l'equivalence de catégories de Beilinson et Bernstein. C.R. Acad. Sci. Paris (I) **313** (1991) 705–709

Kac, V.G.

[Ka1] Infinite dimensional Lie algebras. Cambridge Univ. Press 1985

Kac, V.G., Kazhdan, D.A.

[KK1] Structure of representations with highest weight of infinite dimensional Lie algebras. Adv. Math. **34** (1979) 97–108

Kashiwara, M.

[K1] Kazhdan-Lusztig conjecture for symmetrizable Kac-Moody Lie algebra. In: Progress in Mathematics **87**, Birkhäuser, Boston 1991, pp. 407–433

[K2] Crystalizing the q-Analogue of Universal Enveloping Algebras. Comm. Math. Phys. **133** (1990) 249–260

[K3] On crystal bases of the q-analogue of the universal enveloping algebra. Duke Math. J. **63** (1991) 465–516

[K4] Global crystal bases of quantum groups. Duke Math. J. **69** (1993) 455–485

[K5] Crystal base and Littelmann's refined Demazure character formula. Duke Math. J. **71** (1993) 839–858

[K6] Crystal bases of modified quantized enveloping algebra. Duke Math. J. **73** (1994) 383–413

Kashiwara, M., Tanisaki, T.

[KT1] Kazhdan-Lusztig conjecture for symmetrizable Kac-Moody Lie algebra, II. Intersection cohomologies of Schubert varieties. In: Progress in Mathematics **92**, Birkhäuser, Boston 1990, pp. 159–195

Kassel, C.

[Ks1] Algèbres de Hopf – Catégories tensorielles. Mimeographed Lecture Notes, Strasbourg 1993

Kazhdan, D.A., Lusztig, G.

[KL1] Representations of Coexter groups and Hecke algebras. Invent. Math. **53** (1979) 165–184

[KL2] Tensor structures arising from affine Lie algebras, I–IV. J. Amer. Math. Soc. **6** (1993) 905–947, 949–1011; **7** (1994) 335–381, 383–453

Khoroshkin, S.M., Tolstoy, V.N.

[KT1] Uniqueness theorem for the universal R-matrix. Lett. Math. Phys. **24** (1992) 231–244

Kirillov, A.N., Reshetikhin, N.

[KR1] q-Weyl group and a multiplicative formula for universal R-matrices. Comm. Math. Phys. 184 (1990) 421–431

Kostant, B.

[Ko1] Lie group representations on polynomial rings. Amer. J. Math. **85** (1963) 327–404

[Ko2] Lie algebra cohomology and generalized Schubert cells. Ann. Math. **77** (1963) 72–144

[Ko3] Groups over $\mathbb{Z}$. In: A. Borel, G.D. Mostow (eds.), Algebraic Groups and Their Discontinuous Subgroups, Proc. Symp. Pure Math. **8**, Amer. Math. Soc. 1966, pp. 90–98

Kostant, B., Kumar, S.

[KoK1] T-equivarient K-theory of generalized flag varieties. J. Diff. Geom. **32** (1990) 549–603

Krause, G., Lenagen, T.H.

[KrL1] Growth of Algebras and Gelfand-Kirillov Dimension. Research Notes in Mathematics 116. Pitman, London 1985

Kulish, P.P., Sklyanin, E.K.

[KS1] On the solutions of the Yang-Baxter equations. Zap. Nauchn. Sem. LOMI **95** (1980) 126–160

Kumar, S.

[Ku1] Demazure character formula in arbitrary Kac-Moody setting. Invent. Math. **89** (1987) 395–423

[Ku2] Proof of the Parthasarathy-Ranga Rao-Varadarajan conjecture. Invent. Math. **93** (1988) 117–130

[Ku3] Bernstein-Gelfand-Gelfand resolution for arbitrary Kac-Moody Lie algebra. Math. Ann. **286** (1990) 709–729

Lakshmibai, V.

[La1] Bases for Demazure modules for symmetrizable Kac-Moody algebras. Contemp. Math. **153** (1993) 59–78

[La2] Bases for quantum Demazure modules – I. (To appear in volume dedicated to M.S. Narasimhan and C.S. Seshadri on their 60th birthday)

[La3] Bases for quantum Demazure modules – II. Proc. Symp. Pure Math. **56** (1994) 149–168

Lakshmibai, V., Seshadri, C.S.

[LS1] Standard monomial theory. In: Hyderabad Conference on Algebraic Groups, Manoj Prakashan, Madras 1991, pp. 279–323

Lancaster, G., Towber, J.

[LT1] Representation-Functors and Flag-algebras for the Classical Groups I. J. Algebra **59** (1979) 16–38

Lascoux, A., Schützenberger,

[LaS1] Le monoïde plaxique. La ricerca scientifica, **109** (1981) 129–156

Levendorskiĭ, S.Z., Soibelman, Ya. S.

[LeS1] Some applications of quantum Weyl groups. J. Geom. Phys. **7** (1990) 241–254

Levasseur, T., Stafford, J.T.

[LevS1] The quantum coordinate ring of the special linear group. J. Pure Appl. Algebra **86** (1993) 181–186

Littelmann, P.

[Li1] Crystal graphs and Young Tableaux. J. Algebra (in press)

[Li2] A Littlewood-Richardson rule for symmetrizable Kac-Moody algebras. Invent. Math. **116** (1994) 329–346

[Li3] Path and root operators in representation theory. Ann. Math. (in press)

Liu, C.

[Liu1] Non-semisimplicity of the adjoint representation of quantized enveloping algebras. Preprint 1993

[Liu2] Projective functors for quantized eveloping algebras. Comm. Algebra **22** (**6**) (1994) 2125–2172

Lu, J.-H., Weinstein, A.

[LW1] Poisson Lie groups, Dressing transformations, and Bruhat decompositions. J. Diff. Geom. **31** (1990) 501–526

Lusztig, G.

[L1] Characters of reductive groups over a finite field. Annals of Mathematics Studies 107, Princeton University Press, New Jersey 1984

[L2] Quantum deformations of certain simple modules over enveloping algebras. Adv. Math. **70** (1988) 237–249

[L3] Modular representations and quantum groups. Contemp. Math. **82** (1989) 59–77

[L4] On Quantum Groups. J. Algebra **131** (1990) 466–475

[L5] Finite dimensional Hopf algebras arising from quantized enveloping algebras. J. Amer. Math. Soc. **3** (1990) 257–296

[L6] Quantum groups at roots of 1. Geom. Dedicata **35** (1990) 89–114

[L7] Canonical bases arising from quantized enveloping algebras. J. Amer. Math. Soc. **3** (1990) 447–498

[L8] Canonical bases arising from quantized enveloping algebras II. Progr. Theor. Phys. Suppl. **102** (1990) 175–201

[L9] Quivers, perverse sheaves and quantized enveloping algebras. J. Amer. Math. Soc. **4** (1991) 365–421

[L10] Problems on canonical bases. Amer. Math. Soc., Proc. Symp. (in press)

[L11] Tight monomials in quantized enveloping algebras. In: A. Joseph and S. Shnider (eds.), Quantum Deformations of Algebras and their representations, Israel Math. Conf. Proc. **7** (1993), pp. 117–132

[L12] Introduction to quantum groups. Progress in Mathematics **110**, Birkhäuser, Boston 1993

[L13] Cells in affine Weyl groups II. J. Algebra **109** (1987) 536–548

[L14] Cells in affine Weyl groups III. J. Fac. Sci. Tokyo **34** (1987) 223–243

[L15] Total positivity in reductive groups. In: Progress in Mathematics **123**, Birkhäuser, Boston 1994, pp. 531–568

McConnell, J.C.

[Mc1] Quantum groups, filtered rings and Gelfand-Kirillov dimension. In: Lecture Notes in Mathematics, vol. 1448, Springer, Berlin Heidelberg New York 1991, pp. 139–149

McConnell, J.C., Robson, J.C.

[MR1] Noncommutative Noetherian Rings. John Wiley, New York 1987

McConnell, J.C., Stafford, T.T.

[MS1] Gelfand-Kirillov dimension and associated graded modules. J. Algebra **125** (1989) 197–214

Macdonald, I.G.

[Mcd1] Symmetric functions and Hall polynomials. Clarendon Press, Oxford 1979

McGovern, W.M.

[McG1] Completely prime maximal ideals and quantization. Memoirs of the AMS, vol. 108, no. 519 (1994)

[McG2] Unipotent representations and Dixmier algebras. Compositio Math. **69** (1989) 241–276

MacLane, S.

[Mac1] Natural associativity and commutativity. Rice Univ. Studies **49** (1963) 28–46

[Mac2] Categories for the working mathematician. Springer, Berlin Heidelberg New York 1971

Majid, S.

[Maj1] Braided groups. J. Pure Appl. Alg. **86** (1993) 187–221

Majid, S., Soibelman, Ya. S.

[MS1] Bicrossproduct structure of the quantum Weyl group. J. Algebra **163** (1994) 68–87

Makar-Limanov, L.

[ML1] A version of the Poincaré-Birkhoff-Witt theorem. Bull. Lond. Math. Soc. (in press)

Malliavin, M.-P.

[Ma1] L'algèbre d'Heisenberg quantique. C.R. Acad. Sci. Paris (I) **317** (1993) 1099–1102

Manin, Y.I.

[Man1] Quantum groups and noncommutative geometry. Univ. de Montreal, Quebec 1989

[Man2] Some remarks on Koszul algebras and quantum groups. Ann. Inst. Fourier **37** (1987) 191–205

Mathieu, O.

[M1] Formule de caractères pour les algèbres de Kac-Moody générales. Astérisque **159–160** (1989)

[M2] Filtrations of B-Modules. Duke Math. J. **59** (1989) 421–442

[M3] Good Bases for G-Modules. Geom. Dedicata **36** (1990) 51–66

[M4] Filtrations of G-Modules. Ann. Ec. Norm. Sup. **23** (1990) 625–644

[M5] Bicontinuity of the Dixmier map. J. Amer. Math. Soc. **4** (1991) 837–863

Melnikov, A.

[Mel] Robinson-Schensted procedure and combinatorial properties of geometric order $\mathfrak{sl}(n)$. C.R. Acad. Sci. Paris **315** (1992) 709–714

Moeglin, C.

[Mo1] Ideaux primitifs des algèbres enveloppantes. J. Math. Pure et Appl. **59** (1980) 265–336

[Mo2] Ideaux complètement premiers de l'algèbre enveloppante de $\mathfrak{gl}_n(\mathbb{C})$. J. Algebra **106** (1987) 287–366

Moeglin, C., Rentschler, R.

[MoR1] Orbites d'un groupe algébrique dans l'espaces des idéaux rationnels d'une algèbre enveloppante. Bull. Soc. Math. France **109** (1981) 403–426

[MoR2] Sur la classification des idéaux primitifs des algèbres enveloppantes. Bull. Soc. Math. France **112** (1984)

Montgomery, S.

[Mon1] Bi-invertible actions of Hopf algebras. Israel J. Math. **83** (1993) 45–71

[Mon2] Hopf algebras and their actions on rings. CBMS Lecture Notes, vol. 82, AMS 1993

[Mon3] Indecomposable coalgebras, simple comodules and Hopf algebras. Proc. Amer. Math. Soc. (in press)

Neidhardt, W.

[N1] Verma Module Imbeddings and the Bruhat Order for Kac-Moody Algebras. J. Algebra **109** (1987) 430–438

Nuomi, M., Yamada, H., Mimachi, K.

[NYM1] Finite dimensional representations of the quantum group $GL_q(n+1,\mathbb{C})$ and zonal spherical functions on $U_q(n)\backslash U_q(n+1)$. Japanese J. Math. **19** (1993) 31–80

Parthasarathy, K.P., Ranga Rao, R., Varadarajan, V.S.

[PRV1] Representations of complex semisimple Lie groups and Lie algebras. Ann. Math. **85** (1967) 383–429

[PRV2] The Dirac operator and the discrete series. Ann. Math. **96** (1972) 1–30

Polo, P.

[P1] Variétés de Schubert et Excellentes Filtrations. Astérisque **173–174** (1989) 281–311

Radford, D.E.

[Ra1] Irreducible representations of $U_q(\mathfrak{g})$ arising from $Mod^{\bullet}_{C^{\frac{1}{2}}}$. In: A. Joseph and S. Shnider (eds.), Quantum Deformations of Algebras and their Representations, Israel Math. Conf. Proc. **7** (1993), pp. 143–170

Reshetikhin, N.

[Re1] Quantized universal enveloping algebras, the Yang-Baxter equation and invariants of links. Comm. Math. Phys. **127** (1990) 1–20

[Re2] Quantization of Lie bialgebras. Internat. Math. Res. Notices, Duke Math. J. **7** (1992) 143–151

Reshetikhin, N., Semenov-Tian-Shansky, M.A.

[RS1] Quantum R-matrices and factorization problems. J. Geom. Phys. **5** (1988) 533–550

Reshetikhin, N., Turaev, V.

[RT1] Invariants of 3-manifolds via link polynomials and quantum groups. Invent. Math. **103** (1991) 547–597

Ringel, C.M.

[Ri1] Hall algebras and quantum groups. Invent. Math. **101** (1990) 583–592

Rossmann, W.

[Ro1] Nilpotent Orbital Integrals in a Real Semisimple Lie Algebra and Representations of Weyl Groups. In: Progress in Mathematics **92**, Birkhäuser, Boston 1990, pp. 263–287

Rosso, M.

[R1] Finite Dimensional Representations of the Quantum Analog of the Enveloping Algebra of a Complex Simple Lie Algebra. Comm. Math. Phys. **117** (1988) 581–593

[R2] Analogues de la forme de Killing et du théorème d'Harish-Chandra pour les groupes quantiques. Ann. Ec. Norm. Sup. **23** (1990) 445–467

[R3] Certaines formes bilinéaires sur les groupes quantiques et une conjecture de Schechtman et Varchenko. C.R. Acad. Sci. Paris (I) **314** (1992) 5–8

Semenov-Tian-Shansky, M.A.

[Sem1] Dressing transformations and Poisson group actions. Publ. Res. Inst. Math. Sci. **21** (1986)

Serre, J.-P.

[Se1] Algèbres de Lie semi-simples complexes. W.A. Benjamin, New York 1966

Shafarevich, I.R.

[Sh1] Basic Algebraic Geometry. Springer, Berlin Heidelberg New York 1977

Shapovalev, N.N.

[Sha1] On a bilinear form on the universal enveloping algebra of a complex semisimple Lie algebra. Funct. Anal. Appl. **6** (1972) 65–70

Shnider, S., Sternberg, S.

[ShS1] Quantum Groups: From Coalgebras to Drinfeld Algebras. International Press Inc., Boston 1993

Smith, S.P.

[Sm1] Quantum Groups: An Introduction and Survey for Ring Theorists. In: S. Montgomery and L. Small (eds.), Noncommutative Rings, MSRI Publ. 24, Springer, Berlin Heidelberg New York 1992, pp. 131–178

Soergel, W.

[So1] Équivalences de certaines catégories de $\mathfrak{g}$-modules. C.R. Acad. Sci. Paris (I) **303** (1986) 725–728

[So2] The Prime Spectrum of the Enveloping Algebra of a Reductive Lie Algebra. Math. Z. **204** (1990) 559–581

Soibelman, Ya. S.

[S1] The algebra of functions on a compact quantum group, and its representations. Leningrad Math. J. **2** (1991) 161–178

[S2] Irreducible representations of the function algebra on the quantum group $SU(n)$, and Schubert cells. Sov. Math. Dokl. **40** (1990) 34–38

Springer, T.

[Sp1] Linear Algebraic Groups. Progress in Mathematics **9**, Birkhäuser, Boston 1981

Stasheff, J.

[Sta1] H-spaces from a Homotopy Point of View. Lecture Notes in Mathematics, Vol. 161. Springer, Berlin Heidelberg New York 1970

Steinberg, R.

[St1] On a theorem of Pittie. Topology **14** (1975) 173–177

Stolin A.

[Sto1] On rational solutions of Yang-Baxter equation for $\mathfrak{sl}(n)$. Math. Scand. **69** (1991) 57–80

Sweedler, M.E.

[Sw1] Hopf Algebras. W.A. Benjamin, New York 1969

Taft, E., Wilson, R.L.

[TW1] On antipodes in pointed Hopf algebras. J. Algebra **29** (1974) 27–32

Tanisaki, T.

[Ta1] Killing forms, Harish-Chandra isomorphisms and universal R-matrices for quantum algebras. Infinite Analysis, World Scientific, 1992, pp. 941–961

Turaev, V.

[T1] The Yang-Baxter equation and invariants of links. Invent. Math. **92** (1988) 527–553

[T2] Skein quantization of Poisson algebras of loops on surfaces. Ann. Ec. Norm. Sup. **24** (1991) 635–704

Vergne, M.

[Ve1] Polynômes de Joseph et représentations de Springer. Ann. Ec. Norm. Sup. **23** (1990) 543–562

Vogan, D.A.

[V1] Ordering in the primitive spectrum of a semisimple Lie algebra. Math. Ann. **248** (1980) 195–203

[V2] Gelfand-Kirillov dimension for Harish-Chandra modules. Invent. Math. **48** (1978) 75–98

[V3] Unitary Representations of Reductive Lie Groups. Annals of Mathematics Studies **118**, Princeton University Press, New Jersey 1987

[V4] Dixmier Algebras, Sheets and Representation Theory. In: Progress in Mathematics **92**, Birkhäuser, Boston 1990, pp. 333–395

Zelobenko, D.P.

[Z1] Compact Lie groups and their representations. Transl. Math. Monographs **40** Amer. Math. Soc. Providence 1973

Index of Notation

Symbols occurring frequently are noted below together with the place at which they are first defined. The notation $C_1, C_2, \cdots$ (resp. $\hat{C}_1, \hat{C}_2, \cdots$) signifies that the meaning given is only valid (resp. *not* valid) in the chapter or section noted. Diminutive forms (in which say subscripts are omitted) are given in brackets. It should be noted that μ, η, ξ are also used as general symbols for weights and φ, ψ, θ as general symbols for maps. Sometimes the same symbol denotes a different, but equivalent object. Throughout q denotes an indeterminate.

k	1.1.1, $\hat{C}_5, \hat{C}_6$	$A^\wedge$	1.3.4
$1d_V$ $(1d)$	1.1.1, 1.1.7	θ_χ	1.3.4
μ_A (μ)	1.1.1, 1.1.7	$TF(A)$	1.3.6
η_A (η)	1.1.1, 1.1.7	A^*	1.4.1, $\hat{C}_5, \hat{C}_6$
Δ_A (Δ)	1.1.2, 1.1.7	$c^V_{\xi,v}$ $(c_{\xi,v})$	1.4.4
ε_A (ε)	1.1.2, 1.1.7	C^V	1.4.4
τ_{ij}	1.1.3, C_1	$\mathrm{Mod}_f\, A$	1.4.5
μ^{opp}	1.1.3	V^σ	1.4.8
Δ^{opp}	1.1.3	$(A^*)^B$	1.4.15
V^*	1.1.5	$\mathbf{m}$	2.1.1, C_2
σ_A (σ)	1.1.7	$T_{x,X}$ (T_x)	2.1.2, C_2
$B\#A$	1.1.8	$A^\times$	2.1.3, C_2
$R[G]$	1.2.2	$\mathbf{m}_g$	2.1.7, C_2
$T(V)$	1.2.5	$\mathbf{m}_0$	2.1.10, C_2
$\mathfrak{g}$	1.2.6	A_V	2.2.2, C_2
$\mathfrak{gl}(V)$	1.2.6	$\mathbf{m}_V$	2.2.2, C_2
$\mathfrak{sl}(V)$	1.2.6	$\mathfrak{g}_V$	2.2.2, C_2
$U(\mathfrak{g})$	1.2.6	Lie G	2.3.3
$S(\mathfrak{g})$	1.2.6	$[H, H']$	2.3.6
$[n]_q$ $([n])$	1.2.12	$GL(V)$	2.3.7
$[n!]_q$ $([n!])$	1.2.12	ℓ	3.1.1
ad	1.3.1	$\mathfrak{h}_\mathbb{Q}$	3.1.1
$F(A)$	1.3.1	α_i	3.1.1
$Z(A)$	1.3.3	$\mathfrak{h}^*_\mathbb{Q}$	3.1.1

Index

Ergebnisse der Mathematik und ihrer Grenzgebiete, 3. Folge

A Series of Modern Surveys in Mathematics

Volume 1: **A. Fröhlich**

Galois Module Structure of Algebraic Integers

1983. X, 262 pp. ISBN 3-540-11920-5

Volume 2: **W. Fulton**

Intersection Theory

1984. XI, 470 pp. ISBN 3-540-12176-5

Volume 3: **J. C. Jantzen**

Einhüllende Algebren halbeinfacher Lie-Algebren

1983. V, 298 S. ISBN 3-540-12178-1

Volume 4: **W. Barth, C. Peters, A. van de Ven**

Compact Complex Surfaces

1984. X, 304 pp. ISBN 3-540-12172-2

Volume 5: **K. Strebel**

Quadratic Differentials

1984. XII, 184 pp. 74 figs. ISBN 3-540-13035-7

Volume 6: **M. J. Beeson**

Foundations of Constructive Mathematics

Metamathematical Studies

1985. XXIII, 466 pp. ISBN 3-540-12173-0

Volume 8: **R. Mañé**

Ergodic Theory and Differentiable Dynamics

Translated from the Portuguese by Silvio Levy

1987. XII, 317 pp. 32 figs. ISBN 3-540-15278-4

Volume 9: **M. Gromov**

Partial Differential Relations

1986. IX, 363 pp. ISBN 3-540-12177-3

Volume 10: **A. L. Besse**

Einstein Manifolds

1986. XII, 510 pp. 22 figs. ISBN 3-540-15279-2

Volume 11: **M. D. Fried, M. Jarden**

Field Arithmetic

1986. XVII, 458 pp. ISBN 3-540-16640-8

Volume 12: **J. Bochnak, M. Coste, M.-F. Roy**

Géométrie algébrique réelle

1987. X, 373 pp. 44 figs. ISBN 3-540-16951-2

Volume 13: **E. Freitag, R. Kiehl**

Etale Cohomology and the Weil Conjecture

With an Historical Introduction by J. A. Dieudonné

1987. XVIII, 317 pp. ISBN 3-540-12175-7

Volume 14: **M. R. Goresky, R. D. MacPherson**

Stratified Morse Theory

1988. XIV, 272 pp. 84 figs. ISBN 3-540-17300-5

Volume 15: **T. Oda**

Convex Bodies and Algebraic Geometry

An Introduction to the Theory of Toric Varieties

1987. VIII, 212 pp. 42 figs. ISBN 3-540-17600-4

Springer-Verlag and the Environment

We at Springer-Verlag firmly believe that an international science publisher has a special obligation to the environment, and our corporate policies consistently reflect this conviction.

We also expect our business partners – paper mills, printers, packaging manufacturers, etc. – to commit themselves to using environmentally friendly materials and production processes.

The paper in this book is made from low- or no-chlorine pulp and is acid free, in conformance with international standards for paper permanency.